STUDENT'S SOLUTIONS MANUAL

JEFFERY A. COLE
Anoka-Ramsey Community College

INTERMEDIATE ALGEBRA
TENTH EDITION

Margaret L. Lial
American River College

John Hornsby
University of New Orleans

Terry McGinnis

PEARSON

Addison
Wesley

Boston San Francisco New York
London Toronto Sydney Tokyo Singapore Madrid
Mexico City Munich Paris Cape Town Hong Kong Montreal

Reproduced by Pearson Addison-Wesley from electronic files supplied by the author.

Copyright © 2008 Pearson Education, Inc.
Publishing as Pearson Addison-Wesley, 75 Arlington Street, Boston, MA 02116.

ISBN-13: 978-0-321-44115-7
ISBN-10: 0-321-44115-X

8 9 10 11 BRR 10

Preface

This *Student's Solutions Manual* contains solutions to selected exercises in the text *Intermediate Algebra, Tenth Edition* by Margaret L. Lial, John Hornsby, and Terry McGinnis. It contains solutions to the odd-numbered exercises in each section, all Relating Concepts exercises, as well as solutions to all the exercises in the review sections, the chapter tests, and the cumulative review sections.

This manual is a text supplement and should be read along *with* the text. You should read all exercise solutions in this manual because many concept explanations are given and then used in subsequent solutions. All concepts necessary to solve a particular problem are not reviewed for every exercise. If you are having difficulty with a previously covered concept, refer back to the section where it was covered for more complete help.

A significant number of today's students are involved in various outside activities, and find it difficult, if not impossible, to attend all class sessions; this manual should help meet the needs of these students. In addition, it is my hope that this manual's solutions will enhance the understanding of all readers of the material and provide insights to solving other exercises.

I appreciate feedback concerning errors, solution correctness or style, and manual style. Any comments may be sent directly to me at the address below, at jeff.cole@anokaramsey.edu, or in care of the publisher, Pearson Addison-Wesley.

I would like to thank Ken Grace, of Anoka-Ramsey Community College, and Jeannine Grace, for typesetting the manuscript and providing assistance with many features of the manual; Marv Riedesel and Mary Johnson, for their careful accuracy checking and valuable suggestions; Jim McLaughlin, for his help with the entire art package; and the authors and Maureen O'Connor and Lauren Morse, of Pearson Addison-Wesley, for entrusting me with this project.

Jeffery A. Cole
Anoka-Ramsey Community College
11200 Mississippi Blvd. NW
Coon Rapids, MN 55433

Table of Contents

CHAPTER 1   REVIEW OF THE REAL NUMBER SYSTEM

1.1   Basic Concepts

1. $\{x \mid x \text{ is a natural number less than } 6\}$
The set of natural numbers is $\{1, 2, 3, \dots\}$, so the set of natural numbers less than 6 is $\{1, 2, 3, 4, 5\}$.

3. $\{z \mid z \text{ is an integer greater than } 4\}$
The set of integers is
$\{\dots, -3, -2, -1, 0, 1, 2, 3, \dots\}$, so the set of integers greater than 4 is $\{5, 6, 7, 8, \dots\}$.

5. $\{z \mid z \text{ is an integer less than or equal to } 4\}$
The set of integers is
$\{\dots, -3, -2, -1, 0, 1, 2, 3, \dots\}$, so the set of integers less than or equal to 4 is
$\{\dots, -1, 0, 1, 2, 3, 4\}$.

7. $\{a \mid a \text{ is an even integer greater than } 8\}$
The set of even integers is $\{\dots, -2, 0, 2, 4, \dots\}$, so the set of even integers greater than 8 is
$\{10, 12, 14, 16, \dots\}$.

9. $\{x \mid x \text{ is an irrational number that is also rational}\}$
Irrational numbers cannot also be rational numbers. The set of irrational numbers that are also rational is $\emptyset$.

11. $\{p \mid p \text{ is a number whose absolute value is } 4\}$
This is the set of numbers that lie a distance of 4 units from 0 on the number line. Thus, the set of numbers whose absolute value is 4 is $\{-4, 4\}$.

In Exercises 13–16, we give one possible answer.

13. $\{2, 4, 6, 8\}$ can be described by $\{x \mid x \text{ is an even natural number less than or equal to } 8\}$.

15. $\{4, 8, 12, 16, \dots\}$ can be described by $\{x \mid x \text{ is a multiple of 4 greater than } 0\}$.

17. Yes, the choice of variables is not important. The sets have the same description so they represent the same set.

19. Graph $\{-4, -2, 0, 3, 5\}$.
Place dots for $-4, -2, 0, 3,$ and 5 on a number line.

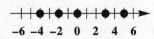

21. Graph $\left\{-\frac{6}{5}, -\frac{1}{4}, 0, \frac{5}{6}, \frac{13}{4}, 5.2, \frac{11}{2}\right\}$
Place dots for
$-\frac{6}{5} = -1.2, -\frac{1}{4} = -0.25, 0, \frac{5}{6} = 0.8\overline{3},$
$\frac{13}{4} = 3.25, 5.2,$ and $\frac{11}{2} = 5.5,$ on a number line.

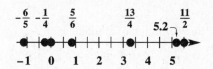

23. $\left\{-9, -\sqrt{6}, -0.7, 0, \frac{6}{7}, \sqrt{7}, 4.\overline{6}, 8, \frac{21}{2}, 13, \frac{75}{5}\right\}$

(a) The elements $8, 13,$ and $\frac{75}{5}$ (or 15) are natural numbers.

(b) The elements $0, 8, 13,$ and $\frac{75}{5}$ are whole numbers.

(c) The elements $-9, 0, 8, 13,$ and $\frac{75}{5}$ are integers.

(d) The elements $-9, -0.7, 0, \frac{6}{7}, 4.\overline{6}, 8, \frac{21}{2}, 13,$ and $\frac{75}{5}$ are rational numbers.

(e) The elements $-\sqrt{6}$ and $\sqrt{7}$ are irrational numbers.

(f) All the elements are real numbers.

25. The statement "Every integer is a whole number" is *false*. Some integers are whole numbers, but the negative integers are not whole numbers.

27. The statement "Every irrational number is an integer" is *false*. Irrational numbers have decimal representations that neither terminate nor repeat, so no irrational numbers are integers.

29. The statement "Every natural number is a whole number" is *true*. The whole numbers consist of the natural numbers and zero.

31. The statement "Some rational numbers are whole numbers" is *true*. Every whole number is rational.

33. The statement "The absolute value of any number is the same as the absolute value of its additive inverse" is *true*. The distance on a number line from 0 to a number is the same as the distance from 0 to its additive inverse.

35. **(a)** $-(-4) = 4$ (Choice **A**)

(b) $|-4| = 4$ (Choice **A**)

(c) $-|-4| = -(4) = -4$ (Choice **B**)

(d) $-|-(-4)| = -|4| = -4$ (Choice **B**)

37. **(a)** The additive inverse of 6 is -6.

(b) $6 > 0$, so $|6| = 6$.

39. **(a)** The additive inverse of -12 is $-(-12) = 12$.

(b) $-12 < 0$, so $|-12| = -(-12) = 12$.

41. **(a)** The additive inverse of $\frac{6}{5}$ is $-\frac{6}{5}$.

(b) $\frac{6}{5} > 0$, so $\left|\frac{6}{5}\right| = \frac{6}{5}$.

43. $|-8|$

Use the definition of absolute value.

$$-8 < 0, \text{ so } |-8| = -(-8) = 8$$

45. $\frac{3}{2} > 0$, so $\left|\frac{3}{2}\right| = \frac{3}{2}$.

47. $-|5| = -(5) = -5$

49. $-|-2| = -[-(-2)] = -(2) = -2$

51. $-|4.5| = -(4.5) = -4.5$

53. $|-2| + |3| = 2 + 3 = 5$

55. $|-9| - |-3| = 9 - 3 = 6$

57. $|-1| + |-2| - |-3| = 1 + 2 - 3$
$$= 3 - 3$$
$$= 0$$

59. **(a)** The greatest absolute value is $|11.5| = 11.5$. Therefore, Las Vegas had the greatest change in population. The population increased 11.5%.

(b) The least absolute value is $|-1.2| = 1.2$. Therefore, Chicago had the least change in population. The population decreased 1.2%.

61. Compare the depths of the bodies of water. The deepest, that is, the body of water whose depth has the greatest absolute value, is the Pacific Ocean ($|-12{,}925| = 12{,}925$) followed by the Indian Ocean, the Caribbean Sea, the South China Sea, and the Gulf of California.

63. True; the absolute value of the depth of the Pacific Ocean is

$$|-12{,}925| = 12{,}925$$

which is greater than the absolute value of the depth of the Indian Ocean,

$$|-12{,}598| = 12{,}598.$$

65. True; since -6 is to the left of -2 on a number line, -6 is less than -2.

67. False; since -4 is to the left of -3 on a number line, -4 is *less* than -3, not greater.

69. True; since 3 is to the right of -2 on a number line, 3 is greater than -2.

71. True; since $-3 = -3$, -3 is greater than *or equal to* -3.

73. The inequality $6 > 2$ can also be written $2 < 6$. In each case, the inequality symbol points toward the smaller number so both inequalities are true.

75. $-9 < 4$ is equivalent to $4 > -9$.

77. $-5 > -10$ is equivalent to $-10 < -5$.

79. $0 < x$ is equivalent to $x > 0$.

81. "7 is greater than y" can be written as $7 > y$.

83. "5 is greater than or equal to 5" can be written as $5 \geq 5$.

85. "$3t - 4$ is less than or equal to 10" can be written as $3t - 4 \leq 10$.

87. "$5x + 3$ is not equal to 0" can be written as $5x + 3 \neq 0$.

89. "t is between -3 and 5" can be written as $-3 < t < 5$.

91. "$3x$ is between -3 and 4, including -3 and excluding 4" can be written as $-3 \leq 3x < 4$.

93. $-6 < 7 + 3$?
$$-6 < 10 \qquad \textit{True}$$

The last statement is true since -6 is to the left of 10 on a number line.

95. $2 \cdot 5 \geq 4 + 6$?
$$10 \geq 10 \qquad \textit{True}$$

97. $-|-3| \geq -3$?
$$-3 \geq -3 \qquad \textit{True}$$

99. $-8 > -|-6|$?
$$-8 > -6 \qquad \textit{False}$$

101. $\{x \mid x > -1\}$ includes all numbers greater than -1, written $(-1, \infty)$. Place a parenthesis at -1 since -1 is not an element of the set. The graph extends from -1 to the right.

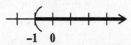

103. $\{x \mid x \leq 6\}$ includes all numbers less than or equal to 6, written $(-\infty, 6]$. Place a bracket at 6 since 6 is an element of the set. The graph extends from 6 to the left.

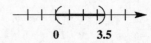

105. $\{x \mid 0 < x < 3.5\}$ includes all numbers between 0 and 3.5, written $(0, 3.5)$. Place parentheses at 0 and 3.5 since these numbers are not elements of the set. The graph goes from 0 to 3.5.

107. $\{x \mid 2 \leq x \leq 7\}$ includes all numbers from 2 to 7, written $[2, 7]$. Place brackets at 2 and 7 since these numbers are elements of the set. The graph goes from 2 to 7.

109. $\{x \mid -4 < x \le 3\}$ includes all numbers between -4 and 3, excluding -4, but including 3, written $(-4, 3]$. Place a parenthesis at -4 and a bracket at 3 to show that -4 is not included, but 3 is. The graph goes from -4 to 3.

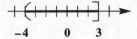

111. $\{x \mid 0 < x \le 3\}$ includes all numbers between 0 and 3, excluding 0, but including 3, written $(0, 3]$. Place a parenthesis at 0 and a bracket at 3 to show that 0 is not included, but 3 is. The graph goes from 0 to 3.

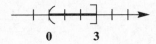

113. 2005: IA $= 12{,}978$, OH $= 7506$, PA $= 6608$, IN $= 6256$
In 2005, Iowa (IA), Ohio (OH), Pennsylvania (PA), and Indiana (IN) had production greater than 6000 million eggs.

115. 2005: TX $= x = 4684$, OH $= y = 7506$
Since $4684 < 7506$, $x < y$ is true.

117. The *natural numbers* are the numbers with which we count. Some examples are 1, 2, and 3.

The *whole numbers* are formed by including 0 with the natural numbers, such as 0, 2, and 17.

The *integers* are formed by including the negatives of the natural numbers with the whole numbers, such as -1, -2, and -3.

The *rational numbers*, such as $\frac{1}{2}$, 0.75, and $-\frac{3}{4}$, are formed by quotients of integers (nonzero denominator).

The *irrational numbers* include positive or negative numbers that are not rational, such as π, $\sqrt{2}$, and $-\sqrt{5}$.

The *real numbers* include all positive numbers, negative numbers, and 0, such as $-\pi$, 0, and $\sqrt{2}$.

1.2   Operations on Real Numbers

1. The sum of a positive number and a negative number is 0 if <u>the numbers are additive inverses</u>. For example, $4 + (-4) = 0$.

3. The sum of two negative numbers is a <u>negative</u> number. For example, $-7 + (-21) = -28$.

5. The sum of a positive number and a negative number is positive if <u>the positive number has a larger absolute value</u>. For example, $15 + (-2) = 13$.

7. The difference between two negative numbers is negative if <u>the number with smaller absolute value is subtracted from the one with larger absolute value</u>. For example, $-15 - (-3) = -12$.

9. The product of two numbers with different signs is <u>negative</u>. For example, $-5(15) = -75$.

11. $-6 + (-13) = -(6 + 13) = -19$

13. $13 + (-4) = 13 - 4 = 9$

15. $-\frac{7}{3} + \frac{3}{4} = -\frac{28}{12} + \frac{9}{12} = -\frac{19}{12}$

17. The difference between 2.3 and 0.45 is 1.85. The number with the larger absolute value, -2.3, is negative, so the answer is negative. Thus, $-2.3 + 0.45 = -1.85$.

19. $-6 - 5 = -6 + (-5) = -(6 + 5) = -11$

21. $8 - (-13) = 8 + 13 = 21$

23. $-16 - (-3) = -16 + 3 = -13$

25. $\begin{aligned} -12.31 - (-2.13) &= -12.31 + 2.13 \\ &= -(12.31 - 2.13) \\ &= -10.18 \end{aligned}$

27. $\frac{9}{10} - \left(-\frac{4}{3}\right) = \frac{9}{10} + \frac{4}{3} = \frac{27}{30} + \frac{40}{30} = \frac{67}{30}$

29. $|-8 - 6| = |-14| = -(-14) = 14$

31. $-|-4 + 9| = -|5| = -5$

33. $-2 - |-4| = -2 - 4 = -2 + (-4) = -6$

35. $-7 + 5 - 9 = (-7 + 5) - 9 = -2 - 9 = -11$

37. $6 - (-2) + 8 = 6 + 2 + 8 = 8 + 8 = 16$

39. $\begin{aligned} -9 - 4 - (-3) + 6 &= (-9 - 4) + 3 + 6 \\ &= -13 + 3 + 6 \\ &= -10 + 6 \\ &= -4 \end{aligned}$

41. $\begin{aligned} -8 - (-12) - (2 - 6) &= -8 + 12 - (-4) \\ &= -8 + 12 + 4 \\ &= 4 + 4 \\ &= 8 \end{aligned}$

43. $\begin{aligned} -0.382 + 4 - 0.6 &= 3.618 - 0.6 \\ &= 3.018 \end{aligned}$

45. $\begin{aligned} \left(-\frac{5}{4} - \frac{2}{3}\right) + \frac{1}{6} &= \left(-\frac{15}{12} - \frac{8}{12}\right) + \left(\frac{2}{12}\right) \\ &= -\frac{23}{12} + \frac{2}{12} \\ &= -\frac{21}{12}, \text{ or } -\frac{7}{4} \end{aligned}$

47. $\begin{aligned} -\frac{3}{4} - \left(\frac{1}{2} - \frac{3}{8}\right) &= -\frac{6}{8} - \left(\frac{4}{8} - \frac{3}{8}\right) \\ &= -\frac{6}{8} - \frac{1}{8} \\ &= -\frac{7}{8} \end{aligned}$

49. $|-11| - |-5| - |7| + |-2|$
$= 11 - 5 - 7 + 2$
$= 6 - 7 + 2$
$= -1 + 2 = 1$

51. $A = -4$ and $B = 2$. The distance between A and B is the absolute value of their difference.
$$|-4 - 2| = |-6| = 6$$

53. $D = -6$ and $F = \frac{1}{2}$. The distance between D and F is the absolute value of their difference.
$$\left|-6 - \frac{1}{2}\right| = \left|-\frac{12}{2} - \frac{1}{2}\right|$$
$$= \left|-\frac{13}{2}\right|$$
$$= \frac{13}{2}, \text{ or } 6\frac{1}{2}$$

55. It is true for multiplication (and division). It is false for addition and for subtraction when the number to be subtracted has the smaller absolute value. A more precise statement is, "The product or quotient of two negative numbers is positive."

57. The product of two numbers with *different* signs is *negative*, so
$$5(-7) = -35.$$

59. The product of two numbers with the *same* sign is *positive*, so
$$-8(-5) = 40.$$

61. The product of two numbers with the *same* sign is *positive*, so
$$-10\left(-\frac{1}{5}\right) = 2.$$

63. The product of two numbers with *different* signs is *negative*, so
$$\frac{3}{4}(-16) = -\frac{3}{4} \cdot 4 \cdot 4 = -12.$$

65. The product of two numbers with the *same* sign is *positive*, so
$$-\frac{5}{2}\left(-\frac{12}{25}\right) = \frac{5 \cdot 2 \cdot 6}{2 \cdot 5 \cdot 5} = \frac{6}{5}.$$

67. The product of two numbers with the *same* sign is *positive*, so
$$-\frac{3}{8}\left(-\frac{24}{9}\right) = \frac{3 \cdot 3 \cdot 8}{8 \cdot 9} = 1.$$

69. $-2.4(-2.45) = 5.88$

71. $3.4(-3.14) = -10.676$

73. The quotient of two nonzero real numbers with *different* signs is *negative*, so
$$\frac{-14}{2} = -14 \cdot \frac{1}{2} = -2 \cdot 7 \cdot \frac{1}{2} = -7.$$

75. The quotient of two nonzero real numbers with the *same* sign is *positive*, so
$$\frac{-24}{-4} = 24 \cdot \frac{1}{4} = 4 \cdot 6 \cdot \frac{1}{4} = 6.$$

77. The quotient of two nonzero real numbers with *different* signs is *negative*, so
$$\frac{100}{-25} = -100 \cdot \frac{1}{25} = -4 \cdot 25 \cdot \frac{1}{25} = -4.$$

79. $\frac{0}{-8} = 0 \cdot \left(-\frac{1}{8}\right) = 0$

81. Division by 0 is undefined, so $\frac{5}{0}$ is undefined.

83. The quotient of two nonzero real numbers with the *same* sign is *positive*, so
$$-\frac{10}{17} \div \left(-\frac{12}{5}\right) = \frac{10}{17} \cdot \frac{5}{12} = \frac{2 \cdot 5 \cdot 5}{17 \cdot 2 \cdot 6} = \frac{25}{102}.$$

85. $\frac{\frac{12}{13}}{-\frac{4}{3}} = \frac{12}{13} \div \left(-\frac{4}{3}\right) = \frac{12}{13}\left(-\frac{3}{4}\right)$
$$= -\frac{3 \cdot 4 \cdot 3}{13 \cdot 4} = -\frac{9}{13}$$

87. $\frac{-27.72}{13.2} = -2.1$

89. $\frac{-100}{-0.01} = \frac{100}{\frac{1}{100}} = 100 \div \frac{1}{100} = 100 \cdot 100$
$$= 10,000$$

91. $\frac{1}{6} - \left(-\frac{7}{9}\right) = \frac{1}{6} + \frac{7}{9} = \frac{3}{18} + \frac{14}{18} = \frac{17}{18}$

93. $-\frac{1}{9} + \frac{7}{12} = -\frac{4}{36} + \frac{21}{36} = \frac{17}{36}$

95. $-\frac{3}{8} - \frac{5}{12} = -\frac{9}{24} - \frac{10}{24} = -\frac{19}{24}$

97. $-\frac{7}{30} + \frac{2}{45} - \frac{3}{10} = -\frac{21}{90} + \frac{4}{90} - \frac{27}{90} = -\frac{44}{90} = -\frac{22}{45}$

99. $\frac{8}{25}\left(-\frac{5}{12}\right) = -\frac{2 \cdot 4 \cdot 5}{5 \cdot 5 \cdot 3 \cdot 4} = -\frac{2}{5 \cdot 3} = -\frac{2}{15}$

101. $\frac{5}{6}\left(-\frac{9}{10}\right)\left(-\frac{4}{5}\right) = \frac{5 \cdot 3 \cdot 3 \cdot 2 \cdot 2}{2 \cdot 3 \cdot 2 \cdot 5 \cdot 5} = \frac{3}{5}$

103. $\frac{7}{6} \div \left(-\frac{9}{10}\right) = \frac{7}{6} \cdot \left(-\frac{10}{9}\right) = -\frac{7 \cdot 2 \cdot 5}{2 \cdot 3 \cdot 9}$
$$= -\frac{7 \cdot 5}{3 \cdot 9} = -\frac{35}{27}, \text{ or } -1\frac{8}{27}$$

105. $\frac{-\frac{8}{9}}{2} = -\frac{8}{9} \div \frac{2}{1} = -\frac{8}{9} \cdot \frac{1}{2} = -\frac{2 \cdot 4}{9 \cdot 2} = -\frac{4}{9}$

107. $-8.6 - 3.751 = -(8.6 + 3.751) = -12.351$

109. $(-4.2)(1.4)(2.7) = (-5.88)(2.7) = -15.876$

111. $-24.84 \div 6 = -4.14$

113. $-2496 \div (-0.52) = 4800$

115. $-14.23 + 9.81 + 74.63 - 18.715$
$$= -4.42 + 74.63 - 18.715$$
$$= 70.21 - 18.715$$
$$= 51.495$$

117. To find the difference between these two temperatures, subtract the lowest temperature from the highest temperature.

$$90° - (-22°) = 90° + 22°$$
$$= 112°$$

The difference is $112°F$.

119. $48.35 - 35.99 - 20.00 - 28.50 + 66.27$
$= 12.36 - 20.00 - 28.50 + 66.27$
$= -7.64 - 28.50 + 66.27$
$= -36.14 + 66.27$
$= 30.13$

His balance is $30.13.

121. $-382.45 + 25.10 + 34.50 - 45.00 - 98.17$
$= -466.02$

His balance is $-$466.02.

(a) To pay off the balance, his payment should be $466.02.

(b) $-466.02 + 300 - 24.66 = -190.68$

His balance is $-$190.68.

123. (a) $-142 - 225 - 185 + 77 = -475$

The total loss for 2003–2006 was $475 thousand.

(b) $77 - (-185) = 77 + 185 = 262$

The difference between the profit or loss from 2005 to 2006 was $262 thousand.

(c) $-225 - (-142) = -225 + 142 = -83$

The difference between the profit or loss from 2003 to 2004 was $-$83 thousand.

125. (a)

Year	Difference (in billions)
2000	$538 - $409 = $129
2010	$916 - $710 = $206
2020	$1479 - $1405 = $74
2030	$2041 - $2542 = -$501

(b) The cost of Social Security will exceed revenue in 2030 by $501 billion.

1.3 Exponents, Roots, and Order of Operations

1. $-4^6 = (-4)^6$ is *false*.
$-4^6 = -(4 \cdot 4 \cdot 4 \cdot 4 \cdot 4 \cdot 4) = -4096$
whereas
$(-4)^6 = (-4)(-4)(-4)(-4)(-4)(-4)$
$= 4096.$

3. The statement "$\sqrt{16}$ is a positive number" is *true*. The symbol $\sqrt{}$ always gives a positive square root.

5. The statement "$(-2)^7$ is a negative number" is *true*. $(-2)^7$ gives an odd number of negative factors, so the product is negative.

7. The statement "The product of 8 positive factors and 8 negative factors is positive" is *true*. The product of an even number of negative factors is positive.

9. The statement "In the exponential -3^5, -3 is the base" is *false*. The base is 3, not -3. If the problem were written $(-3)^5$, then -3 would be the base.

11. (a) $8^2 = 64$

(b) $-8^2 = -(8 \cdot 8) = -64$

(c) $(-8)^2 = (-8)(-8) = 64$

(d) $-(-8)^2 = -(64) = -64$

13. $10 \cdot 10 \cdot 10 \cdot 10 = 10^4$

15. $\frac{3}{4} \cdot \frac{3}{4} \cdot \frac{3}{4} \cdot \frac{3}{4} \cdot \frac{3}{4} = \left(\frac{3}{4}\right)^5$

17. $(-9)(-9)(-9) = (-9)^3$

19. $z \cdot z \cdot z \cdot z \cdot z \cdot z \cdot z = z^7$

21. $4^2 = 4 \cdot 4 = 16$

23. $0.28^3 = (0.28)(0.28)(0.28) = 0.021952$

25. $\left(\frac{1}{5}\right)^3 = \frac{1}{5} \cdot \frac{1}{5} \cdot \frac{1}{5} = \frac{1}{125}$

27. $\left(\frac{4}{5}\right)^4 = \left(\frac{4}{5}\right)\left(\frac{4}{5}\right)\left(\frac{4}{5}\right)\left(\frac{4}{5}\right)$
$= \frac{4 \cdot 4 \cdot 4 \cdot 4}{5 \cdot 5 \cdot 5 \cdot 5} = \frac{256}{625}$

29. $(-5)^3 = (-5)(-5)(-5) = -125$

31. $(-2)^8 = (-2)(-2)(-2)(-2)(-2)(-2)(-2)(-2)$
$= 256$

33. $-3^6 = -(3 \cdot 3 \cdot 3 \cdot 3 \cdot 3 \cdot 3) = -729$

35. $-8^4 = -(8 \cdot 8 \cdot 8 \cdot 8) = -4096$

37. $\sqrt{81} = 9$ since 9 is positive and $9^2 = 81$.

39. $\sqrt{169} = 13$ since 13 is positive and $13^2 = 169$.

41. $-\sqrt{400} = -\left(\sqrt{400}\right) = -(20) = -20$

43. $\sqrt{\frac{100}{121}} = \frac{10}{11}$ since $\frac{10}{11}$ is positive and $\left(\frac{10}{11}\right)^2 = \frac{100}{121}$.

45. $-\sqrt{0.49} = -\sqrt{(0.7)^2} = -(0.7) = -0.7$

47. There is no real number whose square is negative, so $\sqrt{-36}$ is not a real number.

49. (a) $\sqrt{144} = 12$; choice **B**

(b) $\sqrt{-144}$ is not a real number; choice **C**

(c) $-\sqrt{144} = -12$; choice **A**

51. If a is a positive number, then $-a$ is a negative number. Therefore, $\sqrt{-a}$ is not a real number, so $-\sqrt{-a}$ is not a real number.

53. $12 + 3 \cdot 4 = 12 + 12$ *Multiply.*
$$= 24 \qquad \textit{Add.}$$

55. $6 \cdot 3 - 12 \div 4 = 18 - 12 \div 4$ *Multiply.*
$$= 18 - 3 \qquad \textit{Divide.}$$
$$= 15 \qquad \textit{Subtract.}$$

57. $10 + 30 \div 2 \cdot 3 = 10 + 15 \cdot 3$ *Divide.*
$$= 10 + 45 \qquad \textit{Multiply.}$$
$$= 55 \qquad \textit{Add.}$$

59. $-3(5)^2 - (-2)(-8)$
$$= -3(25) - (-2)(-8) \qquad \textit{Evaluate power.}$$
$$= -75 - 16 \qquad \textit{Multiply.}$$
$$= -91 \qquad \textit{Subtract.}$$

61. $5 - 7 \cdot 3 - (-2)^3$
$$= 5 - 7 \cdot 3 - (-8) \qquad \textit{Evaluate power.}$$
$$= 5 - 21 + 8 \qquad \textit{Multiply and subtract.}$$
$$= -16 + 8 \qquad \textit{Subtract.}$$
$$= -8 \qquad \textit{Add.}$$

63. $-7\left(\sqrt{36}\right) - (-2)(-3)$
$$= -7(6) - (-2)(-3) \qquad \textit{Evaluate root.}$$
$$= -42 - 6 \qquad \textit{Multiply.}$$
$$= -48 \qquad \textit{Subtract.}$$

65. $6|4 - 5| - 24 \div 3$
$$= 6|-1| - 24 \div 3 \qquad \textit{Simplify within absolute value bars.}$$
$$= 6(1) - 24 \div 3 \qquad \textit{Take absolute value.}$$
$$= 6 - 8 \qquad \textit{Multiply and divide.}$$
$$= -2 \qquad \textit{Subtract.}$$

67. $|-6 - 5|(-8) + 3^2$
$$= |-11|(-8) + 3^2 \qquad \textit{Simplify within absolute value bars.}$$
$$= 11(-8) + 3^2 \qquad \textit{Take absolute value.}$$
$$= 11(-8) + 9 \qquad \textit{Evaluate power.}$$
$$= -88 + 9 \qquad \textit{Multiply.}$$
$$= -79 \qquad \textit{Add.}$$

69. $6 + \frac{2}{3}(-9) - \frac{5}{8} \cdot 16 = 6 + (-6) - 10$ *Multiply.*
$$= 0 - 10 \qquad \textit{Add.}$$
$$= -10 \qquad \textit{Subtract.}$$

71. $-14\left(-\frac{2}{7}\right) \div (2 \cdot 6 - 10)$
$$= 4 \div (12 - 10) \qquad \textit{Multiply.}$$
$$= 4 \div 2 \qquad \textit{Work inside parentheses.}$$
$$= 2 \qquad \textit{Divide.}$$

73. $\dfrac{\left(-5 + \sqrt{4}\right)(-2^2)}{-5 - 1}$
$$= \frac{(-5 + 2)(-4)}{-6} \qquad \textit{Evaluate root and power; subtract.}$$
$$= \frac{(-3)(-4)}{-6} \qquad \textit{Work inside parentheses.}$$
$$= \frac{12}{-6} \qquad \textit{Multiply.}$$
$$= -2 \qquad \textit{Divide.}$$

75. $\dfrac{2(-5) + (-3)(-2)}{-8 + 3^2 - 1}$
$$= \frac{-10 + 6}{-8 + 9 - 1} \qquad \textit{Evaluate power; multiply.}$$
$$= \frac{-4}{1 - 1} \qquad \textit{Add.}$$
$$= \frac{-4}{0} \qquad \textit{Subtract.}$$

Since division by 0 is undefined, the given expression is *undefined*.

77. $\dfrac{5 - 3\left(\dfrac{-5 - 9}{-7}\right) - 6}{-9 - 11 + 3 \cdot 7}$
$$= \frac{5 - 3\left(\dfrac{-14}{-7}\right) - 6}{-9 - 11 + 21} \qquad \textit{Work in numerator and denominator separately.}$$
$$= \frac{5 - 3(2) - 6}{-20 + 21}$$
$$= \frac{5 - 6 - 6}{1}$$
$$= \frac{-7}{1}$$
$$= -7 \qquad \textit{Write in lowest terms.}$$

In Exercises 79–86, $a = -3$, $b = 64$, and $c = 6$.

79. $3a + \sqrt{b} = 3(-3) + \sqrt{64}$
$$= 3(-3) + 8$$
$$= -9 + 8$$
$$= -1$$

81. $\sqrt{b} + c - a = \sqrt{64} + 6 - (-3)$
$$= 8 + 6 + 3$$
$$= 14 + 3 = 17$$

83. $4a^3 + 2c = 4(-3)^3 + 2(6)$
$$= 4(-27) + 12$$
$$= -108 + 12 = -96$$

85. $\dfrac{2c + a^3}{4b + 6a} = \dfrac{2(6) + (-3)^3}{4(64) + 6(-3)}$

$\qquad = \dfrac{12 + (-27)}{256 + (-18)}$

$\qquad = \dfrac{-15}{238} = -\dfrac{15}{238}$

In Exercises 87–94, $w = 4$, $x = -\frac{3}{4}$, $y = \frac{1}{2}$, and $z = 1.25$.

87. $wy - 8x = 4\left(\frac{1}{2}\right) - 8\left(-\frac{3}{4}\right)$

$\qquad = 2 + 6$

$\qquad = 8$

89. $xy + y^4 = -\frac{3}{4}\left(\frac{1}{2}\right) + \left(\frac{1}{2}\right)^4$

$\qquad = -\frac{3}{8} + \frac{1}{16}$

$\qquad = -\frac{6}{16} + \frac{1}{16} = -\frac{5}{16}$

91. $-w + 2x + 3y + z$

$\qquad = -4 + 2\left(-\frac{3}{4}\right) + 3\left(\frac{1}{2}\right) + 1.25$

$\qquad = -4 - \frac{3}{2} + \frac{3}{2} + 1.25$

$\qquad = -4 + 1.25$

$\qquad = -2.75$

93. $\dfrac{7x + 9y}{w} = \dfrac{7\left(-\frac{3}{4}\right) + 9\left(\frac{1}{2}\right)}{4}$

$\qquad = \dfrac{-\frac{21}{4} + \frac{9}{2}}{4} = \dfrac{-\frac{21}{4} + \frac{18}{4}}{4}$

$\qquad = \dfrac{-\frac{3}{4}}{4} = -\frac{3}{4} \div \frac{4}{1} = -\frac{3}{4} \cdot \frac{1}{4} = -\frac{3}{16}$

95. $(v \times 0.5485 - 4850) \div 1000 \times 31.44$

$\qquad = (100{,}000 \times 0.5485 - 4850) \div 1000 \times 31.44$

$\qquad = (54{,}850 - 4850) \div 1000 \times 31.44$

$\qquad = 50{,}000 \div 1000 \times 31.44$

$\qquad = 50 \times 31.44$

$\qquad = 1572$

The owner would pay \$1572 in property taxes.

97. $(v \times 0.5485 - 4850) \div 1000 \times 31.44$

$\qquad = (200{,}000 \times 0.5485 - 4850) \div 1000 \times 31.44$

$\qquad = (109{,}700 - 4850) \div 1000 \times 31.44$

$\qquad = 104{,}850 \div 1000 \times 31.44$

$\qquad = 104.85 \times 31.44$

$\qquad = 3296.484 \approx 3296$

The owner would pay \$3296 in property taxes.

99. number of oz $\times$ % alcohol $\times 0.075 \div$ body weight in lb $-$ hr of drinking $\times 0.015$

$\qquad = 36 \times 4.0 \times 0.075 \div 135 - 3 \times 0.015$

$\qquad = 144 \times 0.075 \div 135 - 3 \times 0.015$

$\qquad = 10.8 \div 135 - 3 \times 0.015$

$\qquad = 0.08 - 3 \times 0.015$

$\qquad = 0.08 - 0.045$

$\qquad = 0.035$

101. $1.718x - 3409$

(a) $1.718(1994) - 3409 \approx \$16.7$ billion

(b) $1.718(2000) - 3409 = \$27$ billion

(c) $1.718(2005) - 3409 \approx \$35.6$ billion

(d) \$35.6 billion is more than twice \$16.7 billion, so the amount spent on pets more than doubled from 1994 to 2005.

1.4 Properties of Real Numbers

1. The identity element for addition is 0 since, for any real number a, $a + 0 = 0 + a = a$. Choice **B** is correct.

3. The additive inverse of a is $-a$ since, for any real number a, $a + (-a) = 0$ and $-a + a = 0$. Choice **A** is correct.

5. The multiplication property of 0 states that the product of 0 and any real number is $\underline{0}$.

7. The associative property is used to change the grouping of three terms or factors.

9. When simplifying an expression, only like terms can be combined.

11. Using the distributive property,
$$2(m + p) = 2m + 2p.$$

13. Using the distributive property,
$$-12(x - y) = -12[x + (-y)]$$
$$= -12(x) + (-12)(-y)$$
$$= -12x + 12y.$$

15. Using the second form of the distributive property,
$$5k + 3k = (5 + 3)k$$
$$= 8k.$$

17. $7r - 9r = 7r + (-9r)$
$$= [7 + (-9)]r$$
$$= -2r$$

19. $-8z + 4w$
Since there is no common variable factor here, we cannot use the distributive property to simplify the expression.

21. Using the identity property, then the distributive property,
$$a + 7a = 1a + 7a$$
$$= (1 + 7)a$$
$$= 8a.$$

23. $-(2d - f) = -1(2d - f)$
$$= -1(2d) + (-1)(-f)$$
$$= -2d + f$$

25. $-12y + 4y + 3 + 2y$
$= -12y + 4y + 2y + 3$
$= (-12 + 4 + 2)y + 3$
$= -6y + 3$

27. $-6p + 5 - 4p + 6 + 11p$
$= -6p - 4p + 11p + 5 + 6$
$= (-6 - 4 + 11)p + 11$
$= 1p + 11 \quad \text{or} \quad p + 11$

29. $3(k + 2) - 5k + 6 + 3$
$= 3k + 6 - 5k + 6 + 3$
$= 3k - 5k + 6 + 6 + 3$
$= (3 - 5)k + 6 + 6 + 3$
$= -2k + 15$

31. $-2(m + 1) - (m - 4)$
$= -2m - 2 - m + 4$
$= -2m - m - 2 + 4$
$= (-2 - 1)m + 2$
$= -3m + 2$

33. $0.25(8 + 4p) - 0.5(6 + 2p)$
$= 0.25(8) + 0.25(4p) + (-0.5)(6) + (-0.5)(2p)$
$= 2 + p - 3 - p$
$= p - p + 2 - 3$
$= (1 - 1)p + 2 - 3$
$= 0p - 1$
$= -1$

35. $-(2p + 5) + 3(2p + 4) - 2p$
$= -2p - 5 + 6p + 12 - 2p$
$= (-2 + 6 - 2)p + (-5 + 12)$
$= 2p + 7$

37. $2 + 3(2z - 5) - 3(4z + 6) - 8$
$= 2 + 6z - 15 - 12z - 18 - 8$
$= 6z - 12z + 2 - 15 - 18 - 8$
$= (6 - 12)z - 13 - 18 - 8$
$= -6z - 31 - 8$
$= -6z - 39$

39. $5x + 8x = (5 + 8)x = 13x$ *Distributive property*

41. $5(9r) = (5 \cdot 9)r = 45r$ *Associative property*

43. $5x + 9y = 9y + 5x$ *Commutative property*

45. $1 \cdot 7 = 7$ *Identity property*

47. $-\dfrac{1}{4}ty + \dfrac{1}{4}ty = 0$ *Inverse property*
A number plus its opposite equals 0.

49. $8(-4 + x) = 8(-4) + 8x$ *Distributive property*
$= -32 + 8x$

51. $0(0.875x + 9y - 88z) = 0$ *Multiplication property of 0*
Zero times any quantity equals 0.

53. Answers will vary. Commutative: one example is washing your face and brushing your teeth. The activities can be carried out in either order. Not commutative: one example is putting on your socks and putting on your shoes.

55. $96 \cdot 19 + 4 \cdot 19 = (96 + 4)19$
$= (100)19$
$= 1900$

57. $58 \cdot \dfrac{3}{2} - 8 \cdot \dfrac{3}{2} = (58 - 8)\dfrac{3}{2}$
$= (50)\dfrac{3}{2} = \dfrac{50}{1} \cdot \dfrac{3}{2}$
$= \dfrac{150}{2} = 75$

59. $4.31(69) + 4.31(31) = 4.31(69 + 31)$
$= 4.31(100)$
$= 431$

61. The terms have been grouped using the associative property of addition.

62. The terms have been regrouped using the associative property of addition.

63. The order of the terms inside the parentheses has been changed using the commutative property of addition.

64. The terms have been regrouped using the associative property of addition.

65. The common factor, x, has been distributed over the expression in parentheses using the distributive property.

66. The numbers in parentheses have been added to simplify the expression.

67. Is $a + (b \cdot c) = (a + b)(a + c)$?
No. One example is
$$7 + (5 \cdot 3) = (7 + 5)(7 + 3),$$
which is false since
$$7 + (5 \cdot 3) = 7 + 15 = 22$$
and
$$(7 + 5)(7 + 3) = 12(10) = 120.$$

Chapter 1 Review Exercises

1. $\left\{-4, -1, 2, \dfrac{9}{4}, 4\right\}$
Place dots for $-4, -1, 2, \dfrac{9}{4} = 2.25$, and 4 on a number line.

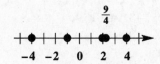

2. $\left\{-5, -\frac{11}{4}, -0.5, 0, 3, \frac{13}{3}\right\}$

Place dots for -5, $-\frac{11}{4} = -2.75$, -0.5, 0, 3, and $\frac{13}{3} = 4.\overline{3}$ on a number line.

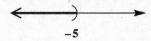

3. $|-16| = -(-16) = 16$

4. $-|-4| = -[-(-4)] = -(4) = -4$

5. $|-8| - |-3| = -(-8) - [-(-3)]$
$= 8 - [3]$
$= 5$

In Exercises 6–9,

$$S = \left\{-9, -\frac{4}{3}, -\sqrt{4}, -0.25, 0, 0.\overline{35}, \frac{5}{3}, \sqrt{7}, \sqrt{-9}, \frac{12}{3}\right\}$$

6. The elements 0 and $\frac{12}{3}$ (or 4) are whole numbers.

7. The elements -9, $-\sqrt{4}$ (or -2), 0, and $\frac{12}{3}$ (or 4) are integers.

8. The elements $-9, -\frac{4}{3}, -\sqrt{4}$ (or -2), -0.25, 0, $0.\overline{35}$, $\frac{5}{3}$, and $\frac{12}{3}$ (or 4) are rational numbers. (Remember that terminating and repeating decimals are rational numbers.)

9. All the elements in the set are real numbers except $\sqrt{-9}$.

10. $\{x \mid x$ is a natural number between 3 and 9$\}$
The natural numbers between 3 and 9 are 4, 5, 6, 7, and 8. Therefore, the set is $\{4, 5, 6, 7, 8\}$.

11. $\{y \mid y$ is a whole number less than 4$\}$
The whole numbers less than 4 are 0, 1, 2, and 3. Therefore, the set is $\{0, 1, 2, 3\}$.

12. $4 \cdot 2 \le |12 - 4|$?
$8 \le |8|$?
$8 \le 8$ *True*

13. $2 + |-2| > 4$?
$2 + 2 > 4$?
$4 > 4$ *False*

14. $4(3 + 7) > -|40|$?
$4(10) > -40$?
$40 > -40$ *True*

15. The longest bar represents the greatest change. Chrysler had the greatest change of 13.7%.

16. The shortest bar represents the least change. Honda had the least change of -2.5%.

17. Ford: $|5.2| = 5.2$

General Motors: $|-5.6| = 5.6$

The absolute value of the percent change for Ford was *less* than the absolute value of the percent change for General Motors, so the statement is *false*.

18. Toyota > 4(Mazda) ?
$13.4 > 4(3.0)$?
$13.4 > 12.0$ *True*

19. $\{x \mid x < -5\}$
In interval notation, $x < -5$ is written as $(-\infty, -5)$.

The parenthesis at -5 indicates that -5 is not included. The graph extends from -5 to the left.

-5

20. $\{x \mid -2 < x \le 3\}$
In interval notation, $-2 < x \le 3$ is written as $(-2, 3]$. The parenthesis indicates that -2 is not included, while the bracket indicates that 3 is included. The graph goes from -2 to 3.

-2 3

21. $-\frac{5}{8} - \left(-\frac{7}{3}\right) = -\frac{5}{8} + \frac{7}{3}$
$= -\frac{15}{24} + \frac{56}{24} = \frac{41}{24}$

22. $-\frac{4}{5} - \left(-\frac{3}{10}\right) = -\frac{4}{5} + \frac{3}{10}$
$= -\frac{8}{10} + \frac{3}{10}$
$= -\frac{5}{10} = -\frac{1}{2}$

23. $-5 + (-11) + 20 - 7$
$= -16 + 20 - 7$
$= 4 - 7$
$= -3$

24. $-9.42 + 1.83 - 7.6 - 1.9$
$= -7.59 - 7.6 - 1.9$
$= -15.19 - 1.9$
$= -17.09$

25. $-15 + (-13) + (-11)$
$= -28 + (-11)$
$= -39$

26. $-1 - 3 - (-10) + (-7)$
$= -4 + 10 + (-7)$
$= 6 + (-7)$
$= -1$

27. $\frac{3}{4} - \left(\frac{1}{2} - \frac{9}{10}\right) = \frac{3}{4} - \left(\frac{5}{10} - \frac{9}{10}\right)$

$\qquad\qquad\qquad = \frac{3}{4} - \left(-\frac{4}{10}\right)$

$\qquad\qquad\qquad = \frac{3}{4} + \frac{4}{10}$

$\qquad\qquad\qquad = \frac{15}{20} + \frac{8}{20} = \frac{23}{20}$

28. $-|-12| - |-9| + (-4) - |10|$

$\qquad = -(12) - 9 + (-4) - 10$

$\qquad = -21 + (-4) - 10$

$\qquad = -25 - 10$

$\qquad = -35$

29. Represent the loss of \$24.4 million as -24.4. Subtract the \$13.1 million profit from the previous year.

$\qquad -24.4 - 13.1 = -(24.4 + 13.1) = -37.5$

Expressed as a positive amount, the difference is \$37.5 million.

30. $2(-5)(-3)(-3) = (-10)(-3)(-3)$

$\qquad\qquad\qquad\quad = (30)(-3)$

$\qquad\qquad\qquad\quad = -90$

31. $-\frac{3}{7}\left(-\frac{14}{9}\right) = \frac{3}{7} \cdot \frac{2 \cdot 7}{3 \cdot 3} = \frac{2}{3}$

32. $\frac{75}{-5} = 15 \cdot 5\left(\frac{1}{-5}\right) = -15$

33. $\frac{-2.3754}{-0.74} = 3.21$

34. $\frac{5}{7 - 7} = \frac{5}{0}$, which is undefined.

$\qquad \frac{7 - 7}{5} = \frac{0}{5} = 0$

35. $10^4 = 10 \cdot 10 \cdot 10 \cdot 10 = 10{,}000$

36. $\left(\frac{3}{7}\right)^3 = \frac{3}{7} \cdot \frac{3}{7} \cdot \frac{3}{7} = \frac{27}{343}$

37. $(-5)^3 = (-5)(-5)(-5) = -125$

38. $-5^3 = -(5 \cdot 5 \cdot 5) = -125$

39. $\sqrt{400} = 20$, because 20 is positive and $20^2 = 400$.

40. $\sqrt{\frac{64}{121}} = \frac{8}{11}$ since $\frac{8}{11}$ is positive and $\left(\frac{8}{11}\right)^2 = \frac{64}{121}$.

41. $-\sqrt{0.81} = -0.9$ since $-0.9^2 = -0.81$.

42. $\sqrt{-64}$ is not a real number.

43. $-14\left(\frac{3}{7}\right) + 6 \div 3 = -2(3) + (6 \div 3)$

$\qquad\qquad\qquad\qquad\quad = -6 + 2 = -4$

44. $-\frac{2}{3}[5(-2) + 8 - 4^3]$

$\qquad = -\frac{2}{3}[5(-2) + 8 - 64]$ *Evaluate the power.*

$\qquad = -\frac{2}{3}[-10 + 8 - 64]$

$\qquad = -\frac{2}{3}[-66] = \frac{2}{3}(22 \cdot 3) = 44$

45. $\dfrac{-5(3^2) + 9\left(\sqrt{4}\right) - 5}{6 - 5(-2)}$

$\qquad = \dfrac{-5(9) + 9(2) - 5}{6 + 10}$

$\qquad = \dfrac{-45 + 18 - 5}{16}$

$\qquad = \dfrac{-32}{16} = -2$

In Exercises 46–48, let $k = -4$, $m = 2$, and $n = 16$.

46. $4k - 7m = 4(-4) - 7(2)$

$\qquad\qquad\quad = -16 - 14$

$\qquad\qquad\quad = -30$

47. $-3\sqrt{n} + m + 5k$

$\qquad = -3\left(\sqrt{16}\right) + 2 + 5(-4)$

$\qquad = -3(4) + 2 - 20$

$\qquad = -12 + 2 - 20$

$\qquad = -10 - 20$

$\qquad = -30$

48. $\dfrac{4m^3 - 3n}{7k^2 - 10} = \dfrac{4(2)^3 - 3(16)}{7(-4)^2 - 10}$

$\qquad\qquad\quad = \dfrac{4(8) - 3(16)}{7(16) - 10}$

$\qquad\qquad\quad = \dfrac{32 - 48}{112 - 10}$

$\qquad\qquad\quad = \dfrac{-16}{102} = -\dfrac{8}{51}$

49. **(a)** 6 ft 1 in. $= (6 \cdot 12 + 1)$ in. $= 73$ in.

$704 \times$ (weight in pounds) $\div$ (height in inches)^2

$\qquad = 704 \times 190 \div 73^2$

$\qquad = 704 \times 190 \div 5329$

$\qquad = 133{,}760 \div 5329$

$\qquad \approx 25$

Carlos Beltran's BMI is 25.

(b) Answers will vary.

50. $2q + 19q$

$\qquad = (2 + 19)q$ *Distributive property*

$\qquad = 21q$

51. $13z - 17z$

$\qquad = (13 - 17)z$ *Distributive property*

$\qquad = -4z$

52. $-m + 6m$
$= -1m + 6m$ *Identity property*
$= (-1 + 6)m$ *Distributive property*
$= 5m$

53. $5p - p$
$= 5p + (-1)p$ *Identity property*
$= [5 + (-1)]p$ *Distributive property*
$= 4p$

54. $-2(k + 3)$
$= -2(k) + (-2)(3)$ *Distributive property*
$= -2k - 6$

55. $6(r + 3)$
$= 6(r) + 6(3)$ *Distributive property*
$= 6r + 18$

56. $9(2m + 3n)$
$= 9(2m) + 9(3n)$ *Distributive property*
$= 18m + 27n$

57. $-(-p + 6q) - (2p - 3q)$
$= -1(-p + 6q) + (-1)(2p - 3q)$
$= -1(-p) + (-1)(6q) + (-1)(2p) + (-1)(-3q)$
$= p - 6q - 2p + 3q$
$= p - 2p - 6q + 3q$
$= -p - 3q$

58. $-3y + 6 - 5 + 4y$
$= -3y + 4y + 6 - 5$
$= y + 1$

59. $2a + 3 - a - 1 - a - 2$
$= 2a - a - a + 3 - 1 - 2$
$= 0$

60. $-3(4m - 2) + 2(3m - 1) - 4(3m + 1)$
$= -12m + 6 + 6m - 2 - 12m - 4$
$= -12m + 6m - 12m + 6 - 2 - 4$
$= -18m$

61. $2x + 3x = (2 + 3)x = 5x$ *Distributive prop.*

62. $-4 \cdot 1 = -4$ *Identity property*

63. $2(4x) = (2 \cdot 4)x = 8x$ *Associative property*

64. $-3 + 13 = 13 + (-3) = 10$ *Commutative prop.*

65. $-3 + 3 = 0$ *Inverse property*

66. $5(x + z) = 5x + 5z$ *Distributive property*

67. $0 + 7 = 7$ *Identity property*

68. $8 \cdot \frac{1}{8} = 1$ *Inverse property*

69. **[1.2]** For 2003 (in millions of dollars):

$|169{,}924 - 221{,}595| = |-51{,}671| = 51{,}671$

The balance of trade (exports minus imports) is *negative* since the amount of money spent on imports is greater than the amount of money received for exports.

70. **[1.2]** For 2004 (in millions of dollars):

$|189{,}880 - 256{,}360| = |-66{,}480| = 66{,}480$

The balance of trade (exports minus imports) is *negative* since the amount of money spent on imports is greater than the amount of money received for exports.

71. **[1.2]** For 2005 (in millions of dollars):

$|211{,}349 - 287{,}870| = |-76{,}521| = 76{,}521$

The balance of trade (exports minus imports) is *negative* since the amount of money spent on imports is greater than the amount of money received for exports.

72. **[1.3]** $\left(-\frac{4}{5}\right)^4 = \left(-\frac{4}{5}\right)\left(-\frac{4}{5}\right)\left(-\frac{4}{5}\right)\left(-\frac{4}{5}\right)$
$= \frac{256}{625}$

73. **[1.2]** $-\frac{5}{8}(-40) = -\frac{5}{8} \cdot \frac{-40}{1}$
$= 5 \cdot 5 = 25$

74. **[1.3]** $-25\left(-\frac{4}{5}\right) + 3^3 - 32 \div \sqrt{4}$
$= -25\left(-\frac{4}{5}\right) + 27 - 32 \div 2$
$= 20 + 27 - 16$
$= 31$

75. **[1.2]** $-8 + |-14| + |-3| = -8 + 14 + 3$
$= 9$

76. **[1.3]** $\dfrac{6 \cdot \sqrt{4} - 3 \cdot \sqrt{16}}{-2 \cdot 5 + 7(-3) - 10}$

$= \dfrac{6 \cdot 2 - 3 \cdot 4}{-2 \cdot 5 + 7(-3) - 10}$

$= \dfrac{12 - 12}{-10 - 21 - 10}$

$= \dfrac{0}{-41} = 0$

77. **[1.3]** $-\sqrt{25} = -(5) = -5$

78. **[1.2]** $-\dfrac{10}{21} \div -\dfrac{5}{14} = -\dfrac{10}{21} \cdot -\dfrac{14}{5}$

$= \dfrac{2 \cdot 5}{3 \cdot 7} \cdot \dfrac{2 \cdot 7}{5}$

$= \dfrac{2 \cdot 2}{3} = \dfrac{4}{3}$

79. **[1.2]** $0.8 - 4.9 - 3.2 + 1.14$
$= -4.1 - 3.2 + 1.14$
$= -7.3 + 1.14$
$= -6.16$

80. **[1.3]** $-3^2 = -(3 \cdot 3) = -9$

81. **[1.2]** $\frac{-38}{-19} = 38\left(\frac{1}{19}\right) = 2 \cdot 19\left(\frac{1}{19}\right) = 2$

82. **[1.4]** $-2(k - 1) + 3k - k$
$= -2k + 2 + 3k - k$
$= -2k + 3k - k + 2$
$= (-2 + 3 - 1)k + 2$
$= 0k + 2 = 2$

83. **[1.3]** Since there is no real number whose square is -100, $-\sqrt{-100}$ is *not a real number*.

84. **[1.4]** $-(3k - 4h)$
$= -1(3k - 4h)$ *Identity property*
$= -1(3k) + (-1)(-4h)$
$= -3k + 4h$

85. **[1.2]** $-4.6(2.48) = -11.408$

86. **[1.3]** $-\frac{2}{3}(-15) + \left(2^4 - 8 \div 4\right)$
$= 10 + [16 - 8 \div 4]$
$= 10 + (16 - 2)$
$= 10 + 14 = 24$

87. **[1.4]** $-2x + 5 - 4x - 1$
$= -2x - 4x + 5 - 1$
$= -6x + 4$

88. **[1.2]** $-\frac{2}{3} - \left(\frac{1}{6} - \frac{5}{9}\right) = -\frac{2}{3} - \left(\frac{3}{18} - \frac{10}{18}\right)$
$= -\frac{2}{3} - \left(-\frac{7}{18}\right)$
$= -\frac{2}{3} + \frac{7}{18}$
$= -\frac{12}{18} + \frac{7}{18} = -\frac{5}{18}$

89. **[1.3]** **(a)** $-m\left(3k^2 + 5m\right) = -2\left[3(-4)^2 + 5(2)\right]$
    Let $k = -4$, $m = 2$.
$= -2[3(16) + 5(2)]$
$= -2[48 + 10]$
$= -2[58]$
$= -116$

    (b) $-m\left(3k^2 + 5m\right) = -\left(-\frac{3}{4}\right)\left[3\left(\frac{1}{2}\right)^2 + 5\left(-\frac{3}{4}\right)\right]$
    Let $k = \frac{1}{2}$, $m = -\frac{3}{4}$.
$= \frac{3}{4}\left[3\left(\frac{1}{4}\right) + 5\left(-\frac{3}{4}\right)\right]$
$= \frac{3}{4}\left[\frac{3}{4} - \frac{15}{4}\right]$
$= \frac{3}{4}\left(-\frac{12}{4}\right)$
$= \frac{3}{4}(-3) = -\frac{9}{4}$

90. **[1.3]** In order to evaluate $(3 + 2)^2$, you should work within the parentheses first.

Chapter 1 Test

1. $\left\{-3, 0.75, \frac{5}{3}, 5, 6.3\right\}$
Place dots at -3, 0.75, $\frac{5}{3} = 1.\overline{6}$, 5, and 6.3.

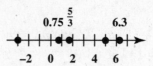

In Exercises 2–5,

$$A = \left\{-\sqrt{6}, -1, -0.5, 0, 3, \sqrt{25}, 7.5, \frac{24}{2}, \sqrt{-4}\right\}.$$

2. The elements 0, 3, $\sqrt{25}$ (or 5), and $\frac{24}{2}$ (or 12) are whole numbers.

3. The elements -1, 0, 3, $\sqrt{25}$ (or 5), and $\frac{24}{2}$ (or 12) are integers.

4. The elements -1, -0.5, 0, 3, $\sqrt{25}$ (or 5), 7.5, and $\frac{24}{2}$ (or 12) are rational numbers.

5. All the elements in the set are real numbers except $\sqrt{-4}$.

6. $\{x \mid x < -3\}$
In interval notation, $x < -3$ is written as $(-\infty, -3)$. The parenthesis at -3 indicates that -3 is not included.. The graph extends from -3 to the left.

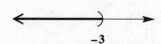

7. $\{y \mid -4 < y \le 2\}$
In interval notation, $-4 < y \le 2$ is written as $(-4, 2]$. The parenthesis indicates that -4 is not included, while the bracket indicates that 2 is included. The graph goes from -4 to 2.

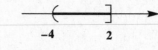

8. $-6 + 14 + (-11) - (-3)$
$= 8 + (-11) + 3$
$= -3 + 3 = 0$

9. $10 - 4 \cdot 3 + 6(-4)$
$= 10 - 12 + (-24)$
$= -2 + (-24) = -26$

10. $7 - 4^2 + 2(6) + (-4)^2$
$= 7 - 16 + 12 + 16$
$= 19$

11. $\dfrac{10 - 24 + (-6)}{\sqrt{16}(-5)}$

$= \dfrac{-14 + (-6)}{4(-5)}$

$= \dfrac{-20}{-20} = 1$

12. $\dfrac{-2[3 - (-1 - 2) + 2]}{\sqrt{9}(-3) - (-2)}$

$= \dfrac{-2[3 - (-3) + 2]}{3(-3) - (-2)}$

$= \dfrac{-2[8]}{-9 - (-2)}$

$= \dfrac{-16}{-7} = \dfrac{16}{7}$

13. $\dfrac{8 \cdot 4 - 3^2 \cdot 5 - 2(-1)}{-3 \cdot 2^3 + 1}$

$= \dfrac{8 \cdot 4 - 9 \cdot 5 - 2(-1)}{-3 \cdot 8 + 1}$

$= \dfrac{32 - 45 + 2}{-24 + 1}$

$= \dfrac{-11}{-23} = \dfrac{11}{23}$

14. $17{,}400 - (-32{,}995) = 17{,}400 + 32{,}995$
$\qquad\qquad\qquad\qquad = 50{,}395$

The difference between the height of Mt. Foraker and the depth of the Philippine Trench is 50,395 feet.

15. $14{,}110 - (-23{,}376) = 14{,}110 + 23{,}376$
$\qquad\qquad\qquad\qquad = 37{,}486$

The difference between the height of Pikes Peak and the depth of the Java Trench is 37,486 feet.

16. $-24{,}721 - (-23{,}376) = -24{,}721 + 23{,}376$
$\qquad\qquad\qquad\qquad\quad = -1345$

The Cayman Trench is 1345 feet deeper than the Java Trench.

17. $\sqrt{196} = 14$, because 14 is positive and $14^2 = 196$.

18. $-\sqrt{225} = -\left(\sqrt{225}\right) = -(15) = -15$

19. Since there is no real number whose square is -16, $\sqrt{-16}$ is not a real number.

20. **(a)** If a is positive, then $\sqrt{a}$ will represent a positive number.

(b) If a is negative, then $\sqrt{a}$ will not represent a real number.

(c) If a is 0, then $\sqrt{a}$ will be 0.

21. $\dfrac{8k + 2m^2}{r - 2} = \dfrac{8(-3) + 2(-3)^2}{25 - 2}$

$\qquad$ *Let $k = -3$, $m = -3$, and $r = 25$.*

$= \dfrac{8(-3) + 2(9)}{23}$

$= \dfrac{-24 + 18}{23}$

$= \dfrac{-6}{23}$ or $-\dfrac{6}{23}$

22. $-3(2k - 4) + 4(3k - 5) - 2 + 4k$
$= -3(2k) + (-3)(-4) + 4(3k)$
$\qquad + 4(-5) - 2 + 4k$
$= -6k + 12 + 12k - 20 - 2 + 4k$
$= -6k + 12k + 4k + 12 - 20 - 2$
$= 10k - 10$

23. When simplifying

$$(3r + 8) - (-4r + 6),$$

the subtraction sign in front of $(-4r + 6)$ changes the sign of the terms $-4r$ and 6.

$(3r + 8) - (-4r + 6)$
$= 3r + 8 - (-4r) - 6$
$= 3r + 8 + 4r - 6$
$= 3r + 4r + 8 - 6$
$= 7r + 2$

24. $6 + (-6) = 0$
The answer is **B**, *Inverse property*.
The sum of 6 and its inverse, -6, equals zero.

25. $-2 + (3 + 6) = (-2 + 3) + 6$
The answer is **D**, *Associative property*.
The order of the terms is the same, but the grouping has changed.

26. $5x + 15x = (5 + 15)x$
The answer is **A**, *Distributive property*.
This is the second form of the distributive property.

27. $13 \cdot 0 = 0$
The answer is **F**, *Multiplication property of* 0.
Multiplication by 0 always equal 0.

28. $-9 + 0 = -9$
The answer is **C**, *Identity property*.
The addition of 0 to any number does not change the number.

29. $4 \cdot 1 = 4$
The answer is **C**, *Identity property*.
Multiplication of any number by 1 does not change the number.

30. $(a + b) + c = (b + a) + c$
The answer is **E**, *Commutative property*.
The order of the terms a and b is reversed.

CHAPTER 2 LINEAR EQUATIONS, INEQUALITIES, AND APPLICATIONS

2.1 Linear Equations in One Variable

1. **A.** $3x + x - 1 = 0$ can be written as $4x = 1$, so it is linear.

    C. $6x + 2 = 9$ is in linear form.

3. $3(x + 4) = 5x$ *Original equation*

    $3(6 + 4) = 5 \cdot 6$? *Let x = 6.*

      $3(10) = 30$? *Add.*

        $30 = 30$ *True*

Since a true statement is obtained, 6 is a solution.

5. **(a)** $3x = 6$ is an *equation* because it contains an equals sign.

    (b) $3x + 6$ is an *expression* because it does not contain an equals sign.

    (c) $5x + 6(x - 3) = 12x + 6$ is an *equation* because it contains an equals sign.

    (d) $5x + 6(x - 3) - (12x + 6)$ is an *expression* because it does not contain an equals sign.

7. The solution contains a sign error when the distributive property was applied. The left side of the second line of the solution should be $8x - 4x + 6$. This gives us $4x + 6 = 3x + 7$ and then $x = 1$. Thus, the correct solution is 1.

In most of the following exercises, we do not show the checks of the solutions. To be sure that your solutions are correct, check them by substituting into the original equations.

9. A conditional equation has one solution, an identity has infinitely many solutions, and a contradiction has no solutions.

11. $7x + 8 = 1$

  $7x + 8 - 8 = 1 - 8$ *Subtract 8.*

      $7x = -7$

      $\dfrac{7x}{7} = \dfrac{-7}{7}$ *Divide by 7.*

        $x = -1$

We will use the following notation to indicate the value of each side of the original equation after we have substituted the proposed solution and simplified.

Check $x = -1$: $-7 + 8 = 1$ *True*

The solution set is $\{-1\}$.

13. $5x + 2 = 3x - 6$

 $5x + 2 - 3x = 3x - 6 - 3x$ *Subtract 3x.*

      $2x + 2 = -6$

  $2x + 2 - 2 = -6 - 2$ *Subtract 2.*

        $2x = -8$

       $\dfrac{2x}{2} = \dfrac{-8}{2}$ *Divide by 2.*

         $x = -4$

Check $x = -4$: $-20 + 2 = -12 - 6$ *True*

The solution set is $\{-4\}$.

15. $7x - 5x + 15 = x + 8$

    $2x + 15 = x + 8$ *Combine terms.*

       $2x = x - 7$ *Subtract 15.*

        $x = -7$ *Subtract x.*

The solution set is $\{-7\}$.

17. $12w + 15w - 9 + 5 = -3w + 5 - 9$

    $27w - 4 = -3w - 4$ *Combine terms.*

    $30w - 4 = -4$ *Add 3w.*

      $30w = 0$ *Add 4.*

        $w = 0$ *Divide by 30.*

The solution set is $\{0\}$.

19. $3(2t - 4) = 20 - 2t$

  $6t - 12 = 20 - 2t$ *Distributive property*

  $8t - 12 = 20$ *Add 2t.*

     $8t = 32$ *Add 12.*

      $t = 4$ *Divide by 8.*

The solution set is $\{4\}$.

21. $-5(x + 1) + 3x + 2 = 6x + 4$

 $-5x - 5 + 3x + 2 = 6x + 4$ *Distributive property*

    $-2x - 3 = 6x + 4$ *Combine terms.*

      $-3 = 8x + 4$ *Add 2x.*

      $-7 = 8x$ *Subtract 4.*

    $-\dfrac{7}{8} = x$ *Divide by 8.*

Check Substitute $-\frac{7}{8}$ for x and show that both sides equal -1.25. The screen shows a typical check on a calculator.

```
-7/8→X
              -.875
-5(X+1)•3X+2
              -1.25
6X+4
              -1.25
```

The solution set is $\left\{-\frac{7}{8}\right\}$.

23.
$$-2x + 5x - 9 = 3(x - 4) - 5$$
$$3x - 9 = 3x - 12 - 5$$
$$3x - 9 = 3x - 17$$
$$-9 = -17 \quad \textit{False}$$

The equation is a *contradiction*.

The solution set is $\emptyset$.

25. $2(x + 3) = -4(x + 1)$

$2x + 6 = -4x - 4$ *Remove parentheses.*

$6x + 6 = -4$ *Add 4x.*

$6x = -10$ *Subtract 6.*

$x = \frac{-10}{6} = -\frac{5}{3}$ *Divide by 6.*

The solution set is $\left\{-\frac{5}{3}\right\}$.

27. $3(2w + 1) - 2(w - 2) = 5$

$6w + 3 - 2w + 4 = 5$ *Remove parentheses.*

$4w + 7 = 5$ *Combine terms.*

$4w = -2$ *Subtract 7.*

$w = \frac{-2}{4} = -\frac{1}{2}$ *Divide by 4.*

The solution set is $\left\{-\frac{1}{2}\right\}$.

29. $2x + 3(x - 4) = 2(x - 3)$
$$2x + 3x - 12 = 2x - 6$$
$$5x - 12 = 2x - 6$$
$$3x = 6$$
$$x = \frac{6}{3} = 2$$

The solution set is $\{2\}$.

31. $6p - 4(3 - 2p) = 5(p - 4) - 10$
$$6p - 12 + 8p = 5p - 20 - 10$$
$$14p - 12 = 5p - 30$$
$$9p = -18$$
$$p = -2$$

The solution set is $\{-2\}$.

33. $-2(t + 3) - t - 4 = -3(t + 4) + 2$
$$-2t - 6 - t - 4 = -3t - 12 + 2$$
$$-3t - 10 = -3t - 10$$

The equation is an *identity*.

The solution set is {all real numbers}.

35. $2[w - (2w + 4) + 3] = 2(w + 1)$
$$2[w - 2w - 4 + 3] = 2(w + 1)$$
$$2[-w - 1] = 2(w + 1)$$
$$-w - 1 = w + 1 \quad \textit{Divide by 2.}$$
$$-1 = 2w + 1 \quad \textit{Add w.}$$
$$-2 = 2w \quad \textit{Subtract 1.}$$
$$-1 = w \quad \textit{Divide by 2.}$$

The solution set is $\{-1\}$.

37. $-[2z - (5z + 2)] = 2 + (2z + 7)$
$$-[2z - 5z - 2] = 2 + 2z + 7$$
$$-[-3z - 2] = 2 + 2z + 7$$
$$3z + 2 = 2z + 9$$
$$z = 7$$

The solution set is $\{7\}$.

39. $-3m + 6 - 5(m - 1) = -5m - (2m - 4) + 5$
$$-3m + 6 - 5m + 5 = -5m - 2m + 4 + 5$$
$$-8m + 11 = -7m + 9$$
$$-m = -2$$
$$m = 2$$

The solution set is $\{2\}$.

41. $7[2 - (3 + 4x)] - 2x = -9 + 2(1 - 15x)$
$$7[2 - 3 - 4x] - 2x = -9 + 2 - 30x$$
$$7[-1 - 4x] - 2x = -7 - 30x$$
$$-7 - 28x - 2x = -7 - 30x$$
$$-7 - 30x = -7 - 30x$$

The equation is an *identity*.

The solution set is {all real numbers}.

43. $-[3x - (2x + 5)] = -4 - [3(2x - 4) - 3x]$
$$-[3x - 2x - 5] = -4 - [6x - 12 - 3x]$$
$$-[x - 5] = -4 - [3x - 12]$$
$$-x + 5 = -4 - 3x + 12$$
$$-x + 5 = -3x + 8$$
$$2x = 3$$
$$x = \frac{3}{2}$$

The solution set is $\left\{\frac{3}{2}\right\}$.

45. The denominators of the fractions are 3, 4, and 1. The least common denominator is $(3)(4)(1) = 12$, since it is the smallest number into which each denominator can divide without a remainder.

47. **(a)** We need to make the coefficient of the first term on the left an integer. Since $0.05 = \frac{5}{100}$, we multiply by 10^2 or 100. This will also take care of the second term.

(b) We need to make 0.006, 0.007, and 0.009 integers. These numbers can be written as $\frac{6}{1000}$, $\frac{7}{1000}$, and $\frac{9}{1000}$. Multiplying by 10^3 or 1000 will eliminate the decimal points (the denominators) so that all the coefficients are integers.

49. $-\frac{5}{9}k = 2$

$-5k = 18$ *Multiply by 9.*

$k = \frac{18}{-5} = -\frac{18}{5}$ *Divide by -5.*

The solution set is $\left\{-\frac{18}{5}\right\}$.

51. $\dfrac{6}{5}x = -1$

$6x = -5 \qquad$ *Multiply by 5.*

$x = \dfrac{-5}{6} = -\dfrac{5}{6} \qquad$ *Divide by 6.*

The solution set is $\left\{-\dfrac{5}{6}\right\}$.

53. $\dfrac{m}{2} + \dfrac{m}{3} = 5$

$\qquad\qquad$ *Multiply by the LCD, 6.*

$6\left(\dfrac{m}{2} + \dfrac{m}{3}\right) = 6(5)$

$6\left(\dfrac{m}{2}\right) + 6\left(\dfrac{m}{3}\right) = 30 \qquad$ *Distributive property*

$3m + 2m = 30$

$5m = 30 \qquad$ *Add.*

$m = 6 \qquad$ *Divide by 5.*

Check $m = 6$: $\ 3 + 2 = 5 \quad$ *True*

The solution set is $\{6\}$.

55. $\dfrac{3x}{4} + \dfrac{5x}{2} = 13$

Multiply each side by the LCD, 4.

$4\left(\dfrac{3x}{4} + \dfrac{5x}{2}\right) = 4(13)$

$4\left(\dfrac{3x}{4}\right) + 4\left(\dfrac{5x}{2}\right) = 4(13) \qquad$ *Distributive property*

$3x + 10x = 52$

$13x = 52 \qquad$ *Combine terms.*

$x = 4 \qquad$ *Divide by 13.*

Check $x = 4$: $\ 3 + 10 = 13 \quad$ *True*

The solution set is $\{4\}$.

57. $\dfrac{x - 10}{5} + \dfrac{2}{5} = -\dfrac{x}{3}$

Multiply both sides by the LCD, 15.

$15\left(\dfrac{x - 10}{5} + \dfrac{2}{5}\right) = 15\left(-\dfrac{x}{3}\right)$

$3(x - 10) + 3(2) = -5x$

$3x - 30 + 6 = -5x$

$8x = 24$

$x = \dfrac{24}{8} = 3$

Check $x = 3$: $\ -\dfrac{7}{5} + \dfrac{2}{5} = -1 \quad$ *True*

The solution set is $\{3\}$.

59. $\dfrac{3x - 1}{4} + \dfrac{x + 3}{6} = 3$

Multiply each side by the LCD, 12.

$12\left(\dfrac{3x - 1}{4} + \dfrac{x + 3}{6}\right) = 12(3)$

$3(3x - 1) + 2(x + 3) = 36$

$9x - 3 + 2x + 6 = 36$

$11x + 3 = 36$

$11x = 33$

$x = 3$

Check $x = 3$: $\ 2 + 1 = 3 \quad$ *True*

The solution set is $\{3\}$.

61. $\dfrac{4t + 1}{3} = \dfrac{t + 5}{6} + \dfrac{t - 3}{6}$

Multiply both sides by the LCD, 6.

$6\left(\dfrac{4t + 1}{3}\right) = 6\left(\dfrac{t + 5}{6} + \dfrac{t - 3}{6}\right)$

$2(4t + 1) = (t + 5) + (t - 3)$

$8t + 2 = 2t + 2$

$6t = 0$

$t = 0$

Check $t = 0$: $\ \dfrac{1}{3} = \dfrac{5}{6} - \dfrac{3}{6} \quad$ *True*

The solution set is $\{0\}$.

63. $0.05x + 0.12(x + 5000) = 940$

$\qquad$ *Multiply both sides by 100.*

$5x + 12(x + 5000) = 100(940)$

$5x + 12x + 60{,}000 = 94{,}000$

$17x = 34{,}000$

$x = 2000$

Check $x = 2000$: $\ 100 + 840 = 940 \quad$ *True*

The solution set is $\{2000\}$.

65. $0.02(50) + 0.08r = 0.04(50 + r)$

$\qquad$ *Multiply both sides by 100.*

$2(50) + 8r = 4(50 + r)$

$100 + 8r = 200 + 4r$

$4r = 100$

$r = 25$

Check $r = 25$: $\ 1 + 2 = 3 \quad$ *True*

The solution set is $\{25\}$.

67. $0.05x + 0.10(200 - x) = 0.45x$

$\qquad$ *Multiply both sides by 100.*

$5x + 10(200 - x) = 45x$

$5x + 2000 - 10x = 45x$

$2000 - 5x = 45x$

$2000 = 50x$

$40 = x$

Check $x = 40$: $\ 2 + 16 = 18 \quad$ *True*

The solution set is $\{40\}$.

69. $0.006(x + 2) = 0.007x + 0.009$

$\qquad$ *Multiply each side by 1000.*

$6(x + 2) = 7x + 9$

$6x + 12 = 7x + 9$

$3 = x$

Check $x = 3$: $\ 0.03 = 0.021 + 0.009 \quad$ *True*

The solution set is $\{3\}$.

71. $2L + 2W$; $L = 10, W = 8$

$$2L + 2W = 2(10) + 2(8)$$
$$= 20 + 16 = 36$$

73. $\frac{1}{3}Bh$; $B = 27, h = 8$

$$\frac{1}{3}Bh = \frac{1}{3}(27)(8)$$
$$= 9(8) = 72$$

75. $\frac{5}{9}(F - 32)$; $F = 122$

$$\frac{5}{9}(F - 32) = \frac{5}{9}(122 - 32)$$
$$= \frac{5}{9}(90) = 50$$

77. $\frac{1}{2}h(b + B)$; $B = 9, b = 4, h = 3$

$$\frac{1}{2}h(b + B) = \frac{1}{2}(3)(4 + 9)$$
$$= \frac{3}{2}(13) = \frac{39}{2}, \text{ or } 19\frac{1}{2}$$

2.2 Formulas

1. (a) $x = \dfrac{5x + 8}{3}$

$$3x = 3\left(\frac{5x + 8}{3}\right)$$
$$3x = 5x + 8$$

(b) $t = \dfrac{bt + k}{c}$ $(c \neq 0)$

$$ct = c\left(\frac{bt + k}{c}\right)$$
$$ct = bt + k$$

2. (a) $3x - 5x = 5x + 8 - 5x$

$$3x - 5x = 8$$

(b) $ct - bt = bt + k - bt$

$$ct - bt = k$$

3. Use the distributive property in each case.

(a) $-2x = 8$ **(b)** $(c - b)t = k$

4. (a) $\dfrac{-2x}{-2} = \dfrac{8}{-2}$ **(b)** $\dfrac{(c - b)t}{c - b} = \dfrac{k}{c - b}$

$$x = -4 \qquad\qquad\qquad t = \frac{k}{c - b}$$

5. The restriction $b \neq c$ must be applied. If $b = c$, the denominator becomes 0 and division by 0 is undefined.

6. To solve an equation for a particular variable, such as solving the second equation for t, go through the same steps as you would in solving for x in the first equation. Treat all other variables as constants.

7. Solve $I = prt$ for r.

$$I = prt$$
$$\frac{I}{pt} = \frac{prt}{pt}$$
$$\frac{I}{pt} = r, \text{ or } r = \frac{I}{pt}$$

9. Solve $P = 2L + 2W$ for L.

$$P = 2L + 2W$$
$$P - 2W = 2L$$
$$\frac{P - 2W}{2} = \frac{2L}{2}$$
$$\frac{P - 2W}{2} = L, \text{ or } L = \frac{P}{2} - W$$

11. (a) Solve for $V = LWH$ for W.

$$V = LWH$$
$$\frac{V}{LH} = \frac{LWH}{LH}$$
$$\frac{V}{LH} = W, \text{ or } W = \frac{V}{LH}$$

(b) Solve for $V = LWH$ for H.

$$V = LWH$$
$$\frac{V}{LW} = \frac{LWH}{LW}$$
$$\frac{V}{LW} = H, \text{ or } H = \frac{V}{LW}$$

13. Solve $C = 2\pi r$ for r.

$$C = 2\pi r$$
$$\frac{C}{2\pi} = \frac{2\pi r}{2\pi} \quad \textit{Divide by } 2\pi.$$
$$\frac{C}{2\pi} = r$$

15. (a) Solve $A = \frac{1}{2}h(b + B)$ for h.

$$2A = h(b + B) \quad \textit{Multiply by 2.}$$
$$\frac{2A}{b + B} = h \qquad \textit{Divide by } b + B.$$

(b) Solve $A = \frac{1}{2}h(b + B)$ for B.

$$2A = h(b + B) \text{ or } \qquad 2A = hb + hB$$
$$\frac{2A}{h} = b + B \qquad\quad 2A - hb = hB$$
$$\frac{2A}{h} - b = B, \qquad \text{or } \frac{2A - hb}{h} = B$$

17. Solve $F = \frac{9}{5}C + 32$ for C.

$$F - 32 = \frac{9}{5}C$$
$$\frac{5}{9}(F - 32) = \frac{5}{9}\left(\frac{9}{5}C\right)$$
$$\frac{5}{9}(F - 32) = C$$

19.
$$A = \tfrac{1}{2}bh$$

$2A = 2\left(\tfrac{1}{2}bh\right)$ *Multiply by 2.*

$2A = bh$

$\dfrac{2A}{b} = \dfrac{bh}{b}$ *Divide by b.*

$\dfrac{2A}{b} = h$

$\dfrac{2A}{b} = \dfrac{2}{1} \cdot \dfrac{A}{b}$

$\phantom{\dfrac{2A}{b}} = 2\left(\dfrac{A}{b}\right)$ *This choice is* **A**.

$\phantom{\dfrac{2A}{b}} = 2A\left(\dfrac{1}{b}\right)$ *This is choice* **B**.

To get choice **C**, divide $A = \tfrac{1}{2}bh$ by $\tfrac{1}{2}b$.

$$\dfrac{A}{\tfrac{1}{2}b} = \dfrac{\tfrac{1}{2}bh}{\tfrac{1}{2}b} \quad \text{gives us} \quad h = \dfrac{A}{\tfrac{1}{2}b}.$$

Choice **D**, $h = \dfrac{\tfrac{1}{2}A}{b}$, can be multiplied by $\dfrac{2}{2}$ on the right side to get $h = \dfrac{A}{2b}$, so it is *not* equivalent to $h = \dfrac{2A}{b}$. Therefore, the correct answer is **D**.

21. Solve $2k + ar = r - 3y$ for r.
Get the "r-terms" on one side and the other terms on the other side.

$ar - r = -2k - 3y$

$(a - 1)r = -2k - 3y$ *Distributive property*

$r = \dfrac{-2k - 3y}{a - 1}$ *Divide by a − 1.*

The answer can also be written as

$$r = \dfrac{2k + 3y}{1 - a},$$

which would occur if you took the r-terms to the right side in your first step.

23. Solve $w = \dfrac{3y - x}{y}$ for y.

$wy = 3y - x$ *Multiply by y.*

$x = 3y - wy$ *Get y-terms on one side.*

$x = (3 - w)y$ *Distributive property*

$\dfrac{x}{3 - w} = y$ *Divide by 3 − w.*

Equivalently, we have

$$y = \dfrac{-x}{w - 3}.$$

25. Solve $d = rt$ for t.

$$t = \dfrac{d}{r}$$

To find t, substitute $d = 500$ and $r = 135.173$.

$$t = \dfrac{500}{135.173} \approx 3.699$$

His time was about 3.699 hours.

27. Solve $d = rt$ for r.

$r = \dfrac{d}{t}$

$r = \dfrac{520}{10} = 52$ *Let d = 520, t = 10.*

Her rate was 52 mph.

29. Use the formula $F = \tfrac{9}{5}C + 32$.

$F = \tfrac{9}{5}(40) + 32$ *Let C = 40.*

$ = 72 + 32$

$ = 104$

The corresponding temperature is 104°F.

31. Solve $P = 4s$ for s.

$$s = \dfrac{P}{4}$$

To find s, substitute 920 for P.

$$s = \dfrac{920}{4} = 230$$

The length of each side is 230 m.

33. Use the formula $C = 2\pi r$.

$480\pi = 2\pi r$ *Let C = 480π.*

$\dfrac{480\pi}{2\pi} = r$ *Divide by 2π.*

So the radius of the circle is 240 inches and the diameter is twice that length, that is, 480 inches.

35. Use $V = LWH$.
Let $V = 187$, $L = 11$, and $W = 8.5$.

$187 = 11(8.5)H$

$187 = 93.5H$

$2 = H$ *Divide by 93.5.*

The ream is 2 inches thick.

37. The mixture is 36 oz and that part which is alcohol is 9 oz. Thus, the percent of alcohol is

$$\dfrac{9}{36} = \dfrac{1}{4} = \dfrac{25}{100} = 25\%.$$

The percent of water is

$$100\% - 25\% = 75\%.$$

39. Find what percent $6300 is of $210,000.

$$\frac{6300}{210,000} = 0.03 = 3\%$$

The agent received a 3% rate of commission.

In Exercises 41–44, use the rule of 78.

$$u = f \cdot \frac{k(k+1)}{n(n+1)}$$

41. Substitute 700 for f, 4 for k, and 36 for n.

$$u = 700 \cdot \frac{4(4+1)}{36(36+1)}$$

$$= 700 \cdot \frac{4(5)}{36(37)} \approx 10.51$$

The unearned interest is $10.51.

43. Substitute 380.50 for f, 8 for k, and 24 for n.

$$u = (380.50) \cdot \frac{8(8+1)}{24(24+1)}$$

$$= (380.50) \cdot \frac{8(9)}{24(25)} \approx 45.66$$

The unearned interest is $45.66.

45. **(a)** Detroit:

$$\text{Pct.} = \frac{W}{W+L} = \frac{19}{19+9} = \frac{19}{28} \approx .679$$

(b) Cleveland:

$$\text{Pct.} = \frac{W}{W+L} = \frac{14}{14+13} = \frac{14}{27} \approx .519$$

(c) Minnesota:

$$\text{Pct.} = \frac{W}{W+L} = \frac{11}{11+16} = \frac{11}{27} \approx .407$$

(d) Kansas City:

$$\text{Pct.} = \frac{W}{W+L} = \frac{5}{5+20} = \frac{5}{25} = .200$$

47. $0.34(242,070) = 82,303.80$

To the nearest dollar, $82,304 will be spent to provide housing.

49. Since 14% is twice as much as 7%, the cost for transportation will be $2(\$16,945) = \$33,890$.

51. In 1982, 3% of the television audience watched basic cable.

$$3\% \text{ of } 50,000$$
$$= 0.03(50,000)$$
$$= 1500$$

53. In 2004, 54% of the television audience watched cable.

$$54\% \text{ of } 35,000$$
$$= 0.54(35,000)$$
$$= 18,900$$

55. $\quad 4x + 4(x + 7) = 124$
$\quad\quad 4x + 4x + 28 = 124$
$\quad\quad\quad\quad 8x + 28 = 124$
$\quad\quad\quad\quad\quad\quad 8x = 96$
$\quad\quad\quad\quad\quad\quad x = \frac{96}{8} = 12$

The solution set is $\{12\}$.

57. $\quad \frac{5}{3} + \frac{2}{3}x = 2$
$\quad\quad 5 + 2x = 6 \quad$ *Multiply by 3.*
$\quad\quad\quad\quad 2x = 1$
$\quad\quad\quad\quad x = \frac{1}{2}$

The solution set is $\left\{\frac{1}{2}\right\}$.

59. $\quad 2.4 + 0.4x = 0.25(6 + x)$
$\quad 240 + 40x = 25(6 + x)$
$\quad\quad\quad\quad$ *Multiply by 100.*
$\quad 240 + 40x = 150 + 25x$
$\quad\quad\quad\quad 15x = -90$
$\quad\quad\quad\quad x = -6$

The solution set is $\{-6\}$.

61. "The product of -3 and 5, divided by 1 less than 6" is translated and evaluated as follows:

$$\frac{-3(5)}{6-1} = \frac{-15}{5} = -3$$

63. "The sum of 6 and -9, multiplied by the additive inverse of 2" is translated and evaluated as follows:

$$[6 + (-9)](-2) = -3(-2) = 6$$

2.3 Applications of Linear Equations

1. **(a)** 12 more than a number $\underline{x + 12}$

(b) 12 is more than a number. $\underline{12 > x}$

3. **(a)** 4 less than a number $\underline{x - 4}$

(b) 4 is less than a number. $\underline{4 < x}$

5. 20% can be written as
$0.20 = 0.2 = \frac{20}{100} = \frac{2}{10} = \frac{1}{5}$, so "20% of a number" can be written as $0.20x$, $0.2x$, or $\frac{x}{5}$. We see that "20% of a number" cannot be written as $20x$, choice **D**.

7. Twice a number, decreased by 13 $\underline{2x - 13}$

9. 12 increased by three times a number $\underline{12 + 3x}$

11. The product of 8 and 12 less than a number
$\underline{8(x - 12)}$

13. The quotient of three times a number and 7 $\frac{3x}{7}$

15. The sentence "the sum of a number and 6 is -31" can be translated as

$$x + 6 = -31.$$
$$x = -37 \qquad \textit{Subtract 6.}$$

The number is -37.

17. The sentence "if the product of a number and -4 is subtracted from the number, the result is 9 more than the number" can be translated as

$$x - (-4x) = x + 9.$$
$$x + 4x = x + 9$$
$$4x = 9$$
$$x = \tfrac{9}{4}$$

The number is $\frac{9}{4}$.

19. The sentence "when $\frac{2}{3}$ of a number is subtracted from 12, the result is 10" can be translated as

$$12 - \tfrac{2}{3}x = 10.$$
$$36 - 2x = 30 \qquad \textit{Multiply by 3.}$$
$$-2x = -6 \qquad \textit{Subtract 36.}$$
$$x = 3 \qquad \textit{Divide by } -2.$$

The number is 3.

21. $5(x + 3) - 8(2x - 6)$ is an *expression* because there is no equals sign.

23. $5(x + 3) - 8(2x - 6) = 12$ has an equals sign, so this represents an *equation*.

25. $\dfrac{r}{2} - \dfrac{r + 9}{6} - 8$ is an *expression* because there is no equals sign.

27. *Step 1*
We are asked to find the number of patents each university secured .

Step 2
Let $x =$ the number of patents MIT secured. Then $x - 38 =$ the number of patents Stanford secured .

Step 3
A total of 230 patents were secured, so

$$\underline{x} + \underline{x - 38} = 230.$$

Step 4 $2x - 38 = 230$
$$2x = 268$$
$$x = \underline{134}$$

Step 5
MIT secured 134 patents and Stanford secured $134 - 38 = \underline{96}$ patents.

Step 6
The number of Stanford patents was 38 fewer than the number of MIT patents and the total number of patents was $134 + \underline{96} = \underline{230}$.

29. *Step 2*
Let $W =$ the width of the base. Then $2W - 65$ is the length of the base.

Step 3
The perimeter of the base is 860 feet. Using $P = 2L + 2W$ gives us

$$2(2W - 65) + 2W = 860.$$

Step 4
$$4W - 130 + 2W = 860$$
$$6W - 130 = 860$$
$$6W = 990$$
$$W = \tfrac{990}{6} = 165$$

Step 5
The width of the base is 165 feet and the length of the base is $2(165) - 65 = 265$ feet.

Step 6
$2L + 2W = 2(265) + 2(165) = 530 + 330 = 860$, which is the perimeter of the base.

31. *Step 2*
Let $x =$ the width of the painting. Then $x + 5.54 =$ the height of the painting.

Step 3
The perimeter of the painting is 108.44 inches. Using $P = 2L + 2W$ gives us

$$2(x + 5.54) + 2x = 108.44.$$

Step 4
$$2x + 11.08 + 2x = 108.44$$
$$4x + 11.08 = 108.44$$
$$4x = 97.36$$
$$x = 24.34$$

Step 5
The width of the painting is 24.34 inches and the height is $24.34 + 5.54 = 29.88$ inches.

Step 6
29.88 is 5.54 more than 24.34 and $2(29.88) + 2(24.34) = 108.44$, as required.

33. *Step 2*

Let x = the length of the middle side. Then the shortest side is $x - 75$ and the longest side is $x + 375$.

Step 3

The perimeter of the Bermuda Triangle is 3075 miles. Using $P = a + b + c$ gives us

$$x + (x - 75) + (x + 375) = 3075.$$

Step 4

$$3x + 300 = 3075$$
$$3x = 2775 \quad \textit{Subtract 300.}$$
$$x = 925 \quad \textit{Divide by 3.}$$

Step 5

The length of the middle side is 925 miles. The length of the shortest side is $x - 75 =$ $925 - 75 = 850$ miles. The length of the longest side is $x + 375 = 925 + 375 = 1300$ miles.

Step 6

The answer checks since $925 + 850 + 1300 =$ 3075 miles, which is the correct perimeter.

35. *Step 2*

Let x = the amount of revenue for Exxon Mobil. Then $x - 24$ is the amount of revenue for Wal-Mart (in billions).

Step 3

The total revenue was \$656 billion, so

$$x + (x - 24) = 656.$$

Step 4

$$2x - 24 = 656$$
$$2x = 680 \quad \textit{Add 24.}$$
$$x = 340 \quad \textit{Divide by 2.}$$

Step 5

The amount of revenue for Exxon Mobil was \$340 billion. The amount of revenue for Wal-Mart was $x - 24 = 340 - 24 = \$316$ billion.

Step 6

The answer checks since $340 + 316 = \$656$ billion, which is the correct total revenue.

37. *Step 2*

Let x = the height of the Eiffel Tower. Then $x - 804$ = the height of the Leaning Tower of Pisa.

Step 3

Together these heights are 1164 ft, so

$$x + (x - 804) = 1164.$$

Step 4 $2x - 804 = 1164$
$$2x = 1968$$
$$x = 984$$

Step 5

The height of the Eiffel Tower is 984 feet and the height of the Leaning Tower of Pisa is $984 - 804 = 180$ feet.

Step 6

180 feet is 804 feet shorter than 984 feet and the sum of 180 feet and 984 feet is 1164 feet.

39. *Step 2*

Let x = votes for Kerry. Then $x + 35$ = votes for Bush.

Step 3

There were 537 total electoral votes, so

$$x + (x + 35) = 537.$$

Step 4

$$2x + 35 = 537$$
$$2x = 502$$
$$x = 251$$

Step 5

Kerry received 251 votes, so Bush received $251 + 35 = 286$ votes.

Step 6

286 is 35 more than 251 and the total is $251 + 286 = 537$.

41. Let x = the percent increase.

$$x = \frac{\text{amount of increase}}{\text{base number}}$$
$$= \frac{1{,}186{,}251 - 817{,}000}{817{,}000} = \frac{369{,}251}{817{,}000}$$
$$\approx 0.452 = 45.2\%$$

The percent increase was approximately 45.2%.

43. Let x = the approximate cost in 2005.
Since x is 95% more than the 1995 cost,

$$x = 2811 + 0.95(2811)$$
$$= 2811 + 2670.45$$
$$= 5481.45.$$

To the nearest dollar, the cost was \$5481.

45. Let x = the 2004 cost. Then

$$x + 3.1\%(x) = 36.78.$$
$$x + 3.1(0.01)(x) = 36.78$$
$$1x + 0.031x = 36.78$$
$$1.031x = 36.78$$
$$x = \frac{36.78}{1.031} \approx 35.67$$

The 2004 cost was \$35.67.

47. Let x = the amount of the receipts excluding tax. Since the sales tax is 9% of x, the total amount is

$$x + 0.09x = 2725$$
$$1x + 0.09x = 2725$$
$$1.09x = 2725$$
$$x = \frac{2725}{1.09} = 2500$$

Thus, the tax was $0.09(2500) = \$225$.

49. Let x = the amount invested at 3%. Then $12{,}000 - x$ = the amount invested at 4%. Complete the table. Use $I = prt$ with $t = 1$.

Principal	Rate (as a decimal)	Interest
x	0.03	$0.03x$
$12{,}000 - x$	0.04	$0.04(12{,}000 - x)$
12,000	← Totals →	440

The last column gives the equation.

$$\begin{array}{ccc} \text{Interest} & \text{interest} & \text{total} \\ \text{at 3\%} & + \quad \text{at 4\%} & = \quad \text{interest.} \end{array}$$
$$0.03x + 0.04(12{,}000 - x) = 440$$

$$3x + 4(12{,}000 - x) = 44{,}000 \quad \textit{Multiply by 100.}$$
$$3x + 48{,}000 - 4x = 44{,}000$$
$$-x = -4000$$
$$x = 4000$$

He should invest $4000 at 3% and $12{,}000 - 4000 = \$8000$ at 4%.

Check \$4000 @ 3% = \$120 and
\$8000 @ 4% = \$320; \$120 + \$320 = \$440.

51. Let x = the amount invested at 5%. Then $2x - 400$ = the amount invested at 6.5%. Use $I = prt$ with $t = 1$. Make a table.

Principal	Rate (as a decimal)	Interest
x	0.05	$0.05x$
$2x - 400$	0.065	$0.065(2x - 400)$
	Total →	298

The last column gives the equation.

$$\begin{array}{ccc} \text{Interest} & \text{interest} & \text{total} \\ \text{at 5\%} & + \quad \text{at 6.5\%} & = \quad \text{interest.} \end{array}$$
$$0.05x + 0.065(2x - 400) = 298$$

$$50x + 65(2x - 400) = 298{,}000 \quad \textit{Multiply by 1000.}$$
$$50x + 130x - 26{,}000 = 298{,}000$$
$$180x - 26{,}000 = 298{,}000$$
$$180x = 324{,}000$$
$$x = \frac{324{,}000}{180} = 1800$$

She invested \$1800 at 5% and
$2x - 400 = 2(1800) - 400 = \$3200$ at 6.5%.

Check \$1800 @ 5% = \$90 and
\$3200 @ 6.5% = \$208; \$90 + \$208 = \$298.

53. Let x = the amount of additional money to be invested at 4%.
Use $I = prt$ with $t = 1$. Make a table.
Use the fact that the total return on the two investments is 6%.

Principal	Rate (as a decimal)	Interest
27,000	0.07	$0.07(27{,}000)$
x	0.04	$0.04x$
$27{,}000 + x$	0.06	$0.06(27{,}000 + x)$

The last column gives the equation.

$$\begin{array}{ccc} \text{Interest} & \text{interest} & \text{interest} \\ \text{at 7\%} & + \quad \text{at 4\%} & = \quad \text{at 6\%} . \end{array}$$
$$0.07(27{,}000) + 0.04x = 0.06(27{,}000 + x)$$

$$7(27{,}000) + 4x = 6(27{,}000 + x) \quad \textit{Multiply by 100.}$$
$$189{,}000 + 4x = 162{,}000 + 6x$$
$$27{,}000 = 2x$$
$$13{,}500 = x$$

They should invest \$13,500 at 4%.

Check \$27,000 @ 7% = \$1890 and
\$13,500 @ 4% = \$540;
\$1890 + \$540 = \$2430, which is the same as
(\$27,000 + \$13,500) @ 6%.

55. Let x = the number of liters of 10% acid solution needed. Make a table.

Liters of Solution	Percent (as a decimal)	Liters of Pure Acid
10	0.04	$0.04(10) = 0.4$
x	0.10	$0.10x$
$x + 10$	0.06	$0.06(x + 10)$

$$\begin{array}{ccc} \text{Acid in 4\%} & \text{acid in 10\%} & \text{acid in 6\%} \\ \text{solution} & + \quad \text{solution} & = \quad \text{solution.} \end{array}$$
$$0.4 + 0.10x = 0.06(x + 10)$$

$$0.4 + 0.10x = 0.06x + 0.6$$
$$\textit{Distributive property}$$
$$0.04x = 0.2$$
$$\textit{Subtract 0.06x and 0.4.}$$
$$x = 5 \quad \textit{Divide by 0.04.}$$

Five liters of the 10% solution are needed.

Check 4% of 10 is 0.4 and 10% of 5 is 0.5;
$0.4 + 0.5 = 0.9$, which is the same as 6% of
$(10 + 5)$.

57. Let $x = $ the number of liters of the 20% alcohol solution. Make a table.

Liters of Solution	Percent (as a decimal)	Liters of Pure Alcohol
12	0.12	$0.12(12) = 1.44$
x	0.20	$0.20x$
$x + 12$	0.14	$0.14(x + 12)$

$$\begin{array}{ccccc} \text{Alcohol in} & & \text{alcohol in} & = & \text{alcohol in} \\ \text{12\% solution} & + & \text{20\% solution} & & \text{14\% solution.} \\ 1.44 & + & 0.20x & = & 0.14(x + 12) \end{array}$$

$$\begin{array}{ll} 144 + 20x = 14(x + 12) & \textit{Multiply by 100.} \\ 144 + 20x = 14x + 168 & \textit{Distributive property} \\ 6x = 24 & \textit{Subtract 14x and 144.} \\ x = 4 & \textit{Divide by 6.} \end{array}$$

4L of 20% alcohol solution are needed.

Check 12% of 12 is 1.44 and 20% of 4 is 0.8; $1.44 + 0.8 = 2.24$, which is the same as 14% of $(12 + 4)$.

59. Let $x = $ the amount of pure dye used (pure dye is 100% dye). Make a table.

Gallons of Solution	Percent (as a decimal)	Gallons of Pure Dye
x	1	$1x = x$
4	0.25	$0.25(4) = 1$
$x + 4$	0.40	$0.40(x + 4)$

Write the equation from the last column in the table.

$$\begin{array}{ll} x + 1 = 0.4(x + 4) & \\ x + 1 = 0.4x + 1.6 & \textit{Distributive property} \\ 0.6x = 0.6 & \textit{Subtract 0.4x and 1.} \\ x = 1 & \textit{Divide by 0.6.} \end{array}$$

One gallon of pure (100%) dye is needed.

Check 100% of 1 is 1 and 25% of 4 is 1; $1 + 1 = 2$, which is the same as 40% of $(1 + 4)$.

61. Let $x = $ the amount of $6 per lb nuts. Make a table.

Pounds of nuts	Cost per lb	Total Cost
50	$2	$2(50) = 100$
x	$6	$6x$
$x + 50$	$5	$5(x + 50)$

The total value of the $2 per lb nuts and the $6 per lb nuts must equal the value of the $5 per lb nuts.

$$\begin{array}{l} 100 + 6x = 5(x + 50) \\ 100 + 6x = 5x + 250 \\ x = 150 \end{array}$$

He should use 150 lb of $6 nuts.

Check 50 pounds of the $2 per lb nuts are worth $100 and 150 pounds of the $6 per lb nuts are worth $900; $100 + $900 = 1000, which is the same as $(50 + 150)$ pounds worth $5 per lb.

63. We cannot expect the final mixture to be worth more than each of the ingredients. Answers will vary.

65. **(a)** Let $\quad x = $ the amount invested at 5%.
$800 - x = $ the amount invested at 10%.

(b) Let $\quad y = $ the amount of 5% acid used.
$800 - y = $ the amount of 10% acid used.

66. Organize the information in a table.

(a)

Principal	Percent (as a decimal)	Interest
x	0.05	$0.05x$
$800 - x$	0.10	$0.10(800 - x)$
800	0.0875	$0.0875(800)$

The amount of interest earned at 5% and 10% is found in the last column of the table, $0.05x$ and $0.10(800 - x)$.

(b)

Liters of Solution	Percent (as a decimal)	Liters of Pure Acid
y	0.05	$0.05y$
$800 - y$	0.10	$0.10(800 - y)$
800	0.0875	$0.0875(800)$

The amount of pure acid in the 5% and 10% mixtures is found in the last column of the table, $0.05y$ and $0.10(800 - y)$.

67. Refer to the tables for Exercise 66. In each case, the last column gives the equation.

(a) $0.05x + 0.10(800 - x) = 0.0875(800)$

(b) $0.05y + 0.10(800 - y) = 0.0875(800)$

68. In both cases, multiply by 10,000 to clear the decimals.

(a) $$\begin{array}{l} 0.05x + 0.10(800 - x) = 0.0875(800) \\ 500x + 1000(800 - x) = 875(800) \\ 500x + 800,000 - 1000x = 700,000 \\ -500x = -100,000 \\ x = 200 \end{array}$$

Jack invested $200 at 5% and $800 - x = 800 - 200 = 600 at 10%.

(b)
$$0.05y + 0.10(800 - y) = 0.0875(800)$$
$$500y + 1000(800 - y) = 875(800)$$
$$500y + 800{,}000 - 1000y = 700{,}000$$
$$-500y = -100{,}000$$
$$y = 200$$

Jill used 200 L of 5% acid solution and $800 - y = 800 - 200 = 600$ L of 10% acid solution.

69. The processes used to solve Problems A and B were virtually the same. Aside from the variables chosen, the problem information was organized in similar tables and the equations solved were the same.

71. $P = 2L + 2W$; $L = 10, W = 6$
$$P = 2(10) + 2(6) = 20 + 12 = 32$$

73. $P = a + b + c$; $b = 13, c = 14, P = 46$
$$46 = a + 13 + 14$$
$$46 = a + 27$$
$$19 = a$$

75. $d = rt$; $d = 75, t = 15$
$$75 = r(15)$$
$$r = \frac{75}{15} = 5$$

2.4 Further Applications of Linear Equations

1. The total amount is
$$38(0.05) + 26(0.10) = 1.90 + 2.60$$
$$= \$4.50.$$

3. Use $d = rt$, or $r = \dfrac{d}{t}$. Substitute 300 for d and 5 for t.
$$r = \frac{300}{5} = 60$$

His rate was 60 mph.

5. The problem asks for the distance Jeff travels to the workplace, so we must multiply the rate, 10 mph, by the time, $\frac{3}{4}$ hr, to get the distance, 7.5 mi.

7. No, the answers must be whole numbers because they represent the number of coins.

9. Let x = the number of pennies. Then x is also the number of dimes, and $44 - 2x$ is the number of quarters.

Number of Coins	Denomination	Value
x	0.01	$0.01x$
x	0.10	$0.10x$
$44 - 2x$	0.25	$0.25(44 - 2x)$
44	← Totals →	4.37

The sum of the values must equal the total value.

$$0.01x + 0.10x + 0.25(44 - 2x) = 4.37$$
$$x + 10x + 25(44 - 2x) = 437$$
Multiply by 100.
$$x + 10x + 1100 - 50x = 437$$
$$-39x + 1100 = 437$$
$$-39x = -663$$
$$x = 17$$

There are 17 pennies, 17 dimes, and $44 - 2(17) = 10$ quarters.

Check The number of coins is $17 + 17 + 10 = 44$ and the value of the coins is $\$.01(17) + \$.10(17) + \$.25(10) = \$4.37$, as required.

11. Let x = the number of loonies. Then $37 - x$ is the number of toonies.

Number of Coins	Denomination	Value
x	1	$1x$
$37 - x$	2	$2(37 - x)$
37	← Totals →	51

The sum of the values must equal the total value.

$$1x + 2(37 - x) = 51$$
$$x + 74 - 2x = 51$$
$$-x + 74 = 51$$
$$23 = x$$

She has 23 loonies and $37 - 23 = 14$ toonies.

Check The value of the coins is $\$1(23) + \$2(14) = \$51$, as required.

13. Let x = the number of \$10 coins. Then $41 - x$ is the number of \$20 coins.

Number of Coins	Denomination	Value
x	10	$10x$
$41 - x$	20	$20(41 - x)$
41	← Totals →	540

The sum of the values must equal the total value.

$$10x + 20(41 - x) = 540$$
$$10x + 820 - 20x = 540$$
$$-10x = -280$$
$$x = 28$$

He has 28 \$10 coins and $41 - 28 = 13$ \$20 coins.

Check The value of the coins is
$\$10(28) + \$20(13) = \$540$, as required.

15. Let x = the number of adult tickets sold. Then $2010 - x$ = the number of senior tickets sold.

Cost of Ticket	Number Sold	Amount Collected
\$12	x	$12x$
\$7	$2010 - x$	$7(2010 - x)$
Totals	2010	\$18,430

Write the equation from the last column of the table.

$$12x + 7(2010 - x) = 18{,}430$$
$$12x + 14{,}070 - 7x = 18{,}430$$
$$5x = 4360$$
$$x = 872$$

872 adult tickets were sold; $2010 - 872 = 1138$ senior tickets were sold.

Check The amount collected was
$\$12(872) + \$7(1138)$
$= \$10{,}464 + \$7966 = \$18{,}430$, as required.

17. $d = rt$, so

$$r = \frac{d}{t} = \frac{100}{12.37} \approx 8.08$$

Her rate was about 8.08 m/sec.

19. $d = rt$, so

$$r = \frac{d}{t} = \frac{400}{47.63} \approx 8.40$$

His rate was about 8.40 m/sec.

21. Let t = the time until they are 110 mi apart. Use the formula $d = rt$. Complete the table.

	Rate	Time	Distance
First Steamer	22	t	$22t$
Second Steamer	22	t	$22t$
			110

The total distance traveled is the sum of the distances traveled by each steamer, since they are traveling in opposite directions. This total is 110 mi.

$$22t + 22t = 110$$
$$44t = 110$$
$$t = \frac{110}{44} = \frac{5}{2} \text{ or } 2\tfrac{1}{2}$$

It will take them $2\tfrac{1}{2}$ hr.

Check Each steamer traveled $22(2.5) = 55$ miles for a total of $2(55) = 110$ miles, as required.

23. Let t = Mulder's time. Then $t - \tfrac{1}{2}$ = Scully's time.

	Rate	Time	Distance
Mulder	65	t	$65t$
Scully	68	$t - \tfrac{1}{2}$	$68\left(t - \tfrac{1}{2}\right)$

The distances are equal.

$$65t = 68\left(t - \tfrac{1}{2}\right)$$
$$65t = 68t - 34$$
$$-3t = -34$$
$$t = \frac{34}{3} \text{ or } 11\tfrac{1}{3}$$

Mulder's time will be $11\tfrac{1}{3}$ hr. Since he left at 8:30 A.M., $11\tfrac{1}{3}$ hr or 11 hr 20 min later is 7:50 P.M.

Check Mulder's distance was $65\left(\frac{34}{3}\right) = 736\tfrac{2}{3}$ miles. Scully's distance was
$68\left(\frac{34}{3} - \frac{1}{2}\right) = 68\left(\frac{65}{6}\right) = 736\tfrac{2}{3}$, as required.

25. Let x = her average speed on Sunday. Then $x + 5$ = her average speed on Saturday.

	Rate	Time	Distance
Saturday	$x + 5$	3.6	$3.6(x + 5)$
Sunday	x	4	$4x$

The distances are equal.

$$3.6(x + 5) = 4x$$
$$3.6x + 18 = 4x$$
$$18 = 0.4x \qquad \textit{Subtract 3.6x.}$$
$$x = \frac{18}{0.4} = 45$$

Her average speed on Sunday was 45 mph.
Check On Sunday, 4 hours @ 45 mph = 180 miles. On Saturday, 3.6 hours @ 50 mph = 180 miles. The distances are equal.

27. Let x = Anne's time. Then $x + \tfrac{1}{2}$ = Johnny's time.

	Rate	Time	Distance
Anne	60	x	$60x$
Johnny	50	$x + \tfrac{1}{2}$	$50\left(x + \tfrac{1}{2}\right)$

The total distance is 80.

$$60x + 50\left(x + \tfrac{1}{2}\right) = 80$$
$$60x + 50x + 25 = 80$$
$$110x = 55$$
$$x = \tfrac{55}{110} = \tfrac{1}{2}$$

They will meet $\frac{1}{2}$ hr after Anne leaves.

Check Anne travels $60\left(\tfrac{1}{2}\right) = 30$ miles. Johnny travels $50\left(\tfrac{1}{2} + \tfrac{1}{2}\right) = 50$ miles. The sum of the distances is 80 miles, as required.

29. The sum of the measures of the three angles of a triangle is 180°.

$$(x - 30) + (2x - 120) + \left(\tfrac{1}{2}x + 15\right) = 180$$
$$\tfrac{7}{2}x - 135 = 180$$
$$7x - 270 = 360 \qquad$$
$$\textit{Multiply by 2.}$$
$$7x = 630$$
$$x = 90$$

With $x = 90$, the three angle measures become

$$(90 - 30)° = 60°,$$
$$[2(90) - 120]° = 60°,$$
$$\text{and} \qquad [\tfrac{1}{2}(90) + 15]° = 60°.$$

31. The sum of the measures of the three angles of a triangle is 180°.

$$(3x + 7) + (9x - 4) + (4x + 1) = 180$$
$$16x + 4 = 180$$
$$16x = 176$$
$$x = 11$$

With $x = 11$, the three angle measures become

$$(3 \cdot 11 + 7)° = 40°,$$
$$(9 \cdot 11 - 4)° = 95°,$$
$$\text{and} \quad (4 \cdot 11 + 1)° = 45°.$$

33. The sum of the measures of the angles of a triangle is 180°.

$$x + 2x + 60 = 180$$
$$3x + 60 = 180$$
$$3x = 120$$
$$x = 40$$

The measures of the unknown angles are 40° and $2x = 80°$.

34. Two angles which form a straight line add to 180°, so $180° - 60° = 120°$. The measure of the unknown angle is 120°.

35. The sum of the measures of the unknown angles in Exercise 33 is $40° + 80° = 120°$. This is equal to the measure of the angle in Exercise 34.

36. The sum of the measures of angles ① and ② is equal to the measure of angle ③ .

37. Vertical angles have equal measure.

$$8x + 2 = 7x + 17$$
$$x = 15$$
$$8 \cdot 15 + 2 = 122 \quad \text{and} \quad 7 \cdot 15 + 17 = 122.$$

The angles are both 122°.

39. The sum of the two angles is 90°.

$$(5x - 1) + 2x = 90$$
$$7x - 1 = 90$$
$$7x = 91$$
$$x = 13$$

The measures of the two angles are $[5(13) - 1]° = 64°$ and $[2(13)]° = 26°$.

41. Let $x =$ the first consecutive integer. Then $x + 1$ will be the second consecutive integer, and $x + 2$ will be the third consecutive integer.

The sum of the first and twice the second is 17 more than twice the third.

$$x + 2(x + 1) = 2(x + 2) + 17$$
$$x + 2x + 2 = 2x + 4 + 17$$
$$3x + 2 = 2x + 21$$
$$x = 19$$

Since $x = 19$, $x + 1 = 20$, and $x + 2 = 21$. The three consecutive integers are 19, 20, and 21.

43. Let $x =$ the current age. Then $x + 1$ will be the age next year.
The sum of these ages will be 103 years.

$$x + (x + 1) = 103$$
$$2x + 1 = 103$$
$$2x = 102$$
$$x = 51$$

If my current age is 51, in 10 years I will be

$$51 + 10 = 61 \text{ years old.}$$

45. $(4, \infty)$ is equivalent to $x > 4$.

47. $(-2, 6)$ is equivalent to $-2 < x < 6$.

49. $[-4, 9)$ is equivalent to $-4 \le x < 9$.

Summary Exercises on Solving Applied Problems

1. Let $x =$ the width of the rectangle. Then $x + 3$ is the length of the rectangle.

If the length were decreased by 2 inches and the width were increased by 1 inch, the perimeter would be 24 inches. Use the formula $P = 2L + 2W$, and substitute 24 for P, $(x + 3) - 2$ or $x + 1$ for L, and $x + 1$ for W.

$$P = 2L + 2W$$
$$24 = 2(x + 1) + 2(x + 1)$$
$$24 = 2x + 2 + 2x + 2$$
$$24 = 4x + 4$$
$$20 = 4x$$
$$5 = x$$

The width of the rectangle is 5 inches, and the length is $5 + 3 = 8$ inches.

3. Let $x =$ the regular price of the item. The sale price after a 37% (or 0.37) discount was $32.09, so an equation is

$$x - 0.37x = 32.09.$$
$$0.63x = 32.09$$
$$x = \tfrac{32.09}{0.63} \approx 50.94$$

To the nearest cent, the regular price was $50.94.

5. Let $x =$ the amount invested at 4%. Then $2x$ is the amount invested at 5%. Use $I = prt$ with $t = 1$ yr. Make a table.

Principal	Rate (as a decimal)	Interest
x	0.04	$0.04x$
$2x$	0.05	$0.05(2x) = 0.10x$
	Total →	112

The last column gives the equation.

$$\begin{matrix} \text{Interest} \\ \text{at 4\%} \end{matrix} + \begin{matrix} \text{interest} \\ \text{at 5\%} \end{matrix} = \begin{matrix} \text{total} \\ \text{interest.} \end{matrix}$$
$$0.04x + 0.10x = 112$$

$$4x + 10x = 11{,}200 \quad \textit{Multiply by 100.}$$
$$14x = 11{,}200$$
$$x = 800$$

$800 is invested at 4% and 2($800) = $1600 at 5%.

Check $800 @ 4% = $32 and $1600 @ 5% = $80; $32 + $80 = $112

7. Let $x =$ the number of Emmy awards won by *The Simpsons*. Then $2x - 9 =$ the number of Emmys won by *Frasier*. The total number of Emmys won by both was 51.

$$x + (2x - 9) = 51$$
$$3x - 9 = 51$$
$$3x = 60$$
$$x = 20$$

The Simpsons won 20 Emmys and *Frasier* won $2(20) - 9 = 31$ Emmys.

9. Let $t =$ the time it will take until John and Pat meet. Use $d = rt$ and make a table.

	Rate	Time	Distance
John	60	t	$60t$
Pat	28	t	$28t$

The total distance is 440 miles.

$$60t + 28t = 440$$
$$88t = 440$$
$$t = 5$$

It will take 5 hours for John and Pat to meet.

Check John traveled $60(5) = 300$ miles and Pat traveled $28(5) = 140$ miles; $300 + 140 = 440$, as required.

11. Let $x =$ the number of liters of the 5% drug solution.

Liters of Solution	Percent (as a decimal)	Liters of Pure Drug
20	0.10	$20(0.10) = 2$
x	0.05	$0.05x$
$20 + x$	0.08	$0.08(20 + x)$

$$\begin{matrix} \text{Drug} \\ \text{in 10\%} \end{matrix} + \begin{matrix} \text{drug} \\ \text{in 5\%} \end{matrix} = \begin{matrix} \text{drug} \\ \text{in 8\%.} \end{matrix}$$
$$2 + 0.05x = 0.08(20 + x)$$

$$200 + 5x = 8(20 + x) \quad \textit{Multiply by 100.}$$
$$200 + 5x = 160 + 8x$$
$$40 = 3x$$
$$x = \tfrac{40}{3} \text{ or } 13\tfrac{1}{3}$$

The pharmacist should add $13\tfrac{1}{3}$ L.

Check 10% of 20 is 2 and 5% of $\tfrac{40}{3}$ is $\tfrac{2}{3}$; $2 + \tfrac{2}{3} = \tfrac{8}{3}$, which is the same as 8% of $\left(20 + \tfrac{40}{3}\right)$.

13. Let $x =$ the number of $5 bills. Then $126 - x$ is the number of $10 bills.

Number of Bills	Denomination	Value
x	5	$5x$
$126 - x$	10	$10(126 - x)$
126	← Totals →	840

The sum of the values must equal the total value.

$$5x + 10(126 - x) = 840$$
$$5x + 1260 - 10x = 840$$
$$-5x = -420$$
$$x = 84$$

There are 84 $5 bills and $126 - 84 = 42$ $10 bills.

Check The value of the bills is
$5(84) + \$10(42) = \$840$, as required.

15. The sum of the measures of the three angles of a triangle is $180°$.

$$x + (6x - 50) + (x - 10) = 180$$
$$8x - 60 = 180$$
$$8x = 240$$
$$x = 30$$

With $x = 30$, the three angle measures become

$$(6 \cdot 30 - 50)° = 130°,$$
$$(30 - 10)° = 20°, \text{ and } 30°.$$

17. Let $x =$ the least integer. Then $x + 1$ is the middle integer and $x + 2$ is the greatest integer.

"The sum of the least and greatest of three consecutive integers is 32 more than the middle integer" translates to

$$x + (x + 2) = 32 + (x + 1).$$
$$2x + 2 = x + 33$$
$$x = 31$$

The three consecutive integers are 31, 32, and 33.

Check The sum of the least and greatest integers is $31 + 33 = 64$, which is the same as 32 more than the middle integer.

19. Let $x =$ the length of the shortest side. Then $2x$ is the length of the middle side and $3x - 2$ is the length of the longest side.

The perimeter is 34 inches. Using $P = a + b + c$ gives us

$$x + 2x + (3x - 2) = 34.$$
$$6x - 2 = 34$$
$$6x = 36$$
$$x = 6$$

The lengths of the three sides are 6 inches, $2(6) = 12$ inches, and $3(6) - 2 = 16$ inches.

Check The sum of the lengths of the three sides is $6 + 12 + 16 = 34$ inches, as required.

2.5 Linear Inequalities in One Variable

1. $x \leq 3$

In interval notation, this inequality is written $(-\infty, 3]$. The bracket indicates that 3 is included. The answer is choice **D**.

3. $x < 3$

In interval notation, this inequality is written $(-\infty, 3)$. The parenthesis indicates that 3 is not included. The graph of this inequality is shown in choice **B**.

5. $-3 \leq x \leq 3$

In interval notation, this inequality is written $[-3, 3]$. The brackets indicates that -3 and 3 are included. The answer is choice **F**.

7. **(a)** The wind speed s of a Category 4 hurricane is between 131 mph and 155 mph [inclusive], which can be described by the three-part inequality $131 \leq s \leq 155$.

(b) The wind speed s of a Category 5 hurricane is greater than 155 mph, which can be described by the inequality $s > 155$.

(c) The storm surge x of a Category 3 hurricane is between 9 ft and 12 ft [inclusive], which can be described by the three-part inequality $9 \leq x \leq 12$.

(d) The storm surge x of a Category 5 hurricane is greater than 18 ft, which can be described by the inequality $x > 18$.

9. Since $4 > 0$, the student should not have reversed the direction of the inequality symbol when dividing by 4. We reverse the inequality symbol only when multiplying or dividing by a *negative* number. The solution set is $[-16, \infty)$.

11. $x - 4 \geq 12$
        $x \geq 16$ *Add 4.*

Check that the solution set is the interval $[16, \infty)$.

13. $3k + 1 > 22$
        $3k > 21$ *Subtract 1.*
        $k > 7$ *Divide by 3.*

Check that the solution set is the interval $(7, \infty)$.

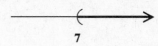

15. $4x < -16$

$\quad\quad x < -4$ *Divide by 4.*

Check that the solution set is the interval $(-\infty, -4)$.

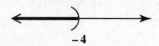

17. $\quad\quad -\frac{3}{4}r \geq 30$

Multiply both sides by $-\frac{4}{3}$, and reverse the inequality symbol.

$-\frac{4}{3}\left(-\frac{3}{4}r\right) \leq -\frac{4}{3}(30)$

$\quad\quad r \leq -40$

Check that the solution set is the interval $(-\infty, -40]$.

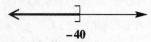

19. $\quad\quad -1.3m \geq -5.2$

Divide both sides by -1.3, and reverse the inequality symbol.

$\dfrac{-1.3m}{-1.3} \leq \dfrac{-5.2}{-1.3}$

$\quad\quad m \leq 4$

Check that the solution set is the interval $(-\infty, 4]$.

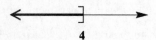

21. $5t + 2 \leq -48$

$\quad 5t \leq -50$ *Subtract 2.*

$\quad\quad t \leq -10$ *Divide by 5.*

Check that the solution set is the interval $(-\infty, -10]$.

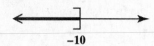

23. $\quad\quad \dfrac{5z - 6}{8} < 8$

$8\left(\dfrac{5z - 6}{8}\right) < 8 \cdot 8$ *Multiply by 8.*

$\quad 5z - 6 < 64$

$\quad\quad 5z < 70$ *Add 6.*

$\quad\quad z < 14$ *Divide by 5.*

Check   Let $z = 14$ in the *equation* $\dfrac{5z - 6}{8} = 8$.

$\dfrac{5(14) - 6}{8} = 8$?

$\dfrac{64}{8} = 8$?

$8 = 8$ *True*

This shows that 14 is the boundary point. Now test a number on each side of 14. We choose 0 and 20.

$$\dfrac{5z - 6}{8} < 8$$

Let z = 0.

$\dfrac{5(0) - 6}{8} < 8$?

$-\dfrac{6}{8} < 8$ *True*

0 is in the solution set.

Let z = 20.

$\dfrac{5(20) - 6}{8} < 8$?

$\dfrac{94}{8}\left(\text{or } 11\tfrac{6}{8}\right) < 8$ *False*

20 is not in the solution set.

The check confirms that $(-\infty, 14)$ is the solution set.

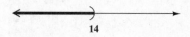

25. $\quad\quad \dfrac{2k - 5}{-4} > 5$

Multiply both sides by -4, and reverse the inequality symbol.

$-4\left(\dfrac{2k - 5}{-4}\right) < -4(5)$

$\quad 2k - 5 < -20$

$\quad\quad 2k < -15$ *Add 5.*

$\quad\quad k < -\dfrac{15}{2}$ *Divide by 2.*

Check that the solution set is the interval $\left(-\infty, -\dfrac{15}{2}\right)$.

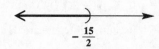

27. $6x - 4 \geq -2x$

$8x - 4 \geq 0$ *Add 2x.*

$\quad 8x \geq 4$ *Add 4.*

$\quad\quad x \geq \dfrac{4}{8} = \dfrac{1}{2}$ *Divide by 8.*

Check that the solution set is the interval $\left[\dfrac{1}{2}, \infty\right)$.

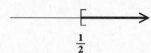

29. $m - 2(m - 4) \leq 3m$

$m - 2m + 8 \leq 3m$

$\quad -m + 8 \leq 3m$

$\quad\quad 8 \leq 4m$ *Add m.*

$\quad\quad 2 \leq m$ or $m \geq 2$

Check that the solution set is the interval $[2, \infty)$.

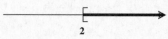

31. $-(4 + r) + 2 - 3r < -14$

$-4 - r + 2 - 3r < -14$ *Distributive property*

Combine terms.

$-4r - 2 < -14$

$-4r < -12$ *Add 2.*

Divide by -4, and reverse the inequality symbol.

$r > 3$

Check that the solution set is the interval $(3, \infty)$.

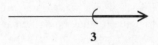

3

33. $-3(z - 6) > 2z - 2$

$-3z + 18 > 2z - 2$ *Distributive property*

$-5z > -20$ *Subtract 2z and 18.*

Divide by -5, and reverse the inequality symbol.

$z < 4$

Check that the solution set is the interval $(-\infty, 4)$.

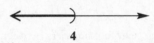

4

35. $\frac{2}{3}(3k - 1) \geq \frac{3}{2}(2k - 3)$

Multiply both sides by 6 to clear the fractions.

$6 \cdot \frac{2}{3}(3k - 1) \geq 6 \cdot \frac{3}{2}(2k - 3)$

$4(3k - 1) \geq 9(2k - 3)$

$12k - 4 \geq 18k - 27$ *Distributive property*

$-6k \geq -23$ *Subtract 18k; add 4.*

Divide by -6, and reverse the inequality symbol.

$k \leq \frac{23}{6}$

Check that the solution set is the interval $\left(-\infty, \frac{23}{6}\right]$.

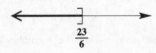

$\frac{23}{6}$

37. $-\frac{1}{4}(p + 6) + \frac{3}{2}(2p - 5) < 10$

Multiply each term by 4 to clear the fractions.

$-1(p + 6) + 6(2p - 5) < 40$

$-p - 6 + 12p - 30 < 40$

$11p - 36 < 40$

$11p < 76$

$p < \frac{76}{11}$

Check that the solution set is the interval $\left(-\infty, \frac{76}{11}\right)$.

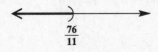

$\frac{76}{11}$

39. $3(2x - 4) - 4x < 2x + 3$

$6x - 12 - 4x < 2x + 3$

$2x - 12 < 2x + 3$

$-12 < 3$ *True*

The statement is true for all values of x. Therefore, the original inequality is true for any real number.

Check that the solution set is the interval $(-\infty, \infty)$.

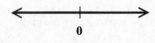

0

41. $8\left(\frac{1}{2}x + 3\right) < 8\left(\frac{1}{2}x - 1\right)$

$4x + 24 < 4x - 8$

$24 < -8$ *False*

This is a false statement, so the inequality is a contradiction.

Check that the solution set is $\emptyset$.

43. $5(x + 3) - 2(x - 4) = 2(x + 7)$

$5x + 15 - 2x + 8 = 2x + 14$

$3x + 23 = 2x + 14$

$x = -9$

Check that the solution set is $\{-9\}$.
The graph is the point -9 on a number line.

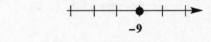

−9

44. $5(x + 3) - 2(x - 4) > 2(x + 7)$

$5x + 15 - 2x + 8 > 2x + 14$

$3x + 23 > 2x + 14$

$x > -9$

Check that the solution set is the interval $(-9, \infty)$.
The graph extends from -9 to the right on a number line; -9 is not included in the graph.

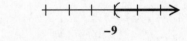

−9

45. $5(x + 3) - 2(x - 4) < 2(x + 7)$

$5x + 15 - 2x + 8 < 2x + 14$

$3x + 23 < 2x + 14$

$x < -9$

Check that the solution set is the interval $(-\infty, -9)$.
The graph extends from -9 to the left on a number line; -9 is not included in the graph.

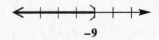

−9

46. If we graph all the solution sets from Exercises 43–45; that is, $\{-9\}$, $(-9, \infty)$, and $(-\infty, -9)$, on the same number line, we will have graphed the set of all real numbers.

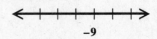

−9

47. The solution set of the given equation is the point -3 on a number line. The solution set of the first inequality extends from -3 to the right (toward ∞) on the same number line. Based on Exercises 43–45, the solution set of the second inequality should then extend from -3 to the left (toward $-\infty$) on the number line. Complete the statement with $\underline{(-\infty, -3)}$.

49. $-4 < x - 5 < 6$

Add 5 to each part of the inequality to isolate the variable x.

$-4 + 5 < x - 5 + 5 < 6 + 5$

$\quad\quad 1 < x < 11$

Check that the solution set is the interval $(1, 11)$.

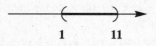

51. $-9 \le k + 5 \le 15$

Subtract 5 from each part.

$-9 - 5 \le k + 5 - 5 \le 15 - 5$

$\quad -14 \le k \le 10$

Check that the solution set is the interval $[-14, 10]$.

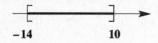

53. $-6 \le 2z + 4 \le 16$

$-10 \le 2z \le 12$ *Subtract 4.*

$\quad -5 \le z \le 6$ *Divide by 2.*

Check that the solution set is the interval $[-5, 6]$.

55. $-19 \le 3x - 5 \le 1$

$-14 \le 3x \le 6$ *Add 5.*

$-\frac{14}{3} \le x \le 2$ *Divide by 3.*

Check that the solution set is the interval $\left[-\frac{14}{3}, 2\right]$.

57. $-1 \le \dfrac{2x - 5}{6} \le 5$

$-6 \le 2x - 5 \le 30$ *Multiply by 6.*

$-1 \le 2x \le 35$ *Add 5.*

$-\frac{1}{2} \le x \le \frac{35}{2}$ *Divide by 2.*

Check that the solution set is the interval $\left[-\frac{1}{2}, \frac{35}{2}\right]$.

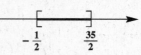

59. $4 \le -9x + 5 < 8$

$-1 \le -9x < 3$ *Subtract 5.*

Divide each part by -9; reverse the inequality symbols.

$\frac{1}{9} \ge x > -\frac{1}{3}$

The last inequality may be written as

$$-\tfrac{1}{3} < x \le \tfrac{1}{9}.$$

Check that the solution set is the interval $\left(-\frac{1}{3}, \frac{1}{9}\right]$.

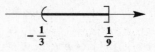

61. Six times a number is between -12 and 12.

$-12 < 6x < 12$

$-2 \;<\; x \;< 2$ *Divide by 6.*

This is the set of all numbers between -2 and 2—that is, $(-2, 2)$.

63. When 1 is added to twice a number, the result is greater than or equal to 7.

$2x + 1 \ge 7$

$2x \ge 6$ *Subtract 1.*

$x \ge 3$ *Divide by 2.*

This is the set of all numbers greater than or equal to 3—that is, $[3, \infty)$.

65. One third of a number is added to 6, giving a result of at least 3.

$6 + \frac{1}{3}x \ge 3$

$\frac{1}{3}x \ge -3$ *Subtract 6.*

$x \ge -9$ *Multiply by 3.*

This is the set of all numbers greater than or equal to -9—that is, $[-9, \infty)$.

67. Draw a horizontal line at the 90°F mark. It intersects the *upper* boundary of the forecasted highs in two places. The temperature is expected to be at least 90°F from about 2:30 P.M. to 6:00 P.M.

69. Locate the dot at 5:39 P.M. and draw a vertical line through it. The line will intersect the lower and upper boundaries of the forecasted highs at about 84°F and 91°F, so the range of predicted temperatures is 84°F–91°F.

71. Let $x =$ his score on the third test.

His average must be at least 84 (≥ 84). To find the average of three numbers, add them and divide by 3.

$$\frac{90 + 82 + x}{3} \geq 84$$

$$\frac{172 + x}{3} \geq 84 \quad \textit{Add.}$$

$$172 + x \geq 252 \quad \textit{Multiply by 3.}$$

$$x \geq 80 \quad \textit{Subtract 172.}$$

He must score at least 80 on his third test.

73. Let $x =$ the number of months. The cost of Plan A is $54.99x$ and the cost of Plan B is $49.99x + 129$. To determine the number of months that would be needed to make Plan B less expensive, solve the following inequality.

Plan B (cost) < Plan A (cost)

$$49.99x + 129 < 54.99x$$

$$129 < 5x \qquad \textit{Subtract 49.99x.}$$

$$5x > 129 \qquad \textit{Equivalent}$$

$$x > \frac{129}{5} \ [= 25.8] \quad \textit{Divide by 5.}$$

It will take 26 months for Plan B to be the better deal.

75. Cost $C = 20x + 100$; Revenue $R = 24x$

The business will show a profit only when $R > C$. Substitute the given expressions for R and C.

$$R > C$$

$$24x > 20x + 100$$

$$4x > 100$$

$$x > 25$$

The company will show a profit upon selling 26 DVDs.

77. $\text{BMI} = \dfrac{704 \times (\text{weight in pounds})}{(\text{height in inches})^2}$

(a) Let the height equal 72.

$$19 \leq \text{BMI} \leq 25$$

$$19 \leq \frac{704w}{72^2} \leq 25$$

$$19(72^2) \leq 704w \leq 25(72^2)$$

$$\frac{19(72^2)}{704} \leq w \leq \frac{25(72^2)}{704}$$

$$(\approx 139.91) \leq w \leq (\approx 184.09)$$

According to the BMI formula, the healthy weight range (rounded to the nearest pound) for a person that is 72 inches tall is 140 to 184 pounds.

(b) Answers will vary.

79. (a) $x > 4$ is equivalent to $(4, \infty)$.

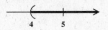

(b) $x < 5$ is equivalent to $(-\infty, 5)$.

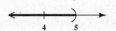

(c) All numbers greater than 4 and less than 5 (that is, $4 < x < 5$) belong to *both* sets.

81. (a) $t < 5$ is equivalent to $(-\infty, 5)$.

(b) $t > 4$ is equivalent to $(4, \infty)$.

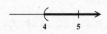

(c) All real numbers belong to *either one or both* of these sets.

2.6 Set Operations and Compound Inequalities

1. This statement is *true*. The solution set of $x + 1 = 5$ is $\{4\}$. The solution set of $x + 1 > 5$ is $(4, \infty)$. The solution set of $x + 1 < 5$ is $(-\infty, 4)$. Taken together we have the set of real numbers. (See Section 2.5, Exercises 43–47, for a discussion of this concept.)

3. This statement is *false*. The union is $(-\infty, 8) \cup (8, \infty)$. The only real number that is *not* in the union is 8.

5. This statement is *false* since 0 is a rational number but not an irrational number. The sets of rational numbers and irrational numbers have no common elements so their intersection is $\emptyset$.

In Exercises 7–14, let

$A = \{1, 2, 3, 4, 5, 6\}, B = \{1, 3, 5\}, C = \{1, 6\}$, and $D = \{4\}$.

7. The intersection of sets B and A contains only those elements in both sets B and A.

$$B \cap A = \{1, 3, 5\} \text{ or set } B$$

9. The intersection of sets A and D is the set of all elements in both set A and D. Therefore,

$$A \cap D = \{4\} \text{ or set } D.$$

11. The intersection of set B and the set of no elements (empty set), $B \cap \emptyset$, is the set of no elements or $\emptyset$.

13. The union of sets A and B is the set of all elements that are in either set A or set B or both sets A and B. Since all numbers in set B are also in set A, the set $A \cup B$ will be the same as set A.

$$A \cup B = \{1, 2, 3, 4, 5, 6\} \text{ or set } A$$

15. The first graph represents the set $(-\infty, 2)$. The second graph represents the set $(-3, \infty)$. The intersection includes the elements common to both sets, that is, $(-3, 2)$.

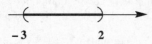

17. The first graph represents the set $(-\infty, 5]$. The second graph represents the set $(-\infty, 2]$. The intersection includes the elements common to both sets, that is, $(-\infty, 2]$.

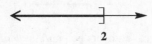

19. $x < 2$ and $x > -3$

The graph of the solution set will be all numbers that are both less than 2 and greater than -3. The solution set is $(-3, 2)$.

21. $x \le 2$ and $x \le 5$

The graph of the solution set will be all numbers that are both less than or equal to 2 and less than or equal to 5. The overlap is the numbers less than or equal to 2. The solution set is $(-\infty, 2]$.

23. $x \le 3$ and $x \ge 6$

The graph of the solution set will be all numbers that are both less than or equal to 3 and greater than or equal to 6. There are no such numbers. The solution set is $\emptyset$.

25. $x - 3 \le 6$ and $x + 2 \ge 7$
   $x \le 9$ and $x \ge 5$
The graph of the solution set is all numbers that are both less than or equal to 9 and greater than or equal to 5. This is the intersection. The elements common to both sets are the numbers between 5 and 9, including the endpoints. The solution set is $[5, 9]$.

27. $-3x > 3$ and $x + 3 > 0$
   $x < -1$ and $x > -3$
The graph of the solution set is all numbers that are both less than -1 and greater than -3. This is the intersection. The elements common to both sets are the numbers between -3 and -1, not including the endpoints. The solution set is $(-3, -1)$.

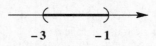

29. $3x - 4 \le 8$ and $-4x + 1 \ge -15$
   $3x \le 12$ and $-4x \ge -16$
   $x \le 4$ and $x \le 4$
Since both inequalities are identical, the graph of the solution set is the same as the graph of one of the inequalities. The solution set is $(-\infty, 4]$.

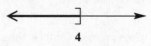

31. The first graph represents the set $(-\infty, 2]$. The second graph represents the set $[4, \infty)$. The union includes all elements in either set, or in both, that is, $(-\infty, 2] \cup [4, \infty)$.

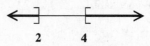

33. The first graph represents the set $[1, \infty)$. The second graph represents the set $(-\infty, 8]$. The union includes all elements in either set, or in both, that is, $(-\infty, \infty)$.

35. $x \le 1$ or $x \le 8$

The word "or" means to take the union of both sets. The graph of the solution set is all numbers that are either less than or equal to 1 *or* less than or equal to 8, or both. This is all numbers less than or equal to 8. The solution set is $(-\infty, 8]$.

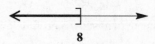

37. $x \ge -2$ or $x \ge 5$

The graph of the solution set will be all numbers that are either greater than or equal to -2 or greater than or equal to 5.
The solution set is $[-2, \infty)$.

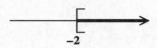

39. $x \ge -2$ or $x \le 4$

The graph of the solution set will be all numbers that are either greater than or equal to -2 or less than or equal to 4. This is the set of all real numbers. The solution set is $(-\infty, \infty)$.

41. $x + 2 > 7$ or $1 - x > 6$
$$-x > 5$$
$$x > 5 \quad \text{or} \quad x < -5$$
The graph of the solution set is all numbers either greater than 5 or less than -5. This is the union. The solution set is $(-\infty, -5) \cup (5, \infty)$.

43. $x + 1 > 3$ or $-4x + 1 > 5$
$$-4x > 4$$
$$x > 2 \quad \text{or} \quad x < -1$$
The graph of the solution set is all numbers either less than -1 or greater than 2. This is the union. The solution set is $(-\infty, -1) \cup (2, \infty)$.

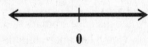

45. $4x + 1 \geq -7$ or $-2x + 3 \geq 5$
$$4x \geq -8 \quad \text{or} \quad -2x \geq 2$$
$$x \geq -2 \quad \text{or} \quad x \leq -1$$
The graph of the solution set is all numbers either greater than or equal to -2 or less than or equal to -1. This is the set of all real numbers. The solution set is $(-\infty, \infty)$.

47. $(-\infty, -1] \cap [-4, \infty)$
The intersection is the set of numbers less than or equal to -1 and greater than or equal to -4. The numbers common to both original sets are between, and including, -4 and -1. The simplest interval form is $[-4, -1]$.

49. $(-\infty, -6] \cap [-9, \infty)$
The intersection is the set of numbers less than or equal to -6 and greater than or equal to -9. The numbers common to both original sets are between, and including, -9 and -6. The simplest interval form is $[-9, -6]$.

51. $(-\infty, 3) \cup (-\infty, -2)$
The union is the set of numbers that are either less than 3 or less than -2, or both. This is all numbers less than 3. The simplest interval form is $(-\infty, 3)$.

53. $[3, 6] \cup (4, 9)$
The union is the set of numbers between, and including, 3 and 6, or between, but not including, 4 and 9. This is the set of numbers greater than or equal to 3 and less than 9. The simplest interval form is $[3, 9)$.

55. $x < -1$ and $x > -5$
The word "and" means to take the intersection of both sets. $x < -1$ and $x > -5$ is true only when
$$-5 < x < -1.$$
The graph of the solution set is all numbers greater than -5 *and* less than -1. This is all numbers between -5 and -1, not including -5 or -1. The solution set is $(-5, -1)$.

57. $x < 4$ or $x < -2$
The word "or" means to take the union of both sets. The graph of the solution set is all numbers that are either less than 4 *or* less than -2, or both. This is all numbers less than 4. The solution set is $(-\infty, 4)$.

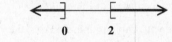

59. $-3x \leq -6$ or $-3x \geq 0$
$$x \geq 2 \quad \text{or} \quad x \leq 0$$
The word "or" means to take the union of both sets. The graph of the solution set is all numbers that are either greater than or equal to 2 *or* less than or equal to 0. The solution set is $(-\infty, 0] \cup [2, \infty)$.

61. $x + 1 \geq 5$ and $x - 2 \leq 10$
$$x \geq 4 \quad \text{and} \quad x \leq 12$$
The word "and" means to take the intersection of both sets. The graph of the solution set is all numbers that are both greater than or equal to 4 *and* less than or equal to 12. This is all numbers between, and including, 4 and 12. The solution set is $[4, 12]$.

63. The set of expenses that are less than $3000 for public schools *and* are greater than $5000 for private schools is {Tuition and fees}.

65. The set of expenses that are greater than $2900 for public schools *or* are greater than $5000 for private schools is {Tuition and fees, Dormitory charges}.

67. Find "the yard can be fenced *and* the yard can be sodded."

A yard that can be fenced has $P \leq 150$. Maria and Joe qualify.

A yard that can be sodded has $A \leq 1400$. Again, Maria and Joe qualify.

Find the intersection. Maria's and Joe's yards are common to both sets, so Maria and Joe can have their yards both fenced and sodded.

68. Find "the yard can be fenced *and* the yard cannot be sodded."

A yard that can be fenced has $P \leq 150$. Maria and Joe qualify.

A yard that cannot be sodded has $A > 1400$. Luigi and Than qualify.

Find the intersection. There are no yards common to both sets, so none of them qualify.

69. Find "the yard cannot be fenced *and* the yard can be sodded."

A yard that cannot be fenced has $P > 150$. Luigi and Than qualify.

A yard that can be sodded has $A \leq 1400$. Maria and Joe qualify.

Find the intersection. There are no yards common to both sets, so none of the qualify.

70. Find "the yard cannot be fenced *and* the yard cannot be sodded."

A yard that cannot be fenced has $P > 150$. Luigi and Than qualify.

A yard that cannot be sodded has $A > 1400$. Again, Luigi and Than qualify.

Find the intersection. Luigi's and Than's yards are common to both sets, so Luigi and Than qualify.

71. Find "the yard can be fenced *or* the yard can be sodded." From Exercise 67, Maria's and Joe's yards qualify for both conditions, so the union is Maria and Joe.

72. Find "the yard cannot be fenced *or* the yard can be sodded." From Exercise 69, Luigi's and Than's yards cannot be fenced, and Maria's and Joe's yards can be sodded. The union includes all of them.

73.
$$2y - 4 \leq 3y + 2$$
$$-y - 4 \leq 2$$
$$-y \leq 6$$
$$y \geq -6$$

The solution set is $[-6, \infty)$.

75.
$$-5 < 2r + 1 \quad < 5$$
$$-6 < 2r \quad\quad < 4$$
$$-3 < r \quad\quad < 2$$

The solution set is $(-3, 2)$.

77.
$$-|6| - |-11| + (-4) = -6 - (11) + (-4)$$
$$= -17 + (-4)$$
$$= -21$$

2.7   Absolute Value Equations and Inequalities

1. $|x| = 5$ has two solutions, $x = 5$ or $x = -5$. The graph is Choice **E**.

$|x| < 5$ is written $-5 < x < 5$. Notice that -5 and 5 are not included. The graph is Choice **C**, which uses parentheses.

$|x| > 5$ is written $x < -5$ or $x > 5$. The graph is Choice **D**, which uses parentheses.

$|x| \leq 5$ is written $-5 \leq x \leq 5$. This time -5 and 5 are included. The graph is Choice **B**, which uses brackets.

$|x| \geq 5$ is written $x \leq -5$ or $x \geq 5$. The graph is Choice **A**, which uses brackets.

3. **(a)** $|ax + b| = k, k = 0$
This means the distance from $ax + b$ to 0 is 0, so $ax + b = 0$, which has one solution.

(b) $|ax + b| = k, k > 0$
This means the distance from $ax + b$ to 0 is a positive number, so $ax + b = k$ or $ax + b = -k$. There are two solutions.

(c) $|ax + b| = k, k < 0$
This means the distance from $ax + b$ to 0 is a negative number, which is impossible because distance is always positive. There are no solutions.

5. $|x| = 12$
$$x = 12 \quad \text{or} \quad x = -12$$
The solution set is $\{-12, 12\}$.

7. $|4x| = 20$
$$4x = 20 \quad \text{or} \quad 4x = -20$$
$$x = 5 \quad \text{or} \quad x = -5$$
The solution set is $\{-5, 5\}$.

9. $|y - 3| = 9$
$$y - 3 = 9 \quad \text{or} \quad y - 3 = -9$$
$$y = 12 \quad \text{or} \quad y = -6$$
The solution set is $\{-6, 12\}$.

11. $|2x - 1| = 11$
$$2x - 1 = 11 \quad \text{or} \quad 2x - 1 = -11$$
$$2x = 12 \quad\quad\quad\quad 2x = -10$$
$$x = 6 \quad \text{or} \quad\quad x = -5$$
The solution set is $\{-5, 6\}$.

13. $|4r - 5| = 17$

$4r - 5 = 17$ or $4r - 5 = -17$

$4r = 22$ $4r = -12$

$r = \frac{22}{4} = \frac{11}{2}$ or $r = -3$

The solution set is $\left\{ -3, \frac{11}{2} \right\}$.

15. $|2y + 5| = 14$

$2y + 5 = 14$ or $2y + 5 = -14$

$2y = 9$ $2y = -19$

$y = \frac{9}{2}$ or $y = -\frac{19}{2}$

The solution set is $\left\{ -\frac{19}{2}, \frac{9}{2} \right\}$.

17. $\left| \frac{1}{2}x + 3 \right| = 2$

$\frac{1}{2}x + 3 = 2$ or $\frac{1}{2}x + 3 = -2$

$\frac{1}{2}x = -1$ $\frac{1}{2}x = -5$

$x = -2$ or $x = -10$

The solution set is $\{ -10, -2 \}$.

19. $\left| 1 + \frac{3}{4}k \right| = 7$

$1 + \frac{3}{4}k = 7$ or $1 + \frac{3}{4}k = -7$

Multiply each side by 4.

$4 + 3k = 28$ or $4 + 3k = -28$

$3k = 24$ $3k = -32$

$k = 8$ or $k = \frac{-32}{3}$

The solution set is $\left\{ -\frac{32}{3}, 8 \right\}$.

21. $|x| > 3$

$x > 3$ or $x < -3$

The solution set is $(-\infty, -3) \cup (3, \infty)$.

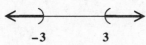

23. $|k| \geq 4$

$k \geq 4$ or $k \leq -4$

The solution set is $(-\infty, -4] \cup [4, \infty)$.

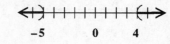

25. $|r + 5| \geq 20$

$r + 5 \leq -20$ or $r + 5 \geq 20$

$r \leq -25$ or $r \geq 15$

The solution set is $(-\infty, -25] \cup [15, \infty)$.

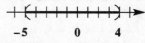

27. $|t + 2| > 10$

$t + 2 > 10$ or $t + 2 < -10$

$t > 8$ or $t < -12$

The solution set is $(-\infty, -12) \cup (8, \infty)$.

29. $|3 - x| > 5$

$3 - x > 5$ or $3 - x < -5$

$-x > 2$ or $-x < -8$

Multiply by -1,

and reverse the inequality symbols.

$x < -2$ or $x > 8$

The solution set is $(-\infty, -2) \cup (8, \infty)$.

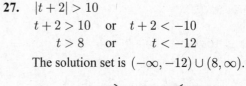

31. $|-5x + 3| \geq 12$

$-5x + 3 \geq 12$ or $-5x + 3 \leq -12$

$-5x \geq 9$ $-5x \leq -15$

$x \leq -\frac{9}{5}$ or $x \geq 3$

The solution set is $\left(-\infty, -\frac{9}{5} \right] \cup [3, \infty)$.

33. **(a)** $|2x + 1| < 9$

The graph of the solution set will be all numbers between -5 and 4, since the absolute value is less than 9.

(b) $|2x + 1| > 9$

The graph of the solution set will be all numbers less than -5 or greater than 4, since the absolute value is greater than 9.

35. $|x| \leq 3$

$-3 \leq x \leq 3$

The solution set is $[-3, 3]$.

37. $|k| < 4$

$-4 < k < 4$

The solution set is $(-4, 4)$.

39. $|r + 5| \leq 20$

$-20 \leq r + 5 \leq 20$

$-25 \leq r \leq 15$ *Subtract 5.*

The solution set is $[-25, 15]$.

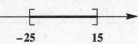

41. $|t + 2| \leq 10$

$-10 \leq t + 2 \leq 10$

$-12 \leq t \leq 8$

The solution set is $[-12, 8]$.

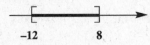

43. $|3 - x| \leq 5$

$-5 \leq 3 - x \leq 5$

$-8 \leq -x \leq 2$

Multiply by -1, and reverse the inequality symbols.

$8 \geq x \geq -2$ or $-2 \leq x \leq 8$

The solution set is $[-2, 8]$.

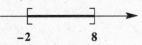

45. $|-5x + 3| \leq 12$

$-12 \leq -5x + 3 \leq 12$

$-15 \leq -5x \leq 9$

Divide by -5, and reverse the inequality symbols.

$3 \geq x \geq -\frac{9}{5}$ or $-\frac{9}{5} \leq x \leq 3$

The solution set is $\left[-\frac{9}{5}, 3\right]$.

47. $|-4 + k| > 9$

$-4 + k > 9$ or $-4 + k < -9$

$k > 13$ or $k < -5$

The solution set is $(-\infty, -5) \cup (13, \infty)$.

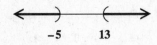

49. $|r + 5| > 20$

$r + 5 > 20$ or $r + 5 < -20$

$r > 15$ or $r < -25$

The solution set is $(-\infty, -25) \cup (15, \infty)$.

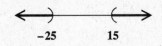

51. $|7 + 2z| = 5$

$7 + 2z = 5$ or $7 + 2z = -5$

$2z = -2$ $2z = -12$

$z = -1$ or $z = -6$

The solution set is $\{-6, -1\}$.

53. $|3r - 1| \leq 11$

$-11 \leq 3r - 1 \leq 11$

$-10 \leq 3r \leq 12$

$-\frac{10}{3} \leq r \leq 4$

The solution set is $\left[-\frac{10}{3}, 4\right]$.

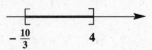

55. $|-6x - 6| \leq 1$

$-1 \leq -6x - 6 \leq 1$

$5 \leq -6x \leq 7$

Divide by -6, and reverse the inequality symbols.

$-\frac{5}{6} \geq x \geq -\frac{7}{6}$ or $-\frac{7}{6} \leq x \leq -\frac{5}{6}$

The solution set is $\left[-\frac{7}{6}, -\frac{5}{6}\right]$.

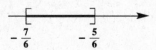

57. $|2x - 1| \geq 7$

$2x - 1 \geq 7$ or $2x - 1 \leq -7$

$2x \geq 8$ or $2x \leq -6$

$x \geq 4$ or $x \leq -3$

The solution set is $(-\infty, -3] \cup [4, \infty)$.

59. $|x + 2| = 3$

$x + 2 = 3$ or $x + 2 = -3$

$x = 1$ or $x = -5$

The solution set is $\{-5, 1\}$.

61. $|x - 6| = 3$

$x - 6 = 3$ or $x - 6 = -3$

$x = 9$ or $x = 3$

The solution set is $\{3, 9\}$.

63. $|x| - 1 = 4$

$|x| = 5$

$x = 5$ or $x = -5$

The solution set is $\{-5, 5\}$.

65. $|x + 4| + 1 = 2$

$|x + 4| = 1$

$x + 4 = 1$ or $x + 4 = -1$

$x = -3$ or $x = -5$

The solution set is $\{-5, -3\}$.

67. $|2x + 1| + 3 > 8$

$|2x + 1| > 5$

$2x + 1 > 5$ or $2x + 1 < -5$

$2x > 4$ $\qquad$ $2x < -6$

$x > 2$ or $\qquad$ $x < -3$

The solution set is $(-\infty, -3) \cup (2, \infty)$.

69. $|x + 5| - 6 \leq -1$

$|x + 5| \leq 5$

$-5 \leq x + 5 \leq 5$

$-10 \leq x \leq 0$

The solution set is $[-10, 0]$.

71. $|3x + 1| = |2x + 4|$

$3x + 1 = 2x + 4$ or $3x + 1 = -(2x + 4)$

$\qquad\qquad\qquad\qquad 3x + 1 = -2x - 4$

$\qquad\qquad\qquad\qquad 5x = -5$

$x = 3$ $\qquad$ or $\qquad$ $x = -1$

The solution set is $\{-1, 3\}$.

73. $\left| m - \frac{1}{2} \right| = \left| \frac{1}{2}m - 2 \right|$

$m - \frac{1}{2} = \frac{1}{2}m - 2$ or $m - \frac{1}{2} = -\left(\frac{1}{2}m - 2 \right)$

Multiply by 2. $\qquad$ $m - \frac{1}{2} = -\frac{1}{2}m + 2$

$2m - 1 = m - 4$ $\qquad$ $2m - 1 = -m + 4$

$\qquad\qquad\qquad\qquad 3m = 5$

$m = -3$ $\qquad$ or $\qquad$ $m = \frac{5}{3}$

The solution set is $\left\{ -3, \frac{5}{3} \right\}$.

75. $|6x| = |9x + 1|$

$6x = 9x + 1$ or $6x = -(9x + 1)$

$-3x = 1$ $\qquad\qquad$ $6x = -9x - 1$

$\qquad\qquad\qquad\qquad 15x = -1$

$x = -\frac{1}{3}$ or $\qquad$ $x = -\frac{1}{15}$

The solution set is $\left\{ -\frac{1}{3}, -\frac{1}{15} \right\}$.

77. $|2p - 6| = |2p + 11|$

$2p - 6 = 2p + 11$ or $2p - 6 = -(2p + 11)$

$-6 = 11$ *False* $\qquad$ $2p - 6 = -2p - 11$

$\qquad\qquad\qquad\qquad 4p = -5$

$\qquad$ *No solution* $\qquad$ or $\qquad$ $p = -\frac{5}{4}$

The solution set is $\left\{ -\frac{5}{4} \right\}$.

79. $|x| \geq -10$

The absolute value of a number is always greater than or equal to 0. Therefore, the inequality is true for all real numbers.

The solution set is $(-\infty, \infty)$.

81. $|12t - 3| = -8$

Since the absolute value of an expression can never be negative, there are no solutions for this equation.

The solution set is $\emptyset$.

83. $|4x + 1| = 0$

The expression $4x + 1$ will equal 0 *only* for the solution of the equation

$$4x + 1 = 0.$$

$$4x = -1$$

$$x = \frac{-1}{4} \text{ or } -\frac{1}{4}$$

The solution set is $\left\{ -\frac{1}{4} \right\}$.

85. $|2q - 1| = -6$

Since the absolute value of an expression can never be negative, there are no solutions for this equation.

The solution set is $\emptyset$.

87. $|x + 5| > -9$

Since the absolute value of an expression is always nonnegative (positive or zero), the inequality is true for any real number x.

The solution set is $(-\infty, \infty)$.

89. $|7x + 3| \leq 0$

The absolute value of an expression is always nonnegative (positive or zero), so this inequality is true only when

$$7x + 3 = 0$$

$$7x = -3$$

$$x = -\frac{3}{7}.$$

The solution set is $\left\{ -\frac{3}{7} \right\}$.

91. $|5x - 2| = 0$

The expression $5x - 2$ will equal 0 *only* for the solution of the equation

$$5x - 2 = 0.$$

$$5x = 2$$

$$x = \frac{2}{5}$$

The solution set is $\left\{ \frac{2}{5} \right\}$.

93. $|x - 2| + 3 \geq 2$

$|x - 2| \geq -1$

Since the absolute value of an expression is always nonnegative (positive or zero), the inequality is true for any real number x.

The solution set is $(-\infty, \infty)$.

95. $|10z + 7| + 3 < 1$
$$|10z + 7| < -2$$

There is no number whose absolute value is less than -2, so this inequality has no solution.

The solution set is $\emptyset$.

97. Let x represent the calcium intake for a specific female. For x to be within 100 mg of 1000 mg, we must have
$$|x - 1000| \le 100.$$
$$-100 \le x - 1000 \le 100$$
$$900 \le \quad x \quad \le 1100$$

99. Add the given heights with a calculator to get 4756. There are 10 numbers, so divide the sum by 10.
$$\frac{4756}{10} = 475.6$$

The average height is 475.6 ft.

100. $|x - k| < 50$

Substitute 475.6 for k and solve the inequality.
$$|x - 475.6| < 50$$
$$-50 < x - 475.6 < 50$$
$$425.6 < \quad x \quad < 525.6$$

The buildings with heights between 425.6 ft and 525.6 ft are the 1201 Walnut, City Hall, Fidelity Bank and Trust Building, Kansas City Power and Light, and the Hyatt Regency Crown Center.

101. $|x - k| < 75$

Substitute 475.6 for k and solve the inequality.
$$|x - 475.6| < 75$$
$$-75 < x - 475.6 < 75$$
$$400.6 < \quad x \quad < 550.6$$

The buildings with heights between 400.6 ft and 550.6 ft are City Center Square, Commerce Tower, Federal Office Building, 1201 Walnut, City Hall, Fidelity Bank and Trust Building, Kansas City Power and Light, and the Hyatt Regency Crown Center.

102. (a) This would be the opposite of the inequality in Exercise 101, that is,
$$|x - 475.6| \ge 75.$$

(b) $|x - 475.6| \ge 75$
$$x - 475.6 \ge 75 \quad \text{or} \quad x - 475.6 \le -75$$
$$x \ge 550.6 \quad \text{or} \quad x \le 400.6$$

(c) The buildings that are not within 75 ft of the average have height less than or equal to 400.6 or greater than or equal to 550.6. They are Town Pavillion and One Kansas City Place.

(d) The answer makes sense because it includes all the buildings *not* listed earlier which had heights within 75 ft of the average.

103. (a) $3x + 2y = 24$
$$3(0) + 2y = 24 \quad \textit{Let x = 0.}$$
$$0 + 2y = 24$$
$$2y = 24$$
$$y = 12$$

(b) $-2x + 5y = 20$
$$-2(0) + 5y = 20 \quad \textit{Let x = 0.}$$
$$0 + 5y = 20$$
$$5y = 20$$
$$y = 4$$

105. (a) $3x + 2y = 24$
$$3(8) + 2y = 24 \quad \textit{Let x = 8.}$$
$$24 + 2y = 24$$
$$2y = 0$$
$$y = 0$$

(b) $-2x + 5y = 20$
$$-2(8) + 5y = 20 \quad \textit{Let x = 8.}$$
$$-16 + 5y = 20$$
$$5y = 36$$
$$y = \frac{36}{5}$$

Summary Exercises on Solving Linear and Absolute Value Equations and Inequalities

1. $4z + 1 = 49$
$$4z = 48$$
$$z = 12$$
The solution set is $\{12\}$.

3. $6q - 9 = 12 + 3q$
$$3q = 21$$
$$q = 7$$
The solution set is $\{7\}$.

5. $|a + 3| = -4$
Since the absolute value of an expression is always nonnegative, there is no number that makes this statement true. Therefore, the solution set is $\emptyset$.

7. $8r + 2 \ge 5r$
$$3r \ge -2$$
$$r \ge -\frac{2}{3}$$
The solution set is $\left[-\frac{2}{3}, \infty\right)$.

9. $2q - 1 = -7$
$$2q = -6$$
$$q = -3$$
The solution set is $\{-3\}$.

11. $6z - 5 \leq 3z + 10$

$\qquad 3z \leq 15$

$\qquad z \leq 5$

The solution set is $(-\infty, 5]$.

13. $9x - 3(x + 1) = 8x - 7$

$\qquad 9x - 3x - 3 = 8x - 7$

$\qquad 6x - 3 = 8x - 7$

$\qquad 4 = 2x$

$\qquad 2 = x$

The solution set is $\{2\}$.

15. $9x - 5 \geq 9x + 3$

$\qquad -5 \geq 3$ *False*

This is a false statement, so the inequality is a contradiction.

The solution set is $\emptyset$.

17. $\qquad |q| < 5.5$

$\qquad -5.5 < q < 5.5$

The solution set is $(-5.5, 5.5)$.

19. $\frac{2}{3}x + 8 = \frac{1}{4}x$

$\qquad 8x + 96 = 3x$ *Multiply by 12.*

$\qquad 5x = -96$

$\qquad x = -\frac{96}{5}$

The solution set is $\left\{-\frac{96}{5}\right\}$.

21. $\qquad \frac{1}{4}p < -6$

$\qquad 4\left(\frac{1}{4}p\right) < 4(-6)$

$\qquad p < -24$

The solution set is $(-\infty, -24)$.

23. $\frac{3}{5}q - \frac{1}{10} = 2$

$\qquad 6q - 1 = 20$ *Multiply by 10.*

$\qquad 6q = 21$

$\qquad q = \frac{21}{6} = \frac{7}{2}$

The solution set is $\left\{\frac{7}{2}\right\}$.

25. $r + 9 + 7r = 4(3 + 2r) - 3$

$\qquad 8r + 9 = 12 + 8r - 3$

$\qquad 8r + 9 = 8r + 9$

$\qquad 0 = 0$ *True*

The last statement is true for any real number r.

The solution set is $(-\infty, \infty)$.

27. $|2p - 3| > 11$

$2p - 3 > 11$ or $2p - 3 < -11$

$\qquad 2p > 14$ $\qquad\qquad 2p < -8$

$\qquad p > 7$ or $\qquad p < -4$

The solution set is $(-\infty, -4) \cup (7, \infty)$.

29. $|5a + 1| \leq 0$

The expression $|5a + 1|$ is never less than 0 since an absolute value expression must be nonnegative. However, $|5a + 1| = 0$ if

$$5a + 1 = 0$$

$$5a = -1$$

$$a = \frac{-1}{5} = -\frac{1}{5}$$

The solution set is $\left\{-\frac{1}{5}\right\}$.

31. $-2 \leq 3x - 1 \leq 8$

$\qquad -1 \leq 3x \leq 9$

$\qquad -\frac{1}{3} \leq x \leq 3$

The solution set is $\left[-\frac{1}{3}, 3\right]$.

33. $|7z - 1| = |5z + 3|$

$7z - 1 = 5z + 3$ or $7z - 1 = -(5z + 3)$

$\qquad 2z = 4$ $\qquad\qquad 7z - 1 = -5z - 3$

$\qquad\qquad\qquad\qquad\qquad 12z = -2$

$\qquad z = 2$ or $\qquad z = \frac{-2}{12} = -\frac{1}{6}$

The solution set is $\left\{-\frac{1}{6}, 2\right\}$.

35. $|1 - 3x| \geq 4$

$1 - 3x \geq 4$ or $1 - 3x \leq -4$

$\qquad -3x \geq 3$ $\qquad\qquad -3x \leq -5$

$\qquad x \leq -1$ or $\qquad x \geq \frac{5}{3}$

The solution set is $\left(-\infty, -1\right] \cup \left[\frac{5}{3}, \infty\right)$.

37. $-(m + 4) + 2 = 3m + 8$

$\qquad -m - 4 + 2 = 3m + 8$

$\qquad -m - 2 = 3m + 8$

$\qquad -10 = 4m$

$\qquad m = \frac{-10}{4} = -\frac{5}{2}$

The solution set is $\left\{-\frac{5}{2}\right\}$.

39. $-6 \leq \frac{3}{2} - x \leq 6$

$\qquad -12 \leq 3 - 2x \leq 12$

$\qquad -15 \leq -2x \leq 9$

$\qquad \frac{15}{2} \geq x \geq -\frac{9}{2}$ or $-\frac{9}{2} \leq x \leq \frac{15}{2}$

The solution set is $\left[-\frac{9}{2}, \frac{15}{2}\right]$.

41. $|x - 1| \geq -6$

The absolute value of an expression is always nonnegative, so the inequality is true for any real number x.

The solution set is $(-\infty, \infty)$.

43. $8q - (1 - q) = 3(1 + 3q) - 4$

$\qquad 8q - 1 + q = 3 + 9q - 4$

$\qquad 9q - 1 = 9q - 1$ *True*

This is an identity.

The solution set is $(-\infty, \infty)$.

45. $|r - 5| = |r + 9|$

$r - 5 = r + 9$ or $r - 5 = -(r + 9)$

$-5 = 9$ *False* $r - 5 = -r - 9$

$2r = -4$

No solution   or $r = -2$

The solution set is $\{-2\}$.

47. $2x + 1 > 5$ or $3x + 4 < 1$

$2x > 4$ $3x < -3$

$x > 2$ or $x < -1$

The solution set is $(-\infty, -1) \cup (2, \infty)$.

Chapter 2 Review Exercises

1. $-(8 + 3z) + 5 = 2z + 6$

$-8 - 3z + 5 = 2z + 6$

$-3z - 3 = 2z + 6$

$-5z = 9$

$z = -\frac{9}{5}$

The solution set is $\left\{-\frac{9}{5}\right\}$.

2. $-\frac{3}{4}x = -12$

$-3x = -48$ *Multiply by 4.*

$x = 16$

The solution set is $\{16\}$.

3. $\dfrac{2q + 1}{3} - \dfrac{q - 1}{4} = 0$

$4(2q + 1) - 3(q - 1) = 0$ *Multiply by 12.*

$8q + 4 - 3q + 3 = 0$

$5q + 7 = 0$

$5q = -7$

$q = -\frac{7}{5}$

The solution set is $\left\{-\frac{7}{5}\right\}$.

4. $5(2x - 3) = 6(x - 1) + 4x$

$10x - 15 = 6x - 6 + 4x$

$10x - 15 = 10x - 6$

$-15 = -6$ *False*

This is a false statement, so the equation is a contradiction.

The solution set is $\emptyset$.

5. $7r - 3(2r - 5) + 5 + 3r = 4r + 20$

$7r - 6r + 15 + 5 + 3r = 4r + 20$

$4r + 20 = 4r + 20$

$20 = 20$ *True*

This equation is an *identity*.

The solution set is $(-\infty, \infty)$.

6. $8p - 4p - (p - 7) + 9p + 6 = 12p - 7$

$8p - 4p - p + 7 + 9p + 6 = 12p - 7$

$12p + 13 = 12p - 7$

$13 = -7$ *False*

This equation is a *contradiction*.

The solution set is $\emptyset$.

7. $-2r + 6(r - 1) + 3r - (4 - r) = -(r + 5) - 5$

$-2r + 6r - 6 + 3r - 4 + r = -r - 5 - 5$

$8r - 10 = -r - 10$

$9r = 0$

$r = 0$

This equation is a *conditional* equation.

The solution set is $\{0\}$.

8. Solve $V = LWH$ for L.

$\dfrac{V}{WH} = \dfrac{LWH}{WH}$

$\dfrac{V}{WH} = L$ or $L = \dfrac{V}{WH}$

9. Solve $A = \frac{1}{2}h(b + B)$ for b.

$2A = h(b + B)$ or $2A = hb + hB$

$\dfrac{2A}{h} = b + B$ or $2A - hB = hb$

$\dfrac{2A}{h} - B = b$ or $\dfrac{2A - hB}{h} = b$

10. Solve $M = -\frac{1}{4}(x + 3y)$ for x.

$-4M = x + 3y$ *Multiply by −4.*

$x = -4M - 3y$

11. Solve $P = \frac{3}{4}x - 12$ for x.

$P + 12 = \frac{3}{4}x$ *Add 12.*

$x = \frac{4}{3}(P + 12)$ *Multiply by 4/3.*

$x = \frac{4}{3}P + 16$

12. Solve $-2x + 5 = 7$.

Begin by subtracting 5 from each side. Then divide each side by -2.

13. Use the formula $V = LWH$ and substitute 180 for V, 6 for L, and 5 for W.

$180 = 6(5)H$

$180 = 30H$

$6 = H$

The height is 6 feet.

14. Divide the amount of decrease by the original amount.

$\dfrac{17{,}849 - 15{,}798}{17{,}849} = \dfrac{2051}{17{,}849} \approx 0.115$

The percent decrease is about 11.5%.

15. Use the formula $I = prt$. Substitute $7800 for I, $30,000 for p, and 4 for t. Solve for r.

$$I = prt$$
$$\$7800 = (\$30{,}000)r(4)$$
$$7800 = 120{,}000r$$
$$r = \frac{7800}{120{,}000} = 0.065$$

The rate is 6.5%.

16. Use the formula $C = \frac{5}{9}(F - 32)$ and substitute 77 for F.

$$C = \frac{5}{9}(77 - 32) = \frac{5}{9}(45) = 25$$

The Celsius temperature is 25°.

17. The amount of money spent on Social Security in 2005 was about

$$0.21(\$2500 \text{ billion}) = \$525 \text{ billion}.$$

18. The amount of money spent on education and social services in 2005 was about

$$0.039(\$2500 \text{ billion}) = \$97.5 \text{ billion}$$

19. "One-third of a number, subtracted from 9" is written

$$9 - \frac{1}{3}x.$$

20. "The product of 4 and a number, divided by 9 more than the number" is written

$$\frac{4x}{x + 9}.$$

21. Let x = the width of the rectangle.
Then $2x - 3$ = the length of the rectangle.

Use the formula $P = 2L + 2W$ with $P = 42$.

$$42 = 2(2x - 3) + 2x$$
$$42 = 4x - 6 + 2x$$
$$48 = 6x$$
$$8 = x$$

The width is 8 meters and the length is $2(8) - 3 = 13$ meters.

22. Let x = the length of each equal side. Then $2x - 15$ = the length of the third side.

Use the formula $P = a + b + c$ with $P = 53$.

$$53 = x + x + (2x - 15)$$
$$53 = 4x - 15$$
$$68 = 4x$$
$$17 = x$$

The lengths of the three sides are 17 inches, 17 inches, and $2(17) - 15 = 19$ inches.

23. Let x = the number of kilograms of peanut clusters. Then $3x$ is the number of kilograms of chocolate creams.
The clerk has a total of 48 kg.

$$x + 3x = 48$$
$$4x = 48$$
$$x = 12$$

The clerk has 12 kilograms of peanut clusters.

24. Let x = the number of liters of the 20% solution. Make a table.

Liters of Solution	Percent (as a decimal)	Liters of Pure Chemical
x	0.20	$0.20x$
15	0.50	$0.50(15) = 7.5$
$x + 15$	0.30	$0.30(x + 15)$

The last column gives the equation.

$$0.20x + 7.5 = 0.30(x + 15)$$
$$0.20x + 7.5 = 0.30x + 4.5$$
$$3 = 0.10x$$
$$30 = x$$

30 L of the 20% solution should be used.

25.

Liters of Solution	Percent (as a decimal)	Liters of Pure Acid
30	0.40	$0.40(30) = 12$
x	0	$0(x) = 0$
$30 + x$	0.30	$0.30(30 + x)$

The last column gives the equation.

$$12 + 0 = 0.30(30 + x)$$
$$12 = 9 + 0.3x$$
$$3 = 0.3x$$
$$10 = x$$

10 L of water should be added.

26. Let x = the amount invested at 6%. Then $x - 4000$ = the amount invested at 4%.

Principal	Rate (as a decimal)	Interest
x	0.06	$0.06x$
$x - 4000$	0.04	$0.04(x - 4000)$
	Total →	$840

The last column gives the equation.

$$0.06x + 0.04(x - 4000) = 840$$
$$6x + 4(x - 4000) = 84{,}000 \quad \textit{Multiply by 100.}$$
$$6x + 4x - 16{,}000 = 84{,}000$$
$$10x = 100{,}000$$
$$x = 10{,}000$$

Jay should invest $10,000 at 6% and $10,000 - $4000 = $6000 at 4%.

27. Let x = the number of quarters. Then $2x - 1$ is the number of dimes.

Number of Coins	Denomination	Value
x	0.25	$0.25x$
$2x - 1$	0.10	$0.10(2x - 1)$
	Total →	3.50

The sum of the values equals the total value.

$$0.25x + 0.10(2x - 1) = 3.50$$
Multiply by 100.
$$25x + 10(2x - 1) = 350$$
$$25x + 20x - 10 = 350$$
$$45x = 360$$
$$x = 8$$

There are 8 quarters and $2(8) - 1 = 15$ dimes.

Check $8(0.25) + 15(0.10) = 3.50$

28. Let x = the number of nickels. Then $19 - x$ is the number of dimes.

Number of Coins	Denomination	Value
x	0.05	$0.05x$
$19 - x$	0.10	$0.10(19 - x)$
	Total →	1.55

The sum of the values equals the total value.

$$0.05x + 0.10(19 - x) = 1.55$$
Multiply by 100.
$$5x + 10(19 - x) = 155$$
$$5x + 190 - 10x = 155$$
$$-5x = -35$$
$$x = 7$$

He had 7 nickels and $19 - 7 = 12$ dimes.

Check $7(0.05) + 12(0.10) = 1.55$

29. Use the formula $d = rt$ or $r = \frac{d}{t}$. Here, d is about 400 mi and t is about 8 hr. Since $\frac{400}{8} = 50$, the best estimate is choice **A**.

30. Use the formula $d = rt$.

(a) Here, $r = 53$ mph and $t = 10$ hr.
$$d = 53(10) = 530$$

The distance is 530 miles.

(b) Here, $r = 164$ mph and $t = 2$ hr.
$$d = 164(2) = 328$$

The distance is 328 miles.

31. Let x = the time it takes for the trains to be 297 mi apart.

Use the formula $d = rt$.

	Rate	Time	Distance
Passenger Train	60	x	$60x$
Freight Train	75	x	$75x$
			297

The total distance traveled is the sum of the distances traveled by each train.

$$60x + 75x = 297$$
$$135x = 297$$
$$x = 2.2$$

It will take the trains 2.2 hours before they are 297 miles apart.

32. Let x = the speed of the faster car and $x - 15$ = the speed of the slower car. Make a table.

	Rate	Time	Distance
Faster Car	x	2	$2x$
Slower Car	$x - 15$	2	$2(x - 15)$
			230

The total distance traveled is the sum of the distances traveled by each car.

$$2x + 2(x - 15) = 230$$
$$2x + 2x - 30 = 230$$
$$4x = 260$$
$$x = 65$$

The faster car travels at 65 km/hr, while the slower car travels at $65 - 15 = 50$ km/hr.

Check $2(65) + 2(50) = 230$

33. Let x = amount of time spent averaging 45 miles per hour. Then $4 - x$ = amount of time at 50 mph.

	Rate	Time	Distance
First Part	45	x	$45x$
Second Part	50	$4 - x$	$50(4 - x)$
Total			195

From the last column:

$$45x + 50(4 - x) = 195$$
$$45x + 200 - 50x = 195$$
$$-5x = -5$$
$$x = 1$$

The automobile averaged 45 mph for 1 hour.

Check 45 mph for 1 hour = 45 miles and 50 mph for 3 hours = 150 miles; $45 + 150 = 195$.

34. Let $x =$ the average speed for the first hour. Then $x - 7 =$ the average speed for the second hour. Using $d = rt$, the distance traveled for the first hour is $x(1)$, for the second hour is $(x - 7)(1)$, and for the whole trip, 85.

$$x + (x - 7) = 85$$
$$2x - 7 = 85$$
$$2x = 92$$
$$x = 46$$

The average speed for the first hour was 46 mph.

Check 46 mph for 1 hour = 46 miles and $46 - 7 = 39$ mph for 1 hour = 39 miles; $46 + 39 = 85$.

35. The sum of the angles in a triangle is $180°$.

$$(3x + 7) + (4x + 1) + (9x - 4) = 180$$
$$16x + 4 = 180$$
$$16x = 176$$
$$x = 11$$

The first angle is $3(11) + 7 = 40°$.
The second angle is $4(11) + 1 = 45°$.
The third angle is $9(11) - 4 = 95°$.

36. The marked angles are supplements which have a sum of $180°$.

$$(15x + 15) + (3x + 3) = 180$$
$$18x + 18 = 180$$
$$18x = 162$$
$$x = 9$$

The angle measures are
$15(9) + 15 = 150°$ and $3(9) + 3 = 30°$.

37. $-\dfrac{2}{3}k < 6$

$-2k < 18$ *Multiply by 3.*
Divide by -2; reverse the inequality symbol.
  $k > -9$
The solution set is $(-9, \infty)$.

38. $-5x - 4 \geq 11$
    $-5x \geq 15$
Divide by -5; reverse the inequality symbol.
    $x \leq -3$
The solution set is $(-\infty, -3]$.

39. $\dfrac{6a + 3}{-4} < -3$

Multiply by -4; reverse the inequality symbol.
$6a + 3 > 12$
  $6a > 9$
  $a > \frac{9}{6} = \frac{3}{2}$
The solution set is $\left(\frac{3}{2}, \infty\right)$.

40. $5 - (6 - 4k) \geq 2k - 7$
$5 - 6 + 4k \geq 2k - 7$
$4k - 1 \geq 2k - 7$
$2k \geq -6$
$k \geq -3$
The solution set is $[-3, \infty)$.

41. $8 \leq 3z - 1 < 14$
$9 \leq 3z < 15$
$3 \leq z < 5$
The solution set is $[3, 5)$.

42. $\frac{5}{3}(m - 2) + \frac{2}{5}(m + 1) > 1$
$25(m - 2) + 6(m + 1) > 15$
        Multiply by 15.
$25m - 50 + 6m + 6 > 15$
$31m - 44 > 15$
$31m > 59$
$m > \frac{59}{31}$
The solution set is $\left(\frac{59}{31}, \infty\right)$.

43. Let $x =$ the other dimension of the rectangle. One dimension of the rectangle is 22 and the perimeter can be no greater than 120.

$$P \leq 120$$
$$2L + 2W \leq 120$$
$$2(x) + 2(22) \leq 120$$
$$2x + 44 \leq 120$$
$$2x \leq 76$$
$$x \leq 38$$

The other dimension must be 38 meters or less.

44. Let $x =$ the number of tickets that can be purchased. The total cost of the tickets is $\$89x$. Including the $\$50$ discount and staying within the available $\$2000$, we have

$$89x - 50 \leq 2000.$$
$$89x \leq 2050$$
$$x \lesssim 23.03$$

The group can purchase 23 tickets or less.

45. Let $x =$ the student's score on the fifth test. The average of the five test scores must be at least 70. The inequality is

$$\frac{75 + 79 + 64 + 71 + x}{5} \geq 70.$$
$$75 + 79 + 64 + 71 + x \geq 350$$
$$289 + x \geq 350$$
$$x \geq 61$$

The student will pass algebra if any score greater than or equal to 61% on the fifth test is achieved.

46. The result, $-8 < -13$, is a false statement. There are no real numbers that make this inequality true. The solution set is $\emptyset$.

For Exercises 47–50, let $A = \{a, b, c, d\}$, $B = \{a, c, e, f\}$, and $C = \{a, e, f, g\}$.

47. $A \cap B = \{a, b, c, d\} \cap \{a, c, e, f\}$
$= \{a, c\}$

48. $A \cap C = \{a, b, c, d\} \cap \{a, e, f, g\}$
$= \{a\}$

49. $B \cup C = \{a, c, e, f\} \cup \{a, e, f, g\}$
$= \{a, c, e, f, g\}$

50. $A \cup C = \{a, b, c, d\} \cup \{a, e, f, g\}$
$= \{a, b, c, d, e, f, g\}$

51. $x > 6$ and $x < 9$

The graph of the solution set will be all numbers which are both greater than 6 and less than 9. The overlap is the numbers between 6 and 9, not including the endpoints.

The solution set is $(6, 9)$.

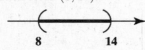

52. $x + 4 > 12$ and $x - 2 < 12$
$x > 8$ and $x < 14$

The graph of the solution set will be all numbers between 8 and 14, not including the endpoints.

The solution set is $(8, 14)$.

53. $x > 5$ or $x \le -3$

The graph of the solution set will be all numbers that are either greater than 5 or less than or equal to -3.

The solution set is $(-\infty, -3] \cup (5, \infty)$.

54. $x \ge -2$ or $x < 2$

The graph of the solution set will be all numbers that are either greater than or equal to -2 or less than 2. All real numbers satisfy these criteria.

The solution set is $(-\infty, \infty)$.

55. $x - 4 > 6$ and $x + 3 \le 10$
$x > 10$ and $x \le 7$

The graph of the solution set will be all numbers that are both greater than 10 and less than or equal to 7. There are no real numbers satisfying these criteria.

The solution set is $\emptyset$.

56. $-5x + 1 \ge 11$ or $3x + 5 \ge 26$
$-5x \ge 10$ $3x \ge 21$
$x \le -2$ or $x \ge 7$

The graph of the solution set will be all numbers that are either less than or equal to -2 or greater than or equal to 7.

The solution set is $(-\infty, -2] \cup [7, \infty)$.

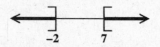

57. $(-3, \infty) \cap (-\infty, 4)$
$(-3, \infty)$ includes all real numbers greater than -3.
$(-\infty, 4)$ includes all real numbers less than 4. Find the intersection. The numbers common to both sets are greater than -3 and less than 4.

$$-3 < x < 4$$

The solution set is $(-3, 4)$.

58. $(-\infty, 6) \cap (-\infty, 2)$
$(-\infty, 6)$ includes all real numbers less than 6.
$(-\infty, 2)$ includes all real numbers less than 2. Find the intersection. The numbers common to both sets are less than 2.

The solution set is $(-\infty, 2)$.

59. $(4, \infty) \cup (9, \infty)$
$(4, \infty)$ includes all real numbers greater than 4.
$(9, \infty)$ includes all real numbers greater than 9. Find the union. The numbers in the first set, the second set, or in both sets are all the real numbers that are greater than 4.

The solution set is $(4, \infty)$.

60. $(1, 2) \cup (1, \infty)$
$(1, 2)$ includes the real numbers between 1 and 2, not including 1 and 2.
$(1, \infty)$ includes all real numbers greater than 1. Find the union. The numbers in the first set, the second set, or in both sets are all real numbers greater than 1.

The solution set is $(1, \infty)$.

61. $|x| = 7$
$x = 7$ or $x = -7$
The solution set is $\{-7, 7\}$.

62. $|x + 2| = 9$

$x + 2 = 9$ or $x + 2 = -9$
$x = 7$ or $x = -11$

The solution set is $\{-11, 7\}$.

63. $|3k - 7| = 8$

$3k - 7 = 8$ or $3k - 7 = -8$

$3k = 15$ $\qquad$ $3k = -1$

$k = 5$ or $\qquad$ $k = -\frac{1}{3}$

The solution set is $\left\{-\frac{1}{3}, 5\right\}$.

64. $|z - 4| = -12$

Since the absolute value of an expression can never be negative, there are no solutions for this equation.

The solution set is $\emptyset$.

65. $|2k - 7| + 4 = 11$

$|2k - 7| = 7$

$2k - 7 = 7$ or $2k - 7 = -7$

$2k = 14$ $\qquad$ $2k = 0$

$k = 7$ or $\qquad$ $k = 0$

The solution set is $\{0, 7\}$.

66. $|4a + 2| - 7 = -3$

$|4a + 2| = 4$

$4a + 2 = 4$ or $4a + 2 = -4$

$4a = 2$ $\qquad$ $4a = -6$

$a = \frac{2}{4}$ $\qquad$ $a = -\frac{6}{4}$

$a = \frac{1}{2}$ or $\qquad$ $a = -\frac{3}{2}$

The solution set is $\left\{-\frac{3}{2}, \frac{1}{2}\right\}$.

67. $|3p + 1| = |p + 2|$

$3p + 1 = p + 2$ or $3p + 1 = -(p + 2)$

$2p = 1$ $\qquad$ $3p + 1 = -p - 2$

$\qquad$ $\qquad$ $4p = -3$

$p = \frac{1}{2}$ or $\qquad$ $p = -\frac{3}{4}$

The solution set is $\left\{-\frac{3}{4}, \frac{1}{2}\right\}$.

68. $|2m - 1| = |2m + 3|$

$2m - 1 = 2m + 3$ or $2m - 1 = -(2m + 3)$

$-1 = 3$ *False* $\qquad$ $2m - 1 = -2m - 3$

$\qquad$ $\qquad$ $4m = -2$

No solution   or $\qquad$ $m = -\frac{2}{4} = -\frac{1}{2}$

The solution set is $\left\{-\frac{1}{2}\right\}$.

69. $|p| < 14$

$-14 < p < 14$

The solution set is $(-14, 14)$.

70. $|-t + 6| \le 7$

$-7 \le -t + 6 \le 7$

$-13 \le -t \le 1$

Multiply by -1; reverse the inequality symbols.

$13 \ge t \ge -1$ or $-1 \le t \le 13$

The solution set is $[-1, 13]$.

71. $|2p + 5| \le 1$

$-1 \le 2p + 5 \le 1$

$-6 \le 2p \le -4$

$-3 \le p \le -2$

The solution set is $[-3, -2]$.

72. $|x + 1| \ge -3$

Since the absolute value of an expression is always nonnegative (positive or zero), the inequality is *true* for any real number x.

The solution set is $(-\infty, \infty)$.

73. $5 - (6 - 4k) > 2k - 5$

$5 - 6 + 4k > 2k - 5$

$-1 + 4k > 2k - 5$

$2k > -4$

$k > -2$

The solution set is $(-2, \infty)$.

74. Solve $ak + bt = 6t - sk$ for k.

Get the "k-terms" on one side.

$$ak + sk = 6t - bt$$
$$(a + s)k = 6t - bt$$
$$k = \frac{6t - bt}{a + s}$$

If we take the k-terms to the right side in the first step, we get

$$k = \frac{bt - 6t}{-a - s}.$$

75. $x < 3$ and $x \ge -2$

The real numbers that are common to both sets are the numbers greater than or equal to -2 and less than 3.

$$-2 \le x < 3$$

The solution set is $[-2, 3)$.

76. $\dfrac{4x + 2}{4} + \dfrac{3x - 1}{8} = \dfrac{x + 6}{16}$

Clear fractions by multiplying by the LCD, 16.

$4(4x + 2) + 2(3x - 1) = x + 6$

$16x + 8 + 6x - 2 = x + 6$

$22x + 6 = x + 6$

$21x = 0$

$x = 0$

The solution set is $\{0\}$.

77. $|3k + 6| \ge 0$

The absolute value of an expression is always nonnegative, so the inequality is true for any real number k.

The solution set is $(-\infty, \infty)$.

78. $-5r \geq -10$

     $r \leq 2$ *Divide by -5; reverse symbol.*

The solution set is $(-\infty, 2]$.

79.   Use the formula $V = LWH$, and solve for H.

$$\frac{V}{LW} = \frac{LWH}{LW}$$

$$\frac{V}{LW} = H \quad \text{or} \quad H = \frac{V}{LW}$$

Substitute 1.5 for W, 5 for L, and 75 for V.

$$H = \frac{75}{5(1.5)} = \frac{75}{7.5} = 10$$

The height of the box is 10 ft.

80.   Let $x =$ the first consecutive integer. Then $x + 1 =$ the second consecutive integer and $x + 2 =$ the third consecutive integer. The sum of the first and third integers is 47 more than the second integer, so an equation is

$$x + (x + 2) = 47 + (x + 1).$$
$$2x + 2 = 48 + x$$
$$x = 46$$

Then $x + 1 = 47$, and $x + 2 = 48$.
The integers are 46, 47, and 48.

81. $|3x + 2| + 4 = 9$

     $|3x + 2| = 5$

$3x + 2 = 5$ or $3x + 2 = -5$

   $3x = 3$ $3x = -7$

    $x = 1$ or $x = -\frac{7}{3}$

The solution set is $\left\{-\frac{7}{3}, 1\right\}$.

82. $0.05x + 0.03(1200 - x) = 42$

    $5x + 3(1200 - x) = 4200$ *Multiply by 100.*

    $5x + 3600 - 3x = 4200$

            $2x = 600$

             $x = 300$

The solution set is $\{300\}$.

83. $|m + 3| \leq 13$

$-13 \leq m + 3 \leq 13$

$-16 \leq \quad m \quad \leq 10$

The solution set is $[-16, 10]$.

84. $\frac{3}{4}(a - 2) - \frac{1}{3}(5 - 2a) < -2$

   $9(a - 2) - 4(5 - 2a) < -24$

                Multiply by 12.

   $9a - 18 - 20 + 8a < -24$

         $17a - 38 < -24$

            $17a < 14$

              $a < \frac{14}{17}$

The solution set is $\left(-\infty, \frac{14}{17}\right)$.

85. $-4 < 3 - 2k < 9$

  $-7 < -2k < 6$

Divide by -2; reverse the inequality symbols.

  $\frac{7}{2} > k > -3$ or $-3 < k < \frac{7}{2}$

The solution set is $\left(-3, \frac{7}{2}\right)$.

86. $-0.3x + 2.1(x - 4) \leq -6.6$

    $-3x + 21(x - 4) \leq -66$

              Multiply by 10.

   $-3x + 21x - 84 \leq -66$

        $18x - 84 \leq -66$

            $18x \leq 18$

              $x \leq 1$

The solution set is $(-\infty, 1]$.

87.   Let $x =$ the angle. Then $90 - x$ is its complement and $180 - x$ is its supplement. The complement of an angle measures $10°$ less than one-fifth of its supplement.

   $90 - x = \frac{1}{5}(180 - x) - 10$

  $450 - 5x = 180 - x - 50$ *Multiply by 5.*

  $450 - 5x = 130 - x$

       $320 = 4x$

        $80 = x$

The measure of the angle is $80°$.

88.   Let $x =$ the employee's earnings during the fifth month. The average of the five months must be at least $1000.

$$\frac{900 + 1200 + 1040 + 760 + x}{5} \geq 1000$$
$$900 + 1200 + 1040 + 760 + x \geq 5000$$
$$3900 + x \geq 5000$$
$$x \geq 1100$$

Any amount greater than or equal to $1100 will qualify the employee for the pension plan.

89. $|5r - 1| > 14$

$5r - 1 > 14$ or $5r - 1 < -14$

  $5r > 15$ $5r < -13$

   $r > 3$ or $r < -\frac{13}{5}$

The solution set is $\left(-\infty, -\frac{13}{5}\right) \cup (3, \infty)$.

90. $x \geq -2$ or $x < 4$

The solution set includes all numbers either greater than or equal to -2 or all numbers less than 4. This is the union and is the set of all real numbers. The solution set is $(-\infty, \infty)$.

91. Let x = the number of liters of the 20% solution. Then $x + 10$ is the number of liters of the resulting 40% solution.

Liters of Solution	Percent (as a decimal)	Liters of Mixture
x	0.20	$0.20x$
10	0.50	$0.50(10) = 5$
$x + 10$	0.40	$0.40(x + 10)$

From the last column:

$$0.20x + 5 = 0.40(x + 10)$$
$$0.20x + 5 = 0.40x + 4$$
$$1 = 0.20x$$
$$5 = x$$

5 L of the 20% solution should be used.

92. $|m - 1| = |2m + 3|$

$m - 1 = 2m + 3$　or　$m - 1 = -(2m + 3)$
$$m - 1 = -2m - 3$$
$$3m = -2$$
$-4 = m$　or　$m = -\frac{2}{3}$

The solution set is $\left\{-4, -\frac{2}{3}\right\}$.

93. $\dfrac{3x}{5} - \dfrac{x}{2} = 3$

$6x - 5x = 30$　*Multiply by 10.*
$$x = 30$$

The solution set is $\{30\}$.

94. $|m + 3| \le 1$
$$-1 \le m + 3 \le 1$$
$$-4 \le m \le -2$$

The solution set is $[-4, -2]$.

95. $|3k - 7| = 4$

$3k - 7 = 4$　or　$3k - 7 = -4$
$3k = 11$　　　　　$3k = 3$
$k = \frac{11}{3}$　or　$k = 1$

The solution set is $\left\{1, \frac{11}{3}\right\}$.

96. $5(2x - 7) = 2(5x + 3)$
$$10x - 35 = 10x + 6$$
$$-35 = 6 \quad \textit{False}$$

This equation is a *contradiction*.
The solution set is $\emptyset$.

97. **(a)** $|5x + 3| < k$

If $k < 0$, then $|5x + 3|$ would be less than a negative number. Since the absolute value of an expression is always nonnegative (positive or zero), the solution set is $\emptyset$.

(b) $|5x + 3| > k$

If $k < 0$, then $|5x + 3|$ would be greater than a negative number. Since the absolute value of an expression is always nonnegative (positive or zero), the solution set is the set of all real numbers, $(-\infty, \infty)$.

(c) $|5x + 3| = k$

If $k < 0$, then $|5x + 3|$ would be equal to a negative number. Since the absolute value of an expression is always nonnegative (positive or zero), the solution set is $\emptyset$.

98. $x > 6$ and $x < 8$

The graph of the solution set is all numbers both greater than 6 *and* less than 8. This is the intersection. The elements common to both sets are the numbers between 6 and 8, not including the endpoints. The solution set is $(6, 8)$.

99. $-5x + 1 \ge 11$　or　$3x + 5 \ge 26$
$-5x \ge 10$　　　　　$3x \ge 21$
$x \le -2$　or　　　　$x \ge 7$

The graph of the solution set is all numbers either less than or equal to -2 *or* greater than or equal to 7. This is the union. The solution set is $(-\infty, -2] \cup [7, \infty)$.

100. **(a)** The set of states with less than 1 million female workers is {Maine, Oregon, Utah}. The set of states with more than 1 million male workers is {Illinois, North Carolina, Oregon, Wisconsin}. Oregon is the only state in both sets, so the set of states with less than 1 million female workers *and* more than 1 million male workers is {Oregon}.

(b) The set of states with less than 1 million female workers *or* more than 2 million male workers is {Illinois, Maine, North Carolina, Oregon, Utah}.

(c) It is easy to see that the sum of the female and male workers for each state doesn't exceed 7 million, so the set of states with a total of more than 7 million civilian workers is { }, or $\emptyset$.

Chapter 2 Test

1. $3(2x - 2) - 4(x + 6) = 3x + 8 + x$
$6x - 6 - 4x - 24 = 4x + 8$
$2x - 30 = 4x + 8$
$-2x = 38$
$x = -19$

The solution set is $\{-19\}$.

2. $0.08x + 0.06(x + 9) = 1.24$
$8x + 6(x + 9) = 124$
Multiply by 100.
$8x + 6x + 54 = 124$
$14x + 54 = 124$
$14x = 70$
$x = 5$

The solution set is $\{5\}$.

3. $\dfrac{x + 6}{10} + \dfrac{x - 4}{15} = \dfrac{x + 2}{6}$
$3(x + 6) + 2(x - 4) = 5(x + 2)$
Multiply by 30.
$3x + 18 + 2x - 8 = 5x + 10$
$5x + 10 = 5x + 10$ *True*

This is an *identity*.

The solution set is $(-\infty, \infty)$.

4. **(a)** $3x - (2 - x) + 4x + 2 = 8x + 3$
$3x - 2 + x + 4x + 2 = 8x + 3$
$8x = 8x + 3$
$0 = 3$ *False*

The false statement indicates that the equation is a *contradiction*.

The solution set is $\emptyset$.

(b) $\dfrac{x}{3} + 7 = \dfrac{5x}{6} - 2 - \dfrac{x}{2} + 9$
Multiply each side by the LCD, 6.
$2x + 42 = 5x - 12 - 3x + 54$
$2x + 42 = 2x + 42$
$0 = 0$ *True*

This equation is an *identity*.

The solution set is $\{$all real numbers$\}$.

(c) $-4(2x - 6) = 5x + 24 - 7x$
$-8x + 24 = -2x + 24$
$24 = 6x + 24$
$0 = 6x$
$0 = x$

This is a *conditional equation*.

The solution set is $\{0\}$.

5. Solve $-16t^2 + vt - S = 0$ for v.
$$vt = S + 16t^2$$
$$v = \frac{S + 16t^2}{t}$$

6. Solve $ar + 2 = 3r - 6t$ for r.
Get the "r-terms" on one side.
$$ar - 3r = -2 - 6t$$
$$r(a - 3) = -2 - 6t$$
$$r = \frac{-2 - 6t}{a - 3}$$

If we take the r-terms to the right side in the first step, we get
$$r = \frac{2 + 6t}{3 - a}.$$

7. Solve $d = rt$ for t and substitute 500 for d and 157.603 for r.
$$t = \frac{d}{r} = \frac{500}{157.603} \approx 3.173$$

Wheldon's time was about 3.173 hr.

8. Use $I = Prt$ and substitute \$2281.25 for I, \$36,500 for P, and 1 for t.
$$2281.25 = 36{,}500r(1)$$
$$r = \frac{2281.25}{36{,}500} = 0.0625$$

The rate of interest is 6.25%.

9. $\dfrac{27{,}385}{37{,}142} \approx 0.737$

73.7% were classified as post offices.

10. Let $x =$ the amount invested at 3%. Then $28{,}000 - x =$ the amount invested at 5%.

Principal	Rate (as a decimal)	Interest
x	0.03	$0.03x$
$28{,}000 - x$	0.05	$0.05(28{,}000 - x)$
\$28,000	← Totals →	\$1240

From the last column:
$$0.03x + 0.05(28{,}000 - x) = 1240$$
$$3x + 5(28{,}000 - x) = 124{,}000$$
Multiply by 100.
$$3x + 140{,}000 - 5x = 124{,}000$$
$$-2x = -16{,}000$$
$$x = 8000$$

He invested \$8000 at 3% and $\$28{,}000 - \$8000 = \$20{,}000$ at 5%.

11. Let $x =$ the speed of the slower car. Then $x + 15 =$ the speed of the faster car.

Use the formula $d = rt$.

	Rate	Time	Distance
Slower Car	x	6	$6x$
Faster Car	$x + 15$	6	$6(x + 15)$
			630

The total distance traveled is the sum of the distances traveled by each car.

$$6x + 6(x + 15) = 630$$
$$6x + 6x + 90 = 630$$
$$12x = 540$$
$$x = 45$$

The slower car traveled at 45 mph, while the faster car traveled at $45 + 15 = 60$ mph.

12. The sum of the three angle measures is $180°$.

$$(2x + 20) + x + x = 180$$
$$4x + 20 = 180$$
$$4x = 160$$
$$x = 40$$

The three angle measures are $40°$, $40°$, and $(2 \cdot 40 + 20)° = 100°$.

13.
$$4 - 6(x + 3) \leq -2 - 3(x + 6) + 3x$$
$$4 - 6x - 18 \leq -2 - 3x - 18 + 3x$$
$$-6x - 14 \leq -20$$
$$-6x \leq -6$$

Divide by -6, and reverse the inequality symbol.
$$x \geq 1$$
The solution set is $[1, \infty)$.

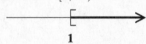

1

14. $-\dfrac{4}{7}x > -16$

$$-4x > -112 \qquad \textit{Multiply by 7.}$$

Divide by -4, and reverse the inequality symbol.
$$x < 28$$

The solution set is $(-\infty, 28)$.

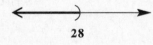

28

15. $-6 \leq \frac{4}{3}x - 2 \leq 2$

$$-18 \leq 4x - 6 \leq 6 \qquad \textit{Multiply by 3.}$$
$$-12 \leq 4x \leq 12 \qquad \textit{Add 6.}$$
$$-3 \leq x \leq 3 \qquad \textit{Divide by 4.}$$

The solution set is $[-3, 3]$.

16. For each inequality, divide both sides by -3 and reverse the direction of the inequality symbol.

A. $\quad -3x < 9$
$\qquad\quad x > -3$

B. $\quad -3x > -9$
$\qquad\quad x < 3$

C. $\quad -3x > 9$
$\qquad\quad x < -3$

D. $\quad -3x < -9$
$\qquad\quad x > 3$

Thus, inequality **C** is equivalent to $x < -3$.

17. Let $x =$ the score on the fourth test.

$$\frac{83 + 76 + 79 + x}{4} \geq 80$$
$$\frac{238 + x}{4} \geq 80$$
$$238 + x \geq 320$$
$$x \geq 82$$

The minimum score must be 82 to guarantee a B.

18. $C = 50x + 5000; \ R = 60x$

$$R \geq C$$
$$60x \geq 50x + 5000$$
$$10x \geq 5000$$
$$x \geq 500$$

The solution set is $[500, \infty)$.

19. **(a)** $A \cap B = \{1, 2, 5, 7\} \cap \{1, 5, 9, 12\}$
$\qquad\qquad = \{1, 5\}$

(b) $A \cup B = \{1, 2, 5, 7\} \cup \{1, 5, 9, 12\}$
$\qquad\qquad = \{1, 2, 5, 7, 9, 12\}$

20. $3k \geq 6 \quad$ and $\quad k - 4 < 5$
$\quad k \geq 2 \quad$ and $\qquad k < 9$
The solution set is all numbers both greater than or equal to 2 *and* less than 9. This is the intersection. The numbers common to both sets are between 2 and 9, including 2 but not 9. The solution set is $[2, 9)$.

21. $-4x \leq -24 \quad$ or $\quad 4x - 2 < 10$
$\qquad\qquad\qquad\qquad\qquad 4x < 12$
$\quad x \geq 6 \qquad$ or $\qquad x < 3$
The solution set is all numbers less than 3 or greater than or equal to 6. This is the union. The solution set is $(-\infty, 3) \cup [6, \infty)$.

22. $|4x + 3| \leq 7$
$$-7 \leq 4x + 3 \leq 7$$
$$-10 \leq 4x \leq 4$$
$$-\frac{10}{4} \leq x \leq \frac{4}{4}$$
$$-\frac{5}{2} \leq x \leq 1$$

The solution set is $\left[-\frac{5}{2}, 1\right]$.

23. $|5 - 6x| > 12$

$$5 - 6x > 12 \quad \text{or} \quad 5 - 6x < -12$$
$$-6x > 7 \qquad\qquad -6x < -17$$
$$x < -\tfrac{7}{6} \quad \text{or} \qquad x > \tfrac{17}{6}$$

The solution set is $\left(-\infty, -\tfrac{7}{6}\right) \cup \left(\tfrac{17}{6}, \infty\right)$.

24. $|7 - x| \le -1$

Since the absolute value of an expression is always nonnegative (positive or zero), the inequality is *false* for any real number x.

The solution set is $\emptyset$.

25. $|-3x + 4| - 4 < -1$

$$|-3x + 4| < 3$$
$$-3 < -3x + 4 < 3$$
$$-7 < -3x < -1$$
$$\tfrac{7}{3} > x > \tfrac{1}{3}, \quad \text{or} \quad \tfrac{1}{3} < x < \tfrac{7}{3}$$

The solution set is $\left(\tfrac{1}{3}, \tfrac{7}{3}\right)$.

26. $|3k - 2| + 1 = 8$

$$|3k - 2| = 7$$
$$3k - 2 = 7 \qquad \text{or} \quad 3k - 2 = -7$$
$$3k = 9 \qquad \text{or} \qquad 3k = -5$$
$$k = \tfrac{9}{3} = 3 \quad \text{or} \qquad k = -\tfrac{5}{3}$$

The solution set is $\left\{-\tfrac{5}{3}, 3\right\}$.

27. $|3 - 5x| = |2x + 8|$

$$3 - 5x = 2x + 8 \quad \text{or} \quad 3 - 5x = -(2x + 8)$$
$$-7x = 5 \qquad\qquad 3 - 5x = -2x - 8$$
$$\qquad\qquad\qquad\qquad -3x = -11$$
$$x = -\tfrac{5}{7} \qquad \text{or} \qquad x = \tfrac{11}{3}$$

The solution set is $\left\{-\tfrac{5}{7}, \tfrac{11}{3}\right\}$.

28. **(a)** $|8x - 5| < k$

If $k < 0$, then $|8x - 5|$ would be less than a negative number. Since the absolute value of an expression is always nonnegative (positive or zero), the solution set is $\emptyset$.

(b) $|8x - 5| > k$

If $k < 0$, then $|8x - 5|$ would be greater than a negative number. Since the absolute value of an expression is always nonnegative (positive or zero), the solution set is the set of all real numbers, $(-\infty, \infty)$.

(c) $|8x - 5| = k$

If $k < 0$, then $|8x - 5|$ would be equal to a negative number. Since the absolute value of an expression is always nonnegative (positive or zero), the solution set is $\emptyset$.

Cumulative Review Exercises (Chapters 1–2)

Exercises 1–6 refer to set A.

Let $A = \left\{-8, -\tfrac{2}{3}, -\sqrt{6}, 0, \tfrac{4}{5}, 9, \sqrt{36}\right\}$.

Simplify $\sqrt{36} = 6$.

1.   The elements 9 and 6 are natural numbers.

2.   The elements 0, 9, and 6 are whole numbers.

3.   The elements $-8, 0, 9,$ and 6 are integers.

4.   The elements $-8, -\tfrac{2}{3}, 0, \tfrac{4}{5}, 9,$ and 6 are rational numbers.

5.   The element $-\sqrt{6}$ is an irrational number.

6.   All the elements in set A are real numbers.

7. $-\tfrac{4}{3} - \left(-\tfrac{2}{7}\right) = -\tfrac{4}{3} + \tfrac{2}{7}$
$$= -\tfrac{28}{21} + \tfrac{6}{21}$$
$$= -\tfrac{22}{21}$$

8. $|-4| - |2| + |-6| = 4 - 2 + 6$
$$= 2 + 6$$
$$= 8$$

9. $(-2)^4 + (-2)^3 = 16 + (-8) = 8$

10. $\sqrt{25} - 5(-1)^0 = 5 - 5(1) = 5 - 5 = 0$

11. $(-3)^5 = (-3)(-3)(-3)(-3)(-3) = -243$

12. $\left(\tfrac{6}{7}\right)^3 = \tfrac{6}{7} \cdot \tfrac{6}{7} \cdot \tfrac{6}{7} = \tfrac{216}{343}$

13. $\left(-\tfrac{2}{3}\right)^3 = \left(-\tfrac{2}{3}\right)\left(-\tfrac{2}{3}\right)\left(-\tfrac{2}{3}\right) = -\tfrac{8}{27}$

14. $-4^6 = -(4 \cdot 4 \cdot 4 \cdot 4 \cdot 4 \cdot 4) = -4096$

15. $-\sqrt{36} = -(6) = -6$
$\sqrt{-36}$ is not a real number.

16. $\dfrac{4 - 4}{4 + 4} = \dfrac{0}{8} = 0$

$\dfrac{4 + 4}{4 - 4} = \dfrac{8}{0}$, which is *undefined*.

For Exercises 17–19, let $a = 2$, $b = -3$, and $c = 4$.

17. $-3a + 2b - c = -3(2) + 2(-3) - 4$
$$= -6 - 6 - 4$$
$$= -16$$

18. $-8\left(a^2 + b^3\right) = -8\left[2^2 + (-3)^3\right]$
$$= -8[4 + (-27)]$$
$$= -8(-23)$$
$$= 184$$

19. $\dfrac{3a^3 - b}{4 + 3c} = \dfrac{3(2)^3 - (-3)}{4 + 3(4)}$

$\qquad = \dfrac{3(8) - (-3)}{4 + 3(4)}$

$\qquad = \dfrac{24 + 3}{4 + 12}$

$\qquad = \dfrac{27}{16}$

20. $-7r + 5 - 13r + 12$

$\qquad = -7r - 13r + 5 + 12$

$\qquad = (-7 - 13)r + (5 + 12)$

$\qquad = -20r + 17$

21. $-(3k + 8) - 2(4k - 7) + 3(8k + 12)$

$\qquad = -3k - 8 - 8k + 14 + 24k + 36$

$\qquad = -3k - 8k + 24k - 8 + 14 + 36$

$\qquad = 13k + 42$

22. $(a + b) + 4 = 4 + (a + b)$

The order of the terms $(a + b)$ and 4 have been reversed. This is an illustration of the commutative property.

23. $4x + 12x = (4 + 12)x$

The common variable, x, has been removed from each term. This is an illustration of the distributive property.

24. $-4x + 7(2x + 3) = 7x + 36$

$\qquad -4x + 14x + 21 = 7x + 36$

$\qquad\qquad 10x + 21 = 7x + 36$

$\qquad\qquad\qquad 3x = 15$

$\qquad\qquad\qquad\; x = 5$

The solution set is $\{5\}$.

25. $\qquad -\tfrac{3}{5}x + \tfrac{2}{3}x = 2$

$\quad 3(-3x) + 5(2x) = 15(2) \quad$ *Multiply by 15.*

$\qquad -9x + 10x = 30$

$\qquad\qquad\qquad x = 30$

The solution set is $\{30\}$.

26. $0.06x + 0.03(100 + x) = 4.35$

$\quad 6x + 3(100 + x) = 435 \quad$ *Multiply by 100.*

$\quad 6x + 300 + 3x = 435$

$\qquad\qquad 9x = 135$

$\qquad\qquad\; x = 15$

The solution set is $\{15\}$.

27. Solve $P = a + b + c$ for b.

$\qquad P - a - c = b \;$ or $\; b = P - a - c$

28. $4(2x - 6) + 3(x - 2) = 11x + 1$

$\quad 8x - 24 + 3x - 6 = 11x + 1$

$\qquad\quad 11x - 30 = 11x + 1$

$\qquad\qquad\quad -30 = 1 \qquad$ *False*

The solution set is $\emptyset$.

29. $\qquad \tfrac{2}{3}x + \tfrac{5}{8}x = \tfrac{31}{24}x$

Multiply by the LCD, 24.

$\qquad 8(2x) + 3(5x) = 31x$

$\qquad\quad 16x + 15x = 31x$

$\qquad\qquad\; 31x = 31x \qquad$ *True*

The solution set is {all real numbers}.

30. $3 - 2(x + 7) \le -x + 3$

$\quad 3 - 2x - 14 \le -x + 3$

$\qquad -2x - 11 \le -x + 3$

$\qquad\qquad -x \le 14$

Multiply by -1, and reverse the inequality symbol.

$\qquad\qquad x \ge -14$

The solution set is $[-14, \infty)$.

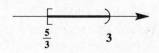

$\qquad\qquad\quad$ **–14**

31. $-4 < 5 - 3x \le 0$

$\quad -9 < -3x \le -5$

Divide by -3, and reverse the inequality symbol.

$\quad 3 > x \ge \tfrac{5}{3} \;$ or $\; \tfrac{5}{3} \le x < 3$

The solution set is $\left[\tfrac{5}{3}, 3\right)$.

$\qquad\qquad \dfrac{5}{3} \qquad\; \mathbf{3}$

32. $2x + 1 > 5 \quad$ or $\quad 2 - x > 2$

$\quad 2x > 4 \qquad\qquad -x > 0$

$\quad\; x > 2 \;$ or $\qquad\; x < 0$

The solution set is $(-\infty, 0) \cup (2, \infty)$.

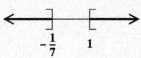

$\qquad\qquad \mathbf{0} \qquad\quad \mathbf{2}$

33. $|-7k + 3| \ge 4$

$\quad -7k + 3 \ge 4 \quad$ or $\quad -7k + 3 \le -4$

$\qquad -7k \ge 1 \qquad\qquad\quad -7k \le -7$

$\qquad\; k \le -\tfrac{1}{7} \;$ or $\qquad\qquad k \ge 1$

The solution set is $\left(-\infty, -\tfrac{1}{7}\right] \cup [1, \infty)$.

$\qquad\qquad -\dfrac{1}{7} \qquad \mathbf{1}$

34. Let x = the amount invested at 5%. Then $x + 2000$ = the amount invested at 6%.

Use $I = prt$ and create a table.

Principal	Rate (as a decimal)	Interest
x	0.05	$0.05x$
$x + 2000$	0.06	$0.06(x + 2000)$
	Total →	$670

From the last column:

$$0.05x + 0.06(x + 2000) = 670$$
$$5x + 6(x + 2000) = 67,000$$

Multiply by 100.

$$5x + 6x + 12,000 = 67,000$$
$$11x = 55,000$$
$$x = 5000$$

She invested $5000 at 5% and $7000 at 6%.

35. Let x = the number of grams of food C. Then $2x$ is the number of grams of food A.
There are 5 grams of food B and the total is at most 24 grams.

$$x + 2x + 5 \le 24$$
$$3x \le 19$$
$$x \le \tfrac{19}{3} \text{ or } 6\tfrac{1}{3}$$

He may use at most $6\tfrac{1}{3}$ grams of food C.

36. Let x = the grade the student must make on the third test.
To find the average of the three tests, add them and divide by 3. This average must be at least 80.

$$\frac{88 + 78 + x}{3} \ge 80$$
$$\frac{166 + x}{3} \ge 80$$
$$166 + x \ge 240$$
$$x \ge 74$$

She must score at least 74 on her third test.

37. Let x = the amount of pure alcohol that should be added.

Liters of Solution	Percent (as a decimal)	Liters of Pure Alcohol
x	1.00	$1.00x$
7	0.10	$0.10(7) = 0.7$
$x + 7$	0.30	$0.30(x + 7)$

From the last column:

$$1.00x + 0.7 = 0.30(x + 7)$$
$$10x + 7 = 3(x + 7) \qquad \textit{Multiply by 10.}$$
$$10x + 7 = 3x + 21$$
$$7x = 14$$
$$x = 2$$

2 L of pure alcohol should be added to the solution.

38. Let x = the number of nickels. Then $x - 4$ = the number of quarters. The collection contains 29 coins, so the number of pennies is

$$29 - x - (x - 4) = 33 - 2x.$$

	Number of Coins	Denomination	Value
Pennies	$33 - 2x$	0.01	$0.01(33 - 2x)$
Nickels	x	0.05	$0.05x$
Quarters	$x - 4$	0.25	$0.25(x - 4)$
Totals	29		$2.69

From the last column:

$$0.01(33 - 2x) + 0.05x + 0.25(x - 4) = 2.69$$
$$1(33 - 2x) + 5x + 25(x - 4) = 269$$

Multiply by 100.

$$33 - 2x + 5x + 25x - 100 = 269$$
$$28x - 67 = 269$$
$$28x = 336$$
$$x = 12$$

There are $33 - 2(12) = 9$ pennies, 12 nickels, and $12 - 4 = 8$ quarters.

39. Clark's rule:

$$\frac{\text{Weight of child in pounds}}{150} \times \frac{\text{adult}}{\text{dose}} = \frac{\text{child's}}{\text{dose}}$$

If the child weighs 55 lb and the adult dosage is 120 mg, then

$$\frac{55}{150} \times 120 = 44.$$

The child's dosage is 44 mg.

40. **(a)** 1990: 1611
2003: 1456

$$1611 - 1456 = 155$$

The number decreased by 155 newspapers.

(b) $\dfrac{155}{1611} \approx 0.096$, or 9.6%.

The number decreased by approximately 9.6%.

CHAPTER 3   GRAPHS, LINEAR EQUATIONS, AND FUNCTIONS

3.1   The Rectangular Coordinate System

1. (a) x represents the year; y represents the revenue in billions of dollars.

    (b) The dot above the year 2002 appears to be at about 1850, so the revenue in 2002 was $1850 billion.

    (c) The ordered pair is $(2002, 1850)$.

    (d) In 2000, federal tax revenues were about $2030 billion.

3. The point with coordinates $(0, 0)$ is called the _origin_ of a rectangular coordinate system.

5. The x-intercept is the point where a line crosses the x-axis. To find the x-intercept of a line, we let _y_ equal 0 and solve for _x_.

    The y-intercept is the point where a line crosses the y-axis. To find the y-intercept of a line, we let _x_ equal 0 and solve for _y_.

7. To graph a straight line, we must find a minimum of _two_ points. A third point is sometimes found to check the accuracy of the first two points.

9. (a) The point $(1, 6)$ is located in quadrant I, since the x- and y-coordinates are both positive.

    (b) The point $(-4, -2)$ is located in quadrant III, since the x- and y-coordinates are both negative.

    (c) The point $(-3, 6)$ is located in quadrant II, since the x-coordinate is negative and the y-coordinate is positive.

    (d) The point $(7, -5)$ is located in quadrant IV, since the x-coordinate is positive and the y-coordinate is negative.

    (e) The point $(-3, 0)$ is located on the x-axis, so it does not belong to any quadrant.

    (f) The point $(0, -0.5)$ is located on the y-axis, so it does not belong to any quadrant.

11. (a) If $xy > 0$, then both x and y have the same sign.
    (x, y) is in quadrant I if x and y are positive.
    (x, y) is in quadrant III if x and y are negative.

    (b) If $xy < 0$, then x and y have different signs.
    (x, y) is in quadrant II if $x < 0$ and $y > 0$.
    (x, y) is in quadrant IV if $x > 0$ and $y < 0$.

    (c) If $\dfrac{x}{y} < 0$, then x and y have different signs.
    (x, y) is in either quadrant II or IV. (See part (b).)

    (d) If $\dfrac{x}{y} > 0$, then x and y have the same sign.
    (x, y) is in either quadrant I or III. (See part (a).)

For Exercises 13–22, see the rectangular coordinate system after Exercise 21.

13. To plot $(2, 3)$, go two units from zero to the right along the x-axis, and then go three units up parallel to the y-axis.

15. To plot $(-3, -2)$, go three units from zero to the left along the x-axis, and then go two units down parallel to the y-axis.

17. To plot $(0, 5)$, do not move along the x-axis at all since the x-coordinate is 0. Move five units up along the y-axis.

19. To plot $(-2, 4)$, go two units from zero to the left along the x-axis, and then go four units up parallel to the y-axis.

21. To plot $(-2, 0)$, go two units to the left along the x-axis. Do not move up or down since the y-coordinate is 0.

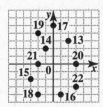

23. $y = x - 4$
    To complete the table, substitute the given values for x and y in the equation.
    For $x = 0$: $y = x - 4$
    $y = 0 - 4$
    $y = -4$ $(0, -4)$

    For $x = 1$: $y = x - 4$
    $y = 1 - 4$
    $y = -3$ $(1, -3)$

    In a similar manner, substitute $x = 2$, 3, and 4 to get $y = -2, -1$, and 0.

x	y
0	-4
1	-3
2	-2
3	-1
4	0

    Plot the ordered pairs and draw the line through them.

continued

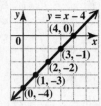

25. $x - y = 3$

To complete the table, substitute the given values for x and y in the equation.

For $x = 0$: $x - y = 3$
$$0 - y = 3$$
$$y = -3 \quad (0, -3)$$

For $y = 0$: $x - y = 3$
$$x - 0 = 3$$
$$x = 3 \quad (3, 0)$$

For $x = 5$: $x - y = 3$
$$5 - y = 3$$
$$-y = -2$$
$$y = 2 \quad (5, 2)$$

For $x = 2$: $x - y = 3$
$$2 - y = 3$$
$$-y = 1$$
$$y = -1 \quad (2, -1)$$

Plot the ordered pairs and draw the line through them.

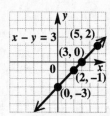

27. $x + 2y = 5$

To complete the table, substitute the given values for x or y in the equation.

For $x = 0$: $x + 2y = 5$
$$0 + 2y = 5$$
$$2y = 5$$
$$y = \tfrac{5}{2} \quad (0, \tfrac{5}{2})$$

For $y = 0$: $x + 2y = 5$
$$x + 2(0) = 5$$
$$x + 0 = 5$$
$$x = 5 \quad (5, 0)$$

For $x = 2$: $x + 2y = 5$
$$2 + 2y = 5$$
$$2y = 3$$
$$y = \tfrac{3}{2} \quad (2, \tfrac{3}{2})$$

For $y = 2$: $x + 2y = 5$
$$x + 2(2) = 5$$
$$x + 4 = 5$$
$$x = 1 \quad (1, 2)$$

Plot the ordered pairs and draw the line through them.

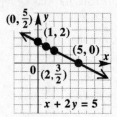

29. $4x - 5y = 20$

For $x = 0$: $4x - 5y = 20$
$$4(0) - 5y = 20$$
$$-5y = 20$$
$$y = -4 \quad (0, -4)$$

For $y = 0$: $4x - 5y = 20$
$$4x - 5(0) = 20$$
$$4x = 20$$
$$x = 5 \quad (5, 0)$$

For $x = 2$: $4x - 5y = 20$
$$4(2) - 5y = 20$$
$$8 - 5y = 20$$
$$-5y = 12$$
$$y = -\tfrac{12}{5} \quad (2, -\tfrac{12}{5})$$

For $y = -3$: $4x - 5y = 20$
$$4x - 5(-3) = 20$$
$$4x + 15 = 20$$
$$4x = 5$$
$$x = \tfrac{5}{4} \quad (\tfrac{5}{4}, -3)$$

Plot the ordered pairs and draw the line through them.

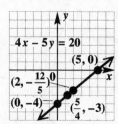

31. $y = -2x + 3$

x	$-2x$	$y = -2x + 3$
0	0	3
1	-2	1
2	-4	-1
3	-6	-3

Notice that as the value of x increases by 1, the value of y decreases by 2.

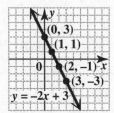

33. **(a)** The y-values are $-4, -3, -2, -1$, and 0. They increase by $\underline{1}$ unit.

(b) The y-values are $3, 1, -1$, and -3. They decrease by $\underline{2}$ units.

(c) It appears that the y-value increases (or decreases) by the value of the coefficient of x. So for $y = 2x + 4$, a conjecture is "For every increase in x by 1 unit, y increases by 2 units."

35. Choose a value *other than* 0 for either x or y and then solve $4x + 5y = 0$ for the other variable. For example, if $x = -5$, then $-20 + 5y = 0$ and $y = 4$. The student should then plot the points $(0, 0)$ and $(-5, 4)$ and draw a line through them.

37. $2x + 3y = 12$
To find the x-intercept, let $y = 0$.

$$2x + 3y = 12$$
$$2x + 3(0) = 12$$
$$2x = 12$$
$$x = 6$$

The x-intercept is $(6, 0)$.
To find the y-intercept, let $x = 0$.

$$2x + 3y = 12$$
$$2(0) + 3y = 12$$
$$3y = 12$$
$$y = 4$$

The y-intercept is $(0, 4)$.
Plot the intercepts and draw the line through them.

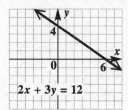

39. $x - 3y = 6$
To find the x-intercept, let $y = 0$.

$$x - 3y = 6$$
$$x - 3(0) = 6$$
$$x - 0 = 6$$
$$x = 6$$

The x-intercept is $(6, 0)$.
To find the y-intercept, let $x = 0$.

$$x - 3y = 6$$
$$0 - 3y = 6$$
$$-3y = 6$$
$$y = -2$$

The y-intercept is $(0, -2)$.
Plot the intercepts and draw the line through them.

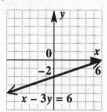

41. $\frac{2}{3}x - 3y = 7$
To find the x-intercept, let $y = 0$.

$$\frac{2}{3}x - 3(0) = 7$$
$$\frac{2}{3}x = 7$$
$$x = \frac{3}{2} \cdot 7 = \frac{21}{2}$$

The x-intercept is $\left(\frac{21}{2}, 0\right)$.
To find the y-intercept, let $x = 0$.

$$\frac{2}{3}(0) - 3y = 7$$
$$-3y = 7$$
$$y = -\frac{7}{3}$$

The y-intercept is $\left(0, -\frac{7}{3}\right)$.
Plot the intercepts and draw the line through them.

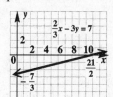

43. $y = 5$
This is a horizontal line. Every point has y-coordinate 5, so no point has y-coordinate 0. There is no x-intercept.
Since every point of the line has y-coordinate 5, the y-intercept is $(0, 5)$. Draw the horizontal line through $(0, 5)$.

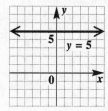

45. $x = 2$

This is a vertical line. Every point has x-coordinate 2, so the x-intercept is $(2, 0)$. Since every point of the line has x-coordinate 2, no point has x-coordinate 0. There is no y-intercept. Draw the vertical line through $(2, 0)$.

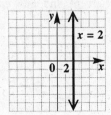

47. $x + 4 = 0$ $(x = -4)$

This is a vertical line. Every point has x-coordinate -4, so the x-intercept is $(-4, 0)$. Since every point of the line has x-coordinate -4, no point has x-coordinate 0. There is no y-intercept. Draw the vertical line through $(-4, 0)$.

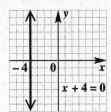

49. $x + 5y = 0$

To find the x-intercept, let $y = 0$.

$$x + 5y = 0$$
$$x + 5(0) = 0$$
$$x = 0$$

The x-intercept is $(0, 0)$, and since $x = 0$, this is also the y-intercept. Since the intercepts are the same, another point is needed to graph the line. Choose any number for y, say $y = -1$, and solve the equation for x.

$$x + 5y = 0$$
$$x + 5(-1) = 0$$
$$x = 5$$

This gives the ordered pair $(5, -1)$. Plot $(5, -1)$ and $(0, 0)$, and draw the line through them.

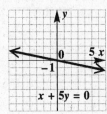

51. $2x = 3y$

If $x = 0$, then $y = 0$, so the x- and y-intercepts are $(0, 0)$. To get another point, let $x = 3$.

$$2(3) = 3y$$
$$2 = y$$

Plot $(3, 2)$ and $(0, 0)$, and draw the line through them.

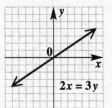

53. $-\frac{2}{3}y = x$

If $x = 0$, then $y = 0$, so the x- and y-intercepts are $(0, 0)$. To get another point, let $y = 3$.

$$-\frac{2}{3}(3) = x$$
$$-2 = x$$

Plot $(-2, 3)$ and $(0, 0)$, and draw the line through them.

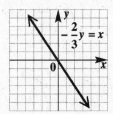

55. By the Midpoint Formula, the midpoint of the segment with endpoints $(-8, 4)$ and $(-2, -6)$ is

$$\left(\frac{-8 + (-2)}{2}, \frac{4 + (-6)}{2} \right) = \left(\frac{-10}{2}, \frac{-2}{2} \right) = (-5, -1).$$

57. By the Midpoint Formula, the midpoint of the segment with endpoints $(3, -6)$ and $(6, 3)$ is

$$\left(\frac{3 + 6}{2}, \frac{-6 + 3}{2} \right) = \left(\frac{9}{2}, \frac{-3}{2} \right) = \left(\frac{9}{2}, -\frac{3}{2} \right).$$

59. By the Midpoint Formula, the midpoint of the segment with endpoints $(-9, 3)$ and $(9, 8)$ is

$$\left(\frac{-9 + 9}{2}, \frac{3 + 8}{2} \right) = \left(\frac{0}{2}, \frac{11}{2} \right) = \left(0, \frac{11}{2} \right).$$

61. By the Midpoint Formula, the midpoint of the segment with endpoints $(2.5, 3.1)$ and $(1.7, -1.3)$ is

$$\left(\frac{2.5 + 1.7}{2}, \frac{3.1 + (-1.3)}{2} \right) = \left(\frac{4.2}{2}, \frac{1.8}{2} \right) = (2.1, 0.9).$$

63. By the Midpoint Formula, the midpoint of the segment with endpoints $\left(\frac{1}{2}, \frac{1}{3}\right)$ and $\left(\frac{3}{2}, \frac{5}{3}\right)$ is

$$\left(\frac{\frac{1}{2} + \frac{3}{2}}{2}, \frac{\frac{1}{3} + \frac{5}{3}}{2}\right) = \left(\frac{\frac{4}{2}}{2}, \frac{\frac{6}{3}}{2}\right) = \left(\frac{2}{2}, \frac{2}{2}\right) = (1, 1).$$

65. By the Midpoint Formula, the midpoint of the segment with endpoints $\left(-\frac{1}{3}, \frac{2}{7}\right)$ and $\left(-\frac{1}{2}, \frac{1}{14}\right)$ is

$$\left(\frac{-\frac{1}{3} + \left(-\frac{1}{2}\right)}{2}, \frac{\frac{2}{7} + \frac{1}{14}}{2}\right) = \left(\frac{-\frac{5}{6}}{2}, \frac{\frac{5}{14}}{2}\right) = \left(-\frac{5}{12}, \frac{5}{28}\right).$$

67. midpoint of $P(5, 8)$ and $Q(x, y) = M(8, 2)$

$$\left(\frac{5 + x}{2}, \frac{8 + y}{2}\right) = (8, 2)$$

The x- and y-coordinates must be equal.

$$\frac{5 + x}{2} = 8 \qquad \frac{8 + y}{2} = 2$$
$$5 + x = 16 \qquad 8 + y = 4$$
$$x = 11 \qquad y = -4$$

Thus, the endpoint Q is $(11, -4)$.

69. midpoint of $P\left(\frac{1}{3}, \frac{1}{5}\right)$ and $Q(x, y) = M\left(\frac{3}{2}, 1\right)$

$$\left(\frac{\frac{1}{3} + x}{2}, \frac{\frac{1}{5} + y}{2}\right) = \left(\frac{3}{2}, 1\right)$$

The x- and y-coordinates must be equal.

$$\frac{\frac{1}{3} + x}{2} = \frac{3}{2} \qquad \frac{\frac{1}{5} + y}{2} = 1$$
$$\frac{1}{3} + x = 3 \qquad \frac{1}{5} + y = 2$$
$$\frac{1}{3} + x = \frac{9}{3} \qquad \frac{1}{5} + y = \frac{10}{5}$$
$$x = \frac{8}{3} \qquad y = \frac{9}{5}$$

Thus, the endpoint Q is $\left(\frac{8}{3}, \frac{9}{5}\right)$.

71. For 2003, $x = 2003 - 2000 = 3$.

$$y = -1237x + 60{,}936$$
$$= -1237(3) + 60{,}936 \quad \textit{Let } x = 3.$$
$$= -3711 + 60{,}936$$
$$y = 57{,}225$$

The approximate number of U.S. travelers to other countries in 2003 was 57,225 thousand (or 57,225,000).

73. The graph goes through the point $(-2, 0)$ which satisfies only equations **B** and **C**. The graph also goes through the point $(0, 3)$ which satisfies only equations **A** and **B**. Therefore, the correct equation is **B**.

75. The screen on the right is more useful because it shows the intercepts.

77. We need to solve the given equation for y.

$$5x + 2y = -10$$
$$2y = -5x - 10 \quad \textit{Subtract 5x.}$$
$$y = -\frac{5}{2}x - 5 \quad \textit{Divide by 2.}$$

Graph $Y_1 = -2.5X - 5$ in a standard viewing window.

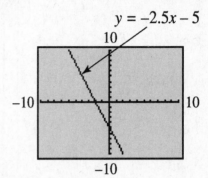

79. $3.6x - y = -5.8$

$$-y = -3.6x - 5.8 \quad \textit{Subtract 3.6x.}$$
$$y = 3.6x + 5.8 \quad \textit{Multiply by } -1.$$

Graph $Y_1 = 3.6X + 5.8$ in a standard viewing window.

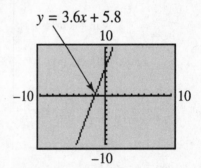

81. $\dfrac{6 - 2}{5 - 3} = \dfrac{4}{2} = 2$

83. $\dfrac{4 - (-1)}{-3 - (-5)} = \dfrac{4 + 1}{-3 + 5} = \dfrac{5}{2}$

85. $\dfrac{-5 - (-5)}{3 - 2} = \dfrac{-5 + 5}{1} = \dfrac{0}{1} = 0$

3.2 The Slope of a Line

1. $\text{slope} = \dfrac{\text{change in vertical position}}{\text{change in horizontal position}}$

$$= \dfrac{30 \text{ feet}}{100 \text{ feet}}$$

Choices **A**, 0.3, **B**, $\frac{3}{10}$, and **D**, $\frac{30}{100}$ are all correct.

3. **(a)** Graph **C** indicates that sales leveled off during the second quarter.

(b) Graph **A** indicates that sales leveled off during the fourth quarter.

(c) Graph **D** indicates that sales rose sharply during the first quarter, and then fell to the original level during the second quarter.

(d) Graph **B** is the only graph that indicates that sales fell during the first two quarters.

5.   slope of $BC = \dfrac{\text{rise}}{\text{run}} = \dfrac{0}{-4} = 0$

7.   slope of $DE = \dfrac{\text{rise}}{\text{run}} = \dfrac{-1}{3} = -\dfrac{1}{3}$

9.   slope of $FG = \dfrac{\text{rise}}{\text{run}} = \dfrac{-4}{1} = -4$

11. $m = \dfrac{6-2}{5-3} = \dfrac{4}{2} = 2$

13. $m = \dfrac{4-(-1)}{-3-(-5)} = \dfrac{4+1}{-3+5} = \dfrac{5}{2}$

15. $m = \dfrac{-5-(-5)}{3-2} = \dfrac{-5+5}{1} = \dfrac{0}{1} = 0$

17. $m = \dfrac{3-8}{-2-(-2)} = \dfrac{-5}{-2+2} = \dfrac{-5}{0}$, which is *undefined*.

19. $m = \dfrac{\frac{4}{3}+\frac{1}{2}}{\frac{1}{6}-\frac{1}{6}} = \dfrac{\frac{11}{6}}{0}$, which is *undefined*.

21.   Let $(x_1, y_1) = (-2, -3)$ and $(x_2, y_2) = (-1, 5)$. Then

$$m = \frac{y_2 - y_1}{x_2 - x_1} = \frac{5-(-3)}{-1-(-2)} = \frac{8}{1} = 8.$$

The slope is 8.

23.   Let $(x_1, y_1) = (-4, 1)$ and $(x_2, y_2) = (2, 6)$. Then

$$m = \frac{y_2 - y_1}{x_2 - x_1} = \frac{6-1}{2-(-4)} = \frac{5}{6}.$$

The slope is $\frac{5}{6}$.

25.   Let $(x_1, y_1) = (2, 4)$ and $(x_2, y_2) = (-4, 4)$. Then

$$m = \frac{y_2 - y_1}{x_2 - x_1} = \frac{4-4}{-4-2} = \frac{0}{-6} = 0.$$

The slope is 0.

27.   Let $(x_1, y_1) = (1.5, 2.6)$ and $(x_2, y_2) = (0.5, 3.6)$. Then

$$m = \frac{y_2 - y_1}{x_2 - x_1} = \frac{3.6-2.6}{0.5-1.5} = \frac{1}{-1} = -1.$$

The slope is -1.

29.   Let $(x_1, y_1) = \left(\frac{1}{6}, \frac{1}{2}\right)$ and $(x_2, y_2) = \left(\frac{5}{6}, \frac{9}{2}\right)$. Then

$$m = \frac{y_2 - y_1}{x_2 - x_1} = \frac{\frac{9}{2}-\frac{1}{2}}{\frac{5}{6}-\frac{1}{6}} = \frac{\frac{8}{2}}{\frac{4}{6}} = 4 \cdot \frac{3}{2} = 6.$$

The slope is 6.

31.   Let $(x_1, y_1) = \left(-\frac{2}{9}, \frac{5}{18}\right)$ and $(x_2, y_2) = \left(\frac{1}{18}, -\frac{5}{9}\right)$. Then

$$m = \frac{y_2 - y_1}{x_2 - x_1} = \frac{-\frac{5}{9}-\frac{5}{18}}{\frac{1}{18}-\left(-\frac{2}{9}\right)} = \frac{-\frac{15}{18}}{\frac{5}{18}}$$
$$= -\frac{15}{18} \cdot \frac{18}{5} = -3.$$

The slope is -3.

33.   The points shown on the line are $(-3, 3)$ and $(-1, -2)$. The slope is

$$m = \frac{-2-3}{-1-(-3)} = \frac{-5}{2} = -\frac{5}{2}.$$

35.   The points shown on the line are $(3, 3)$ and $(3, -3)$. The slope is

$$m = \frac{-3-3}{3-3} = \frac{-6}{0}, \text{ which is } \textit{undefined}.$$

37.   "The line has positive slope" means that the line goes up from left to right. This is line B.

39.   "The line has slope 0" means that there is no vertical change; that is, the line is horizontal. This is line A.

41.   To find the slope of

$$x + 2y = 4,$$

first find the intercepts. Replace y with 0 to find that the x-intercept is $(4, 0)$; replace x with 0 to find that the y-intercept is $(0, 2)$. The slope is then

$$m = \frac{2-0}{0-4} = -\frac{2}{4} = -\frac{1}{2}.$$

To sketch the graph, plot the intercepts and draw the line through them.

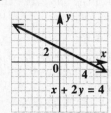

43. To find the slope of

$$5x - 2y = 10,$$

first find the intercepts. Replace y with 0 to find that the x-intercept is $(2, 0)$; replace x with 0 to find that the y-intercept is $(0, -5)$. The slope is then

$$m = \frac{-5 - 0}{0 - 2} = \frac{-5}{-2} = \frac{5}{2}.$$

To sketch the graph, plot the intercepts and draw the line through them.

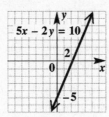

45. In the equation

$$y = 4x,$$

replace x with 0 and then x with 1 to get the ordered pairs $(0, 0)$ and $(1, 4)$, respectively. (There are other possibilities for ordered pairs.) The slope is then

$$m = \frac{4 - 0}{1 - 0} = \frac{4}{1} = 4.$$

To sketch the graph, plot the two points and draw the line through them.

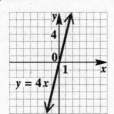

47. $x - 3 = 0 \quad (x = 3)$
The graph of $x = 3$ is the vertical line with x-intercept $(3, 0)$. The slope of a vertical line is undefined.

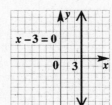

49. $y = -5$
The graph of $y = -5$ is the horizontal line with y-intercept $(0, -5)$. The slope of a horizontal line is 0.

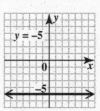

51. $2y = 3 \quad \left(y = \frac{3}{2}\right)$
The graph of $y = \frac{3}{2}$ is the horizontal line with y-intercept $\left(0, \frac{3}{2}\right)$. The slope of a horizontal line is 0.

53. To graph the line through $(-4, 2)$ with slope $m = \frac{1}{2}$, locate $(-4, 2)$ on the graph. To find a second point, use the definition of slope.

$$m = \frac{\text{change in } y}{\text{change in } x} = \frac{1}{2}$$

From $(-4, 2)$, go up 1 unit. Then go 2 units to the right to get to $(-2, 3)$. Draw the line through $(-4, 2)$ and $(-2, 3)$.

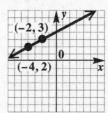

55. To graph the line through $(0, -2)$ with slope $m = -\frac{2}{3}$, locate the point $(0, -2)$ on the graph. To find a second point on the line, use the definition of slope, writing $-\frac{2}{3}$ as $\frac{-2}{3}$.

$$m = \frac{\text{change in } y}{\text{change in } x} = \frac{-2}{3}$$

From $(0, -2)$, move 2 units down and then 3 units to the right. Draw a line through this second point and $(0, -2)$. (Note that the slope could also be written as $\frac{2}{-3}$. In this case, move 2 units up and 3 units to the left to get another point on the same line.)

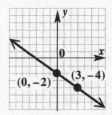

57. Locate $(-1, -2)$. Then use $m = 3 = \frac{3}{1}$ to go 3 units up and 1 unit right to $(0, 1)$.

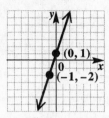

59. Locate $(2, -5)$. A slope of 0 means that the line is horizontal, so $y = -5$ at every point. Draw the horizontal line through $(2, -5)$.

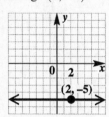

61. Locate $(-3, 1)$. Since the slope is undefined, the line is vertical. The x-value of every point is -3. Draw the vertical line through $(-3, 1)$.

63. If a line has slope $-\frac{4}{9}$, then any line parallel to it has slope $-\frac{4}{9}$ (the slope must be the same), and any line perpendicular to it has slope $\frac{9}{4}$ (the slope must be the negative reciprocal).

65. The slope of the line through $(15, 9)$ and $(12, -7)$ is

$$m = \frac{-7 - 9}{12 - 15} = \frac{-16}{-3} = \frac{16}{3}.$$

The slope of the line through $(8, -4)$ and $(5, -20)$ is

$$m = \frac{-20 - (-4)}{5 - 8} = \frac{-16}{-3} = \frac{16}{3}.$$

Since the slopes are equal, the two lines are *parallel*.

67. $x + 4y = 7$ and $4x - y = 3$
Solve the equations for y.

$4y = -x + 7$ $-y = -4x + 3$

$y = -\frac{1}{4}x + \frac{7}{4}$ $y = 4x - 3$

The slopes, $-\frac{1}{4}$ and 4, are negative reciprocals of one another, so the lines are *perpendicular*.

69. $4x - 3y = 6$ and $3x - 4y = 2$
Solve the equations for y.

$-3y = -4x + 6$ $-4y = -3x + 2$

$y = \frac{4}{3}x - 2$ $y = \frac{3}{4}x - \frac{1}{2}$

The slopes are $\frac{4}{3}$ and $\frac{3}{4}$. The lines are *neither* parallel nor perpendicular.

71. $x = 6$ and $6 - x = 8$
The second equation can be simplified as $x = -2$. Both lines are vertical lines, so they are *parallel*.

73. $4x + y = 0$ and $5x - 8 = 2y$
Solve the equations for y.

$y = -4x$ $\frac{5}{2}x - 4 = y$

The slopes are -4 and $\frac{5}{2}$. The lines are *neither* parallel nor perpendicular.

75. $2x = y + 3$ and $2y + x = 3$
Solve the equations for y.

$2x - 3 = y$ $2y = -x + 3$

          $y = -\frac{1}{2}x + \frac{3}{2}$

The slopes, 2 and $-\frac{1}{2}$, are negative reciprocals of one another, so the lines are *perpendicular*.

77. Use the points $(0, 20)$ and $(4, 4)$.

average rate of change

$$= \frac{\text{change in } y}{\text{change in } x} = \frac{4 - 20}{4 - 0} = \frac{-16}{4} = -4$$

The average rate of change is $-\$4000$ per year, that is, the value of the machine is decreasing $\$4000$ each year during these years.

79. We can see that there is no change in the percent of pay raise. Thus, the average rate of change is 0% per year, that is, the percent of pay raise is not changing—it is 3% each year during these years.

81. Let y be the vertical rise.
Since the slope is the vertical rise divided by the horizontal run,

$$0.13 = \frac{y}{150}.$$

Solving for y gives

$$y = 0.13(150) = 19.5.$$

The vertical rise could be a maximum of 19.5 ft.

83. **(a)** For 1999–2000:

$$m = \frac{109,478 - 86,047}{2000 - 1999} = 23,431$$

For 2000–2001:

$$m = \frac{128,375 - 109,478}{2001 - 2000} = 18,897$$

For 2001–2002:

$$m = \frac{140{,}767 - 128{,}375}{2002 - 2001} = 12{,}392$$

For 2002–2003:

$$m = \frac{158{,}722 - 140{,}767}{2003 - 2002} = 17{,}955$$

For 2003–2004:

$$m = \frac{182{,}140 - 158{,}722}{2004 - 2003} = 23{,}418$$

The average rates of change, m, are measured in thousands.

(b) The average rate of change in successive years is *not* approximately the same. Therefore, an approximately straight line could not be drawn through the plotted ordered pairs.

85. **(a)** Use $(1998, 229.3)$ and $(2004, 338.8)$.

$$m = \frac{338.8 - 229.3}{2004 - 1998} = \frac{109.5}{6} = 18.25$$

The average rate of change is $18.25 billion per year.

(b) The positive slope means that personal spending on recreation in the United States *increased* by an average of $18.25 billion each year.

87. Use $(1997, 500)$ and $(2002, 155)$.

$$m = \frac{155 - 500}{2002 - 1997} = \frac{-345}{5} = -69$$

The average rate of change in price is $-$69 per year, that is, the price decreased an average of $69 each year from 1997 to 2002.

89. Label the points as shown in the figure.

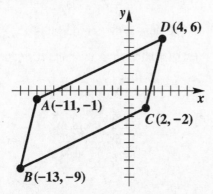

In order to determine whether $ABCD$ is a parallelogram, we need to show that the slope of $\overline{AB}$ equals the slope of $\overline{CD}$ and that the slope of $\overline{AD}$ equals the slope of $\overline{BC}$.

Slope of $\overline{AB} = \dfrac{-9 - (-1)}{-13 - (-11)} = \dfrac{-8}{-2} = 4$

Slope of $\overline{CD} = \dfrac{6 - (-2)}{4 - 2} = \dfrac{8}{2} = 4$

Slope of $\overline{AD} = \dfrac{6 - (-1)}{4 - (-11)} = \dfrac{7}{15}$

Slope of $\overline{BC} = \dfrac{-2 - (-9)}{2 - (-13)} = \dfrac{7}{15}$

Thus, the figure is a parallelogram.

91. Line A has negative slope and line B has positive slope, so line A must be $y_1 = -2x + 3$ and line B must be $y_2 = 3x - 4$.

93. For $A(3, 1)$ and $B(6, 2)$, the slope of $\overline{AB}$ is

$$m = \frac{2 - 1}{6 - 3} = \frac{1}{3}.$$

94. For $B(6, 2)$ and $C(9, 3)$, the slope of $\overline{BC}$ is

$$m = \frac{3 - 2}{9 - 6} = \frac{1}{3}.$$

95. For $A(3, 1)$ and $C(9, 3)$, the slope of $\overline{AC}$ is

$$m = \frac{3 - 1}{9 - 3} = \frac{2}{6} = \frac{1}{3}.$$

96. The slope of $\overline{AB} =$ slope of $\overline{BC}$
$= $ slope of $\overline{AC}$
$= \frac{1}{3}.$

97. For $A(1, -2)$ and $B(3, -1)$, the slope of $\overline{AB}$ is

$$m = \frac{-1 - (-2)}{3 - 1} = \frac{1}{2}.$$

For $B(3, -1)$ and $C(5, 0)$, the slope of $\overline{BC}$ is

$$m = \frac{0 - (-1)}{5 - 3} = \frac{1}{2}.$$

For $A(1, -2)$ and $C(5, 0)$, the slope of $\overline{AC}$ is

$$m = \frac{0 - (-2)}{5 - 1} = \frac{2}{4} = \frac{1}{2}.$$

Since the three slopes are the same, the three points are collinear.

98. For $A(0, 6)$ and $B(4, -5)$, the slope of $\overline{AB}$ is

$$m = \frac{-5 - 6}{4 - 0} = \frac{-11}{4} = -\frac{11}{4}.$$

For $B(4, -5)$ and $C(-2, 12)$, the slope of $\overline{BC}$ is

$$m = \frac{12 - (-5)}{-2 - 4} = \frac{17}{-6} = -\frac{17}{6}.$$

Since these two slopes are not the same, the three points are not collinear.

99. $3x + 2y = 8$
$$2y = -3x + 8$$
$$y = -\frac{3}{2}x + 4$$

101. $y - 2 = 4(x + 3)$
$$y - 2 = 4x + 12$$
$$y = 4x + 14$$

103. $y - (-1) = \frac{5}{3}[x - (-4)]$
$$3(y + 1) = 5(x + 4)$$
$$3y + 3 = 5x + 20$$
$$-17 = 5x - 3y$$

105. $y - (-1) = -\frac{1}{2}[x - (-2)]$
$$y + 1 = -\frac{1}{2}(x + 2)$$
$$2(y + 1) = -1(x + 2)$$
$$2y + 2 = -x - 2$$
$$x + 2y = -4$$

3.3   Linear Equations in Two Variables

1. Choice **A**, $3x - 2y = 5$, is in the form $Ax + By = C$ with $A \geq 0$ and integers A, B, and C having no common factor (except 1).

3. Choice **A**, $y = 6x + 2$, is in the form $y = mx + b$.

5. $y + 2 = -3(x - 4)$
$$y + 2 = -3x + 12$$
$$3x + y = 10 \qquad \textit{Standard form}$$

7. $y = 2x + 3$
This line is in slope-intercept form with slope $m = 2$ and y-intercept $(0, b) = (0, 3)$. The only graph with positive slope and with a positive y-coordinate of its y-intercept is **A**.

9. $y = -2x - 3$
This line is in slope-intercept form with slope $m = -2$ and y-intercept $(0, b) = (0, -3)$. The only graph with negative slope and with a negative y-coordinate of its y-intercept is **C**.

11. $y = 2x$
This line has slope $m = 2$ and y-intercept $(0, b) = (0, 0)$. The only graph with positive slope and with y-intercept $(0, 0)$ is **H**.

13. $y = 3$
This line is a horizontal line with y-intercept $(0, 3)$. Its y-coordinate is positive. The only graph that has these characteristics is **B**.

15. $m = 5$; $b = 15$
Substitute these values in the slope-intercept form.
$$y = mx + b$$
$$y = 5x + 15$$

17. $m = -\frac{2}{3}$; $b = \frac{4}{5}$
Substitute these values in the slope-intercept form.
$$y = mx + b$$
$$y = -\frac{2}{3}x + \frac{4}{5}$$

19. Slope $\frac{2}{5}$; y-intercept $(0, 5)$
Here, $m = \frac{2}{5}$ and $b = 5$. Substitute these values in the slope-intercept form.
$$y = mx + b$$
$$y = \frac{2}{5}x + 5$$

21. To get to the point $(3, 3)$ from the y-intercept $(0, 1)$, we must go up 2 units and to the right 3 units, so the slope is $\frac{2}{3}$. The slope-intercept form is
$$y = \frac{2}{3}x + 1.$$

23. $-x + y = 4$
(a) Solve for y to get the equation in slope-intercept form.
$$-x + y = 4$$
$$y = x + 4$$

(b) The slope is the coefficient of x, 1.

(c) The y-intercept is the point $(0, b)$, or $(0, 4)$.

(d)
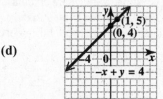

25. $6x + 5y = 30$
(a) Solve for y to get the equation in slope-intercept form.
$$6x + 5y = 30$$
$$5y = -6x + 30$$
$$y = -\frac{6}{5}x + 6$$

(b) The slope is the coefficient of x, $-\frac{6}{5}$.

(c) The y-intercept is the point $(0, b)$, or $(0, 6)$.

(d)
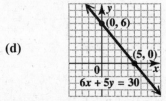

27. $4x - 5y = 20$
(a) Solve for y to get the equation in slope-intercept form.
$$4x - 5y = 20$$
$$-5y = -4x + 20$$
$$y = \frac{4}{5}x - 4$$

(b) The slope is the coefficient of x, $\frac{4}{5}$.

(c) The y-intercept is the point $(0, b)$, or $(0, -4)$.

(d)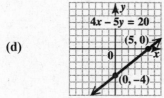

29. $x + 2y = -4$

(a) Solve for y to get the equation in slope-intercept form.

$$x + 2y = -4$$
$$2y = -x - 4$$
$$y = -\frac{1}{2}x - 2$$

(b) The slope is the coefficient of x, $-\frac{1}{2}$.

(c) The y-intercept is the point $(0, b)$, or $(0, -2)$.

(d)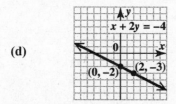

31. $-4x + 3y = 12$

(a) Solve for y to get the equation in slope-intercept form.

$$-4x + 3y = 12$$
$$3y = 4x + 12$$
$$y = \frac{4}{3}x + 4$$

(b) The slope is the coefficient of x, $\frac{4}{3}$.

(c) The y-intercept is the point $(0, b)$, or $(0, 4)$.

(d)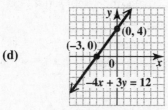

33. **(a)** Through $(5, 8)$; slope -2
Use the point-slope form with $(x_1, y_1) = (5, 8)$ and $m = -2$.

$$y - y_1 = m(x - x_1)$$
$$y - 8 = -2(x - 5)$$
$$y - 8 = -2x + 10$$
$$2x + y = 18$$

(b) Solve the last equation from part (a) for y.

$$2x + y = 18$$
$$y = -2x + 18$$

35. **(a)** Through $(-2, 4)$; slope $-\frac{3}{4}$
Use the point-slope form with $(x_1, y_1) = (-2, 4)$ and $m = -\frac{3}{4}$.

$$y - y_1 = m(x - x_1)$$
$$y - 4 = -\frac{3}{4}[x - (-2)]$$
$$4(y - 4) = -3(x + 2)$$
$$4y - 16 = -3x - 6$$
$$3x + 4y = 10$$

(b) Solve the last equation from part (a) for y.

$$3x + 4y = 10$$
$$4y = -3x + 10$$
$$y = -\frac{3}{4}x + \frac{10}{4}$$
$$y = -\frac{3}{4}x + \frac{5}{2}$$

37. **(a)** Through $(-5, 4)$; slope $\frac{1}{2}$
Use the point-slope form with $(x_1, y_1) = (-5, 4)$ and $m = \frac{1}{2}$.

$$y - y_1 = m(x - x_1)$$
$$y - 4 = \frac{1}{2}[x - (-5)]$$
$$2(y - 4) = 1(x + 5)$$
$$2y - 8 = x + 5$$
$$-x + 2y = 13$$
$$x - 2y = -13$$

(b) Solve the last equation from part (a) for y.

$$x - 2y = -13$$
$$-2y = -x - 13$$
$$y = \frac{1}{2}x + \frac{13}{2}$$

39. **(a)** Through $(3, 0)$; slope 4
Use the point-slope form with $(x_1, y_1) = (3, 0)$ and $m = 4$.

$$y - y_1 = m(x - x_1)$$
$$y - 0 = 4(x - 3)$$
$$y = 4x - 12$$
$$-4x + y = -12$$
$$4x - y = 12$$

(b) Solve the last equation from part (a) for y.

$$4x - y = 12$$
$$-y = -4x + 12$$
$$y = 4x - 12$$

41. (a) Through $(2, 6.8)$; slope 1.4

Use the point-slope form with $(x_1, y_1) = (2, 6.8)$ and $m = 1.4$.

$$y - y_1 = m(x - x_1)$$
$$y - 6.8 = 1.4(x - 2)$$
$$y - 6.8 = \tfrac{7}{5}(x - 2)$$
$$5(y - 6.8) = 7(x - 2)$$
$$5y - 34 = 7x - 14$$
$$-7x + 5y = 20$$
$$7x - 5y = -20$$

(b) Solve the last equation from part (a) for y.

$$7x - 5y = -20$$
$$-5y = -7x - 20$$
$$y = \tfrac{7}{5}x + 4 \quad \text{or} \quad y = 1.4x + 4$$

43. Through $(9, 5)$; slope 0

A line with slope 0 is a horizontal line. A horizontal line through the point (x, k) has equation $y = k$. Here $k = 5$, so an equation is $y = 5$.

45. Through $(9, 10)$; undefined slope

A vertical line has undefined slope and equation $x = c$. Since the x-value in $(9, 10)$ is 9, the equation is $x = 9$.

47. Through $(0.5, 0.2)$; vertical

A vertical line through the point (k, y) has equation $x = k$. Here $k = 0.5$, so the equation is $x = 0.5$.

49. Through $(-7, 8)$; horizontal

A horizontal line through the point (x, k) has equation $y = k$, so the equation is $y = 8$.

51. (a) $(3, 4)$ and $(5, 8)$

Find the slope.

$$m = \frac{8 - 4}{5 - 3} = \frac{4}{2} = 2$$

Use the point-slope form with $(x_1, y_1) = (3, 4)$ and $m = 2$.

$$y - y_1 = m(x - x_1)$$
$$y - 4 = 2(x - 3)$$
$$y - 4 = 2x - 6$$
$$-2x + y = -2$$
$$2x - y = 2$$

(b) Solve the last equation from part (a) for y.

$$2x - y = 2$$
$$-y = -2x + 2$$
$$y = 2x - 2$$

Note: You could use *any* of the equations in part (a) to solve for y. In this exercise, choosing

$$y - 4 = 2x - 6 \quad \text{or} \quad -2x + y = -2,$$

easily leads to the equation $y = 2x - 2$.

53. (a) $(6, 1)$ and $(-2, 5)$

Find the slope.

$$m = \frac{5 - 1}{-2 - 6} = \frac{4}{-8} = -\frac{1}{2}$$

Use the point-slope form with $(x_1, y_1) = (6, 1)$ and $m = -\frac{1}{2}$.

$$y - y_1 = m(x - x_1)$$
$$y - 1 = -\tfrac{1}{2}(x - 6)$$
$$2(y - 1) = -1(x - 6)$$
$$2y - 2 = -x + 6$$
$$x + 2y = 8$$

(b) Solve the last equation from part (a) for y.

$$x + 2y = 8$$
$$2y = -x + 8$$
$$y = -\tfrac{1}{2}x + 4$$

55. (a) $\left(-\tfrac{2}{5}, \tfrac{2}{5}\right)$ and $\left(\tfrac{4}{3}, \tfrac{2}{3}\right)$

Find the slope.

$$m = \frac{\tfrac{2}{3} - \tfrac{2}{5}}{\tfrac{4}{3} - \left(-\tfrac{2}{5}\right)} = \frac{\tfrac{10-6}{15}}{\tfrac{20+6}{15}}$$
$$= \frac{\tfrac{4}{15}}{\tfrac{26}{15}} = \frac{4}{26} = \frac{2}{13}$$

Use the point-slope form with $(x_1, y_1) = \left(-\tfrac{2}{5}, \tfrac{2}{5}\right)$ and $m = \tfrac{2}{13}$.

$$y - \tfrac{2}{5} = \tfrac{2}{13}\left[x - \left(-\tfrac{2}{5}\right)\right]$$
$$13\left(y - \tfrac{2}{5}\right) = 2\left(x + \tfrac{2}{5}\right)$$
$$13y - \tfrac{26}{5} = 2x + \tfrac{4}{5}$$
$$-2x + 13y = \tfrac{30}{5}$$
$$2x - 13y = -6$$

(b) Solve the last equation from part (a) for y.

$$2x - 13y = -6$$
$$-13y = -2x - 6$$
$$y = \tfrac{2}{13}x + \tfrac{6}{13}$$

57. (a) $(2, 5)$ and $(1, 5)$

Find the slope.

$$m = \frac{5 - 5}{1 - 2} = \frac{0}{-1} = 0$$

A line with slope 0 is horizontal. A horizontal line through the point (x, k) has equation $y = k$, so the equation is $y = 5$.

(b) $y = 5$ is already in the slope-intercept form.

59. **(a)** $(7,6)$ and $(7,-8)$
Find the slope.

$$m = \frac{-8-6}{7-7} = \frac{-14}{0} \quad \textit{Undefined}$$

A line with undefined slope is a vertical line. The equation of a vertical line is $x = k$, where k is the common x-value. So the equation is $x = 7$.

(b) It is not possible to write $x = 7$ in slope-intercept form.

61. **(a)** $\left(\frac{1}{2}, -3\right)$ and $\left(-\frac{2}{3}, -3\right)$
Find the slope.

$$m = \frac{-3-(-3)}{-\frac{2}{3}-\frac{1}{2}} = \frac{0}{-\frac{7}{6}} = 0$$

A line with slope 0 is horizontal. A horizontal line through the point (x, k) has equation $y = k$, so the equation is $y = -3$.

(b) $y = -3$ is already in the slope-intercept form.

63. **(a)** Through $(7, 2)$; parallel to the graph of the line having equation $3x - y = 8$
Find the slope of $3x - y = 8$.

$$-y = -3x + 8$$
$$y = 3x - 8$$

The slope is 3, so a line parallel to it also has slope 3. Use $m = 3$ and $(x_1, y_1) = (7, 2)$ in the point-slope form.

$$y - y_1 = m(x - x_1)$$
$$y - 2 = 3(x - 7)$$
$$y - 2 = 3x - 21$$
$$y = 3x - 19$$

(b) $$y = 3x - 19$$
$$-3x + y = -19$$
$$3x - y = 19$$

65. **(a)** Through $(-2, -2)$; parallel to $-x + 2y = 10$
Find the slope of $-x + 2y = 10$.

$$2y = x + 10$$
$$y = \frac{1}{2}x + 5$$

The slope is $\frac{1}{2}$, so a line parallel to it also has slope $\frac{1}{2}$. Use $m = \frac{1}{2}$ and $(x_1, y_1) = (-2, -2)$ in the point-slope form.

$$y - y_1 = m(x - x_1)$$
$$y - (-2) = \frac{1}{2}[x - (-2)]$$
$$y + 2 = \frac{1}{2}(x + 2)$$
$$y + 2 = \frac{1}{2}x + 1$$
$$y = \frac{1}{2}x - 1$$

(b) $$y = \frac{1}{2}x - 1$$
$$2y = x - 2 \quad \textit{Multiply by 2.}$$
$$-x + 2y = -2$$
$$x - 2y = 2$$

67. **(a)** Through $(8, 5)$; perpendicular to $2x - y = 7$
Find the slope of $2x - y = 7$.

$$-y = -2x + 7$$
$$y = 2x - 7$$

The slope of the line is 2. Therefore, the slope of the line perpendicular to it is $-\frac{1}{2}$ since $2\left(-\frac{1}{2}\right) = -1$. Use $m = -\frac{1}{2}$ and $(x_1, y_1) = (8, 5)$ in the point-slope form.

$$y - y_1 = m(x - x_1)$$
$$y - 5 = -\frac{1}{2}(x - 8)$$
$$y - 5 = -\frac{1}{2}x + 4$$
$$y = -\frac{1}{2}x + 9$$

(b) $$y = -\frac{1}{2}x + 9$$
$$2y = -x + 18 \quad \textit{Multiply by 2.}$$
$$x + 2y = 18$$

69. **(a)** Through $(-2, 7)$; perpendicular to $x = 9$
$x = 9$ is a vertical line so a line perpendicular to it will be a horizontal line. It goes through $(-2, 7)$ so its equation is

$$y = 7.$$

(b) $y = 7$ is already in standard form.

71. Distance = (rate)(time), so

$$y = 45x.$$

x	$y = 45x$	Ordered Pair
0	$45(0) = 0$	$(0, 0)$
5	$45(5) = 225$	$(5, 225)$
10	$45(10) = 450$	$(10, 450)$

73. Total cost = (cost/gal)(number of gallons), so

$$y = 3.01x.$$

x	$y = 3.01x$	Ordered Pair
0	$3.01(0) = 0$	$(0, 0)$
5	$3.01(5) = 15.05$	$(5, 15.05)$
10	$3.01(10) = 30.10$	$(10, 30.10)$

75. **(a)** The fixed cost is \$99, so that is the value of b. The variable cost is \$39, so

$$y = mx + b = 39x + 99.$$

(b) If $x = 5$, $y = 39(5) + 99 = 294$. The ordered pair is $(5, 294)$. The cost of a 5-month membership is \$294.

(c) If $x = 12$, $y = 39(12) + 99 = 567$. The cost of the first year's membership is \$567.

77. (a) The fixed cost is $19.95 + $25 = $44.95, so that is the value of b. The variable cost is $35, so

$$y = mx + b = 35x + 44.95.$$

(b) If $x = 5$, $y = 35(5) + 44.95 = 219.95$. The ordered pair is $(5, 219.95)$. The cost of the plan for 5 months is $219.95.

(c) For a 1-year contract, $x = 12$, so $y = 35(12) + 44.95 = 464.95$. The cost of the plan for 1 year is $464.95.

79. (a) The fixed cost is $30, so that is the value of b. The variable cost is $6, so

$$y = mx + b = 6x + 30.$$

(b) If $x = 5$, $y = 6(5) + 30 = 60$. The ordered pair is $(5, 60)$. It costs $60 to rent the saw for 5 days.

(c) $138 = 6x + 30$ *Let y = 138.*

$$108 = 6x$$
$$x = \frac{108}{6} = 18$$

The saw is rented for 18 days.

81. (a) Use $(0, 91)$ and $(5, 63)$.

$$m = \frac{63 - 91}{5 - 0} = \frac{-28}{5} = -5.6$$

The equation is $y = -5.6x + 91$. The slope tells us that the percent of households accessing the Internet by dial-up is decreasing 5.6% per year.

(b) The year 2006 corresponds to $x = 6$, so $y = -5.6(6) + 91 = 57.4 \approx 57\%$.

83. (a) Use $(5, 22{,}393)$ and $(13, 29{,}645)$.

$$m = \frac{29{,}645 - 22{,}393}{13 - 5} = \frac{7252}{8} = 906.5$$

Now use the point-slope form.

$$y - 22{,}393 = 906.5(x - 5)$$
$$y - 22{,}393 = 906.5x - 4532.5$$
$$y = 906.5x + 17{,}860.5$$

(b) The year 1999 corresponds to $x = 9$, so $y = 906.5(9) + 17{,}860.5 = 26{,}019$. This value is slightly lower than the actual value of $27{,}910.

85. (a) $2x + 7 - x = 4x - 2$

$$x + 7 = 4x - 2$$
$$-3x + 9 = 0$$
$$-3x + 9 = y$$

(b) From the screen, we see that $x = 3$ is the solution.

(c) $2x + 7 - x = 4x - 2$

$$x + 7 = 4x - 2$$
$$9 = 3x$$
$$3 = x$$

The solution set is $\{3\}$.

87. (a) $3(2x + 1) - 2(x - 2) = 5$

$$6x + 3 - 2x + 4 - 5 = 0$$
$$4x + 2 = 0$$
$$4x + 2 = y$$

(b) From the screen, we see that $x = -0.5$ is the solution.

(c) $3(2x + 1) - 2(x - 2) = 5$

$$6x + 3 - 2x + 4 = 5$$
$$4x + 7 = 5$$
$$4x = -2$$
$$x = -\tfrac{1}{2} \text{ or } -0.5$$

The solution set is $\{-0.5\}$.

89. The solution to the equation $y_1 = 0$ is the x-coordinate of the x-intercept. In this case, the x-intercept is greater than 10, so the correct choice must be **D**.

91. When $C = 0°$, $F = \underline{32°}$, and when $C = 100°$, $F = \underline{212°}$. These are the freezing and boiling temperatures for water.

92. The two points of the form (C, F) would be $(0, 32)$ and $(100, 212)$.

93. $m = \dfrac{212 - 32}{100 - 0} = \dfrac{180}{100} = \dfrac{9}{5}$

94. Let $m = \frac{9}{5}$ and $(x_1, y_1) = (0, 32)$.

$$y - y_1 = m(x - x_1)$$
$$F - 32 = \tfrac{9}{5}(C - 0)$$
$$F - 32 = \tfrac{9}{5}C$$
$$F = \tfrac{9}{5}C + 32$$

95. $$F = \tfrac{9}{5}C + 32$$
$$F - 32 = \tfrac{9}{5}C$$
$$\tfrac{5}{9}(F - 32) = C$$

96. A temperature of $50°C$ corresponds to a temperature of $122°F$.

97. $2x + 5 < 9$

$$2x < 4$$
$$x < 2$$

The solution set is $(-\infty, 2)$.

99. $5 - 3x \geq 9$

$$-3x \geq 4$$
$$x \leq -\tfrac{4}{3}$$

The solution set is $\left(-\infty, -\tfrac{4}{3}\right]$.

Summary Exercises on Slopes and Equations of Lines

1. For $3x + 5y = 9$, slope $= -\dfrac{A}{B} = -\dfrac{3}{5}$.

3. For $2x - y = 5$, slope $= -\dfrac{A}{B} = -\dfrac{2}{-1} = 2$.

5. For $0.2x + 0.8y = 0$,
$$\text{slope} = -\dfrac{A}{B} = -\dfrac{0.2}{0.8} = -0.25.$$

7. Through $(-2, 6)$ and $(4, 1)$

(a) The slope is
$$m = \dfrac{1 - 6}{4 - (-2)} = \dfrac{-5}{6} = -\dfrac{5}{6}.$$

Use the point-slope form.
$$y - y_1 = m(x - x_1)$$
$$y - 6 = -\tfrac{5}{6}[x - (-2)]$$
$$y - 6 = -\tfrac{5}{6}x - \tfrac{5}{3}$$
$$y = -\tfrac{5}{6}x - \tfrac{5}{3} + \tfrac{18}{3}$$
$$y = -\tfrac{5}{6}x + \tfrac{13}{3}$$

(b)
$$y = -\tfrac{5}{6}x + \tfrac{13}{3}$$
$$6y = -5x + 26 \quad \textit{Multiply by 6.}$$
$$5x + 6y = 26$$

9. Through $(0, 0)$; perpendicular to $2x - 5y = 6$

(a) Find the slope of $2x - 5y = 6$.
$$-5y = -2x + 6$$
$$y = \tfrac{2}{5}x - \tfrac{6}{5}$$

The slope of the line is $\tfrac{2}{5}$. Therefore, the slope of the line perpendicular to it is $-\tfrac{5}{2}$ since $\tfrac{2}{5}\left(-\tfrac{5}{2}\right) = -1$. Use $m = -\tfrac{5}{2}$ and $(x_1, y_1) = (0, 0)$ in the point-slope form.
$$y - y_1 = m(x - x_1)$$
$$y - 0 = -\tfrac{5}{2}(x - 0)$$
$$y = -\tfrac{5}{2}x$$

(b)
$$y = -\tfrac{5}{2}x$$
$$2y = -5x$$
$$5x + 2y = 0$$

11. Through $\left(\tfrac{3}{4}, -\tfrac{7}{9}\right)$; perpendicular to $x = \tfrac{2}{3}$

(a) $x = \tfrac{2}{3}$ is a vertical line so a line perpendicular to it will be a horizontal line. It goes through $\left(\tfrac{3}{4}, -\tfrac{7}{9}\right)$ so its equation is
$$y = -\tfrac{7}{9}.$$

(b) $y = -\tfrac{7}{9}$
$$9y = -7 \quad \textit{Multiply by 9.}$$

13. Through $(-4, 2)$; parallel to the line through $(3, 9)$ and $(6, 11)$

(a) The slope of the line through $(3, 9)$ and $(6, 11)$ is
$$m = \dfrac{11 - 9}{6 - 3} = \dfrac{2}{3}.$$

Use the point-slope form with $(x_1, y_1) = (-4, 2)$ and $m = \tfrac{2}{3}$ (since the slope of the desired line must equal the slope of the given line).
$$y - y_1 = m(x - x_1)$$
$$y - 2 = \tfrac{2}{3}[x - (-4)]$$
$$y - 2 = \tfrac{2}{3}(x + 4)$$
$$y - 2 = \tfrac{2}{3}x + \tfrac{8}{3}$$
$$y = \tfrac{2}{3}x + \tfrac{8}{3} + \tfrac{6}{3}$$
$$y = \tfrac{2}{3}x + \tfrac{14}{3}$$

(b)
$$y = \tfrac{2}{3}x + \tfrac{14}{3}$$
$$3y = 2x + 14$$
$$-2x + 3y = 14$$
$$2x - 3y = -14$$

15. Through $(-4, 12)$ and the midpoint of the segment with endpoints $(5, 8)$ and $(-3, 2)$

(a) The midpoint of the segment with endpoints $(5, 8)$ and $(-3, 2)$ is
$$\left(\dfrac{5 + (-3)}{2}, \dfrac{8 + 2}{2}\right) = \left(\dfrac{2}{2}, \dfrac{10}{2}\right) = (1, 5).$$

The slope of the line through $(-4, 12)$ and $(1, 5)$ is
$$m = \dfrac{5 - 12}{1 - (-4)} = \dfrac{-7}{5} = -\dfrac{7}{5}.$$

Use the point-slope form with $(x_1, y_1) = (-4, 12)$ and $m = -\tfrac{7}{5}$.
$$y - y_1 = m(x - x_1)$$
$$y - 12 = -\tfrac{7}{5}[x - (-4)]$$
$$y - 12 = -\tfrac{7}{5}(x + 4)$$
$$y - 12 = -\tfrac{7}{5}x - \tfrac{28}{5}$$
$$y = -\tfrac{7}{5}x - \tfrac{28}{5} + \tfrac{60}{5}$$
$$y = -\tfrac{7}{5}x + \tfrac{32}{5}$$

(b)
$$y = -\tfrac{7}{5}x + \tfrac{32}{5}$$
$$5y = -7x + 32$$
$$7x + 5y = 32$$

17. Through $\left(\frac{3}{7}, \frac{1}{6}\right)$; parallel to $y = \frac{1}{5}x + \frac{7}{4}$

(a) The slope of the desired line is the same as the slope of the given line, so use the point-slope form with $(x_1, y_1) = \left(\frac{3}{7}, \frac{1}{6}\right)$ and $m = \frac{1}{5}$.

$$y - y_1 = m(x - x_1)$$
$$y - \frac{1}{6} = \frac{1}{5}\left(x - \frac{3}{7}\right)$$
$$y - \frac{1}{6} = \frac{1}{5}x - \frac{3}{35}$$
$$y = \frac{1}{5}x - \frac{18}{210} + \frac{35}{210}$$
$$y = \frac{1}{5}x + \frac{17}{210}$$

(b)
$$y = \frac{1}{5}x + \frac{17}{210}$$
$$210y = 42x + 17 \quad \textit{Multiply by 210.}$$
$$-42x + 210y = 17$$
$$42x - 210y = -17$$

19. Through $(0.3, 1.5)$ and $(0.4, 1.7)$

(a) $m = \dfrac{1.7 - 1.5}{0.4 - 0.3} = \dfrac{0.2}{0.1} = 2$

Use the point-slope form with $(x_1, y_1) = (0.3, 1.5)$ and $m = 2$.

$$y - y_1 = m(x - x_1)$$
$$y - 1.5 = 2(x - 0.3)$$
$$y - 1.5 = 2x - 0.6$$
$$y = 2x + 0.9$$

(b)
$$y = 2x + 0.9$$
$$10y = 20x + 9 \quad \textit{Multiply by 10.}$$
$$-20x + 10y = 9$$
$$20x - 10y = -9$$

21. Slope -0.5, $b = -2$

The slope-intercept form of a line, $y = mx + b$, becomes $y = -0.5x - 2$, or $y = -\frac{1}{2}x - 2$, which is choice **B**.

23. Passes through $(4, -2)$ and $(0, 0)$

$$m = \frac{0 - (-2)}{0 - 4} = \frac{2}{-4} = -\frac{1}{2}$$

Using $m = -\frac{1}{2}$ and a y-intercept of $(0, 0)$, we get $y = -\frac{1}{2}x + 0$, which is choice **A**.

25. $m = \frac{1}{2}$, passes through the origin
Use the point-slope form with $(x_1, y_1) = (0, 0)$ and $m = \frac{1}{2}$.

$$y - y_1 = m(x - x_1)$$
$$y - 0 = \frac{1}{2}(x - 0)$$
$$y = \frac{1}{2}x \quad \text{or} \quad 2y = x$$

This is choice **E**.

3.4 Linear Inequalities in Two Variables

1. The boundary of the graph of $y \leq -x + 2$ will be a _solid_ line (since the inequality involves $\leq$), and the shading will be _below_ the line (since the inequality sign is $\leq$ or $<$).

3. The boundary of the graph of $y > -x + 2$ will be a _dashed_ line (since the inequality involves $>$), and the shading will be _above_ the line (since the inequality sign is $\geq$ or $>$).

5. The graph of $Ax + By = C$ divides the plane into two regions. In one of these regions, the ordered pairs satisfy $Ax + By < C$; in the other, they satisfy $Ax + By > C$.

7. $x + y \leq 2$
Graph the line $x + y = 2$ by drawing a solid line (since the inequality involves $\leq$) through the intercepts $(2, 0)$ and $(0, 2)$.
Test a point not on this line, such as $(0, 0)$.

$$x + y \leq 2$$
$$0 + 0 \leq 2 \quad ?$$
$$0 \leq 2 \quad \textit{True}$$

Shade that side of the line containing the test point $(0, 0)$.

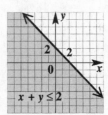

9. $4x - y < 4$
Graph the line $4x - y = 4$ by drawing a dashed line (since the inequality involves $<$) through the intercepts $(1, 0)$ and $(0, -4)$. Instead of using a test point, we will solve the inequality for y.

$$-y < -4x + 4$$
$$y > 4x - 4$$

Since we have "$y >$" in the last inequality, shade the region *above* the boundary line.

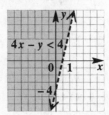

11. $x + 3y \geq -2$

Graph the solid line $x + 3y = -2$ (since the inequality involves $\geq$) through the intercepts $(-2, 0)$ and $\left(0, -\frac{2}{3}\right)$.

Test a point not on this line such as $(0, 0)$.

$$0 + 3(0) \geq -2 \quad ?$$
$$0 \geq -2 \quad \textit{True}$$

Shade that side of the line containing the test point $(0, 0)$.

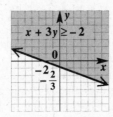

13. $x + y > 0$

Graph the line $x + y = 0$, which includes the points $(0, 0)$ and $(2, -2)$, as a dashed line (since the inequality involves $>$). Solving the inequality for y gives us

$$y > -x,$$

So shade the region above the boundary line.

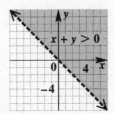

15. $x - 3y \leq 0$

Graph the solid line $x - 3y = 0$ through the points $(0, 0)$ and $(3, 1)$.

Solve the inequality for y.

$$-3y \leq -x$$
$$y \geq \tfrac{1}{3}x$$

Shade the region above the boundary line.

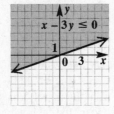

17. $y < x$

Graph the dashed line $y = x$ through $(0, 0)$ and $(2, 2)$. Since we have "$y <$ " in the inequality, shade the region *below* the boundary line.

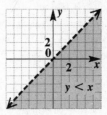

19. $x + y \leq 1$ and $x \geq 1$

Graph the solid line $x + y = 1$ through $(0, 1)$ and $(1, 0)$. The inequality $x + y \leq 1$ can be written as $y \leq -x + 1$, so shade the region below the boundary line.

Graph the solid vertical line $x = 1$ through $(1, 0)$ and shade the region to the right. The required graph is the common shaded area as well as the portions of the lines that bound it.

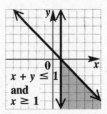

21. $2x - y \geq 2$ and $y < 4$

Graph the solid line $2x - y = 2$ through the intercepts $(1, 0)$ and $(0, -2)$. Test $(0, 0)$ to get $0 \geq 2$, a false statement. Shade that side of the graph not containing $(0, 0)$. To graph $y < 4$ on the same axes, graph the dashed horizontal line through $(0, 4)$. Test $(0, 0)$ to get $0 < 4$, a true statement. Shade that side of the dashed line containing $(0, 0)$.

The word "and" indicates the intersection of the two graphs. The final solution set consists of the region where the two shaded regions overlap.

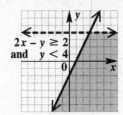

23. $x + y > -5$ and $y < -2$

Graph $x + y = -5$, which has intercepts $(-5, 0)$ and $(0, -5)$, as a dashed line. Test $(0, 0)$, which yields $0 > -5$, a true statement. Shade the region that includes $(0, 0)$.

Graph $y = -2$ as a dashed horizontal line. Shade the region below $y = -2$. The required graph of the intersection is the region common to both graphs.

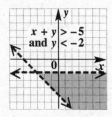

25. $|x| < 3$ can be rewritten as $-3 < x < 3$. The boundaries are the dashed vertical lines $x = -3$ and $x = 3$. Since x is between -3 and 3, the graph includes all points between the lines.

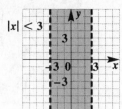

27. $|x + 1| < 2$ can be rewritten as

$$-2 < x + 1 < 2$$
$$-3 < x < 1.$$

The boundaries are the dashed vertical lines $x = -3$ and $x = 1$. Since x is between -3 and 1, the graph includes all points between the lines.

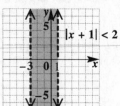

29. $x - y \geq 1$ or $y \geq 2$

Graph the solid line $x - y = 1$, which crosses the y-axis at -1 and the x-axis at 1. Use $(0, 0)$ as a test point, which yields $0 \geq 1$, a false statement. Shade the region that does not include $(0, 0)$. Now graph the solid line $y = 2$. Since the inequality is $y \geq 2$, shade above this line. The required graph of the union includes all the shaded regions, that is, all the points that satisfy either inequality.

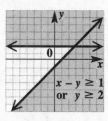

31. $x - 2 > y$ or $x < 1$

Graph $x - 2 = y$, which has intercepts $(2, 0)$ and $(0, -2)$, as a dashed line. Test $(0, 0)$, which yields $-2 > 0$, a false statement. Shade the region that does not include $(0, 0)$.

Graph $x = 1$ as a dashed vertical line. Shade the region to the left of $x = 1$.

The required graph of the union includes all the shaded regions, that is, all the points that satisfy either inequality.

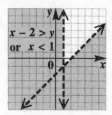

33. $3x + 2y < 6$ or $x - 2y > 2$

Graph $3x + 2y = 6$, which has intercepts $(2, 0)$ and $(0, 3)$, as a dashed line. Test $(0, 0)$, which yields $0 < 6$, a true statement. Shade the region that includes $(0, 0)$.

Graph $x - 2y = 2$, which has intercepts $(2, 0)$ and $(0, -1)$, as a dashed line. Test $(0, 0)$, which yields $0 > 2$, a false statement. Shade the region that does not include $(0, 0)$.

The required graph of the union includes all the shaded regions, that is, all the points that satisfy either inequality.

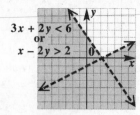

35. $y \leq 3x - 6$

The boundary line, $y = 3x - 6$, has slope 3 and y-intercept -6. This would be graph **B** or graph **C**. Since we want the region less than or equal to $3x - 6$, we want the region on or below the boundary line. The answer is graph **C**.

37. $y \leq -3x - 6$

The slope of the boundary line $y = -3x - 6$ is -3, and the y-intercept is -6. This would be graph **A** or graph **D**. The inequality sign is $\leq$, so we want the region on or below the boundary line. The answer is graph **A**.

39. **(a)** The x-intercept is $(-4, 0)$, so the solution set for $y = 0$ is $\{-4\}$.

(b) The solution set for $y < 0$ is $(-\infty, -4)$, since the graph is below the x-axis for these values of x.

(c) The solution set for $y > 0$ is $(-4, \infty)$, since the graph is above the x-axis for these values of x.

41. **(a)** The x-intercept is $(3.5, 0)$, so the solution set for $y = 0$ is $\{3.5\}$.

(b) The solution set for $y < 0$ is $(3.5, \infty)$, since the graph is below the x-axis for these values of x.

(c) The solution set for $y > 0$ is $(-\infty, 3.5)$, since the graph is above the x-axis for these values of x.

43. **(a)** $5x + 3 = 0$
$$5x = -3$$
$$x = -\tfrac{3}{5} = -0.6$$
The solution set is $\{-0.6\}$.

(b) $5x + 3 > 0$
$$5x > -3$$
$$x > -\tfrac{3}{5} \text{ or } -0.6$$
The solution set is $(-0.6, \infty)$.

(c) $5x + 3 < 0$
$$5x < -3$$
$$x < -\tfrac{3}{5} \text{ or } -0.6$$
The solution set is $(-\infty, -0.6)$.

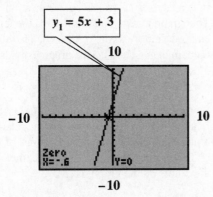

The x-intercept is $(-0.6, 0)$, as in part (a). The graph is above the x-axis for $x > -0.6$, as in part (b), and below the x-axis for $x < -0.6$, as in part (c).

45. **(a)** $-8x - (2x + 12) = 0$
$$-8x - 2x - 12 = 0$$
$$-10x - 12 = 0$$
$$-10x = 12$$
$$x = -1.2$$
The solution set is $\{-1.2\}$.

(b) $-8x - (2x + 12) \geq 0$
$$-8x - 2x - 12 \geq 0$$
$$-10x - 12 \geq 0$$
$$-10x \geq 12$$
$$x \leq -1.2$$
The solution set is $(-\infty, -1.2]$.

(c) $-8x - (2x + 12) \leq 0$
$$-8x - 2x - 12 \leq 0$$
$$-10x - 12 \leq 0$$
$$-10x \leq 12$$
$$x \geq -1.2$$
The solution set is $[-1.2, \infty)$.

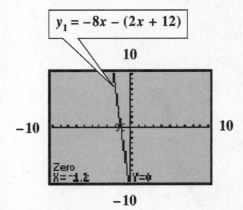

The x-intercept is $(-1.2, 0)$, as in part (a). The graph is on or above the x-axis for $(-\infty, -1.2]$, as in part (b), and on or below the x-axis for $[-1.2, \infty)$, as in part (c).

47. "A factory can have *no more than* 200 workers on a shift, but must have *at least* 100" can be translated as $x \leq 200$ and $x \geq 100$. "Must manufacture *at least* 3000 units" can be translated as $y \geq 3000$.

48.

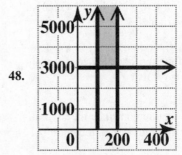

49. The total daily cost C consists of \$50 per worker and \$100 to manufacture 1 unit, so
$$C = 50x + 100y.$$

50. Some examples of points in the shaded region are $(150, 4000)$, $(120, 3500)$, and $(180, 6000)$. Some examples of points on the boundary are $(100, 5000)$, $(150, 3000)$, and $(200, 4000)$. The corner points are $(100, 3000)$ and $(200, 3000)$.

51.

(x, y)	$50x + 100y = C$
$(150, 4000)$	$50(150) + 100(4000) = 407{,}500$
$(120, 3500)$	$50(120) + 100(3500) = 356{,}000$
$(180, 6000)$	$50(180) + 100(6000) = 609{,}000$
$(100, 5000)$	$50(100) + 100(5000) = 505{,}000$
$(150, 3000)$	$50(150) + 100(3000) = 307{,}500$
$(200, 4000)$	$50(200) + 100(4000) = 410{,}000$
$(100, 3000)$	$50(100) + 100(3000) = 305{,}000$
	(least value)
$(200, 3000)$	$50(200) + 100(3000) = 310{,}000$

52. The company should use 100 workers and manufacture 3000 units to achieve the least possible cost.

53. $y = -7x + 12$
$y = -7(3) + 12$ *Let x = 3.*
$y = -9$

55. $y = 3x - 8$
$y = 3(3) - 8$ *Let x = 3.*
$y = 1$

57. $3x - 7y = 8$
$-7y = -3x + 8$
$y = \frac{3}{7}x - \frac{8}{7}$

59. $\frac{1}{2}x - 4y = 5$
$-4y = -\frac{1}{2}x + 5$
$y = \frac{1}{8}x - \frac{5}{4}$

3.5 Introduction to Functions

1. We give one of many possible answers here. A function is a set of ordered pairs in which each first component corresponds to exactly one second component. For example, $\{(0, 1), (1, 2), (2, 3), (3, 4) \ldots\}$ is a function.

3. In an ordered pair of a relation, the first element is the independent variable.

5. $\{(5, 1), (3, 2), (4, 9), (7, 6)\}$
The relation is a function since for each x-value, there is only one y-value.

7. $\{(2, 4), (0, 2), (2, 5)\}$
The relation is not a function since the x-value 2 has two different y-values associated with it, 4 and 5.

9. $\{(-3, 1), (4, 1), (-2, 7)\}$
The relation is a function since for each x-value, there is only one y-value.

11. $\{(1, 1), (1, -1), (0, 0), (2, 4), (2, -4)\}$
The relation is not a function since the x-value 1 has two different y-values associated with it, 1 and -1. (A similar statement can be made for $x = 2$.)

The domain is the set of x-values: $\{0, 1, 2\}$.
The range is the set of y-values: $\{-4, -1, 0, 1, 4\}$.

13. The relation can be described by the set of ordered pairs

$$\{(2, 1), (5, 1), (11, 7), (17, 20), (3, 20)\}.$$

The relation is a function since for each x-value, there is only one y-value.

The domain is the set of x-values: $\{2, 3, 5, 11, 17\}$.
The range is the set of y-values: $\{1, 7, 20\}$.

15. The relation can be described by the set of ordered pairs

$$\{(1, 5), (1, 2), (1, -1), (1, -4)\}.$$

The relation is not a function since the x-value 1 has four different y-values associated with it, 5, 2, -1, and -4.

The domain is the set of x-values: $\{1\}$.
The range is the set of y-values: $\{5, 2, -1, -4\}$.

17. Using the vertical line test, we find any vertical line will intersect the graph at most once. This indicates that the graph represents a function. This graph extends indefinitely to the left $(-\infty)$ and indefinitely to the right (∞). Therefore, the domain is $(-\infty, \infty)$. This graph extends indefinitely downward $(-\infty)$, and indefinitely upward (∞). Thus, the range is $(-\infty, \infty)$.

19. Using the vertical line test, we find any vertical line will intersect the graph at most once. This indicates that the graph represents a function. This graph extends indefinitely to the left $(-\infty)$ and indefinitely to the right (∞). Therefore, the domain is $(-\infty, \infty)$. This graph extends indefinitely downward $(-\infty)$, and reaches a high point at $y = 4$. Therefore, the range is $(-\infty, 4]$.

21. Since a vertical line can intersect the graph of the relation in more than one point, the relation is not a function. The domain, the x-values of the points on the graph, is $[-4, 4]$. The range, the y-values of the points on the graph, is $[-3, 3]$.

23. $y = x^2$
Each value of x corresponds to one y-value. For example, if $x = 3$, then $y = 3^2 = 9$. Therefore, $y = x^2$ defines y as a function of x.
Since any x-value, positive, negative, or zero, can be squared, the domain is $(-\infty, \infty)$.

25. $x = y^6$
The ordered pairs $(64, 2)$ and $(64, -2)$ both satisfy the equation. Since one value of x, 64, corresponds to two values of y, 2 and -2, the relation does not define a function. Because x is equal to the sixth power of y, the values of x must always be nonnegative. The domain is $[0, \infty)$.

27. $y = 2x - 6$

For any value of x, there is exactly one value of y, so this equation defines a function. The domain is the set of all real numbers, $(-\infty, \infty)$.

29. $x + y < 4$

For a particular x-value, more than one y-value can be selected to satisfy $x + y < 4$. For example, if $x = 2$ and $y = 0$, then

$$2 + 0 < 4. \quad \textit{True}$$

Now, if $x = 2$ and $y = 1$, then

$$2 + 1 < 4. \quad \textit{Also true}$$

Therefore, $x + y < 4$ does not define y as a function of x.

The graph of $x + y < 4$ consists of the shaded region below the dashed line $x + y = 4$, which extends indefinitely from left to right. Therefore, the domain is $(-\infty, \infty)$.

31. $y = \sqrt{x}$

For any value of x, there is exactly one corresponding value for y, so this relation defines a function. Since the radicand must be a nonnegative number, x must always be nonnegative. The domain is $[0, \infty)$.

33. $xy = 1$

Rewrite $xy = 1$ as $y = \dfrac{1}{x}$. Note that x can never equal 0, otherwise the denominator would equal 0. The domain is $(-\infty, 0) \cup (0, \infty)$.

Each nonzero x-value gives exactly one y-value. Therefore, $xy = 1$ defines y as a function of x.

35. $y = \sqrt{4x + 2}$

To determine the domain of $y = \sqrt{4x + 2}$, recall that the radicand must be nonnegative. Solve the inequality $4x + 2 \geq 0$, which gives us $x \geq -\frac{1}{2}$. Therefore, the domain is $\left[-\frac{1}{2}, \infty\right)$.

Each x-value from the domain produces exactly one y-value. Therefore, $y = \sqrt{4x + 2}$ defines a function.

37. $y = \dfrac{2}{x - 4}$

Given any value of x, y is found by subtracting 4, then dividing the result into 2. This process produces exactly one value of y for each x-value, so the relation represents a function. The domain includes all real numbers except those that make the denominator 0, namely 4. The domain is $(-\infty, 4) \cup (4, \infty)$.

39. $f(3)$ is the value of the dependent variable when the independent variable is 3—choice **B**.

41. $f(x) = -3x + 4$
$f(0) = -3(0) + 4$
$\quad\;\; = 0 + 4$
$\quad\;\; = 4$

43. $g(x) = -x^2 + 4x + 1$
$g(-2) = -(-2)^2 + 4(-2) + 1$
$\quad\quad\; = -(4) - 8 + 1$
$\quad\quad\; = -11$

45. $f(x) = -3x + 4$
$f\left(\frac{1}{3}\right) = -3\left(\frac{1}{3}\right) + 4$
$\quad\quad = -1 + 4$
$\quad\quad = 3$

47. $g(x) = -x^2 + 4x + 1$
$g(0.5) = -(0.5)^2 + 4(0.5) + 1$
$\quad\quad\; = -0.25 + 2 + 1$
$\quad\quad\; = 2.75$

49. $f(x) = -3x + 4$
$f(p) = -3(p) + 4$
$\quad\;\; = -3p + 4$

51. $f(x) = -3x + 4$
$f(-x) = -3(-x) + 4$
$\quad\quad = 3x + 4$

53. $f(x) = -3x + 4$
$f(x + 2) = -3(x + 2) + 4$
$\quad\quad\quad = -3x - 6 + 4$
$\quad\quad\quad = -3x - 2$

55. $g(x) = -x^2 + 4x + 1$
$g(\pi) = -\pi^2 + 4\pi + 1$

57. $f(x) = -3x + 4$
$f(x + h) = -3(x + h) + 4$
$\quad\quad\quad = -3x - 3h + 4$

59. $f(4) - g(4)$
$\quad = [-3(4) + 4] - [-(4)^2 + 4(4) + 1]$
$\quad = [-8] - [1]$
$\quad = -9$

61. **(a)** When $x = 2$, $y = 2$, so $f(2) = 2$.

(b) When $x = -1$, $y = 3$, so $f(-1) = 3$.

63. **(a)** When $x = 2$, $y = 15$, so $f(2) = 15$.

(b) When $x = -1$, $y = 10$, so $f(-1) = 10$.

65. **(a)** The point $(2, 3)$ is on the graph of f, so $f(2) = 3$.

(b) The point $(-1, -3)$ is on the graph of f, so $f(-1) = -3$.

67. **(a)** Solve the equation for y.

$$x + 3y = 12$$
$$3y = 12 - x$$
$$y = \frac{12 - x}{3}$$

Since $y = f(x)$,

$$f(x) = \frac{12 - x}{3} = -\frac{1}{3}x + 4.$$

(b) $f(3) = \frac{12 - 3}{3} = \frac{9}{3} = 3$

69. **(a)** Solve the equation for y.

$$y + 2x^2 = 3$$
$$y = 3 - 2x^2$$

Since $y = f(x)$,

$$f(x) = 3 - 2x^2.$$

(b) $f(3) = 3 - 2(3)^2$
$$= 3 - 2(9)$$
$$= -15$$

71. **(a)** Solve the equation for y.

$$4x - 3y = 8$$
$$-3y = 8 - 4x$$
$$y = \frac{8 - 4x}{-3}$$

Since $y = f(x)$,

$$f(x) = \frac{8 - 4x}{-3} = \frac{4}{3}x - \frac{8}{3}.$$

(b) $f(3) = \frac{8 - 4(3)}{-3} = \frac{8 - 12}{-3}$
$$= \frac{-4}{-3} = \frac{4}{3}$$

73. The equation $2x + y = 4$ has a straight <u>line</u> as its graph. To find y in $(3, \underline{\ y\ })$, let $x = 3$ in the equation.

$$2x + y = 4$$
$$2(3) + y = 4$$
$$6 + y = 4$$
$$y = -2$$

To use functional notation for $2x + y = 4$, solve for y to get

$$y = -2x + 4.$$

Replace y with $f(x)$ to get

$$f(x) = \underline{\ -2x + 4\ }.$$
$$f(3) = -2(3) + 4 = \underline{\ -2\ }$$

Because $y = -2$ when $x = 3$, the point $\underline{\ (3, -2)\ }$ lies on the graph of the function.

75. $f(x) = -2x + 5$
The graph will be a line. The intercepts are $(0, 5)$ and $\left(\frac{5}{2}, 0\right)$.
The domain is $(-\infty, \infty)$. The range is $(-\infty, \infty)$.

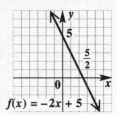

77. $h(x) = \frac{1}{2}x + 2$
The graph will be a line. The intercepts are $(0, 2)$ and $(-4, 0)$.
The domain is $(-\infty, \infty)$. The range is $(-\infty, \infty)$.

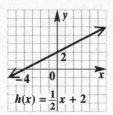

79. $G(x) = 2x$
This line includes the points $(0, 0), (1, 2)$, and $(2, 4)$. The domain is $(-\infty, \infty)$. The range is $(-\infty, \infty)$.

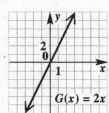

81. $g(x) = -4$
Using a y-intercept of $(0, -4)$ and a slope of $m = 0$, we graph the horizontal line. From the graph we see that the domain is $(-\infty, \infty)$. The range is $\{-4\}$.

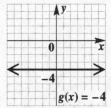

83. $f(x) = 0$
Draw the horizontal line through the point $(0, 0)$. On the horizontal line the value of x can be any real number, so the domain is $(-\infty, \infty)$. The range is $\{0\}$.

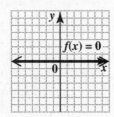

85. **(a)** $f(x) = 2.75x$

$$f(3) = 2.75(3)$$
$$= 8.25 \text{ (dollars)}$$

(b) 3 is the value of the independent variable, which represents a package weight of 3 pounds; $f(3)$ is the value of the dependent variable representing the cost to mail a 3-pound package.

(c) $2.75(5) = \$13.75$, the cost to mail a 5-lb package. Using function notation, we have $f(5) = 13.75$.

87. **(a)** Since the length of a man's femur is given, use the formula $h(r) = 69.09 + 2.24r$.

$$h(56) = 69.09 + 2.24(56) \quad \text{Let } r = 56.$$
$$= 194.53$$

The man is 194.53 cm tall.

(b) Use the formula $h(t) = 81.69 + 2.39t$.

$$h(40) = 81.69 + 2.39(40) \quad \text{Let } t = 40.$$
$$= 177.29$$

The man is 177.29 cm tall.

(c) Since the length of a woman's femur is given, use the formula $h(r) = 61.41 + 2.32r$.

$$h(50) = 61.41 + 2.32(50) \quad \text{Let } r = 50.$$
$$= 177.41$$

The woman is 177.41 cm tall.

(d) Use the formula $h(t) = 72.57 + 2.53t$.

$$h(36) = 72.57 + 2.53(36) \quad \text{Let } t = 36.$$
$$= 163.65$$

The woman is 163.65 cm tall.

89. **(a)** The independent variable is t, the number of hours, and the possible values are in the set $[0, 100]$. The dependent variable is g, the number of gallons, and the possible values are in the set $[0, 3000]$.

(b) The graph rises for the first 25 hours, so the water level increases for 25 hours. The graph falls for $t = 50$ to $t = 75$, so the water level decreases for 25 hours.

(c) There are 2000 gallons in the pool when $t = 90$.

(d) $f(0)$ is the number of gallons in the pool at time $t = 0$. Here, $f(0) = 0$, which means the pool is empty at time 0.

(e) $f(25) = 3000$; After 25 hours, there are 3000 gallons of water in the pool.

91. The graph shows $x = 3$ and $y = 7$. In function notation, this is

$$f(3) = 7.$$

93.
$$6x + 5y = 2$$
$$6(-3) + 5y = 2 \quad \text{Let } x = -3.$$
$$-18 + 5y = 2$$
$$5y = 20$$
$$y = 4$$

95.
$$1.5x + 2.5y = 5.5$$
$$1.5(-3) + 2.5y = 5.5 \quad \text{Let } x = -3.$$
$$-4.5 + 2.5y = 5.5$$
$$2.5y = 10$$
$$y = 4$$

97. $-4\left(\dfrac{x-3}{3}\right) + 2x = 8$

$$-4(x - 3) + 6x = 24 \quad \text{Multiply by 3.}$$
$$-4x + 12 + 6x = 24$$
$$2x + 12 = 24$$
$$2x = 12$$
$$x = 6$$

The solution set is $\{6\}$.

Chapter 3 Review Exercises

1. $3x + 2y = 10$

For $x = 0$:

$$3(0) + 2y = 10$$
$$2y = 10$$
$$y = 5 \quad (0, 5)$$

For $y = 0$:

$$3x + 2(0) = 10$$
$$3x = 10$$
$$x = \tfrac{10}{3} \quad \left(\tfrac{10}{3}, 0\right)$$

For $x = 2$:

$$3(2) + 2y = 10$$
$$6 + 2y = 10$$
$$2y = 4$$
$$y = 2 \quad (2, 2)$$

continued

For $y = -2$:

$$3x + 2(-2) = 10$$
$$3x - 4 = 10$$
$$3x = 14$$
$$x = \frac{14}{3} \quad \left(\frac{14}{3}, -2\right)$$

Plot the ordered pairs, and draw the line through them.

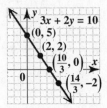

2. $x - y = 8$

For $x = 2$:

$$2 - y = 8$$
$$-y = 6$$
$$y = -6 \quad (2, -6)$$

For $y = -3$:

$$x - (-3) = 8$$
$$x + 3 = 8$$
$$x = 5 \quad (5, -3)$$

For $x = 3$:

$$3 - y = 8$$
$$-y = 5$$
$$y = -5 \quad (3, -5)$$

For $y = -2$:

$$x - (-2) = 8$$
$$x + 2 = 8$$
$$x = 6 \quad (6, -2)$$

Plot the ordered pairs, and draw the line through them.

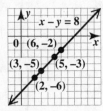

3. $4x - 3y = 12$

To find the x-intercept, let $y = 0$.

$$4x - 3y = 12$$
$$4x - 3(0) = 12$$
$$4x = 12$$
$$x = 3$$

The x-intercept is $(3, 0)$.

To find the y-intercept, let $x = 0$.

$$4x - 3y = 12$$
$$4(0) - 3y = 12$$
$$-3y = 12$$
$$y = -4$$

The y-intercept is $(0, -4)$.

Plot the intercepts and draw the line through them.

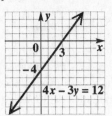

4. $5x + 7y = 28$

To find the x-intercept, let $y = 0$.

$$5x + 7y = 28$$
$$5x + 7(0) = 28$$
$$5x = 28$$
$$x = \frac{28}{5}$$

The x-intercept is $\left(\frac{28}{5}, 0\right)$.

To find the y-intercept, let $x = 0$.

$$5x + 7y = 28$$
$$5(0) + 7y = 28$$
$$7y = 28$$
$$y = 4$$

The y-intercept is $(0, 4)$.

Plot the intercepts and draw the line through them.

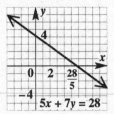

5. $2x + 5y = 20$

To find the x-intercept, let $y = 0$.

$$2x + 5y = 20$$
$$2x + 5(0) = 20$$
$$2x = 20$$
$$x = 10$$

The x-intercept is $(10, 0)$.

To find the y-intercept, let $x = 0$.

$$2x + 5y = 20$$
$$2(0) + 5y = 20$$
$$5y = 20$$
$$y = 4$$

The y-intercept is $(0, 4)$.

Plot the intercepts and draw the line through them.

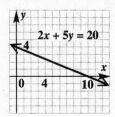

6. $x - 4y = 8$

To find the x-intercept, let $y = 0$.

$$x - 4y = 8$$
$$x - 4(0) = 8$$
$$x = 8$$

The x-intercept is $(8, 0)$.
To find the y-intercept, let $x = 0$.

$$0 - 4y = 8$$
$$-4y = 8$$
$$y = -2$$

The y-intercept is $(0, -2)$.
Plot the intercepts and draw the line through them.

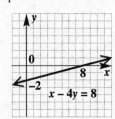

7. If both coordinates are positive, the point lies in quadrant I. If the first coordinate is negative and the second is positive, the point lies in quadrant II. To lie in quadrant III, the point must have both coordinates negative. To lie in quadrant IV, the first coordinate must be positive and the second must be negative.

8. Through $(-1, 2)$ and $(4, -5)$
$$m = \frac{\text{change in } y}{\text{change in } x} = \frac{-5 - 2}{4 - (-1)} = \frac{-7}{5} = -\frac{7}{5}$$

9. Through $(0, 3)$ and $(-2, 4)$
Let $(x_1, y_1) = (0, 3)$ and $(x_2, y_2) = (-2, 4)$.
$$m = \frac{y_2 - y_1}{x_2 - x_1} = \frac{4 - 3}{-2 - 0} = \frac{1}{-2} = -\frac{1}{2}$$

10. The slope of $y = 2x + 3$ is 2, the coefficient of x.

11. $3x - 4y = 5$
Write the equation in slope-intercept form.
$$-4y = -3x + 5$$
$$y = \frac{3}{4}x - \frac{5}{4}$$

The slope is $\frac{3}{4}$.

12. $x = 5$ is a vertical line and has *undefined* slope.

13. Parallel to $3y = 2x + 5$
Write the equation in slope-intercept form.
$$3y = 2x + 5$$
$$y = \frac{2}{3}x + \frac{5}{3}$$

The slope of $3y = 2x + 5$ is $\frac{2}{3}$; all lines parallel to it will also have a slope of $\frac{2}{3}$.

14. Perpendicular to $3x - y = 4$
Solve for y.
$$y = 3x - 4$$

The slope is 3; the slope of a line perpendicular to it is $-\frac{1}{3}$ since
$$3\left(-\frac{1}{3}\right) = -1.$$

15. Through $(-1, 5)$ and $(-1, -4)$
$$m = \frac{\Delta y}{\Delta x} = \frac{-4 - 5}{-1 - (-1)} = \frac{-9}{0} \quad \textit{Undefined}$$

This is a vertical line; it has undefined slope.

16. Through $(3, -1)$ and $(-3, 1)$
$$m = \frac{\Delta y}{\Delta x} = \frac{1 - (-1)}{-3 - 3} = \frac{2}{-6} = -\frac{1}{3}.$$

17. The x-intercept is $(2, 0)$ and the y-intercept is $(0, 2)$. The slope is
$$m = \frac{\text{change in } y}{\text{change in } x} = \frac{2 - 0}{0 - 2} = \frac{2}{-2} = -1.$$

18. The line goes up from left to right, so it has positive slope.

19. The line goes down from left to right, so it has negative slope.

20. The line is vertical, so it has *undefined* slope.

21. The line is horizontal, so it has 0 slope.

22. The slope is $\frac{2}{10}$ which can be written as 0.2, 20%, $\frac{20}{100}$, or $\frac{1}{5}$.

The correct responses are **A**, **B**, **C**, **D**, and **F**.

23. To rise 1 foot, we must move 4 feet in the horizontal direction. To rise 3 feet, we must move $3(4) = 12$ feet in the horizontal direction.

24. Let $(x_1, y_1) = (1980, 21{,}000)$ and $(x_2, y_2) = (2003, 52{,}700)$. Then
$$m = \frac{\Delta y}{\Delta x} = \frac{52{,}700 - 21{,}000}{2003 - 1980} = \frac{31{,}700}{23}$$
$$\approx 1378.$$

The average rate of change is \$1378 per year.

25. **(a)** Slope $-\frac{1}{3}$, y-intercept $(0, -1)$

Use the slope-intercept form with $m = -\frac{1}{3}$ and $b = -1$.

$$y = mx + b$$
$$y = -\frac{1}{3}x - 1$$

(b)
$$y = -\frac{1}{3}x - 1$$
$$3y = -x - 3$$
$$x + 3y = -3$$

26. **(a)** Slope 0, y-intercept $(0, -2)$

Use the slope-intercept form with $m = 0$ and $b = -2$.

$$y = mx + b$$
$$y = (0)x - 2$$
$$y = -2$$

(b) $y = -2$ is already in standard form.

27. **(a)** Slope $-\frac{4}{3}$, through $(2, 7)$

Use the point-slope form with $m = -\frac{4}{3}$ and $(x_1, y_1) = (2, 7)$.

$$y - y_1 = m(x - x_1)$$
$$y - 7 = -\frac{4}{3}(x - 2)$$
$$y - 7 = -\frac{4}{3}x + \frac{8}{3}$$
$$y = -\frac{4}{3}x + \frac{29}{3}$$

(b)
$$y = -\frac{4}{3}x + \frac{29}{3}$$
$$3y = -4x + 29$$
$$4x + 3y = 29$$

28. **(a)** Slope 3, through $(-1, 4)$

Use the point-slope form with $m = 3$ and $(x_1, y_1) = (-1, 4)$.

$$y - y_1 = m(x - x_1)$$
$$y - 4 = 3[x - (-1)]$$
$$y - 4 = 3(x + 1)$$
$$y - 4 = 3x + 3$$
$$y = 3x + 7$$

(b)
$$y = 3x + 7$$
$$-3x + y = 7$$
$$3x - y = -7$$

29. **(a)** Vertical, through $(2, 5)$

The equation of any vertical line is in the form $x = k$. Since the line goes through $(2, 5)$, the equation is $x = 2$. (Slope-intercept form is not possible.)

(b) $x = 2$ is already in standard form.

30. **(a)** Through $(2, -5)$ and $(1, 4)$

Find the slope.

$$m = \frac{\Delta y}{\Delta x} = \frac{4 - (-5)}{1 - 2} = \frac{9}{-1} = -9$$

Use the point-slope form with $m = -9$ and $(x_1, y_1) = (2, -5)$.

$$y - y_1 = m(x - x_1)$$
$$y - (-5) = -9(x - 2)$$
$$y + 5 = -9x + 18$$
$$y = -9x + 13$$

(b)
$$y = -9x + 13$$
$$9x + y = 13$$

31. **(a)** Through $(-3, -1)$ and $(2, 6)$

Find the slope.

$$m = \frac{\Delta y}{\Delta x} = \frac{6 - (-1)}{2 - (-3)} = \frac{7}{5}$$

Use the point-slope form with $m = \frac{7}{5}$ and $(x_1, y_1) = (2, 6)$.

$$y - y_1 = m(x - x_1)$$
$$y - 6 = \frac{7}{5}(x - 2)$$
$$y - 6 = \frac{7}{5}x - \frac{14}{5}$$
$$y = \frac{7}{5}x + \frac{16}{5}$$

(b)
$$y = \frac{7}{5}x + \frac{16}{5}$$
$$5y = 7x + 16$$
$$-7x + 5y = 16$$
$$7x - 5y = -16$$

32. **(a)** From Exercise 17, we have $m = -1$ and a y-intercept of $(0, 2)$. The slope-intercept form is

$$y = -1x + 2 \quad \text{or} \quad y = -x + 2.$$

(b)
$$y = -x + 2$$
$$x + y = 2$$

33. **(a)** Parallel to $4x - y = 3$ and through $(7, -1)$

Writing $4x - y = 3$ in slope-intercept form gives us $y = 4x - 3$, which has slope 4. Lines parallel to it will also have slope 4. The line with slope 4 through $(7, -1)$ is :

$$y - y_1 = m(x - x_1)$$
$$y - (-1) = 4(x - 7)$$
$$y + 1 = 4x - 28$$
$$y = 4x - 29$$

(b)
$$y = 4x - 29$$
$$-4x + y = -29$$
$$4x - y = 29$$

34. **(a)** Perpendicular to $2x - 5y = 7$ and through $(4, 3)$

Write the equation in slope-intercept form.

$$2x - 5y = 7$$
$$-5y = -2x + 7$$
$$y = \frac{2}{5}x - \frac{7}{5}$$

$y = \frac{2}{5}x - \frac{7}{5}$ has slope $\frac{2}{5}$ and is perpendicular to lines with slope $-\frac{5}{2}$.

The line with slope $-\frac{5}{2}$ through $(4, 3)$ is

$$y - y_1 = m(x - x_1)$$
$$y - 3 = -\frac{5}{2}(x - 4)$$
$$y - 3 = -\frac{5}{2}x + 10$$
$$y = -\frac{5}{2}x + 13$$

(b)
$$y = -\frac{5}{2}x + 13$$
$$2y = -5x + 26$$
$$5x + 2y = 26$$

35. (a) The fixed cost is $159, so that is the value of b. The variable cost is $57, so

$$y = mx + b = 57x + 159.$$

The cost of a 1-year membership can be found by substituting 12 for x.

$$y = 57(12) + 159$$
$$= 684 + 159 = 843$$

The cost is $843.

(b) As in part (a),

$$y = 47x + 159.$$
$$y = 47(12) + 159$$
$$= 564 + 159 = 723$$

The cost is $723.

36. (a) Use $(6, 12.6)$ and $(12, 35.0)$.

$$m = \frac{\Delta y}{\Delta x} = \frac{35.0 - 12.6}{12 - 6} = \frac{22.4}{6} \approx 3.73$$

Use the point-slope form of a line.

$$y - y_1 = m(x - x_1)$$
$$y - 12.6 = 3.73(x - 6)$$
$$y - 12.6 = 3.73x - 22.38$$
$$y = 3.73x - 9.78$$
$$\text{or } y = 3.73x - 9.8$$

The slope, 3.73, indicates that the number of e-filing taxpayers increased by 3.73% each year from 1996 to 2002.

(b) The year 2005 corresponds to $x = 15$.

$$y = 3.73(15) - 9.8$$
$$= 55.95 - 9.8 = 46.15$$

According to the equation from part (a), 46.15% of tax returns will be filed electronically in 2005.

37. $3x - 2y \leq 12$
Graph $3x - 2y = 12$ as a solid line through $(0, -6)$ and $(4, 0)$. Use $(0, 0)$ as a test point. Since $(0, 0)$ satisfies the inequality, shade the region on the side of the line containing $(0, 0)$.

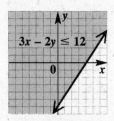

38. $5x - y > 6$
Graph $5x - y = 6$ as a dashed line through $(0, -6)$ and $\left(\frac{6}{5}, 0\right)$. Use $(0, 0)$ as a test point. Since $(0, 0)$ does not satisfy the inequality, shade the region on the side of the line that does not contain $(0, 0)$.

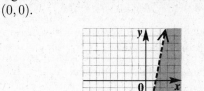

39. $2x + y \leq 1$ and $x \geq 2y$
Graph $2x + y = 1$ as a solid line through $\left(\frac{1}{2}, 0\right)$ and $(0, 1)$, and shade the region on the side containing $(0, 0)$ since it satisfies the inequality. Next, graph $x = 2y$ as a solid line through $(0, 0)$ and $(2, 1)$, and shade the region on the side containing $(2, 0)$ since $2 > 2(0)$ or $2 > 0$ is true. The intersection is the region where the graphs overlap.

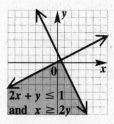

40. $x \geq 2$ or $y \geq 2$

Graph $x = 2$ as a solid vertical line through $(2, 0)$. Shade the region to the right of $x = 2$.

Graph $y = 2$ as a solid horizontal line through $(0, 2)$. Shade the region above $y = 2$. The graph of

$$x \geq 2 \quad \text{or} \quad y \geq 2$$

includes all the shaded regions.

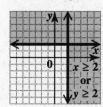

41. In $y < 4x + 3$, the " $<$ " symbol indicates that the graph has a dashed boundary line and that the shading is below the line, so the correct choice is **D**.

42. $\{(-4, 2), (-4, -2), (1, 5), (1, -5)\}$
The domain, the set of x-values, is $\{-4, 1\}$.
The range, the set of y-values, is $\{2, -2, 5, -5\}$.
Since each x-value has more than one y-value, the relation is not a function.

43. The relation can be described by the set of ordered pairs

$\{(9, 32), (11, 47), (4, 47), (17, 69), (25, 14)\}.$

The relation is a function since for each x-value, there is only one y-value.

The domain is the set of x-values:
$\{9, 11, 4, 17, 25\}.$
The range is the set of y-values: $\{32, 47, 69, 14\}.$

44. The domain, the x-values of the points on the graph, is $[-4, 4]$. The range, the y-values of the points on the graph, is $[0, 2]$. Since a vertical line intersects the graph of the relation in at most one point, the relation is a function.

45. The x-values are negative or zero, so the domain is $(-\infty, 0]$. The y-values can be any real number, so the range is $(-\infty, \infty)$. A vertical line, such as $x = -3$, will intersect the graph twice, so by the vertical line test, the relation is not a function.

46. $y = 3x - 3$
For any value of x, there is exactly one value of y, so the equation defines a function, actually a linear function. The domain is the set of all real numbers, $(-\infty, \infty)$.

47. $y < x + 2$
For any value of x, there are many values of y. For example, $(1, 0)$ and $(1, 1)$ are both solutions of the inequality that have the same x-value but different y-values. The inequality does not define a function. The domain is the set of all real numbers, $(-\infty, \infty)$.

48. $y = |x|$
For any value of x, there is exactly one value of y, so the equation defines a function. The domain is the set of all real numbers, $(-\infty, \infty)$.

49. $y = \sqrt{4x + 7}$
Given any value of x, y is found by multiplying x by 4, adding 7, and taking the square root of the result. This process produces exactly one value of y for each x-value, so the equation defines a function. Since the radicand must be nonnegative,

$$4x + 7 \geq 0$$
$$4x \geq -7$$
$$x \geq -\tfrac{7}{4}.$$

The domain is $\left[-\tfrac{7}{4}, \infty\right)$.

50. $x = y^2$
The ordered pairs $(4, 2)$ and $(4, -2)$ both satisfy the equation. Since one value of x, 4, corresponds to two values of y, 2 and -2, the equation does not define a function. Because x is equal to the square of y, the values of x must always be nonnegative. The domain is $[0, \infty)$.

51. $y = \dfrac{7}{x - 6}$
Given any value of x, y is found by subtracting 6, then dividing the result into 7. This process produces exactly one value of y for each x-value, so the equation defines a function. The domain includes all real numbers except those that make the denominator 0, namely 6. The domain is $(-\infty, 6) \cup (6, \infty)$.

52. If no vertical line intersects the graph in more than one point, then it is the graph of a function.

In Exercises 53–56, use

$$f(x) = -2x^2 + 3x - 6.$$

53. $f(0) = -2(0)^2 + 3(0) - 6 = -6$

54. $f(2.1) = -2(2.1)^2 + 3(2.1) - 6$
$\qquad = -8.82 + 6.3 - 6 = -8.52$

55. $f\left(-\tfrac{1}{2}\right) = -2\left(-\tfrac{1}{2}\right)^2 + 3\left(-\tfrac{1}{2}\right) - 6$
$\qquad = -\tfrac{1}{2} - \tfrac{3}{2} - 6 = -8$

56. $f(k) = -2k^2 + 3k - 6$

57. **(a)** For each year, there is exactly one life expectancy associated with the year, so the table defines a function.

(b) The domain is the set of years, that is, $\{1943, 1953, 1963, 1973, 1983, 1993, 2003\}$. The range is the set of life expectancies, that is, $\{63.3, 68.8, 69.9, 71.4, 74.6, 75.5, 77.6\}$.

(c) Answers will vary. Two possible answers are $(1943, 63.3)$ and $(1953, 68.8)$.

(d) $f(1973) = 71.4$. In 1973, life expectancy at birth was 71.4 yr.

(e) Since $f(1993) = 75.5$, $x = 1993$.

58. $2x^2 - y = 0$

$$-y = -2x^2$$
$$y = 2x^2$$

Since $y = f(x)$,
$$f(x) = 2x^2,$$
and $f(3) = 2(3)^2 = 2(9) = 18.$

59.   Solve for y in terms of x.

$$2x - 5y = 7$$
$$2x - 7 = 5y$$
$$\tfrac{2}{5}x - \tfrac{7}{5} = y$$

Thus, choice **C** is correct.

60.   No, because the equation of a line with an undefined slope is $x = a$. The ordered pairs have the form (a, y), where a is a constant and y is a variable. Thus, the number a corresponds to an infinite number of values of y.

61.   The slope is negative since the line falls from left to right.

62.   Use the points $(-1, 5)$ and $(3, -1)$.

$$m = \frac{\Delta y}{\Delta x} = \frac{-1 - 5}{3 - (-1)} = \frac{-6}{4} = -\frac{3}{2}$$

63.   Since $m = -\frac{3}{2}$, the slope of any line parallel to this line is also $-\frac{3}{2}$, whereas the slope of any line perpendicular to this line is $\frac{2}{3}$ since $\frac{2}{3}$ is the negative reciprocal of $-\frac{3}{2}$.

64. $2y = -3x + 7$
To find the x-intercept, let $y = 0$.

$$2(0) = -3x + 7$$
$$3x = 7$$
$$x = \tfrac{7}{3}$$

The x-intercept is $\left(\tfrac{7}{3}, 0\right)$.

65. $2y = -3x + 7$
To find the y-intercept, let $x = 0$.

$$2y = -3(0) + 7$$
$$2y = 7$$
$$y = \tfrac{7}{2}$$

The y-intercept is $\left(0, \tfrac{7}{2}\right)$.

66.   Solve $2y = -3x + 7$ for y.
$$y = -\tfrac{3}{2}x + \tfrac{7}{2}$$

Since $y = f(x)$,
$$f(x) = -\tfrac{3}{2}x + \tfrac{7}{2}.$$

67. $f(x) = -\frac{3}{2}x + \frac{7}{2}$
$$f(8) = -\tfrac{3}{2}(8) + \tfrac{7}{2}$$
$$= -\tfrac{24}{2} + \tfrac{7}{2} = -\tfrac{17}{2}$$

68. $f(x) = -\frac{3}{2}x + \frac{7}{2}$

$$-8 = -\tfrac{3}{2}x + \tfrac{7}{2} \quad \textit{Let } f(x) = -8.$$
$$-16 = -3x + 7 \quad \textit{Multiply by 2.}$$
$$-23 = -3x \quad \textit{Subtract 7.}$$
$$x = \tfrac{23}{3} \quad \textit{Divide by } -3.$$

69.
$$f(x) \geq 0$$
$$-\tfrac{3}{2}x + \tfrac{7}{2} \geq 0$$
$$-\tfrac{3}{2}x \geq -\tfrac{7}{2}$$
$$x \leq \left(-\tfrac{7}{2}\right)\left(-\tfrac{2}{3}\right)$$
$$x \leq \tfrac{7}{3}$$

$$\tfrac{7}{3}$$

70. $f(x) = 0$ is equivalent to $y = 0$, which is the equation we solved in Exercise 64 to find the x-intercept.
The solution set is $\left\{\tfrac{7}{3}\right\}$.

71.   The graph is below the x-axis for $x > \frac{7}{3}$, so the solution set of $f(x) < 0$ is $\left(\tfrac{7}{3}, \infty\right)$.

72.   The graph is above the x-axis for $x < \frac{7}{3}$, so the solution set of $f(x) > 0$ is $\left(-\infty, \tfrac{7}{3}\right)$.

Chapter 3 Test

1. $2x - 3y = 12$
For $x = 1$:

$$2(1) - 3y = 12$$
$$2 - 3y = 12$$
$$-3y = 10$$
$$y = -\tfrac{10}{3} \quad \left(1, -\tfrac{10}{3}\right)$$

For $x = 3$:

$$2(3) - 3y = 12$$
$$6 - 3y = 12$$
$$-3y = 6$$
$$y = -2 \quad (3, -2)$$

For $y = -4$:

$$2x - 3(-4) = 12$$
$$2x + 12 = 12$$
$$2x = 0$$
$$x = 0 \quad (0, -4)$$

2. $3x - 2y = 20$
To find the x-intercept, let $y = 0$.

$$3x - 2(0) = 20$$
$$3x = 20$$
$$x = \frac{20}{3}$$

The x-intercept is $\left(\frac{20}{3}, 0\right)$.
To find the y-intercept, let $x = 0$.

$$3(0) - 2y = 20$$
$$-2y = 20$$
$$y = -10$$

The y-intercept is $(0, -10)$.
Draw the line through these two points.

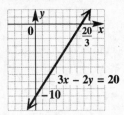

3. The graph of $y = 5$ is the horizontal line with slope 0 and y-intercept $(0, 5)$. There is no x-intercept.

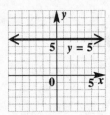

4. The graph of $x = 2$ is the vertical line with x-intercept at $(2, 0)$. There is no y-intercept.

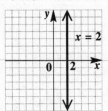

5. Through $(6, 4)$ and $(-4, -1)$

$$m = \frac{\Delta y}{\Delta x} = \frac{-1 - 4}{-4 - 6} = \frac{-5}{-10} = \frac{1}{2}$$

The slope of the line is $\frac{1}{2}$.

6. The graph of a line with undefined slope is the graph of a vertical line.

7. Find the slope of each line.

$$5x - y = 8$$
$$-y = -5x + 8$$
$$y = 5x - 8$$

The slope is 5.

$$5y = -x + 3$$
$$y = -\frac{1}{5}x + \frac{3}{5}$$

The slope is $-\frac{1}{5}$.
Since $5\left(-\frac{1}{5}\right) = -1$, the two slopes are negative reciprocals and the lines are perpendicular.

8. Find the slope of each line.

$$2y = 3x + 12$$
$$y = \frac{3}{2}x + 6$$

The slope is $\frac{3}{2}$.

$$3y = 2x - 5$$
$$y = \frac{2}{3}x - \frac{5}{3}$$

The slope is $\frac{2}{3}$.
The lines are neither parallel nor perpendicular.

9. Use the points $(1980, 119{,}000)$ and $(2005, 89{,}000)$.

average rate of change

$$= \frac{\text{change in } y}{\text{change in } x} = \frac{89{,}000 - 119{,}000}{2005 - 1980}$$
$$= \frac{-30{,}000}{25} = -1200$$

The average rate of change is about -1200 farms per year, that is, the number of farms decreased by about 1200 each year from 1980 to 2005.

10. Through $(4, -1)$; $m = -5$

(a) Let $m = -5$ and $(x_1, y_1) = (4, -1)$ in the point-slope form.

$$y - y_1 = m(x - x_1)$$
$$y - (-1) = -5(x - 4)$$
$$y + 1 = -5x + 20$$
$$y = -5x + 19$$

(b) $\quad y = -5x + 19 \quad$ *From part (a)*
$\quad 5x + y = 19 \quad\quad$ *Standard form*

11. Through $(-3, 14)$; horizontal

(a) A horizontal line has equation $y = k$. Here $k = 14$, so the line has equation $y = 14$.

(b) $y = 14$ is already in standard form.

12. Through $(-7, 2)$ and parallel to $3x + 5y = 6$

(a) To find the slope of $3x + 5y = 6$, write the equation in slope-intercept form by solving for y.

$$3x + 5y = 6$$
$$5y = -3x + 6$$
$$y = -\frac{3}{5}x + \frac{6}{5}$$

The slope is $-\frac{3}{5}$, so a line parallel to it also has slope $-\frac{3}{5}$. Let $m = -\frac{3}{5}$ and $(x_1, y_1) = (-7, 2)$ in the point-slope form.

$$y - y_1 = m(x - x_1)$$
$$y - 2 = -\tfrac{3}{5}[x - (-7)]$$
$$y - 2 = -\tfrac{3}{5}(x + 7)$$
$$y - 2 = -\tfrac{3}{5}x - \tfrac{21}{5}$$
$$y = -\tfrac{3}{5}x - \tfrac{11}{5}$$

(b) $y = -\tfrac{3}{5}x - \tfrac{11}{5}$ *From part (a)*

$\tfrac{3}{5}x + y = -\tfrac{11}{5}$ *Variable terms on one side*

$5\left(\tfrac{3}{5}x + y\right) = 5\left(-\tfrac{11}{5}\right)$ *Multiply by 5.*

$3x + 5y = -11$ *Standard form*

13. Through $(-7, 2)$ and perpendicular to $y = 2x$

(a) Since $y = 2x$ is in slope-intercept form ($b = 0$), the slope, m, of $y = 2x$ is 2. A line perpendicular to it has a slope that is the negative reciprocal of 2, that is, $-\tfrac{1}{2}$. Let $m = -\tfrac{1}{2}$ and $(x_1, y_1) = (-7, 2)$ in the point-slope form.

$$y - y_1 = m(x - x_1)$$
$$y - 2 = -\tfrac{1}{2}(x + 7)$$
$$y - 2 = -\tfrac{1}{2}x - \tfrac{7}{2}$$
$$y = -\tfrac{1}{2}x - \tfrac{3}{2}$$

(b) $y = -\tfrac{1}{2}x - \tfrac{3}{2}$ *From part (a)*

$\tfrac{1}{2}x + y = -\tfrac{3}{2}$ *Variable terms on one side*

$2\left(\tfrac{1}{2}x + y\right) = 2\left(-\tfrac{3}{2}\right)$ *Multiply by 2.*

$x + 2y = -3$ *Standard form*

14. Through $(-2, 3)$ and $(6, -1)$

(a) First find the slope.

$$m = \frac{\Delta y}{\Delta x} = \frac{-1 - 3}{6 - (-2)} = \frac{-4}{8} = -\frac{1}{2}$$

Use $m = -\tfrac{1}{2}$ and $(x_1, y_1) = (-2, 3)$ in the point-slope form.

$$y - y_1 = m(x - x_1)$$
$$y - 3 = -\tfrac{1}{2}[x - (-2)]$$
$$y - 3 = -\tfrac{1}{2}(x + 2)$$
$$y - 3 = -\tfrac{1}{2}x - 1$$
$$y = -\tfrac{1}{2}x + 2$$

(b) $y = -\tfrac{1}{2}x + 2$ *From part (a)*

$\tfrac{1}{2}x + y = 2$ *Variable terms on one side*

$2\left(\tfrac{1}{2}x + y\right) = 2(2)$ *Multiply by 2.*

$x + 2y = 4$ *Standard form*

15. Through $(5, -6)$; vertical

(a) The equation of any vertical line is in the form $x = k$. Since the line goes through $(5, -6)$, the equation is $x = 5$. Writing $x = 5$ in slope-

intercept form is *not possible* since there is no y-term.

(b) From part (a), the standard form is $x = 5$.

16. Positive slope means that the line goes up from left to right. The only line that has positive slope and a negative y-coordinate for its y-intercept is choice **B**.

17. **(a)** Use the points $(5, 22{,}860)$ and $(11, 33{,}565)$.

$$m = \frac{\Delta y}{\Delta x} = \frac{33{,}565 - 22{,}860}{11 - 5} = \frac{10{,}705}{6}$$
$$\approx 1784.17$$

Use the point-slope form with $m = \tfrac{10{,}705}{6}$ and $(x_1, y_1) = (5, 22{,}860)$.

$$y - 22{,}860 = \tfrac{10{,}705}{6}(x - 5)$$
$$y - 22{,}860 = \tfrac{10{,}705}{6}x - \tfrac{53{,}525}{6}$$
$$y \approx 1784.17x + 13{,}939.17$$

(b) $y = 1784.17(9) + 13{,}939.17$ *Let x = 9.*
  $= 29{,}996.7 \approx \$29{,}997,$
which is slightly less than the actual value.

18. $3x - 2y > 6$
Graph the line $3x - 2y = 6$, which has intercepts $(2, 0)$ and $(0, -3)$, as a dashed line since the inequality involves $>$. Test $(0, 0)$, which yields $0 > 6$, a false statement. Shade the region that does not include $(0, 0)$.

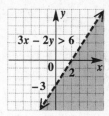

19. $y < 2x - 1$ and $x - y < 3$
First graph $y = 2x - 1$ as a dashed line through $(2, 3)$ and $(0, -1)$. Test $(0, 0)$, which yields $0 < -1$, a false statement. Shade the side of the line not containing $(0, 0)$.
Next, graph $x - y = 3$ as a dashed line through $(3, 0)$ and $(0, -3)$. Test $(0, 0)$, which yields $0 < 3$, a true statement. Shade the side of the line containing $(0, 0)$. The intersection is the region where the graphs overlap.

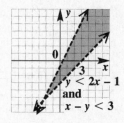

20. Choice **D** is the only graph that passes the vertical line test.

21. Choice **D** does not define a function, since its domain (input) element 0 is paired with two different range (output) elements, 1 and 2.

22. The x-values are greater than or equal to zero, so the domain is $[0, \infty)$. Since y can be any value, the range is $(-\infty, \infty)$.

23. The domain is the set of x-values: $\{0, -2, 4\}$. The range is the set of y-values: $\{1, 3, 8\}$.

24. $f(x) = -x^2 + 2x - 1$

   (a) $f(1) = -(1)^2 + 2(1) - 1$
$$= -1 + 2 - 1$$
$$= 0$$

   (b) $f(a) = -a^2 + 2a - 1$

25. $f(x) = \frac{2}{3}x - 1$

This function represents a line with y-intercept $(0, -1)$ and x-intercept $\left(\frac{3}{2}, 0\right)$.
Draw the line through these two points.
The domain is $(-\infty, \infty)$, and the range is $(-\infty, \infty)$.

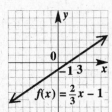

Cumulative Review Exercises (Chapters 1–3)

1. The absolute value of a negative number is a positive number and the additive inverse of the same negative number is the same positive number. For example, suppose the negative number is -5:

$$|-5| = -(-5) = 5$$
$$\text{and} \quad -(-5) = 5$$

The statement is *always true*.

2. The statement is *always true*; in fact, it is the definition of a rational number.

3. The sum of two negative numbers is another negative number, so the statement is *never true*.

4. The statement is *sometimes true*. For example,

$$3 + (-3) = 0,$$
$$\text{but} \quad 3 + (-1) = 2 \neq 0.$$

5. $-|-2| - 4 + |-3| + 7 = -2 - 4 + 3 + 7$
$$= -6 + 3 + 7$$
$$= -3 + 7$$
$$= 4$$

6. $(-0.8)^2 = (-0.8)(-0.8) = 0.64$

7. $\sqrt{-64}$ is not a real number.

8. The product of two numbers that have the same sign is positive.

$$-\frac{2}{3}\left(-\frac{12}{5}\right) = \frac{2 \cdot 12}{3 \cdot 5} = \frac{2 \cdot 4}{5} = \frac{8}{5}$$

9. $-(-4m + 3) = -(-4m) - 3$
$$= 4m - 3$$

10. $3x^2 - 4x + 4 + 9x - x^2$
$$= 3x^2 - x^2 - 4x + 9x + 4$$
$$= 2x^2 + 5x + 4$$

11. $\dfrac{(4^2 - 4) - (-1)7}{4 + (-6)} = \dfrac{(16 - 4) - (-7)}{-2}$
$$= \frac{12 + 7}{-2} = -\frac{19}{2}$$

12. $-3 < x \leq 5$

This is the set of numbers between -3 and 5, not including -3 (use a parenthesis), but including 5 (use a bracket). In interval notation, the set is $(-3, 5]$.

13. $\sqrt{\dfrac{-2 + 4}{-5}} = \sqrt{\dfrac{2}{-5}} = \sqrt{-\dfrac{2}{5}}$

This is not a real number since the number under the radical sign is negative.

For Exercises 14–16, let $p = -4$, $q = \frac{1}{2}$, and $r = 16$.

14. $-3(2q - 3p) = -3\left[2\left(\frac{1}{2}\right) - 3(-4)\right]$
$$= -3(1 + 12)$$
$$= -3(13)$$
$$= -39$$

15. $|p|^3 - |16q^3| = |-4|^3 - \left|16\left(\frac{1}{2}\right)^3\right|$
$$= 4^3 - \left|16\left(\frac{1}{8}\right)\right|$$
$$= 64 - |2|$$
$$= 64 - 2 = 62$$

16. $\dfrac{\sqrt{r}}{8p + 2r} = \dfrac{\sqrt{16}}{8(-4) + 2(16)}$
$$= \frac{4}{-32 + 32}$$
$$= \frac{4}{0}, \text{ which is } \textit{undefined.}$$

17. $2z - 5 + 3z = 2 - z$
$5z - 5 = 2 - z$
$6z = 7$
$z = \frac{7}{6}$

The solution set is $\left\{\frac{7}{6}\right\}$.

18. $\dfrac{3a - 1}{5} + \dfrac{a + 2}{2} = -\dfrac{3}{10}$

Multiply both sides by the LCD, 10.

$10\left(\dfrac{3a - 1}{5} + \dfrac{a + 2}{2}\right) = 10\left(-\dfrac{3}{10}\right)$

$2(3a - 1) + 5(a + 2) = -3$
$6a - 2 + 5a + 10 = -3$
$11a + 8 = -3$
$11a = -11$
$a = -1$

The solution set is $\{-1\}$.

19. Solve $V = \dfrac{1}{3}\pi r^2 h$ for h.

$3V = \pi r^2 h$ *Multiply by 3.*

$\dfrac{3V}{\pi r^2} = h$ *Divide by πr^2.*

20. $F = \frac{9}{5}C + 32$

$F = \frac{9}{5}(-50) + 32$ *Let C= –50.*

$= -90 + 32 = -58$

$-50°C$ is equivalent to $-58°F$.

21. Let x denote the side of the original square and $4x$ the perimeter. Now $x + 4$ is the side of the new square and $4(x + 4)$ is its perimeter.
"The perimeter would be 8 inches less than twice the perimeter of the original square " translates as

$4(x + 4) = 2(4x) - 8.$
$4x + 16 = 8x - 8$
$24 = 4x$
$6 = x$

The length of a side of the original square is 6 inches.

22. Let $x =$ the time it takes for the planes to be 2100 miles apart.
Make a table. Use the formula $d = rt$.

	r	t	d
Eastbound Plane	550	x	$550x$
Westbound Plane	500	x	$500x$

The total distance is 2100 miles.

$550x + 500x = 2100$
$1050x = 2100$
$x = 2$

It will take 2 hr for the planes to be 2100 mi apart.

23. $-4 < 3 - 2k < 9$
$-7 < -2k < 6$
Divide by -2; reverse the inequality symbols.
$\frac{7}{2} > k > -3$ or $-3 < k < \frac{7}{2}$
The solution set is $\left(-3, \frac{7}{2}\right)$.

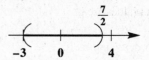

24. $-0.3x + 2.1(x - 4) \le -6.6$
$-3x + 21(x - 4) \le -66$
Multiply by 10.
$-3x + 21x - 84 \le -66$
$18x - 84 \le -66$
$18x \le 18$
$x \le 1$
The solution set is $(-\infty, 1]$.

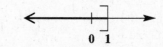

25. $\frac{1}{2}x > 3$ and $\frac{1}{3}x < \frac{8}{3}$
$x > 6$ and $x < 8$
The graph of the solution set is all numbers both greater than 6 *and* less than 8. This is the intersection. The elements common to both sets are the numbers between 6 and 8, not including the endpoints. The solution set is $(6, 8)$.

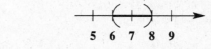

26. $-5x + 1 \ge 11$ or $3x + 5 > 26$
$-5x \ge 10$ $3x > 21$
$x \le -2$ or $x > 7$
The graph of the solution set is all numbers either less than or equal to -2 *or* greater than 7. This is the union. The solution set is $(-\infty, -2] \cup (7, \infty)$.

27. $|2k - 7| + 4 = 11$
$|2k - 7| = 7$
$2k - 7 = 7$ or $2k - 7 = -7$
$2k = 14$ $2k = 0$
$k = 7$ or $k = 0$
The solution set is $\{0, 7\}$.

28. $|3m + 6| \ge 0$
The absolute value of an expression is always nonnegative, so the inequality is true for any real number m.
The solution set is $(-\infty, \infty)$.

29. The union of the three solution sets is $(-\infty, \infty)$; that is, the set of all real numbers.

30. To complete the table of ordered pairs, substitute the given values for x or y in the equation $3x - 4y = 12$.

For $x = 0$:
$$3x - 4y = 12$$
$$3(0) - 4y = 12$$
$$-4y = 12$$
$$y = -3$$

The ordered pair is $(0, -3)$.

For $y = 0$:
$$3x - 4y = 12$$
$$3x - 4(0) = 12$$
$$3x = 12$$
$$x = 4$$

The ordered pair is $(4, 0)$.

For $x = 2$:
$$3x - 4y = 12$$
$$3(2) - 4y = 12$$
$$6 - 4y = 12$$
$$-4y = 6$$
$$y = \frac{6}{-4} = -\frac{3}{2}$$

The ordered pair is $\left(2, -\frac{3}{2}\right)$.

31. $3x + 5y = 12$
To find the x-intercept, let $y = 0$.
$$3x + 5(0) = 12$$
$$3x = 12$$
$$x = 4$$

The x-intercept is $(4, 0)$.
To find the y-intercept, let $x = 0$.
$$3(0) + 5y = 12$$
$$5y = 12$$
$$y = \frac{12}{5}$$

The y-intercept is $\left(0, \frac{12}{5}\right)$.
Plot the intercepts and draw the line through them.

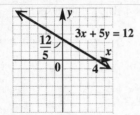

32. $A(-2, 1)$ and $B(3, -5)$

(a) The slope of line AB is
$$m = \frac{\Delta y}{\Delta x} = \frac{-5 - 1}{3 - (-2)} = \frac{-6}{5} = -\frac{6}{5}.$$

(b) The slope of a line perpendicular to line AB is the negative reciprocal of $-\frac{6}{5}$, which is $\frac{5}{6}$.

33. $-2x + y < -6$
Graph the line $-2x + y = -6$, which has intercepts $(3, 0)$ and $(0, -6)$, as a dashed line since the inequality involves $<$. Test $(0, 0)$, which yields $0 < -6$, a false statement. Shade the region that does not include $(0, 0)$.

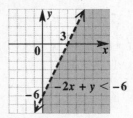

34. (a) Slope $-\frac{3}{4}$; y-intercept $(0, -1)$
To write an equation of this line, let $m = -\frac{3}{4}$ and $b = -1$ in the slope-intercept form.
$$y = mx + b$$
$$y = -\frac{3}{4}x - 1$$

(b)
$$y = -\frac{3}{4}x - 1$$
$$4y = -3x - 4$$
$$3x + 4y = -4$$

35. (a) Horizontal; through $(2, -2)$
A horizontal line through the point (c, d) has equation $y = d$. Here $d = -2$, so the equation of the line is $y = -2$.

(b) $y = -2$ is already in standard form.

36. (a) Through $(4, -3)$ and $(1, 1)$
First find the slope of the line.
$$m = \frac{\Delta y}{\Delta x} = \frac{1 - (-3)}{1 - 4} = \frac{4}{-3} = -\frac{4}{3}$$

Now substitute $(x_1, y_1) = (4, -3)$ and $m = -\frac{4}{3}$ in the point-slope form. Then solve for y.
$$y - y_1 = m(x - x_1)$$
$$y - (-3) = -\frac{4}{3}(x - 4)$$
$$y + 3 = -\frac{4}{3}x + \frac{16}{3}$$
$$y = -\frac{4}{3}x + \frac{7}{3}$$

(b)
$$y = -\frac{4}{3}x + \frac{7}{3}$$
$$3y = -4x + 7$$
$$4x + 3y = 7$$

37. (a) Vertical; through $(4, -6)$
A vertical line through the point (c, d) has equation $x = c$. Here $c = 4$, so the equation of the line is $x = 4$. It is not possible to write this equation in slope-intercept form.

(b) $x = 4$ is in standard form.

38. The domain of the relation consists of the elements in the leftmost figure; that is, $\{14, 91, 75, 23\}$.

The range of the relation consists of the elements in the rightmost figure; that is, $\{9, 70, 56, 5\}$.

Since the element 75 in the domain is paired with two different values, 70 and 56, in the range, the relation is not a function.

39. $f(x) = -4x + 10$

(a) The variable x can be any real number, so the domain is $(-\infty, \infty)$. The function is a non-constant linear function, so its range is $(-\infty, \infty)$.

(b) $f(-3) = -4(-3) + 10 = 12 + 10 = 22$

40. Use $(1997, 12{,}000)$ and $(2004, 86{,}000)$.

$$m = \frac{\Delta y}{\Delta x} = \frac{86{,}000 - 12{,}000}{2004 - 1997} = \frac{74{,}000}{7} \approx 10{,}571$$

So the average rate of change is 10,571 per year; that is, the number of motor scooters sold in the United States increased by an average of 10,571 per year from 1997 to 2004.

CHAPTER 4 SYSTEMS OF LINEAR EQUATIONS

4.1 Systems of Linear Equations in Two Variables

1. If $(3, -6)$ is a solution of a linear system in two variables, then substituting $\underline{3}$ for x and $\underline{-6}$ for y leads to true statements in *both* equations.

3. If the solution process leads to a false statement such as $0 = 5$, the solution set is $\underline{\emptyset}$.

5. If the two lines forming a system have the same slope and different y-intercepts, the system has $\underline{0}$ solutions. (The lines are parallel.)

7. **D**; The ordered pair solution must be in quadrant IV, since that is where the graphs of the equations intersect.

9. **(a)** $x - y = 0$ implies that $x = y$, so the coordinates of the solution must be equal—this limits our choice to **B** or **C**. Since $x + y = 6$, graph **B** is correct.

(b) As in part (a), we must have **B** or **C**. Since $x + y = -6$, graph **C** is correct.

(c) $x + y = 0$ implies that $x = -y$, so the coordinates of the solution must be opposites—this limits our choice to **A** or **D**. Since $x - y = -6$, graph **A** is correct.

(d) As in part (c), we must have **A** or **D**. Since $x - y = 6$, graph **D** is correct.

11. $x - y = 17$
$x + y = -1$
To decide if $(8, -9)$ is a solution, replace x with 8 and y with -9 in each equation of the system.

$$x - y = 17$$
$$8 - (-9) = 17$$
$$17 = 17 \quad \textit{True}$$

$$x + y = -1$$
$$8 + (-9) = -1$$
$$-1 = -1 \quad \textit{True}$$

Since $(8, -9)$ makes both equations true, $(8, -9)$ is a solution of the system.

13. $3x - 5y = -12$
$x - y = 1$
Replace x with -1 and y with 2.

$$3(-1) - 5(2) = -12$$
$$-3 - 10 = -12$$
$$-13 = -12 \quad \textit{False}$$

It is not necessary to check if $(-1, 2)$ satisfies the second equation since it does not satisfy the first equation. Therefore, $(-1, 2)$ is *not* a solution of the system.

15. $x + y = -5$
$-2x + y = 1$
Graph the line $x + y = -5$ through its intercepts, $(-5, 0)$ and $(0, -5)$, and the line $-2x + y = 1$ through its intercepts, $\left(-\frac{1}{2}, 0\right)$ and $(0, 1)$. The lines appear to intersect at $(-2, -3)$.

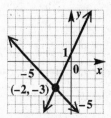

Check this ordered pair in the system.

$$(-2) + (-3) = -5$$
$$-5 = -5 \quad \textit{True}$$

$$-2(-2) + (-3) = 1$$
$$4 + (-3) = 1 \quad \textit{True}$$

The solution set is $\{(-2, -3)\}$.

17. $4x + y = 6 \quad (1)$
$y = 2x \quad (2)$
Since equation (2) is already solved for y, substitute $2x$ for y in equation (1).

$$4x + y = 6 \quad (1)$$
$$4x + 2x = 6 \quad \textit{Let } y = 2x.$$
$$6x = 6$$
$$x = 1$$

Substitute 1 for x in (2).

$$y = 2(1) = 2$$

The solution $(1, 2)$ checks.
The solution set is $\{(1, 2)\}$.

19. $-x - 4y = -14 \quad (1)$
$y = 2x - 1 \quad (2)$

Substitute $2x - 1$ for y in equation (1) and solve for x.

$$-x - 4y = -14 \quad (1)$$
$$-x - 4(2x - 1) = -14$$
$$-x - 8x + 4 = -14$$
$$-9x = -18$$
$$x = 2$$

Substitute 2 for x in (2).

$$y = 2(2) - 1 = 3$$

The solution $(2, 3)$ checks.

The solution set is $\{(2, 3)\}$.

21. $3x - 4y = -22$ (1)
$-3x + y = 0$ (2)

Solve equation (2) for y to get

$$y = 3x.$$

Substitute $3x$ for y in equation (1).

$$3x - 4y = -22 \quad (1)$$
$$3x - 4(3x) = -22 \quad \textit{Let } y = 3x.$$
$$3x - 12x = -22$$
$$-9x = -22$$
$$x = \frac{-22}{-9} = \frac{22}{9}$$

Substitute $\frac{22}{9}$ for x in $y = 3x$ to get

$$y = 3\left(\frac{22}{9}\right) = \frac{22}{3}.$$

The solution $\left(\frac{22}{9}, \frac{22}{3}\right)$ checks.
The solution set is $\left\{\left(\frac{22}{9}, \frac{22}{3}\right)\right\}$.

23. $5x - 4y = 9$ (1)
$3 - 2y = -x$ (2)

Solve equation (2) for x.

$$3 - 2y = -x \quad (2)$$
$$2y - 3 = x \quad (3) \quad -1 \times (2)$$

Substitute $2y - 3$ for x in equation (1).

$$5x - 4y = 9 \quad (1)$$
$$5(2y - 3) - 4y = 9 \quad \textit{Let } x = 2y - 3.$$
$$10y - 15 - 4y = 9$$
$$6y - 15 = 9$$
$$6y = 24$$
$$y = 4$$

Substitute 4 for y in $x = 2y - 3$.

$$x = 2(4) - 3 = 5$$

The solution $(5, 4)$ checks.
The solution set is $\{(5, 4)\}$.

25. $4x - 5y = -11$ (1)
$x + 2y = 7$ (2)

Solve equation (2) for x.

$$x = -2y + 7$$

Substitute $-2y + 7$ for x in (1).

$$4x - 5y = -11 \quad (1)$$
$$4(-2y + 7) - 5y = -11 \quad \textit{Let } x = -2y + 7.$$
$$-8y + 28 - 5y = -11$$
$$-13y + 28 = -11$$
$$-13y = -39$$
$$y = 3$$

Substitute 3 for y in $x = -2y + 7$.

$$x = -2(3) + 7 = -6 + 7 = 1$$

The solution $(1, 3)$ checks.
The solution set is $\{(1, 3)\}$.

27. $x = 3y + 5$ (1)
$x = \frac{3}{2}y$ (2)

Both equations are given in terms of x. Choose equation (2), and substitute $\frac{3}{2}y$ for x in equation (1).

$$x = 3y + 5 \quad (1)$$
$$\tfrac{3}{2}y = 3y + 5 \quad \textit{Let } x = \tfrac{3}{2}y.$$
$$3y = 6y + 10 \quad \textit{Multiply by 2.}$$
$$-3y = 10$$
$$y = -\frac{10}{3}$$

Since $x = \frac{3}{2}y$ and $y = -\frac{10}{3}$,

$$x = \tfrac{3}{2}\left(-\tfrac{10}{3}\right) = -\frac{10}{2} = -5.$$

The solution $\left(-5, -\frac{10}{3}\right)$ checks.
The solution set is $\left\{\left(-5, -\frac{10}{3}\right)\right\}$.

29. $\frac{1}{2}x + \frac{1}{3}y = 3$ (1)
$-3x + y = 0$ (2)

Multiply (1) by its LCD, 6, to eliminate fractions

$$3x + 2y = 18 \quad (3)$$

From (2), we have $y = 3x$, so substitute $3x$ for y in (3).

$$3x + 2(3x) = 18$$
$$3x + 6x = 18$$
$$9x = 18$$
$$x = 2$$

Substitute 2 for x in $y = 3x$.

$$y = 3(2) = 6$$

The solution $(2, 6)$ checks.
The solution set is $\{(2, 6)\}$.

31. $y = 2x$ (1)
$4x - 2y = 0$ (2)

From equation (1), substitute $2x$ for y in equation (2).

$$4x - 2y = 0 \quad (2)$$
$$4x - 2(2x) = 0$$
$$4x - 4x = 0$$
$$0 = 0 \quad \textit{True}$$

The equations are dependent, and the solution is the set of all points on the line.
The solution set is $\{(x, y) \mid y = 2x\}$.

33.
$$x = 5y \quad (1)$$
$$5x - 25y = 5 \quad (2)$$

From equation (1), substitute $5y$ for x in equation (2).

$$5x - 25y = 5 \quad (2)$$
$$5(5y) - 25y = 5$$
$$25y - 25y = 5$$
$$0 = 5 \quad \textit{False}$$

The system is *inconsistent*. Since the graphs of the equations are parallel lines, there are no ordered pairs that satisfy both equations.
The solution set is $\emptyset$.

35.
$$-2x + 3y = -16 \quad (1)$$
$$2x - 5y = 24 \quad (2)$$
To eliminate x, add the equations.

$$\begin{array}{rcr} -2x + 3y &=& -16 \quad (1) \\ 2x - 5y &=& 24 \quad (2) \\ \hline -2y &=& 8 \\ y &=& -4 \end{array}$$

To find x, substitute -4 for y in equation (2).

$$2x - 5y = 24 \quad (2)$$
$$2x - 5(-4) = 24$$
$$2x + 20 = 24$$
$$2x = 4$$
$$x = 2$$

The ordered pair $(2, -4)$ satisfies both equations, so it checks.
The solution set is $\{(2, -4)\}$.

37.
$$2x - 5y = 11 \quad (1)$$
$$3x + y = 8 \quad (2)$$
To eliminate y, multiply equation (2) by 5 and add the result to equation (1).

$$\begin{array}{rclll} 2x - 5y &=& 11 & (1) \\ 15x + 5y &=& 40 & (3) & 5 \times (2) \\ \hline 17x &=& 51 \\ x &=& 3 \end{array}$$

To find y, substitute 3 for x in equation (2).

$$3x + y = 8 \quad (2)$$
$$3(3) + y = 8$$
$$9 + y = 8$$
$$y = -1$$

The ordered pair $(3, -1)$ satisfies both equations, so it checks.
The solution set is $\{(3, -1)\}$.

39.
$$3x + 4y = -6 \quad (1)$$
$$5x + 3y = 1 \quad (2)$$
To eliminate x, multiply equation (1) by 5 and equation (2) by -3. Then add the results.

$$\begin{array}{rclll} 15x + 20y &=& -30 & (3) & 5 \times (1) \\ -15x - 9y &=& -3 & (4) & -3 \times (2) \\ \hline 11y &=& -33 \\ y &=& -3 \end{array}$$

To find x, substitute -3 for y in equation (2).

$$5x + 3y = 1 \quad (2)$$
$$5x + 3(-3) = 1$$
$$5x - 9 = 1$$
$$5x = 10$$
$$x = 2$$

The ordered pair $(2, -3)$ satisfies both equations, so it checks.
The solution set is $\{(2, -3)\}$.

41.
$$7x + 2y = 6 \quad (1)$$
$$-14x - 4y = -12 \quad (2)$$
To eliminate y, multiply equation (1) by 2 and add the result to equation (2).

$$\begin{array}{rclll} 14x + 4y &=& 12 & (3) & 2 \times (1) \\ -14x - 4y &=& -12 & (2) \\ \hline 0 &=& 0 & \textit{True} \end{array}$$

Multiplying equation (1) by -2 gives equation (2). The equations are dependent, and the solution is the set of all points on the line.
The solution set is $\{(x, y) \mid 7x + 2y = 6\}$.

43.
$$3x + 3y = 0 \quad (1)$$
$$4x + 2y = 3 \quad (2)$$
To eliminate y, multiply equation (1) by 2 and equation (2) by -3. Then add the results.

$$\begin{array}{rclll} 6x + 6y &=& 0 & (3) & 2 \times (1) \\ -12x - 6y &=& -9 & (4) & -3 \times (2) \\ \hline -6x &=& -9 \\ x &=& \frac{-9}{-6} = \frac{3}{2} \end{array}$$

To find y, substitute $\frac{3}{2}$ for x in equation (1).

$$3x + 3y = 0 \quad (1)$$
$$3\left(\tfrac{3}{2}\right) + 3y = 0$$
$$\tfrac{9}{2} + 3y = 0$$
$$3y = -\tfrac{9}{2}$$
$$y = \tfrac{1}{3}\left(-\tfrac{9}{2}\right) = -\tfrac{3}{2}$$

The solution $\left(\tfrac{3}{2}, -\tfrac{3}{2}\right)$ checks.
The solution set is $\left\{\left(\tfrac{3}{2}, -\tfrac{3}{2}\right)\right\}$.

continued

When you get a solution that has non-integer components, it is sometimes more difficult to check the problem than it was to solve it. A graphing calculator can be very helpful in this case. Just store the values for x and y in their respective memory locations, and then type the expressions as shown in the following screen. The results 0 and 3 (the right sides of the equations) indicate that we have found the correct solution.

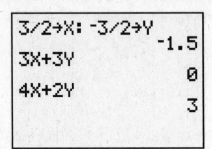

45. $5x - 5y = 3 \quad (1)$
$x - y = 12 \quad (2)$

To eliminate x, multiply equation (2) by -5 and add the result to equation (1).

$$\begin{array}{rl}
5x - 5y = & 3 \quad (1) \\
-5x + 5y = & -60 \quad (3) \quad -5 \times (2) \\
\hline
0 = & -57 \quad \textit{False}
\end{array}$$

The system is *inconsistent*. Since the graphs of the equations are parallel lines, there are no ordered pairs that satisfy both equations. The solution set is $\emptyset$.

47. $x + y = 0 \quad (1)$
$2x - 2y = 0 \quad (2)$

To eliminate y, multiply equation (1) by 2 and add the result to equation (2).

$$\begin{array}{rl}
2x + 2y = 0 & \quad (3) \quad 2 \times (1) \\
2x - 2y = 0 & \quad (2) \\
\hline
4x = 0 & \\
x = 0 &
\end{array}$$

Substitute 0 for x in (1) to get $y = 0$. The solution $(0, 0)$ checks.

The solution set is $\{(0, 0)\}$.

49. $x - \frac{1}{2}y = 2 \quad (1)$
$-x + \frac{2}{5}y = -\frac{8}{5} \quad (2)$

To eliminate x, add equations (1) and (2) to get

$$-\tfrac{1}{2}y + \tfrac{2}{5}y = 2 - \tfrac{8}{5}.$$

Multiply by the LCD, 10.

$$\begin{array}{r}
-5y + 4y = 20 - 16 \\
-y = 4 \\
y = -4
\end{array}$$

Substitute -4 for y in equation (1) to find x.

$$\begin{array}{r}
x - \tfrac{1}{2}(-4) = 2 \\
x + 2 = 2 \\
x = 0
\end{array}$$

The solution $(0, -4)$ checks.

The solution set is $\{(0, -4)\}$.

51. $\frac{1}{2}x + \frac{1}{3}y = -\frac{1}{3} \quad (1)$
$\frac{1}{2}x + 2y = -7 \quad (2)$

Eliminate the fractions by multiplying equation (1) by -6 and equation (2) by 6. Then add the results to eliminate x.

$$\begin{array}{rl}
-3x - 2y = & 2 \quad (3) \quad -6 \times (1) \\
3x + 12y = & -42 \quad (4) \quad 6 \times (2) \\
\hline
10y = & -40 \\
y = & -4
\end{array}$$

To find x, substitute -4 for y in equation (3).

$$\begin{array}{r}
-3x - 2y = 2 \quad (3) \\
-3x - 2(-4) = 2 \\
-3x + 8 = 2 \\
-3x = -6 \\
x = 2
\end{array}$$

The solution $(2, -4)$ checks.
The solution set is $\{(2, -4)\}$.

53. $3x + 7y = 4 \quad (1)$
$6x + 14y = 3 \quad (2)$

Write each equation in slope-intercept form by solving for y.

$$\begin{array}{r}
3x + 7y = 4 \quad\quad (1) \\
7y = -3x + 4 \\
y = -\tfrac{3}{7}x + \tfrac{4}{7}
\end{array}$$

$$\begin{array}{r}
6x + 14y = 3 \quad\quad (2) \\
14y = -6x + 3 \\
y = -\tfrac{6}{14}x + \tfrac{3}{14} \\
y = -\tfrac{3}{7}x + \tfrac{3}{14}
\end{array}$$

Since the equations have the same slope, $-\frac{3}{7}$, but different y-intercepts, $\frac{4}{7}$ and $\frac{3}{14}$, the lines when graphed are parallel. The system is inconsistent and has no solution.

55. $2x = -3y + 1 \quad (1)$
$6x = -9y + 3 \quad (2)$

Write each equation in slope-intercept form by solving for y.

$$\begin{array}{r}
2x = -3y + 1 \quad\quad (1) \\
3y = -2x + 1 \\
y = -\tfrac{2}{3}x + \tfrac{1}{3}
\end{array}$$

$$6x = -9y + 3 \quad (2)$$
$$9y = -6x + 3$$
$$y = -\frac{6}{9}x + \frac{3}{9}$$
$$y = -\frac{2}{3}x + \frac{1}{3}$$

Since both equations are the same, the solution set is all points on the line $y = -\frac{2}{3}x + \frac{1}{3}$. The system has infinitely many solutions.

57. **(a)** $6x - y = 5 \quad (1)$
$\qquad\quad y = 11x \quad (2)$

Use substitution, since the second equation is solved for y.

(b) $3x + y = -7 \quad (1)$
$\qquad x - y = -5 \quad (2)$

Use elimination, since the coefficients of the y-terms are opposites.

(c) $3x - 2y = 0 \quad (1)$
$\qquad 9x + 8y = 7 \quad (2)$

Use elimination, since the equations are in standard form with no coefficients of 1 or −1. Solving by substitution would involve fractions.

59. $3x + y = -7 \quad (1)$
$\quad\; x - y = -5 \quad (2)$

Add the equations to eliminate y.

$$
\begin{array}{rcl}
3x \;+\; y &=& -7 \\
x \;-\; y &=& -5 \\
\hline
4x &=& -12 \\
x &=& -3
\end{array}
$$

Substitute −3 for x in (2).

$$-3 - y = -5$$
$$-y = -2$$
$$y = 2$$

The solution $(-3, 2)$ checks.

The solution set is $\{(-3, 2)\}$.

61. $2x + 3y = 10 \quad (1)$
$\quad -3x + y = 18 \quad (2)$

Solve equation (2) for y.

$$y = 3x + 18 \quad (3)$$

Substitute $3x + 18$ for y in (1).

$$2x + 3(3x + 18) = 10$$
$$2x + 9x + 54 = 10$$
$$11x + 54 = 10$$
$$11x = -44$$
$$x = -4$$

Substitute −4 for x in (3).

$$y = 3(-4) + 18 = 6$$

The solution $(-4, 6)$ checks.

The solution set is $\{(-4, 6)\}$.

63. $\frac{1}{2}x - \frac{1}{8}y = -\frac{1}{4} \quad (1)$
$\quad 4x - y = -2 \quad (2)$

Multiply equation (1) by −8 and add to equation (2).

$$
\begin{array}{rcll}
-4x \;+\; y &=& 2 & (3) \quad -8 \times (1) \\
4x \;-\; y &=& -2 & (2) \\
\hline
0 &=& 0 &
\end{array}
$$

The equations are dependent.

The solution set is $\{(x, y) \mid 4x - y = -2\}$.

65. $0.3x + 0.2y = 0.4 \quad (1)$
$\quad 0.5x + 0.4y = 0.7 \quad (2)$

To eliminate y, multiply equation (1) by −20 and equation (2) by 10, then add.

$$
\begin{array}{rcll}
-6x \;-\; 4y &=& -8 & -20 \times (1) \\
5x \;+\; 4y &=& 7 & (3) \quad 10 \times (2) \\
\hline
-x &=& -1 & \\
x &=& 1 &
\end{array}
$$

Substitute 1 for x in equation (3).

$$5(1) + 4y = 7$$
$$4y = 2$$
$$y = \frac{1}{2}$$

Check $\left(1, \frac{1}{2}\right)$: $0.3 + 0.1 = 0.4$; $0.5 + 0.2 = 0.7$

The solution set is $\left\{\left(1, \frac{1}{2}\right)\right\}$.

67. The table shows that when $X = 3$, $Y_1 = -4$ and $Y_2 = -4$. Since no other values of X in the table give the same values for Y_1 and Y_2, and since the functions are linear, the point $(3, -4)$ is the only point of intersection for the two graphs.

69. $y_1 = 3x - 5$
$\quad y_2 = -4x + 2$
Both of the graphs in **B** have negative y-intercepts. But the y-intercept of y_2 is positive, so the graph in **B** is not acceptable. The slope of y_1 is positive, and y_1 has a negative y-intercept. The slope of y_2 is negative, and y_2 has a positive y-intercept. This fits graph **A**.

71. **(a)**
$$
\begin{array}{rcll}
x \;+\; y &=& 10 & (1) \\
2x \;-\; y &=& 5 & (2) \\
\hline
3x &=& 15 & \\
x &=& 5 &
\end{array}
$$

Substitute 5 for x in equation (1).

$$x + y = 10 \quad (1)$$
$$5 + y = 10$$
$$y = 5$$

The solution set is $\{(5, 5)\}$.

(b) Solve (1) and (2) for y.

$$x + y = 10 \qquad (1)$$
$$y = -x + 10 \qquad (3)$$

$$2x - y = 5 \qquad (2)$$
$$2x - 5 = y \qquad (4)$$

Now graph (3) and (4).

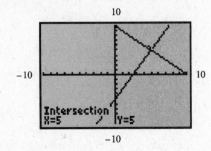

73. (a)
$$3x - 2y = 4 \qquad (1)$$
$$3x + y = -2 \qquad (2)$$

Multiply equation (2) by -1 and add the result to equation (1).

$$
\begin{array}{rl}
3x - 2y = & 4 \quad (1) \\
\underline{-3x - y = 2} & \quad (3) \quad -1 \times (2) \\
-3y = & 6 \\
y = & -2
\end{array}
$$

Substitute -2 for y in equation (1).

$$3x - 2y = 4 \quad (1)$$
$$3x - 2(-2) = 4$$
$$3x + 4 = 4$$
$$3x = 0$$
$$x = 0$$

The solution set is $\{(0, -2)\}$.

(b) Solving (1) and (2) for y gives us

$$y = \frac{3}{2}x - 2 \qquad (3)$$
$$\text{and} \quad y = -3x - 2. \qquad (4)$$

Now graph (3) and (4).

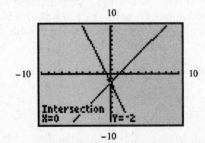

75. (a) The supply and demand graphs intersect at 4, so supply equals demand at a price of $4 per half-gallon.

(b) At a price of $4 per half-gallon, the supply and demand are both about 300 half-gallons.

(c) At a price of $2 per half-gallon, the supply is 200 half-gallons and the demand is 400 half-gallons.

77. (a) The graph for ABC is above the others for years 94–96. Therefore, ABC dominated between 1994 and 1996.

(b) ABC's dominance ended in 1997. ABC and CBS both had a 16% share in 1997.

(c) In 1998 and in 1999, ABC and NBC both had 16% share.

(d) The first year in which two networks had an equal share is 1994. CBS and NBC both had a share of 18%, so the ordered pair is $(1994, 18)$.

(e) Viewership has generally declined during these years.

79. The graphs intersect at about 3.5 (years since 2000). So the sales of digital cameras were less than the sales of conventional cameras for 2000, 2001, 2002, and the first half of 2003.

81.
$$2.5x + y = 19.4 \quad (1)$$
$$-1.7x + y = 4.4 \quad (2)$$

Solve equation (2) for y.

$$y = 1.7x + 4.4 \quad (3)$$

Substitute $1.7x + 4.4$ for y in equation (1).

$$2.5x + (1.7x + 4.4) = 19.4$$
$$4.2x + 4.4 = 19.4$$
$$4.2x = 15$$
$$x = \frac{15}{4.2} = \frac{150}{42} = \frac{25}{7}$$

Substitute $\frac{25}{7}$ for x in (3).

$$y = 1.7\left(\frac{25}{7}\right) + 4.4 = \frac{17}{10}\left(\frac{25}{7}\right) + \frac{44}{10} \cdot \frac{7}{7} = \frac{733}{70}$$

Approximating these values to the nearest tenth, we get $\frac{25}{7} \approx 3.6$ and $\frac{733}{70} \approx 10.5$. Written as an ordered pair, the solution is $(3.6, 10.5)$. (Values may vary slightly based on the method of solution used.)

83.
$$\frac{3}{x} + \frac{4}{y} = \frac{5}{2} \qquad (1)$$
$$\frac{5}{x} - \frac{3}{y} = \frac{7}{4} \qquad (2)$$

If $p = \frac{1}{x}$ and $q = \frac{1}{y}$, equations (1) and (2) can be written as

$$3p + 4q = \tfrac{5}{2} \qquad (3)$$
$$5p - 3q = \tfrac{7}{4}. \qquad (4)$$

To eliminate q, multiply equation (3) by 3 and equation (4) by 4. Then add the results.

$$
\begin{array}{rll}
9p + 12q = \frac{15}{2} & (5) & 3 \times (3) \\
\underline{20p - 12q = 7} & (6) & 4 \times (4) \\
29p \qquad\;\; = \frac{29}{2} & & \\
p = \frac{1}{2} & &
\end{array}
$$

To find q, substitute $\frac{1}{2}$ for p in equation (3).

$$
\begin{array}{rl}
3p + 4q = \frac{5}{2} & (3) \\
3\left(\frac{1}{2}\right) + 4q = \frac{5}{2} & \text{Let } p = \frac{1}{2}. \\
\frac{3}{2} + 4q = \frac{5}{2} & \\
4q = 1 & \\
q = \frac{1}{4} &
\end{array}
$$

Since $p = \dfrac{1}{x}$ and $p = \dfrac{1}{2}$, $\dfrac{1}{2} = \dfrac{1}{x}$ and $x = 2$.

Since $q = \dfrac{1}{y}$ and $q = \dfrac{1}{4}$, $\dfrac{1}{4} = \dfrac{1}{y}$ and $y = 4$.

The solution $(2, 4)$ checks.

The solution set is $\{(2, 4)\}$.

85.
$$
\begin{array}{rl}
\dfrac{2}{x} + \dfrac{3}{y} = \dfrac{11}{2} & (1) \\[2mm]
-\dfrac{1}{x} + \dfrac{2}{y} = -1 & (2)
\end{array}
$$

Let $p = \dfrac{1}{x}$ and $q = \dfrac{1}{y}$. Rewrite the system as

$$
\begin{array}{rl}
2p + 3q = \frac{11}{2} & (3) \\
-p + 2q = -1. & (4)
\end{array}
$$

To eliminate p, multiply (3) by 2 and (4) by 4. Then add the results.

$$
\begin{array}{rll}
4p + 6q = 11 & (5) & 2 \times (3) \\
\underline{-4p + 8q = -4} & (6) & 4 \times (4) \\
14q = 7 & & \\
q = \frac{1}{2} & &
\end{array}
$$

Substitute $\frac{1}{2}$ for q in (4).

$$
\begin{array}{rl}
-p + 2\left(\frac{1}{2}\right) = -1 & \\
-p + 1 = -1 & \\
-p = -2 & \\
p = 2 &
\end{array}
$$

Since $p = \dfrac{1}{x}$ and $p = 2$,

$2 = \dfrac{1}{x}$, so $2x = 1$ and $x = \dfrac{1}{2}$.

Since $q = \dfrac{1}{y}$ and $q = \dfrac{1}{2}$,

$\dfrac{1}{2} = \dfrac{1}{y}$ and $y = 2$.

The solution $\left(\frac{1}{2}, 2\right)$ checks.

The solution set is $\left\{\left(\frac{1}{2}, 2\right)\right\}$.

87.
$$
\begin{array}{rl}
ax + by = c & (1) \\
ax - 2by = c & (2)
\end{array}
$$

To eliminate y, multiply (1) by 2 and add the result to equation (2).

$$
\begin{array}{rll}
2ax + 2by = 2c & (3) & 2 \times (1) \\
\underline{ax - 2by = c} & (2) & \\
3ax \qquad\;\; = 3c & & \\
x = \dfrac{3c}{3a} = \dfrac{c}{a} & &
\end{array}
$$

Substitute $\dfrac{c}{a}$ for x in (1).

$$
\begin{array}{r}
a\left(\dfrac{c}{a}\right) + by = c \\
c + by = c \\
by = 0 \\
y = 0
\end{array}
$$

The solution $\left(\dfrac{c}{a}, 0\right)$ checks.

The solution set is $\left\{\left(\dfrac{c}{a}, 0\right)\right\}$.

89.
$$
\begin{array}{rl}
2ax - y = 3 & (1) \\
y = 5ax & (2)
\end{array}
$$

Substitute $5ax$ for y in (1).

$$
\begin{array}{r}
2ax - 5ax = 3 \\
-3ax = 3 \\
x = \dfrac{3}{-3a} = -\dfrac{1}{a}
\end{array}
$$

Substitute $-\dfrac{1}{a}$ for x in (2).

$$
y = 5a\left(-\dfrac{1}{a}\right) = -5
$$

The solution $\left(-\dfrac{1}{a}, -5\right)$ checks.

The solution set is $\left\{\left(-\dfrac{1}{a}, -5\right)\right\}$.

91.
$$
\begin{array}{rl}
3x + y = 6 & (1) \\
-2x + 3y = 7 & (2)
\end{array}
$$

Multiply equation (1) by -3 and add the result to equation (2).

$$
\begin{array}{rll}
-9x - 3y = -18 & (3) & -3 \times (1) \\
\underline{-2x + 3y = 7} & (2) & \\
-11x \qquad\;\; = -11 & & \\
x = 1 & &
\end{array}
$$

To find y, substitute 1 for x in equation (1).

$$
\begin{array}{rl}
3x + y = 6 & (1) \\
3(1) + y = 6 & \\
y = 3 &
\end{array}
$$

The solution $(1, 3)$ checks.

The solution set is $\{(1, 3)\}$.

92. $3x + y = 6$

$y = -3x + 6$

Replace y with $f(x)$.

$$f(x) = -3x + 6$$

Since f is in the form $f(x) = mx + b$, it is a linear function.

93. $-2x + 3y = 7$

$3y = 2x + 7$

$y = \frac{2}{3}x + \frac{7}{3}$

Replace y with $g(x)$.

$$g(x) = \frac{2}{3}x + \frac{7}{3}$$

Since g is in the form $g(x) = mx + b$, it is a linear function.

94. Because the graphs of f and g are straight lines that are neither parallel nor coincide, they intersect in exactly <u>one</u> point. The coordinates of the point are (<u>1</u> , <u>3</u>). Using functional notation, this is given by $f(\underline{1}) = \underline{3}$ and $g(\underline{1}) = \underline{3}$.

95. $4(2x - 3y + z) = 4(5)$

$8x - 12y + 4z = 20$

97. $x + 2y + 3z = 9$

$1 + 2(-2) + 3z = 9$ *Let $x = 1$, $y = -2$.*

$1 - 4 + 3z = 9$

$3z = 12$

$z = 4$

99. Multiplying the first equation by -3 will give us $-3x$, which when added to $3x$ in the second equation, gives us 0.

4.2 Systems of Linear Equations in Three Variables

1. Substitute 1 for x, 2 for y, and 3 for z in $3x + 2y - z$, which is the left side of each equation.

$$3(1) + 2(2) - (3) = 3 + 4 - 3$$
$$= 7 - 3$$
$$= 4$$

Choice **B** is correct since its right side is 4.

3. $2x - 5y + 3z = -1$ (1)

$x + 4y - 2z = 9$ (2)

$x - 2y - 4z = -5$ (3)

Eliminate x by adding equation (1) to -2 times equation (2).

$$
\begin{array}{rrrrr}
2x & - & 5y & + & 3z & = & -1 & (1) \\
-2x & - & 8y & + & 4z & = & -18 & \quad -2 \times (2) \\
\hline
 & - & 13y & + & 7z & = & -19 & (4)
\end{array}
$$

Now eliminate x by adding equation (2) to -1 times equation (3).

$$
\begin{array}{rrrrr}
x & + & 4y & - & 2z & = & 9 & (2) \\
-x & + & 2y & + & 4z & = & 5 & \quad -1 \times (3) \\
\hline
 & & 6y & + & 2z & = & 14 & (5)
\end{array}
$$

Use equations (4) and (5) to eliminate z. Multiply equation (4) by -2 and add the result to 7 times equation (5).

$$
\begin{array}{rrrrr}
26y & - & 14z & = & 38 & \quad -2 \times (4) \\
42y & + & 14z & = & 98 & \quad 7 \times (5) \\
\hline
68y & & & = & 136 & \\
 & & y & = & 2 &
\end{array}
$$

Substitute 2 for y in equation (5) to find z.

$$6y + 2z = 14 \quad (5)$$
$$6(2) + 2z = 14$$
$$12 + 2z = 14$$
$$2z = 2$$
$$z = 1$$

Substitute 2 for y and 1 for z in equation (3) to find x.

$$x - 2y - 4z = -5 \quad (3)$$
$$x - 2(2) - 4(1) = -5$$
$$x - 4 - 4 = -5$$
$$x - 8 = -5$$
$$x = 3$$

The solution $(3, 2, 1)$ checks in all three of the original equations.

The solution set is $\{(3, 2, 1)\}$.

5. $3x + 2y + z = 8$ (1)

$2x - 3y + 2z = -16$ (2)

$x + 4y - z = 20$ (3)

Eliminate z by adding equations (1) and (3).

$$
\begin{array}{rrrrr}
3x & + & 2y & + & z & = & 8 & (1) \\
x & + & 4y & - & z & = & 20 & (3) \\
\hline
4x & + & 6y & & & = & 28 & (4)
\end{array}
$$

To get another equation without z, multiply equation (3) by 2 and add the result to equation (2).

$$
\begin{array}{rrrrr}
2x & + & 8y & - & 2z & = & 40 & \quad 2 \times (3) \\
2x & - & 3y & + & 2z & = & -16 & (2) \\
\hline
4x & + & 5y & & & = & 24 & (5)
\end{array}
$$

Use equations (4) and (5) to eliminate x. Multiply equation (4) by -1 and add the result to equation (5).

$$
\begin{array}{rrrr}
-4x & - & 6y & = & -28 & \quad -1 \times (4) \\
4x & + & 5y & = & 24 & (5) \\
\hline
 & - & y & = & -4 & \\
 & & y & = & 4 &
\end{array}
$$

Substitute 4 for y in equation (5) to find x.

$$4x + 5y = 24 \quad (5)$$
$$4x + 5(4) = 24$$
$$4x + 20 = 24$$
$$4x = 4$$
$$x = 1$$

Substitute 1 for x and 4 for y in equation (3) to find z.

$$x + 4y - z = 20 \quad (3)$$
$$1 + 4(4) - z = 20$$
$$1 + 16 - z = 20$$
$$17 - z = 20$$
$$-z = 3$$
$$z = -3$$

The solution $(1, 4, -3)$ checks in all three of the original equations.
The solution set is $\{(1, 4, -3)\}$.

7.
$$2x + 5y + 2z = 0 \quad (1)$$
$$4x - 7y - 3z = 1 \quad (2)$$
$$3x - 8y - 2z = -6 \quad (3)$$

Add equations (1) and (3) to eliminate z.

$$\begin{array}{rcll}
2x + 5y + 2z &=& 0 & (1) \\
3x - 8y - 2z &=& -6 & (3) \\
\hline
5x - 3y &=& -6 & (4)
\end{array}$$

Multiply equation (1) by 3 and equation (2) by 2. Then add the results to eliminate z again.

$$\begin{array}{rcll}
6x + 15y + 6z &=& 0 & 3 \times (1) \\
8x - 14y - 6z &=& 2 & 2 \times (2) \\
\hline
14x + y &=& 2 & (5)
\end{array}$$

Solve the system

$$\begin{array}{rcll}
5x - 3y &=& -6 & (4) \\
14x + y &=& 2. & (5)
\end{array}$$

Multiply equation (5) by 3 then add this result to (4).

$$\begin{array}{rcll}
5x - 3y &=& -6 & (4) \\
42x + 3y &=& 6 & 3 \times (5) \\
\hline
47x &=& 0 \\
x &=& 0
\end{array}$$

To find y, substitute $x = 0$ into equation (4).

$$5x - 3y = -6 \quad (4)$$
$$5(0) - 3y = -6$$
$$y = 2$$

To find z, substitute $x = 0$ and $y = 2$ in equation (1).

The solution set is $\{(0, 2, -5)\}$.

$$2x + 5y + 2z = 0 \quad (1)$$
$$2(0) + 5(2) + 2z = 0$$
$$10 + 2z = 0$$
$$2z = -10$$
$$z = -5$$

The solution set is $\{(0, 2, -5)\}$.

9.
$$\begin{array}{rcll}
x + 2y + z &=& 4 & (1) \\
2x + y - z &=& -1 & (2) \\
x - y - z &=& -2 & (3)
\end{array}$$

Add (1) and (2) to eliminate z.

$$\begin{array}{rcll}
x + 2y + z &=& 4 & (1) \\
2x + y - z &=& -1 & (2) \\
\hline
3x + 3y &=& 3 & (4)
\end{array}$$

Add (1) and (3) to eliminate z.

$$\begin{array}{rcll}
x + 2y + z &=& 4 & (1) \\
x - y - z &=& -2 & (3) \\
\hline
2x + y &=& 2 & (5)
\end{array}$$

Multiply equation (5) by -3.

$$\begin{array}{rcll}
-6x - 3y &=& -6 & (6) \; -3 \times (5) \\
3x + 3y &=& 3 & (4) \\
\hline
-3x &=& -3 \\
x &=& 1
\end{array}$$

Substitute 1 for x in equation (5) and solve for y.

$$2(1) + y = 2$$
$$y = 0$$

Substitute 1 for x and 0 for y in (1).

$$(1) + 2(0) + z = 4$$
$$z = 3$$

The solution set is $\{(1, 0, 3)\}$.

11.
$$\begin{array}{rcll}
\frac{1}{3}x + \frac{1}{6}y - \frac{2}{3}z &=& -1 & (1) \\
-\frac{3}{4}x - \frac{1}{3}y - \frac{1}{4}z &=& 3 & (2) \\
\frac{1}{2}x + \frac{3}{2}y + \frac{3}{4}z &=& 21 & (3)
\end{array}$$

Eliminate all fractions.

$$\begin{array}{rcll}
2x + y - 4z &=& -6 & (4) \quad 6 \times (1) \\
-9x - 4y - 3z &=& 36 & (5) \quad 12 \times (2) \\
2x + 6y + 3z &=& 84 & (6) \quad 4 \times (3)
\end{array}$$

Multiply (4) by 4.

$$\begin{array}{rcll}
8x + 4y - 16z &=& -24 \\
-9x - 4y - 3z &=& 36 & (5) \\
\hline
-x - 19z &=& 12 & (7)
\end{array}$$

continued

Multiply (4) by -6.

$$
\begin{array}{rcrcrcr}
-12x & - & 6y & + & 24z & = & 36 \\
2x & + & 6y & + & 3z & = & 84 \quad (6) \\
\hline
-10x & & & + & 27z & = & 120 \quad (8)
\end{array}
$$

Multiply (7) by -10.

$$
\begin{array}{rcrcr}
10x & + & 190z & = & -120 \\
-10x & + & 27z & = & 120 \quad (8) \\
\hline
& & 217z & = & 0 \\
& & z & = & 0
\end{array}
$$

Substituting 0 for z in (7) gives us $x = -12$. Substitute -12 for x and 0 for z in (4).

$$
\begin{aligned}
2(-12) + y - 4(0) &= -6 \\
-24 + y &= -6 \\
y &= 18
\end{aligned}
$$

The solution set is $\{(-12, 18, 0)\}$.

13.
$$
\begin{array}{rcrcrcr}
-x & + & 2y & + & 6z & = & 2 \quad (1) \\
3x & + & 2y & + & 6z & = & 6 \quad (2) \\
x & + & 4y & - & 3z & = & 1 \quad (3)
\end{array}
$$

Eliminate y and z by adding equation (1) to -1 times equation (2).

$$
\begin{array}{rcrcrcrl}
-x & + & 2y & + & 6z & = & 2 & (1) \\
-3x & - & 2y & - & 6z & = & -6 & \quad -1 \times (2) \\
\hline
-4x & & & & & = & -4 & \\
& & & & x & = & 1 &
\end{array}
$$

Eliminate z by adding equation (2) to 2 times equation (3).

$$
\begin{array}{rcrcrcrl}
3x & + & 2y & + & 6z & = & 6 & (2) \\
2x & + & 8y & - & 6z & = & 2 & \quad 2 \times (3) \\
\hline
5x & + & 10y & & & = & 8 & (4)
\end{array}
$$

Substitute 1 for x in equation (4).

$$
\begin{aligned}
5x + 10y &= 8 \quad (4) \\
5(1) + 10y &= 8 \\
10y &= 3 \\
y &= \tfrac{3}{10}
\end{aligned}
$$

Substitute 1 for x and $\tfrac{3}{10}$ for y in equation (1).

$$
\begin{aligned}
-x + 2y + 6z &= 2 \quad (1) \\
-1 + 2\left(\tfrac{3}{10}\right) + 6z &= 2 \\
\tfrac{3}{5} + 6z &= 3 \\
6z &= \tfrac{12}{5} \\
z &= \tfrac{2}{5}
\end{aligned}
$$

The solution set is $\left\{\left(1, \tfrac{3}{10}, \tfrac{2}{5}\right)\right\}$.

15.
$$
\begin{array}{rcrcrcrl}
x & + & y & - & z & = & -2 & (1) \\
2x & - & y & + & z & = & -5 & (2) \\
-x & + & 2y & - & 3z & = & -4 & (3)
\end{array}
$$

Eliminate y and z by adding equations (1) and (2).

$$
\begin{array}{rcrcrcrl}
x & + & y & - & z & = & -2 & (1) \\
2x & - & y & + & z & = & -5 & (2) \\
\hline
3x & & & & & = & -7 & \\
& & & & x & = & -\tfrac{7}{3} &
\end{array}
$$

To get another equation without y, multiply equation (2) by 2 and add the result to equation (3).

$$
\begin{array}{rcrcrcrl}
4x & - & 2y & + & 2z & = & -10 & \quad 2 \times (2) \\
-x & + & 2y & - & 3z & = & -4 & (3) \\
\hline
3x & & & - & z & = & -14 & (4)
\end{array}
$$

Substitute $-\tfrac{7}{3}$ for x in equation (4) to find z.

$$
\begin{aligned}
3x - z &= -14 \quad (4) \\
3\left(-\tfrac{7}{3}\right) - z &= -14 \\
-7 - z &= -14 \\
-z &= -7 \\
z &= 7
\end{aligned}
$$

Substitute $-\tfrac{7}{3}$ for x and 7 for z in equation (1) to find y.

$$
\begin{aligned}
x + y - z &= -2 \quad (1) \\
-\tfrac{7}{3} + y - 7 &= -2 \\
-7 + 3y - 21 &= -6 \quad \textit{Multiply by 3.} \\
3y - 28 &= -6 \\
3y &= 22 \\
y &= \tfrac{22}{3}
\end{aligned}
$$

The solution set is $\left\{\left(-\tfrac{7}{3}, \tfrac{22}{3}, 7\right)\right\}$.

A calculator check reduces the probability of making any arithmetic errors and is highly recommended. The following screen shows the substitution of the solution for x, y, and z along with the left sides of the three original equations. The evaluation of the three expressions, -2, -5, and -4 (the right sides of the three equations), indicates that we have found the correct solution.

```
-7/3→X:22/3→Y:7→
Z:X+Y-Z
                    -2
2X-Y+Z
                    -5
-X+2Y-3Z
                    -4
```

17.
$$2x - 3y + 2z = -1 \quad (1)$$
$$x + 2y + z = 17 \quad (2)$$
$$2y - z = 7 \quad (3)$$

Multiply equation (2) by -2, and add the result to equation (1).

$$2x - 3y + 2z = -1 \quad (1)$$
$$\underline{-2x - 4y - 2z = -34} \quad -2 \times (2)$$
$$-7y = -35$$
$$y = 5$$

To find z, substitute 5 for y in equation (3).

$$2y - z = 7 \quad (3)$$
$$2(5) - z = 7$$
$$10 - z = 7$$
$$-z = -3$$
$$z = 3$$

To find x, substitute $y = 5$ and $z = 3$ into equation (1).

$$2x - 3y + 2z = -1 \quad (1)$$
$$2x - 3(5) + 2(3) = -1$$
$$2x - 9 = -1$$
$$2x = 8$$
$$x = 4$$

The solution set is $\{(4, 5, 3)\}$.

19.
$$4x + 2y - 3z = 6 \quad (1)$$
$$x - 4y + z = -4 \quad (2)$$
$$-x + 2z = 2 \quad (3)$$

Equation (3) is missing y. Eliminate y again by multiplying equation (1) by 2 and adding the result to equation (2).

$$8x + 4y - 6z = 12 \quad 2 \times (1)$$
$$\underline{x - 4y + z = -4} \quad (2)$$
$$9x - 5z = 8 \quad (4)$$

Use equations (3) and (4) to eliminate x. Multiply equation (3) by 9 and add the result to equation (4).

$$-9x + 18z = 18 \quad 9 \times (3)$$
$$\underline{9x - 5z = 8} \quad (4)$$
$$13z = 26$$
$$z = 2$$

Substitute 2 for z in equation (3) to find x.

$$-x + 2z = 2 \quad (3)$$
$$-x + 2(2) = 2$$
$$-x + 4 = 2$$
$$-x = -2$$
$$x = 2$$

Substitute 2 for x and 2 for z in equation (2) to find y.

$$x - 4y + z = -4 \quad (2)$$
$$2 - 4y + 2 = -4$$
$$-4y + 4 = -4$$
$$-4y = -8$$
$$y = 2$$

The solution set is $\{(2, 2, 2)\}$.

21.
$$2x + y = 6 \quad (1)$$
$$3y - 2z = -4 \quad (2)$$
$$3x - 5z = -7 \quad (3)$$

To eliminate y, multiply equation (1) by -3 and add the result to equation (2).

$$-6x - 3y = -18 \quad -3 \times (1)$$
$$\underline{3y - 2z = -4} \quad (2)$$
$$-6x - 2z = -22 \quad (4)$$

Since equation (3) does not have a y-term, we can multiply equation (3) by 2 and add the result to equation (4) to eliminate x and solve for z.

$$6x - 10z = -14 \quad 2 \times (3)$$
$$\underline{-6x - 2z = -22} \quad (4)$$
$$-12z = -36$$
$$z = 3$$

To find x, substitute 3 for z into equation (3).

$$3x - 5z = -7 \quad (3)$$
$$3x - 5(3) = -7$$
$$3x = 8$$
$$x = \frac{8}{3}$$

To find y, substitute 3 for z into equation (2).

$$3y - 2z = -4 \quad (2)$$
$$3y - 2(3) = -4$$
$$3y = 2$$
$$y = \frac{2}{3}$$

The solution set is $\left\{ \left(\frac{8}{3}, \frac{2}{3}, 3 \right) \right\}$.

23.
$$-5x + 2y + z = 5 \quad (1)$$
$$-3x - 2y - z = 3 \quad (2)$$
$$-x + 6y = 1 \quad (3)$$

Add (1) and (2) to eliminate y and z.

$$-8x = 8$$
$$x = -1$$

Substitute -1 for x in equation (3).

$$-x + 6y = 1 \quad (3)$$
$$-(-1) + 6y = 1$$
$$6y = 0$$
$$y = 0$$

continued

Substitute -1 for x and 0 for y in (1).

$$-5x + 2y + z = 5 \quad (1)$$
$$-5(-1) + 2(0) + z = 5$$
$$5 + z = 5$$
$$z = 0$$

The solution set is $\{(-1, 0, 0)\}$.

25.
$$4x \quad - \quad z = -6 \quad (1)$$
$$\tfrac{3}{5}y + \tfrac{1}{2}z = \quad 0 \quad (2)$$
$$\tfrac{1}{3}x \quad + \tfrac{2}{3}z = -5 \quad (3)$$

Eliminate fractions first.

$$6y + 5z = \quad 0 \quad (4) \quad 10 \times (2)$$
$$x \quad + 2z = -15 \quad (5) \quad 3 \times (3)$$

Eliminate z by adding (5) to 2 times (1).

$$8x - 2z = -12 \quad 2 \times (1)$$
$$\underline{x + 2z = -15 \quad (5)}$$
$$9x \qquad = -27$$
$$x = -3$$

Substitute -3 for x in (5).

$$x + 2z = -15 \quad (5)$$
$$-3 + 2z = -15$$
$$2z = -12$$
$$z = -6$$

Substitute -6 for z in (4).

$$6y + 5z = 0 \quad (4)$$
$$6y + 5(-6) = 0$$
$$6y = 30$$
$$y = 5$$

The solution set is $\{(-3, 5, -6)\}$.

27.
$$2x + 2y - 6z = \quad 5 \quad (1)$$
$$-3x + y - z = -2 \quad (2)$$
$$-x - y + 3z = \quad 4 \quad (3)$$

Multiply equation (3) by 2 and add the result to equation (1).

$$2x + 2y - 6z = 5 \quad (1)$$
$$\underline{-2x - 2y + 6z = 8 \qquad 2 \times (3)}$$
$$0 = 13 \quad \textit{False}$$

The solution set is $\emptyset$; inconsistent system.

29.
$$-5x + 5y - 20z = -40 \quad (1)$$
$$x - y + 4z = \quad 8 \quad (2)$$
$$3x - 3y + 12z = \quad 24 \quad (3)$$

Dividing equation (1) by -5 gives equation (2). Dividing equation (3) by 3 also gives equation (2). The resulting equations are the same, so the three equations are dependent.
The solution set is $\{(x, y, z) \mid x - y + 4z = 8\}$.

31.
$$x + 5y - 2z = -1 \quad (1)$$
$$-2x + 8y + z = -4 \quad (2)$$
$$3x - y + 5z = 19 \quad (3)$$

Eliminate x by adding (2) to 2 times (1).

$$2x + 10y - 4z = -2 \qquad 2 \times (1)$$
$$\underline{-2x + 8y + z = -4 \quad (2)}$$
$$18y - 3z = -6 \quad (4)$$

Eliminate x by adding (3) to -3 times (1).

$$-3x - 15y + 6z = \quad 3 \qquad -3 \times (1)$$
$$\underline{3x - y + 5z = 19 \quad (3)}$$
$$-16y + 11z = 22 \quad (5)$$

To eliminate z from equations (4) and (5), we could first divide equation (4) by 3 and then multiply the resulting equation by 11, or we could simply multiply equation (4) by $\frac{11}{3}$ and add it to equation (5).

$$66y - 11z = -22 \qquad \tfrac{11}{3} \times (4)$$
$$\underline{-16y + 11z = \quad 22 \quad (5)}$$
$$50y \qquad = \quad 0$$
$$y = 0$$

Substitute 0 for y in equation (5).

$$-16y + 11z = 22 \quad (5)$$
$$-16(0) + 11z = 22$$
$$11z = 22$$
$$z = 2$$

Substitute 0 for y and 2 for z in equation (1).

$$x + 5y - 2z = -1 \quad (1)$$
$$x + 5(0) - 2(2) = -1$$
$$x - 4 = -1$$
$$x = 3$$

The solution set is $\{(3, 0, 2)\}$.

33.
$$2x + y - z = \quad 6 \quad (1)$$
$$4x + 2y - 2z = 12 \quad (2)$$
$$-x - \tfrac{1}{2}y + \tfrac{1}{2}z = -3 \quad (3)$$

Multiplying equation (1) by 2 gives equation (2). Multiplying equation (3) by -4 also gives equation (2). The resulting equations are the same, so the three equations are dependent.
The solution set is $\{(x, y, z) \mid 2x + y - z = 6\}$.

35.
$$x + y - 2z = 0 \quad (1)$$
$$3x - y + z = 0 \quad (2)$$
$$4x + 2y - z = 0 \quad (3)$$

Eliminate z by adding equations (2) and (3).

$$3x - y + z = 0 \quad (2)$$
$$\underline{4x + 2y - z = 0 \quad (3)}$$
$$7x + y \qquad = 0 \quad (4)$$

To get another equation without z, multiply equation (2) by 2 and add the result to equation (1).

$$
\begin{array}{rl}
6x - 2y + 2z = 0 & 2 \times (2) \\
\underline{x + y - 2z = 0} & (1) \\
7x - y = 0 & (5)
\end{array}
$$

Add equations (4) and (5) to find x.

$$
\begin{array}{rl}
7x + y = 0 & (4) \\
\underline{7x - y = 0} & (5) \\
14x = 0 & \\
x = 0 &
\end{array}
$$

Substitute 0 for x in equation (4) to find y.

$$
\begin{aligned}
7x + y &= 0 \quad (4) \\
7(0) + y &= 0 \\
0 + y &= 0 \\
y &= 0
\end{aligned}
$$

Substitute 0 for x and 0 for y in equation (1) to find z.

$$
\begin{aligned}
x + y - 2z &= 0 \quad (1) \\
0 + 0 - 2z &= 0 \\
-2z &= 0 \\
z &= 0
\end{aligned}
$$

The solution set is $\{(0, 0, 0)\}$.

37.
$$
\begin{array}{rl}
x - 2y + \frac{1}{3}z = 4 & (1) \\
3x - 6y + z = 12 & (2) \\
-6x + 12y - 2z = -3 & (3)
\end{array}
$$

The coefficients of z are $\frac{1}{3}$, 1, and -2. We'll multiply the equations by values that will eliminate fractions and make the coefficients of z easy to compare.

$$
\begin{array}{rl}
-6x + 12y - 2z = -24 & (4) \ -6 \times (1) \\
-6x + 12y - 2z = -24 & (5) \ -2 \times (2) \\
-6x + 12y - 2z = -3 & (3)
\end{array}
$$

We can now easily see that (4) and (5) are dependent equations (they have the same graph—in fact, their graph is the same plane). Equation (3) has the same coefficients, but a different constant term, so its graph is a plane parallel to the other plane—that is, there are no points in common. Thus, the system is inconsistent and the solution set is $\emptyset$.

39.
$$
\begin{array}{rl}
x + y + z - w = 5 & (1) \\
2x + y - z + w = 3 & (2) \\
x - 2y + 3z + w = 18 & (3) \\
-x - y + z + 2w = 8 & (4)
\end{array}
$$

Eliminate w. Add equations (1) and (2).

$$
\begin{array}{rl}
x + y + z - w = 5 & (1) \\
\underline{2x + y - z + w = 3} & (2) \\
3x + 2y = 8 & (5)
\end{array}
$$

Eliminate w again. Add equations (1) and (3).

$$
\begin{array}{rl}
x + y + z - w = 5 & (1) \\
\underline{x - 2y + 3z + w = 18} & (3) \\
2x - y + 4z = 23 & (6)
\end{array}
$$

Eliminate w again. Multiply equation (2) by -2. Add the result to equation (4).

$$
\begin{array}{rl}
-4x - 2y + 2z - 2w = -6 & -2 \times (2) \\
\underline{-x - y + z + 2w = 8} & (4) \\
-5x - 3y + 3z = 2 & (7)
\end{array}
$$

Equations (5), (6), and (7) do not contain a w-term. Since (5) does not have a z-term, we will find another equation without a z-term.

Eliminate z. Multiply equation (6) by 3 and equation (7) by -4. Then add the results.

$$
\begin{array}{rl}
6x - 3y + 12z = 69 & 3 \times (6) \\
\underline{20x + 12y - 12z = -8} & -4 \times (7) \\
26x + 9y = 61 & (8)
\end{array}
$$

Eliminate z again. Multiply equation (5) by 9 and equation (8) by -2. Then add the results.

$$
\begin{array}{rl}
27x + 18y = 72 & 9 \times (5) \\
\underline{-52x - 18y = -122} & -2 \times (8) \\
-25x = -50 & \\
x = 2 &
\end{array}
$$

To find y, substitute $x = 2$ into equation (5).

$$
\begin{aligned}
3x + 2y &= 8 \quad (5) \\
3(2) + 2y &= 8 \\
2y &= 2 \\
y &= 1
\end{aligned}
$$

To find z, substitute $x = 2$ and $y = 1$ into equation (6).

$$
\begin{aligned}
2x - y + 4z &= 23 \quad (6) \\
2(2) - 1 + 4z &= 23 \\
4z &= 20 \\
z &= 5
\end{aligned}
$$

To find w, substitute $x = 2$, $y = 1$, and $z = 5$ into equation (1).

$$
\begin{aligned}
x + y + z - w &= 5 \quad (1) \\
2 + 1 + 5 - w &= 5 \\
-w &= -3 \\
w &= 3
\end{aligned}
$$

The solution set is $\{(2, 1, 5, 3)\}$.

41.
$$\begin{array}{rrrrrr} 3x & + & y & - & z & + & w & = & -3 & (1) \\ 2x & + & 4y & + & z & - & w & = & -7 & (2) \\ -2x & + & 3y & - & 5z & + & w & = & 3 & (3) \\ 5x & + & 4y & - & 5z & + & 2w & = & -7 & (4) \end{array}$$

Eliminate w. Add equations (1) and (2).

$$\begin{array}{rrrrrr} 3x & + & y & - & z & + & w & = & -3 & (1) \\ 2x & + & 4y & + & z & - & w & = & -7 & (2) \\ \hline 5x & + & 5y & & & & & = & -10 & (5) \end{array}$$

Eliminate w again. Add equations (2) and (3).

$$\begin{array}{rrrrrr} 2x & + & 4y & + & z & - & w & = & -7 & (2) \\ -2x & + & 3y & - & 5z & + & w & = & 3 & (3) \\ \hline & & 7y & - & 4z & & & = & -4 & (6) \end{array}$$

Eliminate w again. Multiply equation (2) by 2. Add the result to equation (4).

$$\begin{array}{rrrrrrl} 4x & + & 8y & + & 2z & - & 2w & = & -14 & \quad 2 \times (2) \\ 5x & + & 4y & - & 5z & + & 2w & = & -7 & (4) \\ \hline 9x & + & 12y & - & 3z & & & = & -21 & (7) \end{array}$$

Equations (5), (6), and (7) do not contain a w-term. Since (5) does not have a z-term, we will find another equation without a z-term.

Eliminate z. Multiply equation (6) by 3 and equation (7) by -4. Then add the results.

$$\begin{array}{rrrrl} 21y & - & 12z & = & -12 & \quad 3 \times (6) \\ -36x & - & 48y & + & 12z & = & 84 & \quad -4 \times (7) \\ \hline -36x & - & 27y & & & = & 72 & (8) \end{array}$$

Eliminate y using (5) and (8).

$$\begin{array}{rrrl} 3x & + & 3y & = & -6 & \quad \frac{3}{5} \times (5) \\ -4x & - & 3y & = & 8 & \quad (8) \div 9 \\ \hline -x & & & = & 2 \\ & & x & = & -2 \end{array}$$

Substitute -2 for x in (5).

$$5(-2) + 5y = -10$$
$$5y = 0$$
$$y = 0$$

Substitute 0 for y in (6).

$$7(0) - 4z = -4$$
$$-4z = -4$$
$$z = 1$$

Substitute -2 for x, 0 for y, and 1 for z in (1).

$$3(-2) + (0) - (1) + w = -3$$
$$-6 - 1 + w = -3$$
$$w = 4$$

The solution set is $\{(-2, 0, 1, 4)\}$.

43. Let $x =$ the length of the longest side.
Then $\frac{5}{6}x =$ the length of the shortest side,
and $x - 17 =$ the length of the medium side.

The perimeter is 323, so

$$x + \tfrac{5}{6}x + (x - 17) = 323.$$
$$6x + 5x + 6(x - 17) = 6(323) \quad \textit{Multiply by 6.}$$
$$6x + 5x + 6x - 102 = 1938$$
$$17x = 2040$$
$$x = 120$$

Since $x = 120$, $\frac{5}{6}x = 100$, and $x - 17 = 103$. The lengths of the sides are 100 inches, 103 inches, and 120 inches.

45. Let $x =$ the least number.
Then $-3x =$ the greatest number,
and $-3x - 4 =$ the middle number.

The sum is 16, so

$$x + (-3x) + (-3x - 4) = 16.$$
$$-5x - 4 = 16$$
$$-5x = 20$$
$$x = -4$$

Since $x = -4$, $-3x = 12$, and $-3x - 4 = 8$. The three numbers are -4, 8, and 12.

4.3 Applications of Systems of Linear Equations

1. *Step 2*
Let $x =$ the number of games that the White Sox won and let $y =$ the number of games that they lost.

Step 3
They played 162 games, so

$$x + y = 162. \quad (1)$$

They won 36 more games than they lost, so

$$x = 36 + y. \quad (2)$$

Step 4
Substitute $36 + y$ for x in (1).

$$(36 + y) + y = 162$$
$$36 + 2y = 162$$
$$2y = 126$$
$$y = 63$$

Substitute 63 for y in (2).

$$x = 36 + 63 = 99$$

Step 5

The White Sox win-loss record was 99 wins and 63 losses.

Step 6

99 is 36 more than 63 and the sum of 99 and 63 is 162.

3. Let $W =$ the width of the tennis court and $L =$ the length of the court.

Since the length is 42 ft more than the width,

$$L = W + 42. \quad (1)$$

The perimeter of a rectangle is given by

$$2W + 2L = P.$$

With perimeter $P = 228$ ft,

$$2W + 2L = 228. \quad (2)$$

Substitute $W + 42$ for L in equation (2).

$$2W + 2(W + 42) = 228$$
$$2W + 2W + 84 = 228$$
$$4W = 144$$
$$W = 36$$

Substitute $W = 36$ into equation (1).

$$L = W + 42 \quad (1)$$
$$L = 36 + 42$$
$$= 78$$

The length is 78 ft and the width is 36 ft.

5. Let $x =$ the revenue for Wal-Mart and $y =$ the revenue for ExxonMobil (both in billions of dollars).

The total revenue was $656 billion.

$$x + y = 656 \quad (1)$$

ExxonMobil's revenue was $24 billion more than that of Wal-Mart.

$$y = x + 24 \quad (2)$$

Substitute $x + 24$ for y in equation (1).

$$x + (x + 24) = 656$$
$$2x + 24 = 656$$
$$2x = 632$$
$$x = 316$$

Substitute 316 for x in equation (2).

$$y = x + 24 = 316 + 24 = 340$$

Wal-Mart's revenue was $316 billion and ExxonMobil's revenue was $340 billion.

7. From the figure in the text, the angles marked y and $3x + 10$ are supplementary, so

$$(3x + 10) + y = 180. \quad (1)$$

Also, the angles x and y are complementary, so

$$x + y = 90. \quad (2)$$

Solve equation (2) for y to get

$$y = 90 - x. \quad (3)$$

Substitute $90 - x$ for y in equation (1).

$$(3x + 10) + (90 - x) = 180$$
$$2x + 100 = 180$$
$$2x = 80$$
$$x = 40$$

Substitute $x = 40$ into equation (3) to get

$$y = 90 - x = 90 - 40 = 50.$$

The angles measure 40° and 50°.

9. Let $x =$ the hockey FCI and $y =$ the basketball FCI.

The sum is $514.69, so

$$x + y = 514.69. \quad (1)$$

The hockey FCI was $20.05 less than the basketball FCI, so

$$x = y - 20.05. \quad (2)$$

From (2), substitute $y - 20.05$ for x in (1).

$$(y - 20.05) + y = 514.69$$
$$2y - 20.05 = 514.69$$
$$2y = 534.74$$
$$y = 267.37$$

From (2),

$$x = y - 20.05 = 267.37 - 20.05 = 247.32.$$

The hockey FCI was $247.32 and the basketball FCI was $267.37.

11. Let $x =$ the cost of a single Regular Roast Beef sandwich, and $y =$ the cost of a single Large Roast Beef sandwich.

15 Regular Roast Beef sandwiches and 10 Large Roast Beef sandwiches cost $77.75.

$$15x + 10y = 77.75 \quad (1)$$

30 Regular Roast Beef sandwiches and 5 Large Roast Beef sandwiches cost $92.65.

$$30x + 5y = 92.65 \quad (2)$$

Multiply equation (1) by -2 and add to equation (2).

continued

$$
\begin{array}{rrl}
-30x - 20y &= -155.50 & \quad -2 \times (1)\\
\underline{30x + 5y} &= \underline{92.65} & \quad (2)\\
-15y &= -62.85\\
y &= 4.19
\end{array}
$$

Substitute 4.19 for y in equation (1).

$$
\begin{aligned}
15x + 10y &= 77.75\\
15x + 10(4.19) &= 77.75\\
15x + 41.90 &= 77.75\\
15x &= 35.85\\
x &= 2.39
\end{aligned}
$$

A single Regular Roast Beef sandwich costs $2.39 and a single Large Roast Beef sandwich costs $4.19.

13. Use the formula (rate of percent) • (base amount) = amount (percentage) of pure acid to compute parts (a) – (d).

(a) $0.10(60) = 6$ oz

(b) $0.25(60) = 15$ oz

(c) $0.40(60) = 24$ oz

(d) $0.50(60) = 30$ oz

15. The cost is the price per pound, $1.29, times the number of pounds, x, or $1.29x$.

17. Let x = the amount of 25% alcohol solution, and y = the amount of 35% alcohol solution.

Make a table. The percent times the amount of solution gives the amount of pure alcohol in the third column.

Gallons of Solution	Percent (as a decimal)	Gallons of Pure Alcohol
x	25% = 0.25	$0.25x$
y	35% = 0.35	$0.35y$
20	32% = 0.32	$0.32(20) = 6.4$

The third row gives the total amounts of solution and pure alcohol. From the columns in the table, write a system of equations.

$$
\begin{aligned}
x + y &= 20 \quad (1)\\
0.25x + 0.35y &= 6.4 \quad (2)
\end{aligned}
$$

Solve the system. Multiply equation (1) by -25 and equation (2) by 100. Then add the results.

$$
\begin{array}{rrl}
-25x - 25y &= -500 & \quad -25 \times (1)\\
\underline{25x + 35y} &= \underline{640} & \quad 100 \times (2)\\
10y &= 140\\
y &= 14
\end{array}
$$

Substitute $y = 14$ into equation (1).

$$
\begin{aligned}
x + y &= 20 \quad (1)\\
x + 14 &= 20\\
x &= 6
\end{aligned}
$$

Mix 6 gal of 25% solution and 14 gal of 35% solution.

19. Let x = the amount of pure acid and y = the amount of 10% acid.

Make a table.

Liters of Solution	Percent (as a decimal)	Liters of Pure Acid
x	100% = 1	$1.00x = x$
y	10% = 0.10	$0.10y$
54	20% = 0.20	$0.20(54) = 10.8$

Solve the following system.

$$
\begin{aligned}
x + y &= 54 \quad (1)\\
x + 0.10y &= 10.8 \quad (2)
\end{aligned}
$$

Multiply equation (2) by 10 to clear the decimals.

$$
10x + y = 108 \quad (3)
$$

To eliminate y, multiply equation (1) by -1 and add the result to equation (3).

$$
\begin{array}{rrl}
-x - y &= -54 & \quad -1 \times (1)\\
\underline{10x + y} &= \underline{108} & \quad (3)\\
9x &= 54\\
x &= 6
\end{array}
$$

Since $x = 6$,

$$
\begin{aligned}
x + y &= 54 \quad (1)\\
6 + y &= 54\\
y &= 48.
\end{aligned}
$$

Use 6 L of pure acid and 48 L of 10% acid.

21. Complete the table.

	Number of Kilograms	Price per Kilogram	Value
Nuts	x	2.50	$2.50x$
Cereal	y	1.00	$1.00y$
Mixture	30	1.70	$1.70(30) = 51$

From the "Number of Kilograms" column,

$$
x + y = 30. \quad (1)
$$

From the "Value" column,

$$
2.50x + 1.00y = 51. \quad (2)
$$

Solve the system.

$$
\begin{array}{rrl}
-10x - 10y &= -300 & \quad -10 \times (1)\\
\underline{25x + 10y} &= \underline{510} & \quad 10 \times (2)\\
15x &= 210\\
x &= 14
\end{array}
$$

From (1), $14 + y = 30$, so $y = 16$.
The party mix should be made from 14 kg of nuts and 16 kg of cereal.

23. From the "Principal" column in the text,

$$x + y = 3000. \quad (1)$$

From the "Interest" column in the text,

$$0.02x + 0.04y = 100. \quad (2)$$

Multiply equation (2) by 100 to clear the decimals.

$$2x + 4y = 10{,}000 \quad (3)$$

To eliminate x, multiply equation (1) by -2 and add the result to equation (3).

$$
\begin{array}{rrl}
-2x & - 2y & = -6000 \qquad -2 \times (1) \\
2x & + 4y & = 10{,}000 \quad (3) \\
\hline
& 2y & = 4000 \\
& y & = 2000
\end{array}
$$

From (1), $x + 2000 = 3000$, so $x = 1000$. $1000 is invested at 2%, and $2000 is invested at 4%.

25. (a) The speed of the boat going upstream is *decreased* by the speed of the current, so it is $(10 - x)$ mph.

(b) The speed of the boat going downstream is *increased* by the speed of the current, so it is $(10 + x)$ mph.

27. Let x = the speed of the train and y = the speed of the plane.

	r	t	d
Train	x	$\dfrac{150}{x}$	150
Plane	y	$\dfrac{400}{y}$	400

The times are equal, so

$$\frac{150}{x} = \frac{400}{y}. \quad (1)$$

The speed of the plane is 20 km per hour less than 3 times the speed of the train, so

$$3x - 20 = y. \quad (2)$$

Multiply (1) by xy (or use cross products).

$$150y = 400x \quad (3)$$

From (2), substitute $3x - 20$ for y in (3).

$$
\begin{aligned}
150(3x - 20) &= 400x \\
450x - 3000 &= 400x \\
50x &= 3000 \\
x &= 60
\end{aligned}
$$

From (2), $y = 3(60) - 20 = 160$. The speed of the train is 60 km/hr, and the speed of the plane is 160 km/hr.

29. Let x = the speed of the boat in still water and y = the speed of the current.

Furthermore,

$$\text{rate upstream} = x - y$$

and $\quad$ rate downstream $= x + y$.

Use these rates and the information in the problem to make a table.

	r	t	d
Upstream	$x - y$	2	36
Downstream	$x + y$	1.5	36

From the table, use the formula $d = rt$ to write a system of equations.

$$
\begin{aligned}
36 &= 2(x - y) \\
36 &= 1.5(x + y)
\end{aligned}
$$

Remove the parentheses and move the variables to the left side.

$$
\begin{aligned}
2x - 2y &= 36 \quad (1) \\
1.5x + 1.5y &= 36 \quad (2)
\end{aligned}
$$

Solve the system. Multiply equation (1) by -3 and equation (2) by 4. Then add the results.

$$
\begin{array}{rrll}
-6x & + & 6y & = -108 \quad -3 \times (1) \\
6x & + & 6y & = 144 \quad 4 \times (2) \\
\hline
& & 12y & = 36 \\
& & y & = 3
\end{array}
$$

Substitute $y = 3$ into equation (1).

$$
\begin{aligned}
2x - 2y &= 36 \quad (1) \\
2x - 2(3) &= 36 \\
2x &= 42 \\
x &= 21
\end{aligned}
$$

The speed of the boat is 21 mph, and the speed of the current is 3 mph.

31. Let x = the number of pounds of the $0.75-per-lb candy and y = the number of pounds of the $1.25-per-lb candy.

Make a table.

	Price per Pound	Number of Pounds	Value
Less Expensive Candy	$0.75	x	$0.75x
More Expensive Candy	$1.25	y	$1.25y
Mixture	$0.96	9	$0.96(9) = $8.64

continued

From the "Number of Pounds" column,

$$x + y = 9. \qquad (1)$$

From the "Value" column,

$$0.75x + 1.25y = 8.64. \qquad (2)$$

Solve the system.

$$
\begin{array}{rll}
-75x - 75y & = -675 & -75 \times (1) \\
75x + 125y & = 864 & 100 \times (2) \\
\hline
50y & = 189 & \\
y & = 3.78 &
\end{array}
$$

From (1), $x + 3.78 = 9$, so $x = 5.22$.
Mix 5.22 pounds of the \$0.75-per-lb candy with 3.78 pounds of the \$1.25-per-lb candy to obtain 9 pounds of a mixture that sells for \$0.96 per pound.

33. Let $x =$ the number of general admission tickets and $y =$ the number of student tickets.

Make a table.

Ticket	Number	Value of Tickets
General	x	$5 \cdot x = 5x$
Student	y	$4 \cdot y = 4y$
Totals	184	812

Solve the system.

$$
\begin{array}{rl}
x + y = 184 & (1) \\
5x + 4y = 812 & (2)
\end{array}
$$

To eliminate y, multiply equation (1) by -4 and add the result to equation (2).

$$
\begin{array}{rll}
-4x - 4y & = -736 & -4 \times (1) \\
5x + 4y & = 812 & (2) \\
\hline
x & = 76 &
\end{array}
$$

From (1), $76 + y = 184$, so $y = 108$.

76 general admission tickets and 108 student tickets were sold.

35. Let $x =$ the price for a citron and let $y =$ the price for a wood apple.

"9 citrons and 7 fragrant wood apples is 107" gives us

$$9x + 7y = 107. \qquad (1)$$

"7 citrons and 9 fragrant wood apples is 101" gives us

$$7x + 9y = 101. \qquad (2)$$

Multiply equation (1) by -7 and equation (2) by 9. Then add.

$$
\begin{array}{rll}
-63x - 49y & = -749 & -7 \times (1) \\
63x + 81y & = 909 & 9 \times (2) \\
\hline
32y & = 160 & \\
y & = 5 &
\end{array}
$$

Substitute 5 for y in equation (1).

$$
\begin{array}{rl}
9x + 7(5) & = 107 \\
9x + 35 & = 107 \\
9x & = 72 \\
x & = 8
\end{array}
$$

The prices are 8 for a citron and 5 for a wood apple.

37. Let $x =$ the measure of one angle,
$y =$ the measure of another angle,
and $z =$ the measure of the last angle.

Two equations are given, so

$$
\begin{array}{rll}
& z = x + 10 & \\
\text{or} & -x + z = 10 & (1) \\
\text{and} & x + y = 100. & (2)
\end{array}
$$

Since the sum of the measures of the angles of a triangle is $180°$, the third equation of the system is

$$x + y + z = 180. \qquad (3)$$

Equation (1) is missing y. To eliminate y again, multiply equation (2) by -1 and add the result to equation (3).

$$
\begin{array}{rll}
-x - y & = -100 & -1 \times (2) \\
x + y + z & = 180 & (3) \\
\hline
z & = 80 &
\end{array}
$$

Since $z = 80$,

$$
\begin{array}{rl}
-x + z = 10 & (1) \\
-x + 80 = 10 & \\
-x = -70 & \\
x = 70. &
\end{array}
$$

From (2), $70 + y = 100$, so $y = 30$.
The measures of the angles are $70°$, $30°$, and $80°$.

39. Let $x =$ the measure of the first angle,
$y =$ the measure of the second angle, and
$z =$ the measure of the third angle.

The sum of the angles in a triangle equals $180°$, so

$$x + y + z = 180. \qquad (1)$$

The measure of the second angle is $10°$ more than 3 times that of the first angle, so

$$y = 3x + 10. \qquad (2)$$

The third angle is equal to the sum of the other two, so

$$z = x + y. \qquad (3)$$

Solve the system. Substitute z for $x + y$ in equation (1).

$$(x + y) + z = 180 \quad (1)$$
$$z + z = 180$$
$$2z = 180$$
$$z = 90$$

Substitute 90 for z and $3x + 10$ for y in equation (3).

$$z = x + y \quad (3)$$
$$90 = x + (3x + 10)$$
$$80 = 4x$$
$$20 = x$$

Substitute $x = 20$ and $z = 90$ into equation (3).

$$z = x + y \quad (3)$$
$$90 = 20 + y$$
$$70 = y$$

The three angles have measures of 20°, 70°, and 90°.

41. Let $x =$ the length of the longest side,
$y =$ the length of the middle side,
and $z =$ the length of the shortest side.

Perimeter is the sum of the measures of the sides, so

$$x + y + z = 70. \quad (1)$$

The longest side is 4 cm less than the sum of the other sides, so

$$x = y + z - 4$$
$$\text{or} \quad x - y - z = -4. \quad (2)$$

Twice the shortest side is 9 cm less than the longest side, so

$$2z = x - 9$$
$$\text{or} \quad -x + 2z = -9 \quad (3)$$

Add equations (1) and (2) to eliminate y and z.

$$\begin{array}{r} x + y + z = 70 \quad (1) \\ x - y - z = -4 \quad (2) \\ \hline 2x \qquad = 66 \\ x = 33 \end{array}$$

Substitute 33 for x in (3).

$$-x + 2z = -9 \quad (3)$$
$$-33 + 2z = -9$$
$$2z = 24$$
$$z = 12$$

Substitute 33 for x and 12 for z in (1).

$$x + y + z = 70 \quad (1)$$
$$33 + y + 12 = 70$$
$$y + 45 = 70$$
$$y = 25$$

The shortest side is 12 cm long, the middle side is 25 cm long, and the longest side is 33 cm long.

43. Let $x =$ the number of gold medals, $y =$ the number of silver medals, and $z =$ the number of bronze medals.

The U.S. earned 6 more gold medals than bronze, so

$$x = 6 + z. \quad (1)$$

The number of silver medals earned was 19 less than twice the number of bronze medals, so

$$y = 2z - 19. \quad (2)$$

The total number of medals earned was 103, so

$$x + y + z = 103. \quad (3)$$

Substitute $6 + z$ for x and $2z - 19$ for y in (3).

$$(6 + z) + (2z - 17) + z = 103$$
$$4z - 13 = 103$$
$$4z = 116$$
$$z = 29$$

From (1), $x = 6 + 29 = 35$.
From (2), $y = 2(29) - 19 = 58 - 19 = 39$.

The United States earned 35 gold, 39 silver, and 29 bronze medals.

45. Let $x =$ the number of $14 tickets, $y =$ the number of $20 tickets, and $z =$ the number of VIP $50 tickets.

Five times as many $14 tickets have been sold as VIP tickets, so

$$x = 5z. \quad (1)$$

The number of $14 tickets is 15 more than the sum of the number of $20 tickets and the number of VIP tickets, so

$$x = 15 + y + z. \quad (2)$$

Since x is in terms of z in (1), we'll substitute $5z$ for x in (2) and then get y in terms of z.

$$5z = 15 + y + z$$
$$4z - 15 = y \quad (3)$$

Sales of these tickets totaled $11,700.

$$14x + 20y + 50z = 11{,}700$$
$$14(5z) + 20(4z - 15) + 50z = 11{,}700$$
$$70z + 80z - 300 + 50z = 11{,}700$$
$$200z = 12{,}000$$
$$z = 60$$

From (1), $x = 5(60) = 300$.
From (3), $y = 4(60) - 15 = 225$.

There were 300 $14 tickets, 225 $20 tickets, and 60 $50 tickets sold.

47. Let $x =$ the number of T-shirts shipped
to bookstore A,
$y =$ the number of T-shirts shipped
to bookstore B,
and $z =$ the number of T-shirts shipped
to bookstore C.

Twice as many T-shirts were shipped to bookstore B as to bookstore A, so

$$y = 2x. \quad (1)$$

The number shipped to bookstore C was 40 less than the sum of the numbers shipped to the other two bookstores, so

$$z = x + y - 40. \quad (2)$$

Substitute $2x$ for y [from (1)] into equation (2) to get z in terms of x.

$$z = x + y - 40 \qquad (2)$$
$$z = x + (2x) - 40$$
$$z = 3x - 40 \qquad (3)$$

The total number of T-shirts shipped was 800, so

$$x + y + z = 800. \quad (4)$$

Substitute $2x$ for y and $3x - 40$ for z in (4).

$$x + (2x) + (3x - 40) = 800$$
$$6x - 40 = 800$$
$$6x = 840$$
$$x = 140$$

From (1), $y = 2(140) = 280$.
From (3), $z = 3(140) - 40 = 380$.
The number of T-shirts shipped to bookstores A, B, and C was 140, 280, and 380, respectively.

49. Let x, y, and z denote the number of kilograms of the first, second, and third chemicals, respectively. The mix must include 60% of the first and second chemicals, so

$$x + y = 0.60(750) = 450. \quad (1)$$

The second and third chemicals must be in a ratio of 4 to 3 by weight, so

$$y = \tfrac{4}{3}z. \quad (2)$$

From (1),

$$x = 450 - y. \quad (3)$$

From (2),

$$z = \tfrac{3}{4}y. \quad (4)$$

The total is 750, so

$$x + y + z = 750.$$
$$(450 - y) + y + \tfrac{3}{4}y = 750$$
$$\tfrac{3}{4}y = 300$$
$$y = 400$$

From (3), $x = 450 - 400 = 50$.
From (4), $z = \tfrac{3}{4}(400) = 300$.

Use 50 kg of the first chemical, 400 kg of the second chemical, and 300 kg of the third chemical to make the plant food.

51. *Step 2*
Let $x =$ the number of wins,
$y =$ the number of losses,
and $z =$ the number of ties.

Step 3
They played 82 games, so

$$x + y + z = 82. \quad (1)$$

Their wins and losses totaled 71, so

$$x + y = 71. \quad (2)$$

They tied 14 fewer games than they lost, so

$$z = y - 14. \quad (3)$$

Step 4
Multiply (2) by -1 and add to (1).

$$\begin{array}{rcl} x + y + z = & 82 & (1) \\ -x - y = & -71 & -1 \times (2) \\ \hline z = & 11 & \end{array}$$

Substitute 11 for z in (3).

$$11 = y - 14$$
$$25 = y$$

Substitute 25 for y in (2).

$$x + 25 = 71$$
$$x = 46$$

Step 5
The Flames won 46 games, lost 25 games, and tied 11 games.

Step 6
Adding 46, 25, and 11 gives 82 total games. The wins and losses add up to 71, and there were 14 fewer ties than losses. The solution is correct.

53. **(a)** The additive inverse of -6 is $-(-6) = 6$.

(b) The multiplicative inverse (reciprocal) of -6 is $\tfrac{1}{-6} = -\tfrac{1}{6}$.

55. **(a)** The additive inverse of $\tfrac{7}{8}$ is $-\tfrac{7}{8}$.

(b) The multiplicative inverse (reciprocal) of $\tfrac{7}{8}$ is $\tfrac{1}{7/8} = \tfrac{8}{7}$.

4.4   Solving Systems of Linear Equations by Matrix Methods

1.
$$\begin{bmatrix} -2 & 3 & 1 \\ 0 & 5 & -3 \\ 1 & 4 & 8 \end{bmatrix}$$

(a) The elements of the second row are 0, 5, and -3.

(b) The elements of the third column are 1, -3, and 8.

(c) The matrix is square since the number of rows (three) is the same as the number of columns.

(d) The matrix obtained by interchanging the first and third rows is
$$\begin{bmatrix} 1 & 4 & 8 \\ 0 & 5 & -3 \\ -2 & 3 & 1 \end{bmatrix}.$$

(e) The matrix obtained by multiplying the first row by $-\frac{1}{2}$ is
$$\begin{bmatrix} -2\left(-\frac{1}{2}\right) & 3\left(-\frac{1}{2}\right) & 1\left(-\frac{1}{2}\right) \\ 0 & 5 & -3 \\ 1 & 4 & 8 \end{bmatrix} = \begin{bmatrix} 1 & -\frac{3}{2} & -\frac{1}{2} \\ 0 & 5 & -3 \\ 1 & 4 & 8 \end{bmatrix}.$$

(f) The matrix obtained by multiplying the third row by 3 and adding to the first row is
$$\begin{bmatrix} -2+3(1) & 3+3(4) & 1+3(8) \\ 0 & 5 & -3 \\ 1 & 4 & 8 \end{bmatrix} = \begin{bmatrix} 1 & 15 & 25 \\ 0 & 5 & -3 \\ 1 & 4 & 8 \end{bmatrix}.$$

3. $\begin{aligned} 4x + 8y &= 44 \\ 2x - y &= -3 \end{aligned}$

$$\begin{bmatrix} 4 & 8 & | & 44 \\ 2 & -1 & | & -3 \end{bmatrix}$$

$$\begin{bmatrix} 1 & 2 & | & 11 \\ 2 & -1 & | & -3 \end{bmatrix} \qquad \frac{1}{4}R_1$$

$$\begin{bmatrix} 1 & 2 & | & 11 \\ 0 & -5 & | & -25 \end{bmatrix} \qquad -2R_1 + R_2$$

Note: $\begin{cases} -2(2) + (-1) = -5 \\ -2(11) + (-3) = -25 \end{cases}$

$$\begin{bmatrix} 1 & 2 & | & 11 \\ 0 & 1 & | & 5 \end{bmatrix} \qquad -\frac{1}{5}R_2$$

This represents the system
$$\begin{aligned} x + 2y &= 11 \\ y &= 5. \end{aligned}$$

Substitute $y = 5$ in the first equation.
$$\begin{aligned} x + 2y &= 11 \\ x + 2(5) &= 11 \\ x + 10 &= 11 \\ x &= 1 \end{aligned}$$

The solution set is $\{(1, 5)\}$.

5. $\begin{aligned} x + y &= 5 \\ x - y &= 3 \end{aligned}$

Write the augmented matrix for this system.

$$\begin{bmatrix} 1 & 1 & | & 5 \\ 1 & -1 & | & 3 \end{bmatrix}$$

$$\begin{bmatrix} 1 & 1 & | & 5 \\ 0 & -2 & | & -2 \end{bmatrix} \qquad -1R_1 + R_2$$

$$\begin{bmatrix} 1 & 1 & | & 5 \\ 0 & 1 & | & 1 \end{bmatrix} \qquad -\frac{1}{2}R_2$$

This matrix gives the system
$$\begin{aligned} x + y &= 5 \\ y &= 1. \end{aligned}$$

Substitute $y = 1$ in the first equation.
$$\begin{aligned} x + y &= 5 \\ x + 1 &= 5 \\ x &= 4 \end{aligned}$$

The solution set is $\{(4, 1)\}$.

7. $\begin{aligned} 2x + 4y &= 6 \\ 3x - y &= 2 \end{aligned}$

Write the augmented matrix.

$$\begin{bmatrix} 2 & 4 & | & 6 \\ 3 & -1 & | & 2 \end{bmatrix}$$

The easiest way to get a 1 in the first row, first column position is to multiply the elements in the first row by $\frac{1}{2}$.

$$\begin{bmatrix} 1 & 2 & | & 3 \\ 3 & -1 & | & 2 \end{bmatrix} \qquad \frac{1}{2}R_1$$

To get a 0 in row two, column 1, we need to subtract 3 from the 3 that is in that position. To do this we will multiply row 1 by -3 and add the result to row 2.

$$\begin{bmatrix} 1 & 2 & | & 3 \\ 0 & -7 & | & -7 \end{bmatrix} \qquad -3R_1 + R_2$$

$$\begin{bmatrix} 1 & 2 & | & 3 \\ 0 & 1 & | & 1 \end{bmatrix} \qquad -\frac{1}{7}R_2$$

This matrix gives the system
$$\begin{aligned} x + 2y &= 3 \\ y &= 1. \end{aligned}$$

Substitute $y = 1$ in the first equation.
$$\begin{aligned} x + 2y &= 3 \\ x + 2(1) &= 3 \\ x &= 1 \end{aligned}$$

The solution set is $\{(1, 1)\}$.

9. $3x + 4y = 13$
$2x - 3y = -14$

Write the augmented matrix.

$$\begin{bmatrix} 3 & 4 & | & 13 \\ 2 & -3 & | & -14 \end{bmatrix}$$

$$\begin{bmatrix} 1 & 7 & | & 27 \\ 2 & -3 & | & -14 \end{bmatrix} \quad -1R_2 + R_1$$

$$\begin{bmatrix} 1 & 7 & | & 27 \\ 0 & -17 & | & -68 \end{bmatrix} \quad -2R_1 + R_2$$

$$\begin{bmatrix} 1 & 7 & | & 27 \\ 0 & 1 & | & 4 \end{bmatrix} \quad -\frac{1}{17}R_2$$

This matrix gives the system

$$x + 7y = 27$$
$$y = 4.$$

Substitute $y = 4$ in the first equation.

$$x + 7y = 27$$
$$x + 7(4) = 27$$
$$x + 28 = 27$$
$$x = -1$$

The solution set is $\{(-1, 4)\}$.

11. $-4x + 12y = 36$
$x - 3y = 9$

Write the augmented matrix.

$$\begin{bmatrix} -4 & 12 & | & 36 \\ 1 & -3 & | & 9 \end{bmatrix}$$

$$\begin{bmatrix} -1 & 3 & | & 9 \\ 1 & -3 & | & 9 \end{bmatrix} \quad \frac{1}{4}R_1$$

$$\begin{bmatrix} -1 & 3 & | & 9 \\ 0 & 0 & | & 18 \end{bmatrix} \quad R_1 + R_2$$

The corresponding system is

$$-x + 3y = 9$$
$$0 = 18 \quad \textit{False}$$

which is inconsistent and has no solution.

The solution set is $\emptyset$.

13. $2x + y = 4$
$4x + 2y = 8$

Write the augmented matrix.

$$\begin{bmatrix} 2 & 1 & | & 4 \\ 4 & 2 & | & 8 \end{bmatrix}$$

$$\begin{bmatrix} 1 & \frac{1}{2} & | & 2 \\ 4 & 2 & | & 8 \end{bmatrix} \quad \frac{1}{2}R_1$$

$$\begin{bmatrix} 1 & \frac{1}{2} & | & 2 \\ 0 & 0 & | & 0 \end{bmatrix} \quad -4R_1 + R_2$$

Row 2, $0 = 0$, indicates that the system has dependent equations.

The solution set is $\{(x, y) \mid 2x + y = 4\}$.

15. $x + y - z = -3$
$2x + y + z = 4$
$5x - y + 2z = 23$

Write the augmented matrix.

$$\begin{bmatrix} 1 & 1 & -1 & | & -3 \\ 2 & 1 & 1 & | & 4 \\ 5 & -1 & 2 & | & 23 \end{bmatrix}$$

$$\begin{bmatrix} 1 & 1 & -1 & | & -3 \\ 0 & -1 & 3 & | & 10 \\ 0 & -6 & 7 & | & 38 \end{bmatrix} \quad \begin{array}{l} -2R_1 + R_2 \\ -5R_1 + R_3 \end{array}$$

$$\begin{bmatrix} 1 & 1 & -1 & | & -3 \\ 0 & 1 & -3 & | & -10 \\ 0 & -6 & 7 & | & 38 \end{bmatrix} \quad -1R_2$$

$$\begin{bmatrix} 1 & 1 & -1 & | & -3 \\ 0 & 1 & -3 & | & -10 \\ 0 & 0 & -11 & | & -22 \end{bmatrix} \quad 6R_2 + R_3$$

$$\begin{bmatrix} 1 & 1 & -1 & | & -3 \\ 0 & 1 & -3 & | & -10 \\ 0 & 0 & 1 & | & 2 \end{bmatrix} \quad -\frac{1}{11}R_3$$

This matrix gives the system

$$x + y - z = -3$$
$$y - 3z = -10$$
$$z = 2.$$

Substitute $z = 2$ in the second equation.

$$y - 3z = -10$$
$$y - 3(2) = -10$$
$$y - 6 = -10$$
$$y = -4$$

Substitute $y = -4$ and $z = 2$ in the first equation.

$$x + y - z = -3$$
$$x - 4 - 2 = -3$$
$$x - 6 = -3$$
$$x = 3$$

The solution set is $\{(3, -4, 2)\}$.

17. $x + y - 3z = 1$
$2x - y + z = 9$
$3x + y - 4z = 8$

Write the augmented matrix.

$$\begin{bmatrix} 1 & 1 & -3 & | & 1 \\ 2 & -1 & 1 & | & 9 \\ 3 & 1 & -4 & | & 8 \end{bmatrix}$$

$$\begin{bmatrix} 1 & 1 & -3 & | & 1 \\ 0 & -3 & 7 & | & 7 \\ 0 & -2 & 5 & | & 5 \end{bmatrix} \quad \begin{array}{l} -2R_1 + R_2 \\ -3R_1 + R_3 \end{array}$$

$$\begin{bmatrix} 1 & 1 & -3 & | & 1 \\ 0 & 1 & -\frac{7}{3} & | & -\frac{7}{3} \\ 0 & -2 & 5 & | & 5 \end{bmatrix} \quad -\frac{1}{3}R_2$$

$$\begin{bmatrix} 1 & 1 & -3 & | & 1 \\ 0 & 1 & -\frac{7}{3} & | & -\frac{7}{3} \\ 0 & 0 & \frac{1}{3} & | & \frac{1}{3} \end{bmatrix} \quad 2R_2 + R_3$$

$$\begin{bmatrix} 1 & 1 & -3 & | & 1 \\ 0 & 1 & -\frac{7}{3} & | & -\frac{7}{3} \\ 0 & 0 & 1 & | & 1 \end{bmatrix} \quad 3R_3$$

This matrix gives the system

$$x + y - 3z = 1$$
$$y - \frac{7}{3}z = -\frac{7}{3}$$
$$z = 1.$$

Substitute $z = 1$ in the second equation.

$$y - \frac{7}{3}z = -\frac{7}{3}$$
$$y - \frac{7}{3}(1) = -\frac{7}{3}$$
$$y = 0$$

Substitute $y = 0$ and $z = 1$ in the first equation.

$$x + y - 3z = 1$$
$$x + 0 - 3(1) = 1$$
$$x - 3 = 1$$
$$x = 4$$

The solution set is $\{(4, 0, 1)\}$.

19.
$$x + y - z = 6$$
$$2x - y + z = -9$$
$$x - 2y + 3z = 1$$

Write the augmented matrix.

$$\begin{bmatrix} 1 & 1 & -1 & | & 6 \\ 2 & -1 & 1 & | & -9 \\ 1 & -2 & 3 & | & 1 \end{bmatrix}$$

$$\begin{bmatrix} 1 & 1 & -1 & | & 6 \\ 0 & -3 & 3 & | & -21 \\ 0 & -3 & 4 & | & -5 \end{bmatrix} \quad \begin{array}{l} -2R_1 + R_2 \\ -1R_1 + R_3 \end{array}$$

$$\begin{bmatrix} 1 & 1 & -1 & | & 6 \\ 0 & 1 & -1 & | & 7 \\ 0 & -3 & 4 & | & -5 \end{bmatrix} \quad -\frac{1}{3}R_2$$

$$\begin{bmatrix} 1 & 1 & -1 & | & 6 \\ 0 & 1 & -1 & | & 7 \\ 0 & 0 & 1 & | & 16 \end{bmatrix} \quad 3R_2 + R_3$$

This matrix gives the system

$$x + y - z = 6$$
$$y - z = 7$$
$$z = 16.$$

Substitute $z = 16$ in the second equation.

$$y - 16 = 7$$
$$y = 23$$

Substitute $y = 23$ and $z = 16$ in the first equation.

$$x + 23 - 16 = 6$$
$$x + 7 = 6$$
$$x = -1$$

The solution set is $\{(-1, 23, 16)\}$.

21.
$$x - y \quad = 1$$
$$y - z = 6$$
$$x + \quad z = -1$$

Write the augmented matrix.

$$\begin{bmatrix} 1 & -1 & 0 & | & 1 \\ 0 & 1 & -1 & | & 6 \\ 1 & 0 & 1 & | & -1 \end{bmatrix}$$

$$\begin{bmatrix} 1 & -1 & 0 & | & 1 \\ 0 & 1 & -1 & | & 6 \\ 0 & 1 & 1 & | & -2 \end{bmatrix} \quad -1R_1 + R_3$$

$$\begin{bmatrix} 1 & -1 & 0 & | & 1 \\ 0 & 1 & -1 & | & 6 \\ 0 & 0 & 2 & | & -8 \end{bmatrix} \quad -1R_2 + R_3$$

$$\begin{bmatrix} 1 & -1 & 0 & | & 1 \\ 0 & 1 & -1 & | & 6 \\ 0 & 0 & 1 & | & -4 \end{bmatrix} \quad \frac{1}{2}R_3$$

This matrix gives the system

$$x - y \quad = 1$$
$$y - z = 6$$
$$z = -4.$$

Substitute $z = -4$ in the second equation.

$$y - z = 6$$
$$y - (-4) = 6$$
$$y = 2$$

Substitute $y = 2$ in the first equation.

$$x - y = 1$$
$$x - 2 = 1$$
$$x = 3$$

The solution set is $\{(3, 2, -4)\}$.

23.
$$\begin{aligned} x - 2y + z &= 4 \\ 3x - 6y + 3z &= 12 \\ -2x + 4y - 2z &= -8 \end{aligned}$$

Write the augmented matrix.

$$\left[\begin{array}{ccc|c} 1 & -2 & 1 & 4 \\ 3 & -6 & 3 & 12 \\ -2 & 4 & -2 & -8 \end{array}\right]$$

$$\left[\begin{array}{ccc|c} 1 & -2 & 1 & 4 \\ 1 & -2 & 1 & 4 \\ -1 & 2 & -1 & -4 \end{array}\right] \begin{array}{c} \\ \frac{1}{3}R_2 \\ \frac{1}{2}R_3 \end{array}$$

$$\left[\begin{array}{ccc|c} 1 & -2 & 1 & 4 \\ 0 & 0 & 0 & 0 \\ 0 & 0 & 0 & 0 \end{array}\right] \begin{array}{c} \\ -1R_1 + R_2 \\ R_1 + R_3 \end{array}$$

This augmented matrix represents a system of dependent equations.
The solution set is $\{(x, y, z) \mid x - 2y + z = 4\}$.

25.
$$\begin{aligned} x + 2y + 3z &= -2 \\ 2x + 4y + 6z &= -5 \\ x - y + 2z &= 6 \end{aligned}$$

Write the augmented matrix.

$$\left[\begin{array}{ccc|c} 1 & 2 & 3 & -2 \\ 2 & 4 & 6 & -5 \\ 1 & -1 & 2 & 6 \end{array}\right]$$

$$\left[\begin{array}{ccc|c} 1 & 2 & 3 & -2 \\ 0 & 0 & 0 & -1 \\ 0 & -3 & -1 & 8 \end{array}\right] \begin{array}{c} \\ -2R_1 + R_2 \\ -R_1 + R_3 \end{array}$$

From the second row, $0 = -1$, we see that the system is inconsistent.
The solution set is $\emptyset$.

27.
$$\begin{aligned} 4x + y &= 5 \\ 2x + y &= 3 \end{aligned}$$

Enter the augmented matrix as $[A]$.

$$\left[\begin{array}{cc|c} 4 & 1 & 5 \\ 2 & 1 & 3 \end{array}\right]$$

The TI-83 screen for A follows. (Use MATRX EDIT.)

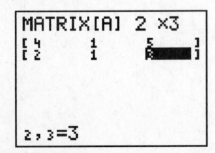

Now use the reduced row echelon form (rref) command to simplify the system. (Use MATRX MATH ALPHA B for rref and MATRX 1 for $[A]$.)

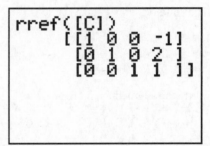

This matrix gives the system
$$\begin{aligned} 1x + 0y &= 1 \\ 0x + 1y &= 1, \end{aligned}$$

or simply, $x = 1$ and $y = 1$.
The solution set is $\{(1, 1)\}$.

29.
$$\begin{aligned} 5x + y - 3z &= -6 \\ 2x + 3y + z &= 5 \\ -3x - 2y + 4z &= 3 \end{aligned}$$

Enter the augmented matrix as [C].

$$\left[\begin{array}{ccc|c} 5 & 1 & -3 & -6 \\ 2 & 3 & 1 & 5 \\ -3 & -2 & 4 & 3 \end{array}\right]$$

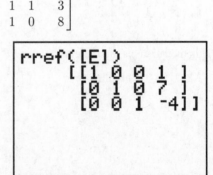

The solution set is $\{(-1, 2, 1)\}$.

31.
$$\begin{aligned} x + z &= -3 \\ y + z &= 3 \\ x + y &= 8 \end{aligned}$$

Enter the augmented matrix as [E].

$$\left[\begin{array}{ccc|c} 1 & 0 & 1 & -3 \\ 0 & 1 & 1 & 3 \\ 1 & 1 & 0 & 8 \end{array}\right]$$

The solution set is $\{(1, 7, -4)\}$.

33. $2^6 = 2 \cdot 2 \cdot 2 \cdot 2 \cdot 2 \cdot 2 = 64$

35. $(-5)^4 = (-5)(-5)(-5)(-5) = 625$

37. $\left(\frac{3}{4}\right)^4 = \frac{3}{4} \cdot \frac{3}{4} \cdot \frac{3}{4} \cdot \frac{3}{4}$

$= \frac{3 \cdot 3 \cdot 3 \cdot 3}{4 \cdot 4 \cdot 4 \cdot 4} = \frac{81}{256}$

Chapter 4 Review Exercises

1. **(a)** The graphs meet between the years of 1980 and 1985. Therefore, the number of degrees for men equal the number of degrees for women between 1980 and 1985.

(b) The number was just less than 500,000.

2. $x + 3y = 8 \quad (1)$
$2x - y = 2 \quad (2)$

Graph the two lines. They appear to intersect at the point $(2, 2)$. Check $(2, 2)$ in both equations.

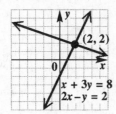

$x + 3y = 8 \qquad (1)$
$2 + 3(2) = 8$
$8 = 8 \quad True$

$2x - y = 2 \qquad (2)$
$2(2) - 2 = 2$
$2 = 2 \quad True$

The solution set is $\{(2, 2)\}$.

3. Checking the ordered pairs in choices **A**, **B**, and **C** yields true statements.

For choice **D**, we have:

$3x + 2y = 6$
$3(3) + 2(-2) = 6 \quad ?$
$9 - 4 = 6 \quad ?$
$5 = 6 \qquad False$

So **D** is the answer.

4. **(a)** The graphs of these two linear equations will intersect once.

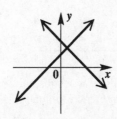

(b) The graphs of these two linear equations will not intersect. They are parallel lines.

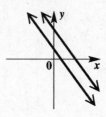

(c) The graphs of these two linear equations will be the same line.

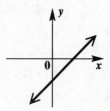

5. $3x + y = -4 \qquad (1)$
$ x = \frac{2}{3}y \qquad (2)$

Substitute $\frac{2}{3}y$ for x in equation (1) and solve for y.

$3x + y = -4 \quad (1)$
$3\left(\frac{2}{3}y\right) + y = 4$
$2y + y = -4$
$3y = -4$
$y = -\frac{4}{3}$

Since $x = \frac{2}{3}y$ and $y = -\frac{4}{3}$,

$x = \frac{2}{3}\left(-\frac{4}{3}\right) = -\frac{8}{9}.$

The solution set is $\left\{\left(-\frac{8}{9}, -\frac{4}{3}\right)\right\}$.

6. $9x - y = -4 \qquad (1)$
$ y = x + 4 \quad (2)$

Substitute $x + 4$ for y in equation (1) and solve for x.

$9x - y = -4 \qquad (1)$
$9x - (x + 4) = -4$
$9x - x - 4 = -4$
$8x = 0$
$x = 0$

Since $x = 0$, $y = x + 4 = 0 + 4 = 4$.
The solution set is $\{(0, 4)\}$.

7. $\quad -5x + 2y = -2 \qquad (1)$
$\qquad\quad x + 6y = 26 \qquad (2)$

Solve equation (2) for x.

$$x = 26 - 6y$$

Substitute $26 - 6y$ for x in equation (1).

$$-5x + 2y = -2 \quad (1)$$
$$-5(26 - 6y) + 2y = -2$$
$$-130 + 30y + 2y = -2$$
$$-130 + 32y = -2$$
$$32y = 128$$
$$y = 4$$

Since $x = 26 - 6y$ and $y = 4$,

$$x = 26 - 6(4) = 26 - 24 = 2.$$

The solution set is $\{(2, 4)\}$.

8. $\quad 5x + \ y = 12 \qquad (1)$
$\qquad 2x - 2y = \ 0 \qquad (2)$

To eliminate y, multiply equation (1) by 2 and add the result to equation (2).

$$
\begin{array}{rcl}
10x \ + \ 2y \ = \ 24 & & 2 \times (1) \\
\underline{2x \ - \ 2y \ = \ \ 0} & & (2) \\
12x \qquad\quad = \ 24 & & \\
x \ = \ 2 & &
\end{array}
$$

To find y, substitute 2 for x in equation (2).

$$2x - 2y = 0 \qquad (2)$$
$$2(2) - 2y = 0$$
$$4 = 2y$$
$$2 = y$$

The ordered pair $(2, 2)$ satisfies both equations, so it checks.
The solution set is $\{(2, 2)\}$.

9. $\quad x - 4y = -4 \qquad (1)$
$\qquad 3x + \ y = \ \ 1 \qquad (2)$

To eliminate y, multiply equation (2) by 4 and add the result to equation (1).

$$
\begin{array}{rcl}
x \ - \ 4y \ = \ -4 & & (1) \\
\underline{12x \ + \ 4y \ = \ \ \ 4} & & 4 \times (2) \\
13x \qquad\quad = \ \ \ 0 & & \\
x \ = \ \ \ 0 & &
\end{array}
$$

Substitute 0 for x in equation (1) to find y.

$$x - 4y = -4 \quad (1)$$
$$0 - 4y = -4$$
$$y = 1$$

The solution set is $\{(0, 1)\}$.

10. $\quad 6x + 5y = 4 \qquad (1)$
$\qquad -4x + 2y = 8 \qquad (2)$

To eliminate x, multiply equation (1) by 2 and equation (2) by 3. Then add the results.

$$
\begin{array}{rcl}
12x \ + \ 10y \ = \ \ 8 & & 2 \times (1) \\
\underline{-12x \ + \ \ 6y \ = \ 24} & & 3 \times (2) \\
16y \ = \ 32 & & \\
y \ = \ 2 & &
\end{array}
$$

Since $y = 2$,

$$-4x + 2y = 8 \qquad (2)$$
$$-4x + 2(2) = 8$$
$$-4x + 4 = 8$$
$$-4x = 4$$
$$x = -1.$$

The solution set is $\{(-1, 2)\}$.

11. $\quad \frac{1}{6}x + \frac{1}{6}y = -\frac{1}{2} \qquad (1)$
$\qquad x - y \ = -9 \qquad (2)$

Multiply equation (1) by 6 to clear the fractions. Add the result to equation (2) to eliminate y.

$$
\begin{array}{rcl}
x \ + \ y \ = \ \ -3 & & 6 \times (1) \\
\underline{x \ - \ y \ = \ \ -9} & & (2) \\
2x \qquad\quad = \ -12 & & \\
x \ = \ -6 & &
\end{array}
$$

Since $x = -6$,

$$x - y = -9 \quad (2)$$
$$-6 - y = -9$$
$$-y = -3$$
$$y = 3.$$

The solution set is $\{(-6, 3)\}$.

12. $\quad -3x + y = 6 \qquad\quad (1)$
$\qquad\qquad\ y = 6 + 3x \qquad (2)$

Since equation (2) can be rewritten as $-3x + y = 6$, the two equations are the same, and hence, dependent.
The solution set is $\{(x, y) \mid -3x + y = 6\}$.

13. $\quad 5x - 4y = 2 \qquad (1)$
$\qquad -10x + 8y = 7 \qquad (2)$

Multiply equation (1) by 2 and add the result to equation (2).

$$
\begin{array}{rcl}
10x \ - \ 8y \ = \ \ 4 & & 2 \times (1) \\
\underline{-10x \ + \ 8y \ = \ \ 7} & & (2) \\
0 \ = \ 11 & & \textit{False}
\end{array}
$$

Since a false statement results, the system is *inconsistent*. The solution set is $\emptyset$.

14. $y = 3x + 2$
$\quad\;\; y = 3x - 4$

Since both equations are in slope-intercept form, their slopes and y-intercepts can be easily determined. The two lines have the same slope, 3, but the y-intercepts, $(0, 2)$ and $(0, -4)$, are different. Therefore, the lines are parallel, do not intersect, and have no common solution.

15. $\quad 2x + 3y - z = -16 \quad\;\;(1)$
$\quad\quad\; x + 2y + 2z = -3 \quad\;\;\;(2)$
$\quad -3x + y + z = -5 \quad\;\;\;(3)$

To eliminate z, add equations (1) and (3).

$$\begin{array}{rcl}
2x + 3y - z = -16 & & (1) \\
\underline{-3x + y + z = -5} & & (3) \\
-x + 4y \quad\quad\;\; = -21 & & (4)
\end{array}$$

To eliminate z again, multiply equation (1) by 2 and add the result to equation (2).

$$\begin{array}{rcl}
4x + 6y - 2z = -32 & & 2 \times (1) \\
\underline{x + 2y + 2z = -3} & & (2) \\
5x + 8y \quad\quad\;\; = -35 & & (5)
\end{array}$$

Use equations (4) and (5) to eliminate x. Multiply equation (4) by 5 and add the result to equation (5).

$$\begin{array}{rcl}
-5x + 20y = -105 & & 5 \times (4) \\
\underline{5x + 8y = -35} & & (5) \\
28y = -140 & & \\
y = -5 & &
\end{array}$$

Substitute -5 for y in equation (4) to find x.

$$-x + 4y = -21 \quad (4)$$
$$-x + 4(-5) = -21$$
$$-x - 20 = -21$$
$$-x = -1$$
$$x = 1$$

Substitute 1 for x and -5 for y in equation (2) to find z.

$$x + 2y + 2z = -3 \quad (2)$$
$$1 + 2(-5) + 2z = -3$$
$$1 - 10 + 2z = -3$$
$$2z = 6$$
$$z = 3$$

The solution set is $\{(1, -5, 3)\}$.

16. $4x - y \quad\quad\;\; = 2 \quad (1)$
$\quad\quad\; 3y + z = 9 \quad (2)$
$\quad x \quad\quad + 2z = 7 \quad (3)$

To eliminate y, multiply equation (1) by 3 and add the result to equation (2).

$$\begin{array}{rcl}
12x - 3y \quad\quad = 6 & & 3 \times (1) \\
\underline{3y + z = 9} & & (2) \\
12x \quad\quad + z = 15 & & (4)
\end{array}$$

To eliminate z, multiply equation (4) by -2 and add the result to equation (3).

$$\begin{array}{rcl}
-24x - 2z = -30 & & -2 \times (4) \\
\underline{x + 2z = 7} & & (3) \\
-23x \quad\quad = -23 & & \\
x = 1 & &
\end{array}$$

Substitute 1 for x in equation (3) to find z.

$$x + 2z = 7 \quad (3)$$
$$1 + 2z = 7$$
$$2z = 6$$
$$z = 3$$

Substitute 1 for x in equation (1) to find y.

$$4x - y = 2 \quad (1)$$
$$4(1) - y = 2$$
$$4 - y = 2$$
$$-y = -2$$
$$y = 2$$

The solution set is $\{(1, 2, 3)\}$.

17. $\quad 3x - y - z = -8 \quad\;\;(1)$
$\quad\quad 4x + 2y + 3z = 15 \quad\;\;(2)$
$\quad -6x + 2y + 2z = 10 \quad\;\;(3)$

To eliminate y, multiply equation (1) by 2 and add the result to equation (3).

$$\begin{array}{rcl}
6x - 2y - 2z = -16 & & 2 \times (1) \\
\underline{-6x + 2y + 2z = 10} & & (3) \\
0 = -6 & & \textit{False}
\end{array}$$

Since a false statement results, equations (1) and (3) have no common solution. The system is *inconsistent*. The solution set is $\emptyset$.

18. Let $x =$ the width of the rink, and
$y =$ the length of the rink.

The length is 30 ft longer than twice the width.

$$y = 2x + 30 \quad (1)$$

The perimeter is 570 ft.

$$2x + 2y = 570 \quad (2)$$

Substitute $2x + 30$ for y in equation (2).

$$2x + 2y = 570 \quad (2)$$
$$2x + 2(2x + 30) = 570$$
$$2x + 4x + 60 = 570$$
$$6x = 510$$
$$x = 85$$

From (1), $y = 2(85) + 30 = 200$.

The width is 85 ft and the length is 200 ft.

19.   Let $x =$ the average price for a Red Sox ticket and $y =$ the average price for a Cubs ticket.

From the given information, we get the following system of equations:

$$4x + 4y = 276.88 \quad (1)$$
$$2x + 6y = 252.24 \quad (2)$$

Divide (1) by 4 and (2) by -2.

$$\begin{array}{rcl} x + y &=& 69.22 \quad (3) \quad (1) \div 4 \\ -x - 3y &=& -126.12 \quad (4) \quad (2) \div (-2) \\ \hline -2y &=& -56.90 \quad \textit{Add (3) and (4).} \\ y &=& 28.45 \end{array}$$

From (3), $x + 28.45 = 69.22$, so $x = 40.77$. The average price for a Red Sox ticket was \$40.77 and the average price for a Cubs ticket was \$28.45.

20.   Let $x =$ the speed of the plane and $y =$ the speed of the wind.

Complete the chart.

	r	t	d
With Wind	$x + y$	1.75	$1.75(x + y)$
Against Wind	$x - y$	2	$2(x - y)$

The distance each way is 560 miles. From the chart,

$$1.75(x + y) = 560.$$

Divide by 1.75.

$$x + y = 320 \quad (1)$$

From the chart,

$$2(x - y) = 560$$
$$x - y = 280. \quad (2)$$

Solve the system by adding equations (1) and (2) to eliminate y.

$$\begin{array}{rcl} x + y &=& 320 \quad (1) \\ x - y &=& 280 \quad (2) \\ \hline 2x &=& 600 \\ x &=& 300 \end{array}$$

From (1), $300 + y = 320$, so $y = 20$.

The speed of the plane was 300 mph, and the speed of the wind was 20 mph.

21.   Let $x =$ the amount of \$2-per-pound nuts and $y =$ the amount of \$1-per-pound chocolate candy.

Complete the chart.

	Number of Pounds	Price per Pound	Value
Nuts	x	2	$2x$
Chocolate	y	1	$1y = y$
Mixture	100	1.30	$1.30(100) = 130$

Solve the system formed from the first and third columns.

$$x + y = 100 \quad (1)$$
$$2x + y = 130 \quad (2)$$

Solve equation (1) for y.

$$y = 100 - x \quad (3)$$

Substitute $100 - x$ for y in equation (2).

$$2x + (100 - x) = 130$$
$$x = 30$$

From (3), $y = 100 - 30 = 70$.

She should use 30 lb of \$2-per-pound nuts and 70 lb of \$1-per-pound chocolate candy.

22.   Let $x =$ the measure of the largest angle, $y =$ the measure of the middle-sized angle, and $z =$ the measure of the smallest angle.

Since the sum of the measures of the angles of a triangle is $180°$,

$$x + y + z = 180. \quad (1)$$

Since the largest angle measures $10°$ less than the sum of the other two,

$$x = y + z - 10$$
$$\text{or} \quad x - y - z = -10. \quad (2)$$

Since the measure of the middle-sized angle is the average of the other two,

$$y = \frac{x + z}{2}$$
$$2y = x + z$$
$$-x + 2y - z = 0. \quad (3)$$

Solve the system.

$$\begin{array}{rcl} x + y + z &=& 180 \quad (1) \\ x - y - z &=& -10 \quad (2) \\ -x + 2y - z &=& 0 \quad (3) \end{array}$$

Add equations (1) and (2) to find x.

$$\begin{array}{rcl} x + y + z &=& 180 \quad (1) \\ x - y - z &=& -10 \quad (2) \\ \hline 2x &=& 170 \\ x &=& 85 \end{array}$$

Add equations (1) and (3), to find y.

$$\begin{array}{rcl} x + y + z &=& 180 \quad (1) \\ -x + 2y - z &=& 0 \quad (3) \\ \hline 3y &=& 180 \\ y &=& 60 \end{array}$$

Substitute 85 for x and 60 for y in equation (1) to find z.

$$x + y + z = 180 \quad (1)$$
$$85 + 60 + z = 180$$
$$145 + z = 180$$
$$z = 35$$

The three angle measures are $85°$, $60°$, and $35°$.

23. Let $x =$ the value of sales at 10%,
    $y =$ the value of sales at 6%,
and $z =$ the value of sales at 5%.

Since her total sales were $280,000,

$$x + y + z = 280{,}000 \quad (1)$$

Since her commissions on the sales totaled $17,000,

$$0.10x + 0.06y + 0.05z = 17{,}000.$$

Multiply by 100 to clear the decimals, so

$$10x + 6y + 5z = 1{,}700{,}000. \quad (2)$$

Since the 5% sale amounted to the sum of the other two sales,

$$z = x + y. \quad (3)$$

Solve the system.

$$x + y + z = 280{,}000 \quad (1)$$
$$10x + 6y + 5z = 1{,}700{,}000 \quad (2)$$
$$z = x + y \quad (3)$$

Since equation (3) is given in terms of z, substitute $x + y$ for z in equations (1) and (2).

$$x + y + z = 280{,}000 \quad (1)$$
$$x + y + (x + y) = 280{,}000$$
$$2x + 2y = 280{,}000$$
$$x + y = 140{,}000 \quad (4)$$

$$10x + 6y + 5z = 1{,}700{,}000 \quad (2)$$
$$10x + 6y + 5(x + y) = 1{,}700{,}000$$
$$10x + 6y + 5x + 5y = 1{,}700{,}000$$
$$15x + 11y = 1{,}700{,}000 \quad (5)$$

To eliminate x, multiply equation (4) by -11 and add the result to equation (5).

$$
\begin{array}{rll}
-11x - 11y = & -1{,}540{,}000 & -11 \times (4) \\
15x + 11y = & 1{,}700{,}000 & (5) \\
\hline
4x \quad\quad = & 160{,}000 & \\
x = & 40{,}000 &
\end{array}
$$

From (4), $y = 100{,}000$.
From (3), $z = 40{,}000 + 100{,}000 = 140{,}000$.
He sold $40,000 at 10%, $100,000 at 6%, and $140,000 at 5%.

24. Let $x =$ the number of liters of 8% solution,
    $y =$ the number of liters of 10% solution,
and $z =$ the number of liters of 20% solution.

Since the amount of the mixture will be 8 L,

$$x + y + z = 8. \quad (1)$$

Since the final solution will be 12.5% hydrogen peroxide,

$$0.08x + 0.10y + 0.20z = 0.125(8).$$

Multiply by 100 to clear the decimals.

$$8x + 10y + 20z = 100 \quad (2)$$

Since the amount of 8% solution used must be 2 L more than the amount of 20% solution,

$$x = z + 2. \quad (3)$$

Solve the system.

$$
\begin{array}{rll}
x + y + z = & 8 & (1) \\
8x + 10y + 20z = & 100 & (2) \\
x = & z + 2 & (3)
\end{array}
$$

Since equation (3) is given in terms of x, substitute $z + 2$ for x in equations (1) and (2).

$$x + y + z = 8 \quad (1)$$
$$(z + 2) + y + z = 8$$
$$y + 2z = 6 \quad (4)$$

$$8x + 10y + 20z = 100 \quad (2)$$
$$8(z + 2) + 10y + 20z = 100$$
$$8z + 16 + 10y + 20z = 100$$
$$10y + 28z = 84 \quad (5)$$

To eliminate y, multiply equation (4) by -10 and add the result to equation (5).

$$
\begin{array}{rll}
-10y - 20z = & -60 & -10 \times (4) \\
10y + 28z = & 84 & (5) \\
\hline
8z = & 24 & \\
z = & 3 &
\end{array}
$$

From (3), $x = z + 2 = 3 + 2 = 5$.
From (4), $y = 6 - 2z = 6 - 2(3) = 0$.
Mix 5 L of 8% solution, none of 10% solution, and 3 L of 20% solution.

25. Let $x =$ the number of home runs hit by Mantle,
    $y =$ the number of home runs hit by Maris,
and $z =$ the number of home runs
        hit by Blanchard.

They combined for 136 home runs, so

$$x + y + z = 136. \quad (1)$$

Mantle hit 7 fewer than Maris, so

$$x = y - 7. \quad (2)$$

continued

Maris hit 40 more than Blanchard, so

$$y = z + 40 \quad \text{or} \quad z = y - 40. \quad (3)$$

Substitute $y - 7$ for x and $y - 40$ for z in (1).

$$
\begin{aligned}
x + y + z &= 136 \quad (1) \\
(y - 7) + y + (y - 40) &= 136 \\
3y - 47 &= 136 \\
3y &= 183 \\
y &= 61
\end{aligned}
$$

From (2), $x = y - 7 = 61 - 7 = 54$.
From (3), $z = y - 40 = 61 - 40 = 21$.
Mantle hit 54 home runs, Maris hit 61 home runs, and Blanchard hit 21 home runs.

26.
$$
\begin{aligned}
2x + 5y &= -4 \\
4x - y &= 14
\end{aligned}
$$

Write the augmented matrix.

$$\begin{bmatrix} 2 & 5 & \vline & -4 \\ 4 & -1 & \vline & 14 \end{bmatrix}$$

$$\begin{bmatrix} 2 & 5 & \vline & -4 \\ 0 & -11 & \vline & 22 \end{bmatrix} \quad -2R_1 + R_2$$

$$\begin{bmatrix} 2 & 5 & \vline & -4 \\ 0 & 1 & \vline & -2 \end{bmatrix} \quad -\tfrac{1}{11}R_2$$

This matrix gives the system

$$
\begin{aligned}
2x + 5y &= -4 \\
y &= -2.
\end{aligned}
$$

Substitute $y = -2$ in the first equation.

$$
\begin{aligned}
2x + 5y &= -4 \\
2x + 5(-2) &= -4 \\
2x - 10 &= -4 \\
2x &= 6 \\
x &= 3
\end{aligned}
$$

The solution set is $\{(3, -2)\}$.

27.
$$
\begin{aligned}
6x + 3y &= 9 \\
-7x + 2y &= 17
\end{aligned}
$$

Write the augmented matrix.

$$\begin{bmatrix} 6 & 3 & \vline & 9 \\ -7 & 2 & \vline & 17 \end{bmatrix}$$

$$\begin{bmatrix} 1 & -5 & \vline & -26 \\ -7 & 2 & \vline & 17 \end{bmatrix} \quad -R_1 - R_2$$

$$\begin{bmatrix} 1 & -5 & \vline & -26 \\ 0 & -33 & \vline & -165 \end{bmatrix} \quad 7R_1 + R_2$$

$$\begin{bmatrix} 1 & -5 & \vline & -26 \\ 0 & 1 & \vline & 5 \end{bmatrix} \quad -\tfrac{1}{33}R_2$$

This matrix gives the system

$$
\begin{aligned}
x - 5y &= -26 \\
y &= 5.
\end{aligned}
$$

Substitute $y = 5$ in the first equation.

$$
\begin{aligned}
x - 5y &= -26 \\
x - 5(5) &= -26 \\
x - 25 &= -26 \\
x &= -1
\end{aligned}
$$

The solution set is $\{(-1, 5)\}$.

28.
$$
\begin{aligned}
x + 2y - z &= 1 \\
3x + 4y + 2z &= -2 \\
-2x - y + z &= -1
\end{aligned}
$$

$$\begin{bmatrix} 1 & 2 & -1 & \vline & 1 \\ 3 & 4 & 2 & \vline & -2 \\ -2 & -1 & 1 & \vline & -1 \end{bmatrix}$$

$$\begin{bmatrix} 1 & 2 & -1 & \vline & 1 \\ 0 & -2 & 5 & \vline & -5 \\ 0 & 3 & -1 & \vline & 1 \end{bmatrix} \quad \begin{matrix} -3R_1 + R_2 \\ 2R_1 + R_3 \end{matrix}$$

$$\begin{bmatrix} 1 & 2 & -1 & \vline & 1 \\ 0 & 1 & 4 & \vline & -4 \\ 0 & 3 & -1 & \vline & 1 \end{bmatrix} \quad R_3 + R_2$$

$$\begin{bmatrix} 1 & 2 & -1 & \vline & 1 \\ 0 & 1 & 4 & \vline & -4 \\ 0 & 0 & -13 & \vline & 13 \end{bmatrix} \quad -3R_2 + R_3$$

$$\begin{bmatrix} 1 & 2 & -1 & \vline & 1 \\ 0 & 1 & 4 & \vline & -4 \\ 0 & 0 & 1 & \vline & -1 \end{bmatrix} \quad -\tfrac{1}{13}R_3$$

This matrix gives the system

$$
\begin{aligned}
x + 2y - z &= 1 \\
y + 4z &= -4 \\
z &= -1.
\end{aligned}
$$

Substitute $z = -1$ in the second equation.

$$
\begin{aligned}
y + 4z &= -4 \\
y + 4(-1) &= -4 \\
y &= 0
\end{aligned}
$$

Substitute $y = 0$ and $z = -1$ in the first equation.

$$
\begin{aligned}
x + 2y - z &= 1 \\
x + 2(0) - (-1) &= 1 \\
x + 1 &= 1 \\
x &= 0
\end{aligned}
$$

The solution set is $\{(0, 0, -1)\}$.

29.
$$
\begin{aligned}
x + 3y &= 7 \\
3x + z &= 2 \\
y - 2z &= 4
\end{aligned}
$$

$$\begin{bmatrix} 1 & 3 & 0 & \vline & 7 \\ 3 & 0 & 1 & \vline & 2 \\ 0 & 1 & -2 & \vline & 4 \end{bmatrix}$$

$$\begin{bmatrix} 1 & 3 & 0 & \vline & 7 \\ 0 & -9 & 1 & \vline & -19 \\ 0 & 1 & -2 & \vline & 4 \end{bmatrix} \quad -3R_1 + R_2$$

$$\begin{bmatrix} 1 & 3 & 0 & | & 7 \\ 0 & 1 & -2 & | & 4 \\ 0 & -9 & 1 & | & -19 \end{bmatrix} \qquad R_2 \leftrightarrow R_3$$

We use $\leftrightarrow$ to represent the interchanging of 2 rows.

$$\begin{bmatrix} 1 & 3 & 0 & | & 7 \\ 0 & 1 & -2 & | & 4 \\ 0 & 0 & -17 & | & 17 \end{bmatrix} \qquad 9R_2 + R_3$$

$$\begin{bmatrix} 1 & 3 & 0 & | & 7 \\ 0 & 1 & -2 & | & 4 \\ 0 & 0 & 1 & | & -1 \end{bmatrix} \qquad -\frac{1}{17}R_3$$

This matrix gives the system

$$x + 3y = 7$$
$$y - 2z = 4$$
$$z = -1.$$

Substitute $z = -1$ in the second equation.

$$y - 2z = 4$$
$$y - 2(-1) = 4$$
$$y + 2 = 4$$
$$y = 2$$

Substitute $y = 2$ in the first equation.

$$x + 3y = 7$$
$$x + 3(2) = 7$$
$$x + 6 = 7$$
$$x = 1$$

The solution set is $\{(1, 2, -1)\}$.

30. **[4.1]** System **B** would be easier to solve using the substitution method because the second equation is already solved for y.

31. **[4.1]** $\frac{2}{3}x + \frac{1}{6}y = \frac{19}{2}$ $\qquad (1)$
$\frac{1}{3}x - \frac{2}{9}y = 2$ $\qquad (2)$

Multiply equation (1) by 6 and equation (2) by 9 to clear the fractions.

$$4x + y = 57 \quad (3)\ 6 \times (1)$$
$$3x - 2y = 18 \quad (4)\ 9 \times (2)$$

To eliminate y, multiply equation (3) by 2 and add the result to equation (4).

$$\begin{array}{rcll} 8x + 2y &=& 114 & 2 \times (3) \\ 3x - 2y &=& 18 & (4) \\ \hline 11x &=& 132 & \\ x &=& 12 & \end{array}$$

Substitute 12 for x in equation (3) to find y.

$$4x + y = 57 \quad (3)$$
$$4(12) + y = 57$$
$$48 + y = 57$$
$$y = 9$$

The solution set is $\{(12, 9)\}$.

32. **[4.2]** $\begin{array}{rcll} 2x + 5y - z &=& 12 & (1) \\ -x + y - 4z &=& -10 & (2) \\ -8x - 20y + 4z &=& 31 & (3) \end{array}$

Multiply equation (1) by 4 and add the result to equation (3).

$$\begin{array}{rcll} 8x + 20y - 4z &=& 48 & 4 \times (1) \\ -8x - 20y + 4z &=& 31 & (3) \\ \hline 0 &=& 79 & \textit{False} \end{array}$$

Since a false statement results, the system is *inconsistent*. The solution set is $\emptyset$.

33. **[4.1]** $x = 7y + 10$ $\qquad (1)$
$2x + 3y = 3$ $\qquad (2)$

Since equation (1) is given in terms of x, substitute $7y + 10$ for x in equation (2) and solve for y.

$$2(7y + 10) + 3y = 3$$
$$14y + 20 + 3y = 3$$
$$17y = -17$$
$$y = -1$$

From (1), $x = 7(-1) + 10 = 3$.

The solution set is $\{(3, -1)\}$.

34. **[4.1]** $\begin{array}{rcll} x + 4y &=& 17 & (1) \\ -3x + 2y &=& -9 & (2) \end{array}$

To eliminate x, multiply equation (1) by 3 and add the result to equation (2).

$$\begin{array}{rcll} 3x + 12y &=& 51 & 3 \times (1) \\ -3x + 2y &=& -9 & (2) \\ \hline 14y &=& 42 & \\ y &=& 3 & \end{array}$$

Substitute 3 for y in equation (1) to find x.

$$x + 4y = 17 \quad (1)$$
$$x + 4(3) = 17$$
$$x + 12 = 17$$
$$x = 5$$

The solution set is $\{(5, 3)\}$.

35. **[4.1]** $\begin{array}{rcll} -7x + 3y &=& 12 & (1) \\ 5x + 2y &=& 8 & (2) \end{array}$

To eliminate y, multiply equation (1) by 2 and equation (2) by -3. Then add the results.

$$\begin{array}{rcll} -14x + 6y &=& 24 & 2 \times (1) \\ -15x - 6y &=& -24 & -3 \times (2) \\ \hline -29x &=& 0 & \\ x &=& 0 & \end{array}$$

continued

Substitute 0 for x in equation (1) to find y.

$$-7x + 3y = 12 \quad (1)$$
$$-7(0) + 3y = 12$$
$$3y = 12$$
$$y = 4$$

The solution set is $\{(0, 4)\}$.

36. **[4.1]** $2x - 5y = 8 \quad (1)$
$ 3x + 4y = 10 \quad (2)$

To eliminate y, multiply equation (1) by 4 and equation (2) by 5 and add the results.

$$8x - 20y = 32 \quad 4 \times (1)$$
$$\underline{15x + 20y = 50 \quad 5 \times (2)}$$
$$23x = 82$$
$$x = \tfrac{82}{23}$$

Instead of substituting to find y, we'll choose different multipliers and eliminate x from the original system.

$$6x - 15y = 24 \quad 3 \times (1)$$
$$\underline{-6x - 8y = -20 \quad -2 \times (2)}$$
$$-23y = 4$$
$$y = -\tfrac{4}{23}$$

The solution set is $\left\{\left(\tfrac{82}{23}, -\tfrac{4}{23}\right)\right\}$.

37. **[4.3]** Let $x =$ the number of liters of 5% solution and $y =$ the number of liters of 10% solution.

Liters of Solution	Percent (as a decimal)	Liters of Pure Acid
x	0.05	$0.05x$
10	0.20	$0.20(10) = 2$
y	0.10	$0.10y$

Solve the system formed from the first and third columns.

$$x + 10 = y \quad (1)$$
$$0.05x + 2 = 0.10y \quad (2)$$

Multiply equation (2) by 100 to clear the decimals.

$$5x + 200 = 10y \quad (3)$$

Substitute $x + 10$ for y in equation (3) and solve for x.

$$5x + 200 = 10y \quad (3)$$
$$5x + 200 = 10(x + 10)$$
$$5x + 200 = 10x + 100$$
$$100 = 5x$$
$$20 = x$$

He should use 20 L of 5% solution.

38. **[4.3]** Let x, y, and z denote the number of medals won by Germany, the United States, and Canada, respectively.

The total number of medals won was 78, so

$$x + y + z = 78. \quad (1)$$

Germany won four more medals than the United States, so

$$x = y + 4. \quad (2)$$

Canada won one fewer medal than the United States, so

$$z = y - 1. \quad (3)$$

Substitute $y + 4$ for x and $y - 1$ for z in (1).

$$(y + 4) + y + (y - 1) = 78$$
$$3y + 3 = 78$$
$$3y = 75$$
$$y = 25$$

From (2), $x = 25 + 4 = 29$.
From (3), $z = 25 - 1 = 24$.

Germany won 29 medals, the United States won 25 medals, and Canada won 24 medals.

Chapter 4 Test

1. **(a)** Rising graphs indicate population growth, so Houston, Phoenix, and Dallas will experience population growth.

(b) Philadelphia's graph indicates that it will experience population decline.

(c) In the year 2000, the city populations from least to greatest are Dallas, Phoenix, Philadelphia, and Houston.

2. **(a)** The graphs for Dallas and Philadelphia intersect in the year 2010. The population for each city will be about 1.45 million.

(b) The graphs for Houston and Phoenix appear to intersect in the year 2025. The population for each city will be about 2.8 million. This can be represented by the ordered pair $(2025, 2.8)$.

3. When each equation of the system

$$x + y = 7$$
$$x - y = 5$$

is graphed, the point of intersection appears to be $(6, 1)$. To check, substitute 6 for x and 1 for y in each of the equations. Since $(6, 1)$ makes both equations true, the solution set of the system is $\{(6, 1)\}$.

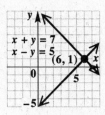

4.
$$2x - 3y = 24 \quad (1)$$
$$y = -\tfrac{2}{3}x \quad (2)$$

Since equation (2) is solved for y, substitute $-\tfrac{2}{3}x$ for y in equation (1) and solve for x.

$$2x - 3y = 24 \quad (1)$$
$$2x - 3\left(-\tfrac{2}{3}x\right) = 24$$
$$2x + 2x = 24$$
$$4x = 24$$
$$x = 6$$

From (2), $y = -\tfrac{2}{3}(6) = -4$.

The solution set is $\{(6, -4)\}$.

5.
$$3x - y = -8 \quad (1)$$
$$2x + 6y = 3 \quad (2)$$

To eliminate x, multiply equation (1) by 6. Then add that equation and equation (2).

$$\begin{array}{rl} 18x - 6y = -48 & 6 \times (1) \\ 2x + 6y = 3 & (2) \\ \hline 20x = -45 & \end{array}$$
$$x = -\tfrac{45}{20} = -\tfrac{9}{4}$$

To find y, substitute $-\tfrac{9}{4}$ for x in equation (2). (We could also use elimination by adding $2 \times (1)$ and $-3 \times (2)$.)

$$2x + 6y = 3 \quad (2)$$
$$2\left(-\tfrac{9}{4}\right) + 6y = 3$$
$$-\tfrac{9}{2} + 6y = 3$$
$$6y = \tfrac{15}{2}$$
$$y = \tfrac{15}{12} = \tfrac{5}{4}$$

The solution set is $\left\{-\tfrac{9}{4}, \tfrac{5}{4}\right\}$.

6.
$$12x - 5y = 8 \quad (1)$$
$$3x = \tfrac{5}{4}y + 2$$
$$\text{or} \quad x = \tfrac{5}{12}y + \tfrac{2}{3} \quad (2)$$

Substitute $\tfrac{5}{12}y + \tfrac{2}{3}$ for x in equation (1) and solve for y.

$$12x - 5y = 8 \quad (1)$$
$$12\left(\tfrac{5}{12}y + \tfrac{2}{3}\right) - 5y = 8$$
$$5y + 8 - 5y = 8$$
$$8 = 8 \quad True$$

Equations (1) and (2) are dependent.
The solution set is $\{(x, y) \mid 12x - 5y = 8\}$.

7.
$$3x + y = 12 \quad (1)$$
$$2x - y = 3 \quad (2)$$

To eliminate y, add equations (1) and (2).

$$\begin{array}{rl} 3x + y = 12 & (1) \\ 2x - y = 3 & (2) \\ \hline 5x = 15 & \end{array}$$
$$x = 3$$

Substitute 3 for x in equation (1) to find y.

$$3x + y = 12 \quad (1)$$
$$3(3) + y = 12$$
$$9 + y = 12$$
$$y = 3$$

The solution set is $\{(3, 3)\}$.

8.
$$-5x + 2y = -4 \quad (1)$$
$$6x + 3y = -6 \quad (2)$$

To eliminate x, multiply equation (1) by 6 and equation (2) by 5. Then add the results.

$$\begin{array}{rl} -30x + 12y = -24 & 6 \times (1) \\ 30x + 15y = -30 & 5 \times (2) \\ \hline 27y = -54 & \end{array}$$
$$y = -2$$

Substitute -2 for y in equation (1) to find x.

$$-5x + 2y = -4 \quad (1)$$
$$-5x + 2(-2) = -4$$
$$-5x - 4 = -4$$
$$-5x = 0$$
$$x = 0$$

The solution set is $\{(0, -2)\}$.

9.
$$3x + 4y = 8 \quad (1)$$
$$8y = 7 - 6x$$
$$\text{or} \quad 6x + 8y = 7 \quad (2)$$

Multiply equation (1) by -2 and add the result to equation (2).

$$\begin{array}{rl} -6x - 8y = -16 & -2 \times (1) \\ 6x + 8y = 7 & (2) \\ \hline 0 = -9 & False \end{array}$$

Since a false statement results, the system is *inconsistent*. The solution set is $\emptyset$.

10.
$$3x + 5y + 3z = 2 \quad (1)$$
$$6x + 5y + z = 0 \quad (2)$$
$$3x + 10y - 2z = 6 \quad (3)$$

To eliminate x, multiply equation (1) by -1 and add the result to equation (3).

$$\begin{array}{rl} -3x - 5y - 3z = -2 & -1 \times (1) \\ 3x + 10y - 2z = 6 & (3) \\ \hline 5y - 5z = 4 & (4) \end{array}$$

continued

To eliminate x again, multiply equation (1) by -2 and add the result to equation (2).

$$\begin{array}{rcl}
-6x - 10y - 6z = -4 & \quad -2 \times (1) \\
\underline{6x + 5y + z = 0} & \quad (2) \\
-5y - 5z = -4 & \quad (5)
\end{array}$$

To eliminate y, add equations (4) and (5).

$$\begin{array}{rcl}
5y - 5z = 4 & (4) \\
\underline{-5y - 5z = -4} & (5) \\
-10z = 0 \\
z = 0
\end{array}$$

Substitute 0 for z in equation (4) to find y.

$$\begin{aligned}
5y - 5z &= 4 \quad (4) \\
5y - 5(0) &= 4 \\
5y - 0 &= 4 \\
5y &= 4 \\
y &= \tfrac{4}{5}
\end{aligned}$$

Substitute $\tfrac{4}{5}$ for y and 0 for z in equation (1) to find x.

$$\begin{aligned}
3x + 5y + 3z &= 2 \quad (1) \\
3x + 5\left(\tfrac{4}{5}\right) + 3(0) &= 2 \\
3x + 4 + 0 &= 2 \\
3x &= -2 \\
x &= -\tfrac{2}{3}
\end{aligned}$$

The solution set is $\left\{ \left(-\tfrac{2}{3}, \tfrac{4}{5}, 0 \right) \right\}$.

11.
$$\begin{array}{rcl}
4x + y + z = 11 & (1) \\
x - y - z = 4 & (2) \\
y + 2z = 0 & (3)
\end{array}$$

To eliminate x, multiply equation (2) by -4 and add the result to equation (1).

$$\begin{array}{rcl}
4x + y + z = 11 & (1) \\
\underline{-4x + 4y + 4z = -16} & \quad -4 \times (2) \\
5y + 5z = -5 & (4)
\end{array}$$

To eliminate y, divide equation (4) by -5 and add the result to equation (3).

$$\begin{array}{rcl}
-y - z = 1 & (4) \div (-5) \\
\underline{y + 2z = 0} & (3) \\
z = 1
\end{array}$$

From (3), $y + 2(1) = 0$, so $y = -2$.
From (2), $x - (-2) - 1 = 4$, so $x = 3$.

The solution set is $\{(3, -2, 1)\}$.

12. Let $x =$ the gross (in millions of dollars) for *Ocean's Eleven*, and $y =$ the gross (in millions of dollars) for *Runaway Bride*.

Together the movies grossed \$335.5 million, so

$$x + y = 335.5. \quad (1)$$

Runaway Bride grossed \$31.3 million less than *Ocean's Eleven*, so

$$y = x - 31.3. \quad (2)$$

Substitute $x - 31.3$ for y in equation (1).

$$\begin{aligned}
x + y &= 335.5 \quad (1) \\
x + (x - 31.3) &= 335.5 \\
2x &= 366.8 \\
x &= 183.4
\end{aligned}$$

From (2), $y = 183.4 - 31.3 = 152.1$.

Ocean's Eleven grossed \$183.4 million and *Runaway Bride* grossed \$152.1 million.

13. Let $x =$ the speed of the faster car and $y =$ the speed of the slower car.

Make a table.

	r	t	d
Faster Car	x	3.5	$3.5x$
Slower Car	y	3.5	$3.5y$

Since the slow car travels 30 mph slower than the fast car,

$$x - y = 30. \quad (1)$$

Since the cars travel a total of 420 miles,

$$3.5x + 3.5y = 420.$$

Multiply by 10 to clear the decimals.

$$35x + 35y = 4200 \quad (2)$$

To eliminate y, multiply equation (1) by 35 and add the result to equation (2).

$$\begin{array}{rcl}
35x - 35y = 1050 & \quad 35 \times (1) \\
\underline{35x + 35y = 4200} & \quad (2) \\
70x = 5250 \\
x = 75
\end{array}$$

Substitute 75 for x in equation (1) to find y.

$$\begin{aligned}
x - y &= 30 \quad (1) \\
75 - y &= 30 \\
-y &= -45 \\
y &= 45
\end{aligned}$$

The faster car is traveling at 75 mph, and the slower car is traveling at 45 mph.

14. Let x = the number of liters of 20% solution and y = the number of liters of 50% solution.

Make a table.

Liters of Solution	Percent (as a decimal)	Liters of Pure Alcohol
x	0.20	$0.20x$
y	0.50	$0.50y$
12	0.40	$0.40(12) = 4.8$

Since 12 L of the mixture are needed,

$$x + y = 12. \quad (1)$$

Since the amount of pure alcohol in the 20% solution plus the amount of pure alcohol in the 50% solution must equal the amount of alcohol in the mixture,

$$0.2x + 0.5y = 4.8.$$

Multiply by 10 to clear the decimals.

$$2x + 5y = 48 \quad (2)$$

Multiply equation (1) by -2 and add the result to equation (2).

$$\begin{array}{rl} -2x - 2y = -24 & -2 \times (1) \\ 2x + 5y = 48 & (2) \\ \hline 3y = 24 & \\ y = 8 & \end{array}$$

From (1), $x + 8 = 12$, so $x = 4$.
4 L of 20% solution and 8 L of 50% solution are needed.

15. Let x = the price of an AC adaptor and y = the price of a rechargeable flashlight.

Since 7 AC adaptors and 2 rechargeable flashlights cost $86,

$$7x + 2y = 86. \quad (1)$$

Since 3 AC adaptors and 4 rechargeable flashlights cost $84,

$$3x + 4y = 84. \quad (2)$$

Solve the system.

$$\begin{array}{rl} 7x + 2y = 86 & (1) \\ 3x + 4y = 84 & (2) \end{array}$$

To eliminate y, multiply equation (1) by -2 and add the result to equation (2).

$$\begin{array}{rl} -14x - 4y = -172 & -2 \times (1) \\ 3x + 4y = 84 & (2) \\ \hline -11x = -88 & \\ x = 8 & \end{array}$$

Substitute 8 for x in equation (1) to find y.

$$\begin{array}{l} 7x + 2y = 86 \quad (1) \\ 7(8) + 2y = 86 \\ 56 + 2y = 86 \\ 2y = 30 \\ y = 15 \end{array}$$

An AC adaptor costs $8, and a rechargeable flashlight costs $15.

16. Let x = the amount of Orange Pekoe, y = the amount of Irish Breakfast, and z = the amount of Earl Grey.

The owner wants 100 oz of tea, so

$$x + y + z = 100. \quad (1)$$

An equation which relates the prices of the tea is

$$0.80x + 0.85y + 0.95z = 0.83(100).$$

Multiply by 100 to clear the decimals.

$$80x + 85y + 95z = 8300 \quad (2)$$

The mixture must contain twice as much Orange Pekoe as Irish Breakfast, so

$$x = 2y. \quad (3)$$

To eliminate z, multiply equation (1) by -95 and add the result to equation (2).

$$\begin{array}{rl} -95x - 95y - 95z = -9500 & -95 \times (1) \\ 80x + 85y + 95z = 8300 & (2) \\ \hline -15x - 10y = -1200 & (4) \end{array}$$

Divide equation (4) by -5.

$$3x + 2y = 240 \quad (5)$$

Substitute $2y$ for x in equation (5) to find y.

$$\begin{array}{l} 3x + 2y = 240 \quad (5) \\ 3(2y) + 2y = 240 \\ 8y = 240 \\ y = 30 \end{array}$$

From (3), $x = 2(30) = 60$.
Substitute 60 for x and 30 for y in equation (1) to find z.

$$\begin{array}{l} x + y + z = 100 \quad (1) \\ 60 + 30 + z = 100 \\ z = 10 \end{array}$$

He should use 60 oz of Orange Pekoe, 30 oz of Irish Breakfast, and 10 oz of Earl Grey.

17. $3x + 2y = 4$
      $5x + 5y = 9$

Write the augmented matrix.

$$\begin{bmatrix} 3 & 2 & | & 4 \\ 5 & 5 & | & 9 \end{bmatrix}$$

We could divide row 1 by 3, but to avoid working with fractions, we'll multiply row 1 by 2 and then multiply row 2 by -1 and add to row 1 to obtain a "1" in the first row.

$$\begin{bmatrix} 6 & 4 & | & 8 \\ 5 & 5 & | & 9 \end{bmatrix} \quad 2R_1$$

$$\begin{bmatrix} 1 & -1 & | & -1 \\ 5 & 5 & | & 9 \end{bmatrix} \quad -R_2 + R_1$$

$$\begin{bmatrix} 1 & -1 & | & -1 \\ 0 & 10 & | & 14 \end{bmatrix} \quad -5R_1 + R_2$$

$$\begin{bmatrix} 1 & -1 & | & -1 \\ 0 & 1 & | & \frac{7}{5} \end{bmatrix} \quad \frac{1}{10}R_2$$

This matrix gives the system

$$x - y = -1$$
$$y = \tfrac{7}{5}.$$

Substitute $y = \frac{7}{5}$ in the first equation.

$$x - \tfrac{7}{5} = -1$$
$$x = -1 + \tfrac{7}{5}$$
$$= -\tfrac{5}{5} + \tfrac{7}{5} = \tfrac{2}{5}$$

The solution set is $\left\{ \left(\frac{2}{5}, \frac{7}{5} \right) \right\}$.

18. $x + 3y + 2z = 11$
      $3x + 7y + 4z = 23$
      $5x + 3y - 5z = -14$

Write the augmented matrix.

$$\begin{bmatrix} 1 & 3 & 2 & | & 11 \\ 3 & 7 & 4 & | & 23 \\ 5 & 3 & -5 & | & -14 \end{bmatrix}$$

$$\begin{bmatrix} 1 & 3 & 2 & | & 11 \\ 0 & -2 & -2 & | & -10 \\ 0 & -12 & -15 & | & -69 \end{bmatrix} \quad \begin{matrix} -3R_1 + R_2 \\ -5R_1 + R_3 \end{matrix}$$

$$\begin{bmatrix} 1 & 3 & 2 & | & 11 \\ 0 & 1 & 1 & | & 5 \\ 0 & -12 & -15 & | & -69 \end{bmatrix} \quad -\tfrac{1}{2}R_2$$

$$\begin{bmatrix} 1 & 3 & 2 & | & 11 \\ 0 & 1 & 1 & | & 5 \\ 0 & 0 & -3 & | & -9 \end{bmatrix} \quad 12R_2 + R_3$$

$$\begin{bmatrix} 1 & 3 & 2 & | & 11 \\ 0 & 1 & 1 & | & 5 \\ 0 & 0 & 1 & | & 3 \end{bmatrix} \quad -\tfrac{1}{3}R_3$$

This matrix gives the system

$$x + 3y + 2z = 11$$
$$y + z = 5$$
$$z = 3.$$

Substitute $z = 3$ in the second equation.

$$y + z = 5$$
$$y + 3 = 5$$
$$y = 2$$

Substitute $y = 2$ and $z = 3$ in the first equation.

$$x + 3y + 2z = 11$$
$$x + 3(2) + 2(3) = 11$$
$$x + 6 + 6 = 11$$
$$x = -1$$

The solution set is $\{(-1, 2, 3)\}$.

Cumulative Review Exercises (Chapters 1–4)

1. $-\dfrac{3}{4} - \dfrac{2}{5} = -\dfrac{15}{20} - \dfrac{8}{20}$

   $= \dfrac{-15 - 8}{20} = \dfrac{-23}{20} = -\dfrac{23}{20}$

2. $\dfrac{8}{15} \div \left(-\dfrac{12}{5} \right) = \dfrac{8}{15} \left(-\dfrac{5}{12} \right)$

   $= -\dfrac{2 \cdot 4 \cdot 5}{3 \cdot 5 \cdot 3 \cdot 4}$

   $= -\dfrac{2}{3 \cdot 3} = -\dfrac{2}{9}$

3. $(-3)^4 = (-3)(-3)(-3)(-3) = 81$

4. $-3^4 = -(3)(3)(3)(3) = -81$

5. $-(-3)^4 = -(-3)(-3)(-3)(-3) = -81$

6. $\sqrt{0.49} = 0.7$, since 0.7 is positive and $(0.7)^2 = 0.49$.

7. $-\sqrt{0.49} = -0.7$, since $(0.7)^2 = 0.49$ and the negative sign in front of the radical must be applied.

8. $\sqrt{-0.49}$ is not a real number because of the negative sign under the radical. No real number squared is negative.

In Exercises 9–10, let $x = -4$, $y = 3$, and $z = 6$.

9. $|2x| + 3y - z^3$

   $= |(2)(-4)| + 3(3) - (6)^3$

   $= |-8| + 9 - 216$

   $= 8 + 9 - 216$

   $= -199$

10. $-5(x^3 - y^3) = -5[(-4)^3 - (3)^3]$
$$= -5(-64 - 27)$$
$$= -5(-91)$$
$$= 455$$

11. The *commutative property* says that $3 \cdot 6 = 6 \cdot 3$, so that is the property that justifies the given statement.

12. $7(2x + 3) - 4(2x + 1) = 2(x + 1)$
$$14x + 21 - 8x - 4 = 2x + 2$$
$$6x + 17 = 2x + 2$$
$$4x = -15$$
$$x = -\frac{15}{4}$$
The solution set is $\left\{-\frac{15}{4}\right\}$.

13. $|6x - 8| = 4$
$6x - 8 = 4$ or $6x - 8 = -4$
$6x = 12$ $\qquad$ $6x = 4$
$x = 2$ or $\qquad$ $x = \frac{4}{6} = \frac{2}{3}$
The solution set is $\left\{\frac{2}{3}, 2\right\}$.

14. $\qquad ax + by = cx + d$
To solve for x, get all terms with x alone on one side of the equals sign.

$$ax - cx = d - by$$
$$x(a - c) = d - by$$
$$x = \frac{d - by}{a - c}$$
or $\qquad x = \frac{by - d}{c - a}$

if the x-terms were put on the right side of the equals sign.

15. $0.04x + 0.06(x - 1) = 1.04$
Multiply both sides by 100 to clear the decimals.
$$4x + 6(x - 1) = 104$$
$$4x + 6x - 6 = 104$$
$$10x - 6 = 104$$
$$10x = 110$$
$$x = 11$$
The solution set is $\{11\}$.

16. $\frac{2}{3}x + \frac{5}{12}x \leq 20$
Multiply both sides by 12.
$$12\left(\frac{2}{3}x + \frac{5}{12}x\right) \leq 12(20)$$
$$8x + 5x \leq 240$$
$$13x \leq 240$$
$$x \leq \frac{240}{13}$$
The solution set is $\left(-\infty, \frac{240}{13}\right]$.

17. $|3x + 2| \leq 4$
$-4 \leq 3x + 2 \leq 4$
$-6 \leq 3x \leq 2$ $\qquad$ *Subtract 2.*
$-2 \leq x \leq \frac{2}{3}$ $\qquad$ *Divide by 3.*
The solution set is $\left[-2, \frac{2}{3}\right]$.

18. $|12t + 7| \geq 0$
The solution set is $(-\infty, \infty)$ since the absolute value of any number is greater than or equal to 0.

19. 80.4% of 2500 is $0.804(2500) = 2010$
72.5% of 2500 is $0.725(2500) = 1812.5 \approx 1813$

$$\frac{1570}{2500} = 0.628 \text{ or } 62.8\%$$

$$\frac{1430}{2500} = 0.572 \text{ or } 57.2\%$$

Product or Company	Percent	Actual Number
Charmin	80.4%	2010
Wheaties	72.5%	1813
Budweiser	62.8%	1570
State Farm	57.2%	1430

20. Let $h =$ the height of the triangle.
Use the formula $A = \frac{1}{2}bh$. Here, $A = 42$ and $b = 14$, so substitute these values in the formula and solve for h.

$$A = \tfrac{1}{2}bh$$
$$42 = \tfrac{1}{2}(14)h$$
$$42 = 7h$$
$$6 = h$$

The height is 6 m.

21. Let $x =$ the number of nickels,
$x + 1 =$ the number of dimes, and
$x + 6 =$ the number of pennies.

The total value is \$4.80, so

$$0.05x + 0.10(x + 1) + 0.01(x + 6) = 4.80.$$

Multiply both sides by 100 to clear the decimals.

$$5x + 10(x + 1) + 1(x + 6) = 480$$
$$5x + 10x + 10 + x + 6 = 480$$
$$16x + 16 = 480$$
$$16x = 464$$
$$x = 29$$

Then, $x + 1 = 29 + 1 = 30$,
and $x + 6 = 29 + 6 = 35$.
There are 35 pennies, 29 nickels, and 30 dimes.

22. Let $x =$ the measure of the equal angles and $2x - 4 =$ the measure of the third angle.

The sum of the measures of the angles in a triangle is 180°, so

$$x + x + (2x - 4) = 180$$
$$4x - 4 = 180$$
$$4x = 184$$
$$x = 46.$$

So, $2x - 4 = 2(46) - 4 = 92 - 4 = 88.$
The measures of the angles are 46°, 46°, and 88°.

23. A horizontal line through the point (x, k) has equation $y = k$. Since point A has coordinates $(-2, 6)$, $k = 6$. The equation of the horizontal line through A is $y = 6$.

24. A vertical line through the point (k, y) has equation $x = k$. Since point B has coordinates $(4, -2)$, $k = 4$. The equation of the vertical line through B is $x = 4$.

25. Let $(x_1, y_1) = (-2, 6)$ and $(x_2, y_2) = (4, -2)$. Then,

$$m = \frac{y_2 - y_1}{x_2 - x_1} = \frac{-2 - 6}{4 - (-2)} = \frac{-8}{6} = -\frac{4}{3}.$$

The slope is $-\frac{4}{3}$.

26. Perpendicular lines have slopes that are negative reciprocals of each other. The slope of line AB is $-\frac{4}{3}$ (from Exercise 25). The negative reciprocal of $-\frac{4}{3}$ is $\frac{3}{4}$, so the slope of a line perpendicular to line AB is $\frac{3}{4}$.

27. Let $m = -\frac{4}{3}$ and $(x_1, y_1) = (4, -2)$ in the point-slope form.

$$y - y_1 = m(x - x_1)$$
$$y - (-2) = -\frac{4}{3}(x - 4)$$
$$y + 2 = -\frac{4}{3}x + \frac{16}{3}$$

Multiply by 3 to clear the fractions, and then write the equation in standard form, $Ax + By = C$.

$$3y + 6 = -4x + 16$$
$$4x + 3y = 10$$

28. First locate the point $(-1, -3)$ on a graph. Then use the definition of slope to find a second point on the line.

$$m = \frac{\text{change in } y}{\text{change in } x} = \frac{2}{3}$$

From $(-1, -3)$, move 2 units up and 3 units to the right. The line through $(-1, -3)$ and the new point, $(2, -1)$, is the graph.

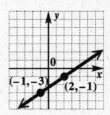

29. $-3x - 2y \leq 6$

Graph the line $-3x - 2y = 6$ through its intercepts, $(-2, 0)$ and $(0, -3)$, as a solid line, since the inequality involves $\leq$.
To determine the region that belongs to the graph, test $(0, 0)$.

$$-3x - 2y \leq 6$$
$$-3(0) - 2(0) \leq 6$$
$$0 \leq 6 \quad \textit{True}$$

Since the result is true, shade the region that includes $(0, 0)$.

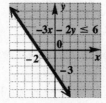

30. $f(x) = x^2 + 3x - 6$

   (a) $f(-3) = (-3)^2 + 3(-3) - 6$
   $$= 9 - 9 - 6 = -6$$

   (b) $f(a) = (a)^2 + 3(a) - 6$
   $$= a^2 + 3a - 6$$

31.
$$-2x + 3y = -15 \quad (1)$$
$$4x - y = 15 \quad (2)$$

To eliminate x, multiply equation (1) by 2 and add the result to equation (2).

$$
\begin{array}{rcll}
-4x + 6y &=& -30 & 2 \times (1) \\
4x - y &=& 15 & (2) \\
\hline
5y &=& -15 & \\
y &=& -3 &
\end{array}
$$

Substitute -3 for y in equation (2) to find x.

$$4x - y = 15 \quad (2)$$
$$4x - (-3) = 15$$
$$4x + 3 = 15$$
$$4x = 12$$
$$x = 3$$

The solution set is $\{(3, -3)\}$.

32.
$$x - 3y = 7 \quad (1)$$
$$2x - 6y = 14 \quad (2)$$

Multiplying equation (1) by 2 gives equation (2). Since those equations are the same, the two equations are dependent.

The solution set is $\{(x, y) \mid x - 3y = 7\}$.

33.
$$x + y + z = 10 \quad (1)$$
$$x - y - z = 0 \quad (2)$$
$$-x + y - z = -4 \quad (3)$$

Add equations (1) and (2) to eliminate y and z. The result is

$$2x = 10$$
$$x = 5.$$

Add equations (2) and (3) to eliminate x and y. The result is

$$-2z = -4$$
$$z = 2.$$

Substitute 5 for x and 2 for z in equation (1) to find y.

$$x + y + z = 10 \quad (1)$$
$$5 + y + 2 = 10$$
$$y + 7 = 10$$
$$y = 3$$

The solution set is $\{(5, 3, 2)\}$.

34. Let $x =$ the number of pounds of oranges
and $y =$ the number of pounds of apples.

Since she bought 6 lb of fruit,

$$x + y = 6. \quad (1)$$

The total cost of x lb of oranges at \$.90/lb and y lb of apples at \$.70/lb is \$5.20, so

$$0.90x + 0.70y = 5.20.$$

Multiply by 10 to clear the decimals.

$$9x + 7y = 52 \quad (2)$$

To solve the system, solve equation (1) for x.

$$x + y = 6 \quad (1)$$
$$x = 6 - y$$

Substitute $6 - y$ for x in equation (2) and solve for y.

$$9x + 7y = 52 \quad (2)$$
$$9(6 - y) + 7y = 52$$
$$54 - 9y + 7y = 52$$
$$-2y = -2$$
$$y = 1$$

Since $x = 6 - y$ and $y = 1$,

$$x = 6 - 1 = 5.$$

She bought 5 lb of oranges and 1 lb of apples.

35. Let $x =$ the average cost of the original Tickle Me Elmo
and $y =$ the recommended cost of T.M.X.

The original's cost was \$12.37 less than T.M.X.'s, so

$$x = y - 12.37. \quad (1)$$

Together they cost \$67.63, so

$$x + y = 67.63. \quad (2)$$

From (1), substitute $y - 12.37$ for x in equation (2).

$$(y - 12.37) + y = 67.63$$
$$2y - 12.37 = 67.63$$
$$2y = 80$$
$$y = 40$$

From (1), $x = y - 12.37 = 40 - 12.37 = 27.63$. The average cost of the original Elmo was \$27.63 and the recommended cost of a T.M.X. is \$40.

36. Let $x =$ the cost of a small box
and $y =$ the cost of a large box.

The cost of 10 small and 20 large boxes is \$65, so

$$10x + 20y = 65. \quad (1)$$

The cost of 6 small and 10 large boxes is \$34, so

$$6x + 10y = 34. \quad (2)$$

To eliminate y, multiply equation (2) by -2 and add the result to equation (1).

$$\begin{array}{rcll} 10x + 20y & = & 65 & (1) \\ -12x - 20y & = & -68 & -2 \times (2) \\ \hline -2x & = & -3 & \\ x & = & \frac{3}{2} & \text{or} \quad 1.5 \end{array}$$

Substitute $\frac{3}{2}$ for x in equation (2) to find y.

$$6x + 10y = 34 \qquad (2)$$
$$6\left(\tfrac{3}{2}\right) + 10y = 34$$
$$9 + 10y = 34$$
$$10y = 25$$
$$y = \tfrac{25}{10} \quad \text{or} \quad \tfrac{5}{2} \quad \text{or} \quad 2.5$$

A small box costs \$1.50 and a large box costs \$2.50.

37. **(a)** Let $x = $ the cost of a pound of peanuts
and $y = $ the cost of a pound of cashews.

The cost of 6 lb of peanuts and 12 lb of cashews is $60, so

$$6x + 12y = 60. \quad (1)$$

The cost of 3 lb of peanuts and 4 lb of cashews is $22, so

$$3x + 4y = 22. \quad (2)$$

To eliminate x, multiply equation (2) by -2 and add the result to equation (1).

$$
\begin{array}{rcll}
6x + 12y &=& 60 & (1) \\
-6x - 8y &=& -44 & -2 \times (2) \\
\hline
4y &=& 16 & \\
y &=& 4 &
\end{array}
$$

Substitute 4 for y in equation (2) to find x.

$$
\begin{aligned}
3x + 4y &= 22 \quad (2) \\
3x + 4(4) &= 22 \\
3x + 16 &= 22 \\
3x &= 6 \\
x &= 2
\end{aligned}
$$

Peanuts cost $2/lb and cashews cost $4/lb.

38. **(a)** The lines intersect at $(8, 3000)$, so the cost equals the revenue at $x = 8$ (which is 800 items). The revenue is $3000.

(b) On the sale of 1100 parts $(x = 11)$, the revenue is about $4100 and the cost is about $3700.

$$
\begin{aligned}
\text{Profit} &= \text{Revenue} - \text{Cost} \\
&\approx 4100 - 3700 \\
&= 400
\end{aligned}
$$

The profit is about $400.

CHAPTER 5 EXPONENTS, POLYNOMIALS, AND POLYNOMIAL FUNCTIONS

5.1 Integer Exponents and Scientific Notation

1. $(ab)^2 = a^2b^2$ by a power rule. Since $a^2b^2 \neq ab^2$, the expression $(ab)^2 = ab^2$ has been simplified incorrectly. The exponent should apply to both a and b.

3. $\left(\dfrac{4}{a}\right)^3 = \dfrac{4^3}{a^3}$

Since $\dfrac{4^3}{a^3} \neq \dfrac{4^3}{a}$, the expression

$$\left(\dfrac{4}{a}\right)^3 = \dfrac{4^3}{a}$$

has been simplified incorrectly.

5. The product rule says that when exponential expressions with like bases are multiplied, the base stays the same and the exponents are added. For example, $x^5 \cdot x^6 = x^{11}$.

7. $13^4 \cdot 13^8 = 13^{4+8} = 13^{12}$

9. $x^3 \cdot x^5 \cdot x^9 = x^{3+5+9} = x^{17}$

11. $(-3w^5)(9w^3) = (-3)(9)w^{5+3} = -27w^8$

13. $(2x^2y^5)(9xy^3) = (2)(9)x^{2+1}y^{5+3} = 18x^3y^8$

15. $r^2 \cdot s^4$ cannot be simplified because the product rule does not apply.

17. (a) $9^0 = 1$ **(B)**

(b) $-9^0 = -(9^0) = -(1) = -1$ **(C)**

(c) $(-9)^0 = 1$ **(B)**

(d) $-(-9)^0 = -1$ **(C)**

19. $25^0 = 1$, since $a^0 = 1$ for any nonzero base a.

21. $-7^0 = -(7^0) = -(1) = -1$

23. $(-15)^0 = 1$ since -15 is in parentheses.

25. $3^0 + (-3)^0 = 1 + 1 = 2$

27. $-3^0 + 3^0 = -1 + 1 = 0$

29. $-4^0 - m^0 = -1 - 1 = -2$

31. (a) $4^{-2} = \dfrac{1}{4^2} = \dfrac{1}{16}$ **(B)**

(b) $-4^{-2} = -\dfrac{1}{4^2} = -\dfrac{1}{16}$ **(D)**

(c) $(-4)^{-2} = \dfrac{1}{(-4)^2} = \dfrac{1}{16}$ **(B)**

(d) $-(-4)^{-2} = -\dfrac{1}{(-4)^2} = -\dfrac{1}{16}$ **(D)**

33. $5^{-4} = \dfrac{1}{5^4}$, or $\dfrac{1}{625}$

35. $8^{-1} = \dfrac{1}{8^1} = \dfrac{1}{8}$

37. $(4x)^{-2} = \dfrac{1}{(4x)^2} = \dfrac{1}{4^2x^2} = \dfrac{1}{16x^2}$

39. $4x^{-2} = \dfrac{4}{x^2}$

41. $-a^{-3} = -\dfrac{1}{a^3}$

43. $(-a)^{-4} = \dfrac{1}{(-a)^4} = \dfrac{1}{a^4}$

45. $5^{-1} + 6^{-1} = \dfrac{1}{5} + \dfrac{1}{6} = \dfrac{6}{30} + \dfrac{5}{30} = \dfrac{11}{30}$

47. $8^{-1} - 3^{-1} = \dfrac{1}{8} - \dfrac{1}{3} = \dfrac{3}{24} - \dfrac{8}{24} = -\dfrac{5}{24}$

49. Consider $-a^n$ and $(-a)^n$. Let $a = 2$.

If $n = 2$, then
$-a^n = -2^2 = -4$ and $(-a)^n = (-2)^2 = 4$.
If $n = 3$, then
$-a^n = -2^3 = -8$ and $(-a)^n = (-2)^3 = -8$.
If $n = 4$, then
$-a^n = -2^4 = -16$ and $(-a)^n = (-2)^4 = 16$.
If $n = 5$, then
$-a^n = -2^5 = -32$ and $(-a)^n = (-2)^5 = -32$.
If $n = 6$, then
$-a^n = -2^6 = -64$ and $(-a)^n = (-2)^6 = 64$.

Based on these cases, when n is even, the expressions are opposites. When n is odd, the expressions are equal.

51. $\dfrac{1}{4^{-2}} = 4^2 = 16$

53. $\dfrac{2^{-2}}{3^{-3}} = \dfrac{3^3}{2^2} = \dfrac{27}{4}$

55. $\left(\dfrac{2}{3}\right)^{-3} = \left(\dfrac{3}{2}\right)^3 = \dfrac{3^3}{2^3} = \dfrac{27}{8}$

57. $\left(\dfrac{4}{5}\right)^{-2} = \left(\dfrac{5}{4}\right)^2 = \dfrac{5^2}{4^2} = \dfrac{25}{16}$

59. (a) $\left(\dfrac{1}{3}\right)^{-1} = \left(\dfrac{3}{1}\right)^1 = 3$ **(B)**

(b) $\left(-\dfrac{1}{3}\right)^{-1} = \left(-\dfrac{3}{1}\right)^{1} = -3$ **(D)**

(c) $-\left(\dfrac{1}{3}\right)^{-1} = -\left(\dfrac{3}{1}\right)^{1} = -3$ **(D)**

(d) $-\left(-\dfrac{1}{3}\right)^{-1} = -\left(-\dfrac{3}{1}\right)^{1} = -(-3) = 3$ **(B)**

61. The quotient rule says that when exponential expressions with like bases are divided, the base stays the same and the exponents are subtracted.

For example, $\dfrac{x^8}{x^5} = x^3$.

63. $\dfrac{4^8}{4^6} = 4^{8-6} = 4^2, \ \text{or} \ 16$

65. $\dfrac{x^{12}}{x^8} = x^{12-8} = x^4$

67. $\dfrac{r^7}{r^{10}} = r^{7-10} = r^{-3} = \dfrac{1}{r^3}$

69. $\dfrac{6^4}{6^{-2}} = 6^{4-(-2)} = 6^{4+2} = 6^6$

71. $\dfrac{6^{-3}}{6^7} = 6^{-3-7} = 6^{-10} = \dfrac{1}{6^{10}}$

73. $\dfrac{7}{7^{-1}} = 7^{1-(-1)} = 7^2, \ \text{or} \ 49$

75. $\dfrac{r^{-3}}{r^{-6}} = r^{-3-(-6)} = r^{-3+6} = r^3$

77. $\dfrac{x^3}{y^2}$ cannot be simplified because the quotient rule does not apply.

79. $(x^3)^6 = x^{3 \cdot 6} = x^{18}$

81. $\left(\dfrac{3}{5}\right)^3 = \dfrac{3^3}{5^3} = \dfrac{27}{125}$

83. $(4t)^3 = 4^3 t^3 = 64t^3$

85. $(-6x^2)^3 = (-6)^3 x^{2 \cdot 3} = -216x^6$

87. $\left(\dfrac{-4m^2}{t}\right)^3 = \dfrac{(-4)^3 m^{2 \cdot 3}}{t^3} = \dfrac{-64m^6}{t^3} = -\dfrac{64m^6}{t^3}$

89. $3^5 \cdot 3^{-6} = 3^{5+(-6)} = 3^{-1} = \dfrac{1}{3^1} = \dfrac{1}{3}$

91. $a^{-3} a^2 a^{-4} = a^{-3+2+(-4)} = a^{-5} = \dfrac{1}{a^5}$

93. $\left(k^2\right)^{-3} k^4 = k^{2(-3)} k^4$
$= k^{-6} k^4$
$= k^{-6+4}$
$= k^{-2}$
$= \dfrac{1}{k^2}$

95. $-4r^{-2}\left(r^4\right)^2 = -4r^{-2}\left(r^8\right)$
$= -4r^{-2+8}$
$= -4r^6$

97. $\left(5a^{-1}\right)^4 \left(a^2\right)^{-3} = 5^4 a^{-1 \cdot 4} a^{2(-3)}$
$= 5^4 a^{-4} a^{-6}$
$= 5^4 a^{-4-6}$
$= 5^4 a^{-10}$
$= \dfrac{5^4}{a^{10}}, \ \text{or} \ \dfrac{625}{a^{10}}$

99. $\left(z^{-4} x^3\right)^{-1} = z^{-4(-1)} x^{3(-1)}$
$= z^4 x^{-3}$
$= \dfrac{z^4}{x^3}$

101. $7k^2(-2k)\left(4k^{-5}\right)^0 = 7(-2)k^2 k \cdot 1$
$= -14k^{2+1}$
$= -14k^3$

103. $\dfrac{(p^{-2})^0}{5p^{-4}} = \dfrac{1 \cdot p^4}{5} = \dfrac{p^4}{5}$

105. $\dfrac{(3pq)q^2}{6p^2 q^4} = \dfrac{3p^1 q^{1+2}}{6p^2 q^4} = \dfrac{3pq^3}{6p^2 q^4}$
$= \dfrac{1}{2} p^{1-2} q^{3-4}$
$= \dfrac{1}{2} p^{-1} q^{-1} = \dfrac{1}{2} \cdot \dfrac{1}{p^1} \cdot \dfrac{1}{q^1}$
$= \dfrac{1}{2pq}$

107. $\dfrac{4a^5(a^{-1})^3}{(a^{-2})^{-2}} = \dfrac{4a^5 a^{-1 \cdot 3}}{a^{-2(-2)}} = \dfrac{4a^5 a^{-3}}{a^4}$
$= 4a^{5-3-4}$
$= 4a^{-2}$
$= \dfrac{4}{a^2}$

109. The first step may be the most confusing.

$$(-y^{-4})^2 = (-1 \cdot y^{-4})^2$$
$$= (-1)^2 \left(y^{-4}\right)^2$$
$$= 1 \cdot y^{-4(2)} = y^{-8}$$

It is not necessary to include these steps once the concept of squaring a negative is committed to memory.

$$\frac{(-y^{-4})^2}{6(y^{-5})^{-1}} = \frac{y^{-4(2)}}{6y^{-5(-1)}}$$

$$= \frac{y^{-8}}{6y^5}$$

$$= \frac{1}{6y^5 y^8}$$

$$= \frac{1}{6y^{13}}$$

111. $\dfrac{(2k)^2 m^{-5}}{(km)^{-3}} = \dfrac{2^2 k^2 m^{-5}}{k^{-3} m^{-3}}$

$$= 2^2 k^{2-(-3)} m^{-5-(-3)}$$

$$= 2^2 k^5 m^{-2}$$

$$= \frac{2^2 k^5}{m^2} = \frac{4k^5}{m^2}$$

113. $\dfrac{(2k)^2 k^3}{k^{-1} k^{-5}} \left(5k^{-2}\right)^{-3} = \dfrac{2^2 k^2 k^3}{k^{-1} k^{-5}} \left(5^{-3} k^{-2(-3)}\right)$

$$= \frac{2^2 k^5}{k^{-6}} \left(5^{-3} k^6\right)$$

$$= 2^2 k^{11} \left(5^{-3} k^6\right)$$

$$= 2^2 \cdot 5^{-3} k^{11+6}$$

$$= \frac{2^2 k^{17}}{5^3} = \frac{4k^{17}}{125}$$

115. $\left(\dfrac{3k^{-2}}{k^4}\right)^{-1} \cdot \dfrac{2}{k} = \dfrac{(3k^{-2})^{-1}}{(k^4)^{-1}} \cdot \dfrac{2}{k}$

$$= \frac{3^{-1} k^2 \cdot 2}{k^{-4} k^1}$$

$$= \frac{3^{-1} \cdot 2k^2}{k^{-3}}$$

$$= \frac{2k^2 k^3}{3}$$

$$= \frac{2k^5}{3}$$

117. $\left(\dfrac{2p}{q^2}\right)^3 \left(\dfrac{3p^4}{q^{-4}}\right)^{-1} = \dfrac{(2p)^3}{(q^2)^3} \cdot \dfrac{(3p^4)^{-1}}{(q^{-4})^{-1}}$

$$= \frac{2^3 p^3}{q^6} \cdot \frac{3^{-1} p^{-4}}{q^4}$$

$$= \frac{2^3 p^{3-4}}{3^1 q^{6+4}}$$

$$= \frac{8p^{-1}}{3q^{10}} = \frac{8}{3pq^{10}}$$

119. $\dfrac{2^2 y^4 (y^{-3})^{-1}}{2^5 y^{-2}} = \dfrac{y^4 y^3 y^2}{2^{5-2}}$

$$= \frac{y^{4+3+2}}{2^3}$$

$$= \frac{y^9}{8}$$

121. $\dfrac{(2m^2 p^3)^2 (4m^2 p)^{-2}}{(-3mp^4)^{-1} (2m^3 p^4)^3}$

$$= \frac{2^2 m^4 p^6 4^{-2} m^{-4} p^{-2}}{(-3)^{-1} m^{-1} p^{-4} 2^3 m^9 p^{12}} \quad \textit{Power rule}$$

$$= \frac{4(-3)^1 \left(m^{4-4}\right) \left(p^{6-2}\right)}{4^2 \cdot 2^3 (m^{-1+9})(p^{-4+12})} \quad \textit{Product rule}$$

$$= \frac{-3m^0 p^4}{4 \cdot 2^3 m^8 p^8}$$

$$= \frac{-3\left(p^{4-8}\right)}{32m^8} \quad \textit{Quotient rule}$$

$$= \frac{-3p^{-4}}{32m^8}$$

$$= -\frac{3}{32m^8 p^4}$$

123. $\dfrac{(-3y^3 x^3)(-4y^4 x^2)(x^2)^{-4}}{18x^3 y^2 (y^3)^3 (x^3)^{-2}}$

$$= \frac{(-3y^3 x^3)(-4y^4 x^2)(x^{-8})}{18x^3 y^2 y^9 x^{-6}} \quad \textit{Power rule}$$

$$= \frac{12x^{3+2-8} y^{3+4}}{18x^{3-6} y^{2+9}} \quad \textit{Product rule}$$

$$= \frac{2x^{-3} y^7}{3x^{-3} y^{11}}$$

$$= \frac{2x^{-3-(-3)} y^{7-11}}{3} \quad \textit{Quotient rule}$$

$$= \frac{2y^{-4}}{3}$$

$$= \frac{2}{3y^4}$$

125. $\left(\dfrac{p^2 q^{-1}}{2p^{-2}}\right)^2 \cdot \left(\dfrac{p^3 \cdot 4q^{-2}}{3q^{-5}}\right)^{-1} \cdot \left(\dfrac{pq^{-5}}{q^{-2}}\right)^3$

$$= \frac{p^4 q^{-2} p^{-3} 4^{-1} q^2 p^3 q^{-15}}{2^2 p^{-4} 3^{-1} q^5 q^{-6}} \quad \textit{Power rule}$$

$$= \frac{4^{-1} p^{4-3+3} q^{-2+2-15}}{2^2 3^{-1} p^{-4} q^{5-6}} \quad \textit{Product rule}$$

$$= \frac{3p^4 q^{-15}}{2^2 \cdot 4p^{-4} q^{-1}}$$

$$= \frac{3p^{4-(-4)} q^{-15-(-1)}}{4 \cdot 4} \quad \textit{Quotient rule}$$

$$= \frac{3p^8 q^{-14}}{16}$$

$$= \frac{3p^8}{16q^{14}}$$

127. $530 = 5\!\overset{\frown}{}\!30.\ \leftarrow$ Decimal point

Count 2 places.

Since the number 5.3 is to be made larger, the exponent on 10 is positive.

$$530 = 5.3 \times 10^2$$

129. $0.830 = 0.8_\wedge 30$

Count 1 place.

Since the number 8.3 is to be made smaller, the exponent on 10 is negative.

$$0.830 = 8.3 \times 10^{-1}$$

131. $0.000\,006\,92 = 0.00\,0\,0\,0\,6_\wedge 92$

Count 6 places.

Since the number 6.92 is to be made smaller, the exponent on 10 is negative.

$$0.00000692 = 6.92 \times 10^{-6}$$

133. $-38,500 = -3_\wedge 85\,00.$

Count 4 places.

Since the number 3.85 is to be made larger, the exponent on 10 is positive. Also, affix a negative sign in front of the number.

$$-38,500 = -3.85 \times 10^4$$

135. $7.2 \times 10^4 = 72,000$

Move the decimal point 4 places to the *right* because of the *positive* exponent. Attach extra zeros.

137. $2.54 \times 10^{-3} = 0.002\,54$

Since the exponent is *negative*, move the decimal point 3 places to the *left*.

139. $-6 \times 10^4 = -60,000$

Move the decimal point 4 places to the *right* because of the *positive* exponent. Attach extra zeros.

141. $1.2 \times 10^{-5} = 0.000\,012$

Since the exponent is *negative*, move the decimal point 5 places to the *left*.

143.
$$\frac{12 \times 10^4}{2 \times 10^6} = \frac{12}{2} \times \frac{10^4}{10^6}$$
$$= 6 \times 10^{4-6}$$
$$= 6 \times 10^{-2}$$
$$= 0.06$$

145.
$$\frac{3 \times 10^{-2}}{12 \times 10^3} = \frac{3 \times 10^{-2}}{1.2 \times 10^4}$$
$$= \frac{3}{1.2} \times \frac{10^{-2}}{10^4}$$
$$= 2.5 \times 10^{-6}$$
$$= 0.000\,002\,5$$

147.
$$\frac{0.05 \times 1600}{0.0004} = \frac{5 \times 10^{-2} \times 1.6 \times 10^3}{4 \times 10^{-4}}$$
$$= \frac{5(1.6)}{4} \times \frac{10^{-2} \times 10^3}{10^{-4}}$$
$$= 2 \times 10^{-2+3-(-4)}$$
$$= 2 \times 10^5$$
$$= 200,000$$

149.
$$\frac{20,000 \times 0.018}{300 \times 0.0004} = \frac{2 \times 10^4 \times 1.8 \times 10^{-2}}{3 \times 10^2 \times 4 \times 10^{-4}}$$
$$= \frac{2 \times 1.8}{3 \times 4} \times \frac{10^4 \times 10^{-2}}{10^2 \times 10^{-4}}$$
$$= 0.3 \times 10^{4-2-2-(-4)}$$
$$= 0.3 \times 10^4$$
$$= 3000$$

151.
$$\$1,000,000,000 = \$1 \times 10^9$$
$$\$1,000,000,000,000 = \$1 \times 10^{12}$$
$$\$2,128,000,000,000 = \$2.128 \times 10^{12}$$
$$144,419 = 1.44419 \times 10^5$$

153. (a) 281.4 million
$$= 281,400,000$$
$$= 2.814 \times 10^8$$

(b) $\$1,000,000,000,000 = \$1 \times 10^{12}$

(c) Divide the amount by the number of people to determine how much each person would have to contribute.

$$\frac{\$1 \times 10^{12}}{2.814 \times 10^8} \approx \$.3554 \times 10^4$$
$$= \$3554$$

In 2000, each person in the United States would have had to contribute about \$3554 in order to make someone a trillionaire.

155. Since $d = rt$, $t = \dfrac{d}{r}$. Divide the distance traveled by the speed of light.

$$\frac{9 \times 10^{12}}{3 \times 10^{10}} = \frac{9}{3} \times 10^{12-10} = 3 \times 10^2$$

It will take 300 seconds.

157. First find the number of seconds in a year.

$$\begin{aligned}
1 \text{ year} &= 365 \text{ days}\\
&= 365(24) \text{ hr}\\
&= 8760 \text{ hr}\\
&= 8760(60) \text{ min}\\
&= 525,600 \text{ min}\\
&= 525,600(60) \text{ sec}\\
&= 31,536,000 \text{ sec}\\
&= 3.1536 \times 10^7 \text{ sec}
\end{aligned}$$

Now use $d = rt$ and multiply the rate light travels by the number of seconds in a year.

$$1 \text{ light year} = \left(1.86 \times 10^5 \text{ mi/sec}\right)$$
$$\cdot \left(3.1536 \times 10^7 \text{ sec}\right)$$
$$= 1.86(3.1536) \times 10^{5+7}$$
$$\approx 5.87 \times 10^{12} \text{ mi}$$

There are about 5.87×10^{12} miles in a light year.

159. (a) The distance from Mercury to the sun is 3.6×10^7 mi and the distance from Venus to the sun is 6.7×10^7 mi, so the distance between Mercury and Venus in miles is

$$\left(6.7 \times 10^7\right) - \left(3.6 \times 10^7\right)$$
$$= (6.7 - 3.6) \times 10^7$$
$$= 3.1 \times 10^7.$$

Use $d = rt$, or $\dfrac{d}{r} = t$, where $d = 3.1 \times 10^7$ and $r = 1.55 \times 10^3$.

$$\frac{3.1 \times 10^7}{1.55 \times 10^3} = \frac{3.1}{1.55} \times 10^{7-3}$$
$$= 2 \times 10^4$$
$$= 20{,}000$$

It would take 20,000 hr.

(b) From part (a), it takes 20,000 hr for a spacecraft to travel from Venus to Mercury. Convert this to days (24 hours = 1 day).

$$20{,}000 \text{ hr} = \frac{20{,}000}{24} \text{ days}$$
$$\approx 833 \text{ days}$$

It would take about 833 days.

161. $(1.5 \,\text{E}\, 12) * (5 \,\text{E}\, {}^-3)$
$$= \left(1.5 \times 10^{12}\right)\left(5 \times 10^{-3}\right)$$
$$= (1.5 \times 5)\left(10^{12-3}\right)$$
$$= 7.5 \times 10^9$$

163. $(8.4 \,\text{E}\, 14)/(2.1 \,\text{E}\, {}^-3)$
$$= \frac{8.4 \times 10^{14}}{2.1 \times 10^{-3}}$$
$$= \frac{8.4}{2.1} \times \frac{10^{14}}{10^{-3}}$$
$$= 4 \times 10^{14-(-3)}$$
$$= 4 \times 10^{17}$$

165. $9x + 5x - x + 8x - 12x$
$$= (9 + 5 - 1 + 8 - 12)x$$
$$= 9x$$

167. $6 - 4(3 - z) + 5(2 - 3z)$
$$= 6 - 12 + 4z + 10 - 15z$$
$$= (6 - 12 + 10) + (4z - 15z)$$
$$= 4 - 11z$$

169. $7x - (5 + 5x) + 3$
$$= 7x - 5 - 5x + 3$$
$$= (7x - 5x) + (-5 + 3)$$
$$= 2x - 2$$

5.2 Adding and Subtracting Polynomials

1. $2x^3 + x - 3x^2 + 4$

The polynomial is written in descending powers of the variable if the exponents on the terms of the polynomial decrease from left to right.

$$2x^3 - 3x^2 + x + 4$$

3. $4p^3 - 8p^5 + p^7 = p^7 - 8p^5 + 4p^3$

5. $-m^3 + 5m^2 + 3m^4 + 10$
$$= 3m^4 - m^3 + 5m^2 + 10$$

7. In $7z$, the coefficient is 7 and, since $7z = 7z^1$, the degree is 1.

9. In $-15p^2$, the coefficient is -15 and the degree is 2.

11. In x^4, since $x^4 = 1x^4$, the coefficient is 1 and the degree is 4.

13. In $\dfrac{t}{6} = \dfrac{1}{6}t$, the coefficient is $\dfrac{1}{6}$ and since $\dfrac{t}{6} = \dfrac{1}{6}t^1$, the degree is 1.

15. In $-mn^5$, the coefficient is -1, since $-mn^5 = -1mn^5$, and the degree is 6, since the sum of the exponents on m and n is $1 + 5 = 6$.

17. 25 is one term, so it's a *monomial*. 25 is a nonzero constant, so it has degree zero.

19. $7m - 22$ has two terms, so it's a *binomial*. The exponent on m is 1, so $7m - 22$ has degree 1.

21. $-7y^6 + 11y^8$ is a *binomial* of degree 8.

23. $-5m^3 + 6m - 9m^2$ has three terms, so it's a *trinomial*. The greatest exponent is 3, so the degree is 3.

25. $-6p^4q - 3p^3q^2 + 2pq^3 - q^4$ has four terms, so it is classified as *none of these*. The greatest sum of exponents on any term is 5, so the polynomial has degree 5.

27. Only choice **A** is a trinomial (it has three terms) with descending powers and having degree 6.

29. $5z^4 + 3z^4 = (5 + 3)z^4 = 8z^4$

31. $-m^3 + 2m^3 + 6m^3 = (-1 + 2 + 6)m^3 = 7m^3$

33. $x + x + x + x + x$
$$= (1 + 1 + 1 + 1 + 1)x$$
$$= 5x$$

35. $m^4 - 3m^2 + m$ is *already simplified* since there are no like terms to be combined.

37. $5t + 4s - 6t + 9s = (5t - 6t) + (4s + 9s)$
$$= -t + 13s$$

39. $2k + 3k^2 + 5k^2 - 7$
$$= (3k^2 + 5k^2) + 2k - 7$$
$$= (3 + 5)k^2 + 2k - 7$$
$$= 8k^2 + 2k - 7$$

41. $n^4 - 2n^3 + n^2 - 3n^4 + n^3$
$$= n^4 - 3n^4 - 2n^3 + n^3 + n^2$$
$$= (1 - 3)n^4 + (-2 + 1)n^3 + n^2$$
$$= -2n^4 - n^3 + n^2$$

43. $3ab^2 + 7a^2b - 5ab^2 + 13a^2b$
$$= (3 - 5)ab^2 + (7 + 13)a^2b$$
$$= -2ab^2 + 20a^2b$$

45. $4 - (2 + 3m) + 6m + 9$
$$= 4 - 2 - 3m + 6m + 9$$
$$= (-3 + 6)m + (4 - 2 + 9)$$
$$= 3m + 11$$

47. $(6 + 3p) - (2p + 1) - (2p + 9)$
$$= 6 + 3p - 2p - 1 - 2p - 9$$
$$= (3 - 2 - 2)p + (6 - 1 - 9)$$
$$= -p - 4$$

49. A *monomial* (or *term*) is a numeral, a variable, or a product of numerals and variables raised to positive integer powers. Some examples of monomials are 6, x, and $-4x^2y^3$.

A *binomial* is a sum or difference of exactly two terms, such as $x^2 + y^2$ and $x^2 - y^2$.

A *trinomial* consists of exactly three terms, such as $x^2 - 3x + 8$. These are all examples of *polynomials*.

51. $(5x^2 + 7x - 4) + (3x^2 - 6x + 2)$
$$= 5x^2 + 3x^2 + 7x - 6x - 4 + 2$$
$$= 8x^2 + x - 2$$

53. $(6t^2 - 4t^4 - t) + (3t^4 - 4t^2 + 5)$
$$= -4t^4 + 3t^4 + 6t^2 - 4t^2 - t + 5$$
$$= (-4 + 3)t^4 + (6 - 4)t^2 - t + 5$$
$$= -t^4 + 2t^2 - t + 5$$

55. $(y^3 + 3y + 2) + (4y^3 - 3y^2 + 2y - 1)$
$$= y^3 + 4y^3 - 3y^2 + 3y + 2y + 2 - 1$$
$$= (1 + 4)y^3 - 3y^2 + (3 + 2)y + (2 - 1)$$
$$= 5y^3 - 3y^2 + 5y + 1$$

57. $(3r + 8) - (2r - 5)$

Change all signs in the second polynomial and add.
$$= (3r + 8) + (-2r + 5)$$
$$= 3r + 8 - 2r + 5$$
$$= 3r - 2r + 8 + 5$$
$$= r + 13$$

59. $(2a^2 + 3a - 1) - (4a^2 + 5a + 6)$
$$= (2a^2 + 3a - 1) + (-4a^2 - 5a - 6)$$
$$= 2a^2 - 4a^2 + 3a - 5a - 1 - 6$$
$$= -2a^2 - 2a - 7$$

61. $(z^5 + 3z^2 + 2z) - (4z^5 + 2z^2 - 5z)$
$$= z^5 + 3z^2 + 2z - 4z^5 - 2z^2 + 5z$$
$$= z^5 - 4z^5 + 3z^2 - 2z^2 + 2z + 5z$$
$$= -3z^5 + z^2 + 7z$$

63.
$$\begin{array}{r} 21p - 8 \\ -9p + 4 \\ \hline 12p - 4 \end{array}$$ *Add vertically.*

65.
$$\begin{array}{r} -12p^2 + 4p - 1 \\ 3p^2 + 7p - 8 \\ \hline -9p^2 + 11p - 9 \end{array}$$ *Add vertically.*

67. Subtract.
$$\begin{array}{r} 12a + 15 \\ 7a - 3 \\ \hline \end{array}$$

Change all the signs in the second polynomial, and add.
$$\begin{array}{r} 12a + 15 \\ -7a + 3 \\ \hline 5a + 18 \end{array}$$

69. Subtract.
$$\begin{array}{r} 6m^2 - 11m + 5 \\ -8m^2 + 2m - 1 \\ \hline \end{array}$$

Change all the signs in the second polynomial, and add.
$$\begin{array}{r} 6m^2 - 11m + 5 \\ 8m^2 - 2m + 1 \\ \hline 14m^2 - 13m + 6 \end{array}$$

71. Add column by column to obtain the result on the bottom line.
$$\begin{array}{r} 12z^2 - 11z + 8 \\ 5z^2 + 16z - 2 \\ -4z^2 + 5z - 9 \\ \hline 13z^2 + 10z - 3 \end{array}$$

73.

$$6y^3 - 9y^2 \qquad + 8$$
$$\underline{4y^3 + 2y^2 + 5y}$$
$$10y^3 - 7y^2 + 5y + 8$$

75. Subtract.

$$\underline{\begin{array}{l} -5a^4 \quad + 8a^2 - 9 \\ 6a^3 - \ a^2 + 2 \end{array}}$$

Change all the signs in the second polynomial, and add.

$$\underline{\begin{array}{l} -5a^4 \qquad + 8a^2 - \ 9 \\ -6a^3 + \ a^2 - \ 2 \end{array}}$$
$$-5a^4 - 6a^3 + 9a^2 - 11$$

77. $7y^2 - 6y + 5 - \left(4y^2 - 2y + 3\right)$

$$= 7y^2 - 6y + 5 - 4y^2 + 2y - 3$$
$$= 7y^2 - 4y^2 - 6y + 2y + 5 - 3$$
$$= 3y^2 - 4y + 2$$

79. Simplify the expression in brackets first.

$$\left(3m^2 - 5n^2 + 2n\right) + \left(-3m^2\right) + 4n^2$$
$$= 3m^2 - 5n^2 + 2n - 3m^2 + 4n^2$$
$$= 3m^2 - 3m^2 - 5n^2 + 4n^2 + 2n$$
$$= -n^2 + 2n$$

Now perform the subtraction.

$$\left(-4m^2 + 3n^2 - 5n\right) - \left(-n^2 + 2n\right)$$
$$= -4m^2 + 3n^2 - 5n + n^2 - 2n$$
$$= -4m^2 + 3n^2 + n^2 - 5n - 2n$$
$$= -4m^2 + 4n^2 - 7n$$

81. $\left[-\left(y^4 - y^2 + 1\right) - \left(y^4 + 2y^2 + 1\right)\right]$
$\qquad + \left(3y^4 - 3y^2 - 2\right)$

$$= -y^4 + y^2 - 1 - y^4 - 2y^2 - 1$$
$$\qquad + 3y^4 - 3y^2 - 2$$
$$= -y^4 - y^4 + 3y^4 + y^2 - 2y^2 - 3y^2$$
$$\qquad - 1 - 1 - 2$$
$$= y^4 - 4y^2 - 4$$

83. $-\left[3z^2 + 5z - \left(2z^2 - 6z\right)\right]$
$\qquad + \left[\left(8z^2 - \left[5z - z^2\right]\right) + 2z^2\right]$

$$= -\left(3z^2 + 5z - 2z^2 + 6z\right)$$
$$\qquad + \left(8z^2 - 5z + z^2 + 2z^2\right)$$
$$= -\left(3z^2 - 2z^2 + 5z + 6z\right)$$
$$\qquad + \left(8z^2 + z^2 + 2z^2 - 5z\right)$$
$$= -\left(z^2 + 11z\right) + \left(11z^2 - 5z\right)$$
$$= -z^2 - 11z + 11z^2 - 5z$$
$$= -z^2 + 11z^2 - 11z - 5z$$
$$= 10z^2 - 16z$$

85. $f(x) = 3x + 1$

(a) $f(-1) = 3(-1) + 1$
$$\qquad = -3 + 1 = -2$$

(b) $f(2) = 3(2) + 1$
$$\qquad = 6 + 1 = 7$$

87. $f(x) = x^3 - 8$

(a) $f(-1) = (-1)^3 - 8$
$$\qquad = -1 - 8 = -9$$

(b) $f(2) = 2^3 - 8$
$$\qquad = 8 - 8 = 0$$

89. Using the vertical line test, we find any vertical line will intersect the graph at most once. This indicates that the graph represents a function.

91. Using the vertical line test, we find any vertical line will intersect the graph at most once. This indicates that the graph represents a function.

5.3 Polynomial Functions, Graphs, and Composition

1. $f(x) = 6x - 4$

(a) $f(-1) = 6(-1) - 4$
$$\qquad = -6 - 4 = -10$$

(b) $f(2) = 6(2) - 4$
$$\qquad = 12 - 4 = 8$$

3. $f(x) = x^2 - 3x + 4$

(a) $f(-1) = (-1)^2 - 3(-1) + 4$
$$\qquad = 1 + 3 + 4 = 8$$

(b) $f(2) = (2)^2 - 3(2) + 4$
$$\qquad = 4 - 6 + 4 = 2$$

5. $f(x) = 5x^4 - 3x^2 + 6$

(a) $f(-1) = 5(-1)^4 - 3(-1)^2 + 6$
$$\qquad = 5 \cdot 1 - 3 \cdot 1 + 6$$
$$\qquad = 5 - 3 + 6 = 8$$

(b) $f(2) = 5(2)^4 - 3(2)^2 + 6$
$$\qquad = 5 \cdot 16 - 3 \cdot 4 + 6$$
$$\qquad = 80 - 12 + 6 = 74$$

7. $f(x) = -x^2 + 2x^3 - 8$

(a) $f(-1) = -(-1)^2 + 2(-1)^3 - 8$
$$\qquad = -(1) + 2(-1) - 8$$
$$\qquad = -1 - 2 - 8 = -11$$

(b) $f(2) = -(2)^2 + 2(2)^3 - 8$
$$\qquad = -(4) + 2 \cdot 8 - 8$$
$$\qquad = -4 + 16 - 8 = 4$$

9. $f(x) = -2.19x^2 + 245.7x + 15{,}163$

(a) The year 1980 corresponds to $x = 0$.

$f(0) = -2.19(0)^2 + 245.7(0) + 15{,}163$
$= 15{,}163$ airports

(b) $x = 1990 - 1980 = 10$

$f(10) = -2.19(10)^2 + 245.7(10) + 15{,}163$
$= 17{,}401$ airports

(c) $x = 2003 - 1980 = 23$

$f(23) = -2.19(23)^2 + 245.7(23) + 15{,}163$
$\approx 19{,}655.59$
$\approx 19{,}655$ (rounded down) airports

11. $P(x) = -0.31x^3 + 5.8x^2 - 15x + 9$

(a) The year 1990 corresponds to $x = 0$.

$P(0) = -0.31(0)^3 + 5.8(0)^2 - 15(0) + 9$
$= 9$ million bank debit cards

(b) $x = 1996 - 1990 = 6$

$P(6) = -0.31(6)^3 + 5.8(6)^2 - 15(6) + 9$
$= 60.84 \approx 61$ million bank debit cards

(c) $x = 1999 - 1990 = 9$

$P(9) = -0.31(9)^3 + 5.8(9)^2 - 15(9) + 9$
$= 117.81 \approx 118$ million bank debit cards

13. (a) $(f + g)(x) = f(x) + g(x)$
$= (5x - 10) + (3x + 7)$
$= 8x - 3$

(b) $(f - g)(x) = f(x) - g(x)$
$= (5x - 10) - (3x + 7)$
$= (5x - 10) + (-3x - 7)$
$= 2x - 17$

15. (a) $(f + g)(x)$
$= f(x) + g(x)$
$= (4x^2 + 8x - 3) + (-5x^2 + 4x - 9)$
$= -x^2 + 12x - 12$

(b) $(f - g)(x)$
$= f(x) - g(x)$
$= (4x^2 + 8x - 3) - (-5x^2 + 4x - 9)$
$= (4x^2 + 8x - 3) + (5x^2 - 4x + 9)$
$= 9x^2 + 4x + 6$

For Exercises 17–31, let $f(x) = x^2 - 9$, $g(x) = 2x$, and $h(x) = x - 3$.

17. $(f + g)(x) = f(x) + g(x)$
$= (x^2 - 9) + (2x)$
$= x^2 + 2x - 9$

19. $(f + g)(3) = f(3) + g(3)$
$= (3^2 - 9) + 2(3)$
$= 0 + 6 = 6$

Alternatively, we could evaluate the polynomial in Exercise 17, $x^2 + 2x - 9$, using $x = 3$.

21. $(f - h)(x) = f(x) - h(x)$
$= (x^2 - 9) - (x - 3)$
$= x^2 - 9 - x + 3$
$= x^2 - x - 6$

23. $(f - h)(-3) = f(-3) - h(-3)$
$= [(-3)^2 - 9] - [(-3) - 3]$
$= (9 - 9) - (-6)$
$= 0 + 6 = 6$

25. $(g + h)(-10) = g(-10) + h(-10)$
$= 2(-10) + [(-10) - 3]$
$= -20 + (-13)$
$= -33$

27. $(g - h)(-3) = g(-3) - h(-3)$
$= 2(-3) - [(-3) - 3]$
$= -6 - (-6)$
$= -6 + 6 = 0$

29. $(g + h)\left(\frac{1}{4}\right) = g\left(\frac{1}{4}\right) + h\left(\frac{1}{4}\right)$
$= 2\left(\frac{1}{4}\right) + \left(\frac{1}{4} - 3\right)$
$= \frac{1}{2} + \left(-\frac{11}{4}\right)$
$= -\frac{9}{4}$

31. $(g + h)\left(-\frac{1}{2}\right) = g\left(-\frac{1}{2}\right) + h\left(-\frac{1}{2}\right)$
$= 2\left(-\frac{1}{2}\right) + \left(-\frac{1}{2} - 3\right)$
$= -1 + \left(-\frac{7}{2}\right)$
$= -\frac{9}{2}$

33. Answers will vary. Let $f(x) = x^3$ and $g(x) = x^4$.

$(f - g)(x) = f(x) - g(x)$
$= x^3 - x^4$
$(g - f)(x) = g(x) - f(x)$
$= x^4 - x^3$

Because the two differences are not equal, subtraction of polynomial functions is not commutative.

For Exercises 35–50,
$f(x) = x^2 + 4$, $g(x) = 2x + 3$, and $h(x) = x + 5$.

35. $(h \circ g)(4) = h(g(4))$
$= h(2 \cdot 4 + 3)$
$= h(11)$
$= 11 + 5$
$= 16$

37. $(g \circ f)(6) = g(f(6))$
$$= g(6^2 + 4)$$
$$= g(40)$$
$$= 2 \cdot 40 + 3$$
$$= 83$$

39. $(f \circ h)(-2) = f(h(-2))$
$$= f(-2 + 5)$$
$$= f(3)$$
$$= 3^2 + 4$$
$$= 13$$

41. $(f \circ g)(x) = f(g(x))$
$$= f(2x + 3)$$
$$= (2x + 3)^2 + 4$$
$$= 4x^2 + 12x + 9 + 4$$
$$= 4x^2 + 12x + 13$$

43. $(f \circ h)(x) = f(h(x))$
$$= f(x + 5)$$
$$= (x + 5)^2 + 4$$
$$= x^2 + 10x + 25 + 4$$
$$= x^2 + 10x + 29$$

45. $(h \circ g)(x) = h(g(x))$
$$= h(2x + 3)$$
$$= 2x + 3 + 5$$
$$= 2x + 8$$

47. $(f \circ h)\left(\frac{1}{2}\right) = f\left(h\left(\frac{1}{2}\right)\right)$
$$= f\left(\frac{1}{2} + 5\right)$$
$$= f\left(\frac{11}{2}\right)$$
$$= \left(\frac{11}{2}\right)^2 + 4$$
$$= \frac{121}{4} + \frac{16}{4} = \frac{137}{4}$$

49. $(f \circ g)\left(-\frac{1}{2}\right) = f\left(g\left(-\frac{1}{2}\right)\right)$
$$= f\left(2\left(-\frac{1}{2}\right) + 3\right)$$
$$= f(2)$$
$$= 2^2 + 4$$
$$= 4 + 4 = 8$$

51. $f(x) = 12x, \ g(x) = 5280x$

$(f \circ g)(x) = f[g(x)]$
$$= f(5280x)$$
$$= 12(5280x)$$
$$= 63{,}360x$$

$(f \circ g)(x)$ computes the number of inches in x mi.

53. $r(t) = 2t, \ A(r) = \pi r^2$

$(A \circ r)(t) = A[r(t)]$
$$= A(2t)$$
$$= \pi(2t)^2$$
$$= 4\pi t^2$$

This is the area of the circular layer as a function of time.

55.

x	$f(x) = -2x + 1$
-2	$-2(-2) + 1 = 5$
-1	$-2(-1) + 1 = 3$
0	$-2(0) + 1 = 1$
1	$-2(1) + 1 = -1$
2	$-2(2) + 1 = -3$

This is a linear function, so plot the points and draw a line through them.

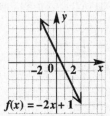

Any x-value can be used, so the domain is $(-\infty, \infty)$. From the graph, we see that any y-value can be obtained from the function, so the range is $(-\infty, \infty)$.

57.

x	$f(x) = -3x^2$
-2	$-3(-2)^2 = -12$
-1	$-3(-1)^2 = -3$
0	$-3(0)^2 = 0$
1	$-3(1)^2 = -3$
2	$-3(2)^2 = -12$

Since the greatest exponent is 2, the graph of f is a parabola.

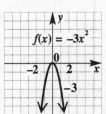

Any x-value can be used, so the domain is $(-\infty, \infty)$. From the graph, we see that the y-values are at most 0, so the range is $(-\infty, 0]$.

59.

x	$f(x) = x^3 + 1$
-2	$(-2)^3 + 1 = -7$
-1	$(-1)^3 + 1 = 0$
0	$(0)^3 + 1 = 1$
1	$(1)^3 + 1 = 2$
2	$(2)^3 + 1 = 9$

The greatest exponent is 3, so the graph of f is s-shaped.

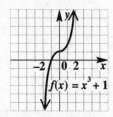

Any x-value can be used, so the domain is $(-\infty, \infty)$. From the graph, we see that any y-value can be obtained from the function, so the range is $(-\infty, \infty)$.

61. $3m^3(4m^2) = 3(4)m^3m^2$
$$= 12m^{3+2}$$
$$= 12m^5$$

63. $-3b^5(2a^3b^4) = -3(2)a^3b^5b^4$
$$= -6a^3b^{5+4}$$
$$= -6a^3b^9$$

65. $12x^2y(5xy^3) = 12(5)x^2xyy^3$
$$= 60x^{2+1}y^{1+3}$$
$$= 60x^3y^4$$

5.4 Multiplying Polynomials

1. $(2x - 5)(3x + 4)$

$\qquad$ **F** $\qquad$ **O** $\qquad$ **I** $\qquad$ **L**
$$= 2x(3x) + 2x(4) + (-5)(3x) + (-5)(4)$$
$$= 6x^2 + 8x - 15x - 20$$
$$= 6x^2 - 7x - 20 \quad \text{(Choice \textbf{C})}$$

3. $(2x - 5)(3x - 4)$

$\qquad$ **F** $\qquad$ **O** $\qquad$ **I** $\qquad$ **L**
$$= 2x(3x) + 2x(-4) + (-5)(3x) + (-5)(-4)$$
$$= 6x^2 - 8x - 15x + 20$$
$$= 6x^2 - 23x + 20 \quad \text{(Choice \textbf{D})}$$

5. $-8m^3(3m^2) = -8(3)m^{3+2} = -24m^5$

7. $(14x^2y^3)(-2x^5y) = 14(-2)x^{2+5}y^{3+1}$
$$= -28x^7y^4$$

9. $3x(-2x + 5) = 3x(-2x) + 3x(5) = -6x^2 + 15x$

11. $-q^3(2 + 3q) = -q^3(2) - q^3(3q) = -2q^3 - 3q^4$

13. $6k^2(3k^2 + 2k + 1)$
$$= 6k^2(3k^2) + 6k^2(2k) + 6k^2(1)$$
$$= 18k^4 + 12k^3 + 6k^2$$

15. $(2m + 3)(3m^2 - 4m - 1)$
$$= 2m(3m^2 - 4m - 1) + 3(3m^2 - 4m - 1)$$
$$= 2m(3m^2) + 2m(-4m) + 2m(-1)$$
$$\qquad + 3(3m^2) + 3(-4m) + 3(-1)$$
$$= 6m^3 - 8m^2 - 2m + 9m^2 - 12m - 3$$
$$= 6m^3 - 8m^2 + 9m^2 - 2m - 12m - 3$$
$$= 6m^3 + m^2 - 14m - 3$$

17. $m(m + 5)(m - 8)$
$$= m(m^2 - 8m + 5m - 40)$$
$$= m(m^2 - 3m - 40)$$
$$= m(m^2) + m(-3m) + m(-40)$$
$$= m^3 - 3m^2 - 40m$$

19. $4z(2z + 1)(3z - 4)$
$$= 4z(6z^2 - 8z + 3z - 4)$$
$$= 4z(6z^2 - 5z - 4)$$
$$= 4z(6z^2) + 4z(-5z) + 4z(-4)$$
$$= 24z^3 - 20z^2 - 16z$$

21. $4x^3(x - 3)(x + 2)$
$$= 4x^3(x^2 + 2x - 3x - 6)$$
$$= 4x^3(x^2 - x - 6)$$
$$= 4x^3(x^2) + 4x^3(-x) + 4x^3(-6)$$
$$= 4x^5 - 4x^4 - 24x^3$$

23. $(2y + 3)(3y - 4)$
Rewrite vertically and multiply.

$$
\begin{array}{r}
2y + 3 \\
3y - 4 \\
\hline
-8y - 12 \quad \leftarrow -4(2y+3) \\
6y^2 + 9y \qquad\quad \leftarrow 3y(2y+3) \\
\hline
6y^2 + y - 12 \qquad
\end{array}
$$
Combine like terms.

25.
$$
\begin{array}{r}
-b^2 + 3b + 3 \\
2b + 4 \\
\hline
-4b^2 + 12b + 12 \\
-2b^3 + 6b^2 + 6b \\
\hline
-2b^3 + 2b^2 + 18b + 12
\end{array}
$$

27.
$$
\begin{array}{r}
5m - 3n \\
5m + 3n \\
\hline
15mn - 9n^2 \\
25m^2 - 15mn \\
\hline
25m^2 \qquad\quad - 9n^2
\end{array}
$$

29.
$$
\begin{array}{r}
2z^3 - 5z^2 + 8z - 1 \\
4z + 3 \\
\hline
6z^3 - 15z^2 + 24z - 3 \\
8z^4 - 20z^3 + 32z^2 - 4z \\
\hline
8z^4 - 14z^3 + 17z^2 + 20z - 3
\end{array}
$$

31.
$$
\begin{array}{r}
2p^2 + 3p + 6 \\
3p^2 - 4p - 1 \\
\hline
-2p^2 - 3p - 6 \\
-8p^3 - 12p^2 - 24p \\
6p^4 + 9p^3 + 18p^2 \\
\hline
6p^4 + p^3 + 4p^2 - 27p - 6
\end{array}
$$

33. $(m+5)(m-8)$

$$\qquad \textbf{F} \quad \textbf{O} \quad \textbf{I} \quad \textbf{L}$$
$$= m^2 - 8m + 5m - 40$$
$$= m^2 - 3m - 40$$

35. $(4k+3)(3k-2)$

$$\qquad \textbf{F} \quad \textbf{O} \quad \textbf{I} \quad \textbf{L}$$
$$= 12k^2 - 8k + 9k - 6$$
$$= 12k^2 + k - 6$$

37. $(z-w)(3z+4w)$

$$\qquad \textbf{F} \quad \textbf{O} \quad \textbf{I} \quad \textbf{L}$$
$$= 3z^2 + 4zw - 3zw - 4w^2$$
$$= 3z^2 + zw - 4w^2$$

39. $(6c-d)(2c+3d)$

$$\qquad \textbf{F} \quad \textbf{O} \quad \textbf{I} \quad \textbf{L}$$
$$= 12c^2 + 18cd - 2cd - 3d^2$$
$$= 12c^2 + 16cd - 3d^2$$

41. $(0.2x+1.3)(0.5x-0.1)$

$$\qquad \textbf{F} \qquad \textbf{O} \qquad \textbf{I} \qquad \textbf{L}$$
$$= 0.1x^2 - 0.02x + 0.65x - 0.13$$
$$= 0.1x^2 + 0.63x - 0.13$$

43. $\left(3w + \frac{1}{4}z\right)(w - 2z)$

$$\qquad \textbf{F} \quad \textbf{O} \quad \textbf{I} \quad \textbf{L}$$
$$= 3w^2 - 6wz + \frac{1}{4}wz - \frac{1}{2}z^2$$
$$= 3w^2 - \frac{23}{4}wz - \frac{1}{2}z^2$$

45. The product of two binomials is the sum of the product of the first terms, the product of the outer terms, the product of the inner terms, and the product of the last terms.

47. Use the formula for the product of the sum and difference of two terms.

$$(2p-3)(2p+3) = (2p)^2 - (3)^2$$
$$= 4p^2 - 9$$

49. $(5m-1)(5m+1) = (5m)^2 - (1)^2$
$$= 25m^2 - 1$$

51. $(3a+2c)(3a-2c) = (3a)^2 - (2c)^2$
$$= 9a^2 - 4c^2$$

53. $\left(4x - \frac{2}{3}\right)\left(4x + \frac{2}{3}\right) = (4x)^2 - \left(\frac{2}{3}\right)^2$
$$= 16x^2 - \frac{4}{9}$$

55. $\left(4m + 7n^2\right)\left(4m - 7n^2\right) = (4m)^2 - \left(7n^2\right)^2$
$$= 16m^2 - 49n^4$$

57. $3y(5y^3 + 2)(5y^3 - 2)$
First multiply $(5y^3 + 2)(5y^3 - 2)$.

$$\left(5y^3 + 2\right)\left(5y^3 - 2\right) = \left(5y^3\right)^2 - (2)^2$$
$$= 25y^6 - 4$$

Now multiply the last polynomial times $3y$.

$$3y(25y^6 - 4) = 75y^7 - 12y$$

59. Use the formula for the square of a binomial.

$$(y-5)^2 = y^2 - 2(y)(5) + 5^2$$
$$= y^2 - 10y + 25$$

61. $(2p+7)^2 = (2p)^2 + 2(2p)(7) + 7^2$
$$= 4p^2 + 28p + 49$$

63. $(4n+3m)^2$
$$= (4n)^2 + 2(4n)(3m) + (3m)^2$$
$$= 16n^2 + 24nm + 9m^2$$

65. $\left(k - \frac{5}{7}p\right)^2 = k^2 - 2(k)\left(\frac{5}{7}p\right) + \left(\frac{5}{7}p\right)^2$
$$= k^2 - \frac{10}{7}kp + \frac{25}{49}p^2$$

67. $(0.2x - 1.4y)^2$
$$= (0.2x)^2 - 2(0.2x)(1.4y) + (1.4y)^2$$
$$= 0.04x^2 - 0.56xy + 1.96y^2$$

69. To find the product $101 \cdot 99$ using the special product rule

$$(x+y)(x-y) = x^2 - y^2,$$

let $x = 100$ and $y = 1$. Then

$$101 \cdot 99 = (100+1)(100-1)$$
$$= 100^2 - 1^2$$
$$= 10,000 - 1$$
$$= 9999.$$

71. $[(5x+1) + 6y]^2$
$$= (5x+1)^2 + 2(5x+1)(6y) + (6y)^2$$
$$\qquad \textit{Square of a binomial}$$
$$= (5x+1)^2 + 12y(5x+1) + 36y^2$$
$$= \left[(5x)^2 + 2(5x)(1) + 1^2\right]$$
$$\quad + 60xy + 12y + 36y^2$$
$$\qquad \textit{Square of a binomial}$$
$$= 25x^2 + 10x + 1 + 60xy + 12y + 36y^2$$

73. $[(2a + b) - 3]^2$
$$= (2a + b)^2 - 2(2a + b)(3) + 3^2$$
$$= [(2a)^2 + 2(2a)(b) + b^2]$$
$$\quad - 6(2a + b) + 3^2$$
$$= 4a^2 + 4ab + b^2 - 12a - 6b + 9$$

75. $[(2a + b) - 3][(2a + b) + 3]$
$$= (2a + b)^2 - 3^2$$
$$\qquad \textit{Product of the sum and}$$
$$\qquad \textit{difference of two terms}$$
$$= [(2a)^2 + 2(2a)(b) + b^2] - 9$$
$$\qquad \textit{Square of a binomial}$$
$$= 4a^2 + 4ab + b^2 - 9$$

77. $[(2h - k) + j][(2h - k) - j]$
$$= (2h - k)^2 - j^2$$
$$= (2h)^2 - 2(2h)(k) + k^2 - j^2$$
$$= 4h^2 - 4hk + k^2 - j^2$$

79. $(y + 2)^3 = (y + 2)^2(y + 2)$
$$= [y^2 + 2(y)(2) + 2^2](y + 2)$$
$$= (y^2 + 4y + 4)(y + 2)$$

$$
\begin{array}{r}
y^2 + 4y + 4 \\
y + 2 \\
\hline
2y^2 + 8y + 8 \\
y^3 + 4y^2 + 4y \\
\hline
y^3 + 6y^2 + 12y + 8
\end{array}
$$

81. $(5r - s)^3$
$$= (5r - s)^2(5r - s)$$
$$= (25r^2 - 10rs + s^2)(5r - s)$$
$$= 125r^3 - 50r^2s + 5rs^2 - 25r^2s + 10rs^2 - s^3$$
$$= 125r^3 - 75r^2s + 15rs^2 - s^3$$

83. $(q - 2)^4 = (q - 2)^2(q - 2)^2$
$$= (q^2 - 4q + 4)(q^2 - 4q + 4)$$

$$
\begin{array}{r}
q^2 - 4q + 4 \\
q^2 - 4q + 4 \\
\hline
4q^2 - 16q + 16 \\
-4q^3 + 16q^2 - 16q \\
q^4 - 4q^3 + 4q^2 \\
\hline
q^4 - 8q^3 + 24q^2 - 32q + 16
\end{array}
$$

85. $(2a + b)(3a^2 + 2ab + b^2)$
Rewrite vertically and multiply.

$$
\begin{array}{r}
3a^2 + 2ab + b^2 \\
2a + b \\
\hline
3a^2b + 2ab^2 + b^3 \\
6a^3 + 4a^2b + 2ab^2 \\
\hline
6a^3 + 7a^2b + 4ab^2 + b^3
\end{array}
$$

87. $(4z - x)(z^3 - 4z^2x + 2zx^2 - x^3)$
Rewrite vertically and multiply.

$$
\begin{array}{r}
z^3 - 4z^2x + 2zx^2 - x^3 \\
4z - x \\
\hline
-z^3x + 4z^2x^2 - 2zx^3 + x^4 \\
4z^4 - 16z^3x + 8z^2x^2 - 4zx^3 \\
\hline
4z^4 - 17z^3x + 12z^2x^2 - 6zx^3 + x^4
\end{array}
$$

89. $(m^2 - 2mp + p^2)(m^2 + 2mp - p^2)$
Rewrite vertically and multiply.

$$
\begin{array}{r}
m^2 - 2mp + p^2 \\
m^2 + 2mp - p^2 \\
\hline
-m^2p^2 + 2mp^3 - p^4 \\
2m^3p - 4m^2p^2 + 2mp^3 \\
m^4 - 2m^3p + m^2p^2 \\
\hline
m^4 \qquad - 4m^2p^2 + 4mp^3 - p^4
\end{array}
$$

91. $ab(a + b)(a + 2b)(a - 3b)$
First multiply $(a + b)(a + 2b)(a - 3b)$.

$$(a + b)(a + 2b)(a - 3b)$$
$$= [(a + b)(a + 2b)](a - 3b)$$
$$= [a^2 + 3ab + 2b^2](a - 3b)$$

$$
\begin{array}{r}
a^2 + 3ab + 2b^2 \\
a - 3b \\
\hline
-3a^2b - 9ab^2 - 6b^3 \\
a^3 + 3a^2b + 2ab^2 \\
\hline
a^3 \qquad - 7ab^2 - 6b^3
\end{array}
$$

Now multiply the last polynomial times ab.

$$ab(a^3 - 7ab^2 - 6b^3)$$
$$= a^4b - 7a^2b^3 - 6ab^4$$

In Exercises 93–96, substitute 3 for x and 4 for y.

93. $(x + y)^2 = (3 + 4)^2 = 7^2 = 49$
$x^2 + y^2 = 3^2 + 4^2 = 9 + 6 = 25$
Since $49 \neq 25$, $(x + y)^2 \neq x^2 + y^2$.

95. $(x + y)^4 = (3 + 4)^4 = 7^4 = 2401$
$x^4 + y^4 = 3^4 + 4^4 = 81 + 256 = 337$
Since $2401 \neq 337$, $(x + y)^4 \neq x^4 + y^4$.

97. The formula for the area of a triangle is $A = \frac{1}{2}bh$.
Use $b = 3x + 2y$ and $h = 3x - 2y$.

$$A = \tfrac{1}{2}(3x + 2y)(3x - 2y)$$
$$= \tfrac{1}{2}(9x^2 - 4y^2)$$
$$= \tfrac{9}{2}x^2 - 2y^2$$

99. The formula for the area of a parallelogram is $A = bh$. Use $b = 5x + 6$ and $h = 3x - 4$.

$$A = (5x + 6)(3x - 4)$$
$$= 15x^2 - 20x + 18x - 24$$
$$= 15x^2 - 2x - 24$$

101. The length of each side of the entire square is a. To find the length of each side of the blue square, subtract b, the length of a side of the green rectangle, that is, $a - b$.

102. The formula for the area A of a square with side s is $A = s^2$. Since $s = a - b$, the formula for the area of the blue square would be $A = (a - b)^2$.

103. The formula for the area of a rectangle is $A = LW$. The green rectangle has length $a - b$ and width b, so each green rectangle has an area of $\underline{(a - b)b \text{ or } ab - b^2}$.

Since there are two green rectangles, the total area in green is $\underline{2(ab - b^2) \text{ or } 2ab - 2b^2}$.

104. The length of each side of the yellow square is b, so the yellow square has an area of $\underline{b^2}$.

105. The area of the entire colored region is $\underline{a^2}$, because each side of the entire colored region has length $\underline{a}$.

106. Using the results from Exercises 101–105, the area of the blue square equals

$$a^2 - (2ab - 2b^2) - b^2 = a^2 - 2ab + b^2.$$

107. **(a)** Both expressions for the area of the blue square must be equal to each other; that is, $(a - b)^2$ from Exercise 102 and $a^2 - 2ab + b^2$ from Exercise 106.

(b) From Exercise 102, the area of the blue square is $(a - b)^2$. From Exercise 106, the area of the blue square is $a^2 - 2ab + b^2$. Since these expressions must be equal

$$(a - b)^2 = a^2 - 2ab + b^2.$$

This equation reinforces the special product for the square of a binomial difference.

108.

The large square is made up of two smaller squares and two congruent rectangles. The sum of the areas is

$$a^2 + 2ab + b^2.$$

Since these expressions represent the same quantity, they must be equal. Thus,

$$(a + b)^2 = a^2 + 2ab + b^2.$$

109. $f(x) = 2x, g(x) = 5x - 1$

$$\begin{aligned}(fg)(x) &= f(x) \cdot g(x) \\ &= 2x(5x - 1) \\ &= 10x^2 - 2x\end{aligned}$$

111. $f(x) = x + 1, g(x) = 2x - 3$

$$\begin{aligned}(fg)(x) &= f(x) \cdot g(x) \\ &= (x + 1)(2x - 3) \\ &= 2x^2 - 3x + 2x - 3 \\ &= 2x^2 - x - 3\end{aligned}$$

113. $f(x) = 2x - 3, g(x) = 4x^2 + 6x + 9$

$$\begin{aligned}(fg)(x) &= f(x) \cdot g(x) \\ &= (2x - 3)(4x^2 + 6x + 9)\end{aligned}$$

Multiply vertically.

$$\begin{array}{r} 4x^2 + 6x + 9 \\ 2x - 3 \\ \hline -12x^2 - 18x - 27 \\ 8x^3 + 12x^2 + 18x \\ \hline 8x^3 \qquad\qquad - 27 \end{array}$$

For Exercises 115–126, let $f(x) = x^2 - 9, g(x) = 2x$, and $h(x) = x - 3$.

115. $\begin{aligned}(fg)(x) &= f(x) \cdot g(x) \\ &= (x^2 - 9)(2x) \\ &= 2x^3 - 18x\end{aligned}$

117. $\begin{aligned}(fg)(2) &= f(2) \cdot g(2) \\ &= (2^2 - 9)[2(2)] \\ &= -5 \cdot 4 = -20\end{aligned}$

119. $\begin{aligned}(gh)(x) &= g(x) \cdot h(x) \\ &= (2x)(x - 3) \\ &= 2x^2 - 6x\end{aligned}$

121. $\begin{aligned}(gh)(-3) &= g(-3) \cdot h(-3) \\ &= [2(-3)] \cdot [(-3) - 3] \\ &= -6(-6) = 36\end{aligned}$

123. $\begin{aligned}(fg)\left(-\tfrac{1}{2}\right) &= f\left(-\tfrac{1}{2}\right) \cdot g\left(-\tfrac{1}{2}\right) \\ &= \left[\left(-\tfrac{1}{2}\right)^2 - 9\right] \cdot \left[2\left(-\tfrac{1}{2}\right)\right] \\ &= \left(\tfrac{1}{4} - \tfrac{36}{4}\right)(-1) \\ &= -\tfrac{35}{4}(-1) = \tfrac{35}{4}\end{aligned}$

125. $\begin{aligned}(fh)\left(-\tfrac{1}{4}\right) &= f\left(-\tfrac{1}{4}\right) \cdot h\left(-\tfrac{1}{4}\right) \\ &= \left[\left(-\tfrac{1}{4}\right)^2 - 9\right] \cdot \left[-\tfrac{1}{4} - 3\right] \\ &= \left(\tfrac{1}{16} - \tfrac{144}{16}\right)\left(-\tfrac{1}{4} - \tfrac{12}{4}\right) \\ &= -\tfrac{143}{16}\left(-\tfrac{13}{4}\right) = \tfrac{1859}{64}\end{aligned}$

127. $\dfrac{12p^7}{6p^3} = \dfrac{12}{6}p^{7-3} = 2p^4$

129. $\dfrac{-8a^3b^7}{6a^5b} = \dfrac{-8}{6}a^{3-5}b^{7-1} = \dfrac{-4}{3}a^{-2}b^6 = \dfrac{-4b^6}{3a^2}$

131. Subtract.

$$-3a^2 + 4a - 5$$
$$\underline{5a^2 + 3a - 9}$$

Change all the signs in the second row, then add.

$$-3a^2 + 4a - 5$$
$$\underline{-5a^2 - 3a + 9}$$
$$-8a^2 + a + 4$$

5.5   Dividing Polynomials

1. We find the quotient of two monomials by using the _quotient_ rule for _exponents_ .

3. When dividing polynomials that are not monomials, first write them in _descending_ powers.

5. $\dfrac{15x^3 - 10x^2 + 5}{5} = \dfrac{15x^3}{5} - \dfrac{10x^2}{5} + \dfrac{5}{5}$

$\phantom{\dfrac{15x^3 - 10x^2 + 5}{5}} = 3x^3 - 2x^2 + 1$

7. $\dfrac{9y^2 + 12y - 15}{3y} = \dfrac{9y^2}{3y} + \dfrac{12y}{3y} - \dfrac{15}{3y}$

$\phantom{\dfrac{9y^2 + 12y - 15}{3y}} = 3y + 4 - \dfrac{5}{y}$

9. $\dfrac{15m^3 + 25m^2 + 30m}{5m^2}$

$= \dfrac{15m^3}{5m^2} + \dfrac{25m^2}{5m^2} + \dfrac{30m}{5m^2}$

$= 3m + 5 + \dfrac{6}{m}$

11. $\dfrac{14m^2n^2 - 21mn^3 + 28m^2n}{14m^2n}$

$= \dfrac{14m^2n^2}{14m^2n} - \dfrac{21mn^3}{14m^2n} + \dfrac{28m^2n}{14m^2n}$

$= n - \dfrac{3n^2}{2m} + 2$

13. $\dfrac{8wxy^2 + 3wx^2y + 12w^2xy}{4wx^2y}$

$= \dfrac{8wxy^2}{4wx^2y} + \dfrac{3wx^2y}{4wx^2y} + \dfrac{12w^2xy}{4wx^2y}$

$= \dfrac{2y}{x} + \dfrac{3}{4} + \dfrac{3w}{x}$

15.

$$
\require{enclose}
\begin{array}{r}
r^2 - 7r + 6 \\
3r - 1 \enclose{longdiv}{3r^3 - 22r^2 + 25r - 6} \\
\underline{3r^3 - r^2} \\
-21r^2 + 25r \\
\underline{-21r^2 + 7r} \\
18r - 6 \\
\underline{18r - 6} \\
0
\end{array}
$$

Answer: $r^2 - 7r + 6$

17.

$$
\begin{array}{r}
y - 4 \\
y + 5 \enclose{longdiv}{y^2 + y - 20} \\
\underline{y^2 + 5y} \\
-4y - 20 \\
\underline{-4y - 20} \\
0
\end{array}
$$

Answer: $y - 4$

19.

$$
\begin{array}{r}
q + 8 \\
q - 4 \enclose{longdiv}{q^2 + 4q - 32} \\
\underline{q^2 - 4q} \\
8q - 32 \\
\underline{8q - 32} \\
0
\end{array}
$$

Answer: $q + 8$

21.

$$
\begin{array}{r}
t + 5 \\
3t + 2 \enclose{longdiv}{3t^2 + 17t + 10} \\
\underline{3t^2 + 2t} \\
15t + 10 \\
\underline{15t + 10} \\
0
\end{array}
$$

Answer: $t + 5$

23.

$$
\begin{array}{r}
p - 4 \\
p + 6 \enclose{longdiv}{p^2 + 2p + 20} \\
\underline{p^2 + 6p} \\
-4p + 20 \\
\underline{-4p - 24} \\
44
\end{array}
$$

Remainder

Answer: $p - 4 + \dfrac{44}{p + 6}$

25.

$$
\begin{array}{r}
m^2 + 2m - 1 \\
3m - 1 \enclose{longdiv}{3m^3 + 5m^2 - 5m + 1} \\
\underline{3m^3 - m^2} \\
6m^2 - 5m \\
\underline{6m^2 - 2m} \\
-3m + 1 \\
\underline{-3m + 1} \\
0
\end{array}
$$

Answer: $m^2 + 2m - 1$

27.
$$
\begin{array}{r}
m^2 + m + 3 \\
m - 3 \overline{\smash{\big)}\, m^3 - 2m^2 + 0m - 9} \\
\underline{m^3 - 3m^2} \\
m^2 + 0m \\
\underline{m^2 - 3m} \\
3m - 9 \\
\underline{3m - 9} \\
0
\end{array}
$$

Answer: $m^2 + m + 3$

29.
$$
\begin{array}{r}
z^2 \qquad + 3 \\
2z - 5 \overline{\smash{\big)}\, 2z^3 - 5z^2 + 6z - 15} \\
\underline{2z^3 - 5z^2} \\
6z - 15 \\
\underline{6z - 15} \\
0
\end{array}
$$

Answer: $z^2 + 3$

31.
$$
\begin{array}{r}
x^2 + 2x - 3 \\
4x + 1 \overline{\smash{\big)}\, 4x^3 + 9x^2 - 10x + 3} \\
\underline{4x^3 + x^2} \\
8x^2 - 10x \\
\underline{8x^2 + 2x} \\
-12x + 3 \\
\underline{-12x - 3} \\
6
\end{array}
$$
Remainder

Answer: $x^2 + 2x - 3 + \dfrac{6}{4x + 1}$

33.
$$
\begin{array}{r}
2x - 5 \\
3x^2 - 2x + 4 \overline{\smash{\big)}\, 6x^3 - 19x^2 + 14x - 15} \\
\underline{6x^3 - 4x^2 + 8x} \\
-15x^2 + 6x - 15 \\
\underline{-15x^2 + 10x - 20} \\
-4x + 5
\end{array}
$$
Remainder

Answer: $2x - 5 + \dfrac{-4x + 5}{3x^2 - 2x + 4}$

35.
$$
\begin{array}{r}
x^2 + x + 3 \\
x - 1 \overline{\smash{\big)}\, x^3 + 0x^2 + 2x - 3} \\
\underline{x^3 - x^2} \\
x^2 + 2x \\
\underline{x^2 - x} \\
3x - 3 \\
\underline{3x - 3} \\
0
\end{array}
$$

Answer: $x^2 + x + 3$

37.
$$
\begin{array}{r}
3x^2 + 6x + 11 \\
x - 2 \overline{\smash{\big)}\, 3x^3 + 0x^2 - x + 4} \\
\underline{3x^3 - 6x^2} \\
6x^2 - x \\
\underline{6x^2 - 12x} \\
11x + 4 \\
\underline{11x - 22} \\
26
\end{array}
$$

Answer: $3x^2 + 6x + 11 + \dfrac{26}{x - 2}$

39.
$$
\begin{array}{r}
2k^2 + 3k - 1 \\
2k^2 + 1 \overline{\smash{\big)}\, 4k^4 + 6k^3 + 0k^2 + 3k - 1} \\
\underline{4k^4 + 2k^2} \\
6k^3 - 2k^2 + 3k \\
\underline{6k^3 + 3k} \\
-2k^2 - 1 \\
\underline{-2k^2 - 1} \\
0
\end{array}
$$

Answer: $2k^2 + 3k - 1$

41.
$$
\begin{array}{r}
2y^2 \qquad + 2 \\
3y^2 + 2y - 3 \overline{\smash{\big)}\, 6y^4 + 4y^3 + 0y^2 + 4y - 6} \\
\underline{6y^4 + 4y^3 - 6y^2} \\
6y^2 + 4y - 6 \\
\underline{6y^2 + 4y - 6} \\
0
\end{array}
$$

Answer: $2y^2 + 2$

43.
$$
\begin{array}{r}
x^2 - 4x + 2 \\
x^2 + 3 \overline{\smash{\big)}\, x^4 - 4x^3 + 5x^2 - 3x + 2} \\
\underline{x^4 + 3x^2} \\
-4x^3 + 2x^2 - 3x \\
\underline{-4x^3 - 12x} \\
2x^2 + 9x + 2 \\
\underline{2x^2 + 6} \\
9x - 4
\end{array}
$$

Answer: $x^2 - 4x + 2 + \dfrac{9x - 4}{x^2 + 3}$

45.
$$
\begin{array}{r}
p^2 + \frac{5}{2}p + 2 \\
2p + 2 \overline{\smash{\big)}\, 2p^3 + 7p^2 + 9p + 3} \\
\underline{2p^3 + 2p^2} \\
5p^2 + 9p \\
\underline{5p^2 + 5p} \\
4p + 3 \\
\underline{4p + 4} \\
-1
\end{array}
$$

Answer: $p^2 + \dfrac{5}{2}p + 2 + \dfrac{-1}{2p + 2}$

47.

$$\begin{array}{r} \frac{3}{2}a \ - \ 10 \\ 2a \ + \ 6\overline{)\ 3a^2 \ - \ \ 11a \ + \ 17\ } \\ \underline{3a^2 \ + \ \ \ 9a\ \ \ \ \ \ \ \ \ } \\ -20a \ + \ 17 \\ \underline{-20a \ - \ 60} \\ 77 \end{array}$$

Answer: $\dfrac{3}{2}a - 10 + \dfrac{77}{2a + 6}$

49.

$$\begin{array}{r} p^2 \ + \ \ p \ + \ 1 \\ p \ - \ 1\overline{)\ p^3 \ + \ 0p^2 \ + \ 0p \ - \ 1\ } \\ \underline{p^3 \ - \ \ p^2\ \ \ \ \ \ \ \ \ \ \ \ \ } \\ p^2 \ + \ 0p \\ \underline{p^2 \ - \ \ p} \\ p \ - \ 1 \\ \underline{p \ - \ 1} \\ 0 \end{array}$$

Answer: $p^2 + p + 1$

51. To start: $\dfrac{2x^2}{3x} = \dfrac{2}{3}x$

$$\begin{array}{r} \frac{2}{3}x \ - \ 1 \\ 3x \ + \ 1\overline{)\ 2x^2 \ - \ \frac{7}{3}x \ - \ 1\ } \\ \underline{2x^2 \ + \ \frac{2}{3}x\ \ \ \ \ \ \ } \\ -3x \ - \ 1 \\ \underline{-3x \ - \ 1} \\ 0 \end{array}$$

Answer: $\dfrac{2}{3}x - 1$

53.

$$\begin{array}{r} \frac{3}{4}a \ - \ 2 \\ 4a \ + \ 3\overline{)\ 3a^2 \ - \ \frac{23}{4}a \ - \ 5\ } \\ \underline{3a^2 \ + \ \frac{9}{4}a\ \ \ \ \ \ \ } \\ -8a \ - \ 5 \\ \underline{-8a \ - \ 6} \\ 1 \end{array}$$

Remainder

Answer: $\dfrac{3}{4}a - 2 + \dfrac{1}{4a + 3}$

55. The volume of a box is the product of the height, length, and width. Use the formula $V = LWH$.

$$V = LWH$$

$$\frac{V}{LH} = W$$

Here,

$L \cdot H = (p + 4)p = p^2 + 4p$, so

$W = \dfrac{V}{LH} = \dfrac{2p^3 + 15p^2 + 28p}{p^2 + 4p}$.

$$\begin{array}{r} 2p \ + \ 7 \\ p^2 \ + \ 4p\overline{)\ 2p^3 \ + \ 15p^2 \ + \ 28p\ } \\ \underline{2p^3 \ + \ \ 8p^2\ \ \ \ \ \ \ \ \ } \\ 7p^2 \ + \ 28p \\ \underline{7p^2 \ + \ 28p} \\ 0 \end{array}$$

The width is $2p + 7$.

57. $P(x) = x^3 - 4x^2 + 3x - 5$

$P(-1) = (-1)^3 - 4(-1)^2 + 3(-1) - 5$

$\quad\quad\quad = -1 - 4 - 3 - 5 = -13$

Now divide the given polynomial by $x + 1$.

$$\begin{array}{r} x^2 \ - \ 5x \ + \ \ \ 8 \\ x \ + \ 1\overline{)\ x^3 \ - \ \ \ 4x^2 \ + \ 3x \ - \ \ \ 5\ } \\ \underline{x^3 \ + \ \ \ \ x^2\ \ \ \ \ \ \ \ \ \ \ \ \ \ \ \ } \\ -5x^2 \ + \ 3x \\ \underline{-5x^2 \ - \ 5x} \\ 8x \ - \ \ \ 5 \\ \underline{8x \ + \ \ \ 8} \\ -13 \end{array}$$

Remainder

The remainder in the division is the same as $P(-1)$, -13. This suggests that if a polynomial is divided by $x - r$, in this case $x - (-1)$ or $x + 1$, then the remainder is equal to $P(r)$, in this case $P(-1)$.

59. $\left(\dfrac{f}{g}\right)(x) = \dfrac{f(x)}{g(x)} = \dfrac{10x^2 - 2x}{2x}$

$\quad\quad\quad\quad = \dfrac{10x^2}{2x} - \dfrac{2x}{2x}$

$\quad\quad\quad\quad = 5x - 1$

The x-values that are not in the domain of the quotient function are found by solving $g(x) = 0$.

$$2x = 0$$
$$x = 0$$

61.

$$\begin{array}{r} 2x \ - \ 3 \\ x \ + \ 1\overline{)\ 2x^2 \ - \ \ \ x \ - \ 3\ } \\ \underline{2x^2 \ + \ \ 2x\ \ \ \ \ \ \ \ } \\ -3x \ - \ 3 \\ \underline{-3x \ - \ 3} \\ 0 \end{array}$$

Quotient: $2x - 3$

$$g(x) = 0$$
$$x + 1 = 0$$
$$x = -1$$

63.

$$
\begin{array}{r}
4x^2 + 6x + 9 \\
2x - 3\overline{\smash{\big)}\,8x^3 + 0x^2 + 0x - 27} \\
\underline{8x^3 - 12x^2} \\
12x^2 + 0x \\
\underline{12x^2 - 18x} \\
18x - 27 \\
\underline{18x - 27} \\
0
\end{array}
$$

Quotient: $4x^2 + 6x + 9$

$$
\begin{aligned}
g(x) &= 0 \\
2x - 3 &= 0 \\
2x &= 3 \\
x &= \tfrac{3}{2}
\end{aligned}
$$

For Exercises 65–76, let $f(x) = x^2 - 9$, $g(x) = 2x$, and $h(x) = x - 3$.

65. $\left(\dfrac{f}{g}\right)(x) = \dfrac{f(x)}{g(x)} = \dfrac{x^2 - 9}{2x}$

We must exclude any values of x that make the denominator equal to zero, that is, $x \neq 0$.

67. $\left(\dfrac{f}{g}\right)(2) = \dfrac{f(2)}{g(2)} = \dfrac{2^2 - 9}{2(2)} = \dfrac{-5}{4} = -\dfrac{5}{4}$

69. $\left(\dfrac{h}{g}\right)(x) = \dfrac{h(x)}{g(x)} = \dfrac{x - 3}{2x},\ x \neq 0$

71. $\left(\dfrac{h}{g}\right)(3) = \dfrac{h(3)}{g(3)} = \dfrac{(3) - 3}{2(3)} = \dfrac{0}{6} = 0$

73. $\left(\dfrac{f}{g}\right)\left(\dfrac{1}{2}\right) = \dfrac{f\left(\frac{1}{2}\right)}{g\left(\frac{1}{2}\right)} = \dfrac{\left(\frac{1}{2}\right)^2 - 9}{2\left(\frac{1}{2}\right)} = \dfrac{\frac{1}{4} - \frac{36}{4}}{1}$
$$= -\dfrac{35}{4}$$

75. $\left(\dfrac{h}{g}\right)\left(-\dfrac{1}{2}\right) = \dfrac{h\left(-\frac{1}{2}\right)}{g\left(-\frac{1}{2}\right)} = \dfrac{-\frac{1}{2} - 3}{2\left(-\frac{1}{2}\right)} = \dfrac{-\frac{7}{2}}{-1} = \dfrac{7}{2}$

77. $9 \cdot 6 + 9 \cdot r^2 = 9(6 + r^2)$

79. $7(2x) - 7(3z) = 7(2x - 3z)$

81. $3x(x + 1) + 4(x + 1) = (x + 1)(3x + 4)$

83. $18z^3 w\left(zw^2\right)^4 = 18z^3 wz^4\left(w^2\right)^4$
$$= 18z^3 z^4 w^1 \cdot w^{2 \cdot 4}$$
$$= 18z^{3+4} w^{1+8}$$
$$= 18z^7 w^9$$

85. $12p^4 q^{-2}(5pq)^{-1} = 12p^4 q^{-2} 5^{-1} p^{-1} q^{-1}$
$$= 12 \cdot 5^{-1} p^{4+(-1)} q^{-2+(-1)}$$
$$= 12 \cdot 5^{-1} p^3 q^{-3}$$
$$= \dfrac{12p^3}{5q^3}$$

Chapter 5 Review Exercises

1. $4^3 = 4 \cdot 4 \cdot 4 = 64$

2. $\left(\dfrac{1}{3}\right)^4 = \dfrac{1}{3} \cdot \dfrac{1}{3} \cdot \dfrac{1}{3} \cdot \dfrac{1}{3} = \dfrac{1}{81}$

3. $(-5)^3 = (-5)(-5)(-5) = -125$

4. $\dfrac{2}{(-3)^{-2}} = \dfrac{2}{\dfrac{1}{(-3)^2}} = \dfrac{2}{1} \cdot \dfrac{(-3)^2}{1}$
$$= 2 \cdot (-3)^2$$
$$= 2 \cdot (-3)(-3)$$
$$= 18$$

5. $\left(\dfrac{2}{3}\right)^{-4} = \left(\dfrac{3}{2}\right)^4$
$$= \dfrac{3}{2} \cdot \dfrac{3}{2} \cdot \dfrac{3}{2} \cdot \dfrac{3}{2}$$
$$= \dfrac{81}{16}$$

6. $\left(\dfrac{5}{4}\right)^{-2} = \left(\dfrac{4}{5}\right)^2 = \dfrac{4}{5} \cdot \dfrac{4}{5} = \dfrac{16}{25}$

7. $5^{-1} + 6^{-1} = \dfrac{1}{5} + \dfrac{1}{6} = \dfrac{6}{30} + \dfrac{5}{30} = \dfrac{11}{30}$

8. $(5 + 6)^{-1} = 11^{-1} = \dfrac{1}{11}$

9. $-3^0 + 3^0 = -1 + 1 = 0$

10. $\left(3^{-4}\right)^2 = 3^{(-4) \cdot 2} = 3^{-8} = \dfrac{1}{3^8}$

11. $\left(x^{-4}\right)^{-2} = x^{-4(-2)} = x^8$

12. $\left(xy^{-3}\right)^{-2} = x^{1(-2)} y^{(-3)(-2)}$
$$= x^{-2} y^6$$
$$= \dfrac{1}{x^2} \cdot y^6 = \dfrac{y^6}{x^2}$$

13. $\left(z^{-3}\right)^3 z^{-6} = z^{-9} z^{-6}$
$$= z^{-9+(-6)}$$
$$= z^{-15} = \dfrac{1}{z^{15}}$$

14. $\left(5m^{-3}\right)^2 \left(m^4\right)^{-3} = 5^2 m^{-6} m^{-12}$
$$= 25m^{-6-12}$$
$$= 25m^{-18} = \dfrac{25}{m^{18}}$$

15. $\dfrac{(3r)^2 r^4}{r^{-2} r^{-3}} \left(9r^{-3}\right)^{-2} = \dfrac{3^2 r^2 r^4 9^{-2} r^6}{r^{-2} r^{-3}}$
$$= \dfrac{9^1 9^{-2} r^{2+4+6}}{r^{-2-3}}$$
$$= \dfrac{9^{-1} r^{12}}{r^{-5}}$$
$$= \dfrac{r^{12-(-5)}}{9} = \dfrac{r^{17}}{9}$$

16. $\left(\dfrac{5z^{-3}}{z^{-1}}\right)\dfrac{5}{z^2} = \dfrac{25z^{-3}}{z^{-1+2}}$

$\qquad = \dfrac{25}{z^3 z^1}$

$\qquad = \dfrac{25}{z^{3+1}} = \dfrac{25}{z^4}$

17. $\left(\dfrac{6m^{-4}}{m^{-9}}\right)^{-1}\left(\dfrac{m^{-2}}{16}\right) = \dfrac{6^{-1}m^4}{m^9} \cdot \dfrac{m^{-2}}{16}$

$\qquad = \dfrac{1}{6 \cdot 16}m^{4+(-2)-9}$

$\qquad = \dfrac{1}{96}m^{-7} = \dfrac{1}{96m^7}$

18. $\left(\dfrac{3r^5}{5r^{-3}}\right)^{-2}\left(\dfrac{9r^{-1}}{2r^{-5}}\right)^3$

$\qquad = \dfrac{3^{-2}r^{-10}}{5^{-2}r^6} \cdot \dfrac{9^3 r^{-3}}{2^3 r^{-15}}$

$\qquad = \dfrac{5^2}{3^2}r^{-10-6} \cdot \dfrac{9^3}{2^3}r^{-3-(-15)}$

$\qquad = \dfrac{25}{9}r^{-16} \cdot \dfrac{729}{8}r^{12}$

$\qquad = \dfrac{(25)(729)}{(9)(8)}r^{-16+12}$

$\qquad = \dfrac{2025}{8}r^{-4} = \dfrac{2025}{8r^4}$

19. $\left(-3x^4 y^3\right)\left(4x^{-2}y^5\right) = -3(4)x^4 x^{-2}y^3 y^5$

$\qquad = -12x^{4-2}y^{3+5}$

$\qquad = -12x^2 y^8$

20. $\dfrac{6m^{-4}n^3}{-3mn^2} = -2m^{-4-1}n^{3-2}$

$\qquad = -2m^{-5}n^1$

$\qquad = -\dfrac{2n}{m^5} \ \text{ or } \ \dfrac{-2n}{m^5}$

21. $\dfrac{(5p^{-2}q)(4p^5 q^{-3})}{2p^{-5}q^5} = \dfrac{20p^{-2+5}q^{1-3}}{2p^{-5}q^5}$

$\qquad = \dfrac{10p^3 q^{-2}}{p^{-5}q^5}$

$\qquad = 10p^{3-(-5)}q^{-2-5}$

$\qquad = 10p^8 q^{-7}$

$\qquad = \dfrac{10p^8}{q^7}$

22. $\left(\dfrac{a^{-2}b^{-1}}{3a^2}\right)^{-2}\left(\dfrac{b^{-2} \cdot 3a^4}{2b^{-3}}\right)^{-2}\left(\dfrac{a^{-4}b^5}{a^3}\right)^{-2}$

$\qquad = \dfrac{a^4 b^2 b^4 3^{-2}a^{-8}a^8 b^{-10}}{3^{-2}a^{-4}2^{-2}b^6 a^{-6}}$

$\qquad = \dfrac{4a^{4-8+8}b^{2+4-10}}{a^{-4-6}b^6}$

$\qquad = \dfrac{4a^4 b^{-4}}{a^{-10}b^6}$

$\qquad = 4a^{4-(-10)}b^{-4-6}$

$\qquad = 4a^{14}b^{-10} = \dfrac{4a^{14}}{b^{10}}$

23. In $(-6)^0$, the base is -6 and the expression simplifies to 1. In -6^0, the base is 6 and the expression simplifies to -1.

24. For example, if $a = 4$, then $(2a)^{-3} = (2 \cdot 4)^{-3} = 8^{-3} = \dfrac{1}{512}$, while $\dfrac{2}{a^3} = \dfrac{2}{4^3} = \dfrac{2}{64} = \dfrac{1}{32}$. $\dfrac{1}{512} \neq \dfrac{1}{32}$.

25. Yes, $\left(\dfrac{a}{b}\right)^{-1} = \dfrac{a^{-1}}{b^{-1}}$ for all $a, b \neq 0$.

26. No, $(ab)^{-1} \neq ab^{-1}$ for all $a, b \neq 0$. For example, let $a = 3$ and $b = 4$. Then $(ab)^{-1} = (3 \cdot 4)^{-1} = 12^{-1} = \dfrac{1}{12}$, while $ab^{-1} = 3 \cdot 4^{-1} = 3 \cdot \dfrac{1}{4} = \dfrac{3}{4} \cdot \dfrac{1}{12} \neq \dfrac{3}{4}$.

27. Let $x = 2$ and $y = 3$. Then

$$\left(x^2 + y^2\right)^2 = \left(2^2 + 3^2\right)^2$$
$$= (4 + 9)^2$$
$$= 13^2 = 169,$$

and $\quad x^4 + y^4 = 2^4 + 3^4$
$$= 16 + 81 = 97.$$

Since $169 \neq 97$,

$$\left(x^2 + y^2\right)^2 \neq x^4 + y^4.$$

28. $13{,}450 = 1{\scriptstyle\wedge}3\,4\,5\,0.$

Place a caret to the right of the first nonzero digit. Count 4 places.

Since the number 1.345 is to be made larger, the exponent on 10 is positive.

$$13{,}450 = 1.345 \times 10^4$$

29. $0.000\,000\,076\,5 = 0.0\,0\,0\,0\,0\,0\,0\,7{\scriptstyle\wedge}65$

Count 8 places

Since the number 7.65 is to be made smaller, the exponent on 10 is negative.

$$0.000\,000\,076\,5 = 7.65 \times 10^{-8}$$

30. $0.138 = 0.1{\scriptstyle\wedge}38$

Count 1 place.

Since the number 1.38 is to be made smaller, the exponent on 10 is negative.

$$0.138 = 1.38 \times 10^{-1}$$

31. $281,400,000 = 2.814 \times 10^8$

$50,454 = 5.0454 \times 10^4$

$100 = 1 \times 10^2$

32. $1.21 \times 10^6 = 1,210,000$

Move the decimal point 6 places to the right because the exponent is positive. Attach extra zeros.

33. $5.8 \times 10^{-3} = 0.0058$

Move the decimal point 3 places to the left because the exponent is negative.

34. $\dfrac{16 \times 10^4}{8 \times 10^8} = \dfrac{16}{8} \times 10^{4-8}$

$= 2 \times 10^{-4} \text{ or } 0.0002$

35. $\dfrac{6 \times 10^{-2}}{4 \times 10^{-5}} = \dfrac{6}{4} \times 10^{-2-(-5)}$

$= 1.5 \times 10^3 \text{ or } 1500$

36. $\dfrac{0.000\,000\,016\,4}{0.0004} = \dfrac{1.64 \times 10^{-8}}{4 \times 10^{-4}}$

$= \dfrac{1.64}{4} \times 10^{-8-(-4)}$

$= 0.41 \times 10^{-4}$

$= 4.1 \times 10^{-5} \text{ or } 0.000\,041$

37. $\dfrac{0.0009 \times 12,000,000}{400,000}$

$= \dfrac{9 \times 10^{-4} \times 1.2 \times 10^7}{4 \times 10^5}$

$= \dfrac{9 \times 1.2}{4} \times \dfrac{10^{-4} \times 10^7}{10^5}$

$= \dfrac{10.8}{4} \times 10^{-4+7-5}$

$= 2.7 \times 10^{-2} \text{ or } 0.027$

38. The density D is the population P divided by the area A.

$$D = \frac{P}{A}$$

We want to find the area, so solve the formula for A.

$DA = P$

$A = \dfrac{P}{D}$

$= \dfrac{4.69 \times 10^5}{470}$

$= \dfrac{4.69 \times 10^5}{4.7 \times 10^2}$

$\approx 0.998 \times 10^{5-2}$

$= 0.998 \times 10^3 = 998$

The area is 998 mi^2.

39. (a) $5449 = 5\,{\scriptstyle\wedge}\,4\,4\,9.$

Count 3 places.

$= 5.449 \times 10^3$

(b) As in Exercise 38, use $A = \dfrac{P}{D}$.

$\dfrac{3.45 \times 10^5}{5.449 \times 10^3} = \dfrac{3.45}{5.449} \times \dfrac{10^5}{10^3}$

$\approx 0.63 \times 10^{5-3}$

$= 0.63 \times 10^2$

$= 63$

The area is approximately 63 mi^2.

40. The coefficient of $14p^5$ is 14.

41. The coefficient of $-z$ is -1.

42. The coefficient of $\frac{x}{10} = \frac{1}{10}x$ is $\frac{1}{10}$.

43. The coefficient of $504p^3r^5$ is 504.

44. $9k + 11k^3 - 3k^2$

(a) In descending powers of k, the polynomial is

$$11k^3 - 3k^2 + 9k.$$

(b) The polynomial is a trinomial since it has three terms.

(c) The degree of the polynomial is 3 since the highest power of k is 3.

45. $14m^6 + 9m^7$

(a) In descending powers of m, the polynomial is

$$9m^7 + 14m^6.$$

(b) The polynomial is a binomial since it has two terms.

(c) The degree of the polynomial is 7 since the highest power of m is 7.

46. $-5y^4 + 3y^3 + 7y^2 - 2y$

(a) The polynomial is already written in descending powers of y.

(b) The polynomial has four terms, so it is none of these choices.

(c) The degree of the polynomial is 4 since the highest power of y is 4.

47. $-7q^5r^3$

(a) The polynomial is already written in descending powers.

(b) The polynomial is a monomial since it has just one term.

(c) The degree is $5 + 3 = 8$, the sum of the exponents of this term.

48. One example of a polynomial in the variable x that has degree 5, is lacking a third-degree term, and is written in descending powers of the variable is

$$x^5 + 2x^4 - x^2 + x + 2.$$

49. Add by columns.

$$
\begin{array}{r}
3x^2 - 5x + 6 \\
-4x^2 + 2x - 5 \\
\hline
-1x^2 - 3x + 1
\end{array}
\text{ or } -x^2 - 3x + 1
$$

50. Subtract.

$$
\begin{array}{r}
-5y^3 \phantom{{}+ 8y} + 8y - 3 \\
4y^2 + 2y + 9 \\
\hline
\end{array}
$$

Change the signs in the second polynomial and add.

$$
\begin{array}{r}
-5y^3 \phantom{{}- 4y^2} + 8y - 3 \\
- 4y^2 - 2y - 9 \\
\hline
-5y^3 - 4y^2 + 6y - 12
\end{array}
$$

51. $\left(4a^3 - 9a + 15\right) - \left(-2a^3 + 4a^2 + 7a\right)$
$= 4a^3 - 9a + 15 + 2a^3 - 4a^2 - 7a$
$= 4a^3 + 2a^3 - 4a^2 - 9a - 7a + 15$
$= 6a^3 - 4a^2 - 16a + 15$

52. $\left(3y^2 + 2y - 1\right) + \left(5y^2 - 11y + 6\right)$
$= 3y^2 + 5y^2 + 2y - 11y - 1 + 6$
$= 8y^2 - 9y + 5$

53. To find the perimeter, add the measures of the three sides.

$\left(4x^2 + 2\right) + \left(6x^2 + 5x + 2\right) + \left(2x^2 + 3x + 1\right)$
$= 4x^2 + 6x^2 + 2x^2 + 5x + 3x + 2 + 2 + 1$
$= 12x^2 + 8x + 5$

The perimeter is $12x^2 + 8x + 5$.

54. $f(x) = -2x^2 + 5x + 7$

(a) $f(-2) = -2(-2)^2 + 5(-2) + 7$
$= -2(4) - 10 + 7$
$= -8 - 10 + 7$
$= -18 + 7 = -11$

(b) $f(3) = -2(3)^2 + 5(3) + 7$
$= -2(9) + 15 + 7$
$= -18 + 15 + 7$
$= -3 + 7 = 4$

55. $f(x) = 2x + 3, g(x) = 5x^2 - 3x + 2$

(a) $(f + g)(x) = f(x) + g(x)$
$= (2x + 3) + \left(5x^2 - 3x + 2\right)$
$= 5x^2 + 2x - 3x + 3 + 2$
$= 5x^2 - x + 5$

(b) $(f - g)(x) = f(x) - g(x)$
$= (2x + 3) - \left(5x^2 - 3x + 2\right)$
$= 2x + 3 - 5x^2 + 3x - 2$
$= -5x^2 + 5x + 1$

(c) $(f + g)(-1)$
$= f(-1) + g(-1)$
$= [2(-1) + 3] + \left[5(-1)^2 - 3(-1) + 2\right]$
$= [1] + [10]$
$= 11$

(d) $(f - g)(-1) = f(-1) - g(-1)$
$= 1 - 10 \quad \textit{from part (c)}$
$= -9$

56. $f(x) = 3x^2 + 2x - 1, g(x) = 5x + 7$

(a) $(g \circ f)(3) = g(f(3))$
$= g\left(3 \cdot 3^2 + 2 \cdot 3 - 1\right)$
$= g(32)$
$= 5 \cdot 32 + 7$
$= 167$

(b) $(f \circ g)(3) = f(g(3))$
$= f(5 \cdot 3 + 7)$
$= f(22)$
$= 3 \cdot 22^2 + 2 \cdot 22 - 1$
$= 1495$

(c) $(f \circ g)(-2) = f(g(-2))$
$= f[5(-2) + 7]$
$= f(-3)$
$= 3(-3)^2 + 2(-3) - 1$
$= 20$

(d) $(g \circ f)(-2) = g(f(-2))$
$= g[3(-2)^2 + 2(-2) - 1]$
$= g(7)$
$= 5 \cdot 7 + 7$
$= 42$

(e) $(f \circ g)(x)$
$= f(g(x))$
$= f(5x + 7)$
$= 3(5x + 7)^2 + 2(5x + 7) - 1$
$= 3\left(25x^2 + 70x + 49\right) + 10x + 14 - 1$
$= 75x^2 + 210x + 147 + 10x + 13$
$= 75x^2 + 220x + 160$

(f) $(g \circ f)(x) = g(f(x))$
$= g\left(3x^2 + 2x - 1\right)$
$= 5\left(3x^2 + 2x - 1\right) + 7$
$= 15x^2 + 10x - 5 + 7$
$= 15x^2 + 10x + 2$

57. $f(x) = -0.574x^2 + 6.01x + 15.9$

  (a) The year 1995 corresponds to $x = 0$.

$$f(0) = -0.574(0)^2 + 6.01(0) + 15.9$$
$$= 15.9 \text{ million}$$

  (b) $x = 2000 - 1995 = 5$

$$f(5) = -0.574(5)^2 + 6.01(5) + 15.9$$
$$= 31.6 \text{ million}$$

  (c) $x = 2003 - 1995 = 8$

$$f(8) = -0.574(8)^2 + 6.01(8) + 15.9$$
$$= 27.244 \text{ million}$$

58.

x	$f(x) = -2x + 5$
-2	$-2(-2) + 5 = 9$
-1	$-2(-1) + 5 = 7$
0	$-2(0) + 5 = 5$
1	$-2(1) + 5 = 3$
2	$-2(2) + 5 = 1$

This is a linear function, so plot the points and draw a line through them.

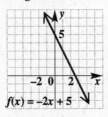

Any x-value can be used, so the domain is $(-\infty, \infty)$. From the graph, we see that any y-value can be obtained from the function, so the range is $(-\infty, \infty)$.

59.

x	$f(x) = x^2 - 6$
-2	$(-2)^2 - 6 = -2$
-1	$(-1)^2 - 6 = -5$
0	$(0)^2 - 6 = -6$
1	$(1)^2 - 6 = -5$
2	$(2)^2 - 6 = -2$

Since the greatest exponent is 2, the graph of f is a parabola.

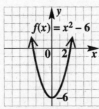

Any x-value can be used, so the domain is $(-\infty, \infty)$. From the graph, we see that the y-values are at least -6, so the range is $[-6, \infty)$.

60.

x	$f(x) = -x^3 + 1$
-2	$-(-2)^3 + 1 = 9$
-1	$-(-1)^3 + 1 = 2$
0	$-(0)^3 + 1 = 1$
1	$-(1)^3 + 1 = 0$
2	$-(2)^3 + 1 = -7$

The greatest exponent is 3, so the graph of f is s-shaped.

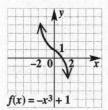

Any x-value can be used, so the domain is $(-\infty, \infty)$. From the graph, we see that any y-value can be obtained from the function, so the range is $(-\infty, \infty)$.

61. $-6k(2k^2 + 7) = -6k(2k^2) - 6k(7)$
$$= -12k^3 - 42k$$

62. $(3m - 2)(5m + 1)$

    F   **O**   **I**   **L**

$$= 15m^2 + 3m - 10m - 2$$
$$= 15m^2 - 7m - 2$$

63. $(3w - 2t)(2w - 3t)$

    F   **O**   **I**   **L**

$$= 6w^2 - 9wt - 4wt + 6t^2$$
$$= 6w^2 - 13wt + 6t^2$$

64. $(2p^2 + 6p)(5p^2 - 4)$

    F   **O**   **I**   **L**

$$= 10p^4 - 8p^2 + 30p^3 - 24p$$
$$= 10p^4 + 30p^3 - 8p^2 - 24p$$

65. $(3q^2 + 2q - 4)(q - 5)$
$$= (3q^2 + 2q - 4)(q) + (3q^2 + 2q - 4)(-5)$$
$$= 3q^3 + 2q^2 - 4q - 15q^2 - 10q + 20$$
$$= 3q^3 + 2q^2 - 15q^2 - 4q - 10q + 20$$
$$= 3q^3 - 13q^2 - 14q + 20$$

66. $(3z^3 - 2z^2 + 4z - 1)(3z - 2)$
$$= (3z^3 - 2z^2 + 4z - 1)(3z)$$
$$\quad + (3z^3 - 2z^2 + 4z - 1)(-2)$$
$$= 9z^4 - 6z^3 + 12z^2 - 3z$$
$$\quad - 6z^3 + 4z^2 - 8z + 2$$
$$= 9z^4 - 6z^3 - 6z^3 + 12z^2 + 4z^2$$
$$\quad - 3z - 8z + 2$$
$$= 9z^4 - 12z^3 + 16z^2 - 11z + 2$$

67. $(6r^2 - 1)(6r^2 + 1) = (6r^2)^2 - 1^2$
$$= 36r^4 - 1$$

68. $\left(z + \dfrac{3}{5}\right)\left(z - \dfrac{3}{5}\right) = z^2 - \left(\dfrac{3}{5}\right)^2$
$$= z^2 - \dfrac{9}{25}$$

69. $(4m + 3)^2 = (4m)^2 + 2(4m)(3) + 3^2$
$$= 16m^2 + 24m + 9$$

70. $(3t + 2)^2 = (3t)^2 + 2(3t)(2) + 2^2$
$$= 9t^2 + 12t + 4$$

Now multiply the last polynomial by t.
$$t(3t + 2)^2 = 9t^3 + 12t^2 + 4t$$

71. $\dfrac{4y^3 - 12y^2 + 5y}{4y} = \dfrac{4y^3}{4y} - \dfrac{12y^2}{4y} + \dfrac{5y}{4y}$
$$= y^2 - 3y + \dfrac{5}{4}$$

72. $\dfrac{x^3 - 9x^2 + 26x - 30}{x - 5}$

$$
\begin{array}{r}
x^2 - 4x + 6 \\
x - 5 \,\overline{\smash{\big)}\, x^3 - 9x^2 + 26x - 30} \\
\underline{x^3 - 5x^2} \\
-4x^2 + 26x \\
\underline{-4x^2 + 20x} \\
6x - 30 \\
\underline{6x - 30} \\
0
\end{array}
$$

Answer: $x^2 - 4x + 6$

73. $\dfrac{2p^3 + 9p^2 + 27}{2p - 3}$

$$
\begin{array}{r}
p^2 + 6p + 9 \\
2p - 3 \,\overline{\smash{\big)}\, 2p^3 + 9p^2 + 0p + 27} \\
\underline{2p^3 - 3p^2} \\
12p^2 + 0p \\
\underline{12p^2 - 18p} \\
18p + 27 \\
\underline{18p - 27} \\
54
\end{array}
$$

Remainder

Answer: $p^2 + 6p + 9 + \dfrac{54}{2p - 3}$

74. $\dfrac{5p^4 + 15p^3 - 33p^2 - 9p + 18}{5p^2 - 3}$

$$
\begin{array}{r}
p^2 + 3p - 6 \\
5p^2 - 3 \,\overline{\smash{\big)}\, 5p^4 + 15p^3 - 33p^2 - 9p + 18} \\
\underline{5p^4 - 3p^2} \\
15p^3 - 30p^2 - 9p \\
\underline{15p^3 - 9p} \\
-30p^2 + 18 \\
\underline{-30p^2 + 18} \\
0
\end{array}
$$

Answer: $p^2 + 3p - 6$

75. **[5.1]** **(a)** $4^{-2} = \dfrac{1}{4^2} = \dfrac{1}{16}$ **(A)**

(b) $-4^2 = -(4^2) = -16$ **(G)**

(c) $4^0 = 1$ **(C)**

(d) $(-4)^0 = 1$ **(C)**

(e) $(-4)^{-2} = \dfrac{1}{(-4)^2} = \dfrac{1}{16}$ **(A)**

(f) $-4^0 = -(4^0) = -1$ **(E)**

(g) $-4^0 + 4^0 = -1 + 1 = 0$ **(B)**

(h) $-4^0 - 4^0 = (-1) - 1 = -2$ **(H)**

(i) $4^{-2} + 4^{-1} = \dfrac{1}{4^2} + \dfrac{1}{4} = \dfrac{1}{16} + \dfrac{4}{16} = \dfrac{5}{16}$ **(F)**

76. **[5.1]** $\dfrac{6^{-1}y^3(y^2)^{-2}}{6y^{-4}(y^{-1})} = \dfrac{y^3 y^{-4}}{6^1 \cdot 6y^{-4}y^{-1}}$
$$= \dfrac{y^{3-4-(-4-1)}}{36}$$
$$= \dfrac{y^{-1+5}}{36} = \dfrac{y^4}{36}$$

77. **[5.1]** $5^{-3} = \dfrac{1}{5^3} = \dfrac{1}{5 \cdot 5 \cdot 5} = \dfrac{1}{125}$

78. **[5.1]** $(y^6)^{-5}(2y^{-3})^{-4}$
$$= y^{-30}(2)^{-4}y^{12}$$
$$= \dfrac{y^{-30+12}}{2^4}$$
$$= \dfrac{y^{-18}}{2^4} = \dfrac{1}{16y^{18}}$$

79. **[5.4]** $7p^5(3p^4 + p^3 + 2p^2)$
$$= 7p^5(3p^4) + 7p^5(p^3) + 7p^5(2p^2)$$
$$= 21p^9 + 7p^8 + 14p^7$$

80. **[5.4]** $(2x - 9)^2 = (2x)^2 - 2(2x)(9) + 9^2$
$$= 4x^2 - 36x + 81$$

81. **[5.1]** $\dfrac{(-z^{-2})^3}{5(z^{-3})^{-1}} = \dfrac{(-1)^3 z^{-2(3)}}{5z^{-3(-1)}}$

$\qquad = \dfrac{-z^{-6}}{5z^3}$

$\qquad = \dfrac{-z^{-6-3}}{5}$

$\qquad = \dfrac{-z^{-9}}{5} = -\dfrac{1}{5z^9}$

82. **[5.1]** $-(-3)^2 = -(9) = -9$

83. **[5.5]**

$$
\begin{array}{r}
8x \;+\; 1 \\
x-3\,\overline{\smash{\big)}\,8x^2 \;-\; 23x \;+\; 2} \\
\underline{8x^2 \;-\; 24x} \\
x \;+\; 2 \\
\underline{x \;-\; 3} \\
5
\end{array}
$$

$\quad$ Answer: $\;8x + 1 + \dfrac{5}{x-3}$

84. **[5.1]** $\dfrac{(5z^2x^3)^2 (2zx^2)^{-1}}{(-10zx^{-3})^{-2}(3z^{-1}x^{-4})^2}$

$\qquad = \dfrac{5^2 z^4 x^6 2^{-1} z^{-1} x^{-2}}{(-10)^{-2} z^{-2} x^6 3^2 z^{-2} x^{-8}}$

$\qquad = \dfrac{25(-10)^2 z^3 x^4}{2 \cdot 9 z^{-4} x^{-2}}$

$\qquad = \dfrac{25(100) z^7 x^6}{2 \cdot 9}$

$\qquad = \dfrac{1250 z^7 x^6}{9}$

85. **[5.4]** $[(3m - 5n) + p][(3m - 5n) - p]$

$\qquad = (3m - 5n)^2 - (p)^2$

$\qquad = (3m)^2 - 2(3m)(5n) + (5n)^2 - p^2$

$\qquad = 9m^2 - 30mn + 25n^2 - p^2$

86. **[5.5]** $\dfrac{20y^3x^3 + 15y^4x + 25yx^4}{10yx^2}$

$\qquad = \dfrac{20y^3x^3}{10yx^2} + \dfrac{15y^4x}{10yx^2} + \dfrac{25yx^4}{10yx^2}$

$\qquad = 2y^2x + \dfrac{3y^3}{2x} + \dfrac{5x^2}{2}$

87. **[5.2]** $(2k - 1) - (3k^2 - 2k + 6)$

$\qquad = 2k - 1 - 3k^2 + 2k - 6$

$\qquad = -3k^2 + 2k + 2k - 1 - 6$

$\qquad = -3k^2 + 4k - 7$

88. **[5.1]** See the solution for Exercise 38.

$$A = \dfrac{P}{D}$$

$\qquad = \dfrac{4.0355 \times 10^6}{38.9}$

$\qquad = \dfrac{4.0355 \times 10^6}{3.89 \times 10^1}$

$\qquad \approx 1.03740 \times 10^5$

$\qquad = 103{,}740$

The area is approximately $103{,}740$ mi^2.

Chapter 5 Test

1. **(a)** $7^{-2} = \dfrac{1}{7^2} = \dfrac{1}{49}$ **(C)**

$\quad$ **(b)** $7^0 = 1$ **(A)**

$\quad$ **(c)** $-7^0 = -(1) = -1$ **(D)**

$\quad$ **(d)** $(-7)^0 = 1$ **(A)**

$\quad$ **(e)** $-7^2 = -49$ **(E)**

$\quad$ **(f)** $7^{-1} + 2^{-1} = \dfrac{1}{7} + \dfrac{1}{2}$

$\qquad\qquad = \dfrac{2}{14} + \dfrac{7}{14} = \dfrac{9}{14}$ **(F)**

$\quad$ **(g)** $(7 + 2)^{-1} = 9^{-1} = \dfrac{1}{9}$ **(B)**

$\quad$ **(h)** $\dfrac{7^{-1}}{2^{-1}} = \dfrac{2^1}{7^1} = \dfrac{2}{7}$ **(G)**

$\quad$ **(i)** $(-7)^{-2} = \dfrac{1}{(-7)^2} = \dfrac{1}{49}$ **(C)**

2. $\quad (3x^{-2}y^3)^{-2} (4x^3y^{-4})$

$\qquad = 3^{-2} x^{-2(-2)} y^{3(-2)} 4x^3 y^{-4}$

$\qquad = 3^{-2} x^4 y^{-6} 4 x^3 y^{-4}$

$\qquad = \dfrac{4x^{4+3} y^{-6-4}}{3^2}$

$\qquad = \dfrac{4x^7 y^{-10}}{9} = \dfrac{4x^7}{9y^{10}}$

3. $\dfrac{36r^{-4}(r^2)^{-3}}{6r^4} = \dfrac{36r^{-4}r^{2(-3)}}{6r^4}$

$\qquad = \dfrac{6r^{-4}r^{-6}}{r^4}$

$\qquad = \dfrac{6r^{-10}}{r^4} = \dfrac{6}{r^{14}}$

4. $\left(\dfrac{4p^2}{q^4}\right)^3 \left(\dfrac{6p^8}{q^{-8}}\right)^{-2}$

$= \dfrac{4^3 p^6}{q^{12}} \cdot \dfrac{6^{-2} p^{-16}}{q^{16}}$

$= \dfrac{4^3 p^{-10}}{6^2 q^{28}}$

$= \dfrac{64}{36 p^{10} q^{28}} = \dfrac{16}{9 p^{10} q^{28}}$

5. $\left(-2x^4 y^{-3}\right)^0 \left(-4x^{-3} y^{-8}\right)^2$

$= 1(-4)^2 x^{-6} y^{-16}$

$= \dfrac{16}{x^6 y^{16}}$

6. $9.1 \times 10^{-7} = 0.000\,000\,91$

Move the decimal point 7 places to the left because the exponent is negative.

7. $\dfrac{(2{,}500{,}000)(0.00003)}{(0.05)(5{,}000{,}000)}$

$= \dfrac{(2.5 \times 10^6)(3 \times 10^{-5})}{(5 \times 10^{-2})(5 \times 10^6)}$

$= \dfrac{7.5 \times 10^1}{25 \times 10^4}$

$= 0.3 \times 10^{1-4}$

$= 0.3 \times 10^{-3}$

$= 3 \times 10^{-4}, \quad \text{or} \quad 0.0003$

8. $f(x) = -2x^2 + 5x - 6,\ g(x) = 7x - 3$

(a) $f(x) = -2x^2 + 5x - 6$

$f(4) = -2(4)^2 + 5(4) - 6$

$= -2 \cdot 16 + 20 - 6$

$= -32 + 20 - 6$

$= -12 - 6 = -18$

(b) $(f + g)(x) = f(x) + g(x)$

$= \left(-2x^2 + 5x - 6\right) + (7x - 3)$

$= -2x^2 + 12x - 9$

(c) $(f - g)(x) = f(x) - g(x)$

$= \left(-2x^2 + 5x - 6\right) - (7x - 3)$

$= -2x^2 + 5x - 6 - 7x + 3$

$= -2x^2 - 2x - 3$

(d) Using the answer in part (c), we have

$(f - g)(-2) = -2(-2)^2 - 2(-2) - 3$

$= -8 + 4 - 3 = -7.$

9. $f(x) = 3x + 5,\ g(x) = x^2 + 2$

(a) $(f \circ g)(-2) = f[g(-2)]$

$= f\left[(-2)^2 + 2\right]$

$= f(6)$

$= 3 \cdot 6 + 5$

$= 23$

(b) $(f \circ g)(x) = f[g(x)]$

$= f\left(x^2 + 2\right)$

$= 3\left(x^2 + 2\right) + 5$

$= 3x^2 + 6 + 5$

$= 3x^2 + 11$

(c) $(g \circ f)(x) = g[f(x)]$

$= g(3x + 5)$

$= (3x + 5)^2 + 2$

$= 9x^2 + 30x + 25 + 2$

$= 9x^2 + 30x + 27$

10.

x	$f(x) = -2x^2 + 3$
-2	$-2(-2)^2 + 3 = -5$
-1	$-2(-1)^2 + 3 = 1$
0	$-2(0)^2 + 3 = 3$
1	$-2(1)^2 + 3 = 1$
2	$-2(2)^2 + 3 = -5$

Since the greatest exponent is 2, the graph of f is a parabola.

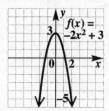

11.

x	$f(x) = -x^3 + 3$
-2	$-(-2)^3 + 3 = 11$
-1	$-(-1)^3 + 3 = 4$
0	$-(0)^3 + 3 = 3$
1	$-(1)^3 + 3 = 2$
2	$-(2)^3 + 3 = -5$

Since the greatest exponent is 3, the graph of f is a cubic (s-shaped).

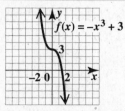

12. $f(x) = -0.141x^2 + 21.5x + 616$

$f(x)$ represents the number of medical doctors, in thousands, in the United States.

	Year	x	$f(x)$
(a)	1990	0	616
(b)	1996	6	$739.924 \approx 740$
(c)	2002	12	$853.696 \approx 854$

13. $(4x^3 - 3x^2 + 2x - 5)$

$\quad - (3x^3 + 11x + 8) + (x^2 - x)$

$= 4x^3 - 3x^2 + 2x - 5 - 3x^3 - 11x$

$\quad - 8 + x^2 - x$

$= x^3 - 2x^2 - 10x - 13$

14. $(5x - 3)(2x + 1)$

$\qquad$ **F** $\quad$ **O** $\quad$ **I** $\quad$ **L**

$= 10x^2 + 5x - 6x - 3$

$= 10x^2 - x - 3$

15. $(2m - 5)(3m^2 + 4m - 5)$

$= 2m(3m^2 + 4m - 5)$

$\quad + (-5)(3m^2 + 4m - 5)$

$= 6m^3 + 8m^2 - 10m$

$\quad - 15m^2 - 20m + 25$

$= 6m^3 - 7m^2 - 30m + 25$

16. $(6x + y)(6x - y) = (6x)^2 - y^2$

$\qquad\qquad\qquad\quad = 36x^2 - y^2$

17. $(3k + q)^2 = (3k)^2 + 2(3k)(q) + q^2$

$\qquad\qquad\quad = 9k^2 + 6kq + q^2$

18. $[2y + (3z - x)][2y - (3z - x)]$

$= (2y)^2 - (3z - x)^2$

$= 4y^2 - (9z^2 - 6zx + x^2)$

$= 4y^2 - 9z^2 + 6zx - x^2$

19. $\dfrac{16p^3 - 32p^2 + 24p}{4p^2}$

$= \dfrac{16p^3}{4p^2} - \dfrac{32p^2}{4p^2} + \dfrac{24p}{4p^2}$

$= 4p - 8 + \dfrac{6}{p}$

20. $(x^3 + 3x^2 - 4) \div (x - 1)$

Insert $0x$ for the missing x-term.

$$
\begin{array}{r}
x^2 + 4x + 4 \\
x - 1 \overline{\smash{\big)}\, x^3 + 3x^2 + 0x - 4} \\
\underline{x^3 - x^2} \\
4x^2 + 0x \\
\underline{4x^2 - 4x} \\
4x - 4 \\
\underline{4x - 4} \\
0
\end{array}
$$

Answer: $x^2 + 4x + 4$

21. $f(x) = x^2 + 3x + 2,\ g(x) = x + 1$

(a) $(fg)(x) = f(x) \cdot g(x)$

$= (x^2 + 3x + 2)(x + 1)$

$= (x^2 + 3x + 2)(x)$

$\quad + (x^2 + 3x + 2)(1)$

$= x^3 + 3x^2 + 2x + x^2 + 3x + 2$

$= x^3 + 4x^2 + 5x + 2$

(b) $(fg)(-2) = f(-2) \cdot g(-2)$

$= [(-2)^2 + 3(-2) + 2] \cdot [(-2) + 1]$

$= [4 - 6 + 2] \cdot [-1]$

$= 0(-1) = 0$

Alternatively, we could have substituted -2 for x into our answer from part (a).

22. **(a)** $\left(\dfrac{f}{g}\right)(x) = \dfrac{f(x)}{g(x)} = \dfrac{x^2 + 3x + 2}{x + 1}$

$$
\begin{array}{r}
x + 2 \\
x + 1 \overline{\smash{\big)}\, x^2 + 3x + 2} \\
\underline{x^2 + x} \\
2x + 2 \\
\underline{2x + 2} \\
0
\end{array}
$$

Thus, $\left(\dfrac{f}{g}\right)(x) = x + 2$ if $x + 1 \neq 0$, that is, $x \neq -1$.

(b) Using our answer from part (a),

$$\left(\dfrac{f}{g}\right)(-2) = (-2) + 2 = 0.$$

Cumulative Review Exercises (Chapters 1–5)

1. 34 is a natural number, so it is also a whole number, an integer, a rational number, and a real number.   **A, B, C, D, F**

2. 0 is a whole number, so it is also an integer, a rational number, and a real number.   **B, C, D, F**

3. 2.16 is a rational number, so it is also a real number.   **D, F**

4. $-\sqrt{36} = -6$ is an integer, so it is also a rational number and a real number.   **C, D, F**

5. $\sqrt{13}$ is an irrational number, so it is also a real number.   **E, F**

6. $-\frac{4}{5}$ is a rational number, so it is also a real number.   **D, F**

7. $9 \cdot 4 - 16 \div 4 = (9 \cdot 4) - (16 \div 4) = 36 - 4 = 32$

8. $\left(\frac{1}{3}\right)^2 - \left(\frac{1}{2}\right)^3 = \frac{1}{9} - \frac{1}{8}$

$\qquad\qquad\qquad = \frac{8}{72} - \frac{9}{72} = -\frac{1}{72}$

9. $-|8 - 13| - |-4| + |-9| = -|-5| - 4 + 9$

$\qquad\qquad\qquad\qquad\quad = -5 - 4 + 9$

$\qquad\qquad\qquad\qquad\quad = -9 + 9 = 0$

10. $-5(8 - 2z) + 4(7 - z) = 7(8 + z) - 3$

$-40 + 10z + 28 - 4z = 56 + 7z - 3$

$\qquad\qquad\qquad\qquad\qquad$ *Distributive property*

$\qquad 6z - 12 = 7z + 53 \quad$ *Combine like terms.*

$\qquad\quad -65 = z \qquad$ *Subtract 6z; 53.*

Thus, the solution set is $\{-65\}$.

11. $3(x + 2) - 5(x + 2) = -2x - 4$

$3x + 6 - 5x - 10 = -2x - 4$

$\qquad\quad -2x - 4 = -2x - 4$

The last statement is true for all real numbers, so the solution set is $(-\infty, \infty)$.

12. Solve $A = p + prt$ for t.

$\qquad\quad A = p + prt$

$\quad A - p = prt \qquad$ *Subtract p.*

$\quad \dfrac{A - p}{pr} = t \qquad$ *Divide by pr.*

13. $2(m + 5) - 3m + 1 > 5$

$2m + 10 - 3m + 1 > 5$

$\qquad\quad -m + 11 > 5$

$\qquad\qquad -m > -6$

$\qquad\qquad\quad m < 6$

The solution set is $(-\infty, 6)$.

14. $|3x - 1| = 2$

$3x - 1 = 2 \quad$ or $\quad 3x - 1 = -2$

$\quad 3x = 3 \qquad\qquad\quad 3x = -1$

$\qquad x = 1 \quad$ or $\qquad x = -\frac{1}{3}$

The solution set is $\left\{-\frac{1}{3}, 1\right\}$.

15. $|3z + 1| \geq 7$

$3z + 1 \geq 7 \quad$ or $\quad 3z + 1 \leq -7$

$\quad 3z \geq 6 \qquad\qquad\quad 3z \leq -8$

$\qquad z \geq 2 \quad$ or $\qquad z \leq -\frac{8}{3}$

The solution set is $\left(-\infty, -\frac{8}{3}\right] \cup [2, \infty)$.

16. $|x + 1| < -3$

There is no number whose absolute value is less than -3, so this inequality has no solution. The solution set is $\emptyset$.

17. Personal computer: $\frac{480}{1500} = 32\%$

Pacemaker:

$\quad 26\%$ of $1500 = 0.26(1500) = 390$

Wireless communication:

$\quad 18\%$ of $1500 = 0.18(1500) = 270$

Television: $\frac{150}{1500} = 10\%$

18. The sum of the measures of the angles of any triangle is $180°$, so

$\quad (x + 15) + (6x + 10) + (x - 5) = 180.$

Solve this equation.

$\qquad\qquad 8x + 20 = 180$

$\qquad\qquad\quad 8x = 160$

$\qquad\qquad\qquad x = 20$

Substitute 20 for x to find the measures of the angles.

$\qquad x - 5 = 20 - 5 = 15$

$\qquad x + 15 = 20 + 15 = 35$

$\qquad 6x + 10 = 6(20) + 10 = 130$

The measures of the angles of the triangle are $15°$, $35°$, and $130°$.

19. Through $(-4, 5)$ and $(2, -3)$

Use the definition of slope with $x_1 = -4$, $y_1 = 5$, $x_2 = 2$, and $y_2 = -3$.

$$m = \frac{y_2 - y_1}{x_2 - x_1} = \frac{-3 - 5}{2 - (-4)} = \frac{-8}{6} = -\frac{4}{3}$$

20. Horizontal, through $(4, 5)$

The slope of every horizontal line is 0.

21. Through $(4, -1)$, $m = -4$

(a) Use the point-slope form with $x_1 = 4$, $y_1 = -1$, and $m = -4$.

$$y - y_1 = m(x - x_1)$$
$$y - (-1) = -4(x - 4)$$
$$y + 1 = -4x + 16$$
$$y = -4x + 15$$

(b) The standard form is $Ax + By = C$.

$$y = -4x + 15$$
$$4x + y = 15$$

22. Through $(0, 0)$ and $(1, 4)$

    (a) Find the slope.

$$m = \frac{4 - 0}{1 - 0} = 4$$

Because the slope is 4 and the y-intercept is 0, the equation of the line in slope-intercept form is

$$y = 4x.$$

    (b) The standard form is $Ax + By = C$.

$$y = 4x$$
$$0 = 4x - y \quad \text{or} \quad 4x - y = 0$$

23. $-3x + 4y = 12$

If $y = 0$, $x = -4$, so the x-intercept is $(-4, 0)$.

If $x = 0$, $y = 3$, so the y-intercept is $(0, 3)$.

Draw a line through these intercepts. A third point may be used as a check.

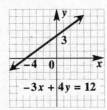

24. $y \leq 2x - 6$

Graph the boundary, $y = 2x - 6$, as a solid line through the intercepts $(3, 0)$ and $(0, -6)$. A third point such as $(1, -4)$ can be used as a check. Using $(0, 0)$ as a test point results in the false inequality $0 \leq -6$, so shade the region *not* containing the origin. This is the region below the line. The solid line shows that the boundary is part of the graph.

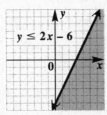

25. $3x + 2y < 0$

Graph the boundary, $3x + 2y = 0$, as a dashed line through $(0, 0)$, $(-2, 3)$, and $(2, -3)$. Choose a test point not on the line. Using $(1, 1)$ results in the false statement $5 < 0$, so shade the region *not* containing $(1, 1)$. This is the region below the line. The dashed line shows that the boundary is not part of the graph.

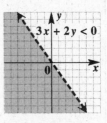

26. **(a)** $m = \dfrac{128{,}665 - 96{,}445}{10 - 0} = \dfrac{32{,}220}{10} = 3222$

The average rate of change is 3222 twin births per year, that is, the number of twin births increased an average of 3222 per year.

    (b) $y = mx + b$

           $y = 3222x + 96{,}445$

    (c) If $x = 2006 - 1993 = 13$, then

$$y = 3222(13) + 96{,}445$$
$$= 138{,}331$$

The model predicts about 138,331 twin births in 2006.

27. $\{(-4, -2), (-1, 0), (2, 0), (5, 2)\}$

The domain is the set of first components, that is, $\{-4, -1, 2, 5\}$.
The range is the set of second components, that is, $\{-2, 0, 2\}$.
The relation is a function since each first component is paired with a unique second component.

28. $g(x) = -x^2 - 2x + 6$

     $g(3) = -3^2 - 2(3) + 6$

           $= -9 - 6 + 6$

           $= -9$

29. $3x - 4y = 1$ (1)

     $2x + 3y = 12$ (2)

To eliminate y, multiply equation (1) by 3 and equation (2) by 4. Then add the results.

$$
\begin{array}{rll}
9x - 12y &= 3 & 3 \times (1) \\
8x + 12y &= 48 & 4 \times (2) \\
\hline
17x &= 51 & \\
x &= 3 &
\end{array}
$$

Since $x = 3$,

$$
\begin{array}{rl}
3x - 4y &= 1 \quad (1) \\
3(3) - 4y &= 1 \\
9 - 4y &= 1 \\
-4y &= -8 \\
y &= 2
\end{array}
$$

The solution set is $\{(3, 2)\}$.

30.

$$3x - 2y = 4 \quad (1)$$
$$-6x + 4y = 7 \quad (2)$$

Multiply equation (1) by 2 and add the result to equation (2).

$$
\begin{array}{rcl}
6x - 4y &=& 8 \qquad 2 \times (1) \\
-6x + 4y &=& 7 \quad (2) \\
\hline
0 &=& 15 \; \textit{False}
\end{array}
$$

Since a false statement results, the system is *inconsistent*. The solution set is $\emptyset$.

31.

$$x + 3y - 6z = 7 \quad (1)$$
$$2x - y + z = 1 \quad (2)$$
$$x + 2y + 2z = -1 \quad (3)$$

To eliminate x, multiply equation (1) by -2 and add the result to equation (2).

$$
\begin{array}{rcl}
-2x - 6y + 12z &=& -14 \qquad -2 \times (1) \\
2x - y + z &=& 1 \quad (2) \\
\hline
-7y + 13z &=& -13 \quad (4)
\end{array}
$$

To eliminate x again, multiply equation (3) by -2 and add the result to equation (2).

$$
\begin{array}{rcl}
2x - y + z &=& 1 \quad (2) \\
-2x - 4y - 4z &=& 2 \qquad -2 \times (3) \\
\hline
-5y - 3z &=& 3 \quad (5)
\end{array}
$$

Use equations (4) and (5) to eliminate z. Multiply equation (4) by 3 and add the result to 13 times equation (5).

$$
\begin{array}{rcl}
-21y + 39z &=& -39 \quad 3 \times (4) \\
-65y - 39z &=& 39 \quad 13 \times (5) \\
\hline
-86y &=& 0 \\
y &=& 0
\end{array}
$$

From (5), $-3z = 3$, so $z = -1$.
From (3), $x - 2 = -1$, so $x = 1$.

The solution set is $\{(1, 0, -1)\}$.

32. The length L of the rectangular flag measured 12 feet more than its width W, so

$$L = W + 12. \quad (1)$$

The perimeter is 144 feet.

$$P = 2L + 2W \quad (2)$$

Substitute $W + 12$ for L into equation (2).

$$144 = 2(W + 12) + 2W$$
$$144 = 2W + 24 + 2W$$
$$120 = 4W$$
$$30 = W$$

From (1), $L = 30 + 12 = 42$.

The length is 42 feet and the width is 30 feet.

33. Make a chart.

Number of Liters of Solution	Percent (as a decimal)	Pure Liters of Alcohol
x	0.15	$0.15x$
y	0.30	$0.30y$
9	0.20	$0.20(9) = 1.8$

From the first and third columns, we have the following system:

$$x + y = 9 \quad (1)$$
$$0.15x + 0.30y = 1.8 \quad (2)$$

To eliminate x, multiply equation (1) by -15 and add the result to 100 times equation (2).

$$
\begin{array}{rcl}
-15x - 15y &=& -135 \\
15x + 30y &=& 180 \\
\hline
15y &=& 45 \\
y &=& 3
\end{array}
$$

From (1), $x + 3 = 9$, so $x = 6$.

She should use 6 L of 15% solution and 3 L of 30% solution.

34. $\left(\dfrac{2m^3 n}{p^2}\right)^3 = \dfrac{2^3 (m^3)^3 n^3}{(p^2)^3} = \dfrac{8m^9 n^3}{p^6}$

35. $\dfrac{x^{-6} y^3 z^{-1}}{x^7 y^{-4} z} = \dfrac{y^4 y^3}{x^6 x^7 z^1 z} = \dfrac{y^7}{x^{13} z^2}$

36. $\left(2m^{-2} n^3\right)^{-3}$

$$= 2^{-3} \left(m^{-2}\right)^{-3} \left(n^3\right)^{-3}$$

$$= 2^{-3} m^{(-2)(-3)} n^{3(-3)}$$

$$= 2^{-3} m^6 n^{-9}$$

$$= \dfrac{m^6}{2^3 n^9} = \dfrac{m^6}{8n^9}$$

37. $2^{-1} - 5^{-1} = \dfrac{1}{2^1} - \dfrac{1}{5^1} = \dfrac{5}{10} - \dfrac{2}{10} = \dfrac{3}{10}$

38. $2\left(3x^2 - 8x + 1\right) - 4\left(x^2 - 3x - 9\right)$

$$= 6x^2 - 16x + 2 - 4x^2 + 12x + 36$$
$$\qquad\qquad\qquad\qquad \textit{Distributive property}$$

$$= \left(6x^2 - 4x^2\right) + (-16x + 12x) + (2 + 36)$$
$$\qquad\qquad\qquad\qquad \textit{Combine like terms.}$$

$$= 2x^2 - 4x + 38$$

39. $(3x + 2y)(5x - y)$

$$\qquad \textbf{F} \qquad \textbf{O} \qquad \textbf{I} \qquad \textbf{L}$$

$$= 3x(5x) + 3x(-y) + 2y(5x) + 2y(-y)$$

$$= 15x^2 - 3xy + 10xy - 2y^2$$

$$= 15x^2 + 7xy - 2y^2$$

40. $(8m + 5n)(8m - 5n) = (8m)^2 - (5n)^2$
$$= 64m^2 - 25n^2$$

41. $(x + 2y)(x^2 - 2xy + 4y^2)$

Multiply vertically.

$$
\begin{array}{r}
x^2 \quad - \quad 2xy \quad + \quad 4y^2 \\
x \quad + \quad 2y \\
\hline
2x^2y \quad - \quad 4xy^2 \quad + \quad 8y^3 \\
x^3 \quad - \quad 2x^2y \quad + \quad 4xy^2 \\
\hline
x^3 \qquad\qquad\qquad\qquad + \quad 8y^3
\end{array}
$$

Thus,

$$(x + 2y)(x^2 - 2xy + 4y^2) = x^3 + 8y^3.$$

42. $\dfrac{16x^3y^5 - 8x^2y^2 + 4}{4x^2y}$

$$= \frac{16x^3y^5}{4x^2y} - \frac{8x^2y^2}{4x^2y} + \frac{4}{4x^2y}$$

$$= 4xy^4 - 2y + \frac{1}{x^2y}$$

43. $\dfrac{m^3 - 3m^2 + 5m - 3}{m - 1}$

$$
\begin{array}{r}
m^2 \quad - \quad 2m \quad + 3 \\
m - 1 \,\overline{\big)\, m^3 \quad - \quad 3m^2 \quad + \quad 5m \quad - 3} \\
\underline{m^3 \quad - \quad m^2 } \\
-2m^2 \quad + \quad 5m \\
\underline{-2m^2 \quad + \quad 2m } \\
3m \quad - 3 \\
\underline{3m \quad - 3} \\
0
\end{array}
$$

The remainder is 0. The answer is the quotient,

$$m^2 - 2m + 3.$$

44. $f(x) = x^2 + 6x + 1, \quad g(x) = 2x$

$$
\begin{aligned}
(f \circ g)(-2) &= f(g(-2)) \\
&= f(2(-2)) \\
&= f(-4) \\
&= (-4)^2 + 6(-4) + 1 \\
&= 16 - 24 + 1 \\
&= -7
\end{aligned}
$$

CHAPTER 6 FACTORING

6.1 Greatest Common Factors; Factoring by Grouping

1. $12m - 60 = 12 \cdot m - 12 \cdot 5$
$\qquad = 12(m - 5)$

3. $4 + 20z = 4 \cdot 1 + 4 \cdot 5z$
$\qquad = 4(1 + 5z)$

5. $8y - 15$ *cannot be factored.*

7. $8k^3 + 24k = 8k \cdot k^2 + 8k \cdot 3$
$\qquad = 8k(k^2 + 3)$

9. $-4p^3q^4 - 2p^2q^5 = -2p^2q^4 \cdot 2p - 2p^2q^4 \cdot q$
$\qquad = -2p^2q^4(2p + q)$

11. $21x^5 + 35x^4 - 14x^3$
$\qquad = 7x^3(3x^2 + 5x - 2)$

13. $10t^5 - 8t^4 - 4t^3$
$\qquad = 2t^3(5t^2 - 4t - 2)$

15. $15a^2c^3 - 25ac^2 + 5ac$
$\qquad = 5ac(3ac^2 - 5c + 1)$

17. $16z^2n^6 + 64zn^7 - 32z^3n^3$
$\qquad = 16zn^3(zn^3 + 4n^4 - 2z^2)$

19. $14a^3b^2 + 7a^2b - 21a^5b^3 + 42ab^4$
$\qquad = 7ab(2a^2b + a - 3a^4b^2 + 6b^3)$

21. $(m - 4)(m + 2) + (m - 4)(m + 3)$
The GCF is $(m - 4)$.

$\qquad = (m - 4)[(m + 2) + (m + 3)]$
$\qquad = (m - 4)(m + 2 + m + 3)$
$\qquad = (m - 4)(2m + 5)$

23. $(2z - 1)(z + 6) - (2z - 1)(z - 5)$
$\qquad = (2z - 1)[(z + 6) - (z - 5)]$
$\qquad = (2z - 1)[z + 6 - z + 5]$
$\qquad = (2z - 1)(11)$
$\qquad = 11(2z - 1)$

25. $5(2 - x)^2 - 2(2 - x)^3$
$\qquad = (2 - x)^2[5 - 2(2 - x)]$
$\qquad = (2 - x)^2(5 - 4 + 2x)$
$\qquad = (2 - x)^2(1 + 2x)$

27. $4(3 - x)^2 - (3 - x)^3 + 3(3 - x)$
$\qquad = (3 - x)[4(3 - x) - (3 - x)^2 + 3]$
$\qquad = (3 - x)[12 - 4x - (9 - 6x + x^2) + 3]$
$\qquad = (3 - x)[12 - 4x - 9 + 6x - x^2 + 3]$
$\qquad = (3 - x)(6 + 2x - x^2)$

29. $15(2z + 1)^3 + 10(2z + 1)^2 - 25(2z + 1)$
The GCF is $5(2z + 1)$.
$\qquad = 5(2z + 1)$
$\qquad \quad \cdot [3(2z + 1)^2 + 2(2z + 1) - 5]$
$\qquad = 5(2z + 1)$
$\qquad \quad \cdot [3(4z^2 + 4z + 1) + 4z + 2 - 5]$
$\qquad = 5(2z + 1)[12z^2 + 12z + 3 + 4z + 2 - 5]$
$\qquad = 5(2z + 1)[12z^2 + 16z]$
$\qquad = 5(2z + 1)[4z(3z + 4)]$
$\qquad = 20z(2z + 1)(3z + 4)$

31. $5(m + p)^3 - 10(m + p)^2 - 15(m + p)^4$
The GCF is $5(m + p)^2$.
$\qquad = 5(m + p)^2[(m + p) - 2 - 3(m + p)^2]$
$\qquad = 5(m + p)^2$
$\qquad \quad \cdot [m + p - 2 - 3(m^2 + 2mp + p^2)]$
$\qquad = 5(m + p)^2$
$\qquad \quad \cdot (m + p - 2 - 3m^2 - 6mp - 3p^2)$

33. $-r^3 + 3r^2 + 5r$
Factor out r.
$\qquad = r(-r^2 + 3r + 5)$
Factor out $-r$.
$\qquad = -r(r^2 - 3r - 5)$

35. $-12s^5 + 48s^4$
Factor out $12s^4$.
$\qquad = 12s^4(-s + 4)$
Factor out $-12s^4$.
$\qquad = -12s^4(s - 4)$

37. $-2x^5 + 6x^3 + 4x^2$
Factor out $2x^2$.
$\qquad = 2x^2(-x^3 + 3x + 2)$
Factor out $-2x^2$.
$\qquad = -2x^2(x^3 - 3x - 2)$

39. $mx + qx + my + qy$
$\qquad = (mx + qx) + (my + qy)$
$\qquad = x(m + q) + y(m + q)$
$\qquad = (m + q)(x + y)$

41. $10m + 2n + 5mk + nk$
$\qquad = (10m + 2n) + (5mk + nk)$
$\qquad = 2(5m + n) + k(5m + n)$
$\qquad = (5m + n)(2 + k)$

43. $4 - 2q - 6p + 3pq$
$\qquad = (4 - 2q) + (-6p + 3pq)$
$\qquad = 2(2 - q) - 3p(2 - q)$
$\qquad = (2 - q)(2 - 3p)$

45. $p^2 - 4zq + pq - 4pz$
$$= (p^2 + pq) + (-4zq - 4pz)$$
$$= p(p + q) - 4z(q + p)$$
$$= (p + q)(p - 4z)$$

47. $2xy + 3y + 2x + 3$
$$= (2xy + 3y) + (2x + 3)$$
$$= y(2x + 3) + 1(2x + 3)$$
$$= (2x + 3)(y + 1)$$

49. $m^3 + 4m^2 - 6m - 24$
$$= (m^3 + 4m^2) + (-6m - 24)$$
$$= m^2(m + 4) - 6(m + 4)$$
$$= (m + 4)(m^2 - 6)$$

51. $-3a^3 - 3ab^2 + 2a^2b + 2b^3$
$$= (-3a^3 - 3ab^2) + (2a^2b + 2b^3)$$
$$= -3a(a^2 + b^2) + 2b(a^2 + b^2)$$
$$= (a^2 + b^2)(-3a + 2b)$$

53. $4 + xy - 2y - 2x$
$$= xy - 2x - 2y + 4$$
$$= (xy - 2x) + (-2y + 4)$$
$$= x(y - 2) - 2(y - 2)$$
$$= (y - 2)(x - 2)$$

55. $8 + 9y^4 - 6y^3 - 12y$
$$= 9y^4 - 6y^3 - 12y + 8$$
$$= (9y^4 - 6y^3) + (-12y + 8)$$
$$= 3y^3(3y - 2) - 4(3y - 2)$$
$$= (3y - 2)(3y^3 - 4)$$

57. $1 - a + ab - b$
$$= 1 - a - b + ab$$
$$= (1 - a) + (-b + ab)$$
$$= 1(1 - a) - b(1 - a)$$
$$= (1 - a)(1 - b)$$

59. $3m^{-5} + m^{-3}$

Factor out m^{-5} since -5 is the smaller exponent.

$$= m^{-5}(3) + m^{-5}(m^{-3-(-5)})$$
$$= m^{-5}(3 + m^2) \quad \text{or} \quad \frac{3 + m^2}{m^5}$$

61. $3p^{-3} + 2p^{-2}$

Factor out p^{-3} since -3 is the smaller exponent.

$$= p^{-3}(3) + p^{-3}(2p^{-2-(-3)})$$
$$= p^{-3}(3 + 2p) \quad \text{or} \quad \frac{3 + 2p}{p^3}$$

63. The directions said that the student was to factor the polynomial *completely*. The completely factored form is $4xy^3(xy^2 - 2)$.

65. Factoring a polynomial involves writing the polynomial as the product of two or more simpler polynomials. Here, choice **A** is the sum of two polynomials; choices **B** and **D** are differences of two polynomials. Only choice **C** involves the product of two polynomials and is an example of a polynomial in factored form. The correct answer is choice **C**.

67. $(k + 7)(k - 1) = k^2 - k + 7k - 7$
$$= k^2 + 6k - 7$$

69. $(4y - 2)(5y + 6) = 20y^2 + 24y - 10y - 12$
$$= 20y^2 + 14y - 12$$

71. $(5x - 2t)(5x + 2t) = (5x)^2 - (2t)^2$
$$= 25x^2 - 4t^2$$

73. $(3y^3 - 4)(2y^3 + 3) = 6y^6 + 9y^3 - 8y^3 - 12$
$$= 6y^6 + y^3 - 12$$

75. $(3t + 2)(t - 8) = 3t^2 - 24t + 2t - 16$
$$= 3t^2 - 22t - 16$$

Thus, $5t(3t + 2)(t - 8) = 15t^3 - 110t^2 - 80t$.

6.2 Factoring Trinomials

1. **D** is not valid.

$$(8x)(4x) = 32x^2 \neq 12x^2$$

3. **B** is not a factored form.

$$(-x - 10)(x + 6) = -x^2 - 16x - 60$$
$$\neq -x^2 + 16x - 60$$

5. To factor $y^2 + 7y - 30$, we need two integer factors whose sum is 7 (coefficient of the middle term) and whose product is -30 (the last term). Since $-3 + 10 = 7$ and $-3 \cdot 10 = -30$, we have

$$y^2 + 7y - 30 = (y - 3)(y + 10).$$

7. $p^2 + 15p + 56$
Two integer factors whose product is 56 and whose sum is 15 are 8 and 7.
$$= (p + 8)(p + 7)$$

9. $m^2 - 11m + 60$

To factor $m^2 - 11m + 60$, we need two integer factors whose product is 60 (the last term) and whose sum is -11 (coefficient of the middle term).

Factors	Sum
$-1, -60$	-61
$-2, -30$	-32
$-3, -20$	-23
$-4, -15$	-19
$-5, -12$	-17
$-6, -10$	-16

No sum is -11, so the trinomial is prime.

11. $a^2 - 2ab - 35b^2$

Two integer factors whose product is -35 and whose sum is -2 are 5 and -7.

$$a^2 - 2ab - 35b^2 = (a + 5b)(a - 7b)$$

13. $y^2 - 3yq - 15q^2$

There are no integer factors of -15 that add up to -3, so this trinomial is prime.

15. $x^2y^2 + 11xy + 18$

Two integer factors whose product is 18 and whose sum is 11 are 9 and 2.

$$x^2y^2 + 11xy + 18 = (xy + 9)(xy + 2)$$

17. $-6m^2 - 13m + 15 = -1(6m^2 + 13m - 15)$

We'll try to factor $6m^2 + 13m - 15$.

Multiply the first and last coefficients to get $6(-15) = -90$.

Two integer factors whose product is -90 and whose sum is 13 are -5 and 18.

Rewrite the trinomial in a form that can be factored by grouping.

$$6m^2 + (+13m) - 15$$
$$= 6m^2 + (-5m + 18m) - 15$$
$$= (6m^2 - 5m) + (18m - 15)$$
$$= m(6m - 5) + 3(6m - 5)$$
$$= (6m - 5)(m + 3)$$

Thus, the final factored form is

$$-1(6m - 5)(m + 3).$$

Note: These exercises can be worked using the alternative method of repeated combinations and FOIL or the grouping method.

19. $10x^2 + 3x - 18$

Two integer factors whose product is $(10)(-18) = -180$ and whose sum is 3 are 15 and -12.

Rewrite the trinomial in a form that can be factored by grouping.
$$10x^2 + 3x - 18$$
$$= 10x^2 + 15x - 12x - 18$$
$$= 5x(2x + 3) - 6(2x + 3)$$
$$= (2x + 3)(5x - 6)$$

21. $20k^2 + 47k + 24$

Two integer factors whose product is $(20)(24) = 480$ and whose sum is 47 are 15 and 32.

Rewrite the trinomial in a form that can be factored by grouping.
$$= 20k^2 + 15k + 32k + 24$$
$$= 5k(4k + 3) + 8(4k + 3)$$
$$= (4k + 3)(5k + 8)$$

23. $15a^2 - 22ab + 8b^2$

Two integer factors whose product is $(15)(8) = 120$ and whose sum is -22 are -10 and -12.

Rewrite the trinomial in a form that can be factored by grouping.
$$= 15a^2 - 10ab - 12ab + 8b^2$$
$$= 5a(3a - 2b) - 4b(3a - 2b)$$
$$= (3a - 2b)(5a - 4b)$$

25. $36m^2 - 60m + 25$

Use the alternative method and write $36m^2$ as $6m \cdot 6m$ and 25 as $5 \cdot 5$. Use these factors in the binomial factors to obtain

$$36m^2 - 60m + 25 = (6m - 5)(6m - 5)$$
$$= (6m - 5)^2.$$

27. $40x^2 + xy + 6y^2$

There are no integer factors of $(40)(6) = 240$ that add up to 1, so this trinomial is prime.

29. $6x^2z^2 + 5xz - 4$

Two integer factors whose product is $(6)(-4) = -24$ and whose sum is 5 are 8 and -3.

Rewrite the trinomial in a form that can be factored by grouping.
$$= 6x^2z^2 + 8xz - 3xz - 4$$
$$= 2xz(3xz + 4) - 1(3xz + 4)$$
$$= (3xz + 4)(2xz - 1)$$

31. $24x^2 + 42x + 15$

Always factor out the GCF first.
$$= 3(8x^2 + 14x + 5)$$
Now factor $8x^2 + 14x + 5$ by the alternative method.

$$8x^2 + 14x + 5 = (4x + 5)(2x + 1)$$

The final factored form is

$$3(4x + 5)(2x + 1).$$

33. $-15a^2 - 70a + 120$
$$= -5(3a^2 + 14a - 24)$$
$$= -5(a + 6)(3a - 4)$$

35. $-11x^3 + 110x^2 - 264x$
$$= -11x(x^2 - 10x + 24)$$
$$= -11x(x - 6)(x - 4)$$

37. $2x^3y^3 - 48x^2y^4 + 288xy^5$
$$= 2xy^3(x^2 - 24xy + 144y^2)$$
$$= 2xy^3(x - 12y)(x - 12y)$$
or $\quad 2xy^3(x - 12y)^2$

39. $6a^3 + 12a^2 - 90a$
$$= 6a(a^2 + 2a - 15)$$
$$= 6a(a - 3)(a + 5)$$

41. $13y^3 + 39y^2 - 52y$
$= 13y(y^2 + 3y - 4)$
$= 13y(y + 4)(y - 1)$

43. $12p^3 - 12p^2 + 3p$
$= 3p(4p^2 - 4p + 1)$
$= 3p(2p - 1)(2p - 1)$
or $3p(2p - 1)^2$

45. There is a GCF of 2. She did not factor the polynomial *completely*. The factor $(4x + 10)$ can be factored further as $2(2x + 5)$, giving the final form as $2(2x + 5)(x - 2)$.

47. In $12p^6 - 32p^3r + 5r^2$, let $x = p^3$ to obtain

$12x^2 - 32xr + 5r^2 = (6x - r)(2x - 5r)$.

Replace x with p^3.

$12p^6 - 32p^3r + 5r^2 = (6p^3 - r)(2p^3 - 5r)$

49. $10(k + 1)^2 - 7(k + 1) + 1$
Let $x = k + 1$ to obtain
$10x^2 - 7x + 1 = (5x - 1)(2x - 1)$.
Replace x with $k + 1$.
$10(k + 1)^2 - 7(k + 1) + 1$
$= [5(k + 1) - 1][2(k + 1) - 1]$
$= (5k + 5 - 1)(2k + 2 - 1)$
$= (5k + 4)(2k + 1)$

51. $3(m + p)^2 - 7(m + p) - 20$
Let $x = m + p$ to obtain
$3x^2 - 7x - 20 = (3x + 5)(x - 4)$.
Replace x with $m + p$.
$3(m + p)^2 - 7(m + p) - 20$
$= [3(m + p) + 5][(m + p) - 4]$
$= (3m + 3p + 5)(m + p - 4)$

53. $a^2(a + b)^2 - ab(a + b)^2 - 6b^2(a + b)^2$
Factor out the GCF, $(a + b)^2$.
$= (a + b)^2(a^2 - ab - 6b^2)$
Factor the trinomial.
$= (a + b)^2(a - 3b)(a + 2b)$

55. $p^2(p + q) + 4pq(p + q) + 3q^2(p + q)$
Factor out the GCF, $p + q$.
$= (p + q)(p^2 + 4pq + 3q^2)$

Factor the trinomial.
$= (p + q)(p + q)(p + 3q)$
$= (p + q)^2(p + 3q)$

57. $z^2(z - x) - zx(x - z) - 2x^2(z - x)$
Factor out -1 from the middle term:
$x - z = -1(z - x)$.
$= z^2(z - x) + zx(z - x) - 2x^2(z - x)$
Factor out the GCF, $z - x$.
$= (z - x)(z^2 + zx - 2x^2)$

Factor the trinomial.
$= (z - x)(z + 2x)(z - x)$
$= (z - x)^2(z + 2x)$

59. In $p^4 - 10p^2 + 16$, let $x = p^2$ to obtain

$x^2 - 10x + 16 = (x - 8)(x - 2)$.

Replace x with p^2.

$p^4 - 10p^2 + 16 = (p^2 - 8)(p^2 - 2)$

61. In $2x^4 - 9x^2 - 18$, let $y = x^2$ to obtain

$2y^2 - 9y - 18 = (2y + 3)(y - 6)$.

Replace y with x^2.

$2x^4 - 9x^2 - 18 = (2x^2 + 3)(x^2 - 6)$

63. In $16x^4 + 16x^2 + 3$, let $m = x^2$ to obtain

$16m^2 + 16m + 3 = (4m + 3)(4m + 1)$.

Replace m with x^2.

$16x^4 + 16x^2 + 3 = (4x^2 + 3)(4x^2 + 1)$

65. $(3x - 5)(3x + 5) = (3x)^2 - 5^2$
$= 9x^2 - 25$

67. $(p + 3q)^2 = p^2 + 2 \cdot p \cdot 3q + (3q)^2$
$= p^2 + 6pq + 9q^2$

69. $(y + 3)(y^2 - 3y + 9)$
$= y(y^2 - 3y + 9) + 3(y^2 - 3y + 9)$
$= y^3 - 3y^2 + 9y + 3y^2 - 9y + 27$
$= y^3 + 27$

6.3 Special Factoring

1. **A.** Yes, 64 and m^2 are squares.

B. No, $2x^2$ is not a square.

C. No, $k^2 + 9$ is a *sum* of squares.

D. Yes, $4z^2$ and 49 are squares.

So, the binomials that are differences of squares are **A** and **D**.

3. **A.** Since $x^2 - 8x - 16$ has a negative third term, it is not a perfect square trinomial.

B. $4m^2 + 20m + 25 = (2m)^2 + 20m + 5^2$

$2(2m)(5) = 20m$, the middle term. Therefore, this trinomial is a perfect square.

C. $9z^4 + 30z^2 + 25 = (3z^2)^2 + 30z^2 + 5^2$

$2(3z^2)(5) = 30z^2$, the middle term. Therefore, this trinomial is a perfect square.

D. $25a^2 - 45a + 81 = (5a)^2 - 45a + (-9)^2$
$2(5a)(-9) = -90a$

This is not the middle term, so the trinomial is not a perfect square.

So, the perfect square trinomials are **B** and **C**.

5. The sum of two squares can be factored only if the binomial has a common factor. For example,

$$9x^2 + 81 = 9(x^2 + 9).$$

7. $p^2 - 16 = p^2 - 4^2$
$= (p + 4)(p - 4)$

9. $25x^2 - 4 = (5x)^2 - 2^2$
$= (5x + 2)(5x - 2)$

11. $18a^2 - 98b^2 = 2(9a^2 - 49b^2)$
$= 2[(3a)^2 - (7b)^2]$
$= 2[(3a + 7b)(3a - 7b)]$
$= 2(3a + 7b)(3a - 7b)$

13. $64m^4 - 4y^4$
$= 4(16m^4 - y^4)$
$= 4[(4m^2)^2 - (y^2)^2]$

Factor the difference of squares.

$= 4(4m^2 + y^2)(4m^2 - y^2)$
$= 4(4m^2 + y^2)[(2m)^2 - y^2]$

Factor the difference of squares again.
$= 4(4m^2 + y^2)(2m + y)(2m - y)$

15. $(y + z)^2 - 81$
$= (y + z)^2 - 9^2$
$= [(y + z) + 9][(y + z) - 9]$
$= (y + z + 9)(y + z - 9)$

17. $16 - (x + 3y)^2$
$= 4^2 - z^2$ Let $z = (x + 3y)$.
$= (4 + z)(4 - z)$
Substitute $x + 3y$ for z.
$= [4 + (x + 3y)][4 - (x + 3y)]$
$= (4 + x + 3y)(4 - x - 3y)$

19. $p^4 - 256 = (p^2)^2 - 16^2$
$= (p^2 + 16)(p^2 - 16)$
$= (p^2 + 16)(p^2 - 4^2)$
$= (p^2 + 16)(p + 4)(p - 4)$

21. $k^2 - 6k + 9 = (k)^2 - 2(k)(3) + 3^2$
$= (k - 3)^2$

23. $4z^2 + 4zw + w^2$
$= (2z)^2 + 2(2z)(w) + w^2$
$= (2z + w)^2$

25. $16m^2 - 8m + 1 - n^2$
Group the first three terms.
$= (16m^2 - 8m + 1) - n^2$
$= [(4m)^2 - 2(4m)(1) + 1^2] - n^2$
$= (4m - 1)^2 - n^2$
$= [(4m - 1) + n][(4m - 1) - n]$
$= (4m - 1 + n)(4m - 1 - n)$

27. $4r^2 - 12r + 9 - s^2$
Group the first three terms.
$= (4r^2 - 12r + 9) - s^2$
$= [(2r)^2 - 2(2r)(3) + 3^2] - s^2$
$= (2r - 3)^2 - s^2$
$= [(2r - 3) + s][(2r - 3) - s]$
$= (2r - 3 + s)(2r - 3 - s)$

29. $x^2 - y^2 + 2y - 1$
Group the last three terms.
$= x^2 - (y^2 - 2y + 1)$
$= x^2 - (y - 1)^2$
$= [x + (y - 1)][x - (y - 1)]$
$= (x + y - 1)(x - y + 1)$

31. $98m^2 + 84mn + 18n^2$
$= 2(49m^2 + 42mn + 9n^2)$
$= 2[(7m)^2 + 2(7m)(3n) + (3n)^2]$
$= 2(7m + 3n)^2$

33. $(p + q)^2 + 2(p + q) + 1$
$= x^2 + 2x + 1$ Let $x = p + q$.
$= (x + 1)^2$
$= (p + q + 1)^2$ Resubstitute.

35. $(a - b)^2 + 8(a - b) + 16$
$= (a - b)^2 + 2(a - b)(4) + 4^2$
$= [(a - b) + 4]^2$
$= (a - b + 4)^2$

37. $x^3 - 27 = x^3 - 3^3$
$= (x - 3)(x^2 + x \cdot 3 + 3^2)$
$= (x - 3)(x^2 + 3x + 9)$

39. $t^3 - 216 = t^3 - 6^3$
$= (t - 6)(t^2 + t \cdot 6 + 6^2)$
$= (t - 6)(t^2 + 6t + 36)$

41. $x^3 + 64 = x^3 + 4^3$
$= (x + 4)(x^2 - x \cdot 4 + 4^2)$
$= (x + 4)(x^2 - 4x + 16)$

43. $1000 + y^3 = 10^3 + y^3$
$= (10 + y)(10^2 - 10 \cdot y + y^2)$
$= (10 + y)(100 - 10y + y^2)$

45. $8x^3 + 1 = (2x)^3 + 1^3$
$$= (2x + 1)\big[(2x)^2 - 2x \cdot 1 + 1^2\big]$$
$$= (2x + 1)\big(4x^2 - 2x + 1\big)$$

47. $125x^3 - 216 = (5x)^3 - 6^3$
$$= (5x - 6)\big[(5x)^2 + 5x \cdot 6 + 6^2\big]$$
$$= (5x - 6)\big(25x^2 + 30x + 36\big)$$

49. $x^3 - 8y^3 = x^3 - (2y)^3$
$$= (x - 2y)\big[x^2 + x \cdot 2y + (2y)^2\big]$$
$$= (x - 2y)\big(x^2 + 2xy + 4y^2\big)$$

51. $64g^3 - 27h^3$
$$= (4g)^3 - (3h)^3$$
$$= (4g - 3h)\big[(4g)^2 + (4g)(3h) + (3h)^2\big]$$
$$= (4g - 3h)\big(16g^2 + 12gh + 9h^2\big)$$

53. $343p^3 + 125q^3$
$$= (7p)^3 + (5q)^3$$
$$= (7p + 5q)\big[(7p)^2 - (7p)(5q) + (5q)^2\big]$$
$$= (7p + 5q)\big(49p^2 - 35pq + 25q^2\big)$$

55. $24n^3 + 81p^3$
$$= 3\big(8n^3 + 27p^3\big)$$
$$= 3\big[(2n)^3 + (3p)^3\big]$$
$$= 3[2n + 3p]\big[(2n)^2 - (2n)(3p) + (3p)^2\big]$$
$$= 3(2n + 3p)\big(4n^2 - 6np + 9p^2\big)$$

57. $(y + z)^3 + 64$
$$= (y + z)^3 + 4^3$$
$$= \big[(y + z) + 4\big]\big[(y + z)^2 - (y + z)(4) + 4^2\big]$$
$$= (y + z + 4)\big(y^2 + 2yz + z^2 - 4y - 4z + 16\big)$$

59. $m^6 - 125 = \big(m^2\big)^3 - (5)^3$
$$= \big[m^2 - 5\big]\big[\big(m^2\big)^2 + \big(m^2\big)(5) + 5^2\big]$$
$$= \big(m^2 - 5\big)\big(m^4 + 5m^2 + 25\big)$$

61. $1000x^9 - 27$
$$= \big(10x^3\big)^3 - 3^3$$
$$= \big(10x^3 - 3\big)\big[\big(10x^3\big)^2 + \big(10x^3\big)(3) + 3^2\big]$$
$$= \big(10x^3 - 3\big)\big(100x^6 + 30x^3 + 9\big)$$

63. $125y^6 + z^3$
$$= \big(5y^2\big)^3 + z^3$$
$$= \big(5y^2 + z\big)\big[\big(5y^2\big)^2 - \big(5y^2\big)(z) + z^2\big]$$
$$= \big(5y^2 + z\big)\big(25y^4 - 5y^2z + z^2\big)$$

64. $x^6 - y^6$
$$= \big(x^3\big)^2 - \big(y^3\big)^2$$
$$= \big(x^3 + y^3\big)\big(x^3 - y^3\big)$$

$$= \big[(x + y)\big(x^2 - xy + y^2\big)\big]$$
$$\cdot \big[(x - y)\big(x^2 + xy + y^2\big)\big]$$
$$= (x + y)\big(x^2 - xy + y^2\big)(x - y)$$
$$\cdot \big(x^2 + xy + y^2\big)$$

65. $x^6 - y^6 = (x - y)(x + y)$
$$\cdot \; \underline{\big(x^2 + xy + y^2\big)\big(x^2 - xy + y^2\big)}$$

66. $x^6 - y^6$
$$= \big(x^2\big)^3 - \big(y^2\big)^3$$
$$= \big(x^2 - y^2\big)\big(x^4 + x^2y^2 + y^4\big)$$
$$= (x + y)(x - y)\big(x^4 + x^2y^2 + y^4\big)$$

67. $x^6 - y^6 = (x - y)(x + y)$
$$\cdot \; \underline{\big(x^4 + x^2y^2 + y^4\big)}$$

68. The product written on the blank in Exercise 65 must equal the product written on the blank in Exercise 67. To verify this, multiply the two factors written in Exercise 65.
$$\big(x^2 + xy + y^2\big)\big(x^2 - xy + y^2\big)$$
$$= x^2\big(x^2 - xy + y^2\big)$$
$$\quad + xy\big(x^2 - xy + y^2\big)$$
$$\quad + y^2\big(x^2 - xy + y^2\big)$$
$$= x^4 - x^3y + x^2y^2 + x^3y - x^2y^2$$
$$\quad + xy^3 + x^2y^2 - xy^3 + y^4$$
$$= x^4 + x^2y^2 + y^4$$

They are equal.

69. Start by factoring as the difference of squares since doing so resulted in the complete factorization more directly.

71. $125p^3 + 25p^2 + 8q^3 - 4q^2$
$$= \big(125p^3 + 8q^3\big) + \big(25p^2 - 4q^2\big)$$
$$= \big[(5p)^3 + (2q)^3\big] + \big[(5p)^2 - (2q)^2\big]$$
Factor within groups.
$$= \big[(5p + 2q)\big(25p^2 - 10pq + 4q^2\big)\big]$$
$$\quad + \big[(5p + 2q)(5p - 2q)\big]$$
Factor out the GCF, $5p + 2q$.
$$= (5p + 2q)$$
$$\quad \cdot \big[\big(25p^2 - 10pq + 4q^2\big) + (5p - 2q)\big]$$
$$= (5p + 2q)$$
$$\quad \cdot \big(25p^2 - 10pq + 4q^2 + 5p - 2q\big)$$

73. $27a^3 + 15a - 64b^3 - 20b$
$$= \big(27a^3 - 64b^3\big) + (15a - 20b)$$
$$= \big[(3a)^3 - (4b)^3\big] + (15a - 20b)$$
Factor within groups.
$$= (3a - 4b)\big(9a^2 + 12ab + 16b^2\big)$$
$$\quad + 5(3a - 4b)$$
Factor out the GCF, $3a - 4b$.
$$= (3a - 4b)\big[\big(9a^2 + 12ab + 16b^2\big) + 5\big]$$
$$= (3a - 4b)\big(9a^2 + 12ab + 16b^2 + 5\big)$$

75. $8t^4 - 24t^3 + t - 3$

$= \left(8t^4 - 24t^3\right) + (t - 3)$

$= 8t^3(t - 3) + 1(t - 3)$

Factor out the GCF, $t - 3$.

$= (t - 3)\left(8t^3 + 1\right)$

Factor the sum of cubes.

$= (t - 3)(2t + 1)\left(4t^2 - 2t + 1\right)$

77. $64m^2 - 512m^3 - 81n^2 + 729n^3$

$= \left(64m^2 - 81n^2\right) - \left(512m^3 - 729n^3\right)$

$= \left[(8m)^2 - (9n)^2\right] - \left[(8m)^3 - (9n)^3\right]$

Factor within groups.

$= (8m + 9n)(8m - 9n)$
$\quad - (8m - 9n)\left(64m^2 + 72mn + 81n^2\right)$

Factor out the GCF, $8m - 9n$.

$= (8m - 9n)$
$\quad \cdot \left[(8m + 9n) - \left(64m^2 + 72mn + 81n^2\right)\right]$

$= (8m - 9n)$
$\quad \cdot \left(8m + 9n - 64m^2 - 72mn - 81n^2\right)$

79. $2ax + ay - 2bx - by$

$= (2ax + ay) + (-2bx - by)$

$= a(2x + y) - b(2x + y)$

$= (2x + y)(a - b)$

81. $p^2 + 4p - 21 = (p + 7)(p - 3)$

6.4 A General Approach to Factoring

1. $100a^2 - 9b^2$

$= (10a)^2 - (3b)^2$ *Difference of squares*

$= (10a + 3b)(10a - 3b)$

3. $3p^4 - 3p^3 - 90p^2$

$= 3p^2\left(p^2 - p - 30\right)$

$= 3p^2(p - 6)(p + 5)$

5. $3a^2pq + 3abpq - 90b^2pq$

$= 3pq\left(a^2 + ab - 30b^2\right)$

$= 3pq(a + 6b)(a - 5b)$

7. $225p^2 + 256$ is the *sum* of squares and cannot be factored. The binomial is prime.

9. $6b^2 - 17b - 3$

Two integer factors whose product is $(6)(-3) = -18$ and whose sum is -17 are -18 and 1.

$= 6b^2 - 18b + b - 3$

$= 6b(b - 3) + 1(b - 3)$

$= (b - 3)(6b + 1)$

11. $x^3 - 1000$

$= (x)^3 - 10^3$ *Difference of cubes*

$= (x - 10)\left(x^2 + 10x + 100\right)$

13. $4(p + 2) + m(p + 2) = (p + 2)(4 + m)$

15. $9m^2 - 45m + 18m^3$

Factor out the GCF, $9m$.

$= 9m(m - 5 + 2m^2)$ or $9m(2m^2 + m - 5)$

There is no pair of integers with a product of -10 and a sum of 1, so this cannot be factored further.

17. $54m^3 - 2000$

Factor out the GCF, 2.

$2\left(27m^3 - 1000\right)$

$= 2\left[(3m)^3 - 10^3\right]$ *Difference of cubes*

$= 2(3m - 10)$
$\quad \cdot \left[(3m)^2 + (3m)(10) + 10^2\right]$

$= 2(3m - 10)\left(9m^2 + 30m + 100\right)$

19. $9m^2 - 30mn + 25n^2$

$= (3m)^2 - 2(3m)(5n) + (5n)^2$

Perfect square trinomial

$= (3m - 5n)^2$

21. $kq - 9q + kr - 9r$

$= q(k - 9) + r(k - 9)$

$= (k - 9)(q + r)$

23. $16z^3x^2 - 32z^2x = 16z^2x(zx - 2)$

25. $x^2 + 2x - 35$

The numbers 7 and -5 have a product of -35 and a sum of 2.

$= (x + 7)(x - 5)$

27. $x^4 - 625$

$= \left(x^2\right)^2 - 25^2$ *Difference of squares*

$= \left(x^2 + 25\right)\left(x^2 - 25\right)$

$= \left(x^2 + 25\right)(x + 5)(x - 5)$ *Difference of squares again*

29. $p^3 + 1 = p^3 + 1^3$ *Sum of cubes*

$= (p + 1)\left(p^2 - p \cdot 1 + 1^2\right)$

$= (p + 1)\left(p^2 - p + 1\right)$

31. $64m^2 - 625$

$= (8m)^2 - 25^2$ *Difference of squares*

$= (8m + 25)(8m - 25)$

33. $12z^3 - 6z^2 + 18z$

Factor out the GCF, $6z$.

$6z\left(2z^2 - z + 3\right)$

Further factoring is not possible. There is no pair of integers whose product is 6 and whose sum is -1.

35. $256b^2 - 400c^2$

$= 16\left(16b^2 - 25c^2\right)$

$= 16\left[(4b)^2 - (5c)^2\right]$ *Difference of squares*

$= 16(4b + 5c)(4b - 5c)$

37. $1000z^3 + 512$

$= 8\left(125z^3 + 64\right)$

$= 8\left[(5z)^3 + 4^3\right]$ *Sum of cubes*

$= 8[5z + 4]\left[(5z)^2 - (5z)(4) + 4^2\right]$

$= 8(5z + 4)\left(25z^2 - 20z + 16\right)$

39. $10r^2 + 23rs - 5s^2$

Two integer factors whose product is $(10)(-5) = -50$ and whose sum is 23 are 25 and -2.

$= 10r^2 + 25rs - 2rs - 5s^2$

$= 5r(2r + 5s) - s(2r + 5s)$

$= (2r + 5s)(5r - s)$

41. $24p^3q + 52p^2q^2 + 20pq^3$

$= 4pq\left(6p^2 + 13pq + 5q^2\right)$

Two integer factors whose product is $(6)(5) = 30$ and whose sum is 13 are 10 and 3.

$= 4pq\left(6p^2 + 10pq + 3pq + 5q^2\right)$

$= 4pq[2p(3p + 5q) + q(3p + 5q)]$

$= 4pq(3p + 5q)(2p + q)$

43. $48k^4 - 243$

$= 3\left(16k^4 - 81\right)$

$= 3\left[\left(4k^2\right)^2 - 9^2\right]$

$= 3\left(4k^2 + 9\right)\left(4k^2 - 9\right)$

$= 3\left(4k^2 + 9\right)\left[(2k)^2 - 3^2\right]$

$= 3\left(4k^2 + 9\right)(2k + 3)(2k - 3)$

45. $m^3 + m^2 - n^3 - n^2$

$= \left(m^3 - n^3\right) + \left(m^2 - n^2\right)$

   Difference   *Difference*

   of cubes;   *of squares*

$= (m - n)\left(m^2 + mn + n^2\right)$

    $+ (m + n)(m - n)$

$= (m - n)\left[\left(m^2 + mn + n^2\right) + (m + n)\right]$

$= (m - n)\left(m^2 + mn + n^2 + m + n\right)$

47. $x^2 - 4m^2 - 4mn - n^2$

$= x^2 - \left(4m^2 + 4mn + n^2\right)$

$= x^2 - \left[(2m)^2 + 2(2m)n + n^2\right]$

     Perfect square trinomial

$= x^2 - (2m + n)^2$

     Difference of squares

$= [x + (2m + n)][x - (2m + n)]$

$= (x + 2m + n)(x - 2m - n)$

49. $18p^5 - 24p^3 + 12p^6$

Factor out the GCF, $6p^3$.

$6p^3\left(3p^2 - 4 + 2p^3\right)$

Further factoring is not possible.

51. $2x^2 - 2x - 40 = 2\left(x^2 - x - 20\right)$

$= 2(x + 4)(x - 5)$

53. $(2m + n)^2 - (2m - n)^2$

     Difference of squares

$= [(2m + n) + (2m - n)]$

   $\cdot [(2m + n) - (2m - n)]$

$= (2m + n + 2m - n)(2m + n - 2m + n)$

$= 4m(2n)$

$= 8mn$

55. $50p^2 - 162$

$= 2\left(25p^2 - 81\right)$

$= 2\left[(5p)^2 - 9^2\right]$

$= 2(5p + 9)(5p - 9)$

57. $12m^2rx + 4mnrx + 40n^2rx$

Factor out the GCF, $4rx$.

$= 4rx\left(3m^2 + mn + 10n^2\right)$

59. $21a^2 - 5ab - 4b^2$

Two integer factors whose product is $(21)(-4) = -84$ and whose sum is -5 are -12 and 7.

$= 21a^2 - 12ab + 7ab - 4b^2$

$= 3a(7a - 4b) + b(7a - 4b)$

$= (7a - 4b)(3a + b)$

61. $x^2 - y^2 - 4$ cannot be factored. The polynomial is *prime*.

63. $(p + 8q)^2 - 10(p + 8q) + 25$

$= x^2 - 10x + 25$ *Let $x = p + 8q$.*

$= (x - 5)^2$

$= [(p + 8q) - 5]^2$ *Resubstitute.*

$= (p + 8q - 5)^2$

65. $21m^4 - 32m^2 - 5$

$= 21x^2 - 32x - 5$ *Let $x = m^2$.*

Two integer factors whose product is $(21)(-5) = -105$ and whose sum is -32 are -35 and 3.

$= 21x^2 - 35x + 3x - 5$

$= 7x(3x - 5) + 1(3x - 5)$

$= (3x - 5)(7x + 1)$

$= \left(3m^2 - 5\right)\left(7m^2 + 1\right)$ *Resubstitute.*

67. $(r + 2t)^3 + (r - 3t)^3$

Let $x = r + 2t$ and $y = r - 3t$.

$= x^3 + y^3$ *Sum of cubes*

$= (x + y)(x^2 - xy + y^2)$

Substitute $r + 2t$ for x and $r - 3t$ for y.

$= [(r + 2t) + (r - 3t)]$
$\quad \cdot [(r + 2t)^2 - (r + 2t)(r - 3t)$
$\quad + (r - 3t)^2]$

$= (2r - t)$
$\quad \cdot (r^2 + 4rt + 4t^2 - r^2 + rt + 6t^2$
$\quad + r^2 - 6rt + 9t^2)$

$= (2r - t)(r^2 - rt + 19t^2)$

69. $x^5 + 3x^4 - x - 3$

$= x^4(x + 3) - 1(x + 3)$

$= (x + 3)(x^4 - 1)$

$= (x + 3)[(x^2)^2 - 1^2]$

$= (x + 3)(x^2 + 1)(x^2 - 1)$

$= (x + 3)(x^2 + 1)[x^2 - 1^2]$

$= (x + 3)(x^2 + 1)(x + 1)(x - 1)$

71. $m^2 - 4m + 4 - n^2 + 6n - 9$

$= (m^2 - 4m + 4) - (n^2 - 6n + 9)$

$= (m^2 - 2(2)m + 2^2)$
$\quad - (n^2 - 2(3)n + 3^2)$

         Perfect square trinomials

$= (m - 2)^2 - (n - 3)^2$

$= [(m - 2) + (n - 3)]$
$\quad \cdot [(m - 2) - (n - 3)]$

         Difference of two squares

$= (m - 2 + n - 3)(m - 2 - n + 3)$

$= (m + n - 5)(m - n + 1)$

73. $3x + 2 = 0$

$3x = -2$

$x = \frac{-2}{3} = -\frac{2}{3}$

The solution set is $\left\{ -\frac{2}{3} \right\}$.

75. $5x = 0$

$x = \frac{0}{5} = 0$

The solution set is $\{0\}$.

77. $\frac{1}{2}t + 5 = 0$

$\frac{1}{2}t = -5$

$2\left(\frac{1}{2}t\right) = 2(-5)$

$t = -10$

The solution set is $\{-10\}$.

6.5   Solving Equations by Factoring

1. First rewrite the equation so that one side is 0. Factor the other side and set each factor equal to 0. The solutions of these linear equations are solutions of the quadratic equation.

In the exercises in this section, check all solutions to the equations by substituting them back in the original equations.

3. $(x + 10)(x - 5) = 0$

$x + 10 = 0$ or $x - 5 = 0$
$\quad x = -10$ or $x = 5$

Check $x = -10$: $0(-15) = 0$ *True*
Check $x = 5$: $15(0) = 0$ *True*

The solution set is $\{-10, 5\}$.

5. $(2k - 5)(3k + 8) = 0$

$2k - 5 = 0$ or $3k + 8 = 0$
$\quad 2k = 5$ $3k = -8$
$\quad k = \frac{5}{2}$ or $k = -\frac{8}{3}$

The solution set is $\left\{ -\frac{8}{3}, \frac{5}{2} \right\}$.

7. $m^2 - 3m - 10 = 0$

$(m + 2)(m - 5) = 0$

Use the zero-factor property.

$m + 2 = 0$ or $m - 5 = 0$
$\quad m = -2$ or $m = 5$

The solution set is $\{-2, 5\}$.

9. $z^2 + 9z + 18 = 0$

$(z + 6)(z + 3) = 0$

$z + 6 = 0$ or $z + 3 = 0$
$\quad z = -6$ or $z = -3$

The solution set is $\{-6, -3\}$.

11. $2x^2 = 7x + 4$

Get 0 on one side.

$2x^2 - 7x - 4 = 0$
$(2x + 1)(x - 4) = 0$

$2x + 1 = 0$ or $x - 4 = 0$
$\quad 2x = -1$ $x = 4$
$\quad x = -\frac{1}{2}$

The solution set is $\left\{ -\frac{1}{2}, 4 \right\}$.

13. $15k^2 - 7k = 4$

$15k^2 - 7k - 4 = 0$
$(3k + 1)(5k - 4) = 0$

$3k + 1 = 0$ or $5k - 4 = 0$
$\quad 3k = -1$ $5k = 4$
$\quad k = -\frac{1}{3}$ or $k = \frac{4}{5}$

The solution set is $\left\{ -\frac{1}{3}, \frac{4}{5} \right\}$.

15. $2x^2 - 12 - 4x = x^2 - 3x$

$x^2 - x - 12 = 0$

$(x + 3)(x - 4) = 0$

$x + 3 = 0 \quad$ or $\quad x - 4 = 0$

$x = -3 \quad$ or $\quad x = 4$

The solution set is $\{-3, 4\}$.

17. $(5z + 1)(z + 3) = -2(5z + 1)$

$(5z + 1)(z + 3) + 2(5z + 1) = 0$

$(5z + 1)[(z + 3) + 2] = 0$

$(5z + 1)(z + 5) = 0$

$5z + 1 = 0 \quad$ or $\quad z + 5 = 0$

$5z = -1 \qquad\qquad z = -5$

$z = -\frac{1}{5}$

The solution set is $\left\{-5, -\frac{1}{5}\right\}$.

19. $4p^2 + 16p = 0$

$4p(p + 4) = 0$

$4p = 0 \quad$ or $\quad p + 4 = 0$

$p = 0 \qquad\qquad p = -4$

The solution set is $\{-4, 0\}$.

21. $6m^2 - 36m = 0$

$6m(m - 6) = 0$

$6m = 0 \quad$ or $\quad m - 6 = 0$

$m = 0 \quad$ or $\quad m = 6$

The solution set is $\{0, 6\}$.

23. $4p^2 - 16 = 0$

$4(p^2 - 4) = 0$

$4(p + 2)(p - 2) = 0$

$p + 2 = 0 \quad$ or $\quad p - 2 = 0$

$p = -2 \quad$ or $\quad p = 2$

The solution set is $\{-2, 2\}$.

25. $-3m^2 + 27 = 0$

$-3(m^2 - 9) = 0$

$-3(m + 3)(m - 3) = 0$

Note that the leading -3 does not affect the solution set of the equation.

$m + 3 = 0 \quad$ or $\quad m - 3 = 0$

$m = -3 \quad$ or $\quad m = 3$

The solution set is $\{-3, 3\}$.

27. $-x^2 = 9 - 6x$

$0 = x^2 - 6x + 9$

$0 = (x - 3)(x - 3)$

$0 = x - 3$

$3 = x$

The solution set is $\{3\}$.

29. $9k^2 + 24k + 16 = 0$

$(3k + 4)(3k + 4) = 0$

$3k + 4 = 0$

$3k = -4$

$k = -\frac{4}{3}$

The solution set is $\left\{-\frac{4}{3}\right\}$.

31. $(x - 3)(x + 5) = -7$

Multiply the factors, and then add 7 on both sides of the equation to get 0 on the right.

$x^2 + 5x - 3x - 15 = -7$

$x^2 + 2x - 8 = 0$

Now factor the polynomial.

$(x + 4)(x - 2) = 0$

$x + 4 = 0 \quad$ or $\quad x - 2 = 0$

$x = -4 \quad$ or $\quad x = 2$

The solution set is $\{-4, 2\}$.

33. $(2x + 1)(x - 3) = 6x + 3$

$2x^2 - 6x + x - 3 = 6x + 3$

$2x^2 - 5x - 3 = 6x + 3$

$2x^2 - 11x - 6 = 0$

$(2x + 1)(x - 6) = 0$

$2x + 1 = 0 \quad$ or $\quad x - 6 = 0$

$2x = -1 \qquad\qquad x = 6$

$x = -\frac{1}{2}$

The solution set is $\left\{-\frac{1}{2}, 6\right\}$.

35. $(x + 3)(x - 6) = (2x + 2)(x - 6)$

$x^2 - 3x - 18 = 2x^2 - 10x - 12$

$0 = x^2 - 7x + 6$

$0 = (x - 1)(x - 6)$

$x - 1 = 0 \quad$ or $\quad x - 6 = 0$

$x = 1 \quad$ or $\quad x = 6$

The solution set is $\{1, 6\}$.

37. $2x^3 - 9x^2 - 5x = 0$

$x(2x^2 - 9x - 5) = 0$

$x(2x + 1)(x - 5) = 0$

$x = 0 \quad$ or $\quad 2x + 1 = 0 \quad$ or $\quad x - 5 = 0$

$2x = -1 \qquad\qquad x = 5$

$x = -\frac{1}{2}$

The solution set is $\left\{-\frac{1}{2}, 0, 5\right\}$.

39. $x^3 - 2x^2 = 3x$

$x^3 - 2x^2 - 3x = 0$

$x(x^2 - 2x - 3) = 0$

$x(x - 3)(x + 1) = 0$

$x = 0 \quad$ or $\quad x - 3 = 0 \quad$ or $\quad x + 1 = 0$

$\qquad\qquad$ or $\qquad x = 3 \qquad\qquad x = -1$

The solution set is $\{-1, 0, 3\}$.

41.

$$9t^3 = 16t$$
$$9t^3 - 16t = 0$$
$$t(9t^2 - 16) = 0$$
$$t(3t + 4)(3t - 4) = 0$$

$t = 0$ or $3t + 4 = 0$ or $3t - 4 = 0$
$\qquad\qquad 3t = -4 \qquad\qquad 3t = 4$
$\qquad\qquad t = -\frac{4}{3}$ or $\qquad t = \frac{4}{3}$

The solution set is $\left\{ -\frac{4}{3}, 0, \frac{4}{3} \right\}$.

43.

$$2r^3 + 5r^2 - 2r - 5 = 0$$

Factor by grouping.

$$(2r^3 - 2r) + (5r^2 - 5) = 0$$
$$2r(r^2 - 1) + 5(r^2 - 1) = 0$$
$$(r^2 - 1)(2r + 5) = 0$$
$$(r + 1)(r - 1)(2r + 5) = 0$$

$r + 1 = 0$ or $r - 1 = 0$ or $2r + 5 = 0$
$\quad r = -1$ or $\quad r = 1$ or $\quad r = -\frac{5}{2}$

The solution set is $\left\{ -\frac{5}{2}, -1, 1 \right\}$.

45.

$$x^3 - 6x^2 - 9x + 54 = 0$$

Factor by grouping.

$$(x^3 - 6x^2) + (-9x + 54) = 0$$
$$x^2(x - 6) - 9(x - 6) = 0$$
$$(x - 6)(x^2 - 9) = 0$$
$$(x - 6)(x + 3)(x - 3) = 0$$

$x - 6 = 0$ or $x + 3 = 0$ or $x - 3 = 0$
$\quad x = 6 \qquad\quad x = -3 \qquad\quad x = 3$

The solution set is $\{-3, 3, 6\}$.

47. By dividing each side by a variable expression, she "lost" the solution 0. The solution set is $\left\{ -\frac{4}{3}, 0, \frac{4}{3} \right\}$.

49. $2(x - 1)^2 - 7(x - 1) - 15 = 0$
Let $y = x - 1$.
$$2y^2 - 7y - 15 = 0$$
$$(2y + 3)(y - 5) = 0$$

$2y + 3 = 0$ or $y - 5 = 0$
$\quad 2y = -3 \qquad\quad y = 5$
$\quad y = -\frac{3}{2}$
Substitute $x - 1$ for y.

$x - 1 = -\frac{3}{2}$ or $x - 1 = 5$
$\quad x = -\frac{1}{2}$ or $\qquad x = 6$

The solution set is $\left\{ -\frac{1}{2}, 6 \right\}$.

51. $5(3a - 1)^2 + 3 = -16(3a - 1)$
Let $x = 3a - 1$.
$$5x^2 + 3 = -16x$$
$$5x^2 + 16x + 3 = 0$$
$$(x + 3)(5x + 1) = 0$$

$x + 3 = 0$ or $5x + 1 = 0$
$\quad x = -3 \qquad\quad 5x = -1$
$\qquad\qquad\qquad\quad x = -\frac{1}{5}$

Substitute $3a - 1$ for x.

$3a - 1 = -3$ or $3a - 1 = -\frac{1}{5}$
$\quad 3a = -2 \qquad\quad 3a = \frac{4}{5}$
$\quad a = -\frac{2}{3} \qquad\quad a = \frac{4}{15}$

The solution set is $\left\{ -\frac{2}{3}, \frac{4}{15} \right\}$.

53.

$$(2k - 3)^2 = 16k^2$$
$$4k^2 - 12k + 9 = 16k^2$$
$$-12k^2 - 12k + 9 = 0$$
$$4k^2 + 4k - 3 = 0 \qquad \textit{Divide by } -3.$$
$$(2k + 3)(2k - 1) = 0$$

$2k + 3 = 0$ or $2k - 1 = 0$
$\quad 2k = -3 \qquad\quad 2k = 1$
$\quad k = -\frac{3}{2}$ or $\quad k = \frac{1}{2}$

The solution set is $\left\{ -\frac{3}{2}, \frac{1}{2} \right\}$.

55. Let $x =$ the width of the garden.
Then $x + 4 =$ the length of the garden.

The area of the rectangular-shaped garden is 320 ft^2, so use the formula $A = LW$ and substitute 320 for A, $x + 4$ for L, and x for W.

$$A = LW$$
$$320 = (x + 4)x$$
$$320 = x^2 + 4x$$
$$0 = x^2 + 4x - 320$$
$$0 = (x - 16)(x + 20)$$

$x - 16 = 0$ or $x + 20 = 0$
$\quad x = 16$ or $\qquad x = -20$

A rectangle cannot have a width that is a negative measure, so reject -20 as a solution. The only possible solution is 16.

The width of the garden is 16 feet, and the length is $16 + 4 = 20$ feet.

57. Let $h =$ the height of the parallelogram.
Then $h + 7 =$ the base of the parallelogram.

The area is 60 ft^2, so use the formula $A = bh$ and substitute 60 for A, and $h + 7$ for b.

$$A = bh$$
$$60 = (h + 7)h$$
$$60 = h^2 + 7h$$
$$0 = h^2 + 7h - 60$$
$$0 = (h + 12)(h - 5)$$

continued

$$h + 12 = 0 \quad \text{or} \quad h - 5 = 0$$
$$h = -12 \quad \text{or} \quad h = 5$$

A parallelogram cannot have a height that is negative, so reject -12 as a solution. The only possible solution is 5.
The height of the parallelogram is 5 feet and the base is $5 + 7 = 12$ feet.

59. Let $L =$ the length of the rectangular area and $W =$ the width.

Use the formula for perimeter, $P = 2L + 2W$, and solve for W in terms of L. The perimeter is 300 ft.

$$300 = 2L + 2W$$
$$300 - 2L = 2W$$
$$150 - L = W$$

Now use the formula for area, $A = LW$, substitute 5000 for A, and solve for L.

$$5000 = L(150 - L)$$
$$5000 = 150L - L^2$$
$$L^2 - 150L + 5000 = 0$$
$$(L - 50)(L - 100) = 0$$

$$L - 50 = 0 \quad \text{or} \quad L - 100 = 0$$
$$L = 50 \quad \text{or} \quad L = 100$$

When $L = 50$, $W = 150 - 50 = 100$.
When $L = 100$, $W = 150 - 100 = 50$.
The dimensions should be 50 feet by 100 feet.

61. Let x and $x + 1$ denote the two consecutive integers.

The sum of their squares is 61, so

$$x^2 + (x + 1)^2 = 61.$$
$$x^2 + x^2 + 2x + 1 = 61$$
$$2x^2 + 2x - 60 = 0 \quad \textit{Divide by 2.}$$
$$x^2 + x - 30 = 0$$
$$(x + 6)(x - 5) = 0$$

$$x + 6 = 0 \quad \text{or} \quad x - 5 = 0$$
$$x = -6 \quad \text{or} \quad x = 5$$

If $x = -6$, then $x + 1 = -5$.
If $x = 5$, then $x + 1 = 6$.

The two possible pairs of consecutive integers are -6 and -5 or 5 and 6.

63. Let $w =$ the width of the cardboard.
Then $w + 6 =$ the length of the cardboard.

If squares that measure 2 inches are cut from each corner of the cardboard, then the width becomes $w - 4$ and the length becomes $(w + 6) - 4 = w + 2$. Use the formula

$V = LWH$ and substitute 110 for V, $w + 2$ for L, $w - 4$ for W, and 2 for H.

$$V = LWH$$
$$110 = (w + 2)(w - 4)2$$
$$110 = (w^2 - 2w - 8)2$$
$$55 = w^2 - 2w - 8$$
$$0 = w^2 - 2w - 63$$
$$0 = (w - 9)(w + 7)$$

$$w - 9 = 0 \quad \text{or} \quad w + 7 = 0$$
$$w = 9 \quad \text{or} \quad w = -7$$

A box cannot have a negative width, so reject -7 as a solution. The only possible solution is 9.
The piece of cardboard has width 9 inches and length $9 + 6 = 15$ inches.

65. When the object hits the ground, its height is given by $f(t) = 0$. Let $f(t) = 0$ in the given equation, and solve for t.

$$-16t^2 + 64t + 80 = f(t)$$
$$-16t^2 + 64t + 80 = 0$$
$$t^2 - 4t - 5 = 0 \quad \textit{Divide by -16.}$$
$$(t - 5)(t + 1) = 0$$

$$t - 5 = 0 \quad \text{or} \quad t + 1 = 0$$
$$t = 5 \quad \text{or} \quad t = -1$$

Time cannot be negative, so reject -1 as a solution. The only possible solution is 5.
The object will hit the ground 5 seconds after it is thrown.

67. Use $f(t) = -16t^2 + 625$ with $f(t) = 0$.

$$0 = -16t^2 + 625$$
$$0 = 16t^2 - 625 \quad \textit{Multiply by -1.}$$
$$0 = (4t + 25)(4t - 25)$$

$$4t + 25 = 0 \quad \text{or} \quad 4t - 25 = 0$$
$$4t = -25 \qquad\qquad 4t = 25$$
$$t = -\frac{25}{4} \quad \text{or} \quad t = \frac{25}{4}$$

Time cannot be negative, so reject $-\frac{25}{4}$ as a solution. The only possible solution is $\frac{25}{4}$ or $6\frac{1}{4}$.
The ball will hit the ground after $6\frac{1}{4}$ seconds.

69.
$$2x^2 - 7x - 4 = 0$$
$$(2x + 1)(x - 4) = 0$$

$$2x + 1 = 0 \quad \text{or} \quad x - 4 = 0$$
$$2x = -1 \qquad\qquad x = 4$$
$$x = -\frac{1}{2}$$

These values correspond to the zeros shown on the screens.

The solution set is $\{-0.5, 4\}$.

71. $-x^2 + 3x = -10$

$$0 = x^2 - 3x - 10$$
$$0 = (x+2)(x-5)$$

$x + 2 = 0$ or $x - 5 = 0$

$x = -2$ $x = 5$

These values correspond to the zeros shown on the screens.

The solution set is $\{-2, 5\}$.

73. $\dfrac{12p^2}{3p} = \dfrac{12}{3}p^{2-1} = 4p$

75. $\dfrac{-27m^2n^5}{36m^6n^8} = -\dfrac{3(9)}{4(9)}m^{2-6}n^{5-8} = -\dfrac{3}{4}m^{-4}n^{-3}$

$$= -\dfrac{3}{4m^4n^3}$$

77. $\dfrac{12}{25} = \dfrac{?}{75}$ $25(3) = 75$

$$\dfrac{12}{25} = \dfrac{12(3)}{25(3)} = \dfrac{36}{75}$$

Chapter 6 Review Exercises

1. $12p^2 - 6p = 6p(2p - 1)$

2. $21x^2 + 35x = 7x(3x + 5)$

3. $12q^2b + 8qb^2 - 20q^3b^2$
$$= 4qb(3q + 2b - 5q^2b)$$

4. $6r^3t - 30r^2t^2 + 18rt^3$
$$= 6rt(r^2 - 5rt + 3t^2)$$

5. $(x + 3)(4x - 1) - (x + 3)(3x + 2)$

The GCF is $(x + 3)$.

$$= (x+3)[(4x-1) - (3x+2)]$$
$$= (x+3)(4x - 1 - 3x - 2)$$
$$= (x+3)(x-3)$$

6. $(z + 1)(z - 4) + (z + 1)(2z + 3)$
$$= (z+1)[(z-4) + (2z+3)]$$
$$= (z+1)(3z-1)$$

7. $4m + nq + mn + 4q$

Rearrange the terms.

$$= 4m + 4q + mn + nq$$
$$= 4(m+q) + n(m+q)$$
$$= (m+q)(4+n)$$

8. $x^2 + 5y + 5x + xy$
$$= x^2 + xy + 5x + 5y$$
$$= x(x+y) + 5(x+y)$$
$$= (x+y)(x+5)$$

9. $2m + 6 - am - 3a$
$$= 2(m+3) - a(m+3)$$
$$= (m+3)(2-a)$$

10. $x^2 + 3x - 3y - xy$
$$= (x^2 + 3x) + (-3y - xy)$$
$$= x(x+3) - y(3+x)$$
$$= x(x+3) - y(x+3)$$
$$= (x+3)(x-y)$$

11. $3p^2 - p - 4$
Two integer factors whose product is $(3)(-4) = -12$ and whose sum is -1 are -4 and 3.
$$= 3p^2 - 4p + 3p - 4$$
$$= p(3p-4) + 1(3p-4)$$
$$= (3p-4)(p+1)$$

12. $6k^2 + 11k - 10$
Two integer factors whose product is $(6)(-10) = -60$ and whose sum is 11 are 15 and -4.
$$= 6k^2 + 15k - 4k - 10$$
$$= 3k(2k+5) - 2(2k+5)$$
$$= (2k+5)(3k-2)$$

13. $12r^2 - 5r - 3$
Two integer factors whose product is $(12)(-3) = -36$ and whose sum is -5 are -9 and 4.
$$= 12r^2 - 9r + 4r - 3$$
$$= 3r(4r-3) + 1(4r-3)$$
$$= (4r-3)(3r+1)$$

14. $10m^2 + 37m + 30$
Two integer factors whose product is $(10)(30) = 300$ and whose sum is 37 are 12 and 25.
$$= 10m^2 + 12m + 25m + 30$$
$$= 2m(5m+6) + 5(5m+6)$$
$$= (5m+6)(2m+5)$$

15. $10k^2 - 11kh + 3h^2$
Two integer factors whose product is $(10)(3) = 30$ and whose sum is -11 are -6 and -5.
$$= 10k^2 - 6kh - 5kh + 3h^2$$
$$= 2k(5k-3h) - h(5k-3h)$$
$$= (5k-3h)(2k-h)$$

16. $9x^2 + 4xy - 2y^2$
There are no integers that have a product of $9(-2) = -18$ and a sum of 4. Therefore, the trinomial cannot be factored and is *prime*.

17. $24x - 2x^2 - 2x^3$
$$= 2x(12 - x - x^2)$$
$$= 2x(4+x)(3-x)$$

18. $6b^3 - 9b^2 - 15b$
$= 3b(2b^2 - 3b - 5)$
$= 3b(2b - 5)(b + 1)$

19. $y^4 + 2y^2 - 8$
$= (y^2)^2 + 2y^2 - 8$
$= (y^2 + 4)(y^2 - 2)$

20. $2k^4 - 5k^2 - 3$
$= 2(k^2)^2 - 5k^2 - 3$
$= (2k^2 + 1)(k^2 - 3)$

21. $p^2(p+2)^2 + p(p+2)^2 - 6(p+2)^2$
Factor out $(p+2)^2$.
$= (p+2)^2(p^2 + p - 6)$
$= (p+2)^2(p+3)(p-2)$

22. $3(r+5)^2 - 11(r+5) - 4$
$= 3x^2 - 11x - 4$ *Let x = r + 5.*
$= (3x + 1)(x - 4)$
$= [3(r+5) + 1][(r+5) - 4]$ *Resubstitute.*
$= (3r + 15 + 1)(r + 1)$
$= (3r + 16)(r + 1)$

23. The student's answer is
$$x^2y^2 - 6x^2 + 5y^2 - 30$$
$$= x^2(y^2 - 6) + 5(y^2 - 6).$$

This is incorrect because the polynomial still has two terms, so it is not factored. The correct answer is
$$(y^2 - 6)(x^2 + 5).$$

24. Since area equals length times width, factor the polynomial given for the area.
$$4p^2 + 3p - 1 = (4p - 1)(p + 1)$$

Since the length is given as $4p - 1$, the width is $p + 1$.

25. $16x^2 - 25 = (4x)^2 - 5^2$ *Difference of squares*
$= (4x + 5)(4x - 5)$

26. $9t^2 - 49 = (3t)^2 - 7^2$
$= (3t + 7)(3t - 7)$

27. $36m^2 - 25n^2 = (6m)^2 - (5n)^2$
$= (6m + 5n)(6m - 5n)$

28. $x^2 + 14x + 49 = x^2 + 2(x)(7) + 7^2$
Perfect square trinomial
$= (x + 7)^2$

29. $9k^2 - 12k + 4 = (3k)^2 - 2(3k)(2) + 2^2$
$= (3k - 2)^2$

30. $r^3 + 27 = r^3 + 3^3$ *Sum of cubes*
$= (r + 3)(r^2 - 3r + 9)$

31. $125x^3 - 1 = (5x)^3 - 1^3$ *Difference of cubes*
$= (5x - 1)(25x^2 + 5x + 1)$

32. $m^6 - 1 = (m^3)^2 - 1^2$
Difference of squares
$= (m^3 + 1)(m^3 - 1)$
$= (m^3 + 1^3)(m^3 - 1^3)$
Sum of   Difference
cubes;   of cubes
$= (m + 1)(m^2 - m + 1)$
$\cdot (m - 1)(m^2 + m + 1)$

33. $x^8 - 1 = (x^4)^2 - 1^2$
Difference of squares
$= (x^4 + 1)(x^4 - 1)$
$= (x^4 + 1)[(x^2)^2 - 1^2]$
Difference of squares again
$= (x^4 + 1)(x^2 + 1)(x^2 - 1)$
Difference of squares again
$= (x^4 + 1)(x^2 + 1)(x + 1)(x - 1)$

34. $x^2 + 6x + 9 - 25y^2$
$= (x^2 + 6x + 9) - 25y^2$
$= [x^2 + 2(x)(3) + 3^2] - 25y^2$
Perfect square trinomial
$= (x + 3)^2 - (5y)^2$
Difference of squares
$= [(x + 3) + 5y][(x + 3) - 5y]$
$= (x + 3 + 5y)(x + 3 - 5y)$

35. $(a + b)^3 - (a - b)^3$
Difference of cubes
$= [(a + b) - (a - b)]$
$\cdot [(a + b)^2 + (a + b)(a - b)$
$+ (a - b)^2]$
$= [2b](a^2 + 2ab + b^2 + a^2 - b^2$
$+ a^2 - 2ab + b^2)$
$= 2b(3a^2 + b^2)$

36. $x^5 - x^3 - 8x^2 + 8$
$= x^3(x^2 - 1) - 8(x^2 - 1)$
$= (x^2 - 1)(x^3 - 8)$
$= (x^2 - 1^2)(x^3 - 2^3)$
Difference   Difference
of squares; of cubes
$= (x + 1)(x - 1)(x - 2)(x^2 + 2x + 4)$

37. $x^2 - 8x + 16 = 0$
$(x - 4)(x - 4) = 0$
$x - 4 = 0$
$x = 4$

The solution set is $\{4\}$.

38. $(5x + 2)(x + 1) = 0$

$5x + 2 = 0 \quad$ or $\quad x + 1 = 0$
$5x = -2 \qquad\qquad x = -1$
$x = -\frac{2}{5}$

The solution set is $\left\{-1, -\frac{2}{5}\right\}$.

39. $p^2 - 5p + 6 = 0$
$(p - 2)(p - 3) = 0$

$p - 2 = 0 \quad$ or $\quad p - 3 = 0$
$p = 2 \quad$ or $\quad p = 3$

The solution set is $\{2, 3\}$.

40. $q^2 + 2q = 8$
$q^2 + 2q - 8 = 0$
$(q + 4)(q - 2) = 0$

$q + 4 = 0 \quad$ or $\quad q - 2 = 0$
$q = -4 \quad$ or $\quad q = 2$

The solution set is $\{-4, 2\}$.

41. $6z^2 = 5z + 50$
$6z^2 - 5z - 50 = 0$
$(3z - 10)(2z + 5) = 0$

$3z - 10 = 0 \quad$ or $\quad 2z + 5 = 0$
$3z = 10 \qquad\qquad 2z = -5$
$z = \frac{10}{3} \quad$ or $\quad z = -\frac{5}{2}$

The solution set is $\left\{-\frac{5}{2}, \frac{10}{3}\right\}$.

42. $6r^2 + 7r = 3$
$6r^2 + 7r - 3 = 0$
$(2r + 3)(3r - 1) = 0$

$2r + 3 = 0 \quad$ or $\quad 3r - 1 = 0$
$2r = -3 \qquad\qquad 3r = 1$
$r = -\frac{3}{2} \quad$ or $\quad r = \frac{1}{3}$

The solution set is $\left\{-\frac{3}{2}, \frac{1}{3}\right\}$.

43. $8k^2 + 14k + 3 = 0$
$(2k + 3)(4k + 1) = 0$

$2k + 3 = 0 \quad$ or $\quad 4k + 1 = 0$
$2k = -3 \qquad\qquad 4k = -1$
$k = -\frac{3}{2} \quad$ or $\quad k = -\frac{1}{4}$

The solution set is $\left\{-\frac{3}{2}, -\frac{1}{4}\right\}$.

44. $-4m^2 + 36 = 0$
$m^2 - 9 = 0 \quad$ *Divide by –4.*
$(m + 3)(m - 3) = 0$

$m + 3 = 0 \quad$ or $\quad m - 3 = 0$
$m = -3 \quad$ or $\quad m = 3$

The solution set is $\{-3, 3\}$.

45. $6x^2 + 9x = 0$
$3x(2x + 3) = 0$

$3x = 0 \quad$ or $\quad 2x + 3 = 0$
$x = 0 \qquad\qquad 2x = -3$
$\qquad\qquad x = -\frac{3}{2}$

The solution set is $\left\{-\frac{3}{2}, 0\right\}$.

46. $(2x + 1)(x - 2) = -3$
$2x^2 - 3x - 2 = -3$
$2x^2 - 3x + 1 = 0$
$(2x - 1)(x - 1) = 0$

$2x - 1 = 0 \quad$ or $\quad x - 1 = 0$
$2x = 1 \qquad\qquad x = 1$
$x = \frac{1}{2}$

The solution set is $\left\{\frac{1}{2}, 1\right\}$.

47. $(r + 2)(r - 2) = (r - 2)(r + 3) - 2$
$r^2 - 4 = r^2 + r - 6 - 2$
$-4 = r - 8$
$4 = r$

The solution set is $\{4\}$.

48. $2x^3 - x^2 - 28x = 0$
$x(2x^2 - x - 28) = 0$
$x(2x + 7)(x - 4) = 0$

$x = 0 \quad$ or $\quad 2x + 7 = 0 \quad$ or $\quad x - 4 = 0$
$\qquad\qquad 2x = -7 \qquad\qquad x = 4$
$\qquad\qquad x = -\frac{7}{2}$

The solution set is $\left\{-\frac{7}{2}, 0, 4\right\}$.

49. $-t^3 - 3t^2 + 4t + 12 = 0$
$t^3 + 3t^2 - 4t - 12 = 0 \quad$ *Multiply by –1.*
$t^2(t + 3) - 4(t + 3) = 0$
$(t + 3)(t^2 - 4) = 0$
$(t + 3)(t + 2)(t - 2) = 0$

$t + 3 = 0 \quad$ or $\quad t + 2 = 0 \quad$ or $\quad t - 2 = 0$
$t = -3 \quad$ or $\quad t = -2 \quad$ or $\quad t = 2$

The solution set is $\{-3, -2, 2\}$.

50. $(r + 2)(5r^2 - 9r - 18) = 0$
$(r + 2)(5r + 6)(r - 3) = 0$

$r + 2 = 0 \quad$ or $\quad 5r + 6 = 0 \quad$ or $\quad r - 3 = 0$
$r = -2 \qquad\qquad 5r = -6 \qquad\qquad r = 3$
$\qquad\qquad r = -\frac{6}{5}$

The solution set is $\left\{-2, -\frac{6}{5}, 3\right\}$.

51. Let x be the length of the shorter side. Then the length of the longer side will be $2x + 1$. The area is 10.5 ft^2. Use the formula for area of a triangle, $A = \frac{1}{2}bh$.

$$\frac{1}{2}(2x + 1)(x) = 10.5$$
$$x(2x + 1) = 21 \qquad \textit{Multiply by 2.}$$
$$2x^2 + x = 21$$
$$2x^2 + x - 21 = 0$$
$$(2x + 7)(x - 3) = 0$$

$$2x + 7 = 0 \quad \text{or} \quad x - 3 = 0$$
$$2x = -7 \qquad\qquad x = 3$$
$$x = -\frac{7}{2}$$

The side cannot have a negative length, so reject $x = -\frac{7}{2}$.
The length of the shorter side is 3 feet.

52. Let w be the width of the lot. Then $w + 20$ will be the length of the lot. The area is 2400 ft^2. Use the formula $LW = A$.

$$(w + 20)w = 2400$$
$$w^2 + 20w = 2400$$
$$w^2 + 20w - 2400 = 0$$
$$(w - 40)(w + 60) = 0$$

$$w - 40 = 0 \quad \text{or} \quad w + 60 = 0$$
$$w = 40 \quad \text{or} \qquad w = -60$$

The lot cannot have a negative width, so reject $w = -60$.
The width of the lot is 40 feet, and the length is $40 + 20 = 60$ feet.

53. The height is 0 when the rock returns to the ground.

$$f(t) = -16t^2 + 256t$$
$$0 = -16t^2 + 256t$$
$$0 = -16t(t - 16)$$

$$-16t = 0 \quad \text{or} \quad t - 16 = 0$$
$$t = 0 \quad \text{or} \qquad t = 16$$

The rock is on the ground when $t = 0$. It will return to the ground again after 16 seconds.

54. $f(t) = -16t^2 + 256t$
$$240 = -16t^2 + 256t \qquad \textit{Let f(t) = 240.}$$
$$0 = -16t^2 + 256t - 240$$
$$0 = -16(t^2 - 16t + 15)$$
$$0 = -16(t - 15)(t - 1)$$

$$t - 15 = 0 \quad \text{or} \quad t - 1 = 0$$
$$t = 15 \quad \text{or} \qquad t = 1$$

The rock will be 240 ft above the ground after 1 second and again after 15 seconds.

55. The question in Exercise 54 has two answers because the rock will be 240 ft above the ground after 1 second on the way up and again after 15 seconds on the way back down.

56. **[6.2]** $30a + am - am^2$
$$= a(30 + m - m^2)$$
$$= a(6 - m)(5 + m)$$

57. **[6.3]** $16 - 81k^2 = 4^2 - (9k)^2$
$$= (4 + 9k)(4 - 9k)$$

58. **[6.3]** $8 - a^3$
$$= 2^3 - a^3$$
$$= (2 - a)(2^2 + 2a + a^2)$$
$$= (2 - a)(4 + 2a + a^2)$$

59. **[6.2]** $9x^2 + 13xy - 3y^2$ is *prime* since it cannot be factored further.

60. **[6.1]** $15y^3 + 20y^2 = 5y^2(3y + 4)$

61. **[6.3]** $25z^2 - 30zm + 9m^2$
$$= (5z)^2 - 2(5z)(3m) + (3m)^2$$
$$= (5z - 3m)^2$$

62. **[6.5]** $\qquad 5x^2 - 17x = 12$
$$5x^2 - 17x - 12 = 0$$
$$(5x + 3)(x - 4) = 0$$

$$5x + 3 = 0 \quad \text{or} \quad x - 4 = 0$$
$$5x = -3 \qquad\qquad x = 4$$
$$x = -\frac{3}{5}$$

The solution set is $\left\{-\frac{3}{5}, 4\right\}$.

63. **[6.5]** $3m^2 - 9m = 0$
$$3m(m - 3) = 0$$

$$3m = 0 \quad \text{or} \quad m - 3 = 0$$
$$m = 0 \qquad\qquad m = 3$$

The solution set is $\{0, 3\}$.

64. **[6.5]** $\qquad\qquad x^3 - x = 0$
$$x(x^2 - 1) = 0$$
$$x(x + 1)(x - 1) = 0$$

$$x = 0 \quad \text{or} \quad x + 1 = 0 \quad \text{or} \quad x - 1 = 0$$
$$x = -1 \qquad\qquad x = 1$$

The solution set is $\{-1, 0, 1\}$.

65. **[6.5]** Let x be the width of the frame. Then $x + 2$ will be the length of the frame. The area is 48 in^2. Use the formula $LW = A$.

$$(x + 2)x = 48$$
$$x^2 + 2x = 48$$
$$x^2 + 2x - 48 = 0$$
$$(x + 8)(x - 6) = 0$$

$$x + 8 = 0 \quad \text{or} \quad x - 6 = 0$$
$$x = -8 \quad \text{or} \quad x = 6$$

The frame cannot have a negative width, so reject $x = -8$.
The width of the frame is 6 inches.

66. **[6.5]** Let x be the width of the floor. Then $x + 85$ will be the length of the floor. The area is 2750 ft². Use the formula $LW = A$.

$$(x + 85)x = 2750$$
$$x^2 + 85x = 2750$$
$$x^2 + 85x - 2750 = 0$$
$$(x + 110)(x - 25) = 0$$

$$x + 110 = 0 \quad \text{or} \quad x - 25 = 0$$
$$x = -110 \quad \text{or} \quad x = 25$$

The floor cannot have a negative width, so reject $x = -110$.
The width of the floor was 25 feet and the length was $25 + 85 = 110$ feet.

Chapter 6 Test

1. $11z^2 - 44z = 11z(z - 4)$

2. $10x^2y^5 - 5x^2y^3 - 25x^5y^3$
Factor out the GCF, $5x^2y^3$.
$$= 5x^2y^3\left(2y^2 - 1 - 5x^3\right)$$

3. $3x + by + bx + 3y$
$$= 3x + 3y + bx + by$$
$$= 3(x + y) + b(x + y)$$
$$= (x + y)(3 + b)$$

4. $-2x^2 - x + 36 = -1(2x^2 + x - 36)$

Two integer factors whose product is $(2)(-36) = -72$ and whose sum is 1 are 9 and -8.

$$2x^2 + x - 36 = 2x^2 + 9x - 8x - 36$$
$$= x(2x + 9) - 4(2x + 9)$$
$$= (2x + 9)(x - 4)$$

Thus, the final factored form is

$$-(2x + 9)(x - 4).$$

5. $6x^2 + 11x - 35$

Two integer factors whose product is $(6)(-35) = -210$ and whose sum is 11 are 21 and -10.

$$= 6x^2 + 21x - 10x - 35$$
$$= 3x(2x + 7) - 5(2x + 7)$$
$$= (2x + 7)(3x - 5)$$

6. $4p^2 + 3pq - q^2$
Two integer factors whose product is $(4)(-1) = -4$ and whose sum is 3 are 4 and -1.

$$= 4p^2 + 4pq - pq - q^2$$
$$= 4p(p + q) - q(p + q)$$
$$= (p + q)(4p - q)$$

7. $16a^2 + 40ab + 25b^2$
$$= (4a)^2 + 2(4a)(5b) + (5b)^2$$
$$= (4a + 5b)^2$$

8. $x^2 + 2x + 1 - 4z^2$
$$= \left(x^2 + 2x + 1\right) - 4z^2$$
$$= (x + 1)^2 - (2z)^2$$
$$= [(x + 1) + 2z][(x + 1) - 2z]$$
$$= (x + 1 + 2z)(x + 1 - 2z)$$

9. $a^3 + 2a^2 - ab^2 - 2b^2$
$$= a^2(a + 2) - b^2(a + 2)$$
$$= (a + 2)\left(a^2 - b^2\right)$$
$$= (a + 2)(a + b)(a - b)$$

10. $9k^2 - 121j^2 = (3k)^2 - (11j)^2$
$$= (3k + 11j)(3k - 11j)$$

11. $y^3 - 216 = y^3 - 6^3$
$$= (y - 6)\left(y^2 + 6y + 6^2\right)$$
$$= (y - 6)\left(y^2 + 6y + 36\right)$$

12. $6k^4 - k^2 - 35 = 6\left(k^2\right)^2 - k^2 - 35$
Two integer factors whose product is $(6)(-35) = -210$ and whose sum is -1 are -15 and 14.

$$= 6\left(k^2\right)^2 - 15k^2 + 14k^2 - 35$$
$$= 3k^2\left(2k^2 - 5\right) + 7\left(2k^2 - 5\right)$$
$$= \left(2k^2 - 5\right)\left(3k^2 + 7\right)$$

13. $27x^6 + 1$
$$= \left(3x^2\right)^3 + (1)^3$$
$$= \left(3x^2 + 1\right)\left[\left(3x^2\right)^2 - \left(3x^2\right)(1) + 1^2\right]$$
$$= \left(3x^2 + 1\right)\left(9x^4 - 3x^2 + 1\right)$$

14. It is not in factored form because there are two terms: $(x^2 + 2y)p$ and $3(x^2 + 2y)$. The common factor is $x^2 + 2y$, and the factored form is $(x^2 + 2y)(p + 3)$.

15. **A** $(3 - x)(x + 4)$
$$= 3x + 12 - x^2 - 4x$$
$$= -x^2 - x + 12$$

B $-(x - 3)(x + 4)$
$$= -\left(x^2 + 4x - 3x - 12\right)$$
$$= -\left(x^2 + x - 12\right)$$
$$= -x^2 - x + 12$$

C $(-x + 3)(x + 4)$
$$= -x^2 - 4x + 3x + 12$$
$$= -x^2 - x + 12$$

D $(x - 3)(-x + 4)$
$$= -x^2 + 4x + 3x - 12$$
$$= -x^2 + 7x - 12$$

Therefore, only **D** is *not* a factored form of $-x^2 - x + 12$.

16.
$$3x^2 + 8x = -4$$
$$3x^2 + 8x + 4 = 0$$
$$(x + 2)(3x + 2) = 0$$

$$x + 2 = 0 \quad \text{or} \quad 3x + 2 = 0$$
$$x = -2 \qquad\qquad 3x = -2$$
$$x = -\tfrac{2}{3}$$

The solution set is $\left\{-2, -\tfrac{2}{3}\right\}$.

17. $3x^2 - 5x = 0$
$$x(3x - 5) = 0$$

$$x = 0 \quad \text{or} \quad 3x - 5 = 0$$
$$3x = 5$$
$$x = \tfrac{5}{3}$$

The solution set is $\left\{0, \tfrac{5}{3}\right\}$.

18.
$$5m(m - 1) = 2(1 - m)$$
$$5m^2 - 5m = 2 - 2m$$
$$5m^2 - 3m - 2 = 0$$
$$(5m + 2)(m - 1) = 0$$

$$5m + 2 = 0 \quad \text{or} \quad m - 1 = 0$$
$$5m = -2 \qquad\qquad m = 1$$
$$m = -\tfrac{2}{5}$$

The solution set is $\left\{-\tfrac{2}{5}, 1\right\}$.

19. Using $A = LW$, substitute 40 for A, $x + 7$ for L, and $2x + 3$ for W.

$$A = LW$$
$$40 = (x + 7)(2x + 3)$$
$$40 = 2x^2 + 3x + 14x + 21$$
$$40 = 2x^2 + 17x + 21$$
$$0 = 2x^2 + 17x - 19$$
$$0 = (2x + 19)(x - 1)$$

$$2x + 19 = 0 \quad \text{or} \quad x - 1 = 0$$
$$2x = -19 \qquad\qquad x = 1$$
$$x = -\tfrac{19}{2}$$

Length and width will be negative if $x = -\tfrac{19}{2}$, so reject it as a possible solution. If $x = 1$, then

$$x + 7 = 1 + 7 = 8$$

and

$$2x + 3 = 2(1) + 3 = 5.$$

The length is 8 inches, and the width is 5 inches.

20. Substitute 128 for $f(t)$ in the equation.

$$f(t) = -16t^2 + 96t$$
$$128 = -16t^2 + 96t$$
$$16t^2 - 96t + 128 = 0$$
$$16\left(t^2 - 6t + 8\right) = 0$$
$$16(t - 4)(t - 2) = 0$$

$$t - 4 = 0 \quad \text{or} \quad t - 2 = 0$$
$$t = 4 \quad \text{or} \qquad t = 2$$

The ball is 128 ft high at 2 seconds (on the way up) and again at 4 seconds (on the way down).

Cumulative Review Exercises (Chapters 1–6)

1. $-2(m - 3) = -2(m) - 2(-3) = -2m + 6$

2. $-(-4m + 3) = -(-4m) - (3) = 4m - 3$

3. $3x^2 - 4x + 4 + 9x - x^2$
$$= 3x^2 - x^2 - 4x + 9x + 4$$
$$= 2x^2 + 5x + 4$$

For Exercises 4–7, let $p = -4, q = -2,$ and $r = 5$.

4. $-3(2q - 3p) = -3[2(-2) - 3(-4)]$
$$= -3(-4 + 12)$$
$$= -3(8) = -24$$

5. $8r^2 + q^2 = 8(5)^2 + (-2)^2$
$$= 8(25) + 4$$
$$= 200 + 4 = 204$$

6.
$$\frac{\sqrt{r}}{-p + 2q} = \frac{\sqrt{5}}{-(-4) + 2(-2)}$$
$$= \frac{\sqrt{5}}{4 - 4} = \frac{\sqrt{5}}{0}$$

This is *undefined* since the denominator is zero.

7.
$$\frac{5p + 6r^2}{p^2 + q - 1} = \frac{5(-4) + 6(5)^2}{(-4)^2 + (-2) - 1}$$
$$= \frac{-20 + 6(25)}{16 - 2 - 1}$$
$$= \frac{-20 + 150}{13}$$
$$= \tfrac{130}{13} = 10$$

8. $2z - 5 + 3z = 4 - (z + 2)$
$$5z - 5 = 4 - z - 2$$
$$5z - 5 = 2 - z$$
$$6z = 7$$
$$z = \tfrac{7}{6}$$

The solution set is $\left\{\tfrac{7}{6}\right\}$.

9.
$$\frac{3a-1}{5} + \frac{a+2}{2} = -\frac{3}{10}$$

$2(3a-1) + 5(a+2) = -3$ *Multiply by 10.*

$6a - 2 + 5a + 10 = -3$

$11a + 8 = -3$

$11a = -11$

$a = -1$

The solution set is $\{-1\}$.

10.
$$-\tfrac{4}{3}d \geq -5$$

$$-\tfrac{3}{4}\left(-\tfrac{4}{3}d\right) \leq -\tfrac{3}{4}(-5)$$

$$d \leq \tfrac{15}{4}$$

The solution set is $\left(-\infty, \tfrac{15}{4}\right]$.

11. $3 - 2(m+3) < 4m$

$3 - 2m - 6 < 4m$

$-2m - 3 < 4m$

$-6m < 3$

$m > -\tfrac{3}{6}$ or $-\tfrac{1}{2}$

The solution set is $\left(-\tfrac{1}{2}, \infty\right)$.

12. $2k + 4 < 10$ and $3k - 1 > 5$

$2k < 6$ $3k > 6$

$k < 3$ and $k > 2$

The overlap of these inequalities is the set of all numbers between 2 and 3.

The solution set is $(2, 3)$.

13. $2k + 4 > 10$ or $3k - 1 < 5$

$2k > 6$ $3k < 6$

$k > 3$ or $k < 2$

The solution set is the set of numbers that are either greater than 3 or less than 2.

The solution set is $(-\infty, 2) \cup (3, \infty)$.

14. $|5x + 3| - 10 = 3$

$|5x + 3| = 13$

$5x + 3 = 13$ or $5x + 3 = -13$

$5x = 10$ $5x = -16$

$x = 2$ or $x = -\tfrac{16}{5}$

The solution set is $\left\{-\tfrac{16}{5}, 2\right\}$.

15. $|x + 2| < 9$

$-9 < x + 2 < 9$

$-11 < x < 7$

The solution set is $(-11, 7)$.

16. $|2x - 5| \geq 9$

$2x - 5 \geq 9$ or $2x - 5 \leq -9$

$2x \geq 14$ $2x \leq -4$

$x \geq 7$ or $x \leq -2$

The solution set is $(-\infty, -2] \cup [7, \infty)$.

17.   Solve $V = lwh$ for h.

$$\frac{V}{lw} = \frac{lwh}{lw}$$ *Divide by lw.*

$$\frac{V}{lw} = h$$

18.   Let x be the time it takes for the planes to be 2100 mi apart. Use the formula $d = rt$ to complete the table.

Plane	r	t	d
Eastbound	550	x	$550x$
Westbound	500	x	$500x$

The total distance is 2100 mi.

$$550x + 500x = 2100$$

$$1050x = 2100$$

$$x = 2$$

It will take 2 hours for the planes to be 2100 miles apart.

19. $4x + 2y = -8$

Draw a line through the x- and y-intercepts, $(-2, 0)$ and $(0, -4)$, respectively.

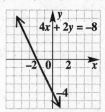

20.   The slope m of the line through $(-4, 8)$ and $(-2, 6)$ is

$$m = \frac{6 - 8}{-2 - (-4)} = \frac{-2}{2} = -1.$$

21. $y = -3$ is an equation of a horizontal line. Its slope is 0.

22.
$$f(x) = 2x + 7$$
$$f(-4) = 2(-4) + 7$$
$$= -8 + 7 = -1$$

23.   To find the x-intercept of the graph of $f(x) = 2x + 7$, let $f(x) = 0$ (which is the same as letting $y = 0$) and solve for x.

$$0 = 2x + 7$$

$$-7 = 2x$$

$$-\tfrac{7}{2} = x$$

The x-intercept is $\left(-\tfrac{7}{2}, 0\right)$.

24. $f(0) = 7$, so the y-intercept is $(0, 7)$.

25. $3x - 2y = -7$ (1)
$2x + 3y = 17$ (2)
To eliminate y, multiply (1) by 3 and (2) by 2, and then add the resulting equations.

$$\begin{array}{rl} 9x - 6y = -21 & 3 \times (1) \\ 4x + 6y = 34 & 2 \times (2) \\ \hline 13x = 13 & \\ x = 1 & \end{array}$$

Substitute 1 for x in (1).

$$\begin{array}{rl} 3x - 2y = -7 & (1) \\ 3(1) - 2y = -7 & \\ -2y = -10 & \\ y = 5 & \end{array}$$

The solution set is $\{(1, 5)\}$.

26. $2x + 3y - 6z = 5$ (1)
$8x - y + 3z = 7$ (2)
$3x + 4y - 3z = 7$ (3)

To eliminate z, add (2) and (3).

$$\begin{array}{rl} 8x - y + 3z = 7 & (2) \\ 3x + 4y - 3z = 7 & (3) \\ \hline 11x + 3y = 14 & (4) \end{array}$$

To eliminate z again, multiply (2) by 2 and add the result to (1).

$$\begin{array}{rl} 2x + 3y - 6z = 5 & (1) \\ 16x - 2y + 6z = 14 & 2 \times (2) \\ \hline 18x + y = 19 & (5) \end{array}$$

To eliminate y, multiply (5) by -3 and add the result to (4).

$$\begin{array}{rl} 11x + 3y = 14 & (4) \\ -54x - 3y = -57 & -3 \times (5) \\ \hline -43x = -43 & \\ x = 1 & \end{array}$$

From (5), $18(1) + y = 19$, so $y = 1$.
To find z, let $x = 1$ and $y = 1$ in (1).

$$\begin{array}{rl} 2x + 3y - 6z = 5 & (1) \\ 2(1) + 3(1) - 6z = 5 & \\ 5 - 6z = 5 & \\ -6z = 0 & \\ z = 0 & \end{array}$$

The solution set is $\{(1, 1, 0)\}$.

27. $\left(3x^2 y^{-1}\right)^{-2} \left(2x^{-3} y\right)^{-1}$
$= 3^{-2} x^{-4} y^2 \cdot 2^{-1} x^3 y^{-1}$
$= 3^{-2} 2^{-1} x^{-1} y$
$= \dfrac{y}{3^2 \cdot 2x} = \dfrac{y}{18x}$

28. $\dfrac{5m^{-2} y^3}{3m^{-3} y^{-1}} = \dfrac{5}{3} \cdot \dfrac{m^{-2}}{m^{-3}} \cdot \dfrac{y^3}{y^{-1}}$
$= \dfrac{5}{3} m^{-2-(-3)} y^{3-(-1)}$
$= \dfrac{5}{3} m^1 y^4$ or $\dfrac{5my^4}{3}$

29. $\left(3x^3 + 4x^2 - 7\right) - \left(2x^3 - 8x^2 + 3x\right)$
$= 3x^3 + 4x^2 - 7 - 2x^3 + 8x^2 - 3x$
$= x^3 + 12x^2 - 3x - 7$

30. $(7x + 3y)^2 = (7x)^2 + 2(7x)(3y) + (3y)^2$
$= 49x^2 + 42xy + 9y^2$

31. $(2p + 3)\left(5p^2 - 4p - 8\right)$
$= 10p^3 - 8p^2 - 16p + 15p^2 - 12p - 24$
$= 10p^3 + 7p^2 - 28p - 24$

32. $16w^2 + 50wz - 21z^2$
Two integer factors whose product is $(16)(-21) = -336$ and whose sum is 50 are 56 and -6.

$= 16w^2 + 56wz - 6wz - 21z^2$
$= 8w(2w + 7z) - 3z(2w + 7z)$
$= (2w + 7z)(8w - 3z)$

33. $4x^2 - 4x + 1 - y^2$
Group the first three terms.
$= \left(4x^2 - 4x + 1\right) - y^2$
$= (2x - 1)^2 - y^2$
$= [(2x - 1) + y][(2x - 1) - y]$
$= (2x - 1 + y)(2x - 1 - y)$

34. $4y^2 - 36y + 81$
$= (2y)^2 - 2(2y)(9) + 9^2$
$= (2y - 9)^2$

35. $100x^4 - 81 = \left(10x^2\right)^2 - 9^2$
$= \left(10x^2 + 9\right)\left(10x^2 - 9\right)$

36. $8p^3 + 27$
$= (2p)^3 + 3^3$
$= (2p + 3)\left[(2p)^2 - (2p)(3) + 3^2\right]$
$= (2p + 3)\left(4p^2 - 6p + 9\right)$

37. $(p - 1)(2p + 3)(p + 4) = 0$

$$\begin{array}{ccccc} p - 1 = 0 & \text{or} & 2p + 3 = 0 & \text{or} & p + 4 = 0 \\ p = 1 & & 2p = -3 & & p = -4 \\ & & p = -\frac{3}{2} & & \end{array}$$

The solution set is $\left\{-4, -\frac{3}{2}, 1\right\}$.

38.
$$9q^2 = 6q - 1$$
$$9q^2 - 6q + 1 = 0$$
$$(3q - 1)^2 = 0$$
$$3q - 1 = 0$$
$$3q = 1$$
$$q = \tfrac{1}{3}$$

The solution set is $\left\{\tfrac{1}{3}\right\}$.

39. Let x be the length of the base. Then $x + 3$ will be the height. The area is 14 square feet. Use the formula $A = \tfrac{1}{2}bh$, and substitute 14 for A, x for b, and $x + 3$ for h.

$$\tfrac{1}{2}bh = A$$
$$\tfrac{1}{2}(x)(x + 3) = 14$$
$$x(x + 3) = 28 \quad \textit{Multiply by 2.}$$
$$x^2 + 3x = 28$$
$$x^2 + 3x - 28 = 0$$
$$(x + 7)(x - 4) = 0$$

$$x + 7 = 0 \quad \text{or} \quad x - 4 = 0$$
$$x = -7 \quad \text{or} \quad x = 4$$

The length cannot be negative, so reject -7 as a solution. The only possible solution is 4. The base is 4 feet long.

40. Let x be the distance between the longer sides. (This is actually the width.) Then $x + 2$ will be the length of the longer side. The area of the rectangle is 288 in^2. Use the formula $LW = A$. Substitute 288 for A, $x + 2$ for L, and x for W.

$$(x + 2)x = 288$$
$$x^2 + 2x = 288$$
$$x^2 + 2x - 288 = 0$$
$$(x + 18)(x - 16) = 0$$

$$x + 18 = 0 \quad \text{or} \quad x - 16 = 0$$
$$x = -18 \quad \text{or} \quad x = 16$$

The distance cannot be negative, so reject -18 as a solution. The only possible solution is 16. The distance between the longer sides is 16 inches, and the length of the longer sides is $16 + 2 = 18$ inches.

CHAPTER 7 RATIONAL EXPRESSIONS AND FUNCTIONS

7.1 Rational Expressions and Functions; Multiplying and Dividing

1. $\dfrac{x-3}{x+4} = \dfrac{(-1)(x-3)}{(-1)(x+4)}$

$\quad = \dfrac{-x+3}{-x-4}$ or $\dfrac{3-x}{-x-4}$ **(C)**

3. $\dfrac{x-3}{x-4} = \dfrac{(-1)(x-3)}{(-1)(x-4)}$

$\quad = \dfrac{-x+3}{-x+4}$ **(D)**

5. $\dfrac{3-x}{x+4} = \dfrac{(-1)(3-x)}{(-1)(x+4)}$

$\quad = \dfrac{-3+x}{-x-4}$ or $\dfrac{x-3}{-x-4}$ **(E)**

7. Replacing x with 2 makes the denominator 0 and the value of the expression undefined. To find the values excluded from the domain, set the denominator equal to 0 and solve the equation. All solutions of the equation are excluded from the domain.

9. $f(x) = \dfrac{x}{x-7}$

Set the denominator equal to zero, and solve the equation.

$$x - 7 = 0$$
$$x = 7$$

The number 7 makes the rational expression undefined, so 7 is not in the domain of the function. In set notation, the domain is $\{x \mid x \neq 7\}$.

11. $f(x) = \dfrac{6x-5}{7x+1}$

Set the denominator equal to zero, and solve the equation.

$$7x + 1 = 0$$
$$7x = -1$$
$$x = -\tfrac{1}{7}$$

The number $-\tfrac{1}{7}$ makes the rational expression undefined, so $-\tfrac{1}{7}$ is not in the domain of the function. In set notation, the domain is $\{x \mid x \neq -\tfrac{1}{7}\}$.

13. $f(x) = \dfrac{12x+3}{x}$

Set the denominator equal to zero and solve.

$$x = 0$$

The number 0 makes the rational expression undefined, so 0 is not in the domain of the function. In set notation, the domain is $\{x \mid x \neq 0\}$.

15. $f(x) = \dfrac{3x+1}{2x^2+x-6}$

Set the denominator equal to zero and solve.

$$2x^2 + x - 6 = 0$$
$$(x+2)(2x-3) = 0$$

$$x + 2 = 0 \quad \text{or} \quad 2x - 3 = 0$$
$$x = -2 \quad \text{or} \quad 2x = 3$$
$$x = \tfrac{3}{2}$$

The numbers -2 and $\tfrac{3}{2}$ are not in the domain of the function. In set notation, the domain is $\{x \mid x \neq -2, \tfrac{3}{2}\}$.

17. $f(x) = \dfrac{x+2}{14}$

The denominator is never zero, so all numbers are in the domain of the function. In set notation, the domain is $(-\infty, \infty)$.

19. $f(x) = \dfrac{2x^2-3x+4}{3x^2+8}$

Set the denominator equal to zero and solve.

$$3x^2 + 8 = 0$$
$$3x^2 = -8$$
$$x^2 = -\tfrac{8}{3}$$

The square of any real number x is positive or zero, so this equation has no solution. There are no real numbers which make this rational expression undefined, so all numbers are in the domain of the function. In set notation, the domain is $(-\infty, \infty)$.

21. **(a)** $\dfrac{x^2+4x}{x+4}$

The two terms in the numerator are x^2 and $4x$. The two terms in the denominator are x and 4.

(b) To express the rational expression in lowest terms, factor the numerator and denominator and replace the quotient of common factors with 1.

$$\frac{x^2+4x}{x+4} = \frac{x(x+4)}{x+4}$$
$$= x \cdot 1 = x$$

23. **A.** $\dfrac{3-x}{x-4} = \dfrac{-1(x-3)}{-1(4-x)} = \dfrac{x-3}{4-x}$

B. $\dfrac{x+3}{4+x}$ cannot be transformed to equal $\dfrac{x-3}{4-x}$.

C. $-\dfrac{3-x}{4-x} = \dfrac{-(3-x)}{4-x} = \dfrac{x-3}{4-x}$

D. $-\dfrac{x-3}{x-4} = \dfrac{x-3}{-(x-4)} = \dfrac{x-3}{4-x}$

Only the expression in **B** is *not* equivalent to $\dfrac{x-3}{4-x}$.

25. $\dfrac{x^2(x+1)}{x(x+1)} = \dfrac{x}{1} \cdot \dfrac{x(x+1)}{x(x+1)} = \dfrac{x}{1} \cdot 1 = x$

27. $\dfrac{(x+4)(x-3)}{(x+5)(x+4)} = \dfrac{x-3}{x+5} \cdot \dfrac{x+4}{x+4} = \dfrac{x-3}{x+5}$

29. $\dfrac{4x(x+3)}{8x^2(x-3)} = \dfrac{(x+3)\cdot 4x}{2x(x-3)\cdot 4x}$

$= \dfrac{x+3}{2x(x-3)}$

31. $\dfrac{3x+7}{3}$ Since the numerator and denominator have no common factors, the expression is already in lowest terms.

33. $\dfrac{6m+18}{7m+21} = \dfrac{6(m+3)}{7(m+3)} = \dfrac{6}{7}$

35. $\dfrac{t^2-9}{3t+9} = \dfrac{(t+3)(t-3)}{3(t+3)}$

$= \dfrac{t-3}{3}$

37. $\dfrac{2t+6}{t^2-9} = \dfrac{2(t+3)}{(t-3)(t+3)}$

$= \dfrac{2}{t-3}$

39. $\dfrac{x^2+2x-15}{x^2+6x+5} = \dfrac{(x+5)(x-3)}{(x+5)(x+1)}$

$= \dfrac{x-3}{x+1}$

41. $\dfrac{8x^2-10x-3}{8x^2-6x-9} = \dfrac{(4x+1)(2x-3)}{(4x+3)(2x-3)}$

$= \dfrac{4x+1}{4x+3}$

43. $\dfrac{a^3+b^3}{a+b} = \dfrac{(a+b)(a^2-ab+b^2)}{a+b}$

$= a^2-ab+b^2$

45. $\dfrac{2c^2+2cd-60d^2}{2c^2-12cd+10d^2}$

$= \dfrac{2(c^2+cd-30d^2)}{2(c^2-6cd+5d^2)}$

$= \dfrac{2(c+6d)(c-5d)}{2(c-d)(c-5d)}$

$= \dfrac{c+6d}{c-d}$

47. $\dfrac{ac-ad+bc-bd}{ac-ad-bc+bd}$

$= \dfrac{a(c-d)+b(c-d)}{a(c-d)-b(c-d)}$

$= \dfrac{(c-d)(a+b)}{(c-d)(a-b)}$ *Factor by grouping.*

$= \dfrac{a+b}{a-b}$

49. $\dfrac{7-b}{b-7} = \dfrac{-1(b-7)}{b-7} = -1$

In Exercises 51–54, there are other acceptable ways to express each answer.

51. $\dfrac{x^2-y^2}{y-x} = \dfrac{(x-y)(x+y)}{y-x}$

$= \dfrac{-1(y-x)(x+y)}{y-x}$

$= -(x+y)$

53. $\dfrac{(a-3)(x+y)}{(3-a)(x-y)} = \dfrac{(a-3)(x+y)}{-1(a-3)(x-y)}$

$= \dfrac{x+y}{-1(x-y)}$

$= -\dfrac{x+y}{x-y}$

55. $\dfrac{5k-10}{20-10k} = \dfrac{-5(2-k)}{10(2-k)} = \dfrac{-5}{10} = -\dfrac{1}{2}$

57. $\dfrac{a^2-b^2}{a^2+b^2} = \dfrac{(a+b)(a-b)}{a^2+b^2}$

The numerator and denominator have no common factors except 1, so the original expression is already in lowest terms.

59. Multiply the numerators, multiply the denominators, and factor each numerator and denominator. (Factoring can be performed first.) Divide the numerator and denominator by any common factors to write the rational expression in lowest terms. For example,

$\dfrac{6r-5s}{3r+2s} \cdot \dfrac{6r+4s}{5s-6r}$

$= \dfrac{(6r-5s)(6r+4s)}{(3r+2s)(5s-6r)}$

$= \dfrac{(6r-5s)2(3r+2s)}{(3r+2s)(-1)(6r-5s)} = \dfrac{2}{-1} = -2.$

61. $\dfrac{x^3}{3y} \cdot \dfrac{9y^2}{x^5} = \dfrac{3y \cdot 3x^3 y}{x^2 \cdot 3x^3 y} = \dfrac{3y}{x^2} \cdot 1 = \dfrac{3y}{x^2}$

63. $\dfrac{5a^4 b^2}{16a^2 b} \div \dfrac{25a^2 b}{60a^3 b^2}$

Multiply by the reciprocal of the divisor.

$= \dfrac{5a^4 b^2}{16a^2 b} \cdot \dfrac{60a^3 b^2}{25a^2 b}$

$= \dfrac{300a^7 b^4}{400a^4 b^2}$

$= \dfrac{100a^4 b^2 \cdot 3a^3 b^2}{100a^4 b^2 \cdot 4}$

$= \dfrac{3a^3 b^2}{4}$

65. $\dfrac{(-3mn)^2 \cdot 64(m^2 n)^3}{16m^2 n^4 (mn^2)^3} \div \dfrac{24(m^2 n^2)^4}{(3m^2 n^3)^2}$

$= \dfrac{9m^2 n^2 \cdot 64m^6 n^3}{16m^2 n^4 \cdot m^3 n^6} \cdot \dfrac{9m^4 n^6}{24m^8 n^8}$

$= \dfrac{9 \cdot 4 \cdot 9m^{12} n^{11}}{3 \cdot 4 \cdot 2m^{13} n^{18}}$

$= \dfrac{3 \cdot 9}{2m^1 n^7} = \dfrac{27}{2mn^7}$

67. $\dfrac{(x+2)(x+1)}{(x+3)(x-2)} \cdot \dfrac{(x+3)(x+4)}{(x+2)(x+1)}$

$= \dfrac{(x+1)(x+2)(x+3)}{(x+1)(x+2)(x+3)} \cdot \dfrac{x+4}{x-2} = \dfrac{x+4}{x-2}$

69. $\dfrac{(2x+3)(x-4)}{(x+8)(x-4)} \div \dfrac{(x-4)(x+2)}{(x-4)(x+8)}$

$= \dfrac{2x+3}{x+8} \div \dfrac{x+2}{x+8}$

$= \dfrac{2x+3}{x+8} \cdot \dfrac{x+8}{x+2}$

$= \dfrac{2x+3}{x+2}$

71. $\dfrac{4x}{8x+4} \cdot \dfrac{14x+7}{6} = \dfrac{4x \cdot 7(2x+1)}{4(2x+1) \cdot 6}$

$= \dfrac{7x}{6}$

For Exercises 73–74, there are several other ways to express the answer.

73. $\dfrac{p^2 - 25}{4p} \cdot \dfrac{2}{5-p} = \dfrac{(p+5)(p-5)2}{2 \cdot 2p(-1)(p-5)}$

$= \dfrac{p+5}{(2p)(-1)} = -\dfrac{p+5}{2p}$

75. $(7k+7) \div \dfrac{4k+4}{5}$

$= \dfrac{7(k+1)}{1} \cdot \dfrac{5}{4(k+1)}$

$= \dfrac{7 \cdot 5}{4} = \dfrac{35}{4}$

77. $(z^2 - 1) \cdot \dfrac{1}{1-z}$

$= \dfrac{(z+1)(z-1)}{1} \cdot \dfrac{1}{-1(z-1)}$

$= \dfrac{z+1}{-1} = -(z+1)$ or $-z-1$

79. $\dfrac{4x-20}{5x} \div \dfrac{2x-10}{7x^3}$

$= \dfrac{4(x-5)}{5x} \cdot \dfrac{7x^3}{2(x-5)}$

$= \dfrac{2 \cdot 7x^2}{5}$

$= \dfrac{14x^2}{5}$

81. $\dfrac{12x-10y}{3x+2y} \cdot \dfrac{6x+4y}{10y-12x}$

$= \dfrac{2(6x-5y) \cdot 2(3x+2y)}{(3x+2y) \cdot 2(5y-6x)}$

$= \dfrac{2(-1)(5y-6x)}{(5y-6x)} = -2$

83. $\dfrac{x^2-25}{x^2+x-20} \cdot \dfrac{x^2+7x+12}{x^2-2x-15}$

$= \dfrac{(x-5)(x+5)}{(x+5)(x-4)} \cdot \dfrac{(x+3)(x+4)}{(x-5)(x+3)}$

$= \dfrac{x+4}{x-4}$

85. $\dfrac{a^3-b^3}{a^2-b^2} \div \dfrac{2a-2b}{2a+2b}$

$= \dfrac{(a-b)(a^2+ab+b^2)}{(a+b)(a-b)} \cdot \dfrac{2(a+b)}{2(a-b)}$

$= \dfrac{a^2+ab+b^2}{a-b}$

87. $\dfrac{8x^3-27}{2x^2-18} \cdot \dfrac{2x+6}{8x^2+12x+18}$

$= \dfrac{(2x)^3-3^3}{2(x^2-9)} \cdot \dfrac{2(x+3)}{2(4x^2+6x+9)}$

$= \dfrac{(2x-3)(4x^2+6x+9)}{2(x+3)(x-3)} \cdot \dfrac{2(x+3)}{2(4x^2+6x+9)}$

$= \dfrac{2x-3}{2(x-3)}$

89. $\dfrac{a^3-8b^3}{a^2-ab-6b^2} \cdot \dfrac{a^2+ab-12b^2}{a^2+2ab-8b^2}$

$= \dfrac{(a-2b)(a^2+2ab+4b^2)}{(a-3b)(a+2b)} \cdot \dfrac{(a-3b)(a+4b)}{(a-2b)(a+4b)}$

$= \dfrac{a^2+2ab+4b^2}{a+2b}$

91. $\dfrac{6x^2 + 5x - 6}{12x^2 - 11x + 2} \div \dfrac{4x^2 - 12x + 9}{8x^2 - 14x + 3}$

$= \dfrac{(3x - 2)(2x + 3)}{(3x - 2)(4x - 1)} \cdot \dfrac{(2x - 3)(4x - 1)}{(2x - 3)(2x - 3)}$

$= \dfrac{2x + 3}{2x - 3}$

93. $\dfrac{3k^2 + 17kp + 10p^2}{6k^2 + 13kp - 5p^2} \div \dfrac{6k^2 + kp - 2p^2}{6k^2 - 5kp + p^2}$

$= \dfrac{(3k + 2p)(k + 5p)}{(3k - p)(2k + 5p)} \cdot \dfrac{(3k - p)(2k - p)}{(3k + 2p)(2k - p)}$

$= \dfrac{k + 5p}{2k + 5p}$

95. $\left(\dfrac{6k^2 - 13k - 5}{k^2 + 7k} \div \dfrac{2k - 5}{k^3 + 6k^2 - 7k} \right)$

$\cdot \dfrac{k^2 - 5k + 6}{3k^2 - 8k - 3}$

Factor k from the denominator of the divisor; multiply by the reciprocal.

$= \left[\dfrac{6k^2 - 13k - 5}{k^2 + 7k} \cdot \dfrac{k(k^2 + 6k - 7)}{2k - 5} \right]$

$\cdot \dfrac{k^2 - 5k + 6}{3k^2 - 8k - 3}$

$= \left[\dfrac{(3k + 1)(2k - 5)}{k(k + 7)} \cdot \dfrac{k(k + 7)(k - 1)}{2k - 5} \right]$

$\cdot \dfrac{(k - 2)(k - 3)}{(3k + 1)(k - 3)}$

$= (k - 1)(k - 2)$

97. $\dfrac{a^2(2a + b) + 6a(2a + b) + 5(2a + b)}{3a^2(a + 2b) - 2a(a + 2b) - (a + 2b)} \div \dfrac{a + 1}{a - 1}$

$= \dfrac{(2a + b)(a^2 + 6a + 5)}{(a + 2b)(3a^2 - 2a - 1)} \cdot \dfrac{a - 1}{a + 1}$

$= \dfrac{(2a + b)(a + 5)(a + 1)(a - 1)}{(a + 2b)(3a + 1)(a - 1)(a + 1)}$

$= \dfrac{(a + 5)(2a + b)}{(3a + 1)(a + 2b)}$

99. $\dfrac{4}{7} + \dfrac{1}{3} - \dfrac{1}{2} = \dfrac{24}{42} + \dfrac{14}{42} - \dfrac{21}{42}$ *LCD = 42*

$= \dfrac{24 + 14 - 21}{42} = \dfrac{17}{42}$

101. $-\dfrac{3}{4} + \dfrac{1}{12} = -\dfrac{9}{12} + \dfrac{1}{12}$ *LCD = 12*

$= \dfrac{-9 + 1}{12} = \dfrac{-8}{12} = -\dfrac{2}{3}$

7.2   Adding and Subtracting Rational Expressions

1. $\dfrac{7}{t} + \dfrac{2}{t} = \dfrac{7 + 2}{t} = \dfrac{9}{t}$

3. $\dfrac{6x}{7} + \dfrac{y}{7} = \dfrac{6x + y}{7}$

5. $\dfrac{11}{5x} - \dfrac{1}{5x} = \dfrac{11 - 1}{5x} = \dfrac{10}{5x} = \dfrac{2}{x}$

7. $\dfrac{9}{4x^3} - \dfrac{17}{4x^3} = \dfrac{9 - 17}{4x^3} = \dfrac{-8}{4x^3} = -\dfrac{2}{x^3}$

9. $\dfrac{5x + 4}{6x + 5} + \dfrac{x + 1}{6x + 5} = \dfrac{5x + 4 + x + 1}{6x + 5}$

$= \dfrac{6x + 5}{6x + 5} = 1$

11. $\dfrac{x^2}{x + 5} - \dfrac{25}{x + 5} = \dfrac{x^2 - 25}{x + 5}$

$= \dfrac{(x + 5)(x - 5)}{x + 5}$

$= x - 5$

13. $\dfrac{-3p + 7}{p^2 + 7p + 12} + \dfrac{8p + 13}{p^2 + 7p + 12}$

$= \dfrac{-3p + 7 + 8p + 13}{p^2 + 7p + 12}$

$= \dfrac{5p + 20}{(p + 3)(p + 4)}$

$= \dfrac{5(p + 4)}{(p + 3)(p + 4)} = \dfrac{5}{p + 3}$

15. $\dfrac{a^3}{a^2 + ab + b^2} - \dfrac{b^3}{a^2 + ab + b^2}$

$= \dfrac{a^3 - b^3}{a^2 + ab + b^2}$

$= \dfrac{(a - b)(a^2 + ab + b^2)}{a^2 + ab + b^2}$

$= a - b$

17. First add or subtract the numerators. Then place the result over the common denominator. Write the answer in lowest terms. We give one example:

$\dfrac{5}{x} - \dfrac{3x + 1}{x} = \dfrac{5 - (3x + 1)}{x}$

$= \dfrac{5 - 3x - 1}{x} = \dfrac{4 - 3x}{x}.$

19. $18x^2y^3,\ 24x^4y^5$

Factor each denominator.

$18x^2y^3 = 2 \cdot 3 \cdot 3 \cdot x^2 \cdot y^3$

$= 2 \cdot 3^2 \cdot x^2 \cdot y^3$

$24x^4y^5 = 2 \cdot 2 \cdot 2 \cdot 3 \cdot x^4 \cdot y^5$

$= 2^3 \cdot 3 \cdot x^4 \cdot y^5$

The least common denominator (LCD) is the product of all the different factors, with each factor raised to the greatest power in any denominator.

$LCD = 2^3 \cdot 3^2 \cdot x^4 \cdot y^5$

$= 8 \cdot 9 \cdot x^4 \cdot y^5$

$= 72x^4y^5$

The LCD is $72x^4y^5$.

21. $z - 2, z$

Both $z - 2$ and z have only 1 and themselves for factors.

LCD $= z(z - 2)$

23. $2y + 8, y + 4$

Factor each denominator.

$$2y + 8 = 2(y + 4)$$

The second denominator, $y + 4$, is already factored. The LCD is

$$2(y + 4).$$

25. $x^2 - 81, x^2 + 18x + 81$

Factor each denominator.

$$x^2 - 81 = (x + 9)(x - 9)$$
$$x^2 + 18x + 81 = (x + 9)(x + 9)$$

LCD $= (x + 9)^2(x - 9)$

27. $m + n, m - n, m^2 - n^2$

Both $m + n$ and $m - n$ have only 1 and themselves for factors, while
$m^2 - n^2 = (m + n)(m - n)$.

LCD $= (m + n)(m - n)$

29. $x^2 - 3x - 4, x + x^2$

Factor each denominator.

$$x^2 - 3x - 4 = (x - 4)(x + 1)$$
$$x + x^2 = x(1 + x) = x(x + 1)$$

The LCD is $x(x - 4)(x + 1)$.

31. $2t^2 + 7t - 15, t^2 + 3t - 10$

Factor each denominator.

$$2t^2 + 7t - 15 = (2t - 3)(t + 5)$$
$$t^2 + 3t - 10 = (t + 5)(t - 2)$$

The LCD is $(2t - 3)(t + 5)(t - 2)$.

33. $2y + 6, y^2 - 9, y$

Factor each denominator.

$$2y + 6 = 2(y + 3)$$
$$y^2 - 9 = (y + 3)(y - 3)$$

Remember the factor y from the third denominator. The LCD is

$$2y(y + 3)(y - 3).$$

35. $2x - 6, x^2 - x - 6, x^2 + 4x + 4$

Factor each denominator.

$$2x - 6 = 2(x - 3)$$
$$x^2 - x - 6 = (x - 3)(x + 2)$$
$$x^2 + 4x + 4 = (x + 2)^2$$

LCD $= 2(x + 2)^2(x - 3)$

37. The first step is incorrect. The third term in the numerator should be $+1$, because the subtraction sign should be distributed to both $4x$ and -1. The correct solution follows.

$$\frac{x}{x + 2} - \frac{4x - 1}{x + 2} = \frac{x - (4x - 1)}{x + 2}$$
$$= \frac{x - 4x + 1}{x + 2}$$
$$= \frac{-3x + 1}{x + 2}$$

39. $\frac{8}{t} + \frac{7}{3t}$ The LCD is $3t$.

$$\frac{8}{t} + \frac{7}{3t} = \frac{8 \cdot 3}{t \cdot 3} + \frac{7}{3t}$$
$$= \frac{24 + 7}{3t} = \frac{31}{3t}$$

41. $\frac{5}{12x^2y} - \frac{11}{6xy}$ The LCD is $12x^2y$.

$$\frac{5}{12x^2y} - \frac{11}{6xy} = \frac{5}{12x^2y} - \frac{11 \cdot 2x}{6xy \cdot 2x}$$
$$= \frac{5}{12x^2y} - \frac{22x}{12x^2y}$$
$$= \frac{5 - 22x}{12x^2y}$$

43. $\frac{4}{15a^4b^5} + \frac{3}{20a^2b^6}$ LCD $= 60a^4b^6$

$$= \frac{4 \cdot 4b}{15a^4b^5 \cdot 4b} + \frac{3 \cdot 3a^2}{20a^2b^6 \cdot 3a^2}$$
$$= \frac{16b + 9a^2}{60a^4b^6}$$

45. $\frac{2r}{7p^3q^4} + \frac{3s}{14p^4q}$ LCD $= 14p^4q^4$

$$= \frac{2r \cdot 2p}{7p^3q^4 \cdot 2p} + \frac{3s \cdot q^3}{14p^4q \cdot q^3}$$
$$= \frac{4pr + 3sq^3}{14p^4q^4}$$

47. $\frac{1}{a^3b^2} - \frac{2}{a^4b} + \frac{3}{a^5b^7}$ LCD $= a^5b^7$

$$= \frac{1 \cdot a^2b^5}{a^3b^2 \cdot a^2b^5} - \frac{2 \cdot ab^6}{a^4b \cdot ab^6} + \frac{3}{a^5b^7}$$
$$= \frac{a^2b^5 - 2ab^6 + 3}{a^5b^7}$$

49. $\dfrac{1}{x-1} - \dfrac{1}{x}$ LCD $= x(x-1)$

$\dfrac{1}{x-1} - \dfrac{1}{x} = \dfrac{1\cdot x}{(x-1)x} - \dfrac{1\cdot(x-1)}{x(x-1)}$

$= \dfrac{x-(x-1)}{x(x-1)}$

$= \dfrac{x-x+1}{x(x-1)}$

$= \dfrac{1}{x(x-1)}$

51. $\dfrac{3a}{a+1} + \dfrac{2a}{a-3}$ LCD $= (a+1)(a-3)$

$= \dfrac{3a(a-3)}{(a+1)(a-3)} + \dfrac{2a(a+1)}{(a-3)(a+1)}$

$= \dfrac{3a(a-3)+2a(a+1)}{(a+1)(a-3)}$

$= \dfrac{3a^2-9a+2a^2+2a}{(a+1)(a-3)}$

$= \dfrac{5a^2-7a}{(a+1)(a-3)}$

53. $\dfrac{17y+3}{9y+7} - \dfrac{-10y-18}{9y+7}$

$= \dfrac{17y+3-(-10y-18)}{9y+7}$

$= \dfrac{17y+3+10y+18}{9y+7}$

$= \dfrac{27y+21}{9y+7}$

$= \dfrac{3(9y+7)}{9y+7}$

$= 3$

55. $\dfrac{2}{4-x} + \dfrac{5}{x-4}$

To get a common denominator of $x-4$, multiply both the numerator and denominator of the first expression by -1.

$= \dfrac{(2)(-1)}{(4-x)(-1)} + \dfrac{5}{x-4}$

$= \dfrac{-2}{x-4} + \dfrac{5}{x-4}$

$= \dfrac{-2+5}{x-4}$

$= \dfrac{3}{x-4}$

If you chose $4-x$ for the LCD, then you should

have obtained the equivalent answer, $\dfrac{-3}{4-x}$.

57. $\dfrac{w}{w-z} - \dfrac{z}{z-w}$

$w-z$ and $z-w$ are opposites, so factor out -1 from $z-w$ to get a common denominator.

$= \dfrac{w}{w-z} - \dfrac{z}{-1(w-z)}$

$= \dfrac{w}{w-z} + \dfrac{z}{w-z}$

$= \dfrac{w+z}{w-z}$, or $\dfrac{-w-z}{z-w}$

59. $\dfrac{1}{x+1} - \dfrac{1}{x-1}$ LCD $= (x+1)(x-1)$

$= \dfrac{1(x-1)}{(x+1)(x-1)} - \dfrac{1(x+1)}{(x-1)(x+1)}$

$= \dfrac{x-1-x-1}{(x+1)(x-1)}$

$= \dfrac{-2}{(x+1)(x-1)}$

61. $\dfrac{4x}{x-1} - \dfrac{2}{x+1} - \dfrac{4}{x^2-1}$

$x^2-1 = (x+1)(x-1)$, the LCD.

$\dfrac{4x}{x-1} - \dfrac{2}{x+1} - \dfrac{4}{x^2-1}$

$= \dfrac{4x(x+1)}{(x-1)(x+1)} - \dfrac{2(x-1)}{(x+1)(x-1)}$

$\quad - \dfrac{4}{(x+1)(x-1)}$

$= \dfrac{4x(x+1)-2(x-1)-4}{(x+1)(x-1)}$

$= \dfrac{4x^2+4x-2x+2-4}{(x-1)(x+1)}$

$= \dfrac{4x^2+2x-2}{(x-1)(x+1)}$

$= \dfrac{2(2x^2+x-1)}{(x-1)(x+1)}$

$= \dfrac{2(2x-1)(x+1)}{(x-1)(x+1)}$

$= \dfrac{2(2x-1)}{x-1}$

63. $\dfrac{15}{y^2+3y} + \dfrac{2}{y} + \dfrac{5}{y+3}$

$y^2+3y = y(y+3)$, the LCD.

$= \dfrac{15}{y(y+3)} + \dfrac{2(y+3)}{y(y+3)} + \dfrac{5y}{(y+3)y}$

$= \dfrac{15+2(y+3)+5y}{y(y+3)}$

$= \dfrac{15+2y+6+5y}{y(y+3)}$

$= \dfrac{7y+21}{y(y+3)}$

$= \dfrac{7(y+3)}{y(y+3)} = \dfrac{7}{y}$

65. $\dfrac{5}{x-2}+\dfrac{1}{x}+\dfrac{2}{x^2-2x}$

$x^2-2x=x(x-2)$, the LCD.

$\dfrac{5}{x-2}+\dfrac{1}{x}+\dfrac{2}{x^2-2x}$

$=\dfrac{5x}{(x-2)x}+\dfrac{1(x-2)}{x(x-2)}+\dfrac{2}{x(x-2)}$

$=\dfrac{5x+x-2+2}{x(x-2)}$

$=\dfrac{6x}{x(x-2)}=\dfrac{6}{x-2}$

67. $\dfrac{3x}{x+1}+\dfrac{4}{x-1}-\dfrac{6}{x^2-1}$

$x^2-1=(x+1)(x-1)$, the LCD.

$=\dfrac{3x(x-1)}{(x+1)(x-1)}+\dfrac{4(x+1)}{(x-1)(x+1)}$

$-\dfrac{6}{(x+1)(x-1)}$

$=\dfrac{3x(x-1)+4(x+1)-6}{(x+1)(x-1)}$

$=\dfrac{3x^2-3x+4x+4-6}{(x+1)(x-1)}$

$=\dfrac{3x^2+x-2}{(x+1)(x-1)}$

$=\dfrac{(3x-2)(x+1)}{(x+1)(x-1)}$

$=\dfrac{3x-2}{x-1}$

69. $\dfrac{4}{x+1}+\dfrac{1}{x^2-x+1}-\dfrac{12}{x^3+1}$

$x^3+1=(x+1)\left(x^2-x+1\right)$, the LCD.

$=\dfrac{4(x^2-x+1)}{(x+1)(x^2-x+1)}$

$+\dfrac{1\cdot(x+1)}{(x^2-x+1)(x+1)}$

$-\dfrac{12}{(x+1)(x^2-x+1)}$

$=\dfrac{4(x^2-x+1)+(x+1)-12}{(x+1)(x^2-x+1)}$

$=\dfrac{4x^2-4x+4+x+1-12}{(x+1)(x^2-x+1)}$

$=\dfrac{4x^2-3x-7}{(x+1)(x^2-x+1)}$

$=\dfrac{(4x-7)(x+1)}{(x+1)(x^2-x+1)}$

$=\dfrac{4x-7}{x^2-x+1}$

71. $\dfrac{2x+4}{x+3}+\dfrac{3}{x}-\dfrac{6}{x^2+3x}$

$x^2+3x=x(x+3)$, the LCD.

$=\dfrac{(2x+4)x}{(x+3)x}+\dfrac{3(x+3)}{x(x+3)}-\dfrac{6}{x(x+3)}$

$=\dfrac{(2x+4)x+3(x+3)-6}{x(x+3)}$

$=\dfrac{2x^2+4x+3x+9-6}{x(x+3)}$

$=\dfrac{2x^2+7x+3}{x(x+3)}$

$=\dfrac{(2x+1)(x+3)}{x(x+3)}=\dfrac{2x+1}{x}$

73. $\dfrac{3}{(p-2)^2}-\dfrac{5}{p-2}+4$

$=\dfrac{3}{(p-2)^2}-\dfrac{5(p-2)}{(p-2)^2}+\dfrac{4(p-2)^2}{(p-2)^2}$

$LCD=(p-2)^2$

$=\dfrac{3-5(p-2)+4(p^2-4p+4)}{(p-2)^2}$

$=\dfrac{3-5p+10+4p^2-16p+16}{(p-2)^2}$

$=\dfrac{4p^2-21p+29}{(p-2)^2}$

75. $\dfrac{3}{x^2-5x+6}-\dfrac{2}{x^2-4x+4}$

$=\dfrac{3}{(x-2)(x-3)}-\dfrac{2}{(x-2)(x-2)}$

$=\dfrac{3(x-2)}{(x-2)(x-3)(x-2)}$

$-\dfrac{2(x-3)}{(x-2)(x-2)(x-3)}$

$LCD=(x-2)^2(x-3)$

$=\dfrac{3x-6-2x+6}{(x-2)^2(x-3)}$

$=\dfrac{x}{(x-2)^2(x-3)}$

77. $\dfrac{5x}{x^2+xy-2y^2}-\dfrac{3x}{x^2+5xy-6y^2}$

Factor each denominator.

$x^2+xy-2y^2=(x+2y)(x-y)$
$x^2+5xy-6y^2=(x+6y)(x-y)$

The LCD is $(x+2y)(x-y)(x+6y)$.

$\dfrac{5x}{(x+2y)(x-y)}-\dfrac{3x}{(x+6y)(x-y)}$

$=\dfrac{(5x)(x+6y)}{(x+2y)(x-y)(x+6y)}$

$-\dfrac{(3x)(x+2y)}{(x+6y)(x-y)(x+2y)}$

continued

$$= \frac{(5x)(x+6y) - (3x)(x+2y)}{(x+6y)(x-y)(x+2y)}$$

$$= \frac{5x^2 + 30xy - (3x^2 + 6xy)}{(x+2y)(x-y)(x+6y)}$$

$$= \frac{2x^2 + 24xy}{(x+2y)(x-y)(x+6y)}$$

$$= \frac{2x(x+12y)}{(x+2y)(x-y)(x+6y)}$$

79. $\dfrac{5x-y}{x^2+xy-2y^2} - \dfrac{3x+2y}{x^2+5xy-6y^2}$

Factor each denominator.

$$x^2 + xy - 2y^2 = (x+2y)(x-y)$$
$$x^2 + 5xy - 6y^2 = (x+6y)(x-y)$$

The LCD is $(x+2y)(x-y)(x+6y)$.

$$\frac{5x-y}{(x+2y)(x-y)} - \frac{3x+2y}{(x+6y)(x-y)}$$

$$= \frac{(5x-y)(x+6y)}{(x+2y)(x-y)(x+6y)}$$

$$- \frac{(3x+2y)(x+2y)}{(x+6y)(x-y)(x+2y)}$$

$$= \frac{(5x-y)(x+6y) - (3x+2y)(x+2y)}{(x+6y)(x-y)(x+2y)}$$

$$= \frac{5x^2 + 29xy - 6y^2 - (3x^2 + 8xy + 4y^2)}{(x+2y)(x-y)(x+6y)}$$

$$= \frac{2x^2 + 21xy - 10y^2}{(x+2y)(x-y)(x+6y)}$$

81. $\dfrac{r+s}{3r^2+2rs-s^2} - \dfrac{s-r}{6r^2-5rs+s^2}$

Factor each denominator.

$$3r^2 + 2rs - s^2 = (3r-s)(r+s)$$
$$6r^2 - 5rs + s^2 = (3r-s)(2r-s)$$

The LCD is $(3r-s)(r+s)(2r-s)$.

$$\frac{r+s}{3r^2+2rs-s^2} - \frac{s-r}{6r^2-5rs+s^2}$$

$$= \frac{r+s}{(3r-s)(r+s)} - \frac{s-r}{(3r-s)(2r-s)}$$

$$= \frac{(r+s)(2r-s)}{(3r-s)(r+s)(2r-s)}$$

$$- \frac{(s-r)(r+s)}{(3r-s)(2r-s)(r+s)}$$

$$= \frac{(r+s)(2r-s) - (s-r)(s+r)}{(3r-s)(r+s)(2r-s)}$$

$$= \frac{2r^2 + rs - s^2 - (s^2 - r^2)}{(3r-s)(r+s)(2r-s)}$$

$$= \frac{2r^2 + rs - s^2 - s^2 + r^2}{(3r-s)(r+s)(2r-s)}$$

$$= \frac{3r^2 + rs - 2s^2}{(3r-s)(r+s)(2r-s)}$$

$$= \frac{(3r-2s)(r+s)}{(3r-s)(r+s)(2r-s)}$$

$$= \frac{3r-2s}{(3r-s)(2r-s)}$$

83. $\dfrac{3}{x^2+4x+4} + \dfrac{7}{x^2+5x+6}$

$$= \frac{3}{(x+2)^2} + \frac{7}{(x+2)(x+3)}$$

$$LCD = (x+2)^2(x+3)$$

$$= \frac{3(x+3)}{(x+2)^2(x+3)} + \frac{7(x+2)}{(x+2)^2(x+3)}$$

$$= \frac{3x+9+7x+14}{(x+2)^2(x+3)}$$

$$= \frac{10x+23}{(x+2)^2(x+3)}$$

85. (a) $c(x) = \dfrac{1010}{49(101-x)} - \dfrac{10}{49}$

$$= \frac{1010}{49(101-x)} - \frac{10(101-x)}{49(101-x)}$$

$$= \frac{1010 - 1010 + 10x}{49(101-x)}$$

$$= \frac{10x}{49(101-x)}$$

(b) $c(95) = \dfrac{10(95)}{49(101-95)}$

$$= \frac{950}{294} \approx 3.23$$

It would cost approximately 3.23 thousand dollars to win 95 points.

87. $\dfrac{3}{7} + \dfrac{5}{9} - \dfrac{6}{63}$

The LCD is $7(9) = 63$.

$$= \frac{3 \cdot 9}{7 \cdot 9} + \frac{5 \cdot 7}{9 \cdot 7} - \frac{6}{63}$$

$$= \frac{27 + 35 - 6}{63}$$

$$= \frac{56}{63} = \frac{8}{9}$$

88. From Example 6, the expression is

$$\frac{3}{x-2} + \frac{5}{x} - \frac{6}{x^2-2x}.$$

Substitute 9 for x.

$$\frac{3}{9-2} + \frac{5}{9} - \frac{6}{9^2-2(9)} = \frac{3}{7} + \frac{5}{9} - \frac{6}{63}$$

The problems in Exercises 87 and 88 are the same.

89. From Exercise 87, the expression is

$$\frac{3}{7} + \frac{5}{9} - \frac{6}{63} = \frac{8}{9}.$$

From Example 6, the answer is $\frac{8}{x}$. If we substitute 9 for x, the answer becomes $\frac{8}{9}$. The answers agree.

90. Answers will vary. For example, suppose the last name is Gore so that $x = 4$. The problem in Example 6 becomes

$$\frac{3}{4-2} + \frac{5}{4} - \frac{6}{4^2 - 2(4)} = \frac{3}{2} + \frac{5}{4} - \frac{6}{8}.$$

The predicted answer is

$$\frac{8}{x}, \text{ or } \frac{8}{4} = 2.$$

Perform the operations to verify our prediction.

$$\frac{3}{2} + \frac{5}{4} - \frac{6}{8} = \frac{3 \cdot 4}{2 \cdot 4} + \frac{5 \cdot 2}{4 \cdot 2} - \frac{6}{8}$$
$$= \frac{12 + 10 - 6}{8}$$
$$= \frac{16}{8} = 2$$

The prediction is correct.

91. If $x = 2$, then the problem from Example 6,

$$\frac{3}{x-2} + \frac{5}{x} - \frac{6}{x^2 - 2x},$$

becomes

$$\frac{3}{2-2} + \frac{5}{2} - \frac{6}{2^2 - 2(2)} = \frac{3}{0} + \frac{5}{2} - \frac{6}{0}.$$

Thus, if $x = 2$, then

$$\frac{3}{x-2} \text{ and } \frac{6}{x^2 - 2x}$$

are undefined.

92. If $x = 0$, then

$$\frac{5}{x} = \frac{5}{0} \text{ and } \frac{6}{x^2 - 2x} = \frac{6}{0},$$

which are undefined. Therefore, 0 is not allowed as a value of x.

93. $\dfrac{\frac{5}{9} - \frac{1}{3}}{\frac{2}{3} + \frac{1}{6}} = \dfrac{\frac{5}{9} - \frac{3}{9}}{\frac{4}{6} + \frac{1}{6}} = \dfrac{\frac{2}{9}}{\frac{5}{6}} = \frac{2}{9} \div \frac{5}{6} = \frac{2}{\cancel{9}_3} \cdot \frac{\cancel{6}^2}{5} = \frac{4}{15}$

95. $\dfrac{2 - \frac{1}{4}}{\frac{5}{4} + 3} = \dfrac{\frac{8}{4} - \frac{1}{4}}{\frac{5}{4} + \frac{12}{4}} = \dfrac{\frac{7}{4}}{\frac{17}{4}} = \frac{7}{4} \div \frac{17}{4} = \frac{7}{\cancel{4}} \cdot \frac{\cancel{4}^1}{17} = \frac{7}{17}$

7.3 Complex Fractions

1. *Method 1*: Begin by simplifying the numerator to a single fraction. Then simplify the denominator to a single fraction. Write as a division problem, and multiply by the reciprocal of the denominator. Simplify the result if possible.

Method 2: Find the LCD of all fractions in the complex fraction. Multiply the numerator and denominator of the complex fraction by this LCD. Simplify the result if possible.

3. $\dfrac{\frac{12}{x-1}}{\frac{6}{x}} = \frac{12}{x-1} \div \frac{6}{x}$

Multiply by the reciprocal of the divisor.

$$= \frac{12}{x-1} \cdot \frac{x}{6}$$
$$= \frac{2x}{x-1}$$

5. $\dfrac{\frac{k+1}{2k}}{\frac{3k-1}{4k}} = \frac{k+1}{2k} \cdot \frac{4k}{3k-1}$

$$= \frac{4k(k+1)}{2k(3k-1)}$$
$$= \frac{2(k+1)}{3k-1}$$

7. $\dfrac{\frac{4z^2 x^4}{9}}{\frac{12x^2 z^5}{15}} = \dfrac{\frac{4z^2 x^4}{9}}{\frac{4x^2 z^5}{5}}$

$$= \frac{4z^2 x^4}{9} \div \frac{4x^2 z^5}{5}$$
$$= \frac{4z^2 x^4}{9} \cdot \frac{5}{4x^2 z^5}$$
$$= \frac{5z^2 x^4}{9x^2 z^5} = \frac{5x^2}{9z^3}$$

9. $\dfrac{6 + \frac{1}{x}}{7 - \frac{3}{x}}$

Multiply the numerator and denominator by x, the LCD of all the fractions.

$$= \frac{x\left(6 + \frac{1}{x}\right)}{x\left(7 - \frac{3}{x}\right)}$$
$$= \frac{x \cdot 6 + x \cdot \frac{1}{x}}{x \cdot 7 + x\left(-\frac{3}{x}\right)}$$
$$= \frac{6x + 1}{7x - 3}$$

11. $\dfrac{\dfrac{3}{x} + \dfrac{3}{y}}{\dfrac{3}{x} - \dfrac{3}{y}}$

Multiply the numerator and denominator by xy, the LCD of all the fractions.

$$= \frac{\left(\dfrac{3}{x} + \dfrac{3}{y}\right)xy}{\left(\dfrac{3}{x} - \dfrac{3}{y}\right)xy}$$

$$= \frac{\dfrac{3}{x} \cdot xy + \dfrac{3}{y} \cdot xy}{\dfrac{3}{x} \cdot xy - \dfrac{3}{y} \cdot xy}$$

$$= \frac{3y + 3x}{3y - 3x}$$

$$= \frac{3(y + x)}{3(y - x)}$$

$$= \frac{y + x}{y - x}$$

13. $\dfrac{\dfrac{8x - 24y}{10}}{\dfrac{x - 3y}{5x}} = \dfrac{8x - 24y}{10} \cdot \dfrac{5x}{x - 3y}$

$$= \frac{8(x - 3y)5x}{10(x - 3y)}$$

$$= \frac{40x}{10} = 4x$$

15. $\dfrac{\dfrac{x^2 - 16y^2}{xy}}{\dfrac{1}{y} - \dfrac{4}{x}}$

Multiply the numerator and denominator by xy, the LCD of all the fractions.

$$= \frac{\left(\dfrac{x^2 - 16y^2}{xy}\right)xy}{\left(\dfrac{1}{y} - \dfrac{4}{x}\right)xy}$$

$$= \frac{x^2 - 16y^2}{\dfrac{1}{y} \cdot xy - \dfrac{4}{x} \cdot xy}$$

$$= \frac{x^2 - 16y^2}{x - 4y}$$

$$= \frac{(x + 4y)(x - 4y)}{x - 4y}$$

$$= x + 4y$$

17. $\dfrac{\dfrac{6}{y - 4}}{\dfrac{12}{y^2 - 16}} = \dfrac{6}{y - 4} \cdot \dfrac{y^2 - 16}{12}$

$$= \frac{6}{y - 4} \cdot \frac{(y + 4)(y - 4)}{12}$$

$$= \frac{y + 4}{2}$$

19. $\dfrac{\dfrac{1}{b^2} - \dfrac{1}{a^2}}{\dfrac{1}{b} - \dfrac{1}{a}} = \dfrac{\dfrac{a^2 - b^2}{b^2 a^2}}{\dfrac{a - b}{ba}}$

$$= \frac{a^2 - b^2}{b^2 a^2} \cdot \frac{ba}{a - b}$$

$$= \frac{(a + b)(a - b)}{b^2 a^2} \cdot \frac{ba}{a - b}$$

$$= \frac{a + b}{ab}$$

21. $\dfrac{x + y}{\dfrac{1}{y} + \dfrac{1}{x}} = \dfrac{x + y}{\dfrac{x + y}{yx}}$

$$= \frac{x + y}{1} \cdot \frac{yx}{x + y}$$

$$= xy$$

23. $\dfrac{y - \dfrac{y - 3}{3}}{\dfrac{4}{9} + \dfrac{2}{3y}}$

Multiply the numerator and denominator by $9y$, the LCD of all the fractions.

$$= \frac{9y\left(y - \dfrac{y - 3}{3}\right)}{9y\left(\dfrac{4}{9} + \dfrac{2}{3y}\right)}$$

$$= \frac{9y^2 - 3y(y - 3)}{4y + 6}$$

$$= \frac{9y^2 - 3y^2 + 9y}{4y + 6}$$

$$= \frac{6y^2 + 9y}{4y + 6}$$

$$= \frac{3y(2y + 3)}{2(2y + 3)} = \frac{3y}{2}$$

25. $\dfrac{\dfrac{x + 2}{x} + \dfrac{1}{x + 2}}{\dfrac{5}{x} + \dfrac{x}{x + 2}}$

Multiply the numerator and denominator by $x(x + 2)$, the LCD of all the fractions.

$$= \frac{x(x + 2)\left(\dfrac{x + 2}{x} + \dfrac{1}{x + 2}\right)}{x(x + 2)\left(\dfrac{5}{x} + \dfrac{x}{x + 2}\right)}$$

$$= \frac{x(x+2)\left(\dfrac{x+2}{x}\right) + x(x+2)\left(\dfrac{1}{x+2}\right)}{x(x+2)\left(\dfrac{5}{x}\right) + x(x+2)\left(\dfrac{x}{x+2}\right)}$$

$$= \frac{(x+2)(x+2) + x}{5(x+2) + x^2}$$

$$= \frac{x^2 + 4x + 4 + x}{5x + 10 + x^2}$$

$$= \frac{x^2 + 5x + 4}{x^2 + 5x + 10}$$

27. To add the fractions in the numerator, use the LCD $m(m-1)$.

$$\frac{4}{m} + \frac{m+2}{m-1} = \frac{4(m-1)}{m(m-1)} + \frac{m(m+2)}{m(m-1)}$$

$$= \frac{4m - 4 + m^2 + 2m}{m(m-1)}$$

$$= \frac{m^2 + 6m - 4}{m(m-1)}$$

28. To subtract the fractions in the denominator, use the same LCD, $m(m-1)$.

$$\frac{m+2}{m} - \frac{2}{m-1}$$

$$= \frac{(m+2)(m-1)}{m(m-1)} - \frac{m \cdot 2}{m(m-1)}$$

$$= \frac{m^2 + m - 2 - 2m}{m(m-1)}$$

$$= \frac{m^2 - m - 2}{m(m-1)}$$

29.
$$\underset{\text{answer}}{\underset{\downarrow}{\text{Exercise 27}}} \qquad \underset{\text{answer}}{\underset{\downarrow}{\text{Exercise 28}}}$$

$$\frac{m^2 + 6m - 4}{m(m-1)} \div \frac{m^2 - m - 2}{m(m-1)}$$

Multiply by the reciprocal.

$$= \frac{m^2 + 6m - 4}{m(m-1)} \cdot \frac{m(m-1)}{m^2 - m - 2}$$

$$= \frac{m^2 + 6m - 4}{m^2 - m - 2}$$

30. The LCD of all the denominators in the complex fraction is $m(m-1)$.

31.
$$\frac{\left(\dfrac{4}{m} + \dfrac{m+2}{m-1}\right) \cdot m(m-1)}{\left(\dfrac{m+2}{m} - \dfrac{2}{m-1}\right) \cdot m(m-1)}$$

$$= \frac{4(m-1) + m(m+2)}{(m+2)(m-1) - 2m}$$

$$= \frac{4m - 4 + m^2 + 2m}{m^2 + m - 2 - 2m}$$

$$= \frac{m^2 + 6m - 4}{m^2 - m - 2}$$

32. Answers will vary. Because of the complicated nature of the numerator and denominator of the complex fraction, using Method 1 takes much longer to simplify the complex fraction. Method 2 is a simpler, more direct means of simplifying and is most likely the preferred method.

33.
$$\frac{1}{x^{-2} + y^{-2}} = \frac{1}{\dfrac{1}{x^2} + \dfrac{1}{y^2}}$$

$$= \frac{x^2 y^2 (1)}{x^2 y^2 \left(\dfrac{1}{x^2} + \dfrac{1}{y^2}\right)} \qquad LCD = x^2 y^2$$

$$= \frac{x^2 y^2}{y^2 + x^2}$$

35.
$$\frac{x^{-2} + y^{-2}}{x^{-1} + y^{-1}}$$

$$= \frac{\dfrac{1}{x^2} + \dfrac{1}{y^2}}{\dfrac{1}{x} + \dfrac{1}{y}}$$

Multiply the numerator and denominator by $x^2 y^2$, the LCD of all the fractions.

$$= \frac{x^2 y^2 \left(\dfrac{1}{x^2} + \dfrac{1}{y^2}\right)}{x^2 y^2 \left(\dfrac{1}{x} + \dfrac{1}{y}\right)}$$

$$= \frac{x^2 y^2 \cdot \dfrac{1}{x^2} + x^2 y^2 \cdot \dfrac{1}{y^2}}{x^2 y^2 \cdot \dfrac{1}{x} + x^2 y^2 \cdot \dfrac{1}{y}}$$

$$= \frac{y^2 + x^2}{xy^2 + x^2 y}, \text{ or } \frac{y^2 + x^2}{xy(y+x)}$$

37.
$$\frac{x^{-1} + 2y^{-1}}{2y + 4x} = \frac{\dfrac{1}{x} + \dfrac{2}{y}}{2y + 4x}$$

Multiply the numerator and denominator by xy, the LCD of all the fractions.

$$= \frac{xy \left(\dfrac{1}{x} + \dfrac{2}{y}\right)}{xy(2y + 4x)}$$

$$= \frac{y + 2x}{2xy(y + 2x)}$$

$$= \frac{1}{2xy}$$

39. (a) $\dfrac{\dfrac{3}{mp} - \dfrac{4}{p} + \dfrac{8}{m}}{2m^{-1} - 3p^{-1}}$

$= \dfrac{\dfrac{3}{mp} - \dfrac{4}{p} + \dfrac{8}{m}}{\dfrac{2}{m} - \dfrac{3}{p}}$

(b) $2m^{-1} = \dfrac{2}{m}$, not $\dfrac{1}{2m}$, since the exponent

applies only to m, not to 2. Likewise, $3p^{-1} = \dfrac{3}{p}$,

not $\dfrac{1}{3p}$.

(c) $\dfrac{\dfrac{3}{mp} - \dfrac{4}{p} + \dfrac{8}{m}}{2m^{-1} - 3p^{-1}} = \dfrac{\dfrac{3}{mp} - \dfrac{4}{p} + \dfrac{8}{m}}{\dfrac{2}{m} - \dfrac{3}{p}}$

Multiply the numerator and denominator by mp, the LCD of all the fractions.

$= \dfrac{mp\left(\dfrac{3}{mp} - \dfrac{4}{p} + \dfrac{8}{m}\right)}{mp\left(\dfrac{2}{m} - \dfrac{3}{p}\right)}$

$= \dfrac{mp \cdot \dfrac{3}{mp} - mp \cdot \dfrac{4}{p} + mp \cdot \dfrac{8}{m}}{mp \cdot \dfrac{2}{m} - mp \cdot \dfrac{3}{p}}$

$= \dfrac{3 - 4m + 8p}{2p - 3m}$

41. $\frac{1}{2}x + \frac{1}{4}x = -9$

$4\left(\frac{1}{2}x + \frac{1}{4}x\right) = 4(-9)$

$2x + x = -36$

$3x = -36$

$x = -12$

The solution set is $\{-12\}$.

43. $\dfrac{x-6}{5} = \dfrac{x+4}{10}$

$10\left(\dfrac{x-6}{5}\right) = 10\left(\dfrac{x+4}{10}\right)$

$2(x-6) = x+4$

$2x - 12 = x + 4$

$x = 16$

The solution set is $\{16\}$.

45. $f(x) = \dfrac{-3x + 2}{x - 6}$

$x - 6 = 0$ only if $x = 6$.
The domain is $\{x \mid x \neq 6\}$.

47. $f(x) = \dfrac{1}{x}$

$x = 0$ is not in the domain.
The domain is $\{x \mid x \neq 0\}$.

7.4 Equations with Rational Expressions and Graphs

1. (a) $\dfrac{1}{3x} + \dfrac{1}{2x} = \dfrac{x}{3}$

Only 0 would make any of the denominators equal to 0, so 0 would have to be rejected as a potential solution.

(b) The domain is $\{x \mid x \neq 0\}$.

3. (a) $\dfrac{1}{x+1} - \dfrac{1}{x-2} = 0$

Set each denominator equal to 0 and solve.

$\begin{array}{ccc} x + 1 = 0 & \text{or} & x - 2 = 0 \\ x = -1 & \text{or} & x = 2 \end{array}$

Solutions of -1 and 2 would be rejected since these values would make a denominator of the original equation equal to 0.

(b) The domain is $\{x \mid x \neq -1, 2\}$.

5. (a) $\dfrac{1}{x^2 - 16} - \dfrac{2}{x - 4} = \dfrac{1}{x + 4}$

$\dfrac{1}{(x+4)(x-4)} - \dfrac{2}{x - 4} = \dfrac{1}{x + 4}$

$x^2 - 16$ equals 0 if $x = \pm 4$.
$x - 4$ equals 0 if $x = 4$.
$x + 4$ equals 0 if $x = -4$.

So ± 4 would have to be rejected as potential solutions.

(b) The domain is $\{x \mid x \neq \pm 4\}$.

7. (a) $\dfrac{2}{x^2 - x} + \dfrac{1}{x + 3} = \dfrac{4}{x - 2}$

$x^2 - x = x(x - 1)$ is 0 if $x = 0$ or $x = 1$.
$x + 3$ is 0 if $x = -3$.
$x - 2$ is 0 if $x = 2$.

So 0, 1, -3, and 2 would have to be rejected as potential solutions.

(b) The domain is $\{x \mid x \neq 0, 1, -3, 2\}$.

9. **(a)** $\dfrac{6}{4x+7} - \dfrac{3}{x} = \dfrac{5}{6x-13}$

$4x + 7$ is 0 if $x = -\frac{7}{4}$.

$6x - 13$ is 0 if $x = \frac{13}{6}$.

So $-\frac{7}{4}, 0,$ and $\frac{13}{6}$ would have to be rejected as potential solutions.

(b) The domain is $\left\{ x \mid x \neq -\frac{7}{4}, 0, \frac{13}{6} \right\}$.

11. **(a)** $\dfrac{3x+1}{x-4} = \dfrac{6x+5}{2x-7}$

$x - 4$ is 0 if $x = 4$.

$2x - 7$ is 0 if $x = \frac{7}{2}$.

So 4 and $\frac{7}{2}$ would have to be rejected as potential solutions.

(b) The domain is $\left\{ x \mid x \neq 4, \frac{7}{2} \right\}$.

13. No, there is no possibility that the proposed solution will be rejected, because there are no variables in the denominators in the original equation.

In Exercises 15–42, check each potential solution in the original equation.

15. $\dfrac{3}{4x} = \dfrac{5}{2x} - \dfrac{7}{4}$

Multiply by the LCD, $4x$. $(x \neq 0)$

$$4x\left(\frac{3}{4x}\right) = 4x\left(\frac{5}{2x} - \frac{7}{4}\right)$$
$$3 = 2(5) - x(7)$$
$$3 = 10 - 7x$$
$$7x = 7$$
$$x = 1$$

Check $x = 1$: $\frac{3}{4} = \frac{10}{4} - \frac{7}{4}$ *True*

The solution set is $\{1\}$.

17. $x - \dfrac{24}{x} = -2$

Multiply by the LCD, x. $(x \neq 0)$

$$x\left(x - \frac{24}{x}\right) = -2 \cdot x$$
$$x^2 - 24 = -2x$$
$$x^2 + 2x - 24 = 0$$
$$(x+6)(x-4) = 0$$

$x + 6 = 0 \quad$ or $\quad x - 4 = 0$
$\quad x = -6 \quad$ or $\quad\quad x = 4$

Check $x = -6$: $-6 + 4 = -2$ *True*
Check $x = 4$: $\quad 4 - 6 = -2$ *True*

The solution set is $\{-6, 4\}$.

19. $\dfrac{x-4}{x+6} = \dfrac{2x+3}{2x-1}$

Multiply by the LCD, $(x+6)(2x-1)$.

Note that $x \neq -6$ and $x \neq \frac{1}{2}$.

$$(x+6)(2x-1)\left(\frac{x-4}{x+6}\right) = (x+6)(2x-1)\left(\frac{2x+3}{2x-1}\right)$$
$$(2x-1)(x-4) = (x+6)(2x+3)$$
$$2x^2 - 9x + 4 = 2x^2 + 15x + 18$$
$$-24x = 14$$
$$x = \frac{14}{-24} = -\frac{7}{12}$$

A calculator check is suggested.

```
-7/12→X:(X-4)/(X
+6)▶Frac
            -11/13
(2X+3)/(2X-1)▶Fr
ac
            -11/13
```

Check $x = -\frac{7}{12}$: $-\frac{11}{13} = -\frac{11}{13}$ *True*

The solution set is $\left\{ -\frac{7}{12} \right\}$.

21. $\dfrac{3x+1}{x-4} = \dfrac{6x+5}{2x-7}$

Multiply by the LCD, $(x-4)(2x-7)$.

Note that $x \neq 4$ and $x \neq \frac{7}{2}$.

$$(x-4)(2x-7)\left(\frac{3x+1}{x-4}\right) = (x-4)(2x-7)\left(\frac{6x+5}{2x-7}\right)$$
$$(2x-7)(3x+1) = (x-4)(6x+5)$$
$$6x^2 - 19x - 7 = 6x^2 - 19x - 20$$
$$-7 = -20 \quad \textit{False}$$

The false statement indicates that the original equation has no solution.

The solution set is $\emptyset$.

23. $\dfrac{1}{y-1} + \dfrac{5}{12} = \dfrac{-2}{3y-3}$

$$\frac{1}{y-1} + \frac{5}{12} = \frac{-2}{3(y-1)}$$

Multiply by the LCD, $12(y-1)$. $(y \neq 1)$

$$12(y-1)\left(\frac{1}{y-1} + \frac{5}{12}\right) = 12(y-1)\left(\frac{-2}{3(y-1)}\right)$$
$$12 + 5(y-1) = -8$$
$$12 + 5y - 5 = -8$$
$$5y + 7 = -8$$
$$5y = -15$$
$$y = -3$$

Check $y = -3$: $-\frac{3}{12} + \frac{5}{12} = \frac{2}{12}$ *True*

The solution set is $\{-3\}$.

25.
$$\frac{7}{6x+3} - \frac{1}{3} = \frac{2}{2x+1}$$
Multiply by the LCD, $3(2x+1)$. $\left(x \neq -\frac{1}{2}\right)$
$$3(2x+1)\left(\frac{7}{6x+3} - \frac{1}{3}\right) = 3(2x+1)\left(\frac{2}{2x+1}\right)$$
$$7 - 1(2x+1) = 3(2)$$
$$7 - 2x - 1 = 6$$
$$-2x = 0$$
$$x = 0$$

Check $x = 0$: $\frac{7}{3} - \frac{1}{3} = 2$ *True*

The solution set is $\{0\}$.

27.
$$\frac{3}{k+2} - \frac{2}{k^2-4} = \frac{1}{k-2}$$
$$\frac{3}{k+2} - \frac{2}{(k+2)(k-2)} = \frac{1}{k-2}$$
Multiply by the LCD, $(k+2)(k-2)$.
$(k \neq -2, 2)$
$$(k+2)(k-2)\left(\frac{3}{k+2} - \frac{2}{(k+2)(k-2)}\right)$$
$$= (k+2)(k-2)\left(\frac{1}{k-2}\right)$$
$$3(k-2) - 2 = k+2$$
$$3k - 6 - 2 = k+2$$
$$3k - 8 = k+2$$
$$2k = 10$$
$$k = 5$$

Check $k = 5$: $\frac{9}{21} - \frac{2}{21} = \frac{1}{3}$ *True*

The solution set is $\{5\}$.

29.
$$\frac{1}{y+2} + \frac{3}{y+7} = \frac{5}{y^2+9y+14}$$
$$\frac{1}{y+2} + \frac{3}{y+7} = \frac{5}{(y+2)(y+7)}$$
Multiply by the LCD, $(y+2)(y+7)$. $(y \neq -2, -7)$
$$(y+2)(y+7)\left(\frac{1}{y+2} + \frac{3}{y+7}\right)$$
$$= (y+2)(y+7)\left(\frac{5}{(y+2)(y+7)}\right)$$
$$(y+7) + 3(y+2) = 5$$
$$y + 7 + 3y + 6 = 5$$
$$4y + 13 = 5$$
$$4y = -8$$
$$y = -2$$

But y cannot equal -2 because that would make the denominator $y + 2$ equal to 0. Since division by 0 is undefined, the equation has no solution.

The solution set is $\emptyset$.

31.
$$\frac{9}{x} + \frac{4}{6x-3} = \frac{2}{6x-3}$$
Multiply by the LCD, $x(6x-3)$. $\left(x \neq 0, \frac{1}{2}\right)$
$$x(6x-3)\left(\frac{9}{x} + \frac{4}{6x-3}\right) = x(6x-3)\left(\frac{2}{6x-3}\right)$$
$$9(6x-3) + 4x = 2x$$
$$54x - 27 + 4x = 2x$$
$$56x = 27$$
$$x = \frac{27}{56}$$

Check $x = \frac{27}{56}$: $\frac{56}{3} + \left(-\frac{112}{3}\right) = -\frac{56}{3}$ *True*

The solution set is $\left\{\frac{27}{56}\right\}$.

33.
$$\frac{6}{w+3} + \frac{-7}{w-5} = \frac{-48}{w^2-2w-15}$$
$$\frac{6}{w+3} + \frac{-7}{w-5} = \frac{-48}{(w+3)(w-5)}$$
Multiply by the LCD, $(w+3)(w-5)$.
$(w \neq -3, 5)$
$$(w+3)(w-5)\left(\frac{6}{w+3} + \frac{-7}{w-5}\right)$$
$$= (w+3)(w-5)\left[\frac{-48}{(w+3)(w-5)}\right]$$
$$6(w-5) - 7(w+3) = -48$$
$$6w - 30 - 7w - 21 = -48$$
$$-w - 51 = -48$$
$$-w = 3$$
$$w = -3$$

But w cannot equal -3 because that would make the denominator $w + 3$ equal to 0. Since division by 0 is undefined, the equation has no solution.

The solution set is $\emptyset$.

35.
$$\frac{x}{x-3} + \frac{4}{x+3} = \frac{18}{x^2-9}$$
$$\frac{x}{x-3} + \frac{4}{x+3} = \frac{18}{(x-3)(x+3)}$$
Multiply by the LCD, $(x-3)(x+3)$. $(x \neq 3, -3)$
$$(x-3)(x+3)\left(\frac{x}{x-3} + \frac{4}{x+3}\right)$$
$$= (x-3)(x+3)\left(\frac{18}{(x-3)(x+3)}\right)$$
$$x(x+3) + 4(x-3) = 18$$
$$x^2 + 3x + 4x - 12 = 18$$
$$x^2 + 7x - 30 = 0$$
$$(x-3)(x+10) = 0$$
$$x - 3 = 0 \quad \text{or} \quad x + 10 = 0$$
$$x = 3 \quad \text{or} \qquad x = -10$$

But $x \neq 3$ since a denominator of 0 results. The only solution to check is -10.

Check $x = -10$: $\frac{10}{13} + \left(-\frac{4}{7}\right) = \frac{18}{91}$ *True*

The solution set is $\{-10\}$.

37.
$$\frac{1}{x+4} + \frac{x}{x-4} = \frac{-8}{x^2-16}$$
$$\frac{1}{x+4} + \frac{x}{x-4} = \frac{-8}{(x+4)(x-4)}$$
Multiply by the LCD, $(x+4)(x-4)$. $(x \neq -4, 4)$
$$(x+4)(x-4)\left(\frac{1}{x+4} + \frac{x}{x-4}\right)$$
$$= (x+4)(x-4)\left(\frac{-8}{(x+4)(x-4)}\right)$$
$$(x-4) + x(x+4) = -8$$
$$x - 4 + x^2 + 4x = -8$$
$$x^2 + 5x + 4 = 0$$
$$(x+4)(x+1) = 0$$

$x + 4 = 0 \quad$ or $\quad x + 1 = 0$
$\quad x = -4 \qquad\qquad x = -1$

But x cannot equal -4, so we only need to check -1.
Check $x = -1$: $\quad \frac{1}{3} + \frac{1}{5} = \frac{8}{15} \quad$ *True*

The solution set is $\{-1\}$.

39.
$$\frac{2}{k^2+k-6} + \frac{1}{k^2-k-2} = \frac{4}{k^2+4k+3}$$
$$\frac{2}{(k+3)(k-2)} + \frac{1}{(k-2)(k+1)} = \frac{4}{(k+1)(k+3)}$$
Multiply by the LCD, $(k+3)(k-2)(k+1)$.
$(k \neq -3, 2, -1)$
$$2(k+1) + 1(k+3) = 4(k-2)$$
$$2k + 2 + k + 3 = 4k - 8$$
$$13 = k$$

Check $k = 13$: $\quad \frac{1}{88} + \frac{1}{154} = \frac{1}{56} \quad$ *True*

The solution set is $\{13\}$.

41.
$$\frac{5x+14}{x^2-9} = \frac{-2x^2-5x+2}{x^2-9} + \frac{2x+4}{x-3}$$
$$\frac{5x+14}{(x+3)(x-3)}$$
$$= \frac{-2x^2-5x+2}{(x+3)(x-3)} + \frac{2x+4}{x-3}$$
Multiply by the LCD, $(x+3)(x-3)$.
$(x \neq -3, 3)$
$$5x + 14 = -2x^2 - 5x + 2 + (2x+4)(x+3)$$
$$5x + 14 = -2x^2 - 5x + 2 + 2x^2 + 10x + 12$$
$$5x + 14 = 5x + 14 \quad \textit{True}$$

This equation is true for every real number value of x, but we have already determined that $x \neq -3$ or $x \neq 3$. So every real number except -3 and 3 is a solution.

The solution set is $\{x \mid x \neq \pm 3\}$ or $(-\infty, -3) \cup (-3, 3) \cup (3, \infty)$.

43. (a) $\frac{x+3}{x+3} = 1$

The left side of the equation is equal to 1 except for $x = -3$, when it is undefined. Hence, the solution set is $\{x \mid x \neq -3\}$.

(b) The solution set is not {all real numbers} because -3 is not in the domain.

45. $f(x) = \frac{2}{x}$ is not defined when $x = 0$, so an equation of the vertical asymptote is $x = 0$.

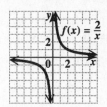

47. $f(x) = \frac{1}{x-2}$ is not defined when $x = 2$, so an equation of the vertical asymptote is $x = 2$.

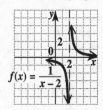

49. $w(x) = \frac{x^2}{2(1-x)}$

(a) $w(0.1) = \frac{(0.1)^2}{2(1-0.1)}$
$$= \frac{0.01}{2(0.9)} \approx 0.006$$

To the nearest tenth, $w(0.1)$ is 0.

(b) $w(0.8) = \frac{(0.8)^2}{2(1-0.8)}$
$$= \frac{0.64}{2(0.2)} = 1.6$$

(c) $w(0.9) = \frac{(0.9)^2}{2(1-0.9)}$
$$= \frac{0.81}{2(0.1)} = 4.05 \approx 4.1$$

(d) Based on the answers in (a), (b), and (c), we see that as the traffic intensity increases, the waiting time also increases.

51. **(a)** $F(r) = \dfrac{225{,}000}{r}$

$\quad\quad 450 = \dfrac{225{,}000}{r}$ *Let F(r) = 450.*

$\quad\quad 450r = 225{,}000$

$\quad\quad\quad r = \dfrac{225{,}000}{450} = 500$

The radius must be 500 feet.

(b) As r increases, the fraction $\dfrac{225{,}000}{r}$ gets

smaller, so the force, $f(r)$, decreases.

53. The number of solutions of the equation $f(x) = 0$ is the same as the number of x-intercepts, four.

55. The x-intercepts are -2, 0, and 3, so the solution set is $\{-2, 0, 3\}$.

57. $d = rt$

$\quad \dfrac{d}{r} = \dfrac{rt}{r}$ *Divide by r.*

$\quad \dfrac{d}{r} = t$

59. $\quad\quad P = a + b + c$

$\quad\quad P - a = b + c$ *Subtract a.*

$\quad P - a - b = c$ *Subtract b.*

Summary Exercises on Rational Expressions and Equations

1. $\dfrac{x}{2} - \dfrac{x}{4} = 5$

There is an equals sign, so this is an *equation*.

$4\left(\dfrac{x}{2} - \dfrac{x}{4}\right) = 4(5)$ *Multiply by 4.*

$\quad\quad 2x - x = 20$

$\quad\quad\quad\quad x = 20$

Check $x = 20$: $10 - 5 = 5$ *True*

The solution set is $\{20\}$.

3. $\dfrac{6}{7x} - \dfrac{4}{x}$

No equals sign appears so this is an *expression*.

$\quad = \dfrac{6}{7x} - \dfrac{4 \cdot 7}{x \cdot 7}$ *LCD = 7x*

$\quad = \dfrac{6 - 28}{7x} = \dfrac{-22}{7x}$

5. $\dfrac{5}{7t} = \dfrac{52}{7} - \dfrac{3}{t}$

There is an equals sign, so this is an *equation*.

Multiply by the LCD, $7t$. $(t \neq 0)$

$7t\left(\dfrac{5}{7t}\right) = 7t\left(\dfrac{52}{7} - \dfrac{3}{t}\right)$

$\quad\quad 5 = 52t - 3(7)$

$\quad\quad 5 = 52t - 21$

$\quad 26 = 52t$

$\quad\quad t = \dfrac{26}{52} = \dfrac{1}{2}$

Check $t = \frac{1}{2}$: $\frac{10}{7} = \frac{52}{7} - 6$ *True*

The solution set is $\left\{\frac{1}{2}\right\}$.

7. $\dfrac{7}{6x} + \dfrac{5}{8x}$

No equals sign appears so this is an *expression*.

$\quad = \dfrac{7}{3 \cdot 2x} + \dfrac{5}{4 \cdot 2x}$

$\quad = \dfrac{7(4)}{3 \cdot 2x \cdot 4} + \dfrac{5(3)}{4 \cdot 2x \cdot 3}$ *LCD = 24x*

$\quad = \dfrac{28 + 15}{3 \cdot 4 \cdot 2x} = \dfrac{43}{24x}$

9. $\dfrac{\dfrac{6}{x+1} - \dfrac{1}{x}}{\dfrac{2}{x} - \dfrac{4}{x+1}}$

No equals sign appears so this is an *expression*.
Multiply the numerator and denominator by the LCD of all the fractions, $x(x + 1)$.

$\quad = \dfrac{6(x) - 1(x+1)}{2(x+1) - 4(x)}$

$\quad = \dfrac{6x - x - 1}{2x + 2 - 4x}$

$\quad = \dfrac{5x - 1}{-2x + 2}$, or $\dfrac{5x - 1}{-2(x - 1)}$

11. $\dfrac{x}{x+y} + \dfrac{2y}{x-y}$

No equals sign appears so this is an *expression*.

$\quad \dfrac{x(x-y)}{(x+y)(x-y)} + \dfrac{2y(x+y)}{(x-y)(x+y)}$

$\quad\quad\quad\quad LCD = (x+y)(x-y)$

$\quad = \dfrac{x^2 - xy + 2xy + 2y^2}{(x+y)(x-y)}$

$\quad = \dfrac{x^2 + xy + 2y^2}{(x+y)(x-y)}$

13. $\dfrac{x-2}{9} \cdot \dfrac{5}{8-4x}$

No equals sign appears so this is an *expression*.

$\quad = \dfrac{x-2}{9} \cdot \dfrac{5}{-4(x-2)}$

$\quad = \dfrac{5}{-36} = -\dfrac{5}{36}$

15. $\dfrac{b^2 + b - 6}{b^2 + 2b - 8} \cdot \dfrac{b^2 + 8b + 16}{3b + 12}$

No equals sign appears so this is an *expression*.

$\quad = \dfrac{(b+3)(b-2)}{(b+4)(b-2)} \cdot \dfrac{(b+4)(b+4)}{3(b+4)}$

$\quad = \dfrac{b+3}{3}$

17. $\dfrac{5}{x^2 - 2x} - \dfrac{3}{x^2 - 4}$

No equals sign appears so this is an *expression*.

$$= \dfrac{5}{x(x-2)} - \dfrac{3}{(x+2)(x-2)}$$

$$= \dfrac{5(x+2)}{x(x-2)(x+2)} - \dfrac{3x}{(x-2)(x+2)x}$$

$$= \dfrac{5x + 10 - 3x}{x(x-2)(x+2)}$$

$$= \dfrac{2x + 10}{x(x-2)(x+2)}$$

19. $\dfrac{\dfrac{5}{x} - \dfrac{3}{y}}{\dfrac{9x^2 - 25y^2}{x^2 y}}$

No equals sign appears so this is an *expression*. Multiply the numerator and denominator by the LCD of all the fractions, $x^2 y$.

$$= \dfrac{x^2 y \left(\dfrac{5}{x} - \dfrac{3}{y} \right)}{x^2 y \left(\dfrac{9x^2 - 25y^2}{x^2 y} \right)}$$

$$= \dfrac{5xy - 3x^2}{9x^2 - 25y^2}$$

$$= \dfrac{-x(3x - 5y)}{(3x + 5y)(3x - 5y)}$$

$$= \dfrac{-x}{3x + 5y}$$

21. $\dfrac{4y^2 - 13y + 3}{2y^2 - 9y + 9} \div \dfrac{4y^2 + 11y - 3}{6y^2 - 5y - 6}$

No equals sign appears so this is an *expression*.

$$= \dfrac{(4y - 1)(y - 3)}{(2y - 3)(y - 3)} \cdot \dfrac{(2y - 3)(3y + 2)}{(4y - 1)(y + 3)}$$

$$= \dfrac{3y + 2}{y + 3}$$

23. $\dfrac{3r}{r - 2} = 1 + \dfrac{6}{r - 2}$

There is an equals sign, so this is an *equation*. Multiply by the LCD, $r - 2$. $(r \neq 2)$

$$3r = r - 2 + 6$$
$$2r = 4$$
$$r = 2$$

But $r \neq 2$.
The solution set is $\emptyset$.

25. $\dfrac{-1}{3 - x} - \dfrac{2}{x - 3} = \dfrac{-1}{-(x - 3)} - \dfrac{2}{x - 3}$

No equals sign appears so this is an *expression*.

$$= \dfrac{1}{x - 3} - \dfrac{2}{x - 3}$$

$$= \dfrac{-1}{x - 3},$$

$$\text{or} \quad \dfrac{-1}{-(3 - x)} = \dfrac{1}{3 - x}$$

27. $\dfrac{2}{y + 1} - \dfrac{3}{y^2 - y - 2} = \dfrac{3}{y - 2}$

$$\dfrac{2}{y + 1} - \dfrac{3}{(y - 2)(y + 1)} = \dfrac{3}{y - 2}$$

There is an equals sign, so this is an *equation*. Multiply by the LCD, $(y - 2)(y + 1)$. $(y \neq -1, 2)$

$$2(y - 2) - 3 = 3(y + 1)$$
$$2y - 4 - 3 = 3y + 3$$
$$2y - 7 = 3y + 3$$
$$-10 = y$$

Check $y = -10$: $-\dfrac{2}{9} - \dfrac{1}{36} = -\dfrac{1}{4}$ *True*

The solution set is $\{-10\}$.

29. $\dfrac{3}{y - 3} - \dfrac{3}{y^2 - 5y + 6} = \dfrac{2}{y - 2}$

$$\dfrac{3}{y - 3} - \dfrac{3}{(y - 3)(y - 2)} = \dfrac{2}{y - 2}$$

There is an equals sign, so this is an *equation*. Multiply by the LCD, $(y - 3)(y - 2)$. $(y \neq 2, 3)$

$$3(y - 2) - 3 = 2(y - 3)$$
$$3y - 6 - 3 = 2y - 6$$
$$3y - 9 = 2y - 6$$
$$y = 3$$

But $y \neq 3$.
The solution set is $\emptyset$.

7.5 Applications of Rational Expressions

1. **A.** $b = \dfrac{p}{r}$ is the same as $p = br$.

B. $r = \dfrac{b}{p}$ is the same as $b = pr$.

C. $b = \dfrac{r}{p}$ is the same as $r = bp$.

D. $p = \dfrac{r}{b}$ is the same as $r = bp$.

Choice **A** is correct.

3. **A.** $a = mF$ is the same as $m = \dfrac{a}{F}$.

B. $F = \dfrac{m}{a}$ is the same as $m = Fa$.

C. $F = \dfrac{a}{m}$ is the same as $Fm = a$, which is the same as $m = \dfrac{a}{F}$.

D. $F = ma$ is the same as $m = \dfrac{F}{a}$.

Choice **D** is correct.

5.
$$\frac{1}{a} = \frac{1}{b} + \frac{1}{c}$$

Let $a = 8$ and $c = 12$.
$$\frac{1}{8} = \frac{1}{b} + \frac{1}{12}$$

Multiply by the LCD, $24b$.
$$24b\left(\frac{1}{8}\right) = 24b\left(\frac{1}{b} + \frac{1}{12}\right)$$
$$3b = 24 + 2b$$
$$b = 24$$

7.
$$c = \frac{100b}{L}$$
$$80 = \frac{100 \cdot 5}{L} \qquad \textit{Let } c = 80, b = 5.$$
$$80L = 500 \qquad \textit{Multiply by L.}$$
$$L = \frac{500}{80} = \frac{25}{4} \text{ or } 6.25$$

9. Solve $F = \dfrac{GMm}{d^2}$ for G.
$$Fd^2 = GMm \qquad \textit{Multiply by } d^2.$$
$$\frac{Fd^2}{Mm} = G \qquad \textit{Divide by Mm.}$$

11. Solve $\dfrac{1}{a} = \dfrac{1}{b} + \dfrac{1}{c}$ for a.

Multiply by the LCD, abc.
$$abc\left(\frac{1}{a}\right) = abc\left(\frac{1}{b} + \frac{1}{c}\right)$$
$$bc = ac + ab$$
$$bc = a(c + b) \qquad \textit{Factor out a.}$$
$$\frac{bc}{c + b} = a \qquad \textit{Divide by } c + b.$$

13. Solve $\dfrac{PV}{T} = \dfrac{pv}{t}$ for v.
$$\frac{PVt}{T} = pv \qquad \textit{Multiply by t.}$$
$$\frac{PVt}{pT} = v \qquad \textit{Divide by p.}$$

15. Solve $I = \dfrac{nE}{R + nr}$ for r.
$$I(R + nr) = nE$$
$$IR + Inr = nE$$
$$Inr = nE - IR$$
$$r = \frac{nE - IR}{In}, \quad \text{or} \quad r = \frac{IR - nE}{-In}$$

17. Solve $A = \dfrac{1}{2}h(b + B)$ for b.
$$\frac{2}{h}(A) = \frac{2}{h}\left[\frac{1}{2}h(b + B)\right] \qquad \textit{Multiply by } \frac{2}{h}.$$
$$\frac{2A}{h} = b + B$$
$$\frac{2A}{h} - B = b, \quad \text{or} \quad b = \frac{2A - hB}{h}$$

19.
$$\frac{E}{e} = \frac{R + r}{r} \text{ for } r$$
$$Er = e(R + r) \qquad \textit{Multiply by er.}$$
$$Er = eR + er$$
$$Er - er = eR \qquad \textit{Subtract er.}$$
$$r(E - e) = eR$$
$$r = \frac{eR}{E - e} \qquad \textit{Divide by } E - e.$$

21. To solve the equation $m = \dfrac{ab}{a - b}$ for a, the first step is to multiply both sides of the equation by the LCD, $a - b$.

23. Let $x =$ the number of girls in the class. Write and solve a proportion.
$$\frac{3}{4} = \frac{x}{20}$$

Multiply by the LCD, 20.
$$20\left(\frac{3}{4}\right) = 20\left(\frac{x}{20}\right)$$
$$15 = x$$

There are 15 girls and $20 - 15 = 5$ boys in the class.

25.
$$\text{Marin's rate} = \frac{1 \text{ job}}{\text{time to complete 1 job}}$$
$$= \frac{1 \text{ job}}{2 \text{ hours}}$$
$$= \tfrac{1}{2} \text{ job per hour}$$

27. Let $x =$ the distance between Chicago and El Paso on the map (in inches).

Set up a proportion with one ratio involving map distances and the other involving actual distances.
$$\frac{x \text{ inches}}{4.125 \text{ inches}} = \frac{1606 \text{ miles}}{1238 \text{ miles}}$$
$$1238x = 4.125(1606)$$
$$1238x = 6624.75$$
$$x \approx 5.351$$

The distance on the map between Chicago and El Paso would be about 5.351 inches.

29. Let $x =$ the distance between Madrid and Rio de Janeiro on the map (in inches).

Set up a proportion with one ratio involving map distances and the other involving actual distances.
$$\frac{x \text{ inches}}{8.5 \text{ inches}} = \frac{5045 \text{ miles}}{5619 \text{ miles}}$$
$$5619x = 8.5(5045)$$
$$5619x = 42{,}882.5$$
$$x \approx 7.632 \approx 7.6$$

The distance on the map between Madrid and Rio de Janeiro would be about 7.6 inches.

31. Let $x =$ the number of teachers.

Write a proportion.

$$\frac{1}{14} = \frac{x}{554}$$

Multiply each side by $554 \cdot 14$.

$$554 \cdot 14 \cdot \frac{1}{14} = 554 \cdot 14 \cdot \frac{x}{554}$$
$$554 = 14x$$
$$x = \frac{554}{14} \approx 40$$

There would be 40 teachers.

33. Let $x =$ the number of deer in the forest preserve. Write and solve a proportion.

$$\frac{\text{total in forest}}{\text{tagged in forest}} = \frac{\text{total in sample}}{\text{tagged in sample}}$$
$$\frac{x}{42} = \frac{75}{15}$$
$$\frac{x}{42} = \frac{5}{1} \qquad \textit{Reduce.}$$
$$x = 5(42)$$
$$x = 210$$

There are approximately 210 deer in the forest preserve.

35. Let $x =$ the number of fish in the lake. Write and solve a proportion.

$$\frac{\text{total in lake}}{\text{tagged in lake}} = \frac{\text{total in sample}}{\text{tagged in sample}}$$
$$\frac{x}{500} = \frac{400}{8}$$
$$\frac{x}{500} = 50$$
$$x = 500(50)$$
$$= 25{,}000$$

There are approximately 25,000 fish in the lake.

37. *Step 2*
Let $x =$ the additional number of gallons of gasoline needed.

Step 3
He knows that he can drive 156 miles with 5 gallons of gasoline. He wants to drive 300 miles using $(3 + x)$ gallons of gasoline. Set up a proportion.

$$\frac{156}{5} = \frac{300}{3 + x}$$

Step 4
Find the cross products and solve for x.

$$156(3 + x) = 5(300)$$
$$468 + 156x = 1500$$
$$156x = 1032$$
$$x = \frac{1032}{156}$$
$$x \approx 6.6$$

Step 5
He will need about 6.6 more gallons of gasoline.

Step 6
Check The 3 gallons plus the 6.6 gallons equals 9.6 gallons. We'll check the rates (miles/gallon). Note that we could also use gallons/mile.

$$\frac{156}{5} = 31.2 \text{ mpg} \qquad \frac{300}{9.6} = 31.25 \text{ mpg}$$

The rates are approximately equal, so the solution is correct. Note that we could have used our exact value for x to get the exact rate

$$\frac{300}{3 + (1032/156)} = 31.2.$$

39. Since $\frac{4}{6} = \frac{6}{9} = \frac{2}{3}$, use the proportion

$$\frac{2}{3} = \frac{2x + 1}{2x + 5}.$$
$$2(2x + 5) = 3(2x + 1)$$
$$4x + 10 = 6x + 3$$
$$7 = 2x$$
$$\frac{7}{2} = x$$

Since $x = \frac{7}{2}$,

$$AC = 2x + 1 = 2\left(\frac{7}{2}\right) + 1 = 8$$
$$\text{and} \quad DF = 2x + 5 = 2\left(\frac{7}{2}\right) + 5 = 12.$$

41. Let x represent the amount to administer in milliliters.

$$\frac{100 \text{ mg}}{2 \text{ mL}} = \frac{120 \text{ mg}}{x \text{ mL}}$$

Multiply each side by the LCD, $2x$.

$$2x\left(\frac{100}{2}\right) = 2x\left(\frac{120}{x}\right)$$
$$100x = 240$$
$$x = 2.4$$

The correct dose is 2.4 mL.

43. *Step 2*
Let x represent the speed of the current of the river. The boat goes 12 mph, so the downstream speed is $12 + x$ and the upstream speed is $12 - x$.

Use $t = \dfrac{d}{r}$ and make a table.

	Distance	Rate	Time
Downstream	10	$12 + x$	$\dfrac{10}{12 + x}$
Upstream	6	$12 - x$	$\dfrac{6}{12 - x}$

Step 3
Because the time upstream equals the time downstream,

$$\frac{6}{12 - x} = \frac{10}{12 + x}$$

Step 4
Multiply by the LCD, $(12 - x)(12 + x)$.

$$(12 - x)(12 + x)\left(\frac{6}{12 - x}\right)$$
$$= (12 - x)(12 + x)\left(\frac{10}{12 + x}\right)$$
$$6(12 + x) = 10(12 - x)$$
$$72 + 6x = 120 - 10x$$
$$16x = 48$$
$$x = 3$$

Step 5
The speed of the current of the river is 3 mph.

Step 6
Check The rate downstream is $12 + 3 = 15$ mph, so she can go 10 miles in $\frac{10}{15} = \frac{2}{3}$ hour. The rate upstream is $12 - 3 = 9$ mph, so she can go 6 miles in $\frac{6}{9} = \frac{2}{3}$ hour. The times are the same, as required.

45.   *Step 2*
Find the distance from Montpelier to Columbia. Let x represent that distance.

Complete the table.

	d	r	t
Actual Trip	x	51	$\frac{x}{51}$
Alternative Trip	x	60	$\frac{x}{60}$

Step 3
At 60 mph, his time at 51 mph would be decreased 3 hr.

$$\frac{x}{60} = \frac{x}{51} - 3$$

Step 4
Multiply by 1020.

$$17x = 20x - 3060$$
$$3060 = 3x$$
$$1020 = x$$

Step 5
The distance from Montpelier to Columbia is 1020 miles.

Step 6
Check 1020 miles at 51 mph takes $\frac{1020}{51}$ or 20 hours; 1020 miles at 60 mph takes $\frac{1020}{60}$ or 17 hours; $17 = 20 - 3$ as required.

47.   *Step 2*
Let $x =$ the one-way distance.

Make a table.

	d	r	t
Trip Going East	x	500	$\frac{x}{500}$
Return Trip	x	350	$\frac{x}{350}$

Step 3
The total flying time in both directions was 8.5 hours, so

$$\frac{x}{500} + \frac{x}{350} = 8.5.$$

Step 4
Multiply by the LCD, 3500.

$$7x + 10x = 29{,}750$$
$$17x = 29{,}750$$
$$x = 1750$$

Step 5
The one-way distance was 1750 miles.

Step 6
Check The time for the trip going east was $\frac{1750}{500} = 3.5$ hours. The time for the return trip was $\frac{1750}{350} = 5$ hours. The total time was $3.5 + 5 = 8.5$ hours.

49.   *Step 2*
Let $x =$ the distance on the first part of the trip.

Make a table.

	d	r	t
First Part	x	60	$\frac{x}{60}$
Second Part	$x + 10$	50	$\frac{x + 10}{50}$

Step 3
From the problem, the equation is stated in words: Time for the second part = time for the first part $+ \frac{1}{2}$ (Note that 30 min $= \frac{1}{2}$ hr.). Use the times given in the table to write the equation.

$$\frac{x + 10}{50} = \frac{x}{60} + \frac{1}{2}$$

Step 4
Multiply by the LCD, 300.

$$300\left(\frac{x + 10}{50}\right) = 300\left(\frac{x}{60} + \frac{1}{2}\right)$$
$$6(x + 10) = 5x + 150$$
$$6x + 60 = 5x + 150$$
$$x = 90$$

Step 5
The distance for both parts of the trip is given by

$$x + (x + 10) = 90 + (90 + 10) = 190.$$

The distance is 190 miles.

Step 6
Check 90 miles at 60 mph takes $\frac{90}{60}$ or $1\frac{1}{2}$ hours; 100 miles at 50 mph takes 2 hours. The second part of the trip takes $\frac{1}{2}$ hour more than the first part, as required.

51. Let $x =$ the time it would take them working together.

Complete the table.

Worker	Rate	Time Working Together	Fractional Part of the Job Done
Butch	$\frac{1}{15}$	x	$\frac{1}{15}x$
Peggy	$\frac{1}{12}$	x	$\frac{1}{12}x$

Part done by Butch	+	part done by Peggy	=	1 whole job.
$\frac{1}{15}x$	+	$\frac{1}{12}x$	=	1

Multiply by the LCD, 60.

$$60\left(\tfrac{1}{15}x + \tfrac{1}{12}x\right) = 60 \cdot 1$$
$$4x + 5x = 60$$
$$9x = 60$$
$$x = \frac{60}{9} = \frac{20}{3} \ \text{ or } \ 6\tfrac{2}{3}$$

Together they could do the job in $\frac{20}{3}$ or $6\frac{2}{3}$ minutes.

53. Let $x =$ the time it would take Kuba working alone.

Worker	Rate	Time Working Together	Fractional Part of the Job Done
Jerry	$\frac{1}{20}$	12	$\frac{1}{20}(12) = \frac{3}{5}$
Kuba	$\frac{1}{x}$	12	$\frac{1}{x}(12) = \frac{12}{x}$

Part done by Jerry	+	part done by Kuba	=	1 whole job.
$\frac{3}{5}$	+	$\frac{12}{x}$	=	1

Multiply by the LCD, $5x$.

$$3x + 60 = 5x$$
$$60 = 2x$$
$$30 = x$$

It would take Kuba 30 hours to do the job alone.

55. Let $x =$ the time needed for Dixie and Trixie to do the job together.
Make a table.

Worker	Rate	Time Working Together	Fractional Part of the Job Done
Dixie	$\frac{1}{3}$	x	$\frac{1}{3}x$
Trixie	$\frac{1}{6}$	x	$\frac{1}{6}x$

Since Dixie has been painting for one hour, $\frac{1}{3}$ of the room is already painted. Thus, the sum of the fractional parts equals $\frac{2}{3}$, not 1.

$$\tfrac{1}{3}x + \tfrac{1}{6}x = \tfrac{2}{3}$$
$$2x + x = 4 \quad \textit{Multiply by 6.}$$
$$3x = 4$$
$$x = \tfrac{4}{3}$$

In summary, Dixie will paint for 1 hour alone, Dixie and Trixie will paint for $1\frac{1}{3}$ hours together, so it will take $2\frac{1}{3}$ hours, after Dixie starts, to finish painting the room.

57. Let $x =$ the time it will take to fill the vat if both pipes are open.

	Rate	Time to Fill the Vat	Fractional Part of the Job Done
Inlet Pipe	$\frac{1}{10}$	x	$\frac{1}{10}x$
Outlet Pipe	$-\frac{1}{20}$	x	$-\frac{1}{20}x$

Notice that the rate of the outlet pipe is negative because it will empty the vat, not fill it.

Part done with the inlet pipe open	+	Part done with the outlet pipe open	=	1 whole job.
$\frac{1}{10}x$	+	$\left(-\frac{1}{20}x\right)$	=	1

Multiply by the LCD, 20.

$$2x - x = 20$$
$$x = 20$$

It will take 20 hours to fill the vat.

59. Let $x =$ the time from Mimi's arrival home to the time the place is a shambles.

	Rate	Time to Mess up House	Fractional Part of the Job Done
Hortense and Mort	$-\frac{1}{7}$	x	$-\frac{1}{7}x$
Mimi	$\frac{1}{2}$	x	$\frac{1}{2}x$

continued

Notice that Hortense and Mort's rate is negative since they are opposing the messing up by cleaning the house.

Part done by Hortense and Mort	+	Part done by Mimi	=	1 whole job of messing up.
$-\frac{1}{7}x$	+	$\frac{1}{2}x$	=	1

Multiply by the LCD, 14.

$$-2x + 7x = 14$$
$$5x = 14$$
$$x = \frac{14}{5} \text{ or } 2\frac{4}{5}$$

It would take $\frac{14}{5}$ or $2\frac{4}{5}$ hours after Mimi got home for the house to be a shambles.

61. $y = kx$
$1 = k(3)$ *Let y = 1, x = 3.*
$\frac{1}{3} = k$

63. $y = \dfrac{k}{x}$
$1 = \dfrac{k}{3}$ *Let y = 1, x = 3.*
$1(3) = k$
$3 = k$

7.6 Variation

1. This suggests *direct* variation.
As the number of tickets you buy increases, so does the probability that you will win.

3. This suggests *direct* variation.
As the pressure put on the accelerator increases, so does the speed of the car.

5. This suggests *inverse* variation.
As your age gets larger, the probability that you believe in Santa Claus gets smaller.

7. This suggests *inverse* variation.
As the number of days gets smaller, the number of home runs increases.

9. The equation $y = \dfrac{3}{x}$ represents *inverse* variation.
y varies inversely as x because x is in the denominator.

11. The equation $y = 10x^2$ represents *direct* variation. The number 10 is the constant of variation, and y varies directly as the square of x.

13. The equation $y = 3xz^4$ represents *joint* variation. y varies directly as x and z^4.

15. The equation $y = \dfrac{4x}{wz}$ represents *combined* variation. In the numerator, 4 is the constant of

variation, and y varies directly as x. In the denominator, y varies inversely as w and z.

17. For $k > 0$, if y varies directly as x (then $y = kx$), when x increases, y *increases*, and when x decreases, y *decreases*.

19. "x varies directly as y" means

$$x = ky$$

for some constant k.
Substitute $x = 9$ and $y = 3$ in the equation and solve for k.

$$x = ky$$
$$9 = k(3)$$
$$k = \frac{9}{3} = 3$$

So $x = 3y$.
To find x when $y = 12$, substitute 12 for y in the equation.

$$x = 3y$$
$$x = 3(12)$$
$$x = 36$$

21. "a varies directly as the square of b" means

$$a = kb^2$$

for some constant k.
Substitute $a = 4$ and $b = 3$ in the equation and solve for k.

$$a = kb^2$$
$$4 = k(3)^2$$
$$k = \frac{4}{9}$$

So $a = \frac{4}{9}b^2$.
To find a when $b = 2$, substitute 2 for b in the equation.

$$a = \frac{4}{9}b^2$$
$$a = \frac{4}{9}(2)^2$$
$$a = \frac{4 \cdot 4}{9} = \frac{16}{9}$$

23. "z varies inversely as w" means

$$z = \frac{k}{w}$$

for some constant k. Since $z = 10$ when $w = 0.5$, substitute these values in the equation and solve for k.

$$z = \frac{k}{w}$$
$$10 = \frac{k}{0.5}$$
$$k = 10(0.5) = 5$$

So $z = \dfrac{5}{w}$.

To find z when $w = 8$, substitute 8 for w in the equation.

$$z = \frac{5}{w}$$
$$z = \frac{5}{8} \text{ or } 0.625$$

25. "m varies inversely as p^2" means

$$m = \frac{k}{p^2}$$

for some constant k. Since $m = 20$ when $p = 2$, substitute these values in the equation and solve for k.

$$m = \frac{k}{p^2}$$
$$20 = \frac{k}{2^2}$$
$$k = 20(4) = 80$$

So $m = \frac{80}{p^2}$. Now let $p = 5$.

$$m = \frac{80}{p^2}$$
$$m = \frac{80}{5^2} = \frac{16}{5}$$

27. "p varies jointly as q and r^2" means

$$p = kqr^2$$

for some constant k. Given that $p = 200$ when $q = 2$ and $r = 3$, solve for k.

$$p = kqr^2$$
$$200 = k(2)(3)^2$$
$$200 = 18k$$
$$k = \frac{200}{18} = \frac{100}{9}$$

So $p = \frac{100}{9}qr^2$. Now let $q = 5$ and $r = 2$.

$$p = \frac{100}{9}qr^2$$
$$p = \frac{100}{9}(5)(2)^2$$
$$= \frac{100}{9}(20)$$
$$= \frac{2000}{9} \text{ or } 222\frac{2}{9}$$

29. Let x = the number of gallons he bought and let C = the cost.
C varies directly as x, so

$$C = kx.$$

Since $C = 43.79$ when $x = 15$,

$$43.79 = k(15)$$
$$k = \frac{43.79}{15} \approx 2.919.$$

The price per gallon is $\$2.91\frac{9}{10}$.

31. Let y = the weight of an object on earth and x = the weight of the object on the moon. y varies directly as x, so

$$y = kx$$

for some constant k. Since $y = 200$ when $x = 32$, substitute these values in the equation and solve for k.

$$y = kx$$
$$200 = k(32)$$
$$k = \frac{200}{32} = 6.25$$

So $y = 6.25x$.
To find x when $y = 50$, substitute 50 for y in the equation

$$y = 6.25x.$$
$$50 = 6.25x$$
$$x = \frac{50}{6.25} = 8$$

The dog would weigh 8 lb on the moon.

33. Let V = the volume of the can and let h = the height of the can.
V varies directly as h, so

$$V = kh.$$

Since $V = 300$ when $h = 10.62$,

$$300 = k(10.62)$$
$$k = \frac{300}{10.62} \approx 28.25.$$

So $V = 28.25h$. Now let $h = 15.92$.

$$V = 28.25h$$
$$V = 28.25(15.92) = 449.74$$

The volume is about 450 cubic inches.

35. Let d = the distance and t = the time.
d varies directly as the square of t, so $d = kt^2$.
Let $d = -576$ and $t = 6$. (You could also use $d = 576$, but the negative sign indicates the direction of the body.)

$$-576 = k(6)^2$$
$$-576 = 36k$$
$$-16 = k$$

So $d = -16t^2$. Now let $t = 4$.

$$d = -16(4)^2 = -256$$

The object fell 256 feet in the first 4 seconds.

37. Let s = the speed and t = the time.
The speed varies inversely with time, so there is a constant k such that $s = k/t$. Find the value of k by replacing s with 160 and t with $\frac{1}{2}$.

$$s = \frac{k}{t}$$
$$160 = \frac{k}{\frac{1}{2}}$$
$$k = 160\left(\tfrac{1}{2}\right) = 80$$

So $s = \dfrac{80}{t}$. Now let $t = \frac{3}{4}$.

$$s = \frac{80}{\frac{3}{4}}$$
$$s = \frac{80}{1} \cdot \frac{4}{3} = \frac{320}{3} \quad \text{or} \quad 106\frac{2}{3}$$

A speed $106\frac{2}{3}$ miles per hour is needed to go the same distance in $\frac{3}{4}$ minute.

39. Let f = the frequency of a string in cycles per second and s = the length in feet.
f varies inversely as s, so

$$f = \frac{k}{s}$$

for some constant k. Since $f = 250$ when $s = 2$, substitute these values in the equation and solve for k.

$$f = \frac{k}{s}$$
$$250 = \frac{k}{2}$$
$$k = 250(2) = 500$$

So $f = \dfrac{500}{s}$. Now let $s = 5$.

$$f = \tfrac{500}{5} = 100.$$

The string would have a frequency of 100 cycles per second.

41. Let I = the illumination produced by a light source and d = the distance from the source.
I varies inversely as d^2, so

$$I = \frac{k}{d^2}$$

for some constant k. Since $I = 768$ when $d = 1$, substitute these values in the equation and solve for k.

$$I = \frac{k}{d^2}$$
$$768 = \frac{k}{1^2}$$
$$768 = k$$

So $I = \dfrac{768}{d^2}$. Now let $d = 6$.

$$I = \frac{768}{d^2}$$
$$I = \frac{768}{6^2} = \frac{768}{36} = \frac{64}{3} \quad \text{or} \quad 21\frac{1}{3}$$

The illumination produced by the light source is $21\frac{1}{3}$ foot-candles.

43. Let I = the simple interest, P the principal, and t the time.
Since I varies jointly as the principal and time, there is a constant k such that $I = kPt$. Find k by replacing I with 280, P with 2000, and t with 4.

$$I = kPt$$
$$280 = k(2000)(4)$$
$$k = \tfrac{280}{8000} = 0.035$$

So $I = 0.035Pt$. Now let $t = 6$.

$$I = 0.035(2000)(6)$$
$$= 420$$

The interest would be \$420.

45. The weight W of a bass varies jointly as its girth G and the square of its length L, so

$$W = kGL^2$$

for some constant k. Substitute 22.7 for W, 21 for G, and 36 for L.

$$22.7 = k(21)(36)^2$$
$$k = \frac{22.7}{27,216} \approx 0.000834$$

So $W = 0.000834GL^2$. Now let $G = 18$ and $L = 28$.

$$W = 0.000834GL^2$$
$$= 0.000834(18)(28)^2$$
$$\approx 11.8$$

The bass would weigh about 11.8 pounds.

47. Let F = the force, w = the weight of the car, s = the speed, and r = the radius.
The force varies inversely as the radius and jointly as the weight and the square of the speed, so

$$F = \frac{kws^2}{r}.$$

Let $F = 242$, $w = 2000$, $r = 500$, and $s = 30$.

$$242 = \frac{k(2000)(30)^2}{500}$$
$$k = \frac{242(500)}{2000(900)} = \frac{121}{1800}$$

So $F = \dfrac{121ws^2}{1800r}$.

Let $r = 750$, $s = 50$, and $w = 2000$.

$$F = \frac{121(2000)(50)^2}{1800(750)} \approx 448.1$$

Approximately 448.1 pounds of force would be needed.

49. Let N = the number of long distance calls,

p_1 = the population of City 1,

p_2 = the population of City 2,

and d = the distance between them.

$$N = \frac{kp_1p_2}{d}$$

Let $N = 80{,}000$, $p_1 = 70{,}000$, $p_2 = 100{,}000$, and $d = 400$.

$$80{,}000 = \frac{k(70{,}000)(100{,}000)}{400}$$
$$80{,}000 = 17{,}500{,}000k$$
$$k = \frac{80{,}000}{17{,}500{,}000} = \frac{4}{875}$$
$$N = \frac{4}{875}\left(\frac{p_1p_2}{d}\right)$$

Let $p_1 = 50{,}000$, $p_2 = 75{,}000$, and $d = 250$.

$$N = \frac{4}{875}\left(\frac{50{,}000 \cdot 75{,}000}{250}\right)$$
$$= \frac{480{,}000}{7} = 68{,}571\tfrac{3}{7}$$

Rounded to the nearest hundred, there are approximately 68,600 calls.

51. Use the BMI from Example 7, with $k = 694$.

$$B = \frac{694w}{h^2}$$

Substitute your weight in pounds for w and your height in inches for h to determine your BMI. Answers will vary. A BMI from 19 to 25 is considered desirable.

53. If y varies inversely as x, then x is the denominator; however, if y varies directly as x, then x is in the numerator. If $k > 0$, then, with inverse variation, as x increases, y decreases. With direct variation, y increases as x increases.

55.
$$x^2 = 81$$
$$x^2 - 81 = 0$$
$$(x+9)(x-9) = 0$$
$$x+9=0 \quad \text{or} \quad x-9=0$$
$$x=-9 \quad \text{or} \quad x=9$$

The solution set is $\{\pm 9\}$.

57.
$$t^2 = 0.25$$
$$t^2 - 0.25 = 0$$
$$(t+0.5)(t-0.5) = 0$$
$$t+0.5=0 \quad \text{or} \quad t-0.5=0$$
$$t=-0.5 \quad \text{or} \quad t=0.5$$

The solution set is $\{\pm 0.5\}$.

59.
$$s^4 = 81$$
$$s^4 - 81 = 0$$
$$(s^2+9)(s^2-9) = 0$$
$$(s^2+9)(s+3)(s-3) = 0$$

$s^2 + 9$ is never equal to 0 for any real number s.

$$s+3=0 \quad \text{or} \quad s-3=0$$
$$s=-3 \quad \text{or} \quad s=3$$

The solution set is $\{\pm 3\}$.

Chapter 7 Review Exercises

1. (a) $f(x) = \dfrac{-7}{3x+18}$

Set the denominator equal to zero and solve.

$$3x+18=0$$
$$3x=-18$$
$$x=-6$$

The number -6 makes the expression undefined, so it is excluded from the domain.

(b) The domain is $\{x \mid x \neq -6\}$.

2. (a) $f(x) = \dfrac{5x+17}{x^2-7x+10}$

Set the denominator equal to zero and solve.

$$x^2-7x+10=0$$
$$(x-5)(x-2)=0$$
$$x-5=0 \quad \text{or} \quad x-2=0$$
$$x=5 \quad \text{or} \quad x=2$$

The numbers 2 and 5 make the expression undefined, so they are excluded from the domain.

(b) The domain is $\{x \mid x \neq 2, 5\}$.

3. (a) $f(x) = \dfrac{9}{x^2-18x+81}$

Set the denominator equal to zero and solve.

$$x^2-18x+81=0$$
$$(x-9)^2=0$$
$$x-9=0$$
$$x=9$$

The number 9 makes the expression undefined, so it is excluded from the domain.

(b) The domain is $\{x \mid x \neq 9\}$.

4. $\dfrac{12x^2+6x}{24x+12} = \dfrac{6x(2x+1)}{12(2x+1)} = \dfrac{x}{2}$

5.
$$\frac{25m^2 - n^2}{25m^2 - 10mn + n^2} = \frac{(5m+n)(5m-n)}{(5m-n)(5m-n)}$$
$$= \frac{5m+n}{5m-n}$$

6.
$$\frac{r-2}{4-r^2} = \frac{r-2}{(2+r)(2-r)}$$
$$= \frac{(-1)(2-r)}{(2+r)(2-r)}$$
$$= \frac{-1}{2+r}$$

7. The reciprocal of a rational expression is another rational expression such that the two rational expressions have a product of 1.

8.
$$\frac{(2y+3)^2}{5y} \cdot \frac{15y^3}{4y^2 - 9}$$
$$= \frac{15y^3(2y+3)^2}{5y(2y+3)(2y-3)}$$
$$= \frac{3y^2(2y+3)}{2y-3}$$

9.
$$\frac{w^2 - 16}{w} \cdot \frac{3}{4-w}$$
$$= \frac{(w-4)(w+4)}{w} \cdot \frac{3}{4-w}$$
$$= \frac{(-1)(4-w)(w+4)}{w} \cdot \frac{3}{4-w}$$
$$= \frac{-3(w+4)}{w}$$

10.
$$\frac{z^2 - z - 6}{z-6} \cdot \frac{z^2 - 6z}{z^2 + 2z - 15}$$
$$= \frac{(z-3)(z+2)}{z-6} \cdot \frac{z(z-6)}{(z-3)(z+5)}$$
$$= \frac{z(z+2)}{z+5}$$

11.
$$\frac{m^3 - n^3}{m^2 - n^2} \div \frac{m^2 + mn + n^2}{m+n}$$
Multiply by the reciprocal.
$$= \frac{m^3 - n^3}{m^2 - n^2} \cdot \frac{m+n}{m^2 + mn + n^2}$$
$$= \frac{(m-n)(m^2 + mn + n^2)}{(m-n)(m+n)} \cdot \frac{m+n}{m^2 + mn + n^2}$$
$$= 1$$

12. $32b^3, 24b^5$

Factor each denominator.
$$32b^3 = 2 \cdot 2 \cdot 2 \cdot 2 \cdot 2 \cdot b^3 = 2^5 \cdot b^3$$
$$24b^5 = 2 \cdot 2 \cdot 2 \cdot 3 \cdot b^5 = 2^3 \cdot 3 \cdot b^5$$
$$\text{LCD} = 2^5 \cdot 3 \cdot b^5 = 96b^5$$

13. $9r^2, 3r+1, 9$

Factor each denominator.
$$9r^2 = 3^2 \cdot r^2$$

The second denominator is already in factored form. The third denominator is $9 = 3^2$. The LCD is

$$3^2 \cdot r^2 \cdot (3r+1) \quad \text{or} \quad 9r^2(3r+1).$$

14. $6x^2 + 13x - 5, 9x^2 + 9x - 4$

Factor each denominator.
$$6x^2 + 13x - 5 = (3x-1)(2x+5)$$
$$9x^2 + 9x - 4 = (3x-1)(3x+4)$$

The LCD is $(3x-1)(2x+5)(3x+4)$.

15. $3x - 12, x^2 - 2x - 8, x^2 - 8x + 16$

Factor each denominator.
$$3x - 12 = 3(x-4)$$
$$x^2 - 2x - 8 = (x-4)(x+2)$$
$$x^2 - 8x + 16 = (x-4)^2$$

The LCD is $3(x-4)^2(x+2)$.

16.
$$\frac{5}{3x^6y^5} - \frac{8}{9x^4y^7}$$
The LCD is $9x^6y^7$.
$$= \frac{5 \cdot 3y^2}{3x^6y^5 \cdot 3y^2} - \frac{8 \cdot x^2}{9x^4y^7 \cdot x^2}$$
$$= \frac{15y^2}{9x^6y^7} - \frac{8x^2}{9x^6y^7}$$
$$= \frac{15y^2 - 8x^2}{9x^6y^7}$$

17.
$$\frac{5y+13}{y+1} - \frac{1-7y}{y+1}$$
$$= \frac{5y+13-(1-7y)}{y+1}$$
$$= \frac{5y+13-1+7y}{y+1}$$
$$= \frac{12y+12}{y+1}$$
$$= \frac{12(y+1)}{y+1} = 12$$

18.
$$\frac{6}{5a+10} + \frac{7}{6a+12}$$
$$= \frac{6}{5(a+2)} + \frac{7}{6(a+2)}$$
The LCD is $30(a+2)$.
$$= \frac{6 \cdot 6}{5(a+2) \cdot 6} + \frac{7 \cdot 5}{6(a+2) \cdot 5}$$

$$= \frac{36}{30(a+2)} + \frac{35}{30(a+2)}$$

$$= \frac{36+35}{30(a+2)} = \frac{71}{30(a+2)}$$

19. $\dfrac{3r}{10r^2 - 3rs - s^2} + \dfrac{2r}{2r^2 + rs - s^2}$

$$= \frac{3r}{(5r+s)(2r-s)} + \frac{2r}{(2r-s)(r+s)}$$

The LCD is $(5r+s)(2r-s)(r+s)$.

$$= \frac{3r(r+s)}{(5r+s)(2r-s)(r+s)}$$

$$+ \frac{2r(5r+s)}{(2r-s)(r+s)(5r+s)}$$

$$= \frac{3r^2 + 3rs + 10r^2 + 2rs}{(5r+s)(2r-s)(r+s)}$$

$$= \frac{13r^2 + 5rs}{(5r+s)(2r-s)(r+s)}$$

20. $\dfrac{1}{y-x} = \dfrac{1}{y-x} \cdot \dfrac{-1}{-1} = \dfrac{-1}{-1(y-x)} = \dfrac{-1}{x-y}$

Both students got the correct answer. The two expressions obtained are equivalent.

21. $\dfrac{\dfrac{3}{t} + 2}{\dfrac{4}{t} - 7}$

Multiply the numerator and denominator by the LCD of all the fractions, t.

$$= \frac{t\left(\dfrac{3}{t} + 2\right)}{t\left(\dfrac{4}{t} - 7\right)} = \frac{3 + 2t}{4 - 7t}$$

22. $\dfrac{\dfrac{2}{m-3n}}{\dfrac{1}{3n-m}} = \dfrac{2}{m-3n} \div \dfrac{1}{3n-m}$

$$= \frac{2}{m-3n} \cdot \frac{3n-m}{1}$$

$$= \frac{2}{m-3n} \cdot \frac{-1(m-3n)}{1}$$

$$= \frac{2 \cdot (-1)}{1} = -2$$

23. $\dfrac{\dfrac{3}{p} - \dfrac{2}{q}}{\dfrac{9q^2 - 4p^2}{qp}}$

Multiply the numerator and denominator by the LCD of all the fractions, qp.

$$= \frac{qp\left(\dfrac{3}{p} - \dfrac{2}{q}\right)}{qp\left(\dfrac{9q^2 - 4p^2}{qp}\right)}$$

$$= \frac{3q - 2p}{9q^2 - 4p^2}$$

$$= \frac{3q - 2p}{(3q+2p)(3q-2p)}$$

$$= \frac{1}{3q + 2p}$$

24. $\dfrac{x^{-2} - y^{-2}}{x^{-1} - y^{-1}} = \dfrac{\dfrac{1}{x^2} - \dfrac{1}{y^2}}{\dfrac{1}{x} - \dfrac{1}{y}}$

Multiply the numerator and denominator by the LCD of all the fractions, $x^2 y^2$.

$$= \frac{x^2 y^2\left(\dfrac{1}{x^2} - \dfrac{1}{y^2}\right)}{x^2 y^2\left(\dfrac{1}{x} - \dfrac{1}{y}\right)}$$

$$= \frac{y^2 - x^2}{xy^2 - x^2 y}$$

$$= \frac{(y+x)(y-x)}{xy(y-x)}$$

$$= \frac{y+x}{xy}$$

25. $\dfrac{1}{t+4} + \dfrac{1}{2} = \dfrac{3}{2t+8}$

$$\frac{1}{t+4} + \frac{1}{2} = \frac{3}{2(t+4)}$$

Multiply by the LCD, $2(t+4)$. $(t \neq -4)$

$$2(t+4)\left(\frac{1}{t+4} + \frac{1}{2}\right) = 2(t+4)\left(\frac{3}{2(t+4)}\right)$$

$$2 + (t+4) = 3$$

$$t + 6 = 3$$

$$t = -3$$

Check $t = -3$: $1 + \frac{1}{2} = \frac{3}{2}$ *True*

The solution set is $\{-3\}$.

26. $\dfrac{-5m}{m+1} + \dfrac{m}{3m+3} = \dfrac{56}{6m+6}$

$$\frac{-5m}{m+1} + \frac{m}{3(m+1)} = \frac{56}{6(m+1)}$$

$$\frac{-5m}{m+1} + \frac{m}{3(m+1)} = \frac{28}{3(m+1)}$$

Multiply by the LCD, $3(m+1)$. $(m \neq -1)$

$$3(m+1)\left(\frac{-5m}{m+1} + \frac{m}{3(m+1)}\right)$$

$$= 3(m+1)\left(\frac{28}{3(m+1)}\right)$$

continued

$$-15m + m = 28$$
$$-14m = 28$$
$$m = -2$$

Check $m = -2$: $-10 + \frac{2}{3} = -\frac{56}{6}$ *True*
The solution set is $\{-2\}$.

27.
$$\frac{2}{k-1} - \frac{4k+1}{k^2-1} = \frac{-1}{k+1}$$
$$\frac{2}{k-1} - \frac{4k+1}{(k+1)(k-1)} = \frac{-1}{k+1}$$

Multiply by the LCD, $(k+1)(k-1)$.
$$(k+1)(k-1)\left(\frac{2}{k-1} - \frac{4k+1}{(k+1)(k-1)}\right)$$
$$= (k+1)(k-1)\left(\frac{-1}{k+1}\right)$$
$$2(k+1) - (4k+1) = -1(k-1)$$
$$2k + 2 - 4k - 1 = -k + 1$$
$$-2k + 1 = -k + 1$$
$$0 = k$$

Check $k = 0$: $-2 + 1 = -1$ *True*
The solution set is $\{0\}$.

28.
$$\frac{5}{x+2} + \frac{3}{x+3} = \frac{x}{x^2+5x+6}$$
$$\frac{5}{x+2} + \frac{3}{x+3} = \frac{x}{(x+2)(x+3)}$$

Multiply by the LCD, $(x+2)(x+3)$.
$(x \neq -3, -2)$
$$(x+2)(x+3)\left(\frac{5}{x+2} + \frac{3}{x+3}\right)$$
$$= (x+2)(x+3)\left(\frac{x}{(x+2)(x+3)}\right)$$
$$5(x+3) + 3(x+2) = x$$
$$5x + 15 + 3x + 6 = x$$
$$8x + 21 = x$$
$$7x = -21$$
$$x = -3$$

Substituting -3 in the original equation results in
division by 0, so -3 is not a solution.
The solution set is $\emptyset$.

29. Although her algebra was correct, 3 is not a
solution because it is not in the domain of the
variable. Thus, $\emptyset$ is correct.

30. In simplifying the expression, we are combining
terms to get a single fraction with a denominator
of $6x$. In solving the equation, we are finding a
value for x that makes the equation true.

31. The graph in choice **C** has a vertical asymptote. Its
equation is $x = 0$.

32.
$$\frac{1}{A} = \frac{1}{B} + \frac{1}{C}$$
Let $B = 30$ and $C = 10$.
$$\frac{1}{A} = \frac{1}{30} + \frac{1}{10}$$
To solve for A, multiply both sides
by the LCD, $30A$.
$$30A\left(\frac{1}{A}\right) = 30A\left(\frac{1}{30} + \frac{1}{10}\right)$$
$$30 = A + 3A$$
$$30 = 4A$$
$$A = \frac{30}{4} = \frac{15}{2}$$

33. Solve $F = \frac{GMm}{d^2}$ for m.
$$Fd^2 = GMm \qquad \textit{Multiply by } d^2.$$
$$\frac{Fd^2}{GM} = m \qquad \textit{Divide by } GM.$$

34. Solve $\mu = \frac{Mv}{M+m}$ for M.
$$\mu(M+m) = Mv \qquad \textit{Multiply by } M+m.$$
$$\mu M + \mu m = Mv$$
$$\mu m = Mv - \mu M$$
$$m\mu = M(v - \mu)$$
$$M = \frac{m\mu}{v - \mu}$$

35. Let $x =$ the number of passenger-kilometers per
day provided by high-speed trains.
Write a proportion.
$$\frac{x}{15,000} = \frac{23,200}{58,000}$$
$$x = \frac{23,200(15,000)}{58,000}$$
$$= 6000$$

The high-speed train would provide 6000
passenger-km per day in that region.

36. Let $x =$ the speed of the boat in still water.

Use $d = rt$, or $t = \frac{d}{r}$, to make a table.

	Distance	Rate	Time
Downstream	40	$x + 4$	$\dfrac{40}{x+4}$
Upstream	24	$x - 4$	$\dfrac{24}{x-4}$

Because the times are equal,
$$\frac{40}{x+4} = \frac{24}{x-4}.$$
Multiply by the LCD, $(x+4)(x-4)$. $(x \neq -4, 4)$

$$(x+4)(x-4)\left(\frac{40}{x+4}\right) = (x+4)(x-4)\left(\frac{24}{x-4}\right)$$
$$40(x-4) = 24(x+4)$$
$$40x - 160 = 24x + 96$$
$$16x = 256$$
$$x = 16$$

The speed of the boat in still water is 16 km/hr.

37. Let $x =$ the time it takes to fill the sink with both taps open.

Make a table.

	Rate	Time Working Together	Fractional Part of the Job Done
Cold	$\frac{1}{8}$	x	$\frac{x}{8}$
Hot	$\frac{1}{12}$	x	$\frac{x}{12}$

Part done by cold	plus	part done by hot	equals	1 whole job.
$\frac{x}{8}$	$+$	$\frac{x}{12}$	$=$	1

Multiply by the LCD, 24.

$$24\left(\frac{x}{8} + \frac{x}{12}\right) = 24 \cdot 1$$
$$3x + 2x = 24$$
$$5x = 24$$
$$x = \frac{24}{5} \text{ or } 4\frac{4}{5}$$

The sink will be filled in $\frac{24}{5}$ or $4\frac{4}{5}$ minutes.

38. Let $x =$ the time to do the job working together. Make a table.

Worker	Rate	Time Working Together	Fractional Part of the Job Done
Melena	$\frac{1}{9}$	x	$\frac{x}{9}$
Jeff	$\frac{1}{6}$	x	$\frac{x}{6}$

Part done by Melena	plus	part done by Jeff	equals	1 whole job.
$\frac{x}{9}$	$+$	$\frac{x}{6}$	$=$	1

Multiply by the LCD, 36.

$$36\left(\frac{x}{9} + \frac{x}{6}\right) = 36 \cdot 1$$
$$4x + 6x = 36$$
$$10x = 36$$
$$x = \frac{36}{10}$$
$$x = \frac{18}{5} \text{ or } 3\frac{3}{5}$$

Working together, they can do the job in $\frac{18}{5}$ or $3\frac{3}{5}$ hours.

39. If y varies inversely as x, then $y = \frac{k}{x}$, for some constant k. This form fits choice **C**.

40. Let $v =$ the viewing distance and $e =$ the amount of enlargement.

v varies directly as e, so

$$v = ke$$

for some constant k. Since $v = 250$ when $e = 5$, substitute these values in the equation and solve for k.

$$v = ke$$
$$250 = k(5)$$
$$50 = k$$

So $v = 50e$. Now let $e = 8.6$.

$$v = 50(8.6) = 430$$

It should be viewed from 430 mm.

41. The frequency f of a vibrating guitar string varies inversely as its length L, so

$$f = \frac{k}{L}$$

for some constant k. Substitute 0.65 for L and 4.3 for f.

$$4.3 = \frac{k}{6.5}$$
$$k = 4.3(0.65) = 2.795$$

So $f = \frac{2.795}{L}$. Now let $L = 0.5$.

$$f = \frac{2.795}{0.5} = 5.59$$

The frequency would be 5.59 vibrations per second.

42. The volume V of a rectangular box of a given height is proportional to its width W and length L, so

$$V = kWL$$

for some constant k. Substitute 2 for W, 4 for L, and 12 for V.

$$12 = k(2)(4)$$
$$k = \frac{12}{8} = \frac{3}{2}$$

So $V = \frac{3}{2}WL$. Now let $W = 3$ and $L = 5$.

$$V = \frac{3}{2}(3)(5) = \frac{45}{2}$$

The volume is 22.5 cubic feet.

43. **[7.1]** $\dfrac{x + 2y}{x^2 - 4y^2} = \dfrac{x + 2y}{(x + 2y)(x - 2y)}$

$$= \dfrac{1}{x - 2y}$$

44. **[7.1]** $\dfrac{x^2 + 2x - 15}{x^2 - x - 6} = \dfrac{(x + 5)(x - 3)}{(x - 3)(x + 2)}$

$$= \dfrac{x + 5}{x + 2}$$

45. **[7.2]** $\dfrac{2}{m} + \dfrac{5}{3m^2}$

The LCD is $3m^2$.

$$= \dfrac{2 \cdot 3m}{m \cdot 3m} + \dfrac{5}{3m^2}$$

$$= \dfrac{6m}{3m^2} + \dfrac{5}{3m^2} = \dfrac{6m + 5}{3m^2}$$

46. **[7.1]** $\dfrac{k^2 - 6k + 9}{1 - 216k^3} \cdot \dfrac{6k^2 + 17k - 3}{9 - k^2}$

Factor $1 - 216k^3$ as the difference of cubes, $1^3 - (6k)^3$.

$$= \dfrac{(k - 3)(k - 3)}{(1 - 6k)(1 + 6k + 36k^2)}$$

$$\cdot \dfrac{(6k - 1)(k + 3)}{(3 - k)(3 + k)}$$

$$= \dfrac{(k - 3)(k - 3)}{(-1)(6k - 1)(1 + 6k + 36k^2)}$$

$$\cdot \dfrac{(6k - 1)(k + 3)}{(-1)(k - 3)(k + 3)}$$

$$= \dfrac{k - 3}{1 + 6k + 36k^2} \quad \text{or} \quad \dfrac{k - 3}{36k^2 + 6k + 1}$$

47. **[7.3]** $\dfrac{\dfrac{-3}{x} + \dfrac{x}{2}}{1 + \dfrac{x + 1}{x}}$

Multiply the numerator and denominator by the LCD of all the fractions, $2x$.

$$= \dfrac{2x\left(\dfrac{-3}{x} + \dfrac{x}{2}\right)}{2x\left(1 + \dfrac{x + 1}{x}\right)}$$

$$= \dfrac{-6 + x^2}{2x + 2(x + 1)}$$

$$= \dfrac{x^2 - 6}{2x + 2x + 2}$$

$$= \dfrac{x^2 - 6}{4x + 2} = \dfrac{x^2 - 6}{2(2x + 1)}$$

48. **[7.1]** $\dfrac{9x^2 + 46x + 5}{3x^2 - 2x - 1} \div \dfrac{x^2 + 11x + 30}{x^3 + 5x^2 - 6x}$

Multiply by the reciprocal.

$$= \dfrac{9x^2 + 46x + 5}{3x^2 - 2x - 1} \cdot \dfrac{x(x^2 + 5x - 6)}{x^2 + 11x + 30}$$

$$= \dfrac{(9x + 1)(x + 5)}{(3x + 1)(x - 1)} \cdot \dfrac{x(x + 6)(x - 1)}{(x + 6)(x + 5)}$$

$$= \dfrac{x(9x + 1)}{3x + 1}$$

49. **[7.3]** $\dfrac{\dfrac{3}{x} - 5}{6 + \dfrac{1}{x}}$

Multiply the numerator and denominator by the LCD of all the fractions, x.

$$= \dfrac{x\left(\dfrac{3}{x} - 5\right)}{x\left(6 + \dfrac{1}{x}\right)}$$

$$= \dfrac{3 - 5x}{6x + 1}$$

50. **[7.2]** $\dfrac{9}{3 - x} - \dfrac{2}{x - 3}$

$$= \dfrac{9}{3 - x} - \dfrac{2(-1)}{(x - 3)(-1)}$$

$$= \dfrac{9}{3 - x} - \dfrac{-2}{3 - x}$$

$$= \dfrac{9 - (-2)}{3 - x}$$

$$= \dfrac{11}{3 - x}, \quad \text{or} \quad \dfrac{-11}{x - 3}$$

51. **[7.1]** $\dfrac{4y + 16}{30} \div \dfrac{2y + 8}{5}$

Multiply by the reciprocal.

$$= \dfrac{4y + 16}{30} \cdot \dfrac{5}{2y + 8}$$

$$= \dfrac{4(y + 4)}{30} \cdot \dfrac{5}{2(y + 4)}$$

$$= \dfrac{4 \cdot 5}{2 \cdot 30} = \dfrac{2}{6} = \dfrac{1}{3}$$

52. **[7.3]** $\dfrac{t^{-2} + s^{-2}}{t^{-1} - s^{-1}} = \dfrac{\dfrac{1}{t^2} + \dfrac{1}{s^2}}{\dfrac{1}{t} - \dfrac{1}{s}}$

Multiply the numerator and denominator by the LCD of all the fractions, $t^2 s^2$.

$$= \frac{t^2 s^2 \left(\dfrac{1}{t^2} + \dfrac{1}{s^2} \right)}{t^2 s^2 \left(\dfrac{1}{t} - \dfrac{1}{s} \right)}$$

$$= \frac{s^2 + t^2}{ts^2 - t^2 s}$$

$$= \frac{s^2 + t^2}{st(s - t)}$$

53. **[7.2]** $\dfrac{4a}{a^2 - ab - 2b^2} - \dfrac{6b - a}{a^2 + 4ab + 3b^2}$

$$= \frac{4a}{(a - 2b)(a + b)} - \frac{6b - a}{(a + 3b)(a + b)}$$

The LCD is $(a + 3b)(a - 2b)(a + b)$.

$$= \frac{4a(a + 3b)}{(a - 2b)(a + b)(a + 3b)}$$

$$- \frac{(6b - a)(a - 2b)}{(a + 3b)(a + b)(a - 2b)}$$

$$= \frac{4a(a + 3b) - (6b - a)(a - 2b)}{(a + 3b)(a + b)(a - 2b)}$$

$$= \frac{4a^2 + 12ab - (6ab - 12b^2 - a^2 + 2ab)}{(a + 3b)(a + b)(a - 2b)}$$

$$= \frac{4a^2 + 12ab - 6ab + 12b^2 + a^2 - 2ab}{(a + 3b)(a + b)(a - 2b)}$$

$$= \frac{5a^2 + 4ab + 12b^2}{(a + 3b)(a + b)(a - 2b)}$$

54. **[7.2]** $\dfrac{a}{b} + \dfrac{b}{c} + \dfrac{c}{d}$

The LCD is bcd.

$$= \frac{a \cdot cd}{b \cdot cd} + \frac{b \cdot bd}{c \cdot bd} + \frac{c \cdot bc}{d \cdot bc}$$

$$= \frac{acd + b^2 d + bc^2}{bcd}$$

55. **[7.4]** $\dfrac{x + 3}{x^2 - 5x + 4} - \dfrac{1}{x} = \dfrac{2}{x^2 - 4x}$

$$\frac{x + 3}{(x - 4)(x - 1)} - \frac{1}{x} = \frac{2}{x(x - 4)}$$

Multiply by the LCD, $x(x - 4)(x - 1)$.
$(x \neq 0, 1, 4)$

$$x(x - 4)(x - 1) \left(\frac{x + 3}{(x - 4)(x - 1)} - \frac{1}{x} \right)$$

$$= x(x - 4)(x - 1) \cdot \left(\frac{2}{x(x - 4)} \right)$$

$$x(x + 3) - (x - 4)(x - 1) = 2(x - 1)$$

$$x^2 + 3x - (x^2 - 5x + 4) = 2x - 2$$

$$x^2 + 3x - x^2 + 5x - 4 = 2x - 2$$

$$8x - 4 = 2x - 2$$

$$6x = 2$$

$$x = \tfrac{1}{3}$$

Check $x = \tfrac{1}{3}$: $\tfrac{15}{11} - 3 = -\tfrac{18}{11}$ *True*

The solution set is $\left\{ \tfrac{1}{3} \right\}$.

56. **[7.5]** Solve $A = \dfrac{Rr}{R + r}$ for r.

$$A(R + r) = Rr$$

$$AR + Ar = Rr$$

$$AR = Rr - Ar$$

$$AR = (R - A)r$$

$$\frac{AR}{R - A} = r, \text{ or } r = \frac{-AR}{A - R}$$

57. **[7.4]** $1 - \dfrac{5}{r} = \dfrac{-4}{r^2}$

Multiply by the LCD, r^2. $(r \neq 0)$

$$r^2 \left(1 - \frac{5}{r} \right) = r^2 \left(\frac{-4}{r^2} \right)$$

$$r^2 - 5r = -4$$

$$r^2 - 5r + 4 = 0$$

$$(r - 4)(r - 1) = 0$$

$$r - 4 = 0 \quad \text{or} \quad r - 1 = 0$$

$$r = 4 \quad \text{or} \quad r = 1$$

Check $r = 1$: $1 - 5 = -4$ *True*

Check $r = 4$: $1 - \tfrac{5}{4} = -\tfrac{1}{4}$ *True*

The solution set is $\{1, 4\}$.

58. **[7.4]** $\dfrac{3x}{x - 4} + \dfrac{2}{x} = \dfrac{48}{x^2 - 4x}$

$$\frac{3x}{x - 4} + \frac{2}{x} = \frac{48}{x(x - 4)}$$

Multiply by the LCD, $x(x - 4)$. $(x \neq 0, 4)$

$$x(x - 4) \left(\frac{3x}{x - 4} + \frac{2}{x} \right) = x(x - 4) \left(\frac{48}{x(x - 4)} \right)$$

$$3x^2 + 2(x - 4) = 48$$

$$3x^2 + 2x - 8 = 48$$

$$3x^2 + 2x - 56 = 0$$

$$(3x + 14)(x - 4) = 0$$

$$3x + 14 = 0 \quad \text{or} \quad x - 4 = 0$$

$$3x = -14 \quad \text{or} \quad x = 4$$

$$x = -\tfrac{14}{3}$$

The number 4 is not allowed as a solution because substituting it in the original equation results in division by 0.

Check $x = -\tfrac{14}{3}$: $\tfrac{21}{13} - \tfrac{3}{7} = \tfrac{108}{91}$ *True*

The solution set is $\left\{ -\tfrac{14}{3} \right\}$.

59. **[7.5]** $f(x) = \dfrac{337}{x}$

(a) $f(40.5) = \dfrac{337}{40.5}$ *Let $x = 40.5$.*

$$\approx 8.32$$

(b) $7.51 = \dfrac{337}{x}$ *Let f(x) = 7.51.*

$$7.51x = 337$$

$$x = \frac{337}{7.51} \approx 44.9$$

60. **[7.5]** Let x = the time to fill the tub working together. Make a table.

	Rate	Time Working Together	Fractional Part of the Job Done
Hot	$\dfrac{1}{20}$	x	$\dfrac{x}{20}$
Cold	$\dfrac{1}{15}$	x	$\dfrac{x}{15}$

Part done by hot	plus	part done by cold	equals	1 whole job.
$\dfrac{x}{20}$	$+$	$\dfrac{x}{15}$	$=$	1

Multiply by the LCD, 60.

$$60\left(\frac{x}{20} + \frac{x}{15}\right) = 60 \cdot 1$$

$$3x + 4x = 60$$

$$7x = 60$$

$$x = \frac{60}{7} \text{ or } 8\frac{4}{7}$$

Working together, the tub can be filled in $\frac{60}{7}$ or $8\frac{4}{7}$ minutes.

61. **[7.6]** Let x = the cost of 13 gallons. Write a proportion.

$$\frac{x}{13} = \frac{4.86}{3}$$

$$x = \frac{13(4.86)}{3}$$

$$= 21.06$$

The cost of 13 gallons of unleaded gasoline is $21.06.

62. **[7.6]** The area A of a triangle varies jointly as the lengths of the base b and height h, so

$$A = kbh.$$

When $b = 10$ and $h = 4$, $A = 20$.

$$20 = k(10)(4)$$

$$k = \frac{20}{40} = \frac{1}{2}$$

Thus, $A = \frac{1}{2}bh$. When $b = 3$ and $h = 8$,

$$A = \frac{1}{2}(3)(8) = 12.$$

The area of the triangle is 12 square feet.

63. **[7.5]** Let x = the distance between his office and home. Make a table using the information in the problem and the formula $t = \dfrac{d}{r}$.

	d	r	t
Bike	x	12	$\dfrac{x}{12}$
Car	x	36	$\dfrac{x}{36}$

From the problem, the equation is stated in words: Car driving time = Bike riding time less $\frac{1}{4}$ hour.

Use the car time and the bike time given in the table to write the equation.

$$\frac{x}{36} = \frac{x}{12} - \frac{1}{4}$$

$$36\left(\frac{x}{36}\right) = 36\left(\frac{x}{12}\right) - 36\left(\frac{1}{4}\right)$$

$$x = 3x - 9$$

$$-2x = -9$$

$$x = 4\frac{1}{2} \text{ miles}$$

His office is $4\frac{1}{2}$ miles from home.

64. **[7.5]** *Step 2*

Let x = the distance from San Francisco to the secret rendezvous.

Make a table.

	d	r	t
First Trip	x	200	$\dfrac{x}{200}$
Return Trip	x	300	$\dfrac{x}{300}$

Step 3

Time there	plus	time back	equals	4 hr.
$\dfrac{x}{200}$	$+$	$\dfrac{x}{300}$	$=$	4

Step 4

Multiply by the LCD, 600.

$$600\left(\frac{x}{200} + \frac{x}{300}\right) = 600(4)$$

$$3x + 2x = 2400$$

$$5x = 2400$$

$$x = 480$$

Step 5

The distance is 480 miles.

Step 6

Check 480 miles at 200 mph takes $\frac{480}{200}$ or 2.4 hours; 480 miles at 300 mph takes $\frac{480}{300}$ or 1.6 hours. The total time is 4 hours, as required.

65. [7.5] *Step 2*

Let $x =$ the distance to the fishing hole.

Make a table.

	d	r	t
Old Highway	x	30	$\dfrac{x}{30}$
Interstate	x	50	$\dfrac{x}{50}$

Step 3

The time on the interstate is 2 hr less than the time on the old highway.

$$\frac{x}{50} = \frac{x}{30} - 2$$

Step 4

Multiply by 150.

$$3x = 5x - 300$$
$$300 = 2x$$
$$150 = x$$

Step 5

The distance to the fishing hole is 150 miles.

Step 6

Check 150 miles at 30 mph takes $\frac{150}{30}$ or 5 hours; 150 miles at 50 mph takes $\frac{150}{50}$ or 3 hours. The interstate time is 2 hours less than the highway time, as required.

Chapter 7 Test

1. $f(x) = \dfrac{x+3}{3x^2 + 2x - 8}$

Set the denominator equal to zero and solve.

$$3x^2 + 2x - 8 = 0$$
$$(3x - 4)(x + 2) = 0$$
$$3x - 4 = 0 \quad \text{or} \quad x + 2 = 0$$
$$3x = 4$$
$$x = \tfrac{4}{3} \quad \text{or} \quad x = -2$$

The numbers -2 and $\frac{4}{3}$ make the rational expression undefined and are excluded from the domain of f, which can be written in set notation as $\left\{ x \mid x \neq -2, \frac{4}{3} \right\}$.

2. $\dfrac{6x^2 - 13x - 5}{9x^3 - x} = \dfrac{(3x+1)(2x-5)}{x(9x^2 - 1)}$

$$= \dfrac{(3x+1)(2x-5)}{x(3x+1)(3x-1)}$$
$$= \dfrac{2x - 5}{x(3x - 1)}$$

3. $\dfrac{(x+3)^2}{4} \cdot \dfrac{6}{2x+6} = \dfrac{(x+3)^2}{4} \cdot \dfrac{2 \cdot 3}{2(x+3)}$

$$= \dfrac{3(x+3)}{4}$$

4. $\dfrac{y^2 - 16}{y^2 - 25} \cdot \dfrac{y^2 + 2y - 15}{y^2 - 7y + 12}$

$$= \dfrac{(y+4)(y-4)}{(y+5)(y-5)} \cdot \dfrac{(y+5)(y-3)}{(y-4)(y-3)}$$
$$= \dfrac{y + 4}{y - 5}$$

5. $\dfrac{3 - t}{5} \div \dfrac{t - 3}{10}$

Multiply by the reciprocal of the divisor.

$$= \dfrac{3 - t}{5} \cdot \dfrac{10}{t - 3}$$
$$= \dfrac{-1(t - 3) \cdot 10}{5(t - 3)}$$
$$= -1(2) = -2$$

6. $\dfrac{x^2 - 9}{x^3 + 3x^2} \div \dfrac{x^2 + x - 12}{x^3 + 9x^2 + 20x}$

Multiply by the reciprocal.

$$= \dfrac{x^2 - 9}{x^3 + 3x^2} \cdot \dfrac{x(x^2 + 9x + 20)}{x^2 + x - 12}$$
$$= \dfrac{(x+3)(x-3)}{x^2(x+3)} \cdot \dfrac{x(x+5)(x+4)}{(x+4)(x-3)}$$
$$= \dfrac{x + 5}{x}$$

7. $t^2 + t - 6, \; t^2 + 3t, \; t^2$

Factor each denominator.

$$t^2 + t - 6 = (t + 3)(t - 2)$$
$$t^2 + 3t = t(t + 3)$$

The third denominator is already in factored form. The LCD is

$$t^2(t + 3)(t - 2).$$

8. $\dfrac{7}{6t^2} - \dfrac{1}{3t}$

The LCD is $6t^2$.

$$= \dfrac{7}{6t^2} - \dfrac{1 \cdot 2t}{3t \cdot 2t}$$
$$= \dfrac{7}{6t^2} - \dfrac{2t}{6t^2} = \dfrac{7 - 2t}{6t^2}$$

9. $\dfrac{3}{7a^4 b^3} + \dfrac{5}{21a^5 b^2} \quad$ LCD $= 21a^5 b^3$

$$= \dfrac{3 \cdot 3a}{7a^4 b^3 \cdot 3a} + \dfrac{5 \cdot b}{21a^5 b^2 \cdot b}$$
$$= \dfrac{9a}{21a^5 b^3} + \dfrac{5b}{21a^5 b^3}$$
$$= \dfrac{9a + 5b}{21a^5 b^3}$$

10. $\dfrac{9}{x^2 - 6x + 9} + \dfrac{2}{x^2 - 9}$

Factor each denominator.

$$x^2 - 6x + 9 = (x - 3)^2$$
$$x^2 - 9 = (x + 3)(x - 3)$$

The LCD is $(x + 3)(x - 3)^2$.

$$\dfrac{9}{(x - 3)^2} + \dfrac{2}{(x + 3)(x - 3)}$$

$$= \dfrac{9(x + 3)}{(x - 3)^2(x + 3)} + \dfrac{2(x - 3)}{(x + 3)(x - 3)(x - 3)}$$

$$= \dfrac{9(x + 3) + 2(x - 3)}{(x + 3)(x - 3)^2}$$

$$= \dfrac{9x + 27 + 2x - 6}{(x + 3)(x - 3)^2}$$

$$= \dfrac{11x + 21}{(x + 3)(x - 3)^2}$$

11. $\dfrac{6}{x + 4} + \dfrac{1}{x + 2} - \dfrac{3x}{x^2 + 6x + 8}$

$$= \dfrac{6}{x + 4} + \dfrac{1}{x + 2} - \dfrac{3x}{(x + 4)(x + 2)}$$

The LCD is $(x + 4)(x + 2)$.

$$= \dfrac{6(x + 2)}{(x + 4)(x + 2)} + \dfrac{1(x + 4)}{(x + 2)(x + 4)}$$

$$\quad - \dfrac{3x}{(x + 4)(x + 2)}$$

$$= \dfrac{6(x + 2) + x + 4 - 3x}{(x + 4)(x + 2)}$$

$$= \dfrac{6x + 12 + x + 4 - 3x}{(x + 4)(x + 2)}$$

$$= \dfrac{4x + 16}{(x + 4)(x + 2)}$$

$$= \dfrac{4(x + 4)}{(x + 4)(x + 2)} = \dfrac{4}{x + 2}$$

12. $\dfrac{\dfrac{12}{r + 4}}{\dfrac{11}{6r + 24}} = \dfrac{12}{r + 4} \div \dfrac{11}{6r + 24}$

Multiply by the reciprocal.

$$= \dfrac{12}{r + 4} \cdot \dfrac{6r + 24}{11}$$

$$= \dfrac{12}{r + 4} \cdot \dfrac{6(r + 4)}{11} = \dfrac{72}{11}$$

13. $\dfrac{\dfrac{1}{a} - \dfrac{1}{b}}{\dfrac{a}{b} - \dfrac{b}{a}}$

Multiply the numerator and the denominator
by the LCD of all the fractions, ab.

$$= \dfrac{ab\left(\dfrac{1}{a} - \dfrac{1}{b}\right)}{ab\left(\dfrac{a}{b} - \dfrac{b}{a}\right)} = \dfrac{b - a}{a^2 - b^2}$$

$$= \dfrac{b - a}{(a - b)(a + b)} = \dfrac{(-1)(a - b)}{(a - b)(a + b)}$$

$$= \dfrac{-1}{a + b} \quad \text{or} \quad -\dfrac{1}{a + b}$$

14. $\dfrac{2x^{-2} + y^{-2}}{x^{-1} - y^{-1}} = \dfrac{\dfrac{2}{x^2} + \dfrac{1}{y^2}}{\dfrac{1}{x} - \dfrac{1}{y}}$

Multiply the numerator and denominator by the
LCD of all the fractions, $x^2 y^2$.

$$= \dfrac{x^2 y^2\left(\dfrac{2}{x^2} + \dfrac{1}{y^2}\right)}{x^2 y^2\left(\dfrac{1}{x} - \dfrac{1}{y}\right)}$$

$$= \dfrac{2y^2 + x^2}{xy^2 - x^2 y}$$

$$= \dfrac{2y^2 + x^2}{xy(y - x)}$$

15. (a) Simplify this expression.

$$\dfrac{2x}{3} + \dfrac{x}{4} - \dfrac{11}{2}$$

The LCD is 12.

$$= \dfrac{2x \cdot 4}{3 \cdot 4} + \dfrac{x \cdot 3}{4 \cdot 3} - \dfrac{11 \cdot 6}{2 \cdot 6}$$

$$= \dfrac{8x}{12} + \dfrac{3x}{12} - \dfrac{66}{12}$$

$$= \dfrac{8x + 3x - 66}{12}$$

$$= \dfrac{11x - 66}{12} = \dfrac{11(x - 6)}{12}$$

(b) Solve this equation.

$$\dfrac{2x}{3} + \dfrac{x}{4} = \dfrac{11}{2}$$

Multiply by the LCD, 12.

$$12\left(\dfrac{2x}{3} + \dfrac{x}{4}\right) = 12\left(\dfrac{11}{2}\right)$$

$$8x + 3x = 66$$

$$11x = 66$$

$$x = 6$$

Check $x = 6$: $4 + \frac{3}{2} = \frac{11}{2}$ *True*

The solution set is $\{6\}$.

16.
$$\frac{1}{x} - \frac{4}{3x} = \frac{1}{x-2}.$$
Multiply by the LCD, $3x(x-2)$. $(x \neq 0, 2)$
$$3x(x-2)\left(\frac{1}{x} - \frac{4}{3x}\right) = 3x(x-2)\left(\frac{1}{x-2}\right)$$
$$3(x-2) - 4(x-2) = 3x$$
$$3x - 6 - 4x + 8 = 3x$$
$$-x + 2 = 3x$$
$$-4x = -2$$
$$x = \tfrac{1}{2}$$

Check $x = \tfrac{1}{2}$: $2 - \tfrac{8}{3} = -\tfrac{2}{3}$ *True*

The solution set is $\left\{\tfrac{1}{2}\right\}$.

17.
$$\frac{y}{y+2} - \frac{1}{y-2} = \frac{8}{y^2 - 4}$$
$$\frac{y}{y+2} - \frac{1}{y-2} = \frac{8}{(y+2)(y-2)}$$
Multiply by the LCD, $(y+2)(y-2)$.
$(y \neq -2, 2)$
$$(y+2)(y-2)\left(\frac{y}{y+2} - \frac{1}{y-2}\right)$$
$$= (y+2)(y-2)\left(\frac{8}{(y+2)(y-2)}\right)$$
$$y(y-2) - 1(y+2) = 8$$
$$y^2 - 2y - y - 2 = 8$$
$$y^2 - 3y - 10 = 0$$
$$(y-5)(y+2) = 0$$
$$y - 5 = 0 \quad \text{or} \quad y + 2 = 0$$
$$y = 5 \quad \text{or} \quad y = -2$$

The number -2 is not allowed as a solution because substituting it in the original equation results in division by 0.

Check $y = 5$: $\tfrac{5}{7} - \tfrac{1}{3} = \tfrac{8}{21}$ *True*

The solution set is $\{5\}$.

18. A solution cannot make a denominator 0. This is illustrated in Exercise 17 where -2 is not allowed as a solution for this reason.

19. Solve $S = \frac{n}{2}(a + \ell)$ for ℓ.

To solve for ℓ, multiply both sides by $\frac{2}{n}$.

$$\frac{2}{n} \cdot S = \frac{2}{n} \cdot \frac{n}{2}(a + \ell)$$
$$\frac{2S}{n} = a + \ell$$
$$\frac{2S}{n} - a = \ell, \quad \text{or} \quad \ell = \frac{2S - na}{n}$$

20. $f(x) = \dfrac{-2}{x+1}$ is not defined when $x = -1$, so an equation of the vertical asymptote is $x = -1$.

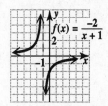

21. Let $x =$ the time to do the job working together. Make a table.

Worker	Rate	Time Working Together	Fractional Part of the Job Done
Wayne	$\frac{1}{9}$	x	$\frac{x}{9}$
Susan	$\frac{1}{5}$	x	$\frac{x}{5}$

Part done by Wayne	plus	part done by Susan	equals	1 whole job.
$\frac{x}{9}$	$+$	$\frac{x}{5}$	$=$	1

Multiply by the LCD, 45.

$$45\left(\frac{x}{9} + \frac{x}{5}\right) = 45 \cdot 1$$
$$5x + 9x = 45$$
$$14x = 45$$
$$x = \tfrac{45}{14} \quad \text{or} \quad 3\tfrac{3}{14}$$

Working together, they can do the job in $\tfrac{45}{14}$ or $3\tfrac{3}{14}$ hours.

22. Let $x =$ the speed of the boat in still water.

Use $d = rt$, or $t = \dfrac{d}{r}$, to make a table.

	d	r	t
Downstream	36	$x+3$	$\frac{36}{x+3}$
Upstream	24	$x-3$	$\frac{24}{x-3}$

Because the times are equal,

$$\frac{36}{x+3} = \frac{24}{x-3}.$$

Multiply by the LCD, $(x+3)(x-3)$. $(x \neq -3, 3)$
$$(x+3)(x-3)\left(\frac{36}{x+3}\right) = (x+3)(x-3)\left(\frac{24}{x-3}\right)$$
$$36(x-3) = 24(x+3)$$
$$36x - 108 = 24x + 72$$
$$12x = 180$$
$$x = 15$$

The speed of the boat in still water is 15 mph.

23. Let $x =$ the number of fish in Lake Linda. Write a proportion.

$$\frac{x}{600} = \frac{800}{10}$$

Multiply by the LCD, 600.

$$600\left(\frac{x}{600}\right) = 600\left(\frac{800}{10}\right)$$
$$x = 60 \cdot 800$$
$$x = 48{,}000$$

There are about 48,000 fish in the lake.

24. $g(x) = \dfrac{5x}{2 + x}$

(a) $3 = \dfrac{5x}{2 + x}$ *Let g(x) = 3.*
$$3(2 + x) = 5x$$
$$6 + 3x = 5x$$
$$6 = 2x$$
$$x = 3 \text{ units}$$

(b) If no food is available, then $x = 0$.

$$g(0) = \frac{5(0)}{2 + 0} = \frac{0}{2} = 0$$

The growth rate is 0.

25. The current I is inversely proportional to the resistance R, so

$$I = \frac{k}{R}$$

for some constant k. Let $I = 80$ and $R = 30$. Find k.

$$80 = \frac{k}{30}$$
$$k = 80(30) = 2400$$

So $I = \dfrac{2400}{R}$. Now let $R = 12$.

$$I = \tfrac{2400}{12} = 200$$

The current is 200 amperes.

26. The force F of the wind blowing on a vertical surface varies jointly as the area A of the surface and the square of the velocity V, so

$$F = kAV^2$$

for some constant k. Let $F = 50$, $A = 500$, and $V = 40$. Find k.

$$50 = k(500)(40)^2$$
$$k = \frac{50}{500(1600)} = \frac{1}{16{,}000}$$

So $F = \frac{1}{16{,}000}AV^2$. Now let $A = 2$ and $V = 80$.

$$F = \frac{1}{16{,}000}(2)(80)^2 = 0.8$$

The force of the wind is 0.8 pounds.

Cumulative Review Exercises (Chapters 1–7)

In Exercises 1–2, $x = -4$, $y = 3$, and $z = 6$.

1. $|2x| + 3y - z^3$
$$= |(2)(-4)| + 3(3) - (6)^3$$
$$= |-8| + 9 - 216$$
$$= 8 + 9 - 216 = -199$$

2. $\dfrac{x(2x - 1)}{3y - z} = \dfrac{-4[2(-4) - 1]}{3(3) - 6}$
$$= \frac{-4[-8 - 1]}{9 - 6}$$
$$= \frac{-4(-9)}{3} = 12$$

3. $7(2x + 3) - 4(2x + 1) = 2(x + 1)$
$$14x + 21 - 8x - 4 = 2x + 2$$
$$6x + 17 = 2x + 2$$
$$4x = -15$$
$$x = -\tfrac{15}{4}$$
The solution set is $\left\{-\tfrac{15}{4}\right\}$.

4. $|6x - 8| - 4 = 0$
$$|6x - 8| = 4$$

$6x - 8 = 4$ or $6x - 8 = -4$
$6x = 12$ $6x = 4$
$x = 2$ or $x = \tfrac{4}{6} = \tfrac{2}{3}$
The solution set is $\left\{\tfrac{2}{3}, 2\right\}$.

5.   Solve $ax + by = cx + d$ for x.
Get the x-terms on one side.

$$ax - cx = d - by$$
$$x(a - c) = d - by$$
$$x = \frac{d - by}{a - c}$$

or $x = \dfrac{by - d}{c - a}$

6. $\tfrac{2}{3}y + \tfrac{5}{12}y \le 20$
Multiply both sides by 12.
$$12\left(\tfrac{2}{3}y + \tfrac{5}{12}y\right) \le 12(20)$$
$$8y + 5y \le 240$$
$$13y \le 240$$
$$y \le \tfrac{240}{13}$$

The solution set is $\left(-\infty, \tfrac{240}{13}\right]$.

7. $|3x + 2| \geq 4$

$$3x + 2 \geq 4 \quad \text{or} \quad 3x + 2 \leq -4$$
$$3x \geq 2 \quad \text{or} \quad 3x \leq -6$$
$$x \geq \tfrac{2}{3} \quad \text{or} \quad x \leq -2$$

The solution set is $(-\infty, -2] \cup \left[\tfrac{2}{3}, \infty\right)$.

8. Let x = the amount of money invested at 4% and $2x$ = the amount of money invested at 3%.

Use $I = prt$ with the time, t, equal to 1 yr.

$$0.04x + 0.03(2x) = 400$$
$$0.04x + 0.06x = 400$$
$$0.10x = 400$$

Multiply by 10 to clear the decimal.

$$x = 4000$$

Then $2x = 2(4000) = 8000$.

He invested \$4000 at 4% and \$8000 at 3%.

9. Let h = the height of the triangle.
Use the formula $A = \tfrac{1}{2}bh$. Here, $A = 42$ and $b = 14$.

$$A = \tfrac{1}{2}bh$$
$$42 = \tfrac{1}{2}(14)h$$
$$42 = 7h$$
$$6 = h$$

The height is 6 meters.

10. $-4x + 2y = 8$
To find the x-intercept, let $y = 0$.

$$-4x + 2(0) = 8$$
$$-4x = 8$$
$$x = -2$$

The x-intercept is $(-2, 0)$.
To find the y-intercept, let $x = 0$.

$$-4(0) + 2y = 8$$
$$2y = 8$$
$$y = 4$$

The y-intercept is $(0, 4)$. Plot the intercepts, and draw the line through them.

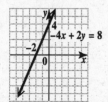

11. Through $(-5, 8)$ and $(-1, 2)$
Let $(x_1, y_1) = (-5, 8)$ and $(x_2, y_2) = (-1, 2)$.
Then,

$$m = \frac{y_2 - y_1}{x_2 - x_1} = \frac{2 - 8}{-1 - (-5)} = \frac{-6}{4} = -\frac{3}{2}.$$

The slope is $-\frac{3}{2}$.

12. Perpendicular to $4x - 3y = 12$

Solve for y to write the equation in slope-intercept form and find the slope.

$$4x - 3y = 12$$
$$-3y = -4x + 12$$
$$y = \tfrac{4}{3}x - 4$$

The slope is $\frac{4}{3}$. Perpendicular lines have slopes that are negative reciprocals of each other. The negative reciprocal of $\frac{4}{3}$ is $-\frac{3}{4}$. The slope of a line perpendicular to the given line is $-\frac{3}{4}$.

13. Use $(x_1, y_1) = (-5, 8)$ and $m = -\frac{3}{2}$ in the point-slope form.

$$y - y_1 = m(x - x_1)$$
$$y - 8 = -\tfrac{3}{2}[x - (-5)]$$
$$y - 8 = -\tfrac{3}{2}(x + 5)$$
$$y - 8 = -\tfrac{3}{2}x - \tfrac{15}{2}$$
$$y = -\tfrac{3}{2}x + \tfrac{1}{2}$$

14. $2x + 5y > 10$
Graph the line $2x + 5y = 10$ by drawing a dashed line (since the inequality involves $>$) through the intercepts $(5, 0)$ and $(0, 2)$.
Test a point not on this line, such as $(0, 0)$.

$$2x + 5y > 10$$
$$2(0) + 5(0) > 10$$
$$0 > 10 \quad \textit{False}$$

Shade the side of the line not containing $(0, 0)$.

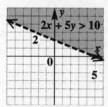

15. $x - y \geq 3$ and $3x + 4y \leq 12$
Graph the solid line $x - y = 3$ through $(3, 0)$ and $(0, -3)$. The inequality $x - y \geq 3$ can be written as $y \leq x - 3$, so shade the region below the boundary line.
Graph the solid line $3x + 4y = 12$ through $(4, 0)$ and $(0, 3)$. The inequality $3x + 4y \leq 12$ can be written as $y \leq -\frac{3}{4}x + 3$, so shade the region below the boundary line.
The required graph is the common shaded area as well as the portions of the lines that bound it.

continued

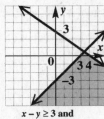

$x - y \geq 3$ and
$3x + 4y \leq 12$

16. Since no year is repeated, the relation defines a function.
The domain is
$\{1990, 1992, 1994, 1996, 1998, 2000, 2002\}$.
The range is
$\{1.25, 1.61, 1.80, 1.21, 1.94, 2.26, 2.60\}$.

17. The relation is not a function since a vertical line may intersect the graph in more than one point. The domain is the set of x-values, $[-2, \infty)$. The range is the set of y-values, $(-\infty, \infty)$.

18. $y = -\sqrt{x + 2}$
The relation is a function since each input value corresponds to exactly one output value. The radicand must be nonnegative; that is, $x + 2 \geq 0$, or $x \geq -2$. The domain is $[-2, \infty)$. The values of $\sqrt{x + 2}$ are nonnegative, so the values of $-\sqrt{x + 2}$ are nonpositive. Thus, the range is $(-\infty, 0]$.

19. (a) Solve the equation for y.
$$5x - 3y = 8$$
$$5x - 8 = 3y$$
$$\frac{5x - 8}{3} = y$$
So $f(x) = \dfrac{5x - 8}{3}$ or $f(x) = \dfrac{5}{3}x - \dfrac{8}{3}$.

(b) $f(1) = \dfrac{5(1) - 8}{3} = \dfrac{-3}{3} = -1$

20. $f(x) = 3x + 6$
$$f(x + 3) = 3(x + 3) + 6$$
$$= 3x + 9 + 6$$
$$= 3x + 15$$

21. $4x - y = -7$ (1)
$5x + 2y = 1$ (2)

To eliminate y, multiply equation (1) by 2 and add the result to (2).

$$\begin{array}{rcl} 8x - 2y &=& -14 \quad 2 \times (1) \\ 5x + 2y &=& 1 \quad (2) \\ \hline 13x &=& -13 \\ x &=& -1 \end{array}$$

To find y, substitute -1 for x in (1).

$$\begin{array}{rcl} 4x - y &=& -7 \quad (1) \\ 4(-1) - y &=& -7 \\ -4 - y &=& -7 \\ -y &=& -3 \\ y &=& 3 \end{array}$$

The solution set is $\{(-1, 3)\}$.

22. $x + y - 2z = -1$ (1)
$2x - y + z = -6$ (2)
$3x + 2y - 3z = -3$ (3)

Add (1) and (2) to eliminate y.

$$\begin{array}{rcl} x + y - 2z &=& -1 \quad (1) \\ 2x - y + z &=& -6 \quad (2) \\ \hline 3x \quad - z &=& -7 \quad (4) \end{array}$$

Multiply (2) by 2 and add it to (3) to eliminate y.

$$\begin{array}{rcl} 4x - 2y + 2z &=& -12 \quad 2 \times (2) \\ 3x + 2y - 3z &=& -3 \quad (3) \\ \hline 7x \quad - z &=& -15 \quad (5) \end{array}$$

Multiply (4) by -1 and add to (5) to eliminate z.

$$\begin{array}{rcl} -3x + z &=& 7 \quad -1 \times (4) \\ 7x - z &=& -15 \quad (5) \\ \hline 4x &=& -8 \\ x &=& -2 \end{array}$$

Substitute -2 for x into (4) and solve for z.

$$\begin{array}{rcl} 3(-2) - z &=& -7 \\ -6 - z &=& -7 \\ -z &=& -1 \\ z &=& 1 \end{array}$$

Substitute -2 for x and 1 for z into (1) and solve for y.

$$\begin{array}{rcl} -2 + y - 2(1) &=& -1 \\ -2 + y - 2 &=& -1 \\ y - 4 &=& -1 \\ y &=& 3 \end{array}$$

The solution set is $\{(-2, 3, 1)\}$.

23. $x + 2y + z = 5$ (1)
$x - y + z = 3$ (2)
$2x + 4y + 2z = 11$ (3)

Multiply (1) by -1 and add to (2).

$$\begin{array}{rcl} -x - 2y - z &=& -5 \quad -1 \times (1) \\ x - y + z &=& 3 \quad (2) \\ \hline - 3y &=& -2 \\ y &=& \dfrac{2}{3} \end{array}$$

Substitute $\frac{2}{3}$ for y in (1) and (3), then add (1) and (3).

$$x + 2\left(\tfrac{2}{3}\right) + z = 5 \quad \text{Let } y = \tfrac{2}{3} \text{ in (1)}.$$
$$x + \tfrac{4}{3} + z = 5$$
$$x + z = \tfrac{11}{3} \quad (4)$$
$$2x + 4\left(\tfrac{2}{3}\right) + 2z = 11 \quad \text{Let } y = \tfrac{2}{3} \text{ in (3)}.$$
$$2x + \tfrac{8}{3} + 2z = 11$$
$$2x + 2z = \tfrac{25}{3} \quad (5)$$

Multiply (4) by -2 and add to (5).

$$\begin{array}{rcll} -2x - 2z &=& -\tfrac{22}{3} & -2 \times (4) \\ 2x + 2z &=& \tfrac{25}{3} & (5) \\ \hline 0 &=& 1 & \textit{False} \end{array}$$

Since this statement is false, the solution is $\emptyset$.

24. Let $x = $ the average speed of the automobile. Then $x + 558 = $ the average speed of the airplane. Write a proportion and solve for x.

$$\frac{7}{100} = \frac{x}{x + 558}$$

Multiply each side by the LCD, $100(x + 558)$.

$$100(x + 558)\left(\frac{7}{100}\right) = 100(x + 558)\left(\frac{x}{x + 558}\right)$$
$$7(x + 558) = 100x$$
$$7x + 3906 = 100x$$
$$3906 = 93x$$
$$42 = x$$

Since $x = 42$, $x + 558 = 600$.

The average speed of the automobile is 42 km per hr and that of the airplane is 600 km per hr.

25. $\left(\dfrac{a^{-3}b^4}{a^2 b^{-1}}\right)^{-2} = \left(\dfrac{b^4 b^1}{a^2 a^3}\right)^{-2} = \left(\dfrac{b^5}{a^5}\right)^{-2}$

$\quad = \left(\dfrac{a^5}{b^5}\right)^2 = \dfrac{a^{10}}{b^{10}}$

26. $\left(\dfrac{m^{-4}n^2}{m^2 n^{-3}}\right) \cdot \left(\dfrac{m^5 n^{-1}}{m^{-2} n^5}\right)$

$\quad = \left(\dfrac{n^2 n^3}{m^2 m^4}\right) \cdot \left(\dfrac{m^5 m^2}{n^1 n^5}\right)$

$\quad = \left(\dfrac{n^5}{m^6}\right) \cdot \left(\dfrac{m^7}{n^6}\right)$

$\quad = \dfrac{n^5 m^7}{n^6 m^6} = \dfrac{m}{n}$

27. $\left(3y^2 - 2y + 6\right) - \left(-y^2 + 5y + 12\right)$
$\quad = 3y^2 - 2y + 6 + y^2 - 5y - 12$
$\quad = 4y^2 - 7y - 6$

28. $-6x^4\left(x^2 - 3x + 2\right)$
$\quad = -6x^6 + 18x^5 - 12x^4$

29. $(4f + 3)(3f - 1) = 12f^2 - 4f + 9f - 3$
$\quad\quad\quad\quad\quad\quad = 12f^2 + 5f - 3$

30. $\left(7t^3 + 8\right)\left(7t^3 - 8\right)$
This is the product of the sum and difference of two terms.
$\quad = \left(7t^3\right)^2 - 8^2$
$\quad = 49t^6 - 64$

31. $\left(\tfrac{1}{4}x + 5\right)^2$
Use the formula for the square of a binomial, $(a + b)^2 = a^2 + 2ab + b^2$.

$\quad = \left(\tfrac{1}{4}x\right)^2 + 2\left(\dfrac{1}{4}x\right)(5) + 5^2$

$\quad = \tfrac{1}{16}x^2 + \tfrac{5}{2}x + 25$

32. $\left(3x^3 + 13x^2 - 17x - 7\right) \div (3x + 1)$

$$\begin{array}{r} x^2 + 4x - 7 \\ 3x + 1 \overline{\smash{\big)}\ 3x^3 + 13x^2 - 17x - 7} \\ \underline{3x^3 + x^2} \\ 12x^2 - 17x \\ \underline{12x^2 + 4x} \\ -21x - 7 \\ \underline{-21x - 7} \\ 0 \end{array}$$

Answer: $x^2 + 4x - 7$

33. **(a)** $(f + g)(x) = f(x) + g(x)$
$\quad\quad\quad\quad = \left(x^2 + 2x - 3\right)$
$\quad\quad\quad\quad\quad + \left(2x^3 - 3x^2 + 4x - 1\right)$
$\quad\quad\quad\quad = 2x^3 - 2x^2 + 6x - 4$

(b) $(g - f)(x) = g(x) - f(x)$
$\quad\quad\quad\quad = \left(2x^3 - 3x^2 + 4x - 1\right)$
$\quad\quad\quad\quad\quad - \left(x^2 + 2x - 3\right)$
$\quad\quad\quad\quad = 2x^3 - 3x^2 + 4x - 1$
$\quad\quad\quad\quad\quad - x^2 - 2x + 3$
$\quad\quad\quad\quad = 2x^3 - 4x^2 + 2x + 2$

(c) Using part (a),
$\quad (f + g)(-1) = 2(-1)^3 - 2(-1)^2 + 6(-1) - 4$
$\quad\quad\quad\quad\quad = -2 - 2 - 6 - 4$
$\quad\quad\quad\quad\quad = -14$

(d) $(f \circ h)(x) = f(h(x))$
$\quad\quad\quad\quad = f\left(x^2\right)$
$\quad\quad\quad\quad = \left(x^2\right)^2 + 2\left(x^2\right) - 3$
$\quad\quad\quad\quad = x^4 + 2x^2 - 3$

34. $2x^2 - 13x - 45 = (2x + 5)(x - 9)$

35. $100t^4 - 25 = 25\left(4t^4 - 1\right)$
$\quad\quad\quad\quad = 25\left[\left(2t^2\right)^2 - 1^2\right]$
$\quad\quad\quad\quad = 25\left(2t^2 + 1\right)\left(2t^2 - 1\right)$

36. Use the sum of cubes formula,

$$x^3 + y^3 = (x+y)(x^2 - xy + y^2).$$

$$8p^3 + 125 = (2p)^3 + 5^3$$
$$= (2p+5)\left[(2p)^2 - (2p)(5) + 5^2\right]$$
$$= (2p+5)\left(4p^2 - 10p + 25\right)$$

37.
$$3x^2 + 4x = 7$$
$$3x^2 + 4x - 7 = 0$$
$$(3x+7)(x-1) = 0$$

$$3x + 7 = 0 \quad \text{or} \quad x - 1 = 0$$
$$3x = -7$$
$$x = -\tfrac{7}{3} \quad \text{or} \quad x = 1$$

The solution set is $\left\{-\tfrac{7}{3}, 1\right\}$.

38.
$$\frac{y^2 - 16}{y^2 - 8y + 16} = \frac{(y+4)(y-4)}{(y-4)(y-4)}$$
$$= \frac{y+4}{y-4}$$

39.
$$\frac{8x^2 - 18}{8x^2 + 4x - 12} = \frac{2(4x^2 - 9)}{4(2x^2 + x - 3)}$$
$$= \frac{2(2x+3)(2x-3)}{4(2x+3)(x-1)}$$
$$= \frac{2x-3}{2(x-1)}$$

40.
$$\frac{2a^2}{a+b} \cdot \frac{a-b}{4a} = \frac{2a^2(a-b)}{4a(a+b)}$$
$$= \frac{a(a-b)}{2(a+b)}$$

41.
$$\frac{x^2 - 9}{2x + 4} \div \frac{x^3 - 27}{4}$$
$$= \frac{x^2 - 9}{2x + 4} \cdot \frac{4}{x^3 - 27}$$
$$= \frac{(x+3)(x-3)}{2(x+2)} \cdot \frac{4}{(x-3)(x^2 + 3x + 9)}$$
$$= \frac{2(x+3)}{(x+2)(x^2 + 3x + 9)}$$

42.
$$\frac{x+4}{x-2} + \frac{2x-10}{x-2} = \frac{x+4+2x-10}{x-2}$$
$$= \frac{3x-6}{x-2}$$
$$= \frac{3(x-2)}{x-2} = 3$$

43.
$$\frac{2x}{2x-1} + \frac{4}{2x+1} + \frac{8}{4x^2 - 1}$$
$$= \frac{2x}{2x-1} + \frac{4}{2x+1} + \frac{8}{(2x+1)(2x-1)}$$

The LCD is $(2x+1)(2x-1)$.

$$= \frac{2x(2x+1)}{(2x-1)(2x+1)} + \frac{4(2x-1)}{(2x+1)(2x-1)}$$
$$+ \frac{8}{(2x+1)(2x-1)}$$
$$= \frac{2x(2x+1) + 4(2x-1) + 8}{(2x+1)(2x-1)}$$
$$= \frac{4x^2 + 2x + 8x - 4 + 8}{(2x+1)(2x-1)}$$
$$= \frac{4x^2 + 10x + 4}{(2x+1)(2x-1)}$$
$$= \frac{2(2x^2 + 5x + 2)}{(2x+1)(2x-1)}$$
$$= \frac{2(2x+1)(x+2)}{(2x+1)(2x-1)}$$
$$= \frac{2(x+2)}{2x-1}$$

44.
$$\frac{-3x}{x+1} + \frac{4x+1}{x} = \frac{-3}{x^2 + x}$$
$$\frac{-3x}{x+1} + \frac{4x+1}{x} = \frac{-3}{x(x+1)}$$

Multiply by the LCD, $x(x+1)$. $(x \neq -1, 0)$

$$x(x+1)\left(\frac{-3x}{x+1} + \frac{4x+1}{x}\right)$$
$$= x(x+1)\left(\frac{-3}{x(x+1)}\right)$$

$$x(-3x) + (x+1)(4x+1) = -3$$
$$-3x^2 + 4x^2 + x + 4x + 1 = -3$$
$$x^2 + 5x + 4 = 0$$
$$(x+4)(x+1) = 0$$

$$x + 4 = 0 \quad \text{or} \quad x + 1 = 0$$
$$x = -4 \quad \text{or} \quad x = -1$$

The number -1 is not allowed as a solution because substituting it in the original equation results in division by 0.

Check $x = -4$: $-4 + \tfrac{15}{4} = -\tfrac{1}{4}$ *True*

The solution set is $\{-4\}$.

45. Solve $\dfrac{1}{f} = \dfrac{1}{p} + \dfrac{1}{q}$ for q.

Multiply by the LCD, fpq.

$$fpq\left(\frac{1}{f}\right) = fpq\left(\frac{1}{p} + \frac{1}{q}\right)$$

$$pq = fq + fp$$

Get all terms with q on one side.

$$pq - fq = fp$$

$$q(p - f) = fp$$

$$q = \frac{fp}{p - f}, \text{ or } q = \frac{-fp}{f - p}$$

46. Let $x =$ the speed of the plane in still air.
Use $d = rt$, or $t = \frac{d}{r}$, to make a table.

	d	r	t
Against the Wind	200	$x - 30$	$\dfrac{200}{x - 30}$
With the Wind	300	$x + 30$	$\dfrac{300}{x + 30}$

The times are the same, so

$$\frac{200}{x - 30} = \frac{300}{x + 30}.$$

Multiply by the LCD, $(x - 30)(x + 30)$.

$$(x - 30)(x + 30)\left(\frac{200}{x - 30}\right)$$

$$= (x - 30)(x + 30)\left(\frac{300}{x + 30}\right)$$

$$200(x + 30) = 300(x - 30)$$

$$200x + 6000 = 300x - 9000$$

$$-100x = -15{,}000$$

$$x = 150$$

The speed of the plane in still air is 150 mph.

47. Let $x =$ the time to complete the job working together.
Make a table.

	Rate	Time Working Together	Fractional Part of the Job Done
Machine A	$\dfrac{1}{2}$	x	$\dfrac{x}{2}$
Machine B	$\dfrac{1}{3}$	x	$\dfrac{x}{3}$

Part done by A	plus	part done by B	equals	1 whole job.
$\dfrac{x}{2}$	$+$	$\dfrac{x}{3}$	$=$	1

Multiply each side by the LCD, 6.

$$6\left(\frac{x}{2} + \frac{x}{3}\right) = 6 \cdot 1$$

$$3x + 2x = 6$$

$$5x = 6$$

$$x = \tfrac{6}{5} \text{ or } 1\tfrac{1}{5}$$

Working together, the machines can complete the job in $\frac{6}{5}$ or $1\frac{1}{5}$ hours.

48. Let $C =$ the cost of a pizza
and $r =$ the radius of the pizza.
C varies directly as r^2, so

$$C = kr^2$$

for some constant k. Since $C = 6$ when $r = 7$, substitute these values in the equation and solve for k.

$$C = kr^2$$

$$6 = k(7)^2$$

$$6 = 49k$$

$$\tfrac{6}{49} = k$$

So $C = \frac{6}{49}r^2$. Now let $r = 9$.

$$C = \tfrac{6}{49}(9)^2 = \tfrac{6}{49}(81) = \tfrac{486}{49} \approx 9.92.$$

A pizza with a 9-inch radius should cost $9.92.

CHAPTER 8 ROOTS, RADICALS, AND ROOT FUNCTIONS

8.1 Radical Expressions and Graphs

1. $-\sqrt{16} = -(4) = -4$ **(E)**

3. $\sqrt[3]{-27} = -3$, because $(-3)^3 = -27$. **(D)**

5. $\sqrt[4]{81} = 3$, because $3^4 = 81$. **(A)**

7. $\sqrt{123.5} \approx \sqrt{121} = 11$, because $11^2 = 121$. **(C)**

9. The length $\sqrt{98}$ is closer to $\sqrt{100} = 10$ than to $\sqrt{81} = 9$. The width $\sqrt{26}$ is closer to $\sqrt{25} = 5$ than to $\sqrt{36} = 6$. Use the estimates $L = 10$ and $W = 5$ in $A = LW$ to find an estimate of the area.

$$A \approx 10 \cdot 5 = 50$$

Choice **C** is the best estimate.

11. **(a)** If $a > 0$, then $\sqrt{-a}$ is not a real number, so $-\sqrt{-a}$ is *not a real number*.

(b) If $a < 0$, then $\sqrt{-a}$ is a positive number and then $-\sqrt{-a}$ is a *negative* number.

(c) If $a = 0$, then $\sqrt{-a} = \sqrt{0} = 0$, so $-\sqrt{-a} = 0$.

13. $\sqrt{81} = 9$, because $9^2 = 81$. Notice that -9 is not an answer because the symbol $\sqrt{81}$ means the *nonnegative* square root of 81. However, the negative in front of the radical does lead to a negative answer since

$$-\sqrt{81} = -(9) = -9.$$

15. $\sqrt[3]{216} = 6$, because $6^3 = 216$.

17. $\sqrt[3]{-64} = -4$, because $(-4)^3 = -64$.

19. $\sqrt[3]{512} = 8$, because $8^3 = 512$, so $-\sqrt[3]{512} = -8$.

21. $\sqrt[4]{1296} = 6$, because $6^4 = 1296$.

23. $\sqrt[4]{16} = 2$, because $2^4 = 16$, so $-\sqrt[4]{16} = -2$.

25. $\sqrt[4]{-625}$ is not a real number since no real number to the fourth power equals -625. Any real number raised to the fourth power is 0 or positive.

27. $\sqrt[6]{64} = 2$, because $2^6 = 64$.

29. $\sqrt[6]{-32}$ is not a real number, because the index, 6, is even and the radicand, -32, is negative.

31. $\sqrt{\frac{64}{81}} = \frac{8}{9}$, because $\left(\frac{8}{9}\right)^2 = \frac{64}{81}$.

33. $\sqrt[3]{\frac{64}{27}} = \frac{4}{3}$, because $\left(\frac{4}{3}\right)^3 = \frac{64}{27}$.

35. $\sqrt[6]{\frac{1}{64}} = \frac{1}{2}$, because $\left(\frac{1}{2}\right)^6 = \frac{1}{64}$, so $-\sqrt[6]{\frac{1}{64}} = -\frac{1}{2}$.

37. $\sqrt{0.49} = 0.7$, because $(0.7)^2 = 0.49$.

39. $\sqrt[3]{0.001} = 0.1$, because $(0.1)^3 = 0.001$.

41. $f(x) = \sqrt{x+3}$
For the radicand to be nonnegative, we must have

$$x + 3 \geq 0 \quad \text{or} \quad x \geq -3.$$

Thus, the domain is $[-3, \infty)$.
The function values are positive or zero (the result of the radical), so the range is $[0, \infty)$.

x	$f(x) = \sqrt{x+3}$
-3	$\sqrt{-3+3} = 0$
-2	$\sqrt{-2+3} = 1$
1	$\sqrt{1+3} = 2$

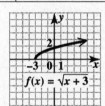

43. $f(x) = \sqrt{x} - 2$
For the radicand to be nonnegative, we must have

$$x \geq 0.$$

Note that the "-2" does not affect the domain, which is $[0, \infty)$.
The result of the radical is positive or zero, but the function values are 2 less than those values, so the range is $[-2, \infty)$.

x	$f(x) = \sqrt{x} - 2$
0	$\sqrt{0} - 2 = -2$
1	$\sqrt{1} - 2 = -1$
4	$\sqrt{4} - 2 = 0$

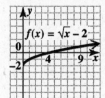

45. $f(x) = \sqrt[3]{x} - 3$

Since we can take the cube root of any real number, the domain is $(-\infty, \infty)$.
The result of a cube root can be any real number, so the range is $(-\infty, \infty)$. (The "-3" does not affect the range.)

x	$f(x) = \sqrt[3]{x} - 3$
-8	$\sqrt[3]{-8} - 3 = -5$
-1	$\sqrt[3]{-1} - 3 = -4$
0	$\sqrt[3]{0} - 3 = -3$
1	$\sqrt[3]{1} - 3 = -2$
8	$\sqrt[3]{8} - 3 = -1$

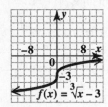

47. $f(x) = \sqrt[3]{x - 3}$

Both the domain and range are $(-\infty, \infty)$.

x	$f(x) = \sqrt[3]{x - 3}$
-5	$\sqrt[3]{-5 - 3} = -2$
2	$\sqrt[3]{2 - 3} = -1$
3	$\sqrt[3]{3 - 3} = 0$
4	$\sqrt[3]{4 - 3} = 1$
11	$\sqrt[3]{11 - 3} = 2$

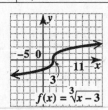

49. $\sqrt{12^2} = |12| = 12$

51. $\sqrt{(-10)^2} = |-10| = -(-10) = 10$

53. Since 6 is an even positive integer, $\sqrt[6]{a^6} = |a|$, so
$\sqrt[6]{(-2)^6} = |-2| = 2$.

55. Since 5 is odd, $\sqrt[5]{a^5} = a$, so
$$\sqrt[5]{(-9)^5} = -9.$$

57. $\sqrt[6]{(-5)^6} = |-5| = 5$, so $-\sqrt[6]{(-5)^6} = -5$.

59. Since the index is even, $\sqrt{x^2} = |x|$.

61. $\sqrt{(-z)^2} = |-z| = |z|$

63. Since the index is odd, $\sqrt[3]{x^3} = x$.

65. $\sqrt[3]{x^{15}} = \sqrt[3]{(x^5)^3} = x^5$ (3 is odd)

67. $\sqrt[6]{x^{30}} = \sqrt[6]{(x^5)^6} = |x^5|$, or $|x|^5$ (6 is even)

In Exercises 69–80, use a calculator and round to three decimal places

69. $\sqrt{9483} \approx 97.381$

71. $\sqrt{284.361} \approx 16.863$

73. $-\sqrt{82} \approx -9.055$

75. $\sqrt[3]{423} \approx 7.507$

77. $\sqrt[4]{100} \approx 3.162$

79. $\sqrt[5]{23.8} \approx 1.885$

81. $f = \dfrac{1}{2\pi\sqrt{LC}}$

$= \dfrac{1}{2\pi\sqrt{(7.237 \times 10^{-5})(2.5 \times 10^{-10})}}$

$\approx 1,183,235$

or about 1,183,000 cycles per second.

83. Since $H = 44 + 6 = 50$ ft, substitute 50 for H in the formula.

$$D = \sqrt{2H} = \sqrt{2 \cdot 50} = \sqrt{100} = 10$$

She will be able to see about 10 miles.

85. Let $a = 850$, $b = 925$, and $c = 1300$.
First find s.

$$s = \tfrac{1}{2}(a + b + c)$$
$$= \tfrac{1}{2}(850 + 925 + 1300)$$
$$= \tfrac{3075}{2} = 1537.5$$

Now find the area using Heron's formula.

$$A = \sqrt{s(s - a)(s - b)(s - c)}$$
$$= \sqrt{1537.5(687.5)(612.5)(237.5)}$$
$$\approx 392,128.8$$

The area of the Bermuda Triangle is about 392,000 square miles.

87. $I = \sqrt{\dfrac{2P}{L}}$

$= \sqrt{\dfrac{2(120)}{80}} = \sqrt{3} \approx 1.732$ amps

89. $x^5 \cdot x^{-1} \cdot x^{-3} = x^{5+(-1)+(-3)} = x^1$, or x

91. $(13x^0 y^5)(13x^4 y^3) = 13^{1+1} x^{0+4} y^{5+3}$
$= 13^2 x^4 y^8$, or $169x^4 y^8$

93. $\dfrac{5}{5^{-1}} = 5^{1-(-1)} = 5^2$, or 25

95. $\left(\dfrac{2}{3}\right)^{-3} = \left(\dfrac{3}{2}\right)^3 = \dfrac{3^3}{2^3}$, or $\dfrac{27}{8}$

8.2 Rational Exponents

1. $2^{1/2} = \sqrt[2]{2^1} = \sqrt{2}$ **(C)**

3. $-16^{1/2} = -\sqrt{16} = -(4) = -4$ **(A)**

5. $(-32)^{1/5} = [(-2)^5]^{1/5} = (-2)^1 = -2$ **(H)**

7. $4^{3/2} = [(2^2)^{1/2}]^3 = 2^3 = 8$ **(B)**

9. $-6^{2/4} = -(6^{1/2}) = -\sqrt{6}$ **(D)**

11. $169^{1/2} = (13^2)^{1/2} = 13^1 = 13$

We could use radical notation as follows:

$$169^{1/2} = \sqrt{169} = 13$$

13. $729^{1/3} = (9^3)^{1/3} = 9$

15. $16^{1/4} = (2^4)^{1/4} = 2$

17. $\left(\frac{64}{81}\right)^{1/2} = \left[\left(\frac{8}{9}\right)^2\right]^{1/2} = \frac{8}{9}$

19. $(-27)^{1/3} = [(-3)^3]^{1/3} = -3$

21. $(-144)^{1/2}$ is not a real number, because no real number squared equals -144.

23. $100^{3/2} = (100^{1/2})^3 = 10^3 = 1000$

25. $81^{3/4} = (81^{1/4})^3 = (\sqrt[4]{81})^3 = 3^3 = 27$

27. $-16^{5/2} = -(16^{1/2})^5 = -(4)^5 = -1024$

29. $(-8)^{4/3} = [(-8)^{1/3}]^4 = (\sqrt[3]{-8})^4$
$= (-2)^4 = 16$

31. $32^{-3/5} = \frac{1}{32^{3/5}} = \frac{1}{(32^{1/5})^3} = \frac{1}{(\sqrt[5]{32})^3}$
$= \frac{1}{2^3} = \frac{1}{8}$

33. $64^{-3/2} = \frac{1}{64^{3/2}} = \frac{1}{(64^{1/2})^3} = \frac{1}{8^3} = \frac{1}{512}$

35. $\left(\frac{125}{27}\right)^{-2/3} = \left(\frac{27}{125}\right)^{2/3}$
$= \left(\left[\left(\frac{3}{5}\right)^3\right]^{1/3}\right)^2$
$= \left(\frac{3}{5}\right)^2 = \frac{9}{25}$

37. $10^{1/2} = (\sqrt[2]{10})^1 = \sqrt{10}$

39. $8^{3/4} = (8^{1/4})^3 = (\sqrt[4]{8})^3$

41. $(9q)^{5/8} - (2x)^{2/3}$
$= [(9q)^{1/8}]^5 - [(2x)^{1/3}]^2$
$= (\sqrt[8]{9q})^5 - (\sqrt[3]{2x})^2$

43. $(2m)^{-3/2} = [(2m)^{1/2}]^{-3}$
$= (\sqrt{2m})^{-3}$
$= \frac{1}{(\sqrt{2m})^3}$

45. $(2y + x)^{2/3} = [(2y+x)^{1/3}]^2$
$= (\sqrt[3]{2y+x})^2$

47. $(3m^4 + 2k^2)^{-2/3}$
$= \frac{1}{(3m^4 + 2k^2)^{2/3}}$
$= \frac{1}{[(3m^4 + 2k^2)^{1/3}]^2}$
$= \frac{1}{(\sqrt[3]{3m^4 + 2k^2})^2}$

49. We are to show that, in general,
$$\sqrt{a^2 + b^2} \neq a + b.$$
When $a = 3$ and $b = 4$,
$$\sqrt{a^2 + b^2} = \sqrt{3^2 + 4^2}$$
$$= \sqrt{9 + 16}$$
$$= \sqrt{25} = 5,$$
but
$$a + b = 3 + 4 = 7.$$
Since $5 \neq 7$, $\sqrt{a^2 + b^2} \neq a + b$.

51. $\sqrt{2^{12}} = (2^{12})^{1/2} = 2^{12/2} = 2^6 = 64$

53. $\sqrt[3]{4^9} = 4^{9/3} = 4^3 = 64$

55. $\sqrt{x^{20}} = x^{20/2} = x^{10}$

57. $\sqrt[3]{x} \cdot \sqrt{x} = x^{1/3} \cdot x^{1/2} = x^{1/3+1/2} = x^{2/6+3/6}$
$= x^{5/6} = \sqrt[6]{x^5}$

59. $\dfrac{\sqrt[3]{t^4}}{\sqrt[5]{t^4}} = \dfrac{t^{4/3}}{t^{4/5}} = \dfrac{t^{20/15}}{t^{12/15}} = t^{20/15-12/15}$
$= t^{8/15} = \sqrt[15]{t^8}$

61. $3^{1/2} \cdot 3^{3/2} = 3^{1/2+3/2} = 3^{4/2} = 3^2 = 9$

63. $\dfrac{64^{5/3}}{64^{4/3}} = 64^{5/3-4/3} = 64^{1/3} = \sqrt[3]{64} = 4$

65. $y^{7/3} \cdot y^{-4/3} = y^{7/3+(-4/3)} = y^{3/3} = y$

67. $x^{2/3} \cdot x^{-1/4} = x^{2/3+(-1/4)} = x^{8/12-3/12} = x^{5/12}$

69. $\dfrac{k^{1/3}}{k^{2/3} \cdot k^{-1}} = \dfrac{k^{1/3}}{k^{-1/3}} = k^{1/3-(-1/3)} = k^{2/3}$

71. $\dfrac{(x^{1/4}y^{2/5})^{20}}{x^2} = \dfrac{x^{(1/4)\cdot 20}y^{(2/5)\cdot 20}}{x^2}$
$= \dfrac{x^5 y^8}{x^2}$
$= x^3 y^8$

73. $\dfrac{\left(x^{2/3}\right)^2}{\left(x^2\right)^{7/3}} = \dfrac{x^{4/3}}{x^{14/3}} = x^{4/3-14/3}$

$= x^{-10/3} = \dfrac{1}{x^{10/3}}$

75. $\dfrac{m^{3/4}n^{-1/4}}{\left(m^2 n\right)^{1/2}} = \dfrac{m^{3/4}n^{-1/4}}{m^1 n^{1/2}}$

$= m^{3/4-1}n^{-1/4-1/2}$

$= m^{3/4-4/4}n^{-1/4-2/4}$

$= m^{-1/4}n^{-3/4}$

$= \dfrac{1}{m^{1/4}n^{3/4}}$

77. $\dfrac{p^{1/5}p^{7/10}p^{1/2}}{\left(p^3\right)^{-1/5}} = \dfrac{p^{2/10+7/10+5/10}}{p^{-3/5}}$

$= \dfrac{p^{14/10}}{p^{-6/10}}$

$= p^{14/10-(-6/10)}$

$= p^{20/10} = p^2$

79. $\left(\dfrac{b^{-3/2}}{c^{-5/3}}\right)^2\left(b^{-1/4}c^{-1/3}\right)^{-1}$

$= \left(\dfrac{c^{5/3}}{b^{3/2}}\right)^2\left(b^{1/4}c^{1/3}\right)$

$= \dfrac{c^{10/3}}{b^3}\left(b^{1/4}c^{1/3}\right)$

$= \dfrac{c^{10/3}b^{1/4}c^{1/3}}{b^3}$

$= c^{10/3+1/3}b^{1/4-3}$

$= c^{11/3}b^{-11/4}$

$= \dfrac{c^{11/3}}{b^{11/4}}$

81. $\left(\dfrac{p^{-1/4}q^{-3/2}}{3^{-1}p^{-2}q^{-2/3}}\right)^{-2}$

$= \dfrac{p^{1/2}q^3}{3^2 p^4 q^{4/3}}$

$= \dfrac{q^{3-4/3}}{9p^{4-1/2}}$

$= \dfrac{q^{5/3}}{9p^{7/2}}$

83. $p^{2/3}\left(p^{1/3} + 2p^{4/3}\right)$

$= p^{2/3}p^{1/3} + p^{2/3}\left(2p^{4/3}\right)$

$= p^{2/3+1/3} + 2p^{2/3+4/3}$

$= p^{3/3} + 2p^{6/3}$

$= p^1 + 2p^2$

$= p + 2p^2$

85. $k^{1/4}\left(k^{3/2} - k^{1/2}\right)$

$= k^{1/4+3/2} - k^{1/4+1/2}$

$= k^{1/4+6/4} - k^{1/4+2/4}$

$= k^{7/4} - k^{3/4}$

87. $6a^{7/4}\left(a^{-7/4} + 3a^{-3/4}\right)$

$= 6a^{7/4+(-7/4)} + 18a^{7/4+(-3/4)}$

$= 6a^0 + 18a^{4/4}$

$= 6(1) + 18a^1 = 6 + 18a$

89. $\sqrt[5]{x^3} \cdot \sqrt[4]{x} = x^{3/5} \cdot x^{1/4}$

$= x^{3/5+1/4}$

$= x^{12/20+5/20}$

$= x^{17/20}$

91. $\dfrac{\sqrt{x^5}}{\sqrt{x^8}} = \dfrac{\left(x^5\right)^{1/2}}{\left(x^8\right)^{1/2}}$

$= x^{5/2-(8/2)}$

$= x^{-3/2} = \dfrac{1}{x^{3/2}}$

93. $\sqrt{y} \cdot \sqrt[3]{yz} = y^{1/2} \cdot (yz)^{1/3}$

$= y^{1/2}y^{1/3}z^{1/3}$

$= y^{1/2+1/3}z^{1/3}$

$= y^{3/6+2/6}z^{1/3}$

$= y^{5/6}z^{1/3}$

95. $\sqrt[4]{\sqrt[3]{m}} = \sqrt[4]{m^{1/3}} = \left(m^{1/3}\right)^{1/4} = m^{1/12}$

97. $\sqrt{\sqrt[3]{\sqrt[4]{x}}} = \sqrt{\sqrt[3]{x^{1/4}}}$

$= \sqrt{\left(x^{1/4}\right)^{1/3}}$

$= \left(x^{1/12}\right)^{1/2}$

$= x^{1/24}$

99. Use $T(D) = 0.07D^{3/2}$ with $D = 16$.

$T(16) = 0.07(16)^{3/2} = 0.07\left(16^{1/2}\right)^3$

$= 0.07(4)^3 = 0.07(64) = 4.48$

To the nearest tenth of an hour, time T is 4.5 hours.

101. Use a calculator and the formula

$W = 35.74 + 0.6215T - 35.75V^{4/25} + 0.4275TV^{4/25}$.

For $T = 30°F$ and $V = 15$ mph, $W \approx 19.0°F$. The table gives 19.0°F.

103. $\sqrt{25} \cdot \sqrt{36} = 5 \cdot 6 = 30$

$\sqrt{25 \cdot 36} = \sqrt{900} = 30$

The results are the same.

105. $\dfrac{\sqrt[3]{27}}{\sqrt[3]{729}} = \dfrac{3}{9} = \dfrac{1}{3}$

$\sqrt[3]{\dfrac{27}{729}} = \sqrt[3]{\dfrac{1}{27}} = \dfrac{1}{3}$

The results are the same.

8.3   Simplifying Radical Expressions

1. Does $2\sqrt{12} = \sqrt{48}$?

$2\sqrt{12} = 2\sqrt{4\cdot 3} = 2\sqrt{4}\cdot\sqrt{3} = 2\cdot 2\cdot\sqrt{3}$
$\qquad = 4\sqrt{3}$

$\sqrt{48} = \sqrt{16\cdot 3} = \sqrt{16}\cdot\sqrt{3} = 4\sqrt{3}$

The calculator approximation for each expression is 6.92820323. The statement is true.

3. Does $3\sqrt{8} = 2\sqrt{18}$?

$3\sqrt{8} = 3\sqrt{4\cdot 2} = 3\sqrt{4}\sqrt{2} = 3\cdot 2\sqrt{2} = 6\sqrt{2}$
$2\sqrt{18} = 2\sqrt{9\cdot 2} = 2\sqrt{9}\sqrt{2} = 2\cdot 3\sqrt{2} = 6\sqrt{2}$

The calculator approximation for each expression is 8.485281374. The statement is true.

5. **A.** $0.5 = \frac{1}{2}$, so $\sqrt{0.5} = \sqrt{\frac{1}{2}}$

B. $\frac{2}{4} = \frac{1}{2}$, so $\sqrt{\frac{2}{4}} = \sqrt{\frac{1}{2}}$

C. $\frac{3}{6} = \frac{1}{2}$, so $\sqrt{\frac{3}{6}} = \sqrt{\frac{1}{2}}$

D. $\dfrac{\sqrt{4}}{\sqrt{16}} = \sqrt{\frac{4}{16}} = \sqrt{\frac{1}{4}} \neq \sqrt{\frac{1}{2}}$

Choice **D** is not equal to $\sqrt{\frac{1}{2}}$.

7. $\sqrt{5}\cdot\sqrt{6} = \sqrt{5\cdot 6} = \sqrt{30}$

9. $\sqrt{14}\cdot\sqrt{x} = \sqrt{14\cdot x} = \sqrt{14x}$

11. $\sqrt{14}\cdot\sqrt{3pqr} = \sqrt{14\cdot 3pqr} = \sqrt{42pqr}$

13. $\sqrt[3]{7x}\cdot\sqrt[3]{2y} = \sqrt[3]{7x\cdot 2y} = \sqrt[3]{14xy}$

15. $\sqrt[4]{11}\cdot\sqrt[4]{3} = \sqrt[4]{11\cdot 3} = \sqrt[4]{33}$

17. $\sqrt[4]{2x}\cdot\sqrt[4]{3x^2} = \sqrt[4]{2x\cdot 3x^2} = \sqrt[4]{6x^3}$

19. $\sqrt[3]{7}\cdot\sqrt[4]{3}$ cannot be multiplied using the product rule, because the indexes (3 and 4) are different.

21. To multiply two radical expressions with the same index, multiply the radicands and keep the same index. For example, $\sqrt[3]{3}\cdot\sqrt[3]{5} = \sqrt[3]{15}$.

23. $\sqrt{\dfrac{64}{121}} = \dfrac{\sqrt{64}}{\sqrt{121}} = \dfrac{8}{11}$

25. $\sqrt{\dfrac{3}{25}} = \dfrac{\sqrt{3}}{\sqrt{25}} = \dfrac{\sqrt{3}}{5}$

27. $\sqrt{\dfrac{x}{25}} = \dfrac{\sqrt{x}}{\sqrt{25}} = \dfrac{\sqrt{x}}{5}$

29. $\sqrt{\dfrac{p^6}{81}} = \dfrac{\sqrt{p^6}}{\sqrt{81}} = \dfrac{\sqrt{(p^3)^2}}{9} = \dfrac{p^3}{9}$

31. $\sqrt[3]{-\dfrac{27}{64}} = \sqrt[3]{\dfrac{-27}{64}} = \dfrac{\sqrt[3]{-27}}{\sqrt[3]{64}} = \dfrac{-3}{4} = -\dfrac{3}{4}$

33. $\sqrt[3]{\dfrac{r^2}{8}} = \dfrac{\sqrt[3]{r^2}}{\sqrt[3]{8}} = \dfrac{\sqrt[3]{r^2}}{2}$

35. $-\sqrt[4]{\dfrac{81}{x^4}} = -\dfrac{\sqrt[4]{3^4}}{\sqrt[4]{x^4}} = -\dfrac{3}{x}$

37. $\sqrt[5]{\dfrac{1}{x^{15}}} = \dfrac{\sqrt[5]{1}}{\sqrt[5]{(x^3)^5}} = \dfrac{1}{x^3}$

39. $\sqrt{12} = \sqrt{4\cdot 3} = \sqrt{4}\cdot\sqrt{3} = 2\sqrt{3}$

41. $\sqrt{288} = \sqrt{144\cdot 2} = \sqrt{144}\cdot\sqrt{2} = 12\sqrt{2}$

43. $-\sqrt{32} = -\sqrt{16\cdot 2} = -\sqrt{16}\cdot\sqrt{2} = -4\sqrt{2}$

45. $-\sqrt{28} = -\sqrt{4\cdot 7} = -\sqrt{4}\cdot\sqrt{7} = -2\sqrt{7}$

47. $\sqrt{30}$ cannot be simplified further.

49. $\sqrt[3]{128} = \sqrt[3]{64\cdot 2} = \sqrt[3]{64}\cdot\sqrt[3]{2} = 4\sqrt[3]{2}$

51. $\sqrt[3]{-16} = \sqrt[3]{-8\cdot 2} = \sqrt[3]{-8}\cdot\sqrt[3]{2} = -2\sqrt[3]{2}$

53. $\sqrt[3]{40} = \sqrt[3]{8\cdot 5} = \sqrt[3]{8}\cdot\sqrt[3]{5} = 2\sqrt[3]{5}$

55. $-\sqrt[4]{512} = -\sqrt[4]{256\cdot 2} = -\sqrt[4]{4^4}\cdot\sqrt[4]{2} = -4\sqrt[4]{2}$

57. $\sqrt[5]{64} = \sqrt[5]{32\cdot 2} = \sqrt[5]{2^5}\cdot\sqrt[5]{2} = 2\sqrt[5]{2}$

59. His reasoning was incorrect. The radicand 14 must be written as a product of two factors (not a sum of two terms) where one of the two factors is a perfect cube.

61. $\sqrt{72k^2} = \sqrt{36k^2\cdot 2} = \sqrt{36k^2}\cdot\sqrt{2} = 6k\sqrt{2}$

63. $\sqrt{144x^3y^9} = \sqrt{144x^2y^8\cdot xy}$
$\qquad = \sqrt{(12xy^4)^2}\cdot\sqrt{xy}$
$\qquad = 12xy^4\sqrt{xy}$

65. $\sqrt{121x^6} = \sqrt{(11x^3)^2} = 11x^3$

67. $-\sqrt[3]{27t^{12}} = -\sqrt[3]{(3t^4)^3} = -3t^4$

69. $-\sqrt{100m^8z^4} = -\sqrt{(10m^4z^2)^2} = -10m^4z^2$

71. $-\sqrt[3]{-125a^6b^9c^{12}} = -\sqrt[3]{(-5a^2b^3c^4)^3}$
$\qquad = -(-5a^2b^3c^4)$
$\qquad = 5a^2b^3c^4$

73. $\sqrt[4]{\dfrac{1}{16}r^8t^{20}} = \sqrt[4]{\left(\dfrac{1}{2}r^2t^5\right)^4} = \dfrac{1}{2}r^2t^5$

75. $\sqrt{50x^3} = \sqrt{25x^2\cdot 2x} = \sqrt{(5x)^2}\cdot\sqrt{2x}$
$\qquad = 5x\sqrt{2x}$

77. $-\sqrt{500r^{11}} = -\sqrt{100r^{10} \cdot 5r}$

$= -\sqrt{(10r^5)^2} \cdot \sqrt{5r} = -10r^5\sqrt{5r}$

79. $\sqrt{13x^7y^8} = \sqrt{(x^6y^8)(13x)}$

$= \sqrt{x^6y^8} \cdot \sqrt{13x}$

$= x^3y^4\sqrt{13x}$

81. $\sqrt[3]{8z^6w^9} = \sqrt[3]{(2z^2w^3)^3} = 2z^2w^3$

83. $\sqrt[3]{-16z^5t^7} = \sqrt[3]{(-2^3z^3t^6)(2z^2t)}$

$= -2zt^2\sqrt[3]{2z^2t}$

85. $\sqrt[4]{81x^{12}y^{16}} = \sqrt[4]{(3x^3y^4)^4} = 3x^3y^4$

87. $-\sqrt[4]{162r^{15}s^{10}} = -\sqrt[4]{81r^{12}s^8(2r^3s^2)}$

$= -\sqrt[4]{81r^{12}s^8} \cdot \sqrt[4]{2r^3s^2}$

$= -3r^3s^2\sqrt[4]{2r^3s^2}$

89. $\sqrt{\dfrac{y^{11}}{36}} = \dfrac{\sqrt{y^{11}}}{\sqrt{36}} = \dfrac{\sqrt{y^{10}\cdot y}}{6} = \dfrac{y^5\sqrt{y}}{6}$

91. $\sqrt[3]{\dfrac{x^{16}}{27}} = \dfrac{\sqrt[3]{x^{15}\cdot x}}{\sqrt[3]{27}} = \dfrac{x^5\sqrt[3]{x}}{3}$

93. $\sqrt[4]{48^2} = 48^{2/4} = 48^{1/2} = \sqrt{48} = \sqrt{16\cdot 3}$

$= \sqrt{16} \cdot \sqrt{3} = 4\sqrt{3}$

95. $\sqrt[4]{25} = 25^{1/4} = (5^2)^{1/4} = 5^{2/4} = 5^{1/2} = \sqrt{5}$

97. $\sqrt[10]{x^{25}} = x^{25/10} = x^{5/2} = \sqrt{x^5}$

$= \sqrt{x^4 \cdot x} = x^2\sqrt{x}$

99. $\sqrt[3]{4} \cdot \sqrt{3}$

The least common index of 3 and 2 is 6. Write each radical as a sixth root.

$$\sqrt[3]{4} = 4^{1/3} = 4^{2/6} = \sqrt[6]{4^2} = \sqrt[6]{16}$$

$$\sqrt{3} = 3^{1/2} = 3^{3/6} = \sqrt[6]{3^3} = \sqrt[6]{27}$$

Therefore,

$$\sqrt[3]{4} \cdot \sqrt{3} = \sqrt[6]{16} \cdot \sqrt[6]{27}$$

$$= \sqrt[6]{16 \cdot 27} = \sqrt[6]{432}.$$

101. $\sqrt[4]{3} \cdot \sqrt[3]{4}$

The least common index of 4 and 3 is 12. Write each radical as a twelfth root.

$$\sqrt[4]{3} = 3^{1/4} = 3^{3/12} = \sqrt[12]{3^3} = \sqrt[12]{27}$$

$$\sqrt[3]{4} = 4^{1/3} = 4^{4/12} = \sqrt[12]{4^4} = \sqrt[12]{256}$$

Therefore,

$$\sqrt[4]{3} \cdot \sqrt[3]{4} = \sqrt[12]{27} \cdot \sqrt[12]{256}$$

$$= \sqrt[12]{27 \cdot 256} = \sqrt[12]{6912}.$$

103. $\sqrt{x} = x^{1/2} = x^{3/6} = \sqrt[6]{x^3}$

$\sqrt[3]{x} = x^{1/3} = x^{2/6} = \sqrt[6]{x^2}$

So $\sqrt{x} \cdot \sqrt[3]{x} = \sqrt[6]{x^3} \cdot \sqrt[6]{x^2}$

$= \sqrt[6]{x^3 \cdot x^2} = \sqrt[6]{x^5}.$

105. Substitute 3 for a and 4 for b in the Pythagorean formula to find c.

$$c^2 = a^2 + b^2$$

$$c = \sqrt{a^2 + b^2} = \sqrt{3^2 + 4^2}$$

$$= \sqrt{9 + 16} = \sqrt{25} = 5$$

The length of the hypotenuse is 5.

107. Substitute 12 for c and 4 for a in the Pythagorean formula to find b.

$$a^2 + b^2 = c^2$$

$$b = \sqrt{c^2 - a^2} = \sqrt{12^2 - 4^2}$$

$$= \sqrt{144 - 16} = \sqrt{128}$$

$$= \sqrt{64}\sqrt{2} = 8\sqrt{2}$$

The length of the unknown leg is $8\sqrt{2}$.

In Exercises 109–120, use the distance formula

$$d = \sqrt{(x_2 - x_1)^2 + (y_2 - y_1)^2}.$$

109. $(6, 13)$ and $(1, 1)$

$$d = \sqrt{(6 - 1)^2 + (13 - 1)^2}$$

$$= \sqrt{5^2 + 12^2} = \sqrt{25 + 144}$$

$$= \sqrt{169} = 13$$

111. $(-6, 5)$ and $(3, -4)$

$$d = \sqrt{(-6 - 3)^2 + [5 - (-4)]^2}$$

$$= \sqrt{(-9)^2 + (9)^2} = \sqrt{81 + 81}$$

$$= \sqrt{162} = \sqrt{81} \cdot \sqrt{2} = 9\sqrt{2}$$

113. $(-8, 2)$ and $(-4, 1)$

$$d = \sqrt{[-8 - (-4)]^2 + (2 - 1)^2}$$

$$= \sqrt{(-4)^2 + 1^2}$$

$$= \sqrt{16 + 1} = \sqrt{17}$$

115. $(4.7, 2.3)$ and $(1.7, -1.7)$

$$d = \sqrt{(4.7 - 1.7)^2 + [2.3 - (-1.7)]^2}$$

$$= \sqrt{3^2 + 4^2} = \sqrt{9 + 16}$$

$$= \sqrt{25} = 5$$

117. $\left(\sqrt{2}, \sqrt{6}\right)$ and $\left(-2\sqrt{2}, 4\sqrt{6}\right)$

$$d = \sqrt{\left[\sqrt{2} - \left(-2\sqrt{2}\right)\right]^2 + \left(\sqrt{6} - 4\sqrt{6}\right)^2}$$
$$= \sqrt{\left(3\sqrt{2}\right)^2 + \left(-3\sqrt{6}\right)^2}$$
$$= \sqrt{9 \cdot 2 + 9 \cdot 6} = \sqrt{18 + 54}$$
$$= \sqrt{72} = \sqrt{36} \cdot \sqrt{2} = 6\sqrt{2}$$

119. $(x + y, y)$ and $(x - y, x)$

$$d = \sqrt{[(x + y) - (x - y)]^2 + (y - x)^2}$$
$$= \sqrt{(2y)^2 + (y - x)^2}$$
$$= \sqrt{4y^2 + y^2 - 2xy + x^2}$$
$$= \sqrt{5y^2 - 2xy + x^2}$$

121. Since $\sqrt{a} = a^{1/2}$, the distance formula

$$d = \sqrt{(x_2 - x_1)^2 + (y_2 - y_1)^2}$$

may be expressed as

$$d = \left[(x_2 - x_1)^2 + (y_2 - y_1)^2\right]^{1/2}.$$

123. To find the lengths of the three sides of the triangle, use the distance formula to find the distance between each pair of points. Then add the distances to find the perimeter.

$$P = \sqrt{(-3 - 2)^2 + (-3 - 6)^2}$$
$$+ \sqrt{(2 - 6)^2 + (6 - 2)^2}$$
$$+ \sqrt{[6 - (-3)]^2 + [2 - (-3)]^2}$$
$$= \sqrt{(-5)^2 + (-9)^2} + \sqrt{(-4)^2 + 4^2}$$
$$+ \sqrt{9^2 + 5^2}$$
$$= \sqrt{25 + 81} + \sqrt{16 + 16} + \sqrt{81 + 25}$$
$$= \sqrt{106} + \sqrt{32} + \sqrt{106}$$
$$= 2\sqrt{106} + \sqrt{16} \cdot \sqrt{2}$$
$$= 2\sqrt{106} + 4\sqrt{2}$$

125. $d = 1.224\sqrt{h}$
$$= 1.224\sqrt{156} \approx 15.3 \text{ miles}$$

127. Substitute 21.7 for a and 16 for b in the Pythagorean formula to find c.

$$c^2 = a^2 + b^2 = (21.7)^2 + 16^2$$
$$c = \sqrt{726.89} \approx 26.96$$

To the nearest tenth of an inch, the length of the diagonal of the screen is 27.0 inches.

129. Use $f_1 = f_2\sqrt{\dfrac{F_1}{F_2}}$ with $F_1 = 300$, $F_2 = 60$, and $f_2 = 260$ to find f_1.

$$f_1 = 260\sqrt{\tfrac{300}{60}} = 260\sqrt{5} \approx 581$$

131. $13x^4 - 12x^3 + 9x^4 + 2x^3$
$$= 13x^4 + 9x^4 - 12x^3 + 2x^3$$
$$= 22x^4 - 10x^3$$

133. $9q^2 + 2q - 5q - q^2$
$$= 9q^2 - q^2 + 2q - 5q$$
$$= 8q^2 - 3q$$

8.4 Adding and Subtracting Radical Expressions

1. Only choice **B** has like radical terms, so it can be simplified without first simplifying the individual radical expressions.

$$3\sqrt{6} + 9\sqrt{6} = 12\sqrt{6}$$

3. $\sqrt{64} + \sqrt[3]{125} + \sqrt[4]{16} = \sqrt{8^2} + \sqrt[3]{5^3} + \sqrt[4]{2^4}$
$$= 8 + 5 + 2 = 15$$

This sum can be found easily since each radicand has a whole number power corresponding to the index of the radical; that is, each radical expression simplifies to a whole number.

5. Simplify each radical and subtract.

$$\sqrt{36} - \sqrt{100} = 6 - 10 = -4$$

7. $-2\sqrt{48} + 3\sqrt{75}$
$$= -2\sqrt{16 \cdot 3} + 3\sqrt{25 \cdot 3}$$
$$= -2 \cdot 4\sqrt{3} + 3 \cdot 5\sqrt{3}$$
$$= -8\sqrt{3} + 15\sqrt{3} = 7\sqrt{3}$$

9. $\sqrt[3]{16} + 4\sqrt[3]{54}$
$$= \sqrt[3]{8 \cdot 2} + 4\sqrt[3]{27 \cdot 2}$$
$$= \sqrt[3]{8}\sqrt[3]{2} + 4\sqrt[3]{27}\sqrt[3]{2}$$
$$= 2\sqrt[3]{2} + 4 \cdot 3\sqrt[3]{2}$$
$$= 2\sqrt[3]{2} + 12\sqrt[3]{2} = 14\sqrt[3]{2}$$

11. $\sqrt[4]{32} + 3\sqrt[4]{2}$
$$= \sqrt[4]{16 \cdot 2} + 3\sqrt[4]{2}$$
$$= \sqrt[4]{16}\sqrt[4]{2} + 3\sqrt[4]{2}$$
$$= 2\sqrt[4]{2} + 3\sqrt[4]{2} = 5\sqrt[4]{2}$$

13. $6\sqrt{18} - \sqrt{32} + 2\sqrt{50}$
$$= 6\sqrt{9 \cdot 2} - \sqrt{16 \cdot 2} + 2\sqrt{25 \cdot 2}$$
$$= 6 \cdot 3\sqrt{2} - 4\sqrt{2} + 2 \cdot 5\sqrt{2}$$
$$= 18\sqrt{2} - 4\sqrt{2} + 10\sqrt{2}$$
$$= 24\sqrt{2}$$

15. $5\sqrt{6} + 2\sqrt{10}$

The radicals differ and are already simplified, so the expression cannot be simplified further.

17. $2\sqrt{5} + 3\sqrt{20} + 4\sqrt{45}$

$\quad = 2\sqrt{5} + 3\sqrt{4 \cdot 5} + 4\sqrt{9 \cdot 5}$

$\quad = 2\sqrt{5} + 3 \cdot 2\sqrt{5} + 4 \cdot 3\sqrt{5}$

$\quad = 2\sqrt{5} + 6\sqrt{5} + 12\sqrt{5}$

$\quad = 20\sqrt{5}$

19. $\sqrt{72x} - \sqrt{8x}$

$\quad = \sqrt{36 \cdot 2x} - \sqrt{4 \cdot 2x}$

$\quad = 6\sqrt{2x} - 2\sqrt{2x}$

$\quad = 4\sqrt{2x}$

21. $3\sqrt{72m^2} - 5\sqrt{32m^2} - 3\sqrt{18m^2}$

$\quad = 3\sqrt{36m^2 \cdot 2} - 5\sqrt{16m^2 \cdot 2} - 3\sqrt{9m^2 \cdot 2}$

$\quad = 3 \cdot 6m\sqrt{2} - 5 \cdot 4m\sqrt{2} - 3 \cdot 3m\sqrt{2}$

$\quad = 18m\sqrt{2} - 20m\sqrt{2} - 9m\sqrt{2}$

$\quad = (18m - 20m - 9m)\sqrt{2} = -11m\sqrt{2}$

23. $2\sqrt[3]{16} - \sqrt[3]{54} = 2\sqrt[3]{8 \cdot 2} - \sqrt[3]{27 \cdot 2}$

$\quad = 2 \cdot 2\sqrt[3]{2} - 3\sqrt[3]{2}$

$\quad = 4\sqrt[3]{2} - 3\sqrt[3]{2}$

$\quad = \sqrt[3]{2}$

25. $2\sqrt[3]{27x} - 2\sqrt[3]{8x}$

$\quad = 2\sqrt[3]{27 \cdot x} - 2\sqrt[3]{8 \cdot x}$

$\quad = 2 \cdot 3\sqrt[3]{x} - 2 \cdot 2\sqrt[3]{x}$

$\quad = 6\sqrt[3]{x} - 4\sqrt[3]{x}$

$\quad = 2\sqrt[3]{x}$

27. $\sqrt[3]{x^2y} - \sqrt[3]{8x^2y} = \sqrt[3]{x^2y} - \sqrt[3]{8}\sqrt[3]{x^2y}$

$\quad = 1\sqrt[3]{x^2y} - 2\sqrt[3]{x^2y}$

$\quad = (1 - 2)\sqrt[3]{x^2y}$

$\quad = -\sqrt[3]{x^2y}$

29. $3x\sqrt[3]{xy^2} - 2\sqrt[3]{8x^4y^2}$

$\quad = 3x\sqrt[3]{xy^2} - 2\sqrt[3]{8x^3} \cdot \sqrt[3]{xy^2}$

$\quad = 3x\sqrt[3]{xy^2} - 2 \cdot 2x \cdot \sqrt[3]{xy^2}$

$\quad = (3x - 4x)\sqrt[3]{xy^2} = -x\sqrt[3]{xy^2}$

31. $5\sqrt[4]{32} + 3\sqrt[4]{162} = 5\sqrt[4]{16 \cdot 2} + 3\sqrt[4]{81 \cdot 2}$

$\quad = 5 \cdot 2\sqrt[4]{2} + 3 \cdot 3\sqrt[4]{2}$

$\quad = 10\sqrt[4]{2} + 9\sqrt[4]{2}$

$\quad = 19\sqrt[4]{2}$

33. $3\sqrt[4]{x^5y} - 2x\sqrt[4]{xy}$

$\quad = 3\sqrt[4]{x^4 \cdot xy} - 2x\sqrt[4]{xy}$

$\quad = 3x\sqrt[4]{xy} - 2x\sqrt[4]{xy}$

$\quad = (3x - 2x)\sqrt[4]{xy} = x\sqrt[4]{xy}$

35. $2\sqrt[4]{32a^3} + 5\sqrt[4]{2a^3}$

$\quad = 2\sqrt[4]{16} \cdot \sqrt[4]{2a^3} + 5\sqrt[4]{2a^3}$

$\quad = 2 \cdot 2 \cdot \sqrt[4]{2a^3} + 5\sqrt[4]{2a^3}$

$\quad = (4 + 5)\sqrt[4]{2a^3} = 9\sqrt[4]{2a^3}$

37. $\sqrt[3]{64xy^2} + \sqrt[3]{27x^4y^5}$

$\quad = \sqrt[3]{64 \cdot xy^2} + \sqrt[3]{27x^3y^3 \cdot xy^2}$

$\quad = 4\sqrt[3]{xy^2} + 3xy\sqrt[3]{xy^2}$

$\quad = (4 + 3xy)\sqrt[3]{xy^2}$

39. $4\sqrt[3]{x} - 6\sqrt{x}$ cannot be simplified further.

41. $2\sqrt[3]{8x^4} + 3\sqrt[4]{16x^5}$

$\quad = 2\sqrt[3]{8x^3 \cdot x} + 3\sqrt[4]{16x^4 \cdot x}$

$\quad = 2 \cdot 2x\sqrt[3]{x} + 3 \cdot 2x\sqrt[4]{x}$

$\quad = 4x\sqrt[3]{x} + 6x\sqrt[4]{x}$

43. $\sqrt{8} - \dfrac{\sqrt{64}}{\sqrt{16}} = \sqrt{4 \cdot 2} - \dfrac{8}{4}$

$\quad = \sqrt{4} \cdot \sqrt{2} - 2$

$\quad = 2\sqrt{2} - 2$

45. $\dfrac{2\sqrt{5}}{3} + \dfrac{\sqrt{5}}{6} = \dfrac{4\sqrt{5}}{6} + \dfrac{1\sqrt{5}}{6}$

$\quad = \dfrac{4\sqrt{5} + 1\sqrt{5}}{6}$

$\quad = \dfrac{5\sqrt{5}}{6}$

47. $\sqrt{\dfrac{8}{9}} + \sqrt{\dfrac{18}{36}} = \dfrac{\sqrt{8}}{\sqrt{9}} + \dfrac{\sqrt{18}}{\sqrt{36}}$

$\quad = \dfrac{\sqrt{4}\sqrt{2}}{3} + \dfrac{\sqrt{9}\sqrt{2}}{6}$

$\quad = \dfrac{2\sqrt{2}}{3} + \dfrac{3\sqrt{2}}{6}$

$\quad = \dfrac{4\sqrt{2}}{6} + \dfrac{3\sqrt{2}}{6}$

$\quad = \dfrac{4\sqrt{2} + 3\sqrt{2}}{6} = \dfrac{7\sqrt{2}}{6}$

49. $\dfrac{\sqrt{32}}{3} + \dfrac{2\sqrt{2}}{3} - \dfrac{\sqrt{2}}{\sqrt{9}}$

$\quad = \dfrac{\sqrt{16}\sqrt{2}}{3} + \dfrac{2\sqrt{2}}{3} - \dfrac{\sqrt{2}}{3}$

$\quad = \dfrac{4\sqrt{2} + 2\sqrt{2} - \sqrt{2}}{3} = \dfrac{5\sqrt{2}}{3}$

51. $3\sqrt{\dfrac{50}{9}} + 8\dfrac{\sqrt{2}}{\sqrt{8}} = 3\dfrac{\sqrt{50}}{\sqrt{9}} + 8\dfrac{\sqrt{2}}{2\sqrt{2}}$

$\quad = 3 \cdot \dfrac{5\sqrt{2}}{3} + 8 \cdot \dfrac{1}{2}$

$\quad = 5\sqrt{2} + 4$

53. $\sqrt{\dfrac{25}{x^8}} - \sqrt{\dfrac{9}{x^6}} = \dfrac{\sqrt{25}}{\sqrt{x^8}} - \dfrac{\sqrt{9}}{\sqrt{x^6}}$

$\qquad = \dfrac{5}{x^4} - \dfrac{3}{x^3}$

$\qquad = \dfrac{5}{x^4} - \dfrac{3 \cdot x}{x^3 \cdot x} \qquad LCD = x^4$

$\qquad = \dfrac{5 - 3x}{x^4}$

55. $3\sqrt[3]{\dfrac{m^5}{27}} - 2m\sqrt[3]{\dfrac{m^2}{64}}$

$\qquad = \dfrac{3\sqrt[3]{m^5}}{\sqrt[3]{27}} - \dfrac{2m\sqrt[3]{m^2}}{\sqrt[3]{64}}$

$\qquad = \dfrac{3\sqrt[3]{m^3}\sqrt[3]{m^2}}{3} - \dfrac{2m\sqrt[3]{m^2}}{4}$

$\qquad = \dfrac{m\sqrt[3]{m^2}}{1} - \dfrac{m\sqrt[3]{m^2}}{2}$

$\qquad = \dfrac{2m\sqrt[3]{m^2} - m\sqrt[3]{m^2}}{2} = \dfrac{m\sqrt[3]{m^2}}{2}$

57. $3\sqrt[3]{\dfrac{2}{x^6}} - 4\sqrt[3]{\dfrac{5}{x^9}} = 3\dfrac{\sqrt[3]{2}}{\sqrt[3]{x^6}} - 4\dfrac{\sqrt[3]{5}}{\sqrt[3]{x^9}}$

$\qquad = 3\dfrac{\sqrt[3]{2}}{x^2} - 4\dfrac{\sqrt[3]{5}}{x^3}$

$\qquad = \dfrac{3 \cdot x \cdot \sqrt[3]{2}}{x^2 \cdot x} - \dfrac{4\sqrt[3]{5}}{x^3}$

$\qquad\qquad\qquad\qquad LCD = x^3$

$\qquad = \dfrac{3x\sqrt[3]{2} - 4\sqrt[3]{5}}{x^3}$

59. $3\sqrt{32} - 2\sqrt{8} \approx 11.3137085$

$\qquad 8\sqrt{2} \approx 11.3137085$

Both calculator approximations are the same, supporting (but not proving) the truth of the statement.

61. Let $L = \sqrt{192} \approx \sqrt{196} = 14$ and $W = \sqrt{48} \approx \sqrt{49} = 7$. An estimate of the perimeter is $2L + 2W = 2(14) + 2(7) = 42$ meters. **(A)**

63. The perimeter, P, of a triangle is the sum of the measures of the sides.

$P = 3\sqrt{20} + 2\sqrt{45} + \sqrt{75}$

$\quad = 3\sqrt{4 \cdot 5} + 2\sqrt{9 \cdot 5} + \sqrt{25 \cdot 3}$

$\quad = 3 \cdot 2\sqrt{5} + 2 \cdot 3\sqrt{5} + 5\sqrt{3}$

$\quad = 6\sqrt{5} + 6\sqrt{5} + 5\sqrt{3}$

$\quad = 12\sqrt{5} + 5\sqrt{3}$

The perimeter is $\left(12\sqrt{5} + 5\sqrt{3}\right)$ inches.

65. To find the perimeter, add the lengths of the sides.

$4\sqrt{18} + \sqrt{108} + 2\sqrt{72} + 3\sqrt{12}$

$\quad = 4\sqrt{9}\sqrt{2} + \sqrt{36}\sqrt{3} + 2\sqrt{36}\sqrt{2} + 3\sqrt{4}\sqrt{3}$

$\quad = 4 \cdot 3\sqrt{2} + 6\sqrt{3} + 2 \cdot 6\sqrt{2} + 3 \cdot 2\sqrt{3}$

$\quad = 12\sqrt{2} + 6\sqrt{3} + 12\sqrt{2} + 6\sqrt{3}$

$\quad = 24\sqrt{2} + 12\sqrt{3}$

The perimeter is $\left(24\sqrt{2} + 12\sqrt{3}\right)$ inches.

67. $5xy\left(2x^2y^3 - 4x\right) = 5xy\left(2x^2y^3\right) - 5xy(4x)$

$\qquad\qquad\qquad\qquad = 10x^3y^4 - 20x^2y$

69. $\left(a^2 + b\right)\left(a^2 - b\right) = \left(a^2\right)^2 - b^2$

$\qquad\qquad\qquad\qquad = a^4 - b^2$

71. $\left(4x^3 + 3\right)^2 = \left(4x^3\right)^2 + 2 \cdot 4x^3 \cdot 3 + 3^2$

$\qquad\qquad\qquad = 16x^6 + 24x^3 + 9$

Now multiply by $4x^3 + 3$.

$$
\begin{array}{r}
16x^6 + 24x^3 + 9 \\
4x^3 + 3 \\
\hline
48x^6 + 72x^3 + 27 \\
64x^9 + 96x^6 + 36x^3 \\
\hline
64x^9 + 144x^6 + 108x^3 + 27
\end{array}
$$

Thus, $\left(4x^3 + 3\right)^3 = 64x^9 + 144x^6 + 108x^3 + 27$.

73. $\dfrac{8x^2 - 10x}{6x^2} = \dfrac{2x(4x - 5)}{2x \cdot 3x}$

$\qquad\qquad = \dfrac{4x - 5}{3x}$

8.5 Multiplying and Dividing Radical Expressions

1. $\left(x + \sqrt{y}\right)\left(x - \sqrt{y}\right)$

$\quad = x^2 - x\sqrt{y} + x\sqrt{y} - \left(\sqrt{y}\right)^2$

$\quad = x^2 - y$ **(E)**

3. $\left(\sqrt{x} + \sqrt{y}\right)\left(\sqrt{x} - \sqrt{y}\right)$

$\quad = \left(\sqrt{x}\right)^2 - \left(\sqrt{y}\right)^2$

$\quad = x - y$ **(A)**

5. $\left(\sqrt{x} - \sqrt{y}\right)^2$

$\quad = \left(\sqrt{x}\right)^2 - 2\sqrt{x}\sqrt{y} + \left(\sqrt{y}\right)^2$

$\quad = x - 2\sqrt{xy} + y$ **(D)**

7. $\sqrt{6}\left(3 + \sqrt{2}\right) = 3\sqrt{6} + \sqrt{6} \cdot \sqrt{2}$

$\qquad\qquad\qquad = 3\sqrt{6} + \sqrt{12}$

$\qquad\qquad\qquad = 3\sqrt{6} + \sqrt{4} \cdot \sqrt{3}$

$\qquad\qquad\qquad = 3\sqrt{6} + 2\sqrt{3}$

9. $5(\sqrt{72} - \sqrt{8}) = 5\sqrt{72} - 5\sqrt{8}$
$= 5\sqrt{36} \cdot \sqrt{2} - 5\sqrt{4} \cdot \sqrt{2}$
$= 5 \cdot 6 \cdot \sqrt{2} - 5 \cdot 2 \cdot \sqrt{2}$
$= 30\sqrt{2} - 10\sqrt{2}$
$= 20\sqrt{2}$

11. $(\sqrt{7} + 3)(\sqrt{7} - 3) = (\sqrt{7})^2 - 3^2$
$= 7 - 9 = -2$

13. $(\sqrt{2} - \sqrt{3})(\sqrt{2} + \sqrt{3}) = (\sqrt{2})^2 - (\sqrt{3})^2$
$= 2 - 3 = -1$

15. $(\sqrt{8} - \sqrt{2})(\sqrt{8} + \sqrt{2}) = (\sqrt{8})^2 - (\sqrt{2})^2$
$= 8 - 2 = 6$

17. $(\sqrt{2} + 1)(\sqrt{3} - 1)$
$\quad\quad\quad$ **F** $\quad\quad$ **O** $\quad\quad$ **I** $\quad\quad$ **L**
$= \sqrt{2} \cdot \sqrt{3} - 1\sqrt{2} + 1\sqrt{3} - 1 \cdot 1$
$= \sqrt{6} - \sqrt{2} + \sqrt{3} - 1$

19. $(\sqrt{11} - \sqrt{7})(\sqrt{2} + \sqrt{5})$
$\quad\quad\quad$ **F** $\quad\quad\quad$ **O** $\quad\quad\quad$ **I** $\quad\quad\quad$ **L**
$= \sqrt{11} \cdot \sqrt{2} + \sqrt{11} \cdot \sqrt{5} - \sqrt{7} \cdot \sqrt{2} - \sqrt{7} \cdot \sqrt{5}$
$= \sqrt{22} + \sqrt{55} - \sqrt{14} - \sqrt{35}$

21. $(2\sqrt{3} + \sqrt{5})(3\sqrt{3} - 2\sqrt{5})$
$= (2\sqrt{3})(3\sqrt{3}) + (2\sqrt{3})(-2\sqrt{5})$
$\quad + (\sqrt{5})(3\sqrt{3}) + (\sqrt{5})(-2\sqrt{5})$
$= 2 \cdot 3 \cdot 3 - 2 \cdot 2\sqrt{3 \cdot 5} + 3\sqrt{5 \cdot 3} - 2 \cdot 5$
$= 18 - 4\sqrt{15} + 3\sqrt{15} - 10$
$= 8 - \sqrt{15}$

23. $(\sqrt{5} + 2)^2 = (\sqrt{5})^2 + 2 \cdot \sqrt{5} \cdot 2 + 2^2$
$= 5 + 4\sqrt{5} + 4$
$= 9 + 4\sqrt{5}$

25. $(\sqrt{21} - \sqrt{5})^2$
$= (\sqrt{21})^2 - 2 \cdot \sqrt{21} \cdot \sqrt{5} + (\sqrt{5})^2$
$= 21 - 2\sqrt{105} + 5$
$= 26 - 2\sqrt{105}$

27. $(2 + \sqrt[3]{6})(2 - \sqrt[3]{6}) = 2^2 - (\sqrt[3]{6})^2$
$= 4 - \sqrt[3]{6} \cdot \sqrt[3]{6}$
$= 4 - \sqrt[3]{36}$

29. $(2 + \sqrt[3]{2})(4 - 2\sqrt[3]{2} + \sqrt[3]{4})$
$= 2 \cdot 4 - 2 \cdot 2\sqrt[3]{2} + 2\sqrt[3]{4}$
$\quad + 4\sqrt[3]{2} - 2\sqrt[3]{2} \cdot \sqrt[3]{2} + \sqrt[3]{2} \cdot \sqrt[3]{4}$
$= 8 - 4\sqrt[3]{2} + 2\sqrt[3]{4} + 4\sqrt[3]{2} - 2\sqrt[3]{4} + \sqrt[3]{8}$
$= 8 + 2$
$= 10$

31. $(3\sqrt{x} - \sqrt{5})(2\sqrt{x} + 1)$
$= (3\sqrt{x})(2\sqrt{x}) + 3\sqrt{x} - 2\sqrt{5}\sqrt{x} - \sqrt{5}$
$= 6x + 3\sqrt{x} - 2\sqrt{5x} - \sqrt{5}$

33. $(3\sqrt{r} - \sqrt{s})(3\sqrt{r} + \sqrt{s})$
$= 3\sqrt{r} \cdot 3\sqrt{r} + 3\sqrt{r} \cdot \sqrt{s}$
$\quad - 3\sqrt{r} \cdot \sqrt{s} - \sqrt{s} \cdot \sqrt{s}$
$= 9r - s$

35. $(\sqrt[3]{2y} - 5)(4\sqrt[3]{2y} + 1)$
$= \sqrt[3]{2y} \cdot 4\sqrt[3]{2y} + \sqrt[3]{2y} - 5 \cdot 4\sqrt[3]{2y} - 5$
$= 4\sqrt[3]{4y^2} + \sqrt[3]{2y} - 20\sqrt[3]{2y} - 5$
$= 4\sqrt[3]{4y^2} - 19\sqrt[3]{2y} - 5$

37. $(\sqrt{3x} + 2)(\sqrt{3x} - 2) = (\sqrt{3x})^2 - 2^2$
$= 3x - 4$

39. $(2\sqrt{x} + \sqrt{y})(2\sqrt{x} - \sqrt{y})$
$= (2\sqrt{x})^2 - (\sqrt{y})^2$
$= 2^2(\sqrt{x})^2 - y$
$= 4x - y$

41. $[(\sqrt{2} + \sqrt{3}) - \sqrt{6}][(\sqrt{2} + \sqrt{3}) + \sqrt{6}]$
$= (\sqrt{2} + \sqrt{3})^2 - (\sqrt{6})^2$
$= [(\sqrt{2})^2 + 2\sqrt{2}\sqrt{3} + (\sqrt{3})^2] - 6$
$= (2 + 2\sqrt{6} + 3) - 6$
$= 2\sqrt{6} - 1$

43. $\dfrac{7}{\sqrt{7}} = \dfrac{7 \cdot \sqrt{7}}{\sqrt{7} \cdot \sqrt{7}} = \dfrac{7\sqrt{7}}{7} = \sqrt{7}$

45. $\dfrac{15}{\sqrt{3}} = \dfrac{15 \cdot \sqrt{3}}{\sqrt{3} \cdot \sqrt{3}} = \dfrac{15\sqrt{3}}{3} = 5\sqrt{3}$

47. $\dfrac{\sqrt{3}}{\sqrt{2}} = \dfrac{\sqrt{3} \cdot \sqrt{2}}{\sqrt{2} \cdot \sqrt{2}} = \dfrac{\sqrt{6}}{2}$

49. $\dfrac{9\sqrt{3}}{\sqrt{5}} = \dfrac{9\sqrt{3} \cdot \sqrt{5}}{\sqrt{5} \cdot \sqrt{5}} = \dfrac{9\sqrt{15}}{5}$

51. $\dfrac{-6}{\sqrt{18}} = \dfrac{-6}{\sqrt{9 \cdot 2}} = \dfrac{-6}{3\sqrt{2}} = \dfrac{-2}{\sqrt{2}} = \dfrac{-2 \cdot \sqrt{2}}{\sqrt{2} \cdot \sqrt{2}}$
$= \dfrac{-2\sqrt{2}}{2} = -\sqrt{2}$

53. $\sqrt{\dfrac{7}{2}} = \dfrac{\sqrt{7}}{\sqrt{2}} = \dfrac{\sqrt{7} \cdot \sqrt{2}}{\sqrt{2} \cdot \sqrt{2}} = \dfrac{\sqrt{14}}{2}$

55. $-\sqrt{\dfrac{7}{50}} = -\dfrac{\sqrt{7}}{\sqrt{25 \cdot 2}} = -\dfrac{\sqrt{7}}{5\sqrt{2}}$
$= -\dfrac{\sqrt{7} \cdot \sqrt{2}}{5\sqrt{2} \cdot \sqrt{2}} = -\dfrac{\sqrt{14}}{5 \cdot 2} = -\dfrac{\sqrt{14}}{10}$

57. $\sqrt{\dfrac{24}{x}} = \dfrac{\sqrt{24}}{\sqrt{x}} = \dfrac{\sqrt{4 \cdot 6}}{\sqrt{x}} = \dfrac{2\sqrt{6}}{\sqrt{x}}$

$\qquad = \dfrac{2\sqrt{6} \cdot \sqrt{x}}{\sqrt{x} \cdot \sqrt{x}} = \dfrac{2\sqrt{6x}}{x}$

59. $\dfrac{-8\sqrt{3}}{\sqrt{k}} = \dfrac{-8\sqrt{3} \cdot \sqrt{k}}{\sqrt{k} \cdot \sqrt{k}} = \dfrac{-8\sqrt{3k}}{k}$

61. $-\sqrt{\dfrac{150m^5}{n^3}} = \dfrac{-\sqrt{150m^5}}{\sqrt{n^3}}$

$\qquad = \dfrac{-\sqrt{25m^4 \cdot 6m}}{\sqrt{n^2 \cdot n}} = \dfrac{-5m^2\sqrt{6m}}{n\sqrt{n}}$

$\qquad = \dfrac{-5m^2\sqrt{6m} \cdot \sqrt{n}}{n\sqrt{n} \cdot \sqrt{n}}$

$\qquad = \dfrac{-5m^2\sqrt{6mn}}{n \cdot n}$

$\qquad = \dfrac{-5m^2\sqrt{6mn}}{n^2}$

63. $\sqrt{\dfrac{288x^7}{y^9}} = \dfrac{\sqrt{288x^7}}{\sqrt{y^9}} = \dfrac{\sqrt{144x^6 \cdot 2x}}{\sqrt{y^8 \cdot y}}$

$\qquad = \dfrac{12x^3\sqrt{2x}}{y^4\sqrt{y}} = \dfrac{12x^3\sqrt{2x} \cdot \sqrt{y}}{y^4\sqrt{y} \cdot \sqrt{y}}$

$\qquad = \dfrac{12x^3\sqrt{2xy}}{y^4 \cdot y} = \dfrac{12x^3\sqrt{2xy}}{y^5}$

65. $\dfrac{5\sqrt{2m}}{\sqrt{y^3}} = \dfrac{5\sqrt{2m}}{\sqrt{y^3}} \cdot \dfrac{\sqrt{y}}{\sqrt{y}}$

$\qquad = \dfrac{5\sqrt{2my}}{\sqrt{y^4}}$

$\qquad = \dfrac{5\sqrt{2my}}{y^2}$

67. $-\sqrt{\dfrac{48k^2}{z}} = -\dfrac{\sqrt{16k^2} \cdot \sqrt{3}}{\sqrt{z}} \cdot \dfrac{\sqrt{z}}{\sqrt{z}}$

$\qquad = -\dfrac{4k\sqrt{3z}}{z}$

69. $\sqrt[3]{\dfrac{2}{3}} = \dfrac{\sqrt[3]{2} \cdot \sqrt[3]{9}}{\sqrt[3]{3} \cdot \sqrt[3]{9}} = \dfrac{\sqrt[3]{18}}{\sqrt[3]{27}} = \dfrac{\sqrt[3]{18}}{3}$

71. $\sqrt[3]{\dfrac{4}{9}} = \dfrac{\sqrt[3]{4}}{\sqrt[3]{9}} = \dfrac{\sqrt[3]{4}}{\sqrt[3]{3^2}} = \dfrac{\sqrt[3]{4} \cdot \sqrt[3]{3}}{\sqrt[3]{3^2} \cdot \sqrt[3]{3}}$

$\qquad = \dfrac{\sqrt[3]{12}}{\sqrt[3]{3^3}} = \dfrac{\sqrt[3]{12}}{3}$

73. $\sqrt[3]{\dfrac{9}{32}} = \dfrac{\sqrt[3]{9}}{\sqrt[3]{32}} = \dfrac{\sqrt[3]{9}}{\sqrt[3]{8} \cdot \sqrt[3]{4}} \cdot \dfrac{\sqrt[3]{2}}{\sqrt[3]{2}}$

$\qquad = \dfrac{\sqrt[3]{9 \cdot 2}}{2 \cdot \sqrt[3]{8}}$

$\qquad = \dfrac{\sqrt[3]{18}}{4}$

75. $-\sqrt[3]{\dfrac{2p}{r^2}} = -\dfrac{\sqrt[3]{2p}}{\sqrt[3]{r^2}} = -\dfrac{\sqrt[3]{2p} \cdot \sqrt[3]{r}}{\sqrt[3]{r^2} \cdot \sqrt[3]{r}}$

$\qquad = -\dfrac{\sqrt[3]{2pr}}{\sqrt[3]{r^3}} = -\dfrac{\sqrt[3]{2pr}}{r}$

77. $\sqrt[3]{\dfrac{x^6}{y}} = \dfrac{\sqrt[3]{x^6}}{\sqrt[3]{y}} = \dfrac{x^2}{\sqrt[3]{y}} \cdot \dfrac{\sqrt[3]{y^2}}{\sqrt[3]{y^2}}$

$\qquad = \dfrac{x^2\sqrt[3]{y^2}}{\sqrt[3]{y^3}}$

$\qquad = \dfrac{x^2\sqrt[3]{y^2}}{y}$

79. $\sqrt[4]{\dfrac{16}{x}} = \dfrac{\sqrt[4]{16}}{\sqrt[4]{x}} = \dfrac{2}{\sqrt[4]{x}} = \dfrac{2 \cdot \sqrt[4]{x^3}}{\sqrt[4]{x} \cdot \sqrt[4]{x^3}}$

$\qquad = \dfrac{2\sqrt[4]{x^3}}{\sqrt[4]{x^4}} = \dfrac{2\sqrt[4]{x^3}}{x}$

81. $\sqrt[4]{\dfrac{2y}{z}} = \dfrac{\sqrt[4]{2y}}{\sqrt[4]{z}} \cdot \dfrac{\sqrt[4]{z^3}}{\sqrt[4]{z^3}}$

$\qquad = \dfrac{\sqrt[4]{2yz^3}}{z}$

83. $\dfrac{3}{4 + \sqrt{5}}$

Multiply both the numerator and denominator by the conjugate of the denominator, $4 - \sqrt{5}$.

$\qquad = \dfrac{3(4 - \sqrt{5})}{(4 + \sqrt{5})(4 - \sqrt{5})}$

$\qquad = \dfrac{3(4 - \sqrt{5})}{16 - 5} = \dfrac{3(4 - \sqrt{5})}{11}$

85. $\dfrac{\sqrt{8}}{3 - \sqrt{2}}$

Multiply both the numerator and denominator by the conjugate of the denominator, $3 + \sqrt{2}$.

$\qquad = \dfrac{\sqrt{4}\sqrt{2}(3 + \sqrt{2})}{(3 - \sqrt{2})(3 + \sqrt{2})}$

$\qquad = \dfrac{2\sqrt{2}(3 + \sqrt{2})}{3^2 - 2}$

$\qquad = \dfrac{2 \cdot 3\sqrt{2} + 2 \cdot 2}{7}$

$\qquad = \dfrac{6\sqrt{2} + 4}{7}$

87. $\dfrac{2}{3\sqrt{5}+2\sqrt{3}}$

Multiply both the numerator and denominator by the conjugate of the denominator, $3\sqrt{5}-2\sqrt{3}$.

$$= \dfrac{2\left(3\sqrt{5}-2\sqrt{3}\right)}{\left(3\sqrt{5}+2\sqrt{3}\right)\left(3\sqrt{5}-2\sqrt{3}\right)}$$

$$= \dfrac{2\left(3\sqrt{5}-2\sqrt{3}\right)}{3^2\cdot5-2^2\cdot3}$$

$$= \dfrac{2\left(3\sqrt{5}-2\sqrt{3}\right)}{45-12} = \dfrac{2\left(3\sqrt{5}-2\sqrt{3}\right)}{33}$$

89. $\dfrac{\sqrt{2}-\sqrt{3}}{\sqrt{6}-\sqrt{5}}$

Multiply both the numerator and denominator by the conjugate of the denominator, $\sqrt{6}+\sqrt{5}$.

$$= \dfrac{\left(\sqrt{2}-\sqrt{3}\right)\left(\sqrt{6}+\sqrt{5}\right)}{\left(\sqrt{6}-\sqrt{5}\right)\left(\sqrt{6}+\sqrt{5}\right)}$$

$$= \dfrac{\sqrt{12}+\sqrt{10}-\sqrt{18}-\sqrt{15}}{\left(\sqrt{6}\right)^2-\left(\sqrt{5}\right)^2}$$

$$= \dfrac{\sqrt{4}\cdot\sqrt{3}+\sqrt{10}-\sqrt{9}\cdot\sqrt{2}-\sqrt{15}}{6-5}$$

$$= 2\sqrt{3}+\sqrt{10}-3\sqrt{2}-\sqrt{15}$$

91. $\dfrac{m-4}{\sqrt{m}+2}$

Multiply both the numerator and denominator by the conjugate of the denominator, $\sqrt{m}-2$.

$$= \dfrac{(m-4)\left(\sqrt{m}-2\right)}{\left(\sqrt{m}+2\right)\left(\sqrt{m}-2\right)}$$

$$= \dfrac{(m-4)\left(\sqrt{m}-2\right)}{m-4}$$

$$= \sqrt{m}-2$$

93. $\dfrac{4}{\sqrt{x}-2\sqrt{y}}$

Multiply both the numerator and denominator by the conjugate of the denominator, $\sqrt{x}+2\sqrt{y}$.

$$= \dfrac{4\left(\sqrt{x}+2\sqrt{y}\right)}{\left(\sqrt{x}-2\sqrt{y}\right)\left(\sqrt{x}+2\sqrt{y}\right)}$$

$$= \dfrac{4\left(\sqrt{x}+2\sqrt{y}\right)}{x-4y}$$

95. $\dfrac{\sqrt{x}-\sqrt{y}}{\sqrt{x}+\sqrt{y}}$

Multiply both the numerator and denominator by the conjugate of the denominator, $\sqrt{x}-\sqrt{y}$.

$$= \dfrac{\left(\sqrt{x}-\sqrt{y}\right)\left(\sqrt{x}-\sqrt{y}\right)}{\left(\sqrt{x}+\sqrt{y}\right)\left(\sqrt{x}-\sqrt{y}\right)}$$

$$= \dfrac{\left(\sqrt{x}\right)^2-2\sqrt{x}\sqrt{y}+\left(\sqrt{y}\right)^2}{\left(\sqrt{x}\right)^2-\left(\sqrt{y}\right)^2}$$

$$= \dfrac{x-2\sqrt{xy}+y}{x-y}$$

97. $\dfrac{5\sqrt{k}}{2\sqrt{k}+\sqrt{q}}$

Multiply both the numerator and denominator by the conjugate of the denominator, $2\sqrt{k}-\sqrt{q}$.

$$= \dfrac{5\sqrt{k}\left(2\sqrt{k}-\sqrt{q}\right)}{\left(2\sqrt{k}+\sqrt{q}\right)\left(2\sqrt{k}-\sqrt{q}\right)}$$

$$= \dfrac{5\sqrt{k}\left(2\sqrt{k}-\sqrt{q}\right)}{4k-q}$$

99. $\dfrac{30-20\sqrt{6}}{10} = \dfrac{10\left(3-2\sqrt{6}\right)}{10} = 3-2\sqrt{6}$

101. $\dfrac{3-3\sqrt{5}}{3} = \dfrac{3\left(1-\sqrt{5}\right)}{3} = 1-\sqrt{5}$

103. $\dfrac{16-4\sqrt{8}}{12} = \dfrac{16-4\left(2\sqrt{2}\right)}{12} = \dfrac{16-8\sqrt{2}}{12}$

$$= \dfrac{4\left(4-2\sqrt{2}\right)}{4\cdot3} = \dfrac{4-2\sqrt{2}}{3}$$

105. $\dfrac{6p+\sqrt{24p^3}}{3p}$

$$= \dfrac{6p+\sqrt{4p^2\cdot6p}}{3p} = \dfrac{6p+2p\sqrt{6p}}{3p}$$

$$= \dfrac{p\left(6+2\sqrt{6p}\right)}{3p} = \dfrac{6+2\sqrt{6p}}{3}$$

107. $\dfrac{1}{\sqrt{x+y}} = \dfrac{1}{\sqrt{x+y}}\cdot\dfrac{\sqrt{x+y}}{\sqrt{x+y}}$

$$= \dfrac{1\cdot\sqrt{x+y}}{\left(\sqrt{x+y}\right)^2}$$

$$= \dfrac{\sqrt{x+y}}{x+y}$$

109. $\dfrac{p}{\sqrt{p+2}} = \dfrac{p}{\sqrt{p+2}}\cdot\dfrac{\sqrt{p+2}}{\sqrt{p+2}}$

$$= \dfrac{p\sqrt{p+2}}{p+2}$$

111. $\dfrac{1}{\sqrt{2}} \cdot \dfrac{\sqrt{3}}{2} - \dfrac{1}{\sqrt{2}} \cdot \dfrac{1}{2} = \dfrac{\sqrt{3}}{2\sqrt{2}} - \dfrac{1}{2\sqrt{2}}$

$\qquad = \dfrac{\sqrt{3}-1}{2\sqrt{2}} = \dfrac{\left(\sqrt{3}-1\right)\sqrt{2}}{\left(2\sqrt{2}\right)\sqrt{2}}$

$\qquad = \dfrac{\sqrt{6}-\sqrt{2}}{2 \cdot 2} = \dfrac{\sqrt{6}-\sqrt{2}}{4}$

Using a calculator,

$\qquad \dfrac{1}{\sqrt{2}} \cdot \dfrac{\sqrt{3}}{2} - \dfrac{1}{\sqrt{2}} \cdot \dfrac{1}{2} \approx 0.2588190451$ and

$\qquad \dfrac{\sqrt{6}-\sqrt{2}}{4} \approx 0.2588190451.$

113. $\dfrac{6-\sqrt{2}}{4} = \dfrac{6-\sqrt{2}}{4} \cdot \dfrac{6+\sqrt{2}}{6+\sqrt{2}}$

$\qquad = \dfrac{6^2 - \left(\sqrt{2}\right)^2}{4\left(6+\sqrt{2}\right)}$

$\qquad = \dfrac{36-2}{4\left(6+\sqrt{2}\right)}$

$\qquad = \dfrac{34}{4\left(6+\sqrt{2}\right)}$

$\qquad = \dfrac{17 \cdot 2}{2 \cdot 2\left(6+\sqrt{2}\right)}$

$\qquad = \dfrac{17}{2\left(6+\sqrt{2}\right)}$

115. $\dfrac{3\sqrt{a}+\sqrt{b}}{b} = \dfrac{3\sqrt{a}+\sqrt{b}}{b} \cdot \dfrac{3\sqrt{a}-\sqrt{b}}{3\sqrt{a}-\sqrt{b}}$

$\qquad = \dfrac{\left(3\sqrt{a}\right)^2 - \left(\sqrt{b}\right)^2}{b\left(3\sqrt{a}-\sqrt{b}\right)}$

$\qquad = \dfrac{9a-b}{b\left(3\sqrt{a}-\sqrt{b}\right)}$

117. $-8t + 7 = 4$

$\qquad -8t = -3$

$\qquad t = \dfrac{3}{8}$

The solution set is $\left\{\dfrac{3}{8}\right\}$.

119. $\qquad 6x^2 - 7x = 3$

$\qquad 6x^2 - 7x - 3 = 0$

$\qquad (3x+1)(2x-3) = 0$

$\quad 3x+1 = 0 \qquad\text{or}\qquad 2x-3 = 0$

$\qquad 3x = -1 \qquad\text{or}\qquad 2x = 3$

$\qquad x = -\dfrac{1}{3} \qquad\text{or}\qquad x = \dfrac{3}{2}$

The solution set is $\left\{-\dfrac{1}{3}, \dfrac{3}{2}\right\}$.

121. $(2x+5)^2 = (2x)^2 + 2 \cdot 2x \cdot 5 + 5^2$

$\qquad = 4x^2 + 20x + 25$

123. $\left(\sqrt{x^4 + 2x^2 + 5}\right)^2 = x^4 + 2x^2 + 5$

125. $\sqrt{4-x} = x+2$

$\quad \sqrt{4-0} = 0+2 \quad ? \quad$ *Let $x = 0$.*

$\qquad \sqrt{4} = 2 \qquad ?$

$\qquad 2 = 2 \qquad\qquad$ *True*

127. $\sqrt{x^2 - 4x + 9} = x - 1$

$\quad \sqrt{4^2 - 4(4) + 9} = 4 - 1 \quad ? \quad$ *Let $x = 4$.*

$\qquad \sqrt{16 - 16 + 9} = 3 \qquad ?$

$\qquad \sqrt{9} = 3 \qquad ?$

$\qquad 3 = 3 \qquad\qquad$ *True*

Summary Exercises on Operations with Radicals and Rational Exponents

1. $6\sqrt{10} - 12\sqrt{10} = (6-12)\sqrt{10}$

$\qquad = -6\sqrt{10}$

3. $\left(1-\sqrt{3}\right)\left(2+\sqrt{6}\right)$

$\qquad$ **F**   **O**   **I**   **L**

$\qquad = 2 + \sqrt{6} - 2\sqrt{3} - \sqrt{18}$

$\qquad = 2 + \sqrt{6} - 2\sqrt{3} - \sqrt{9} \cdot \sqrt{2}$

$\qquad = 2 + \sqrt{6} - 2\sqrt{3} - 3\sqrt{2}$

5. $\left(3\sqrt{5} + 2\sqrt{7}\right)^2$

$\qquad = \left(3\sqrt{5}\right)^2 + 2\left(3\sqrt{5}\right)\left(2\sqrt{7}\right) + \left(2\sqrt{7}\right)^2$

$\qquad = 9 \cdot 5 + 12\sqrt{35} + 4 \cdot 7$

$\qquad = 45 + 12\sqrt{35} + 28$

$\qquad = 73 + 12\sqrt{35}$

7. $\dfrac{8}{\sqrt{7}+\sqrt{5}} = \dfrac{8}{\sqrt{7}+\sqrt{5}} \cdot \dfrac{\sqrt{7}-\sqrt{5}}{\sqrt{7}-\sqrt{5}}$

$\qquad = \dfrac{8\left(\sqrt{7}-\sqrt{5}\right)}{7-5}$

$\qquad = \dfrac{8\left(\sqrt{7}-\sqrt{5}\right)}{2}$

$\qquad = 4\left(\sqrt{7}-\sqrt{5}\right)$

9. $\left(\sqrt{5}+7\right)\left(\sqrt{5}-7\right) = \left(\sqrt{5}\right)^2 - 7^2$

$\qquad = 5 - 49$

$\qquad = -44$

11. $\sqrt[3]{8a^3b^5c^9} = \sqrt[3]{8a^3b^3c^9} \cdot \sqrt[3]{b^2}$

$\qquad = 2abc^3 \sqrt[3]{b^2}$

13. $\dfrac{3}{\sqrt{5}+2} = \dfrac{3}{\sqrt{5}+2} \cdot \dfrac{\sqrt{5}-2}{\sqrt{5}-2}$

$\qquad = \dfrac{3\left(\sqrt{5}-2\right)}{5-4}$

$\qquad = 3\left(\sqrt{5}-2\right)$

15. $\dfrac{16\sqrt{3}}{5\sqrt{12}} = \dfrac{16\sqrt{3}}{5\cdot\sqrt{4}\cdot\sqrt{3}}$

$= \dfrac{16}{5\cdot 2}$

$= \frac{8}{5}$

17. $\dfrac{-10}{\sqrt[3]{10}} = \dfrac{-10}{\sqrt[3]{10}} \cdot \dfrac{\sqrt[3]{100}}{\sqrt[3]{100}}$

$= \dfrac{-10\sqrt[3]{100}}{\sqrt[3]{1000}}$

$= \dfrac{-10\sqrt[3]{100}}{10} = -\sqrt[3]{100}$

19. $\sqrt{12x} - \sqrt{75x} = \sqrt{4}\cdot\sqrt{3x} - \sqrt{25}\cdot\sqrt{3x}$

$= 2\sqrt{3x} - 5\sqrt{3x}$

$= -3\sqrt{3x}$

21. $\sqrt[3]{\dfrac{13}{81}} = \dfrac{\sqrt[3]{13}}{\sqrt[3]{81}}$

$= \dfrac{\sqrt[3]{13}}{\sqrt[3]{27}\cdot\sqrt[3]{3}} \cdot \dfrac{\sqrt[3]{9}}{\sqrt[3]{9}}$

$= \dfrac{\sqrt[3]{13\cdot 9}}{3\cdot\sqrt[3]{27}}$

$= \dfrac{\sqrt[3]{117}}{3\cdot 3} = \dfrac{\sqrt[3]{117}}{9}$

23. $\dfrac{6}{\sqrt[4]{3}} = \dfrac{6}{\sqrt[4]{3}} \cdot \dfrac{\sqrt[4]{3^3}}{\sqrt[4]{3^3}}$

$= \dfrac{6\sqrt[4]{27}}{3} = 2\sqrt[4]{27}$

25. $\sqrt[3]{\dfrac{x^2 y}{x^{-3}y^4}} = \sqrt[3]{x^{2-(-3)}y^{1-4}}$

$= \sqrt[3]{x^5 y^{-3}}$

$= \dfrac{\sqrt[3]{x^5}}{\sqrt[3]{y^3}}$

$= \dfrac{\sqrt[3]{x^3 \cdot x^2}}{y} = \dfrac{x\sqrt[3]{x^2}}{y}$

27. $\dfrac{x^{-2/3}y^{4/5}}{x^{-5/3}y^{-2/5}} = x^{-2/3-(-5/3)}y^{4/5-(-2/5)}$

$= x^{3/3}y^{6/5}$

$= xy^{6/5}$

29. $(125x^3)^{-2/3} = \dfrac{1}{(125x^3)^{2/3}}$

$= \dfrac{1}{\left(\sqrt[3]{125x^3}\right)^2}$

$= \dfrac{1}{(5x)^2}$

$= \dfrac{1}{25x^2}$

31. $\sqrt[3]{16x^2} - \sqrt[3]{54x^2} + \sqrt[3]{128x^2}$

$= \sqrt[3]{8}\cdot\sqrt[3]{2x^2} - \sqrt[3]{27}\cdot\sqrt[3]{2x^2} + \sqrt[3]{64}\cdot\sqrt[3]{2x^2}$

$= 2\sqrt[3]{2x^2} - 3\sqrt[3]{2x^2} + 4\sqrt[3]{2x^2}$

$= (2-3+4)\sqrt[3]{2x^2}$

$= 3\sqrt[3]{2x^2}$

33. $\left(\sqrt{74} - \sqrt{73}\right)\left(\sqrt{74} + \sqrt{73}\right) = 74 - 73 = 1$

35. $(3x^{-2/3}y^{1/2})(-2x^{5/8}y^{-1/3})$

$= 3(-2)x^{-2/3+5/8}y^{1/2+(-1/3)}$

$= -6x^{-16/24+15/24}y^{3/6-2/6}$

$= -6x^{-1/24}y^{1/6}$

$= -\dfrac{6y^{1/6}}{x^{1/24}}$

37. **(a)** $\sqrt{64} = 8$

(b) $x^2 = 64$

$\qquad x = -\sqrt{64} \quad$ or $\quad x = \sqrt{64}$

$\qquad x = -8 \quad\quad$ or $\quad x = 8$

The solution set is $\{-8, 8\}$.

39. **(a)** $x^2 = 16$

$\qquad x = -\sqrt{16} \quad$ or $\quad x = \sqrt{16}$

$\qquad x = -4 \quad\quad$ or $\quad x = 4$

The solution set is $\{-4, 4\}$.

(b) $-\sqrt{16} = -\left(\sqrt{16}\right) = -4$

41. **(a)** Since $\left(\frac{9}{11}\right)^2 = \frac{81}{121}$, $-\sqrt{\frac{81}{121}} = -\frac{9}{11}$.

(b) $x^2 = \dfrac{81}{121}$

$\qquad x = -\sqrt{\frac{81}{121}} \quad$ or $\quad x = \sqrt{\frac{81}{121}}$

$\qquad x = -\frac{9}{11} \quad\quad$ or $\quad x = \frac{9}{11}$

The solution set is $\left\{-\frac{9}{11}, \frac{9}{11}\right\}$.

43. **(a)** $x^2 = 0.04$

$\qquad x = -\sqrt{0.04} \quad$ or $\quad x = \sqrt{0.04}$

$\qquad x = -0.2 \quad\quad$ or $\quad x = 0.2$

The solution set is $\{-0.2, 0.2\}$.

(b) Since $(0.2)^2 = 0.04$, $\sqrt{0.04} = 0.2$.

45. $\qquad\qquad x^2 = 36$

$\qquad\qquad x^2 - 36 = 0$

$\qquad (x+6)(x-6) = 0$

$\qquad x + 6 = 0 \quad$ or $\quad x - 6 = 0$

$\qquad\quad x = -6 \quad$ or $\qquad\quad x = 6$

The solution set is $\{-6, 6\}$.

8.6   Solving Equations with Radicals

1. $\sqrt{3x + 18} = x$

(a) Check $x = 6$.

$$\sqrt{3(6) + 18} \stackrel{?}{=} 6$$
$$\sqrt{18 + 18} \stackrel{?}{=} 6$$
$$\sqrt{36} \stackrel{?}{=} 6$$
$$6 = 6 \quad True$$

The number 6 is a solution.

(b) Check $x = -3$.

$\sqrt{3(-3) + 18} = -3$ is a false statement since the principal square root of a number is nonnegative. The number -3 is not a solution.

3. $\sqrt{x + 2} = \sqrt{9x - 2} - 2\sqrt{x - 1}$

(a) Check $x = 2$.

$$\sqrt{2 + 2} \stackrel{?}{=} \sqrt{9(2) - 2} - 2\sqrt{2 - 1}$$
$$\sqrt{4} \stackrel{?}{=} \sqrt{16} - 2\sqrt{1}$$
$$2 \stackrel{?}{=} 4 - 2$$
$$2 = 2 \quad True$$

The number 2 is a solution.

(b) Check $x = 7$.

$$\sqrt{7 + 2} \stackrel{?}{=} \sqrt{9(7) - 2} - 2\sqrt{7 - 1}$$
$$\sqrt{9} \stackrel{?}{=} \sqrt{61} - 2\sqrt{6}$$
$$3 = \sqrt{61} - 2\sqrt{6} \quad False$$

The number 7 is not a solution.

5. $\sqrt{9} = 3$, not -3. There is no solution of $\sqrt{x} = -3$ since the value of a principal square root cannot equal a negative number.

In Exercises 7–34, check each solution in the original equation.

7. $\sqrt{r - 2} = 3$

Square both sides.

$$\left(\sqrt{r - 2}\right)^2 = 3^2$$
$$r - 2 = 9$$
$$r = 11$$

Check the proposed solution, 11.

Check $r = 11$: $\sqrt{9} = 3$ *True*

The solution set is $\{11\}$.

9. $\sqrt{6k - 1} = 1$

Square both sides.

$$\left(\sqrt{6k - 1}\right)^2 = 1^2$$
$$6k - 1 = 1$$
$$6k = 2$$
$$k = \tfrac{2}{6} = \tfrac{1}{3}$$

Check the proposed solution, $\frac{1}{3}$.

Check $k = \frac{1}{3}$: $\sqrt{1} = 1$ *True*

The solution set is $\left\{\frac{1}{3}\right\}$.

11. $\sqrt{4r + 3} + 1 = 0$

Isolate the radical.

$$\sqrt{4r + 3} = -1$$

This equation has no solution, because $\sqrt{4r + 3}$ cannot be negative.

The solution set is $\emptyset$.

13. $\sqrt{3k + 1} - 4 = 0$

Isolate the radical.

$$\sqrt{3k + 1} = 4$$

Square both sides.

$$\left(\sqrt{3k + 1}\right)^2 = 4^2$$
$$3k + 1 = 16$$
$$3k = 15$$
$$k = 5$$

Check $k = 5$: $\sqrt{16} - 4 = 0$ *True*

The solution set is $\{5\}$.

15. $4 - \sqrt{x - 2} = 0$

$$4 = \sqrt{x - 2} \quad \textit{Isolate.}$$
$$4^2 = \left(\sqrt{x - 2}\right)^2 \quad \textit{Square.}$$
$$16 = x - 2$$
$$18 = x$$

Check $x = 18$: $4 - \sqrt{16} = 0$ *True*

The solution set is $\{18\}$.

17. $\sqrt{9a - 4} = \sqrt{8a + 1}$

$$\left(\sqrt{9a - 4}\right)^2 = \left(\sqrt{8a + 1}\right)^2 \quad \textit{Square.}$$
$$9a - 4 = 8a + 1$$
$$a = 5$$

Check $a = 5$: $\sqrt{41} = \sqrt{41}$ *True*

The solution set is $\{5\}$.

19. $2\sqrt{x} = \sqrt{3x + 4}$

$$\left(2\sqrt{x}\right)^2 = \left(\sqrt{3x + 4}\right)^2 \quad \textit{Square.}$$
$$4x = 3x + 4$$
$$x = 4$$

Check $x = 4$: $4 = \sqrt{16}$ *True*

The solution set is $\{4\}$.

21.
$$3\sqrt{z-1} = 2\sqrt{2z+2}$$
$$(3\sqrt{z-1})^2 = (2\sqrt{2z+2})^2 \quad \text{Square.}$$
$$9(z-1) = 4(2z+2)$$
$$9z - 9 = 8z + 8$$
$$z = 17$$

Check $z = 17$: $3(4) = 2(6)$ *True*
The solution set is $\{17\}$.

23.
$$k = \sqrt{k^2 + 4k - 20}$$
$$k^2 = (\sqrt{k^2 + 4k - 20})^2 \quad \text{Square.}$$
$$k^2 = k^2 + 4k - 20$$
$$20 = 4k$$
$$5 = k$$

Check $k = 5$: $5 = \sqrt{25}$ *True*
The solution set is $\{5\}$.

25.
$$a = \sqrt{a^2 + 3a + 9}$$
$$a^2 = (\sqrt{a^2 + 3a + 9})^2 \quad \text{Square.}$$
$$a^2 = a^2 + 3a + 9$$
$$-3a = 9$$
$$a = -3$$

Substituting -3 for a makes the left side of the original equation negative, but the right side is nonnegative, so the solution set is $\emptyset$.

27.
$$\sqrt{9 - x} = x + 3$$
$$(\sqrt{9 - x})^2 = (x + 3)^2 \quad \text{Square.}$$
$$9 - x = x^2 + 6x + 9$$
$$0 = x^2 + 7x$$
$$0 = x(x + 7)$$
$$x = 0 \quad \text{or} \quad x + 7 = 0$$
$$x = -7$$

Check $x = -7$: $\sqrt{16} = -4$ *False*
Check $x = 0$: $\sqrt{9} = 3$ *True*
The solution set is $\{0\}$.

29.
$$\sqrt{k^2 + 2k + 9} = k + 3$$
$$(\sqrt{k^2 + 2k + 9})^2 = (k + 3)^2 \quad \text{Square.}$$
$$k^2 + 2k + 9 = k^2 + 6k + 9$$
$$0 = 4k$$
$$0 = k$$

Check $k = 0$: $\sqrt{9} = 3$ *True*
The solution set is $\{0\}$.

31.
$$\sqrt{r^2 + 9r + 3} = -r$$
$$(\sqrt{r^2 + 9r + 3})^2 = (-r)^2 \quad \text{Square.}$$
$$r^2 + 9r + 3 = r^2$$
$$9r = -3$$
$$r = -\tfrac{1}{3}$$

Check $r = -\tfrac{1}{3}$: $\sqrt{\tfrac{1}{9}} = \tfrac{1}{3}$ *True*
The solution set is $\{-\tfrac{1}{3}\}$.

33.
$$\sqrt{z^2 + 12z - 4} + 4 - z = 0$$
$$\sqrt{z^2 + 12z - 4} = z - 4 \qquad \text{Isolate.}$$
$$(\sqrt{z^2 + 12z - 4})^2 = (z - 4)^2 \qquad \text{Square.}$$
$$z^2 + 12z - 4 = z^2 - 8z + 16$$
$$20z = 20$$
$$z = 1$$

Substituting 1 for z makes the left side of the original equation positive, but the right side is zero, so the solution set is $\emptyset$.

35.
$$\sqrt{3x + 4} = 8 - x$$

$(8 - x)^2$ equals $64 - 16x + x^2$, not $64 + x^2$. The first step should be
$$3x + 4 = 64 - 16x + x^2.$$
Then we have
$$0 = x^2 - 19x + 60.$$
$$0 = (x - 4)(x - 15)$$
$$x - 4 = 0 \quad \text{or} \quad x - 15 = 0$$
$$x = 4 \quad \text{or} \qquad x = 15$$

Check $x = 4$: $\sqrt{16} = 8 - 4$ *True*
Check $x = 15$: $\sqrt{49} = 8 - 15$ *False*
The solution set is $\{4\}$.

37.
$$\sqrt[3]{2x + 5} = \sqrt[3]{6x + 1}$$
Cube both sides.
$$(\sqrt[3]{2x + 5})^3 = (\sqrt[3]{6x + 1})^3$$
$$2x + 5 = 6x + 1$$
$$4 = 4x$$
$$1 = x$$

Check $x = 1$: $\sqrt[3]{7} = \sqrt[3]{7}$ *True*
The solution set is $\{1\}$.

39.
$$\sqrt[3]{a^2 + 5a + 1} = \sqrt[3]{a^2 + 4a}$$
$$(\sqrt[3]{a^2 + 5a + 1})^3 = (\sqrt[3]{a^2 + 4a})^3 \quad \text{Cube.}$$
$$a^2 + 5a + 1 = a^2 + 4a$$
$$a = -1$$

Check $a = -1$: $\sqrt[3]{-3} = \sqrt[3]{-3}$ *True*
The solution set is $\{-1\}$.

41.
$$\sqrt[3]{2m - 1} = \sqrt[3]{m + 13}$$
$$(\sqrt[3]{2m - 1})^3 = (\sqrt[3]{m + 13})^3 \quad \text{Cube.}$$
$$2m - 1 = m + 13$$
$$m = 14$$

Check $m = 14$: $\sqrt[3]{27} = \sqrt[3]{27}$ *True*
The solution set is $\{14\}$.

43. $\sqrt[4]{a + 8} = \sqrt[4]{2a}$

Raise each side to the fourth power.

$$\left(\sqrt[4]{a + 8}\right)^4 = \left(\sqrt[4]{2a}\right)^4$$
$$a + 8 = 2a$$
$$8 = a$$

Check $a = 8$: $\sqrt[4]{16} = \sqrt[4]{16}$ *True*

The solution set is $\{8\}$.

45. $\sqrt[3]{x - 8} + 2 = 0$

$$\sqrt[3]{x - 8} = -2 \qquad \textit{Isolate.}$$
$$\left(\sqrt[3]{x - 8}\right)^3 = (-2)^3 \quad \textit{Cube.}$$
$$x - 8 = -8$$
$$x = 0$$

Check $x = 0$: $\sqrt[3]{-8} + 2 = 0$ *True*

The solution set is $\{0\}$.

47. $\sqrt[4]{2k - 5} + 4 = 0$

$$\sqrt[4]{2k - 5} = -4 \qquad \textit{Isolate.}$$

This equation has no solution, because $\sqrt[4]{2k - 5}$ cannot be negative.

The solution set is $\emptyset$.

49. $\sqrt{k + 2} - \sqrt{k - 3} = 1$

Get one radical on each side of the equals sign.

$$\sqrt{k + 2} = 1 + \sqrt{k - 3}$$
$$\left(\sqrt{k + 2}\right)^2 = \left(1 + \sqrt{k - 3}\right)^2 \quad \textit{Square.}$$
$$k + 2 = 1 + 2\sqrt{k - 3} + k - 3$$
$$4 = 2\sqrt{k - 3} \qquad \textit{Isolate.}$$
$$2 = \sqrt{k - 3} \qquad \textit{Divide by 2.}$$
$$2^2 = \left(\sqrt{k - 3}\right)^2 \quad \textit{Square again.}$$
$$4 = k - 3$$
$$7 = k$$

Check $k = 7$: $\sqrt{9} - \sqrt{4} = 1$ *True*

The solution set is $\{7\}$.

51. $\sqrt{2r + 11} - \sqrt{5r + 1} = -1$

Get one radical on each side of the equals sign.

$$\sqrt{2r + 11} = -1 + \sqrt{5r + 1}$$

Square both sides.

$$\left(\sqrt{2r + 11}\right)^2 = \left(-1 + \sqrt{5r + 1}\right)^2$$
$$2r + 11 = 1 - 2\sqrt{5r + 1} + 5r + 1$$

Isolate the remaining radical.

$$2\sqrt{5r + 1} = 3r - 9$$

Square both sides again.

$$\left(2\sqrt{5r + 1}\right)^2 = (3r - 9)^2$$
$$4(5r + 1) = 9r^2 - 54r + 81$$
$$20r + 4 = 9r^2 - 54r + 81$$

$$0 = 9r^2 - 74r + 77$$
$$0 = (9r - 11)(r - 7)$$

$$9r - 11 = 0 \quad \text{or} \quad r - 7 = 0$$
$$r = \tfrac{11}{9} \qquad\qquad r = 7$$

Check $r = \frac{11}{9}$: $\frac{11}{3} - \frac{8}{3} = -1$ *False*

Check $r = 7$: $5 - 6 = -1$ *True*

The solution set is $\{7\}$.

53. $\sqrt{3p + 4} - \sqrt{2p - 4} = 2$

Get one radical on each side of the equals sign.

$$\sqrt{3p + 4} = 2 + \sqrt{2p - 4}$$

Square both sides.

$$\left(\sqrt{3p + 4}\right)^2 = \left(2 + \sqrt{2p - 4}\right)^2$$
$$3p + 4 = 4 + 4\sqrt{2p - 4} + 2p - 4$$

Isolate the remaining radical.

$$p + 4 = 4\sqrt{2p - 4}$$

Square both sides again.

$$(p + 4)^2 = \left(4\sqrt{2p - 4}\right)^2$$
$$p^2 + 8p + 16 = 16(2p - 4)$$
$$p^2 + 8p + 16 = 32p - 64$$
$$p^2 - 24p + 80 = 0$$
$$(p - 4)(p - 20) = 0$$

$$p - 4 = 0 \quad \text{or} \quad p - 20 = 0$$
$$p = 4 \qquad\qquad p = 20$$

Check $p = 4$: $\sqrt{16} - \sqrt{4} = 2$ *True*

Check $p = 20$: $\sqrt{64} - \sqrt{36} = 2$ *True*

The solution set is $\{4, 20\}$.

55. $\sqrt{3 - 3p} - 3 = \sqrt{3p + 2}$

Square both sides.

$$\left(\sqrt{3 - 3p} - 3\right)^2 = \left(\sqrt{3p + 2}\right)^2$$
$$3 - 3p - 6\sqrt{3 - 3p} + 9 = 3p + 2$$

Isolate the remaining radical.

$$-6\sqrt{3 - 3p} = 6p - 10$$
$$-3\sqrt{3 - 3p} = 3p - 5$$

Square both sides again.

$$\left(-3\sqrt{3 - 3p}\right)^2 = (3p - 5)^2$$
$$9(3 - 3p) = 9p^2 - 30p + 25$$
$$27 - 27p = 9p^2 - 30p + 25$$
$$0 = 9p^2 - 3p - 2$$
$$0 = (3p + 1)(3p - 2)$$

$$3p + 1 = 0 \quad \text{or} \quad 3p - 2 = 0$$
$$p = -\tfrac{1}{3} \qquad\qquad p = \tfrac{2}{3}$$

Check $p = -\frac{1}{3}$: $\sqrt{4} - 3 = \sqrt{1}$ *False*

Check $p = \frac{2}{3}$: $\sqrt{1} - 3 = \sqrt{4}$ *False*

The solution set is $\emptyset$.

57. $\sqrt{2\sqrt{x+11}} = \sqrt{4x+2}$

$\quad\quad 2\sqrt{x+11} = 4x+2$ *Square.*

$\quad\quad \left(2\sqrt{x+11}\right)^2 = (4x+2)^2$ *Square again.*

$\quad\quad\quad 4(x+11) = 16x^2+16x+4$

$\quad\quad\quad 4x+44 = 16x^2+16x+4$

$\quad\quad\quad\quad 0 = 16x^2+12x-40$

$\quad\quad\quad\quad 0 = 4x^2+3x-10$

$\quad\quad\quad\quad 0 = (x+2)(4x-5)$

$\quad x+2=0 \quad$ or $\quad 4x-5=0$

$\quad\quad x=-2 \quad\quad\quad\quad x=\frac{5}{4}$

Check $x=-2$: $\quad \sqrt{6} = \sqrt{-6}$ *False*

Check $x=\frac{5}{4}$: $\quad \sqrt{7} = \sqrt{7}$ *True*

The solution set is $\left\{\frac{5}{4}\right\}$.

59. Graph the functions

$\quad Y_1 = \sqrt{3-3x}$ and $Y_2 = 3+\sqrt{3x+2}$

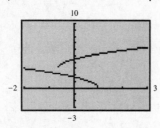

The graphs do not intersect, so the solution set is $\emptyset$.

To find the domain of

$$y = \sqrt{3-3x} - 3 - \sqrt{3x+2}$$

we must have

$\quad 3-3x \geq 0 \quad$ and $\quad 3x+2 \geq 0$

$\quad\quad -3x \geq -3 \quad\quad\quad 3x \geq -2$

$\quad\quad\quad x \leq 1 \quad$ and $\quad x \geq -\frac{2}{3}$.

The domain is $\left[-\frac{2}{3}, 1\right]$. This is the region where the graphs of Y_1 and Y_2 vertically overlap.

61. $\quad (2x-9)^{1/2} = 2+(x-8)^{1/2}$

$\quad\quad \sqrt{2x-9} = 2+\sqrt{x-8}$

$\quad\quad \left(\sqrt{2x-9}\right)^2 = \left(2+\sqrt{x-8}\right)^2$

$\quad\quad\quad 2x-9 = 4+4\sqrt{x-8}+x-8$

$\quad\quad\quad x-5 = 4\sqrt{x-8}$

$\quad\quad\quad (x-5)^2 = \left(4\sqrt{x-8}\right)^2$

$\quad x^2-10x+25 = 16(x-8)$

$\quad x^2-10x+25 = 16x-128$

$\quad x^2-26x+153 = 0$

$\quad (x-9)(x-17) = 0$

$\quad\quad x=9 \quad$ or $\quad x=17$

Check $x=9$: $\quad 9^{1/2} \overset{?}{=} 2+1^{1/2}$

$\quad\quad\quad\quad\quad\quad 3 = 2+1$ *True*

Check $x=17$: $\quad 25^{1/2} \overset{?}{=} 2+9^{1/2}$

$\quad\quad\quad\quad\quad\quad\quad 5 = 2+3$ *True*

The solution set is $\{9, 17\}$.

63. $(2w-1)^{2/3} - w^{1/3} = 0$

$\quad\quad \sqrt[3]{(2w-1)^2} = \sqrt[3]{w}$

$\quad\quad \left[\sqrt[3]{(2w-1)^2}\right]^3 = \left(\sqrt[3]{w}\right)^3$

$\quad\quad\quad (2w-1)^2 = w$

$\quad\quad 4w^2-4w+1 = w$

$\quad\quad 4w^2-5w+1 = 0$

$\quad\quad (4w-1)(w-1) = 0$

$\quad 4w-1=0 \quad$ or $\quad w-1=0$

$\quad\quad w=\frac{1}{4} \quad$ or $\quad\quad w=1$

Check $w=\frac{1}{4}$: $\left(-\frac{1}{2}\right)^{2/3} - \left(\frac{1}{4}\right)^{1/3} \overset{?}{=} 0$

$\quad\quad\quad \left(\frac{1}{4}\right)^{1/3} - \left(\frac{1}{4}\right)^{1/3} = 0$ *True*

Check $w=1$: $\quad 1^{2/3} - 1^{1/3} = 0$ *True*

The solution set is $\left\{\frac{1}{4}, 1\right\}$.

65. Solve $V = \sqrt{\dfrac{2K}{m}}$ for K.

$\quad (V)^2 = \left(\sqrt{\dfrac{2K}{m}}\right)^2$ *Square.*

$\quad\quad V^2 = \dfrac{2K}{m}$ *Multiply by* $\dfrac{m}{2}$.

$\quad\quad \dfrac{V^2 m}{2} = K$

67. Solve $f = \dfrac{1}{2\pi\sqrt{LC}}$ for L.

$\quad 2\pi f\sqrt{LC} = 1$

$\quad \left(2\pi f\sqrt{LC}\right)^2 = 1^2$

$\quad\quad 4\pi^2 f^2 LC = 1$

$\quad\quad\quad L = \dfrac{1}{4\pi^2 f^2 C}$

69. Solve $N = \dfrac{1}{2\pi}\sqrt{\dfrac{a}{r}}$ for r.

$\quad 2\pi N = \sqrt{\dfrac{a}{r}}$

$\quad (2\pi N)^2 = \left(\sqrt{\dfrac{a}{r}}\right)^2$

$\quad 4\pi^2 N^2 = \dfrac{a}{r}$

$\quad 4\pi^2 N^2 r = a$

$\quad\quad r = \dfrac{a}{4\pi^2 N^2}$

71. $(5 + 9x) + (-4 - 8x)$
$= 5 - 4 + 9x - 8x$
$= 1 + x$

73. $(x + 3)(2x - 5) = 2x^2 - 5x + 6x - 15$
$= 2x^2 + x - 15$

75. $\dfrac{-7}{5 - \sqrt{2}} = \dfrac{-7}{5 - \sqrt{2}} \cdot \dfrac{5 + \sqrt{2}}{5 + \sqrt{2}}$

$= \dfrac{-7\left(5 + \sqrt{2}\right)}{5^2 - \left(\sqrt{2}\right)^2}$

$= \dfrac{-7\left(5 + \sqrt{2}\right)}{25 - 2} = \dfrac{-7\left(5 + \sqrt{2}\right)}{23}$

8.7   Complex Numbers

1. $\sqrt{-1} = i$

3. $i^2 = -1$

5. $\dfrac{1}{i} = \dfrac{1}{i} \cdot \dfrac{-i}{-i}$ *–i is the conjugate of i*

$= \dfrac{-i}{-i^2} = \dfrac{i}{i^2} = \dfrac{i}{-1} = -i$

7. $\sqrt{-169} = i\sqrt{169} = 13i$

9. $-\sqrt{-144} = -i\sqrt{144} = -12i$

11. $\sqrt{-5} = i\sqrt{5}$

13. $\sqrt{-48} = i\sqrt{48} = i\sqrt{16 \cdot 3} = 4i\sqrt{3}$

15. $\sqrt{-7} \cdot \sqrt{-15} = i\sqrt{7} \cdot i\sqrt{15} = i^2\sqrt{7 \cdot 15}$
$= -1\sqrt{105} = -\sqrt{105}$

17. $\sqrt{-4} \cdot \sqrt{-25} = i\sqrt{4} \cdot i\sqrt{25} = 2i \cdot 5i = 10i^2$
$= 10(-1) = -10$

19. $\sqrt{-3} \cdot \sqrt{11} = i\sqrt{3} \cdot \sqrt{11}$
$= i\sqrt{33}$

21. $\dfrac{\sqrt{-300}}{\sqrt{-100}} = \dfrac{i\sqrt{300}}{i\sqrt{100}} = \sqrt{\dfrac{300}{100}} = \sqrt{3}$

23. $\dfrac{\sqrt{-75}}{\sqrt{3}} = \dfrac{i\sqrt{75}}{\sqrt{3}} = i\sqrt{\dfrac{75}{3}} = i\sqrt{25} = 5i$

25. **(a)** Since any real number can be written as $a + bi$, where $b = 0$, every real number is also a complex number.

(b) Not every complex number is a real number. For example, any number $a + bi$, $b \neq 0$, such as $3 + 7i$, is a complex number that is not a real number.

27. $(3 + 2i) + (-4 + 5i)$
$= [(3 + (-4)] + (2 + 5)i$
$= -1 + 7i$

29. $(5 - i) + (-5 + i)$
$= (5 - 5) + (-1 + 1)i$
$= 0$

31. $(4 + i) - (-3 - 2i)$
$= [(4 - (-3)] + [(1 - (-2)]i$
$= 7 + 3i$

33. $(-3 - 4i) - (-1 - 4i)$
$= [-3 - (-1)] + [-4 - (-4)]i$
$= -2$

35. $(-4 + 11i) + (-2 - 4i) + (7 + 6i)$
$= (-4 - 2 + 7) + (11 - 4 + 6)i$
$= 1 + 13i$

37. $[(7 + 3i) - (4 - 2i)] + (3 + i)$
Work inside the brackets first.
$= [(7 - 4) + (3 + 2)i] + (3 + i)$
$= (3 + 5i) + (3 + i)$
$= (3 + 3) + (5 + 1)i$
$= 6 + 6i$

39. If $a - c = b$, then $b + c = a$.
So, $(4 + 2i) - (3 + i) = 1 + i$ implies that
$$(1 + i) + (3 + i) = \underline{4 + 2i}.$$

41. $(3i)(27i) = 81i^2 = 81(-1) = -81$

43. $(-8i)(-2i) = 16i^2 = 16(-1) = -16$

45. $5i(-6 + 2i) = (5i)(-6) + (5i)(2i)$
$= -30i + 10i^2$
$= -30i + 10(-1)$
$= -10 - 30i$

47. $(4 + 3i)(1 - 2i)$
     F   **O**   **I**   **L**
$= (4)(1) + 4(-2i) + (3i)(1) + (3i)(-2i)$
$= 4 - 8i + 3i - 6i^2$
$= 4 - 5i - 6(-1)$
$= 4 - 5i + 6 = 10 - 5i$

49. $(4 + 5i)^2 = 4^2 + 2(4)(5i) + (5i)^2$
$= 16 + 40i + 25i^2$
$= 16 + 40i + 25(-1)$
$= 16 + 40i - 25$
$= -9 + 40i$

51. $2i(-4 - i)^2$
$= 2i\left[(-4)^2 - 2(-4)(i) + i^2\right]$
$= 2i(16 + 8i - 1)$
$= 2i(15 + 8i) = 30i + 16i^2$
$= 30i + 16(-1) = -16 + 30i$

53. $(12 + 3i)(12 - 3i)$
$= 12^2 - (3i)^2 = 144 - 9i^2$
$= 144 - 9(-1) = 144 + 9 = 153$

55. $(4 + 9i)(4 - 9i)$

$= 4^2 - (9i)^2 = 16 - 81i^2$

$= 16 - 81(-1) = 16 + 81 = 97$

57. The conjugate of $a + bi$ is $a - bi$.

59. $\dfrac{2}{1 - i}$

Multiply the numerator and the denominator
by the conjugate of the denominator, $1 + i$.

$= \dfrac{2(1 + i)}{(1 - i)(1 + i)} = \dfrac{2(1 + i)}{1^2 - i^2}$

$= \dfrac{2(1 + i)}{1 - (-1)} = \dfrac{2(1 + i)}{2} = 1 + i$

61. $\dfrac{-7 + 4i}{3 + 2i}$

Multiply the numerator and the denominator
by the conjugate of the denominator, $3 - 2i$.

$= \dfrac{(-7 + 4i)(3 - 2i)}{(3 + 2i)(3 - 2i)}$

In the denominator, we make use of the fact
that $(a + bi)(a - bi) = a^2 + b^2$.

$= \dfrac{-21 + 14i + 12i + 8}{3^2 + 2^2}$

$= \dfrac{-13 + 26i}{13} = \dfrac{13(-1 + 2i)}{13} = -1 + 2i$

63. $\dfrac{8i}{2 + 2i}$

Write in lowest terms.

$= \dfrac{2 \cdot 4i}{2(1 + i)} = \dfrac{4i}{1 + i}$

Multiply the numerator and the denominator
by the conjugate of the denominator, $1 - i$.

$= \dfrac{4i(1 - i)}{(1 + i)(1 - i)} = \dfrac{4(i - i^2)}{1^2 + 1^2}$

$= \dfrac{4(i + 1)}{2} = 2(i + 1) = 2 + 2i$

65. $\dfrac{2 - 3i}{2 + 3i}$

Multiply the numerator and the denominator
by the conjugate of the denominator, $2 - 3i$.

$= \dfrac{(2 - 3i)(2 - 3i)}{(2 + 3i)(2 - 3i)} = \dfrac{2^2 - 2(2)(3i) + (3i)^2}{2^2 + 3^2}$

$= \dfrac{4 - 12i + 9i^2}{4 + 9} = \dfrac{4 - 12i - 9}{13}$

$= \dfrac{-5 - 12i}{13} = -\dfrac{5}{13} - \dfrac{12}{13}i$

67. $\dfrac{3 + i}{i}$

Multiply the numerator and the denominator
by the conjugate of the denominator, $-i$.

$= \dfrac{(3 + i)(-i)}{i(-i)} = \dfrac{-3i - i^2}{-i^2}$

$= \dfrac{-3i - (-1)}{-(-1)} = \dfrac{-3i + 1}{1}$

$= 1 - 3i$

69. $i^{18} = i^{16} \cdot i^2 = (i^4)^4 \cdot i^2$

$= 1^4 \cdot (-1) = 1 \cdot (-1) = -1$

71. $i^{89} = i^{88} \cdot i = (i^4)^{22} \cdot i = 1^{22} \cdot i$

$= 1 \cdot i = i$

73. $i^{38} = i^{36} \cdot i^2$

$= (i^4)^9 \cdot i^2 = 1^9(-1) = -1$

75. $i^{43} = i^{40} \cdot i^2 \cdot i$

$= (i^4)^{10} \cdot (-1) \cdot i$

$= 1^{10} \cdot (-i) = -i$

77. $i^{-5} = \dfrac{1}{i^5} = \dfrac{1}{i^4 \cdot i} = \dfrac{1}{1 \cdot i} = \dfrac{1}{i}$

From Exercise 5, $\dfrac{1}{i} = -i$.

79. Since $i^{20} = (i^4)^5 = 1^5 = 1$, the student multiplied
by 1, which is justified by the identity property for
multiplication.

81. $I = \dfrac{E}{R + (X_L - X_c)i}$

Substitute $2 + 3i$ for E, 5 for R, 4 for X_L,
and 3 for X_c.

$I = \dfrac{2 + 3i}{5 + (4 - 3)i} = \dfrac{2 + 3i}{5 + i}$

$= \dfrac{(2 + 3i)(5 - i)}{(5 + i)(5 - i)} = \dfrac{10 - 2i + 15i - 3i^2}{5^2 + 1^2}$

$= \dfrac{10 + 3 + 13i}{25 + 1} = \dfrac{13 + 13i}{26}$

$= \dfrac{13(1 + i)}{13 \cdot 2} = \dfrac{1 + i}{2} = \dfrac{1}{2} + \dfrac{1}{2}i$

83. To check that $1 + 5i$ is a solution of the equation,
substitute $1 + 5i$ for x.

$$x^2 - 2x + 26 = 0$$

$$(1 + 5i)^2 - 2(1 + 5i) + 26 \overset{?}{=} 0$$

$$(1 + 10i + 25i^2) - 2 - 10i + 26 \overset{?}{=} 0$$

$$1 + 10i - 25 - 2 - 10i + 26 \overset{?}{=} 0$$

$$(1 - 25 - 2 + 26) + (10 - 10)i \overset{?}{=} 0$$

$$0 = 0$$

True

Thus, $1 + 5i$ is a solution of

$$x^2 - 2x + 26 = 0.$$

Now substitute $1 - 5i$ for x.

$$(1 - 5i)^2 - 2(1 - 5i) + 26 \overset{?}{=} 0$$

$$\left(1 - 10i + 25i^2\right) - 2 + 10i + 26 \overset{?}{=} 0$$

$$1 - 10i - 25 - 2 + 10i + 26 \overset{?}{=} 0$$

$$(1 - 25 - 2 + 26) + (-10 + 10)i \overset{?}{=} 0$$

$$0 = 0$$

True

Thus, $1 - 5i$ is also a solution of the given equation.

85. $\dfrac{3}{2 - i} + \dfrac{5}{1 + i}$

$$= \frac{3(2 + i)}{(2 - i)(2 + i)} + \frac{5(1 - i)}{(1 + i)(1 - i)}$$

$$= \frac{6 + 3i}{4 + 1} + \frac{5 - 5i}{1 + 1}$$

$$= \frac{2(6 + 3i)}{2 \cdot 5} + \frac{5(5 - 5i)}{5 \cdot 2}$$

$$= \frac{12 + 6i}{10} + \frac{25 - 25i}{10}$$

$$= \frac{37 - 19i}{10} = \frac{37}{10} - \frac{19}{10}i$$

87. $\dfrac{2 + i}{2 - i} + \dfrac{i}{1 + i}$

$$= \frac{(2 + i)(2 + i)}{(2 - i)(2 + i)} + \frac{i(1 - i)}{(1 + i)(1 - i)}$$

$$= \frac{4 + 4i + i^2}{4 + 1} + \frac{i - i^2}{1 + 1}$$

$$= \frac{3 + 4i}{5} + \frac{1 + i}{2}$$

$$= \frac{2(3 + 4i)}{2 \cdot 5} + \frac{5(1 + i)}{5 \cdot 2}$$

$$= \frac{6 + 8i}{10} + \frac{5 + 5i}{10}$$

$$= \frac{11 + 13i}{10} = \frac{11}{10} + \frac{13}{10}i$$

Thus, $\left(\dfrac{2 + i}{2 - i} + \dfrac{i}{1 + i}\right)i$

$$= \left(\frac{11}{10} + \frac{13}{10}i\right)i$$

$$= \frac{11}{10}i + \frac{13}{10}i^2$$

$$= -\frac{13}{10} + \frac{11}{10}i.$$

89. $6x + 13 = 0$

$$6x = -13$$

$$x = -\frac{13}{6}$$

The solution set is $\left\{-\frac{13}{6}\right\}$.

91.
$$x(x + 3) = 40$$
$$x^2 + 3x = 40$$
$$x^2 + 3x - 40 = 0$$
$$(x + 8)(x - 5) = 0$$
$$x + 8 = 0 \qquad \text{or} \qquad x - 5 = 0$$
$$x = -8 \qquad \text{or} \qquad x = 5$$

Check $x = -8$: $-8(-5) = 40$ *True*
Check $x = 5$: $5(8) = 40$ *True*

The solution set is $\{-8, 5\}$.

93.
$$5x^2 - 3x = 2$$
$$5x^2 - 3x - 2 = 0$$
$$(5x + 2)(x - 1) = 0$$
$$5x + 2 = 0 \qquad \text{or} \qquad x - 1 = 0$$
$$x = -\frac{2}{5} \qquad \text{or} \qquad x = 1$$

Check $x = -\frac{2}{5}$: $\frac{4}{5} + \frac{6}{5} = 2$ *True*
Check $x = 1$: $5 - 3 = 2$ *True*

The solution set is $\left\{-\frac{2}{5}, 1\right\}$.

Chapter 8 Review Exercises

1. $\sqrt{1764} = 42$, because $42^2 = 1764$.

2. $-\sqrt{289} = -(17) = -17$, since $17^2 = 289$.

3. $\sqrt[3]{216} = 6$, because $6^3 = 216$.

4. $\sqrt[3]{-125} = -5$, because $(-5)^3 = -125$.

5. $-\sqrt[3]{27} = -(3) = -3$, since $3^3 = 27$.

6. $\sqrt[5]{-32} = -2$, because $(-2)^5 = -32$.

7. $\sqrt[n]{a}$ is not a real number if n is even and a is negative.

8. **(a)** $\sqrt{x^2} = |x|$

    (b) $-\sqrt{x^2} = -|x|$

    (c) $\sqrt[3]{x^3} = x$

9. $-\sqrt{47} \approx -6.856$

10. $\sqrt[3]{-129} \approx -5.053$

11. $\sqrt[4]{605} \approx 4.960$

12. $500^{-3/4} \approx 0.009$

13. $-500^{4/3} \approx -3968.503$

14. $-28^{-1/2} \approx -0.189$

15. $f(x) = \sqrt{x-1}$

For the radicand to be nonnegative, we must have

$$x - 1 \geq 0 \quad \text{or} \quad x \geq 1.$$

Thus, the domain is $[1, \infty)$.

The function values are positive or zero (the result of the radical), so the range is $[0, \infty)$.

x	$f(x) = \sqrt{x-1}$
1	$\sqrt{1-1} = 0$
2	$\sqrt{2-1} = 1$
5	$\sqrt{5-1} = 2$

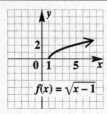

16. $f(x) = \sqrt[3]{x} + 4$

Since we can take the cube root of any real number, the domain is $(-\infty, \infty)$.

The result of a cube root can be any real number, so the range is $(-\infty, \infty)$. (The "+4" does not affect that range.)

x	$f(x) = \sqrt[3]{x} + 4$
-8	$\sqrt[3]{-8} + 4 = 2$
-1	$\sqrt[3]{-1} + 4 = 3$
0	$\sqrt[3]{0} + 4 = 4$
1	$\sqrt[3]{1} + 4 = 5$
8	$\sqrt[3]{8} + 4 = 6$

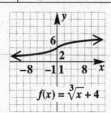

17. The base $\sqrt{38}$ is closest to $\sqrt{36} = 6$. The height $\sqrt{99}$ is closest to $\sqrt{100} = 10$. Use the estimates $b = 6$ and $h = 10$ in $A = \frac{1}{2}bh$ to find an estimate of the area.

$$A \approx \frac{1}{2}(6)(10) = 30$$

Choice **B** is the best estimate.

18. One way to evaluate $8^{2/3}$ is to first find the _cube (or third)_ root of _8_, which is _2_. Then raise that result to the _second_ power, to get an answer of _4_. Therefore, $8^{2/3} = $ _4_.

19. A. $(-27)^{2/3} = \left[(-27)^{1/3}\right]^2$

This number is a square, so it is a positive number.

B. $(-64)^{5/3} = \left[(-64)^{1/3}\right]^5$

This number is the odd power of an odd root of a negative number, so it is a negative number.

C. $(-100)^{1/2}$

This number is the square root of a negative number, so it is not a real number.

D. $(-32)^{1/5}$

This number is an odd root of a negative number, so it is a negative number.

The only positive number is choice **A**.

20. $a^{m/n} = \sqrt[n]{a^m}$

Since n is odd, $\sqrt[n]{a^m}$ is positive if a^m is positive and negative if a^m is negative. Since a is negative, a^m is positive if m is even and negative if m is odd.

(a) If a is negative and n is odd, then $a^{m/n}$ is positive if m is even.

(b) If a is negative and n is odd, then $a^{m/n}$ is negative if m is odd.

21. If a is negative and n is even, then $a^{1/n}$ is not a real number. An example is $(-4)^{1/2}$, which is not a real number.

22. $49^{1/2} = (7^2)^{1/2} = 7$

23. $-121^{1/2} = -(11^2)^{1/2} = -11$

24. $16^{5/4} = \left[(2^4)^{1/4}\right]^5 = 2^5 = 32$

25. $-8^{2/3} = -\left[(2^3)^{1/3}\right]^2 = -2^2 = -4$

26. $-\left(\frac{36}{25}\right)^{3/2} = -\left(\left[\left(\frac{6}{5}\right)^2\right]^{1/2}\right)^3$

$$= -\left(\frac{6}{5}\right)^3 = -\frac{216}{125}$$

27. $\left(-\frac{1}{8}\right)^{-5/3} = (-8)^{5/3} = \left(\left[(-2)^3\right]^{1/3}\right)^5$

$$= (-2)^5 = -32$$

28. $\left(\frac{81}{10{,}000}\right)^{-3/4} = \left(\frac{10{,}000}{81}\right)^{3/4}$

$$= \left(\left[\left(\frac{10}{3}\right)^4\right]^{1/4}\right)^3$$

$$= \left(\frac{10}{3}\right)^3 = \frac{1000}{27}$$

29. The base, -16, is negative and the index, 4, is even, so $(-16)^{3/4}$ is not a real number.

30. Solve $a^2 + b^2 = c^2$ for b. $(b > 0)$

$$b^2 = c^2 - a^2$$

$$b = \sqrt{c^2 - a^2}$$

31. The expression with fractional exponents, $a^{m/n}$, is equivalent to the radical expression, $\sqrt[n]{a^m}$. The denominator of the exponent is the index of the radical. For example, $\sqrt[3]{8^2} = \sqrt[3]{64} = 4$, and $8^{2/3} = (8^{1/3})^2 = 2^2 = 4$.

32. $(m + 3n)^{1/2} = \sqrt{m + 3n}$

33. $(3a + b)^{-5/3} = \dfrac{1}{(3a + b)^{5/3}}$

$= \dfrac{1}{((3a + b)^{1/3})^5}$

$= \dfrac{1}{\left(\sqrt[3]{3a + b}\right)^5}$

or $\dfrac{1}{\sqrt[3]{(3a + b)^5}}$

34. $\sqrt{7^9} = (7^9)^{1/2} = 7^{9/2}$

35. $\sqrt[5]{p^4} = (p^4)^{1/5} = p^{4/5}$

36. $5^{1/4} \cdot 5^{7/4} = 5^{1/4 + 7/4} = 5^{8/4}$

$= 5^2,\ \text{or}\ 25$

37. $\dfrac{96^{2/3}}{96^{-1/3}} = 96^{2/3 - (-1/3)} = 96^1 = 96$

38. $\dfrac{(a^{1/3})^4}{a^{2/3}} = \dfrac{a^{4/3}}{a^{2/3}} = a^{4/3 - 2/3} = a^{2/3}$

39. $\dfrac{y^{-1/3} \cdot y^{5/6}}{y} = \dfrac{y^{-2/6} y^{5/6}}{y^{6/6}}$

$= y^{-2/6 + 5/6 - 6/6}$

$= y^{-3/6} = y^{-1/2} = \dfrac{1}{y^{1/2}}$

40. $\left(\dfrac{z^{-1} x^{-3/5}}{2^{-2} z^{-1/2} x}\right)^{-1} = \dfrac{z^1 x^{3/5}}{2^2 z^{1/2} x^{-1}}$

$= \tfrac{1}{4} z^{1 - 1/2} x^{3/5 - (-1)}$

$= \dfrac{z^{1/2} x^{8/5}}{4}$

41. $r^{-1/2}(r + r^{3/2})$

$= r^{-1/2}(r) + r^{-1/2}(r^{3/2})$

$= r^{-1/2 + 1} + r^{-1/2 + 3/2}$

$= r^{1/2} + r^{2/2}$

$= r^{1/2} + r$

42. $\sqrt[8]{s^4} = (s^4)^{1/8} = s^{4/8} = s^{1/2}$

43. $\sqrt[6]{r^9} = (r^9)^{1/6} = r^{9/6} = r^{3/2}$

44. $\dfrac{\sqrt{p^5}}{p^2} = \dfrac{p^{5/2}}{p^2} = p^{5/2 - 2}$

$= p^{5/2 - 4/2} = p^{1/2}$

45. $\sqrt[4]{k^3} \cdot \sqrt{k^3} = (k^3)^{1/4} (k^3)^{1/2}$

$= k^{3/4} k^{3/2} = k^{3/4 + 3/2}$

$= k^{3/4 + 6/4} = k^{9/4}$

46. $\sqrt[3]{m^5} \cdot \sqrt[3]{m^8} = (m^5)^{1/3} (m^8)^{1/3}$

$= m^{5/3} m^{8/3} = m^{5/3 + 8/3} = m^{13/3}$

47. $\sqrt[4]{\sqrt[3]{z}} = \sqrt[4]{z^{1/3}} = (z^{1/3})^{1/4} = z^{1/12}$

48. $\sqrt{\sqrt{\sqrt{x}}} = \sqrt{\sqrt{x^{1/2}}} = \sqrt{(x^{1/2})^{1/2}}$

$= \sqrt{x^{1/4}} = (x^{1/4})^{1/2} = x^{1/8}$

49. $\sqrt[3]{\sqrt[5]{x}} = \sqrt[3]{x^{1/5}} = (x^{1/5})^{1/3} = x^{1/15}$

50. $\sqrt{\sqrt[6]{\sqrt[3]{x}}} = \sqrt{\sqrt[6]{x^{1/3}}} = \sqrt{(x^{1/3})^{1/6}}$

$= \sqrt{x^{1/18}} = (x^{1/18})^{1/2} = x^{1/36}$

51. The product rule for exponents applies only if the bases are the same.

52. $\sqrt{6} \cdot \sqrt{11} = \sqrt{6 \cdot 11} = \sqrt{66}$

53. $\sqrt{5} \cdot \sqrt{r} = \sqrt{5 \cdot r} = \sqrt{5r}$

54. $\sqrt[3]{6} \cdot \sqrt[3]{5} = \sqrt[3]{6 \cdot 5} = \sqrt[3]{30}$

55. $\sqrt[4]{7} \cdot \sqrt[4]{3} = \sqrt[4]{7 \cdot 3} = \sqrt[4]{21}$

56. $\sqrt{20} = \sqrt{4 \cdot 5} = \sqrt{4}\sqrt{5} = 2\sqrt{5}$

57. $\sqrt{75} = \sqrt{25 \cdot 3} = \sqrt{25}\sqrt{3} = 5\sqrt{3}$

58. $-\sqrt{125} = -\sqrt{25 \cdot 5} = -5\sqrt{5}$

59. $\sqrt[3]{-108} = \sqrt[3]{-27 \cdot 4} = -3\sqrt[3]{4}$

60. $\sqrt{100y^7} = \sqrt{100y^6 \cdot y} = 10y^3\sqrt{y}$

61. $\sqrt[3]{64p^4q^6} = \sqrt[3]{64p^3q^6 \cdot p} = 4pq^2\sqrt[3]{p}$

62. $\sqrt[3]{108a^8b^5} = \sqrt[3]{27a^6b^3 \cdot 4a^2b^2}$

$= 3a^2b\sqrt[3]{4a^2b^2}$

63. $\sqrt[3]{632r^8t^4} = \sqrt[3]{8r^6t^3 \cdot 79r^2t}$

$= 2r^2t\sqrt[3]{79r^2t}$

64. $\sqrt{\dfrac{y^3}{144}} = \dfrac{\sqrt{y^3}}{\sqrt{144}} = \dfrac{\sqrt{y^2 \cdot y}}{12} = \dfrac{y\sqrt{y}}{12}$

65. $\sqrt[3]{\dfrac{m^{15}}{27}} = \dfrac{\sqrt[3]{m^{15}}}{\sqrt[3]{27}} = \dfrac{\sqrt[3]{(m^5)^3}}{\sqrt[3]{3^3}} = \dfrac{m^5}{3}$

66. $\sqrt[3]{\dfrac{r^2}{8}} = \dfrac{\sqrt[3]{r^2}}{\sqrt[3]{8}} = \dfrac{\sqrt[3]{r^2}}{2}$

67. $\sqrt[4]{\dfrac{a^9}{81}} = \dfrac{\sqrt[4]{a^9}}{\sqrt[4]{81}} = \dfrac{\sqrt[4]{a^8 \cdot a}}{3} = \dfrac{a^2\sqrt[4]{a}}{3}$

68. $\sqrt[6]{15^3} = 15^{3/6} = 15^{1/2} = \sqrt{15}$

69. $\sqrt[4]{p^6} = (p^6)^{1/4} = p^{6/4} = p^{3/2}$
$= p^{2/2}p^{1/2} = p\sqrt{p}$

70. $\sqrt[3]{2} \cdot \sqrt[4]{5} = 2^{1/3} \cdot 5^{1/4}$
$= 2^{4/12} \cdot 5^{3/12}$
$= (2^4 \cdot 5^3)^{1/12}$
$= \sqrt[12]{16 \cdot 125} = \sqrt[12]{2000}$

71. $\sqrt{x} \cdot \sqrt[5]{x} = x^{1/2} \cdot x^{1/5}$
$= x^{5/10} \cdot x^{2/10}$
$= x^{7/10} = \sqrt[10]{x^7}$

72. Substitute 8 for a and 6 for b in the Pythagorean formula to find the hypotenuse, c.
$$c^2 = a^2 + b^2$$
$$c = \sqrt{a^2 + b^2} = \sqrt{8^2 + 6^2}$$
$$= \sqrt{64 + 36} = \sqrt{100} = 10$$

The length of the hypotenuse is 10.

73. $(-4, 7)$ and $(10, 6)$
$$d = \sqrt{(x_2 - x_1)^2 + (y_2 - y_1)^2}$$
$$= \sqrt{[10 - (-4)]^2 + (6 - 7)^2}$$
$$= \sqrt{14^2 + (-1)^2}$$
$$= \sqrt{196 + 1} = \sqrt{197}$$

74. $2\sqrt{8} - 3\sqrt{50} = 2\sqrt{4 \cdot 2} - 3\sqrt{25 \cdot 2}$
$= 2 \cdot 2\sqrt{2} - 3 \cdot 5\sqrt{2}$
$= 4\sqrt{2} - 15\sqrt{2} = -11\sqrt{2}$

75. $8\sqrt{80} - 3\sqrt{45} = 8\sqrt{16 \cdot 5} - 3\sqrt{9 \cdot 5}$
$= 8 \cdot 4\sqrt{5} - 3 \cdot 3\sqrt{5}$
$= 32\sqrt{5} - 9\sqrt{5} = 23\sqrt{5}$

76. $-\sqrt{27y} + 2\sqrt{75y} = -\sqrt{9 \cdot 3y} + 2\sqrt{25 \cdot 3y}$
$= -3\sqrt{3y} + 2 \cdot 5\sqrt{3y}$
$= -3\sqrt{3y} + 10\sqrt{3y}$
$= 7\sqrt{3y}$

77. $2\sqrt{54m^3} + 5\sqrt{96m^3}$
$= 2\sqrt{9m^2 \cdot 6m} + 5\sqrt{16m^2 \cdot 6m}$
$= 2 \cdot 3m\sqrt{6m} + 5 \cdot 4m\sqrt{6m}$
$= 6m\sqrt{6m} + 20m\sqrt{6m} = 26m\sqrt{6m}$

78. $3\sqrt[3]{54} + 5\sqrt[3]{16} = 3\sqrt[3]{27 \cdot 2} + 5\sqrt[3]{8 \cdot 2}$
$= 3 \cdot 3\sqrt[3]{2} + 5 \cdot 2\sqrt[3]{2}$
$= 9\sqrt[3]{2} + 10\sqrt[3]{2} = 19\sqrt[3]{2}$

79. $-6\sqrt[4]{32} + \sqrt[4]{512} = -6\sqrt[4]{16 \cdot 2} + \sqrt[4]{256 \cdot 2}$
$= -6 \cdot 2\sqrt[4]{2} + 4\sqrt[4]{2}$
$= -12\sqrt[4]{2} + 4\sqrt[4]{2} = -8\sqrt[4]{2}$

80. $\dfrac{3}{\sqrt{16}} - \dfrac{\sqrt{5}}{2} = \dfrac{3}{4} - \dfrac{2\sqrt{5}}{2 \cdot 2} = \dfrac{3 - 2\sqrt{5}}{4}$

81. $\dfrac{4}{\sqrt{25}} + \dfrac{\sqrt{5}}{4} = \dfrac{4}{5} + \dfrac{\sqrt{5}}{4}$
$= \dfrac{16}{20} + \dfrac{5\sqrt{5}}{20} = \dfrac{16 + 5\sqrt{5}}{20}$

82. Add the measures of the sides.
$$P = a + b + c + d$$
$$P = 4\sqrt{8} + 6\sqrt{12} + 8\sqrt{2} + 3\sqrt{48}$$
$$= 4\sqrt{4 \cdot 2} + 6\sqrt{4 \cdot 3} + 8\sqrt{2} + 3\sqrt{16 \cdot 3}$$
$$= 4 \cdot 2\sqrt{2} + 6 \cdot 2\sqrt{3} + 8\sqrt{2} + 3 \cdot 4\sqrt{3}$$
$$= 8\sqrt{2} + 12\sqrt{3} + 8\sqrt{2} + 12\sqrt{3}$$
$$= 16\sqrt{2} + 24\sqrt{3}$$

The perimeter is $(16\sqrt{2} + 24\sqrt{3})$ feet.

83. Add the measures of the sides.
$$P = a + b + c$$
$$P = 2\sqrt{27} + \sqrt{108} + \sqrt{50}$$
$$= 2\sqrt{9 \cdot 3} + \sqrt{36 \cdot 3} + \sqrt{25 \cdot 2}$$
$$= 2 \cdot 3\sqrt{3} + 6\sqrt{3} + 5\sqrt{2}$$
$$= 6\sqrt{3} + 6\sqrt{3} + 5\sqrt{2}$$
$$= 12\sqrt{3} + 5\sqrt{2}$$

The perimeter is $(12\sqrt{3} + 5\sqrt{2})$ feet.

84. $(\sqrt{3} + 1)(\sqrt{3} - 2) = 3 - 2\sqrt{3} + \sqrt{3} - 2$
$= 1 - \sqrt{3}$

85. $(\sqrt{7} + \sqrt{5})(\sqrt{7} - \sqrt{5}) = (\sqrt{7})^2 - (\sqrt{5})^2$
$= 7 - 5 = 2$

86. $(3\sqrt{2} + 1)(2\sqrt{2} - 3)$
$= 6 \cdot 2 - 9\sqrt{2} + 2\sqrt{2} - 3$
$= 12 - 7\sqrt{2} - 3 = 9 - 7\sqrt{2}$

87. $(\sqrt{13} - \sqrt{2})^2$
$= (\sqrt{13})^2 - 2 \cdot \sqrt{13} \cdot \sqrt{2} + (\sqrt{2})^2$
$= 13 - 2\sqrt{26} + 2 = 15 - 2\sqrt{26}$

88. $(\sqrt[3]{2} + 3)(\sqrt[3]{4} - 3\sqrt[3]{2} + 9)$
$= \sqrt[3]{2} \cdot \sqrt[3]{4} - \sqrt[3]{2} \cdot 3\sqrt[3]{2} + 9\sqrt[3]{2}$
$\quad + 3\sqrt[3]{4} - 3 \cdot 3\sqrt[3]{2} + 27$
$= \sqrt[3]{8} - 3\sqrt[3]{4} + 9\sqrt[3]{2} + 3\sqrt[3]{4} - 9\sqrt[3]{2} + 27$
$= 2 + 27 = 29$

89. $(\sqrt[3]{4y} - 1)(\sqrt[3]{4y} + 3)$
$= \sqrt[3]{16y^2} + 3\sqrt[3]{4y} - \sqrt[3]{4y} - 3$
$= \sqrt[3]{8 \cdot 2y^2} + 2\sqrt[3]{4y} - 3$
$= 2\sqrt[3]{2y^2} + 2\sqrt[3]{4y} - 3$

90. Show that $15 - 2\sqrt{26} \neq 13\sqrt{26}$.

Find a calculator approximation of each term.

$$4.801960973 \neq 66.28725368$$

Therefore, $15 - 2\sqrt{26} \neq 13\sqrt{26}$.

91. Multiplying by $\sqrt[3]{6}$ still results in an expression with a radical in the denominator.

$$\frac{5\sqrt[3]{6}}{\sqrt[3]{6} \cdot \sqrt[3]{6}} = \frac{5\sqrt[3]{6}}{\sqrt[3]{36}}$$

To rationalize the denominator, multiply by $\sqrt[3]{6^2}$ or $\sqrt[3]{36}$.

$$\frac{5\sqrt[3]{6^2}}{\sqrt[3]{6} \cdot \sqrt[3]{6^2}} = \frac{5\sqrt[3]{36}}{\sqrt[3]{6^3}} = \frac{5\sqrt[3]{36}}{6}$$

92. $\dfrac{\sqrt{6}}{\sqrt{5}} = \dfrac{\sqrt{6} \cdot \sqrt{5}}{\sqrt{5} \cdot \sqrt{5}} = \dfrac{\sqrt{30}}{5}$

93. $\dfrac{-6\sqrt{3}}{\sqrt{2}} = \dfrac{-6\sqrt{3} \cdot \sqrt{2}}{\sqrt{2} \cdot \sqrt{2}} = \dfrac{-6\sqrt{6}}{2} = -3\sqrt{6}$

94. $\dfrac{3\sqrt{7p}}{\sqrt{y}} = \dfrac{3\sqrt{7p} \cdot \sqrt{y}}{\sqrt{y} \cdot \sqrt{y}} = \dfrac{3\sqrt{7py}}{y}$

95. $\sqrt{\dfrac{11}{8}} = \dfrac{\sqrt{11}}{\sqrt{8}} = \dfrac{\sqrt{11}}{\sqrt{4 \cdot 2}} = \dfrac{\sqrt{11}}{2\sqrt{2}} = \dfrac{\sqrt{11} \cdot \sqrt{2}}{2\sqrt{2} \cdot \sqrt{2}}$

$$= \dfrac{\sqrt{22}}{2 \cdot 2} = \dfrac{\sqrt{22}}{4}$$

96. $-\sqrt[3]{\dfrac{9}{25}} = -\dfrac{\sqrt[3]{9}}{\sqrt[3]{5^2}} = -\dfrac{\sqrt[3]{9} \cdot \sqrt[3]{5}}{\sqrt[3]{5^2} \cdot \sqrt[3]{5}}$

$$= -\dfrac{\sqrt[3]{45}}{\sqrt[3]{5^3}} = -\dfrac{\sqrt[3]{45}}{5}$$

97. $\sqrt[3]{\dfrac{108m^3}{n^5}} = \dfrac{\sqrt[3]{108m^3}}{\sqrt[3]{n^5}} = \dfrac{\sqrt[3]{27m^3 \cdot 4}}{\sqrt[3]{n^3 \cdot n^2}}$

$$= \dfrac{3m\sqrt[3]{4}}{n\sqrt[3]{n^2}} = \dfrac{3m\sqrt[3]{4} \cdot \sqrt[3]{n}}{n\sqrt[3]{n^2} \cdot \sqrt[3]{n}}$$

$$= \dfrac{3m\sqrt[3]{4n}}{n \cdot n} = \dfrac{3m\sqrt[3]{4n}}{n^2}$$

98. $\dfrac{1}{\sqrt{2} + \sqrt{7}}$

Multiply the numerator and denominator by the conjugate of the denominator, $\sqrt{2} - \sqrt{7}$.

$$= \dfrac{1(\sqrt{2} - \sqrt{7})}{(\sqrt{2} + \sqrt{7})(\sqrt{2} - \sqrt{7})}$$

$$= \dfrac{\sqrt{2} - \sqrt{7}}{2 - 7} = \dfrac{\sqrt{2} - \sqrt{7}}{-5}$$

99. $\dfrac{-5}{\sqrt{6} - 3}$

Multiply the numerator and denominator by the conjugate of the denominator, $\sqrt{6} + 3$.

$$= \dfrac{-5(\sqrt{6} + 3)}{(\sqrt{6} - 3)(\sqrt{6} + 3)} = \dfrac{-5(\sqrt{6} + 3)}{6 - 9}$$

$$= \dfrac{-5(\sqrt{6} + 3)}{-3} = \dfrac{5(\sqrt{6} + 3)}{3}$$

100. $\dfrac{2 - 2\sqrt{5}}{8} = \dfrac{2(1 - \sqrt{5})}{2 \cdot 4} = \dfrac{1 - \sqrt{5}}{4}$

101. $\dfrac{4 - 8\sqrt{8}}{12} = \dfrac{4(1 - 2\sqrt{8})}{3 \cdot 4} = \dfrac{1 - 2\sqrt{8}}{3}$

$$= \dfrac{1 - 2\sqrt{4 \cdot 2}}{3} = \dfrac{1 - 4\sqrt{2}}{3}$$

102. $\dfrac{-18 + \sqrt{27}}{6} = \dfrac{-18 + \sqrt{9} \cdot \sqrt{3}}{6} = \dfrac{-18 + 3\sqrt{3}}{6}$

$$= \dfrac{3(-6 + \sqrt{3})}{3 \cdot 2} = \dfrac{-6 + \sqrt{3}}{2}$$

103.
$$\sqrt{8x + 9} = 5$$
$$(\sqrt{8x + 9})^2 = 5^2 \quad \textit{Square.}$$
$$8x + 9 = 25$$
$$8x = 16$$
$$x = 2$$

Check $x = 2$: $\sqrt{25} = 5$ *True*
The solution set is $\{2\}$.

104.
$$\sqrt{2z - 3} - 3 = 0$$
$$\sqrt{2z - 3} = 3 \quad \textit{Isolate.}$$
$$(\sqrt{2z - 3})^2 = 3^2 \quad \textit{Square.}$$
$$2z - 3 = 9$$
$$2z = 12$$
$$z = 6$$

Check $z = 6$: $\sqrt{9} - 3 = 0$ *True*
The solution set is $\{6\}$.

105.
$$\sqrt{3m + 1} - 2 = -3$$
$$\sqrt{3m + 1} = -1 \quad \textit{Isolate.}$$
This equation has no solution, because $\sqrt{3m + 1}$ cannot be negative.
The solution set is $\emptyset$.

106.
$$\sqrt{7z + 1} = z + 1$$
$$(\sqrt{7z + 1})^2 = (z + 1)^2 \quad \textit{Square.}$$
$$7z + 1 = z^2 + 2z + 1$$
$$0 = z^2 - 5z$$
$$0 = z(z - 5)$$

$z = 0$ or $z = 5$

Check $z = 0$: $\sqrt{1} = 1$ *True*
Check $z = 5$: $\sqrt{36} = 6$ *True*
The solution set is $\{0, 5\}$.

107.
$$3\sqrt{m} = \sqrt{10m - 9}$$
$$\left(3\sqrt{m}\right)^2 = \left(\sqrt{10m - 9}\right)^2 \quad \textit{Square.}$$
$$9m = 10m - 9$$
$$9 = m$$
Check $m = 9$: $3\sqrt{9} = \sqrt{81}$ *True*
The solution set is $\{9\}$.

108.
$$\sqrt{p^2 + 3p + 7} = p + 2$$
$$\left(\sqrt{p^2 + 3p + 7}\right)^2 = (p + 2)^2 \quad \textit{Square.}$$
$$p^2 + 3p + 7 = p^2 + 4p + 4$$
$$3 = p$$
Check $p = 3$: $\sqrt{25} = 5$ *True*
The solution set is $\{3\}$.

109. $\sqrt{a + 2} - \sqrt{a - 3} = 1$
Get one radical on each side of the equals sign.
$$\sqrt{a + 2} = 1 + \sqrt{a - 3}$$
Square both sides.
$$\left(\sqrt{a + 2}\right)^2 = \left(1 + \sqrt{a - 3}\right)^2$$
$$a + 2 = 1 + 2\sqrt{a - 3} + a - 3$$
$$4 = 2\sqrt{a - 3}$$
$$2 = \sqrt{a - 3}$$

Square both sides again.
$$2^2 = \left(\sqrt{a - 3}\right)^2$$
$$4 = a - 3$$
$$7 = a$$
Check $a = 7$: $\sqrt{9} - \sqrt{4} = 1$ *True*
The solution set is $\{7\}$.

110.
$$\sqrt[3]{5m - 1} = \sqrt[3]{3m - 2}$$
Cube both sides.
$$\left(\sqrt[3]{5m - 1}\right)^3 = \left(\sqrt[3]{3m - 2}\right)^3$$
$$5m - 1 = 3m - 2$$
$$2m = -1$$
$$m = -\tfrac{1}{2}$$
Check $m = -\tfrac{1}{2}$: $\sqrt[3]{-\tfrac{7}{2}} = \sqrt[3]{-\tfrac{7}{2}}$ *True*
The solution set is $\left\{-\tfrac{1}{2}\right\}$.

111.
$$\sqrt[3]{2x^2 + 3x - 7} = \sqrt[3]{2x^2 + 4x + 6}$$
Cube both sides.
$$\left(\sqrt[3]{2x^2 + 3x - 7}\right)^3 = \left(\sqrt[3]{2x^2 + 4x + 6}\right)^3$$
$$2x^2 + 3x - 7 = 2x^2 + 4x + 6$$
$$-13 = x$$
Check $x = -13$: $\sqrt[3]{292} = \sqrt[3]{292}$ *True*
The solution set is $\{-13\}$.

112.
$$\sqrt[3]{3y^2 - 4y + 6} = \sqrt[3]{3y^2 - 2y + 8}$$
$$\left(\sqrt[3]{3y^2 - 4y + 6}\right)^3 = \left(\sqrt[3]{3y^2 - 2y + 8}\right)^3$$
$$3y^2 - 4y + 6 = 3y^2 - 2y + 8$$
$$-2 = 2y$$
$$-1 = y$$
Check $y = -1$: $\sqrt[3]{13} = \sqrt[3]{13}$ *True*
The solution set is $\{-1\}$.

113. $\sqrt[3]{1 - 2k} - \sqrt[3]{-k - 13} = 0$
$$\sqrt[3]{1 - 2k} = \sqrt[3]{-k - 13}$$
Cube both sides.
$$\left(\sqrt[3]{1 - 2k}\right)^3 = \left(\sqrt[3]{-k - 13}\right)^3$$
$$1 - 2k = -k - 13$$
$$14 = k$$
Check $k = 14$: $\sqrt[3]{-27} - \sqrt[3]{-27} = 0$ *True*
The solution set is $\{14\}$.

114. $\sqrt[3]{11 - 2t} - \sqrt[3]{-1 - 5t} = 0$
$$\sqrt[3]{11 - 2t} = \sqrt[3]{-1 - 5t}$$
$$\left(\sqrt[3]{11 - 2t}\right)^3 = \left(\sqrt[3]{-1 - 5t}\right)^3$$
$$11 - 2t = -1 - 5t$$
$$3t = -12$$
$$t = -4$$
Check $t = -4$: $\sqrt[3]{19} - \sqrt[3]{19} = 0$ *True*
The solution set is $\{-4\}$.

115. $\sqrt[4]{x - 1} + 2 = 0$
$$\sqrt[4]{x - 1} = -2$$
This equation has no solution, because $\sqrt[4]{x - 1}$ cannot be negative.
The solution set is $\emptyset$.

116. $\sqrt[4]{2k + 3} + 1 = 0$
$$\sqrt[4]{2k + 3} = -1$$
This equation has no solution, because $\sqrt[4]{2k + 3}$ cannot be negative.
The solution set is $\emptyset$.

117. $\sqrt[4]{x + 7} = \sqrt[4]{2x}$
Raise each side to the fourth power.
$$\left(\sqrt[4]{x + 7}\right)^4 = \left(\sqrt[4]{2x}\right)^4$$
$$x + 7 = 2x$$
$$7 = x$$
Check $x = 7$: $\sqrt[4]{14} = \sqrt[4]{14}$ *True*
The solution set is $\{7\}$.

118. $\sqrt[4]{x+8} = \sqrt[4]{3x}$

Raise each side to the fourth power.

$$\left(\sqrt[4]{x+8}\right)^4 = \left(\sqrt[4]{3x}\right)^4$$
$$x + 8 = 3x$$
$$8 = 2x$$
$$4 = x$$

Check $x = 4$: $\sqrt[4]{12} = \sqrt[4]{12}$ *True*

The solution set is $\{4\}$.

119. (a) Solve $L = \sqrt{H^2 + W^2}$ for H.

$$L^2 = H^2 + W^2$$
$$L^2 - W^2 = H^2$$
$$\sqrt{L^2 - W^2} = H$$

(b) Substitute 12 for L and 9 for W in $H = \sqrt{L^2 - W^2}$.

$$H = \sqrt{12^2 - 9^2}$$
$$= \sqrt{144 - 81} = \sqrt{63}$$

To the nearest tenth of a foot, the height is approximately 7.9 feet.

120. $\sqrt{-25} = i\sqrt{25} = 5i$

121. $\sqrt{-200} = i\sqrt{100 \cdot 2} = 10i\sqrt{2}$

122. If a is a positive real number, then $-a$ is negative. So, $\sqrt{-a}$ is not a real number. Therefore, $-\sqrt{-a}$ is not a real number either.

123. $(-2 + 5i) + (-8 - 7i)$
$$= [-2 + (-8)] + [5 + (-7)]i$$
$$= -10 - 2i$$

124. $(5 + 4i) - (-9 - 3i)$
$$= [(5 - (-9)] + [(4 - (-3)]i$$
$$= 14 + 7i$$

125. $\sqrt{-5} \cdot \sqrt{-7} = i\sqrt{5} \cdot i\sqrt{7} = i^2\sqrt{35}$
$$= -1\left(\sqrt{35}\right) = -\sqrt{35}$$

126. $\sqrt{-25} \cdot \sqrt{-81} = 5i \cdot 9i = 45i^2$
$$= 45(-1) = -45$$

127. $\dfrac{\sqrt{-72}}{\sqrt{-8}} = \dfrac{i\sqrt{72}}{i\sqrt{8}} = \sqrt{\dfrac{72}{8}} = \sqrt{9} = 3$

128. $(2 + 3i)(1 - i) = 2 - 2i + 3i - 3i^2$
$$= 2 + i - 3(-1)$$
$$= 2 + i + 3$$
$$= 5 + i$$

129. $(6 - 2i)^2 = 6^2 - 2 \cdot 6 \cdot 2i + (2i)^2$
$$= 36 - 24i + 4i^2$$
$$= 36 - 24i + 4(-1)$$
$$= 36 - 24i - 4$$
$$= 32 - 24i$$

130. $\dfrac{3 - i}{2 + i}$

Multiply by the conjugate of the denominator, $2 - i$.

$$= \frac{(3 - i)(2 - i)}{(2 + i)(2 - i)}$$
$$= \frac{6 - 3i - 2i + i^2}{4 - i^2}$$
$$= \frac{6 - 5i - 1}{4 - (-1)} = \frac{5 - 5i}{5}$$
$$= \frac{5(1 - i)}{5} = 1 - i$$

131. $\dfrac{5 + 14i}{2 + 3i}$

Multiply by the conjugate of the denominator, $2 - 3i$.

$$= \frac{(5 + 14i)(2 - 3i)}{(2 + 3i)(2 - 3i)}$$
$$= \frac{10 - 15i + 28i - 42i^2}{4 - 9i^2}$$
$$= \frac{10 + 13i - 42(-1)}{4 - 9(-1)} = \frac{52 + 13i}{13}$$
$$= \frac{13(4 + i)}{13} = 4 + i$$

132. $i^{11} = i^8 \cdot i^3 = \left(i^4\right)^2 \cdot i^2 \cdot i$
$$= 1^2 \cdot (-1) \cdot i = -i$$

133. $i^{36} = \left(i^4\right)^9 = 1^9 = 1$

134. $i^{-10} = \dfrac{1}{i^{10}} = \dfrac{1}{i^8 \cdot i^2} = \dfrac{1}{(i^4)^2 \cdot (-1)}$
$$= \frac{1}{1^2 \cdot (-1)} = \frac{1}{-1} = -1$$

Another method:
$i^{-10} = i^{-10} \cdot i^{12} = i^2 = -1$

Note that $i^{12} = \left(i^4\right)^3 = 1^3 = 1$.

135. $i^{-8} = \dfrac{1}{i^8} = \dfrac{1}{(i^4)^2} = \dfrac{1}{1^2} = \dfrac{1}{1} = 1$

136. [8.1] $-\sqrt[4]{256} = -\left(\sqrt[4]{4^4}\right) = -(4) = -4$

137. [8.2] $1000^{-2/3} = \dfrac{1}{1000^{2/3}} = \dfrac{1}{\left[(10^3)^{1/3}\right]^2}$
$$= \frac{1}{10^2} = \frac{1}{100}$$

138. [8.2] $\dfrac{z^{-1/5} \cdot z^{3/10}}{z^{7/10}} = z^{-2/10 + 3/10 - 7/10}$
$$= z^{-6/10} = \frac{1}{z^{6/10}} = \frac{1}{z^{3/5}}$$

139. [8.2] $\sqrt[4]{k^{24}} = k^{24/4} = k^6$

140. [8.3] $\sqrt[3]{54z^9t^8} = \sqrt[3]{27z^9t^6 \cdot 2t^2} = 3z^3t^2\sqrt[3]{2t^2}$

141. [8.4]
$$-5\sqrt{18} + 12\sqrt{72} = -5\sqrt{9 \cdot 2} + 12\sqrt{36 \cdot 2}$$
$$= -5 \cdot 3\sqrt{2} + 12 \cdot 6\sqrt{2}$$
$$= -15\sqrt{2} + 72\sqrt{2} = 57\sqrt{2}$$

142. [8.5] $\dfrac{-1}{\sqrt{12}} = \dfrac{-1}{\sqrt{4 \cdot 3}} = \dfrac{-1}{2\sqrt{3}} = \dfrac{-1 \cdot \sqrt{3}}{2\sqrt{3} \cdot \sqrt{3}}$
$$= \dfrac{-\sqrt{3}}{2 \cdot 3} = \dfrac{-\sqrt{3}}{6}$$

143. [8.5] $\sqrt[3]{\dfrac{12}{25}} = \dfrac{\sqrt[3]{12}}{\sqrt[3]{25}} = \dfrac{\sqrt[3]{12}}{\sqrt[3]{5^2}}$
$$= \dfrac{\sqrt[3]{12} \cdot \sqrt[3]{5}}{\sqrt[3]{5^2} \cdot \sqrt[3]{5}} = \dfrac{\sqrt[3]{60}}{\sqrt[3]{5^3}} = \dfrac{\sqrt[3]{60}}{5}$$

144. [8.7] $i^{-1000} = \dfrac{1}{i^{1000}} = \dfrac{1}{(i^4)^{250}}$
$$= \dfrac{1}{1^{250}} = \dfrac{1}{1} = 1$$

145. [8.7] $\sqrt{-49} = i\sqrt{49} = 7i$

146. [8.7] $(4 - 9i) + (-1 + 2i)$
$$= (4 - 1) + (-9 + 2)i$$
$$= 3 - 7i$$

147. [8.7] $\dfrac{\sqrt{50}}{\sqrt{-2}} = \dfrac{\sqrt{25 \cdot 2}}{i\sqrt{2}} = \dfrac{5\sqrt{2}}{i\sqrt{2}} = \dfrac{5}{i}$
The conjugate of i is $-i$.
$$\dfrac{5(-i)}{i(-i)} = \dfrac{-5i}{-i^2} = \dfrac{-5i}{-(-1)} = -5i$$

148. [8.5] $\dfrac{3 + \sqrt{54}}{6} = \dfrac{3 + \sqrt{9 \cdot 6}}{6} = \dfrac{3 + 3\sqrt{6}}{6}$
$$= \dfrac{3(1 + \sqrt{6})}{2 \cdot 3} = \dfrac{1 + \sqrt{6}}{2}$$

149. [8.7] $(3 + 2i)^2 = 3^2 + 2 \cdot 3 \cdot 2i + (2i)^2$
$$= 9 + 12i + 4i^2$$
$$= 9 + 12i + 4(-1)$$
$$= 5 + 12i$$

150. [8.4] $8\sqrt[3]{x^3 y^2} - 2x\sqrt[3]{y^2} = 8x\sqrt[3]{y^2} - 2x\sqrt[3]{y^2}$
$$= 6x\sqrt[3]{y^2}$$

151. [8.4] $9\sqrt{5} - 4\sqrt{15}$ cannot be simplified further.

152. [8.5] $(\sqrt{5} - \sqrt{3})(\sqrt{7} + \sqrt{3})$
$$= \sqrt{35} + \sqrt{15} - \sqrt{21} - 3$$

153. [8.6] $\sqrt{x + 4} = x - 2$
$$(\sqrt{x + 4})^2 = (x - 2)^2 \qquad Square.$$
$$x + 4 = x^2 - 4x + 4$$
$$0 = x^2 - 5x$$
$$0 = x(x - 5)$$
$$x = 0 \quad \text{or} \quad x - 5 = 0$$
$$x = 5$$
Check $x = 0$: $\sqrt{4} = -2$ *False*
Check $x = 5$: $\sqrt{9} = 3$ *True*
The solution set is $\{5\}$.

154. [8.6] $\sqrt[3]{2x - 9} = \sqrt[3]{5x + 3}$
Cube both sides.
$$(\sqrt[3]{2x - 9})^3 = (\sqrt[3]{5x + 3})^3$$
$$2x - 9 = 5x + 3$$
$$-3x = 12$$
$$x = -4$$
Check $x = -4$: $\sqrt[3]{-17} = \sqrt[3]{-17}$ *True*
The solution set is $\{-4\}$.

155. [8.6] $\sqrt{6 + 2x} - 1 = \sqrt{7 - 2x}$
Square both sides.
$$(\sqrt{6 + 2x} - 1)^2 = (\sqrt{7 - 2x})^2$$
$$6 + 2x - 2\sqrt{6 + 2x} + 1 = 7 - 2x$$
$$4x = 2\sqrt{6 + 2x}$$
$$2x = \sqrt{6 + 2x}$$
Square both sides again.
$$(2x)^2 = (\sqrt{6 + 2x})^2$$
$$4x^2 = 6 + 2x$$
$$4x^2 - 2x - 6 = 0$$
$$2x^2 - x - 3 = 0$$
$$(2x - 3)(x + 1) = 0$$
$$2x - 3 = 0 \quad \text{or} \quad x + 1 = 0$$
$$x = \tfrac{3}{2} \quad \text{or} \qquad x = -1$$
Check $x = \tfrac{3}{2}$: $\sqrt{9} - 1 = \sqrt{4}$ *True*
Check $x = -1$: $\sqrt{4} - 1 = \sqrt{9}$ *False*
The solution set is $\left\{\tfrac{3}{2}\right\}$.

156. [8.6] $\sqrt{7x + 11} - 5 = 0$
$$\sqrt{7x + 11} = 5$$
$$(\sqrt{7x + 11})^2 = 5^2$$
$$7x + 11 = 25$$
$$7x = 14$$
$$x = 2$$
Check $x = 2$: $\sqrt{25} - 5 = 0$ *True*
The solution set is $\{2\}$.

157. [8.6]

$$\sqrt{6x+2} - \sqrt{5x+3} = 0$$

Get one radical on each side of the equals sign.

$$\sqrt{6x+2} = \sqrt{5x+3}$$

Square both sides.

$$\left(\sqrt{6x+2}\right)^2 = \left(\sqrt{5x+3}\right)^2$$
$$6x+2 = 5x+3$$
$$x = 1$$

Check $x = 1$: $\sqrt{8} - \sqrt{8} = 0$ *True*
The solution set is $\{1\}$.

158. [8.6] $\sqrt{3+5x} - \sqrt{x+11} = 0$

$$\sqrt{3+5x} = \sqrt{x+11}$$
$$\left(\sqrt{3+5x}\right)^2 = \left(\sqrt{x+11}\right)^2$$
$$3+5x = x+11$$
$$4x = 8$$
$$x = 2$$

Check $x = 2$: $\sqrt{13} - \sqrt{13} = 0$ *True*
The solution set is $\{2\}$.

159. [8.6] $3\sqrt{x} = \sqrt{8x+9}$

$$\left(3\sqrt{x}\right)^2 = \left(\sqrt{8x+9}\right)^2 \quad \textit{Square.}$$
$$9x = 8x+9$$
$$x = 9$$

Check $x = 9$: $3\sqrt{9} = \sqrt{81}$ *True*
The solution set is $\{9\}$.

160. [8.6] $6\sqrt{p} = \sqrt{30p+24}$

$$\left(6\sqrt{p}\right)^2 = \left(\sqrt{30p+24}\right)^2$$
$$36p = 30p+24$$
$$6p = 24$$
$$p = 4$$

Check $p = 4$: $6\sqrt{4} = \sqrt{144}$ *True*
The solution set is $\{4\}$.

161. [8.6]

$$\sqrt{11+2x} + 1 = \sqrt{5x+1}$$

Square both sides.

$$\left(\sqrt{11+2x}+1\right)^2 = \left(\sqrt{5x+1}\right)^2$$
$$\left(\sqrt{11+2x}\right)^2 + 2\cdot\sqrt{11+2x}\cdot 1 + 1$$
$$= 5x+1$$
$$11+2x+2\sqrt{11+2x}+1 = 5x+1$$

Isolate the remaining radical.

$$2\sqrt{11+2x} = 3x-11$$

Square both sides again.

$$\left(2\sqrt{11+2x}\right)^2 = (3x-11)^2$$
$$4(11+2x) = 9x^2 - 66x + 121$$
$$44 + 8x = 9x^2 - 66x + 121$$
$$0 = 9x^2 - 74x + 77$$
$$0 = (9x-11)(x-7)$$

$$9x-11 = 0 \quad \text{or} \quad x-7 = 0$$
$$x = \tfrac{11}{9} \quad \text{or} \quad x = 7$$

Check $x = \tfrac{11}{9}$: $\sqrt{\tfrac{121}{9}} + 1 = \sqrt{\tfrac{64}{9}}$ *False*

Check $x = 7$: $\sqrt{25} + 1 = \sqrt{36}$ *True*
The solution set is $\{7\}$.

162. [8.6]

$$\sqrt{5x+6} - \sqrt{x+3} = 3$$

Get one radical on each side.

$$\sqrt{5x+6} = 3 + \sqrt{x+3}$$
$$\left(\sqrt{5x+6}\right)^2 = \left(3+\sqrt{x+3}\right)^2$$
$$5x+6 = 9 + 6\sqrt{x+3} + x + 3$$

Isolate the remaining radical.

$$4x - 6 = 6\sqrt{x+3}$$
$$2x - 3 = 3\sqrt{x+3}$$
$$(2x-3)^2 = \left(3\sqrt{x+3}\right)^2$$
$$4x^2 - 12x + 9 = 9(x+3)$$
$$4x^2 - 12x + 9 = 9x + 27$$
$$4x^2 - 21x - 18 = 0$$
$$(4x+3)(x-6) = 0$$

$$4x + 3 = 0 \quad \text{or} \quad x - 6 = 0$$
$$x = -\tfrac{3}{4} \qquad\qquad x = 6$$

Check $x = -\tfrac{3}{4}$: $\sqrt{\tfrac{9}{4}} - \sqrt{\tfrac{9}{4}} = 3$ *False*

Check $x = 6$: $\sqrt{36} - \sqrt{9} = 3$ *True*
The solution set is $\{6\}$.

Chapter 8 Test

1. $\sqrt{841} = 29$, because $29^2 = 841$, so
$-\sqrt{841} = -29$.

2. $\sqrt[3]{-512} = -8$, because $(-8)^3 = -512$.

3. $125^{1/3} = \sqrt[3]{125} = 5$, because $5^3 = 125$.

4. $\sqrt{146.25} \approx \sqrt{144} = 12$, because $12^2 = 144$, so choice **C** is the best estimate.

5. $\sqrt{478} \approx 21.863$

6. $\sqrt[3]{-832} \approx -9.405$

7. $f(x) = \sqrt{x+6}$
For the radicand to be nonnegative, we must have

$$x + 6 \geq 0 \quad \text{or} \quad x \geq -6.$$

Thus, the domain is $[-6, \infty)$.
The function values are positive or zero (the result of the radical), so the range is $[0, \infty)$.

continued

x	$f(x) = \sqrt{x+6}$
-6	$\sqrt{-6+6}=0$
-5	$\sqrt{-5+6}=1$
-2	$\sqrt{-2+6}=2$

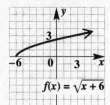

8. $\left(\dfrac{16}{25}\right)^{-3/2} = \left(\dfrac{25}{16}\right)^{3/2} = \dfrac{25^{3/2}}{16^{3/2}}$

$\quad = \dfrac{\left[(5^2)^{1/2}\right]^3}{\left[(4^2)^{1/2}\right]^3} = \dfrac{5^3}{4^3} = \dfrac{125}{64}$

9. $(-64)^{-4/3} = \dfrac{1}{(-64)^{4/3}} = \dfrac{1}{\left(\left[(-4)^3\right]^{1/3}\right)^4}$

$\quad = \dfrac{1}{(-4)^4} = \dfrac{1}{256}$

10. $\dfrac{3^{2/5}x^{-1/4}y^{2/5}}{3^{-8/5}x^{7/4}y^{1/10}}$

$\quad = 3^{2/5-(-8/5)}x^{-1/4-7/4}y^{2/5-1/10}$

$\quad = 3^{10/5}x^{-8/4}y^{4/10-1/10}$

$\quad = 3^2 x^{-2} y^{3/10} = \dfrac{9y^{3/10}}{x^2}$

11. $\left(\dfrac{x^{-4}y^{-6}}{x^{-2}y^3}\right)^{-2/3} = \left(\dfrac{x^2}{x^4y^6y^3}\right)^{-2/3}$

$\quad = \left(\dfrac{1}{x^2y^9}\right)^{-2/3} = (x^2y^9)^{2/3}$

$\quad = (x^2)^{2/3}(y^9)^{2/3} = x^{4/3}y^6$

12. $7^{3/4}\cdot 7^{-1/4} = 7^{3/4+(-1/4)}$

$\quad = 7^{1/2}, \text{ or } \sqrt{7}$

13. $\sqrt[3]{a^4}\cdot\sqrt[3]{a^7} = a^{4/3}\cdot a^{7/3} = a^{4/3+7/3} = a^{11/3}$

$\quad = \sqrt[3]{a^{11}} = \sqrt[3]{(a^3)^3\cdot a^2} = a^3\sqrt[3]{a^2}$

14. $a^2 + b^2 = c^2$

$\quad 12^2 + b^2 = 17^2 \qquad \textit{Let a = 12, c = 17.}$

$\quad 144 + b^2 = 289$

$\qquad b^2 = 145$

$\qquad b = \sqrt{145}$

15. $(-4, 2)$ and $(2, 10)$

$\quad d = \sqrt{(x_2-x_1)^2 + (y_2-y_1)^2}$

$\quad = \sqrt{[2-(-4)]^2 + (10-2)^2}$

$\quad = \sqrt{6^2 + 8^2} = \sqrt{36+64} = \sqrt{100} = 10$

16. $\sqrt{54x^5y^6} = \sqrt{9x^4y^6\cdot 6x} = 3x^2y^3\sqrt{6x}$

17. $\sqrt[4]{32a^7b^{13}} = \sqrt[4]{16a^4b^{12}\cdot 2a^3b}$

$\quad = 2ab^3\sqrt[4]{2a^3b}$

18. $\sqrt{2}\cdot\sqrt[3]{5} = 2^{1/2}\cdot 5^{1/3} = 2^{3/6}\cdot 5^{2/6}$

$\quad = (2^3\cdot 5^2)^{1/6} = \sqrt[6]{2^3\cdot 5^2}$

$\quad = \sqrt[6]{8\cdot 25} = \sqrt[6]{200}$

19. $3\sqrt{20} - 5\sqrt{80} + 4\sqrt{500}$

$\quad = 3\sqrt{4\cdot 5} - 5\sqrt{16\cdot 5} + 4\sqrt{100\cdot 5}$

$\quad = 3\cdot 2\sqrt{5} - 5\cdot 4\sqrt{5} + 4\cdot 10\sqrt{5}$

$\quad = 6\sqrt{5} - 20\sqrt{5} + 40\sqrt{5} = 26\sqrt{5}$

20. $\sqrt[3]{16t^3s^5} - \sqrt[3]{54t^6s^2}$

$\quad = \sqrt[3]{8t^3s^3}\cdot\sqrt[3]{2s^2} - \sqrt[3]{27t^6}\cdot\sqrt[3]{2s^2}$

$\quad = 2ts\sqrt[3]{2s^2} - 3t^2\sqrt[3]{2s^2}$

$\quad = (2ts - 3t^2)\sqrt[3]{2s^2}, \text{ or } t(2s-3t)\sqrt[3]{2s^2}$

21. $(7\sqrt{5}+4)(2\sqrt{5}-1)$

$\quad = 14\cdot 5 - 7\sqrt{5} + 8\sqrt{5} - 4$

$\quad = 70 + \sqrt{5} - 4 = 66 + \sqrt{5}$

22. $(\sqrt{3} - 2\sqrt{5})^2$

$\quad = (\sqrt{3})^2 - 2\cdot\sqrt{3}\cdot 2\sqrt{5} + (2\sqrt{5})^2$

$\quad = 3 - 4\sqrt{15} + 4\cdot 5$

$\quad = 3 - 4\sqrt{15} + 20 = 23 - 4\sqrt{15}$

23. $\dfrac{-5}{\sqrt{40}} = \dfrac{-5}{\sqrt{4\cdot 10}} = \dfrac{-5}{2\sqrt{10}} = \dfrac{-5\cdot\sqrt{10}}{2\sqrt{10}\cdot\sqrt{10}}$

$\quad = \dfrac{-5\sqrt{10}}{2\cdot 10} = \dfrac{-5\sqrt{10}}{20} = -\dfrac{\sqrt{10}}{4}$

24. $\dfrac{2}{\sqrt[3]{5}} = \dfrac{2\cdot\sqrt[3]{5^2}}{\sqrt[3]{5}\sqrt[3]{5^2}} = \dfrac{2\sqrt[3]{25}}{5}$

25. $\dfrac{-4}{\sqrt{7}+\sqrt{5}}$

Multiply the numerator and denominator by the conjugate of the denominator, $\sqrt{7} - \sqrt{5}$.

$\quad = \dfrac{-4(\sqrt{7}-\sqrt{5})}{(\sqrt{7}+\sqrt{5})(\sqrt{7}-\sqrt{5})}$

$\quad = \dfrac{-4(\sqrt{7}-\sqrt{5})}{7-5}$

$\quad = \dfrac{-4(\sqrt{7}-\sqrt{5})}{2} = -2(\sqrt{7}-\sqrt{5})$

26. $\dfrac{6+\sqrt{24}}{2} = \dfrac{6+\sqrt{4\cdot 6}}{2} = \dfrac{6+2\sqrt{6}}{2}$

$\quad = \dfrac{2(3+\sqrt{6})}{2} = 3+\sqrt{6}$

27. **(a)** Substitute 50 for V_0, 0.01 for k, and 30 for T in the formula.

$$V = \frac{V_0}{\sqrt{1 - kT}}$$

$$V = \frac{50}{\sqrt{1 - (0.01)(30)}}$$

$$= \frac{50}{\sqrt{1 - 0.3}} = \frac{50}{\sqrt{0.7}} \approx 59.8$$

The velocity is about 59.8.

(b)
$$V = \frac{V_0}{\sqrt{1 - kT}}$$

$$V^2 = \frac{V_0^2}{1 - kT} \qquad \textit{Square.}$$

$$V^2(1 - kT) = V_0^2$$

$$V^2 - V^2 kT = V_0^2$$

$$V^2 - V_0^2 = V^2 kT$$

$$T = \frac{V^2 - V_0^2}{V^2 k}$$

or $\qquad T = \dfrac{V_0^2 - V^2}{-V^2 k}$

28.
$$\sqrt[3]{5x} = \sqrt[3]{2x - 3}$$

$$\left(\sqrt[3]{5x}\right)^3 = \left(\sqrt[3]{2x - 3}\right)^3$$

$$5x = 2x - 3$$

$$3x = -3$$

$$x = -1$$

Check $x = -1$: $\quad \sqrt[3]{-5} = \sqrt[3]{-5}$ $\quad$ *True*
The solution set is $\{-1\}$.

29. $\quad x + \sqrt{x + 6} = 9 - x$
Isolate the radical.

$$\sqrt{x + 6} = 9 - 2x$$

$$\left(\sqrt{x + 6}\right)^2 = (9 - 2x)^2 \qquad \textit{Square.}$$

$$x + 6 = 81 - 36x + 4x^2$$

$$0 = 4x^2 - 37x + 75$$

$$0 = (x - 3)(4x - 25)$$

$$x - 3 = 0 \quad \text{or} \quad 4x - 25 = 0$$

$$x = 3 \qquad\qquad x = \tfrac{25}{4}$$

Check $x = 3$: $\qquad 3 + \sqrt{9} = 6$ $\quad$ *True*
Check $x = \frac{25}{4}$: $\quad \frac{25}{4} + \sqrt{\frac{49}{4}} = \frac{11}{4}$ $\quad$ *False*
The solution set is $\{3\}$.

30.
$$\sqrt{x + 4} - \sqrt{1 - x} = -1$$

$$\sqrt{x + 4} = -1 + \sqrt{1 - x}$$

$$\left(\sqrt{x + 4}\right)^2 = \left(-1 + \sqrt{1 - x}\right)^2$$

$$x + 4 = 1 - 2\sqrt{1 - x} + 1 - x$$

$$2x + 2 = -2\sqrt{1 - x}$$

$$x + 1 = -\sqrt{1 - x}$$

$$(x + 1)^2 = \left(-\sqrt{1 - x}\right)^2$$

$$x^2 + 2x + 1 = 1 - x$$

$$x^2 + 3x = 0$$

$$x(x + 3) = 0$$

$$x = 0 \quad \text{or} \quad x + 3 = 0$$

$$x = -3$$

Check $x = -3$: $\quad 1 - 2 = -1$ $\quad$ *True*
Check $x = 0$: $\qquad 2 - 1 = -1$ $\quad$ *False*
The solution set is $\{-3\}$.

31. $\quad (-2 + 5i) - (3 + 6i) - 7i$
$$= (-2 - 3) + (5 - 6 - 7)i$$
$$= -5 - 8i$$

32. $\quad (1 + 5i)(3 + i)$
$$= 3 + i + 15i + 5i^2$$
$$= 3 + 16i + 5(-1)$$
$$= -2 + 16i$$

33. $\quad \dfrac{7 + i}{1 - i}$

Multiply the numerator and denominator by the conjugate of the denominator, $1 + i$.

$$= \frac{(7 + i)(1 + i)}{(1 - i)(1 + i)}$$

$$= \frac{7 + 7i + i + i^2}{1 - i^2}$$

$$= \frac{7 + 8i - 1}{1 - (-1)}$$

$$= \frac{6 + 8i}{2} = \frac{2(3 + 4i)}{2} = 3 + 4i$$

34. $\quad i^{37} = i^{36} \cdot i = \left(i^4\right)^9 \cdot i = 1^9 \cdot i = 1 \cdot i = i$

35. **(a)** $i^2 = -1$ is *true.*

(b) $i = \sqrt{-1}$ is *true.*

(c) $i = -1$ is *false; $i = \sqrt{-1}$.*

(d) $\sqrt{-3} = i\sqrt{3}$ is *true.*

Cumulative Review Exercises (Chapters 1–8)

In Exercises 1 and 2, $a = -3$, $b = 5$, and $c = -4$.

1. $\quad \left|2a^2 - 3b + c\right|$
$$= \left|2(-3)^2 - 3(5) + (-4)\right|$$
$$= |2(9) - 15 - 4|$$
$$= |18 - 15 - 4| = |-1| = 1$$

2. $\dfrac{(a+b)(a+c)}{3b-6}$

$= \dfrac{(-3+5)[(-3+(-4)]}{3(5)-6}$

$= \dfrac{(2)(-7)}{15-6} = -\dfrac{14}{9}$

3. $3(x+2)-4(2x+3) = -3x+2$

$3x+6-8x-12 = -3x+2$

$-5x-6 = -3x+2$

$-2x = 8$

$x = -4$

Check $x = -4$: $-6+20 = 14$ *True*

The solution set is $\{-4\}$.

4. $\frac{1}{3}x + \frac{1}{4}(x+8) = x+7$

Multiply by the LCD, 12.

$12\left[\frac{1}{3}x + \frac{1}{4}(x+8)\right] = 12(x+7)$

$4x + 3(x+8) = 12x + 84$

$4x + 3x + 24 = 12x + 84$

$7x + 24 = 12x + 84$

$-5x = 60$

$x = -12$

Check $x = -12$: $-4 - 1 = -5$ *True*

The solution set is $\{-12\}$.

5. $0.04x + 0.06(100-x) = 5.88$

Multiply both sides by 100 to clear decimals.

$4x + 6(100-x) = 588$

$4x + 600 - 6x = 588$

$-2x = -12$

$x = 6$

Check $x = 6$: $0.24 + 5.64 = 5.88$ *True*

The solution set is $\{6\}$.

6. $|6x+7| = 13$

$6x+7 = 13$ or $6x+7 = -13$

$6x = 6$ $6x = -20$

$x = 1$ or $x = -\dfrac{20}{6} = -\dfrac{10}{3}$

Check $x = 1$: $|13| = 13$ *True*

Check $x = -\frac{10}{3}$: $|-13| = 13$ *True*

The solution set is $\left\{-\frac{10}{3}, 1\right\}$.

7. $|-2x+4| = |-2x-3|$

$-2x+4 = -2x-3$ or $-2x+4 = -(-2x-3)$

$4 = -3$ $-2x+4 = 2x+3$

  False $-4x = -1$

$x = \frac{1}{4}$

Check $x = \frac{1}{4}$: $\left|\frac{7}{2}\right| = \left|-\frac{7}{2}\right|$ *True*

The solution set is $\left\{\frac{1}{4}\right\}$.

8. $-5 - 3(m-2) < 11 - 2(m+2)$

$-5 - 3m + 6 < 11 - 2m - 4$

$1 - 3m < 7 - 2m$

$-m < 6$

Multiply by -1; reverse the inequality.

$m > -6$

The solution set is $(-6, \infty)$.

9. The two angles have the same measure, so

$10x - 70 = 7x - 25$

$3x = 45$

$x = 15.$

Then,

$10x - 70 = 10(15) - 70$

$= 150 - 70 = 80.$

Each angle measures $80°$.

10. Let $x =$ the number of nickels and then $50 - x =$ the number of quarters.

Value of nickels	+	value of quarters	=	total value.
$0.05x$	+	$0.25(50-x)$	=	8.90

Multiply by 100 to clear the decimals.

$5x + 25(50-x) = 890$

$5x + 1250 - 25x = 890$

$-20x = -360$

$x = 18$

Since $x = 18$, $50 - x = 50 - 18 = 32$.

There are 18 nickels and 32 quarters.

11. Let $x =$ the amount of pure alcohol.
Make a table.

Number of Liters	Percent (as a decimal)	Liters of Pure Alcohol
x	$100\% = 1$	$1 \cdot x = x$
40	$18\% = 0.18$	$0.18(40) = 7.2$
$x + 40$	$22\% = 0.22$	$0.22(x+40)$

From the last column:

$x + 7.2 = 0.22x + 8.8$

$0.78x = 1.6$

$x = \dfrac{1.6}{0.78} = \dfrac{160}{78} = \dfrac{80}{39}$ or $2\frac{2}{39}$

The required amount is $\frac{80}{39}$ or $2\frac{2}{39}$ liters of pure alcohol.

12. $4x - 3y = 12$
Let $x = 0$ to find the y-intercept, $(0, -4)$.
Let $y = 0$ to find the x-intercept, $(3, 0)$.
Draw a line through the intercepts.

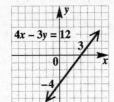

13. $(-4, 6)$ and $(2, -3)$
$$m = \frac{-3 - 6}{2 - (-4)} = \frac{-9}{6} = -\frac{3}{2}$$
Use the point-slope form:
$$y - 6 = -\frac{3}{2}[x - (-4)]$$
$$y - 6 = -\frac{3}{2}(x + 4)$$
$$y - 6 = -\frac{3}{2}x - 6$$
$$y = -\frac{3}{2}x$$

14. $f(x) = 3x - 7$
$f(-10) = 3(-10) - 7 = -30 - 7 = -37$

15. $3x - y = 23 \quad (1)$
$2x + 3y = 8 \quad (2)$
To eliminate y, multiply equation (1) by 3 and add the result to equation (2).

$$\begin{array}{rcll}
9x - 3y &=& 69 & 3 \times (1) \\
2x + 3y &=& 8 & (2) \\
\hline
11x &=& 77 & \\
x &=& 7 &
\end{array}$$

Substitute 7 for x in (2),
$$2(7) + 3y = 8$$
$$14 + 3y = 8$$
$$3y = -6$$
$$y = -2$$

The solution set is $\{(7, -2)\}$.

16. $x + y + z = 1$
$x - y - z = -3$
$x + y - z = -1$

Write the augmented matrix.

$$\begin{bmatrix} 1 & 1 & 1 & | & 1 \\ 1 & -1 & -1 & | & -3 \\ 1 & 1 & -1 & | & -1 \end{bmatrix}$$

$$\begin{bmatrix} 1 & 1 & 1 & | & 1 \\ 0 & -2 & -2 & | & -4 \\ 0 & 0 & -2 & | & -2 \end{bmatrix} \begin{array}{l} \\ -R_1 + R_2 \\ -R_1 + R_3 \end{array}$$

$$\begin{bmatrix} 1 & 1 & 1 & | & 1 \\ 0 & 1 & 1 & | & 2 \\ 0 & 0 & 1 & | & 1 \end{bmatrix} \begin{array}{l} \\ -\frac{1}{2}R_2 \\ -\frac{1}{2}R_3 \end{array}$$

This matrix gives the system

$$x + y + z = 1 \quad (1)$$
$$y + z = 2 \quad (2)$$
$$z = 1.$$

Substitute 1 for z in (2).
$$y + 1 = 2$$
$$y = 1$$

Substitute 1 for y and 1 for z in (1).
$$x + 1 + 1 = 1$$
$$x = -1$$

The solution set is $\{(-1, 1, 1)\}$.

17. Let x = the number of 2-ounce letters and
y = the number of 3-ounce letters.

$$5x + 3y = 5.76 \quad (1)$$
$$3x + 5y = 6.24 \quad (2)$$

To eliminate x, multiply (1) by -3 and (2) by 5 and add the results.

$$\begin{array}{rcll}
-15x - 9y &=& -17.28 & -3 \times (1) \\
15x + 25y &=& 31.20 & 5 \times (2) \\
\hline
16y &=& 13.92 & \\
y &=& 0.87 &
\end{array}$$

Substitute $y = 0.87$ in (1).

$$5x + 3y = 5.76 \quad (1)$$
$$5x + 3(0.87) = 5.76$$
$$5x + 2.61 = 5.76$$
$$5x = 3.15$$
$$x = 0.63$$

The 2006 postage rate for a 2-ounce letter was $0.63 and for a 3-ounce letter, $0.87.

18. $\left(3k^3 - 5k^2 + 8k - 2\right) - \left(4k^3 + 11k + 7\right)$
$+ \left(2k^2 - 5k\right)$
$= 3k^3 - 4k^3 - 5k^2 + 2k^2$
$\quad + 8k - 11k - 5k - 2 - 7$
$= -k^3 - 3k^2 - 8k - 9$

19. $(8x - 7)(x + 3)$
$= 8x^2 + 24x - 7x - 21$
$= 8x^2 + 17x - 21$

20. $\dfrac{8z^3 - 16z^2 + 24z}{8z^2} = \dfrac{8z^3}{8z^2} - \dfrac{16z^2}{8z^2} + \dfrac{24z}{8z^2}$
$= z - 2 + \dfrac{3}{z}$

21.

$$
\begin{array}{r}
3y^3 - 3y^2 + 4y + 1 \\
2y + 1 \overline{\smash{\big)}\ 6y^4 - 3y^3 + 5y^2 + 6y - 9} \\
\underline{6y^4 + 3y^3 } \\
-6y^3 + 5y^2 \\
\underline{-6y^3 - 3y^2 } \\
8y^2 + 6y \\
\underline{8y^2 + 4y } \\
2y - 9 \\
\underline{2y + 1} \\
-10
\end{array}
$$

The answer is

$$3y^3 - 3y^2 + 4y + 1 + \frac{-10}{2y + 1}.$$

22. $2p^2 - 5pq + 3q^2 = (2p - 3q)(p - q)$

23. $3k^4 + k^2 - 4 = \left(3k^2 + 4\right)\left(k^2 - 1\right)$
$$= \left(3k^2 + 4\right)(k + 1)(k - 1)$$

24. $x^3 + 512 = x^3 + 8^3$
$$= (x + 8)\left(x^2 - 8x + 64\right)$$

25. $2x^2 + 11x + 15 = 0$
$(x + 3)(2x + 5) = 0$

$\quad x + 3 = 0 \quad$ or $\quad 2x + 5 = 0$
$\qquad x = -3 \quad$ or $\qquad x = -\frac{5}{2}$
Check $x = -3$: $\quad 18 - 33 + 15 = 0 \quad$ *True*
Check $x = -\frac{5}{2}$: $\quad \frac{25}{2} - \frac{55}{2} + \frac{30}{2} = 0 \quad$ *True*
The solution set is $\left\{-3, -\frac{5}{2}\right\}$.

26. $\qquad 5t(t - 1) = 2(1 - t)$
$\qquad 5t^2 - 5t = 2 - 2t$
$\quad 5t^2 - 3t - 2 = 0$
$(5t + 2)(t - 1) = 0$

$\quad 5t + 2 = 0 \quad$ or $\quad t - 1 = 0$
$\qquad t = -\frac{2}{5} \quad$ or $\qquad t = 1$
Check $x = -\frac{2}{5}$: $\quad -2\left(-\frac{7}{5}\right) = 2\left(\frac{7}{5}\right) \quad$ *True*
Check $x = 1$: $\qquad 5(0) = 2(0) \quad$ *True*
The solution set is $\left\{-\frac{2}{5}, 1\right\}$.

27. $f(x) = \dfrac{2}{x^2 - 9} = \dfrac{2}{(x + 3)(x - 3)}$

The numbers -3 and 3 make the denominator 0 so they must be excluded from the set of all real numbers for the domain of f.
Domain: $\{x \mid x \neq \pm 3\}$

28. $\dfrac{y^2 + y - 12}{y^3 + 9y^2 + 20y} \div \dfrac{y^2 - 9}{y^3 + 3y^2}$

$= \dfrac{y^2 + y - 12}{y(y^2 + 9y + 20)} \cdot \dfrac{y^3 + 3y^2}{y^2 - 9}$

$= \dfrac{(y + 4)(y - 3)}{y(y + 4)(y + 5)} \cdot \dfrac{y^2(y + 3)}{(y + 3)(y - 3)}$

$= \dfrac{y}{y + 5}$

29. $\dfrac{1}{x + y} + \dfrac{3}{x - y} \qquad$ *The LCD is*
$\qquad\qquad\qquad\qquad (x + y)(x - y).$

$= \dfrac{1(x - y)}{(x + y)(x - y)} + \dfrac{3(x + y)}{(x - y)(x + y)}$

$= \dfrac{(x - y) + 3(x + y)}{(x + y)(x - y)}$

$= \dfrac{x - y + 3x + 3y}{(x + y)(x - y)}$

$= \dfrac{4x + 2y}{(x + y)(x - y)}$

30. $\dfrac{\dfrac{-6}{x - 2}}{\dfrac{8}{3x - 6}} = \dfrac{-6}{x - 2} \div \dfrac{8}{3x - 6}$

$= \dfrac{-6}{x - 2} \cdot \dfrac{3x - 6}{8}$

$= \dfrac{-6}{x - 2} \cdot \dfrac{3(x - 2)}{8}$

$= \dfrac{-2 \cdot 3 \cdot 3}{2 \cdot 4} = -\dfrac{9}{4}$

31. $\dfrac{\dfrac{1}{a} - \dfrac{1}{b}}{\dfrac{a}{b} - \dfrac{b}{a}} \qquad$ *The LCD of both the numerator and the denominator is ab.*

$= \dfrac{\dfrac{b - a}{ab}}{\dfrac{a^2 - b^2}{ab}}$

$= \dfrac{b - a}{ab} \div \dfrac{a^2 - b^2}{ab}$

$= \dfrac{b - a}{ab} \cdot \dfrac{ab}{a^2 - b^2}$

$= \dfrac{b - a}{a^2 - b^2}$

$= \dfrac{-(a - b)}{(a - b)(a + b)} = \dfrac{-1}{a + b}$

32. $\dfrac{x^{-1}}{y - x^{-1}} = \dfrac{\dfrac{1}{x}}{y - \dfrac{1}{x}}$

Multiply the numerator and denominator by x.

$$= \dfrac{x\left(\dfrac{1}{x}\right)}{x\left(y - \dfrac{1}{x}\right)} = \dfrac{1}{xy - 1}$$

33. Let x = Mike's speed.
Then $x + 4$ = Cecily's speed.

Use $d = rt$, or $t = \dfrac{d}{r}$, to make a table.

	Distance	Rate	Time
Mike	24	x	$\dfrac{24}{x}$
Cecily	48	$x + 4$	$\dfrac{48}{x + 4}$

Since the times are the same,

$$\dfrac{24}{x} = \dfrac{48}{x + 4}.$$

Multiply by the LCD, $x(x + 4)$.

$$x(x + 4)\left(\dfrac{24}{x}\right) = x(x + 4)\left(\dfrac{48}{x + 4}\right)$$
$$24(x + 4) = 48x$$
$$24x + 96 = 48x$$
$$-24x = -96$$
$$x = 4$$

Since $x = 4$, $x + 4 = 4 + 4 = 8$.
Mike's speed is 4 mph; Cecily's speed is 8 mph.

34. $\dfrac{p + 1}{p - 3} = \dfrac{4}{p - 3} + 6$

Multiply by the LCD, $p - 3$.

$$(p - 3)\left(\dfrac{p + 1}{p - 3}\right) = (p - 3)\left(\dfrac{4}{p - 3} + 6\right)$$
$$p + 1 = 4 + 6(p - 3)$$
$$p + 1 = 4 + 6p - 18$$
$$p + 1 = -14 + 6p$$
$$-5p = -15$$
$$p = 3$$

Substituting 3 in the original equation results in division by zero, so 3 cannot be a solution. The solution set is $\emptyset$.

35. $27^{-2/3} = \dfrac{1}{27^{2/3}} = \dfrac{1}{\left[(3^3)^{1/3}\right]^2} = \dfrac{1}{3^2} = \dfrac{1}{9}$

36. $\sqrt{200x^4} = \sqrt{100x^4 \cdot 2}$
$$= \sqrt{10^2(x^2)^2 \cdot 2} = 10x^2\sqrt{2}$$

37. $\sqrt[3]{16x^2y} \cdot \sqrt[3]{3x^3y}$
$$= \sqrt[3]{8} \cdot \sqrt[3]{2x^2y} \cdot \sqrt[3]{x^3} \cdot \sqrt[3]{3y}$$
$$= 2\sqrt[3]{2x^2y} \cdot x\sqrt[3]{3y}$$
$$= 2x\sqrt[3]{6x^2y^2}$$

38. $\sqrt{50} + \sqrt{8} = \sqrt{25 \cdot 2} + \sqrt{4 \cdot 2}$
$$= 5\sqrt{2} + 2\sqrt{2} = 7\sqrt{2}$$

39. $\dfrac{1}{\sqrt{10} - \sqrt{8}}$

Multiply the numerator and denominator by the conjugate of the denominator, $\sqrt{10} + \sqrt{8}$.

$$= \dfrac{1\left(\sqrt{10} + \sqrt{8}\right)}{\left(\sqrt{10} - \sqrt{8}\right)\left(\sqrt{10} + \sqrt{8}\right)}$$
$$= \dfrac{\sqrt{10} + \sqrt{8}}{10 - 8} = \dfrac{\sqrt{10} + \sqrt{4 \cdot 2}}{2}$$
$$= \dfrac{\sqrt{10} + 2\sqrt{2}}{2}$$

40. $\left(2\sqrt{x} + \sqrt{y}\right)\left(-3\sqrt{x} - 4\sqrt{y}\right)$
$$= -6x - 8\sqrt{xy} - 3\sqrt{xy} - 4y$$
$$= -6x - 11\sqrt{xy} - 4y$$

41. $(-4, 4)$ and $(-2, 9)$
$$d = \sqrt{[(-2) - (-4)]^2 + (9 - 4)^2}$$
$$= \sqrt{2^2 + 5^2}$$
$$= \sqrt{4 + 25} = \sqrt{29}$$

42. $\sqrt{3r - 8} = r - 2$
Square both sides.
$$\left(\sqrt{3r - 8}\right)^2 = (r - 2)^2$$
$$3r - 8 = r^2 - 4r + 4$$
$$0 = r^2 - 7r + 12$$
$$0 = (r - 3)(r - 4)$$

$r - 3 = 0 \quad$ or $\quad r - 4 = 0$
$r = 3 \quad$ or $\qquad r = 4$

Check $r = 3$: $\sqrt{1} = 1$ *True*

Check $r = 4$: $\sqrt{4} = 2$ *True*

The solution set is $\{3, 4\}$.

43. Substitute 32 for D and 5 for h in the given formula.

$$S = \dfrac{2.74D}{\sqrt{h}} = \dfrac{2.74(32)}{\sqrt{5}} = \dfrac{87.68}{\sqrt{5}} \approx 39.2$$

The fall speed is about 39.2 mph.

44. $(5 + 7i) - (3 - 2i)$
$= (5 - 3) + [7 - (-2)]i$
$= 2 + 9i$

45. $\dfrac{6 - 2i}{1 - i}$

Multiply the numerator and denominator by the conjugate of the denominator, $1 + i$.

$= \dfrac{(6 - 2i)(1 + i)}{(1 - i)(1 + i)}$

$= \dfrac{6 + 6i - 2i - 2i^2}{1^2 + 1^2}$

$= \dfrac{6 + 4i + 2}{1 + 1}$

$= \dfrac{8 + 4i}{2} = \dfrac{2(4 + 2i)}{2} = 4 + 2i$

CHAPTER 9 QUADRATIC EQUATIONS, INEQUALITIES, AND FUNCTIONS

9.1 The Square Root Property and Completing the Square

1. By the square root property, if $x^2 = 16$, then

$$x = +\sqrt{16} \quad \text{or} \quad x = -\sqrt{16}$$

Thus, the equation is also true for $x = -4$.

Solution set: $\{-4, 4\}$

3. **(a)** A quadratic equation in standard form has a second-degree polynomial in decreasing powers equal to 0.

(b) The zero-factor property states that if a product equals 0, then at least one of the factors equals 0.

(c) The square root property states that if the square of a quantity equals a number, then the quantity equals the positive or negative square root of the number.

5. $x^2 = 81$
$x = 9 \quad \text{or} \quad x = -9$
Solution set: $\{9, -9\}$

7. $x^2 = 17$
$x = \sqrt{17} \quad \text{or} \quad x = -\sqrt{17}$
Solution set: $\left\{\sqrt{17}, -\sqrt{17}\right\}$

9. $x^2 = 32$
$x = \sqrt{32} \quad \text{or} \quad x = -\sqrt{32}$
$x = 4\sqrt{2} \quad \text{or} \quad x = -4\sqrt{2}$
Solution set: $\left\{4\sqrt{2}, -4\sqrt{2}\right\}$

11. $x^2 - 20 = 0$
$\qquad x^2 = 20$
$x = \sqrt{20} \quad \text{or} \quad x = -\sqrt{20}$
$x = 2\sqrt{5} \quad \text{or} \quad x = -2\sqrt{5}$
Solution set: $\left\{2\sqrt{5}, -2\sqrt{5}\right\}$

13. $3n^2 - 72 = 0$
$\qquad 3n^2 = 72$
$\qquad n = 24$
$n = \sqrt{24} \quad \text{or} \quad n = -\sqrt{24}$
$n = 2\sqrt{6} \quad \text{or} \quad n = -2\sqrt{6}$
Solution set: $\left\{2\sqrt{6}, -2\sqrt{6}\right\}$

15. $(x + 2)^2 = 25$
$x + 2 = \sqrt{25} \quad \text{or} \quad x + 2 = -\sqrt{25}$
$x + 2 = 5 \qquad\qquad x + 2 = -5$
$\quad x = 3 \quad \text{or} \qquad x = -7$
Solution set: $\{-7, 3\}$

17. $(x - 4)^2 = 3$
$x - 4 = \sqrt{3} \quad \text{or} \quad x - 4 = -\sqrt{3}$
$\quad x = 4 + \sqrt{3} \quad \text{or} \quad x = 4 - \sqrt{3}$
Solution set: $\left\{4 + \sqrt{3}, 4 - \sqrt{3}\right\}$

19. $(t + 5)^2 = 48$
$t + 5 = \sqrt{48} \qquad\quad \text{or} \quad t + 5 = -\sqrt{48}$
$t + 5 = 4\sqrt{3} \qquad\qquad\quad t + 5 = -4\sqrt{3}$
$\quad t = -5 + 4\sqrt{3} \quad \text{or} \qquad t = -5 - 4\sqrt{3}$
Solution set: $\left\{-5 + 4\sqrt{3}, -5 - 4\sqrt{3}\right\}$

21. $(3x - 1)^2 = 7$
$3x - 1 = \sqrt{7} \qquad \text{or} \quad 3x - 1 = -\sqrt{7}$
$\quad 3x = 1 + \sqrt{7} \qquad\qquad 3x = 1 - \sqrt{7}$
$\quad x = \dfrac{1 + \sqrt{7}}{3} \quad \text{or} \qquad x = \dfrac{1 - \sqrt{7}}{3}$
Solution set: $\left\{\dfrac{1 + \sqrt{7}}{3}, \dfrac{1 - \sqrt{7}}{3}\right\}$

23. $(4p + 1)^2 = 24$
$4p + 1 = \sqrt{24} \qquad \text{or} \quad 4p + 1 = -\sqrt{24}$
$4p + 1 = 2\sqrt{6} \qquad\qquad 4p + 1 = -2\sqrt{6}$
$\quad 4p = -1 + 2\sqrt{6} \qquad\quad 4p = -1 - 2\sqrt{6}$
$\quad p = \dfrac{-1 + 2\sqrt{6}}{4} \quad \text{or} \qquad p = \dfrac{-1 - 2\sqrt{6}}{4}$
Solution set: $\left\{\dfrac{-1 + 2\sqrt{6}}{4}, \dfrac{-1 - 2\sqrt{6}}{4}\right\}$

25. $(2 - 5t)^2 = 12$
$2 - 5t = \sqrt{12} \qquad\qquad \text{or} \quad 2 - 5t = -\sqrt{12}$
$\quad -5t = -2 + \sqrt{12} \quad \text{or} \qquad -5t = -2 - \sqrt{12}$
$\quad 5t = 2 - \sqrt{12} \qquad \text{or} \qquad 5t = 2 + \sqrt{12}$
$\quad t = \dfrac{2 - 2\sqrt{3}}{5} \qquad \text{or} \qquad t = \dfrac{2 + 2\sqrt{3}}{5}$
Solution set: $\left\{\dfrac{2 + 2\sqrt{3}}{5}, \dfrac{2 - 2\sqrt{3}}{5}\right\}$

27. $\quad d = 16t^2$
$500 = 16t^2$
$\quad t^2 = \dfrac{500}{16} = 31.25$

$t = \sqrt{31.25} \quad \text{or} \quad t = -\sqrt{31.25}$
$t \approx 5.6 \qquad \text{or} \quad t \approx -5.6$

$t \approx 5.6$ seconds (time must be positive)

29. $(2x + 1)^2 = 5$ is more suitable for solving by the square root property. $x^2 + 4x = 12$ is more suitable for solving by completing the square.

31. $x^2 + 6x + \underline{}$

We need to add the square of half the coefficient of x to get a perfect square trinomial.

$$\tfrac{1}{2}(6) = 3 \quad \text{and} \quad 3^2 = 9$$

Add 9 to $x^2 + 6x$ to get a perfect square trinomial.

$$x^2 + 6x + 9 = (x + 3)^2$$

33. $p^2 - 12p + \underline{}$

$$\tfrac{1}{2}(-12) = -6 \quad \text{and} \quad (-6)^2 = 36$$
$$p^2 - 12p + 36 = (p - 6)^2$$

35. $q^2 + 9q + \underline{}$

$$\tfrac{1}{2}(9) = \tfrac{9}{2} \quad \text{and} \quad \left(\tfrac{9}{2}\right)^2 = \tfrac{81}{4}$$
$$q^2 + 9q + \tfrac{81}{4} = \left(q + \tfrac{9}{2}\right)^2$$

37. $x^2 + \tfrac{1}{4}x + \underline{}$

$$\tfrac{1}{2}\left(\tfrac{1}{4}\right) = \tfrac{1}{8} \quad \text{and} \quad \left(\tfrac{1}{8}\right)^2 = \tfrac{1}{64}$$
$$x^2 + \tfrac{1}{4}x + \tfrac{1}{64} = \left(x + \tfrac{1}{8}\right)^2$$

39. $x^2 - 0.8x + \underline{}$

$$\tfrac{1}{2}(-0.8) = -0.4 \quad \text{and} \quad (-0.4)^2 = 0.16$$
$$x^2 - 0.8x + 0.16 = (x - 0.4)^2$$

41. $x^2 + 4x - 2 = 0$
$$x^2 + 4x = 2$$
$$\left[\tfrac{1}{2}(4)\right]^2 = 2^2 = 4$$

43. $x^2 + 10x + 18 = 0$
$$x^2 + 10x = -18$$
$$\left[\tfrac{1}{2}(10)\right]^2 = 5^2 = 25$$

45. $3w^2 - w - 24 = 0$
$$w^2 - \tfrac{1}{3}w - 8 = 0 \quad \textit{Divide by 3.}$$
$$w^2 - \tfrac{1}{3}w = 8$$
$$\left[\tfrac{1}{2}\left(-\tfrac{1}{3}\right)\right]^2 = \left(-\tfrac{1}{6}\right)^2 = \tfrac{1}{36}$$

47. $x^2 - 2x - 24 = 0$
Get the variable terms alone on the left side.

$$x^2 - 2x = 24$$

Complete the square by taking half of -2, the coefficient of x, and squaring the result.

$$\left[\tfrac{1}{2}(-2)\right]^2 = (-1)^2 = 1$$

Add 1 to each side.

$$x^2 - 2x + 1 = 24 + 1$$

Factor the left side.

$$(x - 1)^2 = 25$$

Use the square root property.

$$x - 1 = \sqrt{25} \quad \text{or} \quad x - 1 = -\sqrt{25}$$
$$x - 1 = 5 \quad \text{or} \quad x - 1 = -5$$
$$x = 6 \quad \text{or} \quad x = -4$$

Solution set: $\{-4, 6\}$

49. $x^2 + 4x - 2 = 0$
$$x^2 + 4x = 2$$
$$x^2 + 4x + 4 = 2 + 4 \qquad \left[\tfrac{1}{2}(4)\right]^2 = 4$$
$$(x + 2)^2 = 6$$
$$x + 2 = \sqrt{6} \quad \text{or} \quad x + 2 = -\sqrt{6}$$
$$x = -2 + \sqrt{6} \quad \text{or} \quad x = -2 - \sqrt{6}$$

Solution set: $\left\{-2 + \sqrt{6}, -2 - \sqrt{6}\right\}$

51. $x^2 + 7x - 1 = 0$
$$x^2 + 7x = 1$$
$$x^2 + 7x + \tfrac{49}{4} = 1 + \tfrac{49}{4} \quad \left[\tfrac{1}{2}(7)^2\right] = \tfrac{49}{4}$$
$$\left(x + \tfrac{7}{2}\right)^2 = \tfrac{53}{4}$$
$$x + \tfrac{7}{2} = \sqrt{\tfrac{53}{4}} \quad \text{or} \quad x + \tfrac{7}{2} = -\sqrt{\tfrac{53}{4}}$$
$$x = -\tfrac{7}{2} + \tfrac{\sqrt{53}}{2} \qquad\qquad x = -\tfrac{7}{2} - \tfrac{\sqrt{53}}{2}$$
$$x = \tfrac{-7 + \sqrt{53}}{2} \quad \text{or} \quad x = \tfrac{-7 - \sqrt{53}}{2}$$

Solution set: $\left\{\tfrac{-7 + \sqrt{53}}{2}, \tfrac{-7 - \sqrt{53}}{2}\right\}$

53. $3w^2 - w = 24$
$$w^2 - \tfrac{1}{3}w = 8 \qquad\qquad \textit{Divide by 3.}$$

Complete the square by taking half of $\tfrac{1}{3}$, the coefficient of w, and squaring the result.

$$\left[\tfrac{1}{2}\left(-\tfrac{1}{3}\right)\right]^2 = \left(-\tfrac{1}{6}\right)^2 = \tfrac{1}{36}$$

Add $\tfrac{1}{36}$ to each side.

$$w^2 - \tfrac{1}{3}w + \tfrac{1}{36} = 8 + \tfrac{1}{36}$$
$$\left(w - \tfrac{1}{6}\right)^2 = \tfrac{288}{36} + \tfrac{1}{36}$$
$$\left(w - \tfrac{1}{6}\right)^2 = \tfrac{289}{36}$$
$$w - \tfrac{1}{6} = \sqrt{\tfrac{289}{36}} \quad \text{or} \quad w - \tfrac{1}{6} = -\sqrt{\tfrac{289}{36}}$$
$$w = \tfrac{1}{6} + \tfrac{\sqrt{289}}{\sqrt{36}} \qquad\qquad w = \tfrac{1}{6} - \tfrac{\sqrt{289}}{\sqrt{36}}$$
$$w = \tfrac{1}{6} + \tfrac{17}{6} \qquad\qquad w = \tfrac{1}{6} - \tfrac{17}{6}$$
$$w = \tfrac{18}{6} \qquad\qquad w = -\tfrac{16}{6}$$
$$w = 3 \quad \text{or} \quad w = -\tfrac{8}{3}$$

Solution set: $\left\{-\tfrac{8}{3}, 3\right\}$

55. $2k^2 + 5k - 2 = 0$

$2k^2 + 5k = 2$

$k^2 + \frac{5}{2}k = 1$ *Divide by 2.*

Complete the square.

$$\left(\frac{1}{2} \cdot \frac{5}{2}\right)^2 = \left(\frac{5}{4}\right)^2 = \frac{25}{16}$$

Add $\frac{25}{16}$ to each side.

$k^2 + \frac{5}{2}k + \frac{25}{16} = 1 + \frac{25}{16}$

$\left(k + \frac{5}{4}\right)^2 = \frac{41}{16}$

$k + \frac{5}{4} = \sqrt{\frac{41}{16}}$ or $k + \frac{5}{4} = -\sqrt{\frac{41}{16}}$

$k = -\frac{5}{4} + \frac{\sqrt{41}}{4}$ $k = -\frac{5}{4} - \frac{\sqrt{41}}{4}$

$k = \frac{-5+\sqrt{41}}{4}$ or $k = \frac{-5-\sqrt{41}}{4}$

Solution set: $\left\{\frac{-5+\sqrt{41}}{4}, \frac{-5-\sqrt{41}}{4}\right\}$

57. $5x^2 - 10x + 2 = 0$

$5x^2 - 10x = -2$

$x^2 - 2x = -\frac{2}{5}$ *Divide by 5.*

Complete the square.

$$\left[\frac{1}{2}(-2)\right]^2 = (-1)^2 = 1$$

Add 1 to each side.

$x^2 - 2x + 1 = -\frac{2}{5} + 1$

$(x - 1)^2 = \frac{3}{5}$

$x - 1 = \sqrt{\frac{3}{5}}$ or $x - 1 = -\sqrt{\frac{3}{5}}$

$x - 1 = \frac{\sqrt{3}}{\sqrt{5}} \cdot \frac{\sqrt{5}}{\sqrt{5}}$ $x - 1 = -\frac{\sqrt{3}}{\sqrt{5}} \cdot \frac{\sqrt{5}}{\sqrt{5}}$

$x = \frac{5}{5} + \frac{\sqrt{15}}{5}$ $x = \frac{5}{5} - \frac{\sqrt{15}}{5}$

$x = \frac{5+\sqrt{15}}{5}$ or $x = \frac{5-\sqrt{15}}{5}$

Solution set: $\left\{\frac{5+\sqrt{15}}{5}, \frac{5-\sqrt{15}}{5}\right\}$

59. $9x^2 - 24x = -13$

$x^2 - \frac{24}{9}x = \frac{-13}{9}$ *Divide by 9.*

$x^2 - \frac{8}{3}x = \frac{-13}{9}$

Complete the square.

$$\left[\frac{1}{2}\left(-\frac{8}{3}\right)\right]^2 = \left(-\frac{4}{3}\right)^2 = \frac{16}{9}$$

Add $\frac{16}{9}$ to each side.

$x^2 - \frac{8}{3}x + \frac{16}{9} = \frac{-13}{9} + \frac{16}{9}$

$\left(x - \frac{4}{3}\right)^2 = \frac{3}{9}$

$x - \frac{4}{3} = \sqrt{\frac{3}{9}}$ or $x - \frac{4}{3} = -\sqrt{\frac{3}{9}}$

$x = \frac{4}{3} + \frac{\sqrt{3}}{3}$ $x = \frac{4}{3} - \frac{\sqrt{3}}{3}$

$x = \frac{4+\sqrt{3}}{3}$ or $x = \frac{4-\sqrt{3}}{3}$

Solution set: $\left\{\frac{4+\sqrt{3}}{3}, \frac{4-\sqrt{3}}{3}\right\}$

61. $z^2 - \frac{4}{3}z = -\frac{1}{9}$

Complete the square.

$$\left[\frac{1}{2}\left(-\frac{4}{3}\right)\right]^2 = \left(-\frac{2}{3}\right)^2 = \frac{4}{9}$$

Add $\frac{4}{9}$ to each side.

$z^2 - \frac{4}{3}z + \frac{4}{9} = -\frac{1}{9} + \frac{4}{9}$

$\left(z - \frac{2}{3}\right)^2 = \frac{3}{9}$

$z - \frac{2}{3} = \sqrt{\frac{3}{9}}$ or $z - \frac{2}{3} = -\sqrt{\frac{3}{9}}$

$z = \frac{2}{3} + \frac{\sqrt{3}}{3}$ $z = \frac{2}{3} - \frac{\sqrt{3}}{3}$

$z = \frac{2+\sqrt{3}}{3}$ or $z = \frac{2-\sqrt{3}}{3}$

Solution set: $\left\{\frac{2+\sqrt{3}}{3}, \frac{2-\sqrt{3}}{3}\right\}$

63. $0.1x^2 - 0.2x - 0.1 = 0$

Multiply each side by 10 to clear the decimals.

$x^2 - 2x - 1 = 0$

$x^2 - 2x = 1$

Complete the square.

$$\left[\frac{1}{2}(-2)\right]^2 = (-1)^2 = 1$$

Add 1 to each side.

$x^2 - 2x + 1 = 1 + 1$

$(x - 1)^2 = 2$

$x - 1 = \sqrt{2}$ or $x - 1 = -\sqrt{2}$

$x = 1 + \sqrt{2}$ or $x = 1 - \sqrt{2}$

Solution set: $\left\{1 + \sqrt{2}, 1 - \sqrt{2}\right\}$

65. $x^2 = -12$

$x = \sqrt{-12}$ or $x = -\sqrt{-12}$

$x = i\sqrt{12}$ $x = -i\sqrt{12}$

$x = 2i\sqrt{3}$ or $x = -2i\sqrt{3}$

Solution set: $\left\{2i\sqrt{3}, -2i\sqrt{3}\right\}$

67. $(r - 5)^2 = -4$

$r - 5 = \sqrt{-4}$ or $r - 5 = -\sqrt{-4}$

$r = 5 + 2i$ or $r = 5 - 2i$

Solution set: $\{5 + 2i, 5 - 2i\}$

69. $(6k - 1)^2 = -8$

$6k - 1 = \sqrt{-8}$ or $6k - 1 = -\sqrt{-8}$

$6k - 1 = i\sqrt{8}$ $6k - 1 = -i\sqrt{8}$

$6k - 1 = 2i\sqrt{2}$ $6k - 1 = -2i\sqrt{2}$

$6k = 1 + 2i\sqrt{2}$ $6k = 1 - 2i\sqrt{2}$

$k = \frac{1 + 2i\sqrt{2}}{6}$ or $k = \frac{1 - 2i\sqrt{2}}{6}$

Solution set: $\left\{\frac{1}{6} + \frac{\sqrt{2}}{3}i, \frac{1}{6} - \frac{\sqrt{2}}{3}i\right\}$

71. $m^2 + 4m + 13 = 0$

$m^2 + 4m = -13$

Complete the square.

$\left(\frac{1}{2} \cdot 4\right)^2 = 2^2 = 4$

Add 4 to each side.

$m^2 + 4m + 4 = -13 + 4$

$(m + 2)^2 = -9$

$m + 2 = \sqrt{-9}$ or $m + 2 = -\sqrt{-9}$

$m = -2 + 3i$ or $m = -2 - 3i$

Solution set: $\{-2 + 3i, -2 - 3i\}$

73. $3r^2 + 4r + 4 = 0$

$3r^2 + 4r = -4$

$r^2 + \frac{4}{3}r = \frac{-4}{3}$ *Divide by 3.*

Complete the square.

$\left(\frac{1}{2} \cdot \frac{4}{3}\right)^2 = \left(\frac{2}{3}\right)^2 = \frac{4}{9}$

Add $\frac{4}{9}$ to each side.

$r^2 + \frac{4}{3}r + \frac{4}{9} = \frac{-4}{3} + \frac{4}{9}$

$\left(r + \frac{2}{3}\right)^2 = \frac{-8}{9}$

$r + \frac{2}{3} = \frac{\sqrt{-8}}{\sqrt{9}}$ or $r + \frac{2}{3} = -\frac{\sqrt{-8}}{\sqrt{9}}$

$r = -\frac{2}{3} + \frac{2i\sqrt{2}}{3}$ $\quad r = -\frac{2}{3} - \frac{2i\sqrt{2}}{3}$

Solution set: $\left\{-\frac{2}{3} + \frac{2\sqrt{2}}{3}i, -\frac{2}{3} - \frac{2\sqrt{2}}{3}i\right\}$

75. $-m^2 - 6m - 12 = 0$

Multiply each side by -1.

$m^2 + 6m + 12 = 0$

$m^2 + 6m = -12$

Complete the square.

$\left(\frac{1}{2} \cdot 6\right)^2 = 3^2 = 9$

Add 9 to each side.

$m^2 + 6m + 9 = -12 + 9$

$(m + 3)^2 = -3$

$m + 3 = \sqrt{-3}$ or $m + 3 = -\sqrt{-3}$

$m = -3 + i\sqrt{3}$ or $m = -3 - i\sqrt{3}$

Solution set: $\left\{-3 + i\sqrt{3}, -3 - i\sqrt{3}\right\}$

77. The area of the original square is $x \cdot x$, or x^2.

78. Each rectangular strip has length x and width 1, so each strip has an area of $x \cdot 1$, or x.

79. From Exercise 78, the area of a rectangular strip is x. The area of 6 rectangular strips is $6x$.

80. These are 1 by 1 squares, so each has an area of $1 \cdot 1$, or 1.

81. There are 9 small squares, each with area 1 (from Exercise 80), so the total area is $9 \cdot 1$, or 9.

82. The area of the larger square is $(x + 3)^2$. Using the results from Exercises 77–81,

$$(x + 3)^2, \quad \text{or} \quad x^2 + 6x + 9.$$

83. $x^2 - b = 0$

$x^2 = b$

$x = \sqrt{b}$ or $x = -\sqrt{b}$

Solution set: $\left\{\sqrt{b}, -\sqrt{b}\right\}$

85. $4x^2 = b^2 + 16$

$x^2 = \frac{b^2 + 16}{4}$

$x = \sqrt{\frac{b^2 + 16}{4}}$ or $x = -\sqrt{\frac{b^2 + 16}{4}}$

$x = \frac{\sqrt{b^2 + 16}}{2}$ or $x = -\frac{\sqrt{b^2 + 16}}{2}$

Solution set: $\left\{\frac{\sqrt{b^2 + 16}}{2}, -\frac{\sqrt{b^2 + 16}}{2}\right\}$

87. $(5x - 2b)^2 = 3a$

$5x - 2b = \sqrt{3a}$ or $5x - 2b = -\sqrt{3a}$

$5x = 2b + \sqrt{3a}$ $\quad 5x = 2b - \sqrt{3a}$

$x = \frac{2b + \sqrt{3a}}{5}$ or $x = \frac{2b - \sqrt{3a}}{5}$

Solution set: $\left\{\frac{2b + \sqrt{3a}}{5}, \frac{2b - \sqrt{3a}}{5}\right\}$

89. As in the example, we have $x^2 = 17$ or $x = \pm\sqrt{17} \approx \pm 4.123\,105\,6$.

91. $a = 3, b = 1, c = -1$

$\sqrt{b^2 - 4ac} = \sqrt{1^2 - 4(3)(-1)}$

$= \sqrt{1 + 12}$

$= \sqrt{13}$

93. $a = 6, b = 7, c = 2$

$\sqrt{b^2 - 4ac} = \sqrt{7^2 - 4(6)(2)}$

$= \sqrt{49 - 48}$

$= \sqrt{1} = 1$

95. $a = 3, b = 1, c = -1$ (Exercise 91)

$\frac{-b + \sqrt{b^2 - 4ac}}{2a} = \frac{-1 + \sqrt{13}}{2(3)}$

$= \frac{-1 + \sqrt{13}}{6}$

9.2 The Quadratic Formula

1. The patron forgot the $\pm$ sign in the numerator. The correct formula is
$$x = \frac{-b \pm \sqrt{b^2 - 4ac}}{2a}.$$

3. No, the quadratic formula can be used to solve *any* quadratic equation. Here, the quadratic formula can be used with $a = 2$, $b = 0$, and $c = -5$.

5. $x^2 - 8x + 15 = 0$
Here $a = 1$, $b = -8$, and $c = 15$.
$$x = \frac{-b \pm \sqrt{b^2 - 4ac}}{2a}$$
$$x = \frac{-(-8) \pm \sqrt{(-8)^2 - 4(1)(15)}}{2(1)}$$
$$= \frac{8 \pm \sqrt{64 - 60}}{2}$$
$$= \frac{8 \pm \sqrt{4}}{2} = \frac{8 \pm 2}{2}$$
$$x = \frac{8 + 2}{2} = \frac{10}{2} = 5 \text{ or}$$
$$x = \frac{8 - 2}{2} = \frac{6}{2} = 3$$
Solution set: $\{3, 5\}$

7. $2x^2 + 4x + 1 = 0$
Here $a = 2$, $b = 4$, and $c = 1$.
$$x = \frac{-b \pm \sqrt{b^2 - 4ac}}{2a}$$
$$x = \frac{-4 \pm \sqrt{4^2 - 4(2)(1)}}{2(2)}$$
$$= \frac{-4 \pm \sqrt{16 - 8}}{4}$$
$$= \frac{-4 \pm \sqrt{8}}{4} = \frac{-4 \pm 2\sqrt{2}}{4}$$
$$= \frac{2\left(-2 \pm \sqrt{2}\right)}{2 \cdot 2} = \frac{-2 \pm \sqrt{2}}{2}$$
Solution set: $\left\{\dfrac{-2 + \sqrt{2}}{2}, \dfrac{-2 - \sqrt{2}}{2}\right\}$

9. $2x^2 - 2x = 1$
$2x^2 - 2x - 1 = 0$
Here $a = 2$, $b = -2$, and $c = -1$.
$$x = \frac{-b \pm \sqrt{b^2 - 4ac}}{2a}$$
$$x = \frac{-(-2) \pm \sqrt{(-2)^2 - 4(2)(-1)}}{2(2)}$$
$$= \frac{2 \pm \sqrt{4 + 8}}{4} = \frac{2 \pm \sqrt{12}}{4} = \frac{2 \pm 2\sqrt{3}}{4}$$

$$= \frac{2\left(1 \pm \sqrt{3}\right)}{2 \cdot 2} = \frac{1 \pm \sqrt{3}}{2}$$
Solution set: $\left\{\dfrac{1 + \sqrt{3}}{2}, \dfrac{1 - \sqrt{3}}{2}\right\}$

11. $$x^2 + 18 = 10x$$
$x^2 - 10x + 18 = 0$
Here $a = 1$, $b = -10$, and $c = 18$.
$$x = \frac{-b \pm \sqrt{b^2 - 4ac}}{2a}$$
$$x = \frac{-(-10) \pm \sqrt{(-10)^2 - 4(1)(18)}}{2(1)}$$
$$= \frac{10 \pm \sqrt{100 - 72}}{2} = \frac{10 \pm \sqrt{28}}{2}$$
$$= \frac{10 \pm 2\sqrt{7}}{2} = \frac{2\left(5 \pm \sqrt{7}\right)}{2} = 5 \pm \sqrt{7}$$
Solution set: $\left\{5 + \sqrt{7}, 5 - \sqrt{7}\right\}$

13. $4k^2 + 4k - 1 = 0$
Here $a = 4$, $b = 4$, and $c = -1$.
$$k = \frac{-b \pm \sqrt{b^2 - 4ac}}{2a}$$
$$k = \frac{-4 \pm \sqrt{4^2 - 4(4)(-1)}}{2(4)}$$
$$= \frac{-4 \pm \sqrt{16 + 16}}{8} = \frac{-4 \pm \sqrt{32}}{8}$$
$$= \frac{-4 \pm 4\sqrt{2}}{8} = \frac{4\left(-1 \pm \sqrt{2}\right)}{2 \cdot 4} = \frac{-1 \pm \sqrt{2}}{2}$$
Solution set: $\left\{\dfrac{-1 + \sqrt{2}}{2}, \dfrac{-1 - \sqrt{2}}{2}\right\}$

15. $2 - 2x = 3x^2$
$0 = 3x^2 + 2x - 2$
Here $a = 3$, $b = 2$, and $c = -2$.
$$x = \frac{-b \pm \sqrt{b^2 - 4ac}}{2a}$$
$$x = \frac{-2 \pm \sqrt{2^2 - 4(3)(-2)}}{2(3)}$$
$$= \frac{-2 \pm \sqrt{4 + 24}}{6} = \frac{-2 \pm \sqrt{28}}{6}$$
$$= \frac{-2 \pm 2\sqrt{7}}{6} = \frac{2\left(-1 \pm \sqrt{7}\right)}{2 \cdot 3} = \frac{-1 \pm \sqrt{7}}{3}$$
Solution set: $\left\{\dfrac{-1 + \sqrt{7}}{3}, \dfrac{-1 - \sqrt{7}}{3}\right\}$

17.
$$\frac{x^2}{4} - \frac{x}{2} = 1$$
$$\frac{x^2}{4} - \frac{x}{2} - 1 = 0$$
$$x^2 - 2x - 4 = 0 \quad \textit{Multiply by 4.}$$
Here $a = 1$, $b = -2$, and $c = -4$.

$$x = \frac{-b \pm \sqrt{b^2 - 4ac}}{2a}$$
$$x = \frac{-(-2) \pm \sqrt{(-2)^2 - 4(1)(-4)}}{2(1)}$$
$$= \frac{2 \pm \sqrt{4 + 16}}{2} = \frac{2 \pm \sqrt{20}}{2}$$
$$= \frac{2 \pm 2\sqrt{5}}{2} = 1 \pm \sqrt{5}$$

Solution set: $\left\{ 1 + \sqrt{5}, 1 - \sqrt{5} \right\}$

19.
$$-2t(t + 2) = -3$$
$$-2t^2 - 4t = -3$$
$$-2t^2 - 4t + 3 = 0$$
Here $a = -2$, $b = -4$, and $c = 3$.
$$t = \frac{-b \pm \sqrt{b^2 - 4ac}}{2a}$$
$$t = \frac{-(-4) \pm \sqrt{(-4)^2 - 4(-2)(3)}}{2(-2)}$$
$$= \frac{4 \pm \sqrt{16 + 24}}{-4} = \frac{4 \pm \sqrt{40}}{-4}$$
$$= \frac{4 \pm 2\sqrt{10}}{-4} = \frac{2\left(2 \pm \sqrt{10}\right)}{-2 \cdot 2}$$
$$= \frac{2 \pm \sqrt{10}}{-2} \cdot \frac{-1}{-1} = \frac{-2 \mp \sqrt{10}}{2}$$
$$= \frac{-2 \pm \sqrt{10}}{2}$$

Solution set: $\left\{ \dfrac{-2 + \sqrt{10}}{2}, \dfrac{-2 - \sqrt{10}}{2} \right\}$

21. $(r - 3)(r + 5) = 2$
$$r^2 + 2r - 15 = 2$$
$$r^2 + 2r - 17 = 0$$
Here $a = 1$, $b = 2$, and $c = -17$.
$$r = \frac{-b \pm \sqrt{b^2 - 4ac}}{2a}$$
$$r = \frac{-2 \pm \sqrt{2^2 - 4(1)(-17)}}{2(1)}$$
$$= \frac{-2 \pm \sqrt{4 + 68}}{2} = \frac{-2 \pm \sqrt{72}}{2}$$
$$= \frac{-2 \pm 6\sqrt{2}}{2} = \frac{2\left(-1 \pm 3\sqrt{2}\right)}{2}$$
$$= -1 \pm 3\sqrt{2}$$

Solution set: $\left\{ -1 + 3\sqrt{2}, -1 - 3\sqrt{2} \right\}$

23. $(x + 2)(x - 3) = 1$
$$x^2 - x - 6 = 1$$
$$x^2 - x - 7 = 0$$
Here $a = 1$, $b = -1$, and $c = -7$.
$$x = \frac{-b \pm \sqrt{b^2 - 4ac}}{2a}$$
$$x = \frac{-(-1) \pm \sqrt{(-1)^2 - 4(1)(-7)}}{2(1)}$$
$$= \frac{1 \pm \sqrt{1 + 28}}{2} = \frac{1 \pm \sqrt{29}}{2}$$

Solution set: $\left\{ \dfrac{1 + \sqrt{29}}{2}, \dfrac{1 - \sqrt{29}}{2} \right\}$

25.
$$p = \frac{5(5 - p)}{3(p + 1)}$$
$$3p(p + 1) = 5(5 - p)$$
$$3p^2 + 3p = 25 - 5p$$
$$3p^2 + 8p - 25 = 0$$
Here $a = 3$, $b = 8$, and $c = -25$.

$$p = \frac{-b \pm \sqrt{b^2 - 4ac}}{2a}$$
$$p = \frac{-8 \pm \sqrt{8^2 - 4(3)(-25)}}{2(3)}$$
$$= \frac{-8 \pm \sqrt{64 + 300}}{6} = \frac{-8 \pm \sqrt{364}}{6}$$
$$= \frac{-8 \pm 2\sqrt{91}}{6} = \frac{2\left(-4 \pm \sqrt{91}\right)}{2 \cdot 3} = \frac{-4 \pm \sqrt{91}}{3}$$

Solution set: $\left\{ \dfrac{-4 + \sqrt{91}}{3}, \dfrac{-4 - \sqrt{91}}{3} \right\}$

27. $(2x + 1)^2 = x + 4$
$$4x^2 + 4x + 1 = x + 4$$
$$4x^2 + 3x - 3 = 0$$
Here $a = 4$, $b = 3$, and $c = -3$.

$$x = \frac{-b \pm \sqrt{b^2 - 4ac}}{2a}$$
$$x = \frac{-3 \pm \sqrt{3^2 - 4(4)(-3)}}{2(4)}$$
$$= \frac{-3 \pm \sqrt{9 + 48}}{8}$$
$$= \frac{-3 \pm \sqrt{57}}{8}$$

Solution set: $\left\{ \dfrac{-3 + \sqrt{57}}{8}, \dfrac{-3 - \sqrt{57}}{8} \right\}$

29. $x^2 - 3x + 6 = 0$

Here $a = 1$, $b = -3$, and $c = 6$.

$$x = \frac{-b \pm \sqrt{b^2 - 4ac}}{2a}$$

$$x = \frac{-(-3) \pm \sqrt{(-3)^2 - 4(1)(6)}}{2(1)}$$

$$= \frac{3 \pm \sqrt{9 - 24}}{2}$$

$$= \frac{3 \pm \sqrt{-15}}{2}$$

$$= \frac{3 \pm i\sqrt{15}}{2} = \frac{3}{2} \pm \frac{\sqrt{15}}{2}i$$

Solution set: $\left\{ \frac{3}{2} + \frac{\sqrt{15}}{2}i, \frac{3}{2} - \frac{\sqrt{15}}{2}i \right\}$

31. $r^2 - 6r + 14 = 0$

Here $a = 1$, $b = -6$, and $c = 14$.

$$r = \frac{-b \pm \sqrt{b^2 - 4ac}}{2a}$$

$$r = \frac{-(-6) \pm \sqrt{(-6)^2 - 4(1)(14)}}{2(1)}$$

$$= \frac{6 \pm \sqrt{36 - 56}}{2}$$

$$= \frac{6 \pm \sqrt{-20}}{2} = \frac{6 \pm 2i\sqrt{5}}{2}$$

$$= \frac{2\left(3 \pm i\sqrt{5}\right)}{2} = 3 \pm i\sqrt{5}$$

Solution set: $\left\{ 3 + i\sqrt{5}, 3 - i\sqrt{5} \right\}$

33. $4x^2 - 4x = -7$

$4x^2 - 4x + 7 = 0$

Here $a = 4$, $b = -4$, and $c = 7$.

$$x = \frac{-b \pm \sqrt{b^2 - 4ac}}{2a}$$

$$x = \frac{-(-4) \pm \sqrt{(-4)^2 - 4(4)(7)}}{2(4)}$$

$$= \frac{4 \pm \sqrt{16 - 112}}{8} = \frac{4 \pm \sqrt{-96}}{8}$$

$$= \frac{4 \pm 4i\sqrt{6}}{8} = \frac{4\left(1 \pm i\sqrt{6}\right)}{2 \cdot 4}$$

$$= \frac{1 \pm i\sqrt{6}}{2} = \frac{1}{2} \pm \frac{\sqrt{6}}{2}i$$

Solution set: $\left\{ \frac{1}{2} + \frac{\sqrt{6}}{2}i, \frac{1}{2} - \frac{\sqrt{6}}{2}i \right\}$

35. $x(3x + 4) = -2$

$3x^2 + 4x = -2$

$3x^2 + 4x + 2 = 0$

Here $a = 3$, $b = 4$, and $c = 2$.

$$x = \frac{-b \pm \sqrt{b^2 - 4ac}}{2a}$$

$$x = \frac{-4 \pm \sqrt{4^2 - 4(3)(2)}}{2(3)}$$

$$= \frac{-4 \pm \sqrt{16 - 24}}{6} = \frac{-4 \pm \sqrt{-8}}{6}$$

$$= \frac{-4 \pm 2i\sqrt{2}}{6} = \frac{2\left(-2 \pm i\sqrt{2}\right)}{2 \cdot 3}$$

$$= \frac{-2 \pm i\sqrt{2}}{3} = -\frac{2}{3} \pm \frac{\sqrt{2}}{3}i$$

Solution set: $\left\{ -\frac{2}{3} + \frac{\sqrt{2}}{3}i, -\frac{2}{3} - \frac{\sqrt{2}}{3}i \right\}$

37. $(2x - 1)(8x - 4) = -1$

$16x^2 - 16x + 4 = -1$

$16x^2 - 16x + 5 = 0$

Here $a = 16$, $b = -16$, and $c = 5$.

$$x = \frac{-b \pm \sqrt{b^2 - 4ac}}{2a}$$

$$x = \frac{-(-16) \pm \sqrt{(-16)^2 - 4(16)(5)}}{2(16)}$$

$$= \frac{16 \pm \sqrt{256 - 320}}{32}$$

$$= \frac{16 \pm \sqrt{-64}}{32} = \frac{16 \pm 8i}{32}$$

$$= \frac{16}{32} \pm \frac{8}{32}i = \frac{1}{2} \pm \frac{1}{4}i$$

Solution set: $\left\{ \frac{1}{2} + \frac{1}{4}i, \frac{1}{2} - \frac{1}{4}i \right\}$

Note: We could also solve this equation without the quadratic formula as follows:

$$(2x - 1)(8x - 4) = -1$$
$$(2x - 1)4(2x - 1) = -1$$
$$(2x - 1)^2 = -\frac{1}{4}$$
$$2x - 1 = \pm\sqrt{-\frac{1}{4}}$$
$$2x = 1 \pm \frac{1}{2}i$$
$$x = \frac{1}{2} \pm \frac{1}{4}i$$

39. $25x^2 + 70x + 49 = 0$

Here $a = 25$, $b = 70$, and $c = 49$, so the discriminant is

$$b^2 - 4ac = 70^2 - 4(25)(49)$$
$$= 4900 - 4900$$
$$= 0.$$

Since the discriminant is 0, the quantity under the radical in the quadratic formula is 0, and there is only one rational solution. The answer is **B**.

41. $x^2 + 4x + 2 = 0$

Here $a = 1$, $b = 4$, and $c = 2$, so the discriminant is

$$b^2 - 4ac = 4^2 - 4(1)(2)$$
$$= 16 - 8$$
$$= 8.$$

Since the discriminant is positive, but not a perfect square, there are two distinct irrational number solutions. The answer is **C**.

43. $3x^2 = 5x + 2$
$3x^2 - 5x - 2 = 0$

Here $a = 3$, $b = -5$, and $c = -2$, so the discriminant is

$$b^2 - 4ac = (-5)^2 - 4(3)(-2)$$
$$= 25 + 24$$
$$= 49.$$

Since the discriminant is a perfect square, there are two distinct rational solutions. The answer is **A**.

45. $3m^2 - 10m + 15 = 0$

Here $a = 3$, $b = -10$, and $c = 15$, so the discriminant is

$$b^2 - 4ac = (-10)^2 - 4(3)(15)$$
$$= 100 - 180$$
$$= -80.$$

Since the discriminant is negative, there are two distinct nonreal complex number solutions. The answer is **D**.

47. $0.5x^2 + 10x + 50 = 0$

Here $a = 0.5$, $b = 10$, and $c = 50$, so the discriminant is

$$b^2 - 4ac = 10^2 - 4(0.5)(50)$$
$$= 100 - 100$$
$$= 0.$$

Since the discriminant is 0, the quantity under the radical in the quadratic formula is 0, and there is only one rational solution. The answer is **B**.

49. $25x^2 + 70x + 49 = 0$
$$(5x + 7)^2 = 0$$
$$5x + 7 = 0$$
$$x = -\tfrac{7}{5}$$

Solution set: $\left\{-\tfrac{7}{5}\right\}$

51. $3x^2 = 5x + 2$
$3x^2 - 5x - 2 = 0$
$(3x + 1)(x - 2) = 0$

$3x + 1 = 0$ or $x - 2 = 0$
$x = -\tfrac{1}{3}$ or $x = 2$

Solution set: $\left\{-\tfrac{1}{3}, 2\right\}$

53. **(a)** $3k^2 + 13k = -12$
$3k^2 + 13k + 12 = 0$

Here $a = 3$, $b = 13$, and $c = 12$, so the discriminant is

$$b^2 - 4ac = 13^2 - 4(3)(12)$$
$$= 169 - 144$$
$$= 25.$$

The discriminant is a perfect square, so the equation can be solved by factoring.

$$3k^2 + 13k + 12 = 0$$
$$(3k + 4)(k + 3) = 0$$

$3k + 4 = 0$ or $k + 3 = 0$
$k = -\tfrac{4}{3}$ $k = -3$

Solution set: $\left\{-\tfrac{4}{3}, -3\right\}$

(b) $2x^2 + 19 = 14x$
$2x^2 - 14x + 19 = 0$

Here $a = 2$, $b = -14$, and $c = 19$, so the discriminant is

$$b^2 - 4ac = (-14)^2 - 4(2)(19)$$
$$= 196 - 152$$
$$= 44.$$

The discriminant is not a perfect square, so use the quadratic formula.

$$x = \frac{-b \pm \sqrt{b^2 - 4ac}}{2a}$$

$$x = \frac{-(-14) \pm \sqrt{44}}{2 \cdot 2}$$

$$= \frac{14 \pm 2\sqrt{11}}{2 \cdot 2} = \frac{7 \pm \sqrt{11}}{2}$$

Solution set: $\left\{\dfrac{7 + \sqrt{11}}{2}, \dfrac{7 - \sqrt{11}}{2}\right\}$

55. $p^2 + bp + 25 = 0$
For there to be only one rational solution,
$b^2 - 4ac$ must equal zero.
Since $a = 1$ and $c = 25$,

$$b^2 - 4(1)(25) = 0$$
$$b^2 - 100 = 0$$
$$b^2 = 100$$
$$b = \pm\sqrt{100}$$
$$b = 10 \quad\text{or}\quad b = -10.$$

57. $am^2 + 8m + 1 = 0$

For there to be only one rational solution, $b^2 - 4ac$ must equal zero.

Since $b = 8$ and $c = 1$,

$$8^2 - 4(a)(1) = 0$$
$$64 - 4a = 0$$
$$-4a = -64$$
$$a = 16.$$

59. $9x^2 - 30x + c = 0$

For there to be only one rational solution, $b^2 - 4ac$ must equal zero.

Since $a = 9$ and $b = -30$,

$$(-30)^2 - 4(9)(c) = 0$$
$$900 - 36c = 0$$
$$-36c = -900$$
$$c = 25.$$

61. Substitute $-\frac{5}{2}$ for x and solve for b.

$$4x^2 + bx - 3 = 0$$
$$4\left(-\tfrac{5}{2}\right)^2 + b\left(-\tfrac{5}{2}\right) - 3 = 0$$
$$25 - \tfrac{5}{2}b - 3 = 0$$
$$22 = \tfrac{5}{2}b$$
$$\tfrac{44}{5} = b$$

So the equation is

$$4x^2 + \tfrac{44}{5}x - 3 = 0.$$

Now multiply by 5.

$$20x^2 + 44x - 15 = 0$$

Since $-\frac{5}{2}$ is a solution, $2x + 5$ must be one factor.

$$(2x + 5)(10x - 3) = 0$$

So the other solution is $\frac{3}{10}$.

63. $(7z + 3)^2 + 4(7z + 3) - 5$

$$= u^2 + 4u - 5 \qquad \text{Let } u = 7z + 3.$$
$$= (u + 5)(u - 1)$$

65. $\frac{3}{4}x + \frac{1}{2}x = -10$

$$3x + 2x = -40 \qquad \text{Multiply by 4.}$$
$$5x = -40$$
$$x = -8$$

Solution set: $\{-8\}$.

67. $\sqrt{2x + 6} = x - 1$

$$\left(\sqrt{2x + 6}\right)^2 = (x - 1)^2$$
$$2x + 6 = x^2 - 2x + 1$$
$$0 = x^2 - 4x - 5$$
$$0 = (x - 5)(x + 1)$$
$$x - 5 = 0 \quad \text{or} \quad x + 1 = 0$$
$$x = 5 \quad \text{or} \quad x = -1$$

Check $x = 5$: $\sqrt{16} = 4$ *True*

Check $x = -1$: $\sqrt{4} = -2$ *False*

Solution set: $\{5\}$

9.3 Equations Quadratic in Form

1. $(2x + 3)^2 = 4$

Since the equation has the form $(ax + b)^2 = c$, use the *square root property*.

3. $x^2 + 5x - 8 = 0$

The discriminant is

$$b^2 - 4ac = 5^2 - 4(1)(-8)$$
$$= 25 + 32 = 57.$$

Since the discriminant is not a perfect square, use the *quadratic formula*.

5. $$3x^2 = 2 - 5x$$
$$3x^2 + 5x - 2 = 0$$

The discriminant is

$$b^2 - 4ac = 5^2 - 4(3)(-2)$$
$$= 25 + 24 = 49.$$

Since the discriminant is a perfect square, use *factoring*.

7. $\frac{14}{x} = x - 5$

This is a rational equation, so multiply both sides by the LCD, x.

9. $\left(x^2 + x\right)^2 - 8\left(x^2 + x\right) + 12 = 0$

This is quadratic in form, so substitute a variable for $x^2 + x$.

11. The proposed solution -1 does not check.

Solution set: $\{4\}$

13. $$\frac{14}{x} = x - 5$$

To clear the fraction, multiply each term by the LCD, x.

$$x\left(\frac{14}{x}\right) = x(x) - 5x$$
$$14 = x^2 - 5x$$
$$-x^2 + 5x + 14 = 0$$
$$x^2 - 5x - 14 = 0$$
$$(x + 2)(x - 7) = 0$$

$$x + 2 = 0 \quad \text{or} \quad x - 7 = 0$$
$$x = -2 \quad \text{or} \quad x = 7$$

Check $x = -2$: $-7 = -7$ *True*

Check $x = 7$: $2 = 2$ *True*

Solution set: $\{-2, 7\}$

15.
$$1 - \frac{3}{x} - \frac{28}{x^2} = 0$$

Multiply by the LCD, x^2.

$$x^2(1) - x^2\left(\frac{3}{x}\right) - x^2\left(\frac{28}{x^2}\right) = x^2 \cdot 0$$

$$x^2 - 3x - 28 = 0$$
$$(x + 4)(x - 7) = 0$$
$$x + 4 = 0 \quad \text{or} \quad x - 7 = 0$$
$$x = -4 \quad \text{or} \quad x = 7$$

Check $x = -4$: $1 + \frac{3}{4} - \frac{7}{4} = 0$ *True*

Check $x = 7$: $1 - \frac{3}{7} - \frac{4}{7} = 0$ *True*

Solution set: $\{-4, 7\}$

17. $3 - \frac{1}{t} = \frac{2}{t^2}$

Multiply each term by the LCD, t^2.

$$t^2(3) - t^2\left(\frac{1}{t}\right) = t^2\left(\frac{2}{t^2}\right)$$

$$3t^2 - t = 2$$
$$3t^2 - t - 2 = 0$$
$$(3t + 2)(t - 1) = 0$$

$$3t + 2 = 0 \quad \text{or} \quad t - 1 = 0$$
$$t = -\frac{2}{3} \quad \text{or} \quad t = 1$$

Check $t = -\frac{2}{3}$: $3 + \frac{3}{2} = \frac{9}{2}$ *True*

Check $t = 1$: $3 - 1 = 2$ *True*

Solution set: $\left\{-\frac{2}{3}, 1\right\}$

19. $\frac{1}{x} + \frac{2}{x + 2} = \frac{17}{35}$

Multiply by the LCD, $35x(x + 2)$.

$$35x(x + 2)\left(\frac{1}{x}\right) + 35x(x + 2)\left(\frac{2}{x + 2}\right)$$
$$= 35x(x + 2)\left(\tfrac{17}{35}\right)$$

$$35(x + 2) + 35x(2) = 17x(x + 2)$$
$$35x + 70 + 70x = 17x^2 + 34x$$
$$70 + 105x = 17x^2 + 34x$$
$$0 = 17x^2 - 71x - 70$$
$$0 = (17x + 14)(x - 5)$$

$$17x + 14 = 0 \quad \text{or} \quad x - 5 = 0$$
$$x = -\frac{14}{17} \quad \text{or} \quad x = 5$$

Check $x = -\frac{14}{17}$: $-\frac{17}{14} + \frac{17}{10} = \frac{17}{35}$ *True*

Check $x = 5$: $\frac{1}{5} + \frac{2}{7} = \frac{17}{35}$ *True*

Solution set: $\left\{-\frac{14}{17}, 5\right\}$

21. $\frac{2}{x + 1} + \frac{3}{x + 2} = \frac{7}{2}$

Multiply by the LCD, $2(x + 1)(x + 2)$.

$$2(x + 1)(x + 2)\left(\frac{2}{x + 1} + \frac{3}{x + 2}\right)$$
$$= 2(x + 1)(x + 2)\left(\tfrac{7}{2}\right)$$

$$2(x + 2)(2) + 2(x + 1)(3)$$
$$= (x + 1)(x + 2)(7)$$
$$4x + 8 + 6x + 6 = (x^2 + 3x + 2)(7)$$
$$10x + 14 = 7x^2 + 21x + 14$$
$$0 = 7x^2 + 11x$$
$$0 = x(7x + 11)$$

$$x = 0 \quad \text{or} \quad 7x + 11 = 0$$
$$x = -\frac{11}{7}$$

Check $x = -\frac{11}{7}$: $-\frac{7}{2} + 7 = \frac{7}{2}$ *True*

Check $x = 0$: $2 + \frac{3}{2} = \frac{7}{2}$ *True*

Solution set: $\left\{-\frac{11}{7}, 0\right\}$

23. $\frac{3}{2x} - \frac{1}{2(x + 2)} = 1$

Multiply by the LCD, $2x(x + 2)$.

$$2x(x + 2)\left(\frac{3}{2x} - \frac{1}{2(x + 2)}\right)$$
$$= 2x(x + 2) \cdot 1$$
$$3(x + 2) - x(1) = 2x(x + 2)$$
$$3x + 6 - x = 2x^2 + 4x$$
$$0 = 2x^2 + 2x - 6$$
$$0 = x^2 + x - 3$$

Use $a = 1$, $b = 1$, $c = -3$ in the quadratic formula.

$$x = \frac{-b \pm \sqrt{b^2 - 4ac}}{2a}$$

$$x = \frac{-1 \pm \sqrt{1^2 - 4(1)(-3)}}{2(1)}$$

$$= \frac{-1 \pm \sqrt{1 + 12}}{2} = \frac{-1 \pm \sqrt{13}}{2}$$

Use a calculator to check both proposed solutions. Both solutions check.

Solution set: $\left\{\dfrac{-1 + \sqrt{13}}{2}, \dfrac{-1 - \sqrt{13}}{2}\right\}$

25. $3 = \frac{1}{t + 2} + \frac{2}{(t + 2)^2}$

Multiply by the LCD, $(t + 2)^2$.

$$3(t + 2)^2 = 1(t + 2) + 2$$
$$3(t^2 + 4t + 4) = t + 2 + 2$$
$$3t^2 + 12t + 12 = t + 4$$
$$3t^2 + 11t + 8 = 0$$
$$(3t + 8)(t + 1) = 0$$

$$3t + 8 = 0 \quad \text{or} \quad t + 1 = 0$$
$$t = -\frac{8}{3} \qquad \qquad t = -1$$

Check $t = -\frac{8}{3}$: $3 = -\frac{3}{2} + \frac{9}{2}$ *True*

Check $t = -1$: $3 = 1 + 2$ *True*

Solution set: $\left\{-\frac{8}{3}, -1\right\}$

27.
$$\frac{6}{p} = 2 + \frac{p}{p+1}$$

Multiply by the LCD, $p(p+1)$.
$$6(p+1) = 2p(p+1) + p \cdot p$$
$$6p + 6 = 2p^2 + 2p + p^2$$
$$0 = 3p^2 - 4p - 6$$

Use $a = 3$, $b = -4$, and $c = -6$ in the quadratic formula.

$$p = \frac{-b \pm \sqrt{b^2 - 4ac}}{2a}$$

$$p = \frac{-(-4) \pm \sqrt{(-4)^2 - 4(3)(-6)}}{2(3)}$$

$$= \frac{4 \pm \sqrt{16 + 72}}{2(3)} = \frac{4 \pm \sqrt{88}}{2(3)}$$

$$= \frac{4 \pm 2\sqrt{22}}{2(3)} = \frac{2 \pm \sqrt{22}}{3}$$

Use a calculator to check both proposed solutions. Both solutions check.

Solution set: $\left\{ \dfrac{2 + \sqrt{22}}{3}, \dfrac{2 - \sqrt{22}}{3} \right\}$

29.
$$1 - \frac{1}{2x+1} - \frac{1}{(2x+1)^2} = 0$$

Multiply by the LCD, $(2x+1)^2$.
$$(2x+1)^2 - 1(2x+1) - 1 = 0$$
$$4x^2 + 4x + 1 - 2x - 1 - 1 = 0$$
$$4x^2 + 2x - 1 = 0$$

Use $a = 4$, $b = 2$, $c = -1$ in the quadratic formula.

$$x = \frac{-b \pm \sqrt{b^2 - 4ac}}{2a}$$

$$x = \frac{-2 \pm \sqrt{2^2 - 4(4)(-1)}}{2(4)}$$

$$= \frac{-2 \pm \sqrt{4 + 16}}{2(4)} = \frac{-2 \pm \sqrt{20}}{2(4)}$$

$$= \frac{-2 \pm 2\sqrt{5}}{2(4)} = \frac{-1 \pm \sqrt{5}}{4}$$

Use a calculator to check both proposed solutions. Both solutions check.

Solution set: $\left\{ \dfrac{-1 + \sqrt{5}}{4}, \dfrac{-1 - \sqrt{5}}{4} \right\}$

31. Rate in still water: 20 mph
Rate of current: t mph

(a) When the boat travels upstream, the current works against the rate of the boat in still water, so the rate is $(20 - t)$ mph.

(b) When the boat travels downstream, the current works with the rate of the boat in still water, so the rate is $(20 + t)$ mph.

33. Let $x =$ rate of the boat in still water.
With the speed of the current at 15 mph, then
$x - 15 =$ rate going upstream and
$x + 15 =$ rate going downstream.
Complete a table using the information in the problem, the rates given above, and the formula
$d = rt$ or $t = \dfrac{d}{r}$.

	d	r	t
Upstream	4	$x - 15$	$\dfrac{4}{x-15}$
Downstream	16	$x + 15$	$\dfrac{16}{x+15}$

The time, 48 min, is written as $\frac{48}{60} = \frac{4}{5}$ hr. The time upstream plus the time downstream equals $\frac{4}{5}$.
So, from the table, the equation is written as
$$\frac{4}{x-15} + \frac{16}{x+15} = \frac{4}{5}.$$
Multiply by the LCD, $5(x-15)(x+15)$.

$$5(x-15)(x+15)\left(\frac{4}{x-15} + \frac{16}{x+15} \right)$$
$$= 5(x-15)(x+15) \cdot \frac{4}{5}$$
$$20(x+15) + 80(x-15)$$
$$= 4(x-15)(x+15)$$
$$20x + 300 + 80x - 1200$$
$$= 4(x^2 - 225)$$
$$100x - 900 = 4x^2 - 900$$
$$0 = 4x^2 - 100x$$
$$0 = 4x(x - 25)$$

$4x = 0 \quad$ or $\quad x - 25 = 0$
$x = 0 \quad$ or $\qquad x = 25$

Reject $x = 0$ mph as a possible boat speed.
Eduardo's boat had a top speed of 25 mph.

35. Let $x =$ Rico's speed from Jackson to Lodi.
Then $x + 10 =$ his speed from Lodi to Manteca.

Make a table. Use $t = \frac{d}{r}$.

	d	r	t
Jackson to Lodi	40	x	$\dfrac{40}{x}$
Lodi to Manteca	40	$x + 10$	$\dfrac{40}{x+10}$

Driving time for the entire trip was 88 minutes, or $\frac{88}{60} = \frac{22}{15}$ hours.

continued

$$\frac{40}{x} + \frac{40}{x+10} = \frac{22}{15}$$

Multiply by the LCD, $15x(x+10)$.

$$15x(x+10)\left(\frac{40}{x} + \frac{40}{x+10}\right)$$
$$= 15x(x+10)\left(\tfrac{22}{15}\right)$$

$$600(x+10) + 600x = 22x(x+10)$$
$$600x + 6000 + 600x = 22x^2 + 220x$$
$$0 = 22x^2 - 980x - 6000$$
$$0 = 11x^2 - 490x - 3000$$
$$0 = (11x + 60)(x - 50)$$

$$11x + 60 = 0 \quad \text{or} \quad x - 50 = 0$$
$$x = -\tfrac{60}{11} \quad \text{or} \quad x = 50$$

Reject $-\frac{60}{11}$ since speed cannot be negative. Rico's speed from Jackson to Lodi is 50 mph.

37. Let x be the time in hours required for the faster person to cut the lawn. Then the slower person requires $x + 1$ hours.
Complete the chart.

	Rate	Time Working Together	Fractional Part of the Job Done
Faster Worker	$\frac{1}{x}$	2	$\frac{2}{x}$
Slower Worker	$\frac{1}{x+1}$	2	$\frac{2}{x+1}$

$$\begin{array}{ccc}
\text{Part done by} & & \text{Part done by} & & \text{one whole}\\
\text{faster person} & + & \text{slower person} & = & \text{job.}\\
\frac{2}{x} & + & \frac{2}{x+1} & = & 1
\end{array}$$

Multiply each side by the LCD, $x(x+1)$.

$$x(x+1)\left(\frac{2}{x} + \frac{2}{x+1}\right) = x(x+1)\cdot 1$$
$$2(x+1) + 2x = x^2 + x$$
$$2x + 2 + 2x = x^2 + x$$
$$0 = x^2 - 3x - 2$$

Solve for x using the quadratic formula with $a = 1$, $b = -3$, and $c = -2$.

$$x = \frac{-(-3) \pm \sqrt{(-3)^2 - 4(1)(-2)}}{2(1)}$$
$$= \frac{3 \pm \sqrt{9+8}}{2} = \frac{3 \pm \sqrt{17}}{2}$$
$$x = \frac{3 + \sqrt{17}}{2} \quad \text{or} \quad x = \frac{3 - \sqrt{17}}{2}$$
$$x \approx 3.6 \quad \text{or} \quad x \approx -0.6$$

Discard -0.6 as a solution since time cannot be negative.
It would take the faster person approximately 3.6 hours.

39. Let x represent the time in hours it takes Nancy to plant the flowers. Then $x + 2$ is the time it takes Rusty.
Organize the information in a chart.

Worker	Rate	Time Working Together	Fractional Part of the Job Done
Nancy	$\frac{1}{x}$	12	$\frac{12}{x}$
Rusty	$\frac{1}{x+2}$	12	$\frac{12}{x+2}$

$$\begin{array}{ccc}
\text{Part done} & & \text{part done} & & \text{one whole}\\
\text{by Nancy} & + & \text{by Rusty} & = & \text{job.}\\
\frac{12}{x} & + & \frac{12}{x+2} & = & 1
\end{array}$$

Multiply each side by the LCD, $x(x+2)$.

$$x(x+2)\left(\frac{12}{x} + \frac{12}{x+2}\right) = x(x+2)\cdot 1$$
$$12(x+2) + 12x = x^2 + 2x$$
$$12x + 24 + 12x = x^2 + 2x$$
$$0 = x^2 - 22x - 24$$

Solve for x using the quadratic formula with $a = 1$, $b = -22$, and $c = -24$.

$$x = \frac{-(-22) \pm \sqrt{(-22)^2 - 4(1)(-24)}}{2(1)}$$
$$= \frac{22 \pm \sqrt{580}}{2}$$
$$x = \frac{22 + \sqrt{580}}{2} \approx 23.0 \quad \text{or}$$
$$x = \frac{22 - \sqrt{580}}{2} \approx -1.0$$

Since x represents time, discard the negative solution.
Nancy takes about 23.0 hours planting flowers alone while Rusty takes about 25.0 hours planting alone.

41. Let $x =$ the number of hours it takes for the faster pipe alone to fill the tank.
$x + 3 =$ the number of hours it takes for the slower pipe alone to fill the tank.

Working together, both pipes can fill the tank in 2 hours. Make a chart.

Pipe	Rate	Time	Fractional Part of Tank Filled
Faster	$\dfrac{1}{x}$	2	$\dfrac{2}{x}$
Slower	$\dfrac{1}{x+3}$	2	$\dfrac{2}{x+3}$

Since together the faster and slower pipes fill one tank, the sum of their fractional parts is 1; that is,

$$\frac{2}{x} + \frac{2}{x+3} = 1.$$

Multiply by the LCD, $x(x+3)$.

$$2(x+3) + 2(x) = x(x+3)$$
$$2x + 6 + 2x = x^2 + 3x$$
$$0 = x^2 - x - 6$$
$$0 = (x-3)(x+2)$$

$$x - 3 = 0 \quad \text{or} \quad x + 2 = 0$$
$$x = 3 \quad \text{or} \quad x = -2$$

Reject -2. The faster pipe takes 3 hours to fill the tank alone and the slower pipe takes 6 hours to fill the tank alone.

43.
$$x = \sqrt{7x - 10}$$
$$(x)^2 = \left(\sqrt{7x - 10}\right)^2$$
$$x^2 = 7x - 10$$
$$x^2 - 7x + 10 = 0$$
$$(x-2)(x-5) = 0$$

$$x - 2 = 0 \quad \text{or} \quad x - 5 = 0$$
$$x = 2 \quad \text{or} \quad x = 5$$

Check $x = 2$: $2 = \sqrt{4}$ *True*
Check $x = 5$: $5 = \sqrt{25}$ *True*
Solution set: $\{2, 5\}$

45.
$$2x = \sqrt{11x + 3}$$
$$(2x)^2 = \left(\sqrt{11x + 3}\right)^2$$
$$4x^2 = 11x + 3$$
$$4x^2 - 11x - 3 = 0$$
$$(4x+1)(x-3) = 0$$

$$4x + 1 = 0 \quad \text{or} \quad x - 3 = 0$$
$$x = -\frac{1}{4} \quad \text{or} \quad x = 3$$

Check $x = -\frac{1}{4}$: $-\frac{1}{2} = \sqrt{\frac{1}{4}}$ *False*

Check $x = 3$: $6 = \sqrt{36}$ *True*

Solution set: $\{3\}$

47.
$$3x = \sqrt{16 - 10x}$$
$$(3x)^2 = \left(\sqrt{16 - 10x}\right)^2$$
$$9x^2 = 16 - 10x$$
$$9x^2 + 10x - 16 = 0$$
$$(9x - 8)(x + 2) = 0$$

$$9x - 8 = 0 \quad \text{or} \quad x + 2 = 0$$
$$x = \frac{8}{9} \quad \text{or} \quad x = -2$$

Check $x = \frac{8}{9}$: $\frac{8}{3} = \sqrt{\frac{64}{9}}$ *True*
Check $x = -2$: $-6 = \sqrt{36}$ *False*

Solution set: $\left\{\frac{8}{9}\right\}$

49. $k + \sqrt{k} = 12$
$$\sqrt{k} = 12 - k$$
$$\left(\sqrt{k}\right)^2 = (12 - k)^2$$
$$k = 144 - 24k + k^2$$
$$0 = k^2 - 25k + 144$$
$$0 = (k - 9)(k - 16)$$

$$k - 9 = 0 \quad \text{or} \quad k - 16 = 0$$
$$k = 9 \qquad\qquad k = 16$$

Check $k = 9$: $9 + 3 = 12$ *True*
Check $k = 16$: $16 + 4 = 12$ *False*

Solution set: $\{9\}$

51.
$$m = \sqrt{\frac{6 - 13m}{5}}$$
$$m^2 = \frac{6 - 13m}{5}$$
$$5m^2 = 6 - 13m$$
$$5m^2 + 13m - 6 = 0$$
$$(5m - 2)(m + 3) = 0$$

$$5m - 2 = 0 \quad \text{or} \quad m + 3 = 0$$
$$m = \frac{2}{5} \quad \text{or} \quad m = -3$$

Check $m = \frac{2}{5}$: $\frac{2}{5} = \sqrt{\frac{4}{25}}$ *True*

Check $m = -3$: $-3 = \sqrt{9}$ *False*

Solution set: $\left\{\frac{2}{5}\right\}$

53.
$$-x = \sqrt{\frac{8 - 2x}{3}}$$
$$(-x)^2 = \left(\sqrt{\frac{8 - 2x}{3}}\right)^2$$
$$x^2 = \frac{8 - 2x}{3}$$
$$3x^2 = 8 - 2x$$
$$3x^2 + 2x - 8 = 0$$
$$(3x - 4)(x + 2) = 0$$
$$3x - 4 = 0 \quad \text{or} \quad x + 2 = 0$$
$$x = \frac{4}{3} \quad \text{or} \quad x = -2$$

Check $x = \frac{4}{3}$: $-\frac{4}{3} = \sqrt{\frac{16}{9}}$ *False*

Check $x = -2$: $2 = \sqrt{4}$ *True*

Solution set: $\{-2\}$

55. $x^4 - 29x^2 + 100 = 0$

Let $u = x^2$ and $u^2 = x^4$ to get

$u^2 - 29u + 100 = 0$

$(u - 4)(u - 25) = 0$

$u - 4 = 0$ or $u - 25 = 0$

$u = 4$ or $u = 25$

To find x, substitute x^2 for u.

$x^2 = 4$ or $x^2 = 25$

$x = \pm 2$ or $x = \pm 5$

Check $x = \pm 2$: $16 - 116 + 100 = 0$ *True*

Check $x = \pm 5$: $625 - 725 + 100 = 0$ *True*

Solution set: $\{-5, -2, 2, 5\}$

57. $4k^4 - 13k^2 + 9 = 0$

Let $u = k^2$ and $u^2 = k^4$ to get

$4u^2 - 13u + 9 = 0$

$(4u - 9)(u - 1) = 0$

$4u - 9 = 0$ or $u - 1 = 0$

$u = \frac{9}{4}$ or $u = 1$.

To find k, substitute k^2 for u.

$k^2 = \frac{9}{4}$ or $k^2 = 1$

$k = \pm \frac{3}{2}$ or $k = \pm 1$

Check $k = \pm \frac{3}{2}$: $\frac{81}{4} - \frac{117}{4} + 9 = 0$ *True*

Check $k = \pm 1$: $4 - 13 + 9 = 0$ *True*

Solution set: $\left\{-\frac{3}{2}, -1, 1, \frac{3}{2}\right\}$

59. $\qquad x^4 + 48 = 16x^2$

$x^4 - 16x^2 + 48 = 0$

Let $u = x^2$, so $u^2 = x^4$. The equation becomes

$u^2 - 16u + 48 = 0$.

$(u - 4)(u - 12) = 0$

$u - 4 = 0$ or $u - 12 = 0$

$u = 4$ or $u = 12$

To find x substitute x^2 for u.

$x^2 = 4$ $\qquad$ $x^2 = 12$

$x = \pm \sqrt{4}$ $\qquad$ $x = \pm \sqrt{12}$

$x = \pm 2$ or $x = \pm 2\sqrt{3}$

Check $x = \pm 2$: $\qquad 16 + 48 = 64$ *True*

Check $x = \pm 2\sqrt{3}$: $144 + 48 = 192$ *True*

Solution set: $\left\{-2\sqrt{3}, -2, 2, 2\sqrt{3}\right\}$

61. $(x + 3)^2 + 5(x + 3) + 6 = 0$

Let $u = x + 3$, so $u^2 = (x + 3)^2$.

$u^2 + 5u + 6 = 0$

$(u + 3)(u + 2) = 0$

$u + 3 = 0$ or $u + 2 = 0$

$u = -3$ or $u = -2$

To find x, substitute $x + 3$ for u.

$x + 3 = -3$ or $x + 3 = -2$

$x = -6$ or $x = -5$

Check $x = -6$: $9 - 15 + 6 = 0$ *True*

Check $x = -5$: $4 - 10 + 6 = 0$ *True*

Solution set: $\{-6, -5\}$

63. $3(m + 4)^2 - 8 = 2(m + 4)$

Let $x = m + 4$, so $x^2 = (m + 4)^2$.

$3x^2 - 8 = 2x$

$3x^2 - 2x - 8 = 0$

$(3x + 4)(x - 2) = 0$

$3x + 4 = 0$ or $x - 2 = 0$

$x = -\frac{4}{3}$ $\qquad$ $x = 2$

$m + 4 = -\frac{4}{3}$ or $m + 4 = 2$

$m = -\frac{16}{3}$ $\qquad$ $m = -2$

Check $m = -\frac{16}{3}$: $\frac{16}{3} - 8 = -\frac{8}{3}$ *True*

Check $m = -2$: $12 - 8 = 4$ *True*

Solution set: $\left\{-\frac{16}{3}, -2\right\}$

65. $2 + \dfrac{5}{3k - 1} = \dfrac{-2}{(3k - 1)^2}$

Let $u = 3k - 1$, so $u^2 = (3k - 1)^2$.

$2 + \dfrac{5}{u} = -\dfrac{2}{u^2}$

Multiply by the LCD, u^2.

$u^2\left(2 + \dfrac{5}{u}\right) = u^2\left(-\dfrac{2}{u^2}\right)$

$2u^2 + 5u = -2$

$2u^2 + 5u + 2 = 0$

$(2u + 1)(u + 2) = 0$

$2u + 1 = 0$ or $u + 2 = 0$

$u = -\frac{1}{2}$ $\qquad$ $u = -2$

To find k, substitute $3k - 1$ for u.

$3k - 1 = -\frac{1}{2}$ or $3k - 1 = -2$

$3k = \frac{1}{2}$ $\qquad$ $3k = -1$

$k = \frac{1}{6}$ or $k = -\frac{1}{3}$

Check $k = \frac{1}{6}$: $\qquad 2 - 10 = -8$ *True*

Check $k = -\frac{1}{3}$: $2 - \frac{5}{2} = -\frac{1}{2}$ *True*

Solution set: $\left\{-\frac{1}{3}, \frac{1}{6}\right\}$

67. $2 - 6(m-1)^{-2} = (m-1)^{-1}$

Let $u = m - 1$ to get
$$2 - 6u^{-2} = u^{-1}$$
or $\quad 2 - \dfrac{6}{u^2} = \dfrac{1}{u}.$

Multiply by the LCD, u^2.
$$2u^2 - 6 = u$$
$$2u^2 - u - 6 = 0$$
$$(2u+3)(u-2) = 0$$

$$2u + 3 = 0 \quad \text{or} \quad u - 2 = 0$$
$$u = -\tfrac{3}{2} \quad \text{or} \quad u = 2$$

To find m, substitute $m - 1$ for u.
$$m - 1 = -\tfrac{3}{2} \quad \text{or} \quad m - 1 = 2$$
$$m = -\tfrac{1}{2} \quad \text{or} \quad m = 3$$

Check $m = -\tfrac{1}{2}$: $\quad 2 - \tfrac{8}{3} = -\tfrac{2}{3}$ *True*

Check $m = 3$: $\quad 2 - \tfrac{3}{2} = \tfrac{1}{2}$ *True*

Solution set: $\left\{ -\tfrac{1}{2}, 3 \right\}$

69. $x^{2/3} + x^{1/3} - 2 = 0$

Let $u = x^{1/3}$, so $u^2 = x^{2/3}$.
$$u^2 + u - 2 = 0$$
$$(u+2)(u-1) = 0$$
$$u + 2 = 0 \quad \text{or} \quad u - 1 = 0$$
$$u = -2 \quad \text{or} \quad u = 1$$

To find x, substitute $x^{1/3}$ for u.
$$x^{1/3} = -2 \quad \text{or} \quad x^{1/3} = 1$$

Cube both sides of each equation.
$$\left(x^{1/3}\right)^3 = (-2)^3 \qquad \left(x^{1/3}\right)^3 = 1^3$$
$$x = -8 \quad \text{or} \quad x = 1$$

Check $x = -8$: $\quad 4 - 2 - 2 = 0$ *True*

Check $x = 1$: $\quad 1 + 1 - 2 = 0$ *True*

Solution set: $\{-8, 1\}$

71. $r^{2/3} + r^{1/3} - 12 = 0$

Let $u = r^{1/3}$, so $u^2 = r^{2/3}$. The equation becomes
$$u^2 + u - 12 = 0.$$
$$(u+4)(u-3) = 0$$
$$u + 4 = 0 \quad \text{or} \quad u - 3 = 0$$
$$u = -4 \quad \text{or} \quad u = 3$$

To find r, substitute $r^{1/3}$ for u.
$$r^{1/3} = -4 \quad \text{or} \quad r^{1/3} = 3$$
$$\left(r^{1/3}\right)^3 = (-4)^3 \qquad \left(r^{1/3}\right)^3 = 3^3$$
$$r = -64 \quad \text{or} \quad r = 27$$

Check $r = -64$: $\quad 16 - 4 - 12 = 0$ *True*

Check $r = 27$: $\quad 9 + 3 - 12 = 0$ *True*

Solution set: $\{-64, 27\}$

73. $4k^{4/3} - 13k^{2/3} + 9 = 0$

Let $x = k^{2/3}$, so $x^2 = k^{4/3}$.
$$4x^2 - 13x + 9 = 0$$
$$(4x - 9)(x - 1) = 0$$
$$4x - 9 = 0 \quad \text{or} \quad x - 1 = 0$$
$$x = \tfrac{9}{4} \quad \text{or} \quad x = 1$$
$$k^{2/3} = \tfrac{9}{4} \quad \text{or} \quad k^{2/3} = 1$$
$$\left(k^{2/3}\right)^{1/2} = \left(\tfrac{9}{4}\right)^{1/2} \quad \text{or} \quad \left(k^{2/3}\right)^{1/2} = 1^{1/2}$$
$$k^{1/3} = \pm\tfrac{3}{2} \quad \text{or} \quad k^{1/3} = \pm 1$$
$$\left(k^{1/3}\right)^3 = \left(\pm\tfrac{3}{2}\right)^3 \quad \text{or} \quad \left(k^{1/3}\right)^3 = (\pm 1)^3$$
$$k = \pm\tfrac{27}{8} \quad \text{or} \quad k = \pm 1$$

Check $k = \pm\tfrac{27}{8}$:
$$4\left(\tfrac{81}{16}\right) - 13\left(\tfrac{9}{4}\right) + 9 \overset{?}{=} 0$$
$$\tfrac{81}{4} - \tfrac{117}{4} + \tfrac{36}{4} = 0 \quad \text{*True*}$$

Check $k = \pm 1$:
$$4 - 13 + 9 = 0 \quad \text{*True*}$$

Solution set: $\left\{ -\tfrac{27}{8}, -1, 1, \tfrac{27}{8} \right\}$

75. $2\left(1 + \sqrt{r}\right)^2 = 13\left(1 + \sqrt{r}\right) - 6$

Let $u = 1 + \sqrt{r}$.
$$2u^2 = 13u - 6$$
$$2u^2 - 13u + 6 = 0$$
$$(2u - 1)(u - 6) = 0$$
$$2u - 1 = 0 \quad \text{or} \quad u - 6 = 0$$
$$u = \tfrac{1}{2} \quad \text{or} \quad u = 6$$

Replace u with $1 + \sqrt{r}$.
$$1 + \sqrt{r} = \tfrac{1}{2} \quad \text{or} \quad 1 + \sqrt{r} = 6$$
$$\sqrt{r} = -\tfrac{1}{2} \qquad\qquad \sqrt{r} = 5$$
Not possible, $\qquad\qquad r = 25$
since $\sqrt{r} \geq 0.$

Check $r = 25$: $\quad 72 = 78 - 6$ *True*

Solution set: $\{25\}$

77. $2x^4 + x^2 - 3 = 0$

Let $m = x^2$, so $m^2 = x^4$.
$$2m^2 + m - 3 = 0$$
$$(2m + 3)(m - 1) = 0$$
$$2m + 3 = 0 \quad \text{or} \quad m - 1 = 0$$
$$m = -\tfrac{3}{2} \quad \text{or} \quad m = 1$$

To find x, substitute x^2 for m.
$$x^2 = -\tfrac{3}{2} \quad \text{or} \quad x^2 = 1$$

continued

$$x^2 = -\frac{3}{2} \qquad \text{or} \quad x^2 = 1$$

$$x = \pm\sqrt{-\frac{3}{2}} \qquad x = \pm\sqrt{1}$$

$$x = \pm\frac{\sqrt{3}}{\sqrt{2}} \cdot \frac{\sqrt{2}}{\sqrt{2}}i \qquad x = \pm 1$$

$$x = \pm\frac{\sqrt{6}}{2}i$$

Check $x = \pm\frac{\sqrt{6}}{2}i$: $\frac{9}{2} - \frac{3}{2} - 3 = 0$ *True*

Check $x = \pm 1$: $2 + 1 - 3 = 0$ *True*

Solution set: $\left\{-1, 1, -\frac{\sqrt{6}}{2}i, \frac{\sqrt{6}}{2}i\right\}$

79. $12x^4 - 11x^2 + 2 = 0$

Let $u = x^2$, so $u^2 = x^4$.

$$12u^2 - 11u + 2 = 0$$

$$(4u - 1)(3u - 2) = 0$$

$$4u - 1 = 0 \qquad \text{or} \quad 3u - 2 = 0$$

$$u = \frac{1}{4} \qquad \text{or} \qquad u = \frac{2}{3}$$

$$x^2 = \frac{1}{4} \qquad\qquad x^2 = \frac{2}{3}$$

$$x = \pm\frac{1}{2} \qquad\qquad x = \pm\sqrt{\frac{2}{3}}$$

Note: $x = \pm\sqrt{\frac{2}{3}} = \pm\frac{\sqrt{2}}{\sqrt{3}} \cdot \frac{\sqrt{3}}{\sqrt{3}} = \pm\frac{\sqrt{6}}{3}$

Check $x = \pm\frac{1}{2}$: $\frac{3}{4} - \frac{11}{4} + 2 = 0$ *True*

Check $x = \pm\frac{\sqrt{6}}{3}$: $\frac{16}{3} - \frac{22}{3} + 2 = 0$ *True*

Solution set: $\left\{-\frac{\sqrt{6}}{3}, -\frac{1}{2}, \frac{1}{2}, \frac{\sqrt{6}}{3}\right\}$

81. $\sqrt{2x + 3} = 2 + \sqrt{x - 2}$

Square both sides.

$$\left(\sqrt{2x + 3}\right)^2 = \left(2 + \sqrt{x - 2}\right)^2$$

$$2x + 3 = 4 + 4\sqrt{x - 2} + (x - 2)$$

$$2x + 3 = x + 2 + 4\sqrt{x - 2}$$

Isolate the radical term on one side.

$$x + 1 = 4\sqrt{x - 2}$$

Square both sides again.

$$(x + 1)^2 = \left(4\sqrt{x - 2}\right)^2$$

$$x^2 + 2x + 1 = 16(x - 2)$$

$$x^2 + 2x + 1 = 16x - 32$$

$$x^2 - 14x + 33 = 0$$

$$(x - 11)(x - 3) = 0$$

$$x - 11 = 0 \quad \text{or} \quad x - 3 = 0$$

$$x = 11 \quad \text{or} \qquad x = 3$$

Check $x = 11$: $\sqrt{25} = 2 + \sqrt{9}$ *True*

Check $x = 3$: $\sqrt{9} = 2 + \sqrt{1}$ *True*

Solution set: $\{3, 11\}$

83. $2m^6 + 11m^3 + 5 = 0$

Let $y = m^3$, so $y^2 = m^6$.

$\cdot\ 2y^2 + 11y + 5 = 0$

$(2y + 1)(y + 5) = 0$

$2y + 1 = 0 \qquad \text{or} \quad y + 5 = 0$

$$y = -\frac{1}{2} \quad \text{or} \qquad y = -5$$

To find m, substitute m^3 for y.

$$m^3 = -\frac{1}{2} \quad \text{or} \quad m^3 = -5$$

Take the cube root of both sides of each equation.

$$m = \sqrt[3]{-\frac{1}{2}} \qquad\qquad \text{or} \quad m = \sqrt[3]{-5}$$

$$m = -\sqrt[3]{\frac{1}{2}} \qquad\qquad m = -\sqrt[3]{5}$$

$$= -\frac{\sqrt[3]{1}}{\sqrt[3]{2}} \cdot \frac{\sqrt[3]{2^2}}{\sqrt[3]{2^2}}$$

$$= -\frac{\sqrt[3]{4}}{2}$$

Check $m = -\frac{\sqrt[3]{4}}{2}$: $\frac{1}{2} - \frac{11}{2} + 5 = 0$ *True*

Check $m = -\sqrt[3]{5}$: $50 - 55 + 5 = 0$ *True*

Solution set: $\left\{-\sqrt[3]{5}, -\frac{\sqrt[3]{4}}{2}\right\}$

85. $6 = 7(2w - 3)^{-1} + 3(2w - 3)^{-2}$

Let $x = (2w - 3)^{-1}$, so $x^2 = (2w - 3)^{-2}$.

$6 = 7x + 3x^2$

$0 = 3x^2 + 7x - 6$

$0 = (3x - 2)(x + 3)$

$$3x - 2 = 0 \qquad \text{or} \qquad x + 3 = 0$$

$$x = \frac{2}{3} \qquad \text{or} \qquad\quad x = -3$$

$$(2w - 3)^{-1} = \frac{2}{3} \qquad\quad (2w - 3)^{-1} = -3$$

$$\frac{1}{2w - 3} = \frac{2}{3} \qquad\qquad \frac{1}{2w - 3} = -3$$

$$3 = 2(2w - 3) \qquad\quad 1 = -3(2w - 3)$$

$$3 = 4w - 6 \qquad\qquad 1 = -6w + 9$$

$$9 = 4w \qquad\qquad\quad -8 = -6w$$

$$\frac{9}{4} = w \qquad\qquad\qquad \frac{4}{3} = w$$

Check $w = \frac{4}{3}$: $6 = -21 + 27$ *True*

Check $w = \frac{9}{4}$: $6 = \frac{14}{3} + \frac{4}{3}$ *True*

Solution set: $\left\{\frac{4}{3}, \frac{9}{4}\right\}$

87. $2x^4 - 9x^2 = -2$

$2x^4 - 9x^2 + 2 = 0$

Let $u = x^2$, so $u^2 = x^4$.

$2u^2 - 9u + 2 = 0$

Use $a = 2$, $b = -9$, and $c = 2$ in the quadratic formula.

$$u = \frac{-b \pm \sqrt{b^2 - 4ac}}{2a}$$

$$u = \frac{-(-9) \pm \sqrt{(-9)^2 - 4(2)(2)}}{2(2)}$$

$$= \frac{9 \pm \sqrt{81 - 16}}{4}$$

$$= \frac{9 \pm \sqrt{65}}{4}$$

To find x, substitute x^2 for u.

$$x^2 = \frac{9 \pm \sqrt{65}}{4}$$

$$x = \pm \sqrt{\frac{9 \pm \sqrt{65}}{4}}$$

$$= \pm \frac{\sqrt{9 \pm \sqrt{65}}}{2}$$

Note: the last expression represents four numbers. All four proposed solutions check.

Solution set: $\left\{ \dfrac{\sqrt{9 + \sqrt{65}}}{2}, -\dfrac{\sqrt{9 + \sqrt{65}}}{2}, \dfrac{\sqrt{9 - \sqrt{65}}}{2}, -\dfrac{\sqrt{9 - \sqrt{65}}}{2} \right\}$

89.
$$P = 2L + 2W$$
$$P - 2L = 2W$$
$$\frac{P - 2L}{2} = W, \text{ or } W = \frac{P}{2} - L$$

91.
$$F = \tfrac{9}{5}C + 32$$
$$F - 32 = \tfrac{9}{5}C$$
$$\tfrac{5}{9}(F - 32) = C$$

Summary Exercises on Solving Quadratic Equations

1. $p^2 = 7$
$$p = \sqrt{7} \quad \text{or} \quad p = -\sqrt{7}$$
Solution set: $\left\{ \sqrt{7}, -\sqrt{7} \right\}$

3. $n^2 + 6n + 4 = 0$
$$n^2 + 6n = -4$$
$$n^2 + 6n + 9 = -4 + 9 \quad \left[\tfrac{1}{2}(6) \right]^2 = 9$$
$$(n + 3)^2 = 5$$
$$n + 3 = \sqrt{5} \qquad \text{or} \quad n + 3 = -\sqrt{5}$$
$$n = -3 + \sqrt{5} \quad \text{or} \qquad n = -3 - \sqrt{5}$$
Solution set: $\left\{ -3 + \sqrt{5}, -3 - \sqrt{5} \right\}$

5. $\dfrac{5}{m} + \dfrac{12}{m^2} = 2$
Multiply by the LCD, m^2.
$$5m + 12 = 2m^2$$
$$0 = 2m^2 - 5m - 12$$
$$0 = (2m + 3)(m - 4)$$
$$2m + 3 = 0 \quad \text{or} \quad m - 4 = 0$$
$$m = -\tfrac{3}{2} \quad \text{or} \qquad m = 4$$
Solution set: $\left\{ -\tfrac{3}{2}, 4 \right\}$

7. $2r^2 - 4r + 1 = 0$

Use $a = 2$, $b = -4$, and $c = 1$ in the quadratic formula.

$$r = \frac{-b \pm \sqrt{b^2 - 4ac}}{2a}$$

$$r = \frac{-(-4) \pm \sqrt{(-4)^2 - 4(2)(1)}}{2(2)}$$

$$= \frac{4 \pm \sqrt{16 - 8}}{2(2)}$$

$$= \frac{4 \pm \sqrt{8}}{2(2)} = \frac{4 \pm 2\sqrt{2}}{2(2)}$$

$$= \frac{2 \pm \sqrt{2}}{2}$$

Solution set: $\left\{ \dfrac{2 + \sqrt{2}}{2}, \dfrac{2 - \sqrt{2}}{2} \right\}$

9.
$$x\sqrt{2} = \sqrt{5x - 2}$$
$$\left(x\sqrt{2} \right)^2 = \left(\sqrt{5x - 2} \right)^2$$
$$x^2 \cdot 2 = 5x - 2$$
$$2x^2 - 5x + 2 = 0$$
$$(2x - 1)(x - 2) = 0$$
$$2x - 1 = 0 \quad \text{or} \quad x - 2 = 0$$
$$x = \tfrac{1}{2} \qquad\qquad x = 2$$

Check $x = \tfrac{1}{2}$: $\tfrac{1}{2}\sqrt{2} = \sqrt{\tfrac{1}{2}}$ *True*
Check $x = 2$: $2\sqrt{2} = \sqrt{8}$ *True*
Solution set: $\left\{ \tfrac{1}{2}, 2 \right\}$

11. $(2k + 3)^2 = 8$
$$2k + 3 = \sqrt{8} \qquad \text{or} \quad 2k + 3 = -\sqrt{8}$$
$$2k = -3 + 2\sqrt{2} \qquad\qquad 2k = -3 - 2\sqrt{2}$$
$$k = \frac{-3 + 2\sqrt{2}}{2} \quad \text{or} \qquad k = \frac{-3 - 2\sqrt{2}}{2}$$

Solution set: $\left\{ \dfrac{-3 + 2\sqrt{2}}{2}, \dfrac{-3 - 2\sqrt{2}}{2} \right\}$

13.
$$t^4 + 14 = 9t^2$$
$$t^4 - 9t^2 + 14 = 0$$
$$(t^2 - 2)(t^2 - 7) = 0$$

$$t^2 - 2 = 0 \quad \text{or} \quad t^2 - 7 = 0$$
$$t^2 = 2 \qquad\qquad t^2 = 7$$
$$t = \pm\sqrt{2} \quad \text{or} \qquad t = \pm\sqrt{7}$$

Solution set: $\left\{-\sqrt{7}, -\sqrt{2}, \sqrt{2}, \sqrt{7}\right\}$

15. $z^2 + z + 1 = 0$

Use $a = 1$, $b = 1$, and $c = 1$ in the quadratic formula.

$$z = \frac{-b \pm \sqrt{b^2 - 4ac}}{2a}$$
$$z = \frac{-1 \pm \sqrt{1^2 - 4(1)(1)}}{2(1)}$$
$$= \frac{-1 \pm \sqrt{1 - 4}}{2} = \frac{-1 \pm \sqrt{-3}}{2}$$
$$= \frac{-1 \pm i\sqrt{3}}{2} = -\frac{1}{2} \pm \frac{\sqrt{3}}{2}i$$

Solution set: $\left\{-\frac{1}{2} + \frac{\sqrt{3}}{2}i, -\frac{1}{2} - \frac{\sqrt{3}}{2}i\right\}$

17.
$$4t^2 - 12t + 9 = 0$$
$$(2t - 3)(2t - 3) = 0$$
$$(2t - 3)^2 = 0$$
$$2t - 3 = 0$$
$$t = \frac{3}{2}$$

Solution set: $\left\{\frac{3}{2}\right\}$

19. $r^2 - 72 = 0$
$$r^2 = 72$$
$$r = \pm\sqrt{72} = \pm 6\sqrt{2}$$

Solution set: $\left\{6\sqrt{2}, -6\sqrt{2}\right\}$

21. $x^2 - 5x - 36 = 0$
$$(x + 4)(x - 9) = 0$$
$$x + 4 = 0 \quad \text{or} \quad x - 9 = 0$$
$$x = -4 \quad \text{or} \qquad x = 9$$

Solution set: $\{-4, 9\}$

23.
$$3p^2 = 6p - 4$$
$$3p^2 - 6p + 4 = 0$$

Use $a = 3$, $b = -6$, and $c = 4$ in the quadratic formula.

$$p = \frac{-b \pm \sqrt{b^2 - 4ac}}{2a}$$
$$p = \frac{-(-6) \pm \sqrt{(-6)^2 - 4(3)(4)}}{2(3)}$$

$$= \frac{6 \pm \sqrt{36 - 48}}{2(3)} = \frac{6 \pm \sqrt{-12}}{2(3)}$$
$$= \frac{6 \pm 2i\sqrt{3}}{2(3)} = \frac{3 \pm i\sqrt{3}}{3}$$
$$= 1 \pm \frac{\sqrt{3}}{3}i$$

Solution set: $\left\{1 + \frac{\sqrt{3}}{3}i, 1 - \frac{\sqrt{3}}{3}i\right\}$

25. $\frac{4}{r^2} + 3 = \frac{1}{r}$

Multiply by the LCD, r^2.
$$4 + 3r^2 = r$$
$$3r^2 - r + 4 = 0$$
Use $a = 3$, $b = -1$, and $c = 4$ in the quadratic formula.

$$r = \frac{-b \pm \sqrt{b^2 - 4ac}}{2a}$$
$$r = \frac{-(-1) \pm \sqrt{(-1)^2 - 4(3)(4)}}{2(3)}$$
$$= \frac{1 \pm \sqrt{1 - 48}}{6} = \frac{1 \pm \sqrt{-47}}{6}$$
$$= \frac{1 \pm i\sqrt{47}}{6} = \frac{1}{6} \pm \frac{\sqrt{47}}{6}i$$

Solution set: $\left\{\frac{1}{6} + \frac{\sqrt{47}}{6}i, \frac{1}{6} - \frac{\sqrt{47}}{6}i\right\}$

9.4 Formulas and Further Applications

1. The first step in solving a formula that has the specified variable in the denominator is to multiply both sides by the LCD to clear the equation of fractions.

3. We must recognize that a formula like
$$gw^2 = kw + 24$$
is quadratic in w. So the first step is to write the formula in standard form (with 0 on one side, in decreasing powers of w). This allows us to apply the quadratic formula to solve for w.

5. Since the triangle is a right triangle, use the Pythagorean formula with legs m and n and hypotenuse p.
$$m^2 + n^2 = p^2$$
$$m^2 = p^2 - n^2$$
$$m = \sqrt{p^2 - n^2}$$

Only the positive square root is given since m represents the side of a triangle.

7. Solve $d = kt^2$ for t.

$$kt^2 = d$$

$$t^2 = \frac{d}{k} \qquad \textit{Divide by } k.$$

$$t = \pm\sqrt{\frac{d}{k}} \qquad \begin{array}{l}\textit{Use square root}\\ \textit{property.}\end{array}$$

$$= \frac{\pm\sqrt{d}}{\sqrt{k}} \cdot \frac{\sqrt{k}}{\sqrt{k}} \qquad \begin{array}{l}\textit{Rationalize}\\ \textit{denominator.}\end{array}$$

$$t = \frac{\pm\sqrt{dk}}{k} \qquad \textit{Simplify.}$$

9. Solve $I = \dfrac{ks}{d^2}$ for d.

$$Id^2 = ks \qquad \textit{Multiply by } d^2.$$

$$d^2 = \frac{ks}{I} \qquad \textit{Divide by } I.$$

$$d = \pm\sqrt{\frac{ks}{I}} \qquad \textit{Use square root property.}$$

$$= \pm\frac{\sqrt{ks}}{\sqrt{I}} \cdot \frac{\sqrt{I}}{\sqrt{I}} \qquad \textit{Rationalize denominator.}$$

$$d = \frac{\pm\sqrt{ksI}}{I} \qquad \textit{Simplify.}$$

11. Solve $F = \dfrac{kA}{v^2}$ for v.

$$v^2 F = kA \qquad \textit{Multiply by } v^2.$$

$$v^2 = \frac{kA}{F} \qquad \textit{Divide by } F.$$

$$v = \pm\sqrt{\frac{kA}{F}} \qquad \begin{array}{l}\textit{Use square root}\\ \textit{property.}\end{array}$$

$$= \frac{\pm\sqrt{kA}}{\sqrt{F}} \cdot \frac{\sqrt{F}}{\sqrt{F}} \qquad \begin{array}{l}\textit{Rationalize}\\ \textit{denominator.}\end{array}$$

$$v = \frac{\pm\sqrt{kAF}}{F} \qquad \textit{Simplify.}$$

13. Solve $V = \frac{1}{3}\pi r^2 h$ for r.

$$3V = \pi r^2 h \qquad \textit{Multiply by 3.}$$

$$\frac{3V}{\pi h} = r^2 \qquad \textit{Divide by } \pi h.$$

$$r = \pm\sqrt{\frac{3V}{\pi h}} \qquad \begin{array}{l}\textit{Use square root}\\ \textit{property.}\end{array}$$

$$= \frac{\pm\sqrt{3V} \cdot \sqrt{\pi h}}{\sqrt{\pi h} \cdot \sqrt{\pi h}} \qquad \begin{array}{l}\textit{Rationalize}\\ \textit{denominator.}\end{array}$$

$$r = \frac{\pm\sqrt{3\pi V h}}{\pi h} \qquad \textit{Simplify.}$$

15. Solve $At^2 + Bt = -C$ for t.

$$At^2 + Bt + C = 0$$

Use the quadratic formula.

$$t = \frac{-B \pm \sqrt{B^2 - 4AC}}{2A}$$

17. Solve $D = \sqrt{kh}$ for h.

$$D^2 = kh \qquad \textit{Square both sides.}$$

$$\frac{D^2}{k} = h \qquad \textit{Divide by } k.$$

19. Solve $p = \sqrt{\dfrac{k\ell}{g}}$ for ℓ.

$$p^2 = \frac{k\ell}{g} \qquad \textit{Square both sides.}$$

$$p^2 g = k\ell \qquad \textit{Multiply by } g.$$

$$\frac{p^2 g}{k} = \ell \qquad \textit{Divide by } k.$$

21. Solve $S = 4\pi r^2$ for r.

$$\frac{S}{4\pi} = r^2 \qquad \textit{Divide by } 4\pi.$$

$$r = \pm\sqrt{\frac{S}{4\pi}} \qquad \begin{array}{l}\textit{Use square root}\\ \textit{property.}\end{array}$$

$$= \frac{\pm\sqrt{S} \cdot \sqrt{\pi}}{\sqrt{4\pi} \cdot \sqrt{\pi}} \qquad \begin{array}{l}\textit{Rationalize}\\ \textit{denominator.}\end{array}$$

$$r = \frac{\pm\sqrt{S\pi}}{2\pi} \qquad \textit{Simplify.}$$

23. Solve $p = \dfrac{E^2 R}{(r + R)^2}$ $(E > 0)$ for R.

$$p(r + R)^2 = E^2 R$$

$$p\left(r^2 + 2rR + R^2\right) = E^2 R$$

$$pr^2 + 2prR + pR^2 = E^2 R$$

$$pR^2 + 2prR - E^2 R + pr^2 = 0$$

$$pR^2 + \left(2pr - E^2\right)R + pr^2 = 0$$

Here $a = p$, $b = 2pr - E^2$, and $c = pr^2$.

$$R = \frac{-(2pr - E^2) \pm \sqrt{(2pr - E^2)^2 - 4p \cdot pr^2}}{2p}$$

$$= \frac{E^2 - 2pr \pm \sqrt{4p^2r^2 - 4prE^2 + E^4 - 4p^2r^2}}{2p}$$

$$= \frac{E^2 - 2pr \pm \sqrt{E^4 - 4prE^2}}{2p}$$

$$= \frac{E^2 - 2pr \pm \sqrt{E^2(E^2 - 4pr)}}{2p}$$

$$R = \frac{E^2 - 2pr \pm E\sqrt{E^2 - 4pr}}{2p}$$

25. Solve $10p^2c^2 + 7pcr = 12r^2$ for r.

$0 = 12r^2 - 7pcr - 10p^2c^2$

Here $a = 12$, $b = -7pc$, and $c = -10p^2c^2$.

$r = \dfrac{-(-7pc) \pm \sqrt{(-7pc)^2 - 4(12)(-10p^2c^2)}}{2(12)}$

$= \dfrac{7pc \pm \sqrt{49p^2c^2 + 480p^2c^2}}{24}$

$= \dfrac{7pc \pm \sqrt{529p^2c^2}}{24} = \dfrac{7pc \pm 23pc}{24}$

$r = \dfrac{7pc + 23pc}{24} = \dfrac{30pc}{24} = \dfrac{5pc}{4}$ or

$r = \dfrac{7pc - 23pc}{24} = \dfrac{-16pc}{24} = -\dfrac{2pc}{3}$

27. Solve $LI^2 + RI + \dfrac{1}{c} = 0$ for I.

$cLI^2 + cRI + 1 = 0$ *Multiply by c.*

Here $a = cL$, $b = cR$, and $c = 1$.

$I = \dfrac{-cR \pm \sqrt{(cR)^2 - 4(cL)(1)}}{2(cL)}$

$= \dfrac{-cR \pm \sqrt{c^2R^2 - 4cL}}{2cL}$

29. Apply the Pythagorean formula.

$(x + 4)^2 = x^2 + (x + 1)^2$

$x^2 + 8x + 16 = x^2 + x^2 + 2x + 1$

$0 = x^2 - 6x - 15$

Here $a = 1$, $b = -6$, and $c = -15$.

$x = \dfrac{-(-6) \pm \sqrt{(-6)^2 - 4(1)(-15)}}{2(1)}$

$= \dfrac{6 \pm \sqrt{36 + 60}}{2} = \dfrac{6 \pm \sqrt{96}}{2}$

$x = \dfrac{6 + \sqrt{96}}{2} \approx 7.9$ or

$x = \dfrac{6 - \sqrt{96}}{2} \approx -1.9$

Reject the negative solution.
If $x = 7.9$, then

$x + 4 = 11.9$ and $x + 1 = 8.9$.

The lengths of the sides of the triangle are approximately 7.9, 8.9, and 11.9.

31. Let $x =$ the distance traveled by the eastbound ship. Then $x + 70 =$ the distance traveled by the southbound ship.

Since the ships are traveling at right angles to one another, the distance d between them can be found using the Pythagorean formula.

$c^2 = a^2 + b^2$

$d^2 = x^2 + (x + 70)^2$

Let $d = 170$, and solve for x.

$170^2 = x^2 + (x + 70)^2$

$28{,}900 = x^2 + x^2 + 140x + 4900$

$0 = 2x^2 + 140x - 24{,}000$

$0 = x^2 + 70x - 12{,}000$

$0 = (x + 150)(x - 80)$

$x + 150 = 0$ or $x - 80 = 0$

$x = -150$ or $x = 80$

Distance cannot be negative, so reject -150. If $x = 80$, then $x + 70 = 150$. The eastbound ship traveled 80 miles, and the southbound ship traveled 150 miles.

33. Let $x =$ length of the shorter leg;

$2x - 1 =$ length of the longer leg;

$2x - 1 + 2 =$ length of the hypotenuse.

Use the Pythagorean formula.

$x^2 + (2x - 1)^2 = (2x + 1)^2$

$x^2 + 4x^2 - 4x + 1 = 4x^2 + 4x + 1$

$5x^2 - 4x + 1 = 4x^2 + 4x + 1$

$x^2 - 8x = 0$

$x(x - 8) = 0$

$x = 0$ or $x - 8 = 0$

$x = 8$

Since x represents length, discard 0 as a solution. If $x = 8$, then

$2x - 1 = 2(8) - 1 = 15$ and
$2x - 1 + 2 = 2(8) + 1 = 17.$

The lengths are 8 inches, 15 inches, and 17 inches.

35. Let $x =$ the width of the rug.
Then $2x + 4 =$ the length of the rug.

Use the Pythagorean formula.

$x^2 + (2x + 4)^2 = 26^2$

$x^2 + 4x^2 + 16x + 16 = 676$

$5x^2 + 16x - 660 = 0$

$(5x + 66)(x - 10) = 0$

$5x + 66 = 0$ or $x - 10 = 0$

$x = -\frac{66}{5}$ or $x = 10$

Discard the negative solution.
If $x = 10$, then $2x + 4 = 24$.
The width of the rug is 10 feet, and the length is 24 feet.

37. Let $x =$ the width of the border. Then the width of the pool and the border is $30 + 2x$ ft and the length of the pool and border is $40 + 2x$ ft. Since the area of the pool is $30 \cdot 40 = 1200$ ft^2, we can write an equation using the total area of the pool and the border.

The area of   the area   the area
the pool   is of the plus of the
and border   pool   border.

$$(30 + 2x)(40 + 2x) = 1200 + 296$$
$$1200 + 140x + 4x^2 = 1496$$
$$4x^2 + 140x - 296 = 0$$
$$x^2 + 35x - 74 = 0$$
$$(x - 2)(x + 37) = 0$$
$$x - 2 = 0 \quad \text{or} \quad x + 37 = 0$$
$$x = 2 \quad \text{or} \quad x = -37$$

Discard the negative solution. The strip can be 2 feet wide.

39. Let x = original width of the rectangle. Then $2x - 2$ represents the original length of the rectangle.

Now $x + 5$ is the new width that makes the rectangle a square. Thus, the new width must equal the original length since the sides of a square are equal.

$$x + 5 = 2x - 2$$
$$7 = x$$

The dimensions of the original rectangle are 7 meters and $2(7) - 2 = 12$ meters. Note that the area of the square is $(12)(12) = 144 \text{ m}^2$.

41. Let x be the width of the sheet metal. Then the length is $2x - 4$.

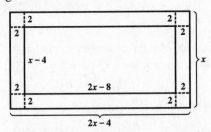

By cutting out 2-inch squares from each corner we get a rectangle with width $x - 4$ and length $(2x - 4) - 4 = 2x - 8$. The uncovered box then has height 2 inches, length $2x - 8$ inches, and width $x - 4$ inches.
Use the formula $V = LWH$ or $V = HLW$.

$$256 = 2(2x - 8)(x - 4)$$
$$256 = 4(x - 4)(x - 4) \qquad \textit{Factor out 2.}$$
$$64 = (x - 4)^2 \qquad \textit{Divide by 4.}$$

Use the square root property.
$$\pm 8 = x - 4$$
$$x - 4 = 8 \quad \text{or} \quad x - 4 = -8$$
$$x = 12 \quad \text{or} \quad x = -4$$

Since x represents width, discard the negative solution.
The width is 12 inches, and the length is $2(12) - 4 = 20$ inches.

43.
$$s = 144t - 16t^2$$
$$128 = 144t - 16t^2 \qquad \textit{Let s = 128.}$$
$$0 = -16t^2 + 144t - 128$$
$$0 = t^2 - 9t + 8 \qquad \textit{Divide by }-16.$$
$$0 = (t - 8)(t - 1)$$
$$t - 8 = 0 \quad \text{or} \quad t - 1 = 0$$
$$t = 8 \quad \text{or} \quad t = 1$$

The object will be 128 feet above the ground at two times, going up and coming down, or at 1 second and at 8 seconds.

45. Let $s = 213$ in the equation.

$$s = -16t^2 + 128t$$
$$213 = -16t^2 + 128t$$
$$0 = -16t^2 + 128t - 213$$

Here $a = -16$, $b = 128$, and $c = -213$.

$$t = \frac{-b \pm \sqrt{b^2 - 4ac}}{2a}$$
$$t = \frac{-128 \pm \sqrt{128^2 - 4(-16)(-213)}}{2(-16)}$$
$$= \frac{-128 \pm \sqrt{16{,}384 - 13{,}632}}{-32}$$
$$= \frac{-128 \pm \sqrt{2752}}{-32}$$
$$t = \frac{-128 + \sqrt{2752}}{-32} \approx 2.4 \quad \text{or}$$
$$t = \frac{-128 - \sqrt{2752}}{-32} \approx 5.6$$

The ball will be 213 feet from the ground after 2.4 seconds and again after 5.6 seconds.

47.
$$D(t) = 13t^2 - 100t$$
$$180 = 13t^2 - 100t \qquad \textit{Let D(t) = 180.}$$
$$0 = 13t^2 - 100t - 180$$

Here $a = 13$, $b = -100$, and $c = -180$.

$$t = \frac{-b \pm \sqrt{b^2 - 4ac}}{2a}$$
$$t = \frac{-(-100) \pm \sqrt{(-100)^2 - 4(13)(-180)}}{2(13)}$$
$$= \frac{100 \pm \sqrt{10{,}000 + 9360}}{2(13)}$$

continued

$$= \frac{100 \pm \sqrt{19{,}360}}{2(13)}$$

$$= \frac{100 \pm 44\sqrt{10}}{2(13)} = \frac{50 \pm 22\sqrt{10}}{13}$$

$$t = \frac{50 + 22\sqrt{10}}{13} \approx 9.2 \quad \text{or}$$

$$t = \frac{50 - 22\sqrt{10}}{13} \approx -1.5$$

Discard the negative solution. The car will skid 180 feet in approximately 9.2 seconds.

49. $s(t) = -16t^2 + 160t$

$$400 = -16t^2 + 160t \qquad \textit{Let s(t) = 400.}$$
$$0 = -16t^2 + 160t - 400$$
$$0 = t^2 - 10t + 25 \qquad \textit{Divide by } -16.$$
$$0 = (t - 5)(t - 5)$$
$$0 = (t - 5)^2$$
$$0 = t - 5$$
$$5 = t$$

The rock reaches a height of 400 feet after 5 seconds. This is its maximum height since this is the only time it reaches 400 feet.

51.
$$V = 3(x - 6)^2$$
$$432 = 3(x - 6)^2 \qquad \textit{Let V = 432.}$$
$$144 = (x - 6)^2 \qquad \textit{Divide by 3.}$$

Use the square root property.
$$\pm\sqrt{144} = x - 6$$
$$6 \pm 12 = x$$
$$x = 6 + 12 = 18$$
or $\quad x = 6 - 12 = -6$

Discard the negative solution since x represents the length.
The original length is 18 inches.

53. Let F denote the Froude number. Solve

$$F = \frac{v^2}{g\ell}$$

for v.
$$v^2 = Fg\ell$$
$$v = \pm\sqrt{Fg\ell}$$
v is positive, so
$$v = \sqrt{Fg\ell}.$$
For the rhinoceros, $\ell = 1.2$ and $F = 2.57$.
$$v = \sqrt{(2.57)(9.8)(1.2)} \approx 5.5$$
or 5.5 meters per second.

55. Write a proportion.

$$\frac{x - 4}{3x - 19} = \frac{4}{x - 3}$$

Multiply by the LCD, $(3x - 19)(x - 3)$.

$$(3x - 19)(x - 3)\left(\frac{x - 4}{3x - 19}\right)$$
$$= (3x - 19)(x - 3)\left(\frac{4}{x - 3}\right)$$
$$(x - 3)(x - 4) = (3x - 19)4$$
$$x^2 - 7x + 12 = 12x - 76$$
$$x^2 - 19x + 88 = 0$$
$$(x - 8)(x - 11) = 0$$

$x - 8 = 0 \quad$ or $\quad x - 11 = 0$
$\quad x = 8 \quad$ or $\qquad x = 11$

If $x = 8$, then
$$3x - 19 = 3(8) - 19 = 5.$$
If $x = 11$, then
$$3x - 19 = 3(11) - 19 = 14.$$

Thus, $AC = 5$ or $AC = 14$.

57. (a) From the graph, the number of miles traveled in 2000 appears to be 2750 billion (to the nearest ten billion).

(b) For 2000, $x = 2000 - 1994 = 6$.

$$f(x) = -1.705x^2 + 75.93x + 2351$$
$$f(6) = -1.705(6)^2 + 75.93(6) + 2351$$
$$= 2745.2$$

To the nearest ten billion, the model gives 2750 billion, the same as the estimate in part (a).

59. Use $f(x) = -1.705x^2 + 75.93x + 2351$ with $f(x) = 2800$.

$$2800 = -1.705x^2 + 75.93x + 2351$$
$$0 = -1.705x^2 + 75.93x - 449$$

Here $a = -1.705$, $b = 75.93$, and $c = -449$.

$$x = \frac{-b \pm \sqrt{b^2 - 4ac}}{2a}$$
$$x = \frac{-75.93 \pm \sqrt{(75.93)^2 - 4(-1.705)(-449)}}{2(-1.705)}$$
$$= \frac{-75.93 \pm \sqrt{2703.1849}}{-3.41}$$
$$\approx 7.02 \text{ or } 37.51$$

The model indicates that the number of miles traveled was 2800 billion in the year $1994 + 7 = 2001$. The other value represents a future year. The graph indicates that vehicle-miles reached 2800 billion in 2001.

61. $f(x) = x^2 + 4x - 3$

$f(2) = 2^2 + 4(2) - 3$

$= 4 + 8 - 3$

$= 9$

63. $f(x) = ax^2 + bx + c$

$f\left(\dfrac{-b}{2a}\right) = a\left(\dfrac{-b}{2a}\right)^2 + b\left(\dfrac{-b}{2a}\right) + c$

$= a\left(\dfrac{b^2}{4a^2}\right) - \dfrac{b^2}{2a} + c$

$= \dfrac{b^2}{4a} - \dfrac{2b^2}{4a} + \dfrac{4ac}{4a}$

$= \dfrac{-b^2 + 4ac}{4a}$

65. $f(x) = (x - 4)^2$

$9 = (x - 4)^2$

$x - 4 = \pm\sqrt{9}$

$x = 4 \pm 3 = 7 \ \text{ or } \ 1$

Solution set: $\{1, 7\}$

9.5 Graphs of Quadratic Functions

1. A parabola with equation $f(x) = a(x - h)^2 + k$ has vertex $V(h, k)$. We'll identify the vertex for each quadratic function.

(a) $f(x) = (x + 2)^2 - 1$

$V(-2, -1)$, choice **B**

(b) $f(x) = (x + 2)^2 + 1$

$V(-2, 1)$, choice **C**

(c) $f(x) = (x - 2)^2 - 1$

$V(2, -1)$, choice **A**

(d) $f(x) = (x - 2)^2 + 1$

$V(2, 1)$, choice **D**

For Exercises 3–11, we write $f(x)$ in the form $f(x) = a(x - h)^2 + k$ and then list the vertex (h, k).

3. $f(x) = -3x^2 = -3(x - 0)^2 + 0$
The vertex (h, k) is $(0, 0)$.

5. $f(x) = x^2 + 4 = 1(x - 0)^2 + 4$
The vertex (h, k) is $(0, 4)$.

7. $f(x) = (x - 1)^2 = 1(x - 1)^2 + 0$
The vertex (h, k) is $(1, 0)$.

9. $f(x) = (x + 3)^2 - 4 = 1[x - (-3)]^2 - 4$
The vertex (h, k) is $(-3, -4)$.

11. $f(x) = -(x - 5)^2 + 6 = -1(x - 5)^2 + 6$
The vertex (h, k) is $(5, 6)$.

13. $f(x) = -\frac{2}{5}x^2$

Since $a = -\frac{2}{5} < 0$, the graph opens down. Since $|a| = \left|-\frac{2}{5}\right| = \frac{2}{5} < 1$, the graph is wider than the graph of $f(x) = x^2$.

15. $f(x) = 3x^2 + 1$

Since $a = 3 > 0$, the graph opens up. Since $|a| = |3| = 3 > 1$, the graph is narrower than the graph of $f(x) = x^2$.

17. Consider $f(x) = a(x - h)^2 + k$.

(a) If $h > 0$ and $k > 0$ in $f(x) = a(x - h)^2 + k$, the shift is to the right and upward, so the vertex is in quadrant I.

(b) If $h > 0$ and $k < 0$, the shift is to the right and downward, so the vertex is in quadrant IV.

(c) If $h < 0$ and $k > 0$, the shift is to the left and upward, so the vertex is in quadrant II.

(d) If $h < 0$ and $k < 0$, the shift is to the left and downward, so the vertex is in quadrant III.

19. **(a)** $f(x) = (x - 4)^2 - 2 = 1(x - 4)^2 - 2$ has vertex $(4, -2)$. Because $a = 1 > 0$, the graph opens up. The correct answer is **D**.

(b) $f(x) = (x - 2)^2 - 4 = 1(x - 2)^2 - 4$ has vertex $(2, -4)$. Because $a = 1 > 0$, the graph opens up. The correct answer is **B**.

(c) $f(x) = -(x - 4)^2 - 2 = -1(x - 4)^2 - 2$ has vertex $(4, -2)$. Because $a = -1 < 0$, the graph opens down. The correct answer is **C**.

(d) $f(x) = -(x - 2)^2 - 4 = -1(x - 2)^2 - 4$ has vertex $(2, -4)$. Because $a = -1 < 0$, the graph opens down. The correct answer is **A**.

21. $f(x) = -2x^2$ written in the form
$f(x) = a(x - h)^2 + k$ is
$f(x) = -2(x - 0)^2 + 0$.

Here, $h = 0$ and $k = 0$, so the vertex (h, k) is $(0, 0)$. Since $a = -2 < 0$, the graph opens down. Since $|a| = |-2| = 2 > 1$, the graph is narrower than the graph of $f(x) = x^2$. By evaluating the function with $x = 2$ and $x = -2$, we see that the points $(2, -8)$ and $(-2, -8)$ are on the graph.

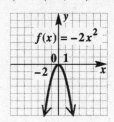

23. $f(x) = x^2 - 1$ written in the form
$f(x) = a(x - h)^2 + k$ is
$f(x) = 1(x - 0)^2 + (-1).$

Here, $h = 0$ and $k = -1$, so the vertex is $(0, -1)$. The graph opens up and has the same shape as $f(x) = x^2$ because $a = 1$. Two other points on the graph are $(-2, 3)$ and $(2, 3)$.

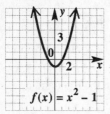

25. $f(x) = -x^2 + 2$ written in the form
$f(x) = a(x - h)^2 + k$ is
$f(x) = -1(x - 0)^2 + 2.$

Here, $h = 0$ and $k = 2$, so the vertex (h, k) is $(0, 2)$. Since $a = -1 < 0$, the graph opens down. Since $|a| = |-1| = 1$, the graph has the same shape as $f(x) = x^2$. The points $(2, -2)$ and $(-2, -2)$ are on the graph.

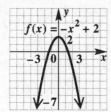

27. $f(x) = (x - 4)^2$ written in the form
$f(x) = a(x - h)^2 + k$ is
$f(x) = 1(x - 4)^2 + 0.$

Here, $h = 4$ and $k = 0$, so the vertex (h, k) is $(4, 0)$ and the axis is $x = 4$. The graph opens up since a is positive and has the same shape as $f(x) = x^2$ because $|a| = 1$. Two other points on the graph are $(2, 4)$ and $(6, 4)$. We can substitute any value for x, so the domain is $(-\infty, \infty)$. The range is $[0, \infty)$ since the smallest y-value is 0.

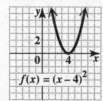

29. $f(x) = (x + 2)^2 - 1$ written in the form
$f(x) = a(x - h)^2 + k$ is
$f(x) = 1[x - (-2)]^2 + (-1).$

Since $h = -2$ and $k = -1$, the vertex (h, k) is $(-2, -1)$ and the axis is $x = -2$. Here, $a = 1$, so the graph opens up and has the same shape as

$f(x) = x^2$. The points $(-1, 0)$ and $(-3, 0)$ are on the graph. The domain is $(-\infty, \infty)$. The range is $[-1, \infty)$ since the smallest y-value is -1.

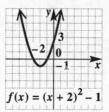

31. $f(x) = 2(x - 2)^2 - 4$ written in the form
$f(x) = a(x - h)^2 + k$ is
$f(x) = 2(x - 2)^2 + (-4).$

Here, $h = 2$ and $k = -4$, so the vertex (h, k) is $(2, -4)$ and the axis is $x = 2$. The graph opens up and is narrower than $f(x) = x^2$ because $|a| = |2| > 1$. Two other points on the graph are $(0, 4)$ and $(4, 4)$.
We can substitute any value for x, so the domain is $(-\infty, \infty)$. The value of y is greater than or equal to -4, so the range is $[-4, \infty)$.

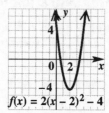

33. $f(x) = -\frac{1}{2}(x + 1)^2 + 2$ written in the form
$f(x) = a(x - h)^2 + k$ is
$f(x) = -\frac{1}{2}[x - (-1)]^2 + 2.$

Since $h = -1$ and $k = 2$, the vertex (h, k) is $(-1, 2)$ and the axis is $x = -1$. Here, $a = -0.5 < 0$, so the graph opens down. Also, $|a| = |-0.5| = 0.5 < 1$, so the graph is wider than the graph of $f(x) = x^2$. The points $(1, 0)$ and $(-3, 0)$ are on the graph.
We can substitute any value for x, so the domain is $(-\infty, \infty)$. The value of y is less than or equal to 2, so the range is $(-\infty, 2]$.

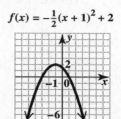

35. $f(x) = 2(x - 2)^2 - 3$ written in the form
$f(x) = a(x - h)^2 + k$ is
$f(x) = 2(x - 2)^2 + (-3).$

Here, $h = 2$ and $k = -3$, so the vertex (h, k) is $(2, -3)$ and the axis is $x = 2$. The graph opens up and is narrower than $f(x) = x^2$ because $|a| = |2| > 1$. Two other points on the graph are $(3, -1)$ and $(1, -1)$.

We can substitute any value for x, so the domain is $(-\infty, \infty)$. The value of y is greater than or equal to -3, so the range is $[-3, \infty)$.

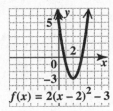

$f(x) = 2(x - 2)^2 - 3$

37. The graph of $F(x) = x^2 + 6$ would be shifted 6 units upward from the graph of $f(x) = x^2$.

38. To graph $G(x) = x + 6$, plot the intercepts $(-6, 0)$ and $(0, 6)$, and draw the line through them.

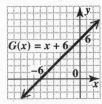

$G(x) = x + 6$

39. When considering the graph of $G(x) = x + 6$, the y-intercept is 6. The graph of $g(x) = x$ has y-intercept 0. Therefore, the graph of $G(x) = x + 6$ is shifted 6 units upward compared to the graph of $g(x) = x$.

40. The graph of $F(x) = (x - 6)^2$ is shifted 6 units to the right compared to the graph of $f(x) = x^2$.

41. To graph $G(x) = x - 6$, plot the intercepts $(6, 0)$ and $(0, -6)$, and draw the line through them.

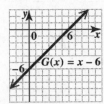

$G(x) = x - 6$

42. When considering the graph of $G(x) = x - 6$, its x-intercept is 6 as compared to the graph of $g(x) = x$ with x-intercept 0. The graph of $G(x) = x - 6$ is shifted 6 units to the right compared to the graph of $g(x) = x$.

43. The points appear to lie on a line, so a *linear* function would be a more appropriate model. The line would rise, so it would have a *positive* slope.

45. The points appear to lie on a parabola, so a *quadratic* function would be a more appropriate model. The parabola would open up, so a would be *positive*.

47. Since the arrangement of the data points is approximately parabolic, a quadratic function would be the more appropriate model for the data set. The coefficient of x^2 should be negative, since the roughly parabolic shape of the graphed data set opens downward.

49. (a)

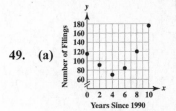

(b) Since the arrangement of the data points is approximately parabolic, a quadratic function would be the more appropriate model for the data set. The coefficient of x^2 should be positive, since the roughly parabolic shape of the graphed data set opens upward.

(c) Use $ax^2 + bx + c = y$ with $(0, 115)$, $(4, 70)$, and $(8, 120)$.

$$0a + 0b + c = 115 \quad (1)$$
$$16a + 4b + c = 70 \quad (2)$$
$$64a + 8b + c = 120 \quad (3)$$

From (1), $c = 115$, so the system becomes

$$16a + 4b = -45 \quad (4)$$
$$64a + 8b = 5 \quad (5)$$

Now eliminate b.

$$-32a - 8b = 90 \qquad -2 \times (4)$$
$$\underline{64a + 8b = 5}$$
$$32a = 95$$
$$a = \tfrac{95}{32} = 2.96875 \approx 2.969$$

From (5) with $a = \tfrac{95}{32}$,

$$64\left(\tfrac{95}{32}\right) + 8b = 5$$
$$190 + 8b = 5$$
$$8b = -185$$
$$b = -\tfrac{185}{8} = -23.125$$

The quadratic function is approximately

$$y = f(x) = 2.969x^2 - 23.125x + 115.$$

(d) $x = 2002 - 1990 = 12$ and $f(12) \approx 265$.

(e) No. About 16 companies filed for bankruptcy each month, so at this rate, filings for 2002 would be about 192. The approximation from the model seems high.

51. (a) $y = f(x) = 0.2455x^2 - 1.856x + 30.7105$
$x = 2002 - 1970 = 32$ and
$f(32) = 222.7105 \approx 222.7$ (per 100,000)

(b) The approximation using the model is high.

53. $x^2 - x - 20 = 0$

From the screens, we see that the x-values of the x-intercepts are -4 and 5, so the solution set is $\{-4, 5\}$.

55. $x^2 + 6x - 3 = 0$

$$x^2 + 6x = 3$$
$$x^2 + 6x + 9 = 3 + 9$$
$$(x + 3)^2 = 12$$
$$x + 3 = \pm\sqrt{12}$$
$$x = -3 \pm 2\sqrt{3}$$

Solution set: $\left\{-3 + 2\sqrt{3}, -3 - 2\sqrt{3}\right\}$

57. $2x^2 - 12x = 5$

$$x^2 - 6x = \tfrac{5}{2}$$
$$x^2 - 6x + 9 = \tfrac{5}{2} + 9$$
$$(x - 3)^2 = \tfrac{23}{2}$$
$$x - 3 = \pm\sqrt{\tfrac{46}{4}}$$
$$x = 3 \pm \tfrac{\sqrt{46}}{2} = \tfrac{6 \pm \sqrt{46}}{2}$$

Solution set: $\left\{\tfrac{6 + \sqrt{46}}{2}, \tfrac{6 - \sqrt{46}}{2}\right\}$

9.6 More About Parabolas and Their Applications

1. If there is an x^2-term in the equation, the axis is vertical. If there is a y^2-term, the axis is horizontal.

3. Use the discriminant, $b^2 - 4ac$, of the function. If it is positive, there are two x-intercepts. If it is zero, there is one x-intercept (at the vertex), and if it is negative, there is no x-intercept.

5. As in Example 1, we'll complete the square to find the vertex.

$$f(x) = x^2 + 8x + 10$$
$$= x^2 + 8x + \underline{16} + 10 - \underline{16} \quad \left[\tfrac{1}{2}(8)\right]^2 = 16$$
$$= (x + 4)^2 - 6$$

The vertex is $(-4, -6)$.

7. As in Example 2, we'll complete the square to find the vertex.

$$f(x) = -2x^2 + 4x - 5$$
$$= -2(x^2 - 2x) - 5$$
$$= -2(x^2 - 2x + 1 - 1) - 5$$
$$= -2(x^2 - 2x + 1) + (-2)(-1) - 5$$
$$= -2(x - 1)^2 - 3$$

The vertex is $(1, -3)$.

9. As in Example 3, we'll use the vertex formula to find the vertex.

$$f(x) = x^2 + x - 7$$

The x-coordinate of the vertex is

$$\frac{-b}{2a} = \frac{-1}{2(1)} = -\frac{1}{2}.$$

The y-coordinate of the vertex is

$$f\left(-\tfrac{1}{2}\right) = \tfrac{1}{4} - \tfrac{1}{2} - 7 = -\tfrac{29}{4}.$$

The vertex is $\left(-\tfrac{1}{2}, -\tfrac{29}{4}\right)$.

11. As in Example 2, we'll complete the square to find the vertex.

$$f(x) = 2x^2 + 4x + 5$$
$$= 2(x^2 + 2x) + 5$$
$$= 2(x^2 + 2x + 1 - 1) + 5$$
$$= 2(x^2 + 2x + 1) + 2(-1) + 5$$
$$= 2(x + 1)^2 - 2 + 5$$
$$f(x) = 2(x + 1)^2 + 3$$

The vertex is $(-1, 3)$.

Because $a = 2 > 1$, the graph opens up and is narrower than the graph of $y = x^2$.

For $y = f(x) = 2x^2 + 4x + 5$, $a = 2$, $b = 4$, and $c = 5$. The discriminant is

$$b^2 - 4ac = 4^2 - 4(2)(5)$$
$$= 16 - 40 = -24.$$

The discriminant is negative, so the parabola has no x-intercepts.

13. $f(x) = -x^2 + 5x + 3$

Use the vertex formula with $a = -1$ and $b = 5$.

The x-coordinate of the vertex is

$$\frac{-b}{2a} = \frac{-5}{2(-1)} = \frac{5}{2}.$$

The y-coordinate of the vertex is

$$f\left(\frac{-b}{2a}\right) = f\left(\frac{5}{2}\right)$$
$$= -\left(\tfrac{5}{2}\right)^2 + 5\left(\tfrac{5}{2}\right) + 3$$
$$= -\tfrac{25}{4} + \tfrac{25}{2} + 3$$
$$= \frac{-25 + 50 + 12}{4} = \frac{37}{4}.$$

The vertex is

$$\left(\frac{-b}{2a}, f\left(\frac{-b}{2a}\right)\right) = \left(\frac{5}{2}, \frac{37}{4}\right).$$

Because $a = -1$, the parabola opens down and has the same shape as the graph of $y = x^2$.

$$b^2 - 4ac = 5^2 - 4(-1)(3)$$
$$= 25 + 12 = 37$$

The discriminant is positive, so the parabola has two x-intercepts.

15. Complete the square on the y-terms to find the vertex.

$$x = \tfrac{1}{3}y^2 + 6y + 24$$
$$= \tfrac{1}{3}(y^2 + 18y) + 24$$
$$= \tfrac{1}{3}(y^2 + 18y + 81 - 81) + 24$$
$$= \tfrac{1}{3}(y^2 + 18y + 81) + \tfrac{1}{3}(-81) + 24$$
$$= \tfrac{1}{3}(y + 9)^2 - 27 + 24$$
$$x = \tfrac{1}{3}(y + 9)^2 - 3$$

The vertex is $(-3, -9)$.
The graph is a horizontal parabola. The graph opens to the right since $a = 0.\overline{3} > 0$ and is wider than the graph of $y = x^2$ since $|a| = |0.\overline{3}| < 1$.

17. The graph of $y = 2x^2 + 4x - 3$ is a vertical parabola opening up, so choice F is correct. **(F)**

19. The graph of $y = -\tfrac{1}{2}x^2 - x + 1$ is a vertical parabola opening down, so choices A and C are possibilities. The graph in C is wider than the graph in A, so it must correspond to $a = -\tfrac{1}{2}$ while the graph in A must correspond to $a = -1$. **(C)**

21. The graph of $x = -y^2 - 2y + 4$ is a horizontal parabola opening to the left, so choice D is correct. **(D)**

23. $y = f(x) = x^2 + 8x + 10$

Step 1
Since $a = 1 > 0$, the graph opens up and is the same shape as the graph of $y = x^2$.

Step 2
From Exercise 5, the vertex is $(-4, -6)$. Since the graph opens up, the axis goes through the x-coordinate of the vertex—its equation is $x = -4$.

Step 3
To find the y-intercept, let $x = 0$.
$f(0) = 10$, so the y-intercept is $(0, 10)$.
To find the x-intercepts, let $y = 0$.

$$0 = x^2 + 8x + 10$$
$$x = \frac{-8 \pm \sqrt{64 - 40}}{2} = \frac{-8 \pm \sqrt{24}}{2}$$
$$= \frac{-8 \pm 2\sqrt{6}}{2} = -4 \pm \sqrt{6}$$

The x-intercepts are approximately $(-6.45, 0)$ and $(-1.55, 0)$.

Step 4
For an additional point on the graph, let $x = -2$ (two units to the right of the axis) to get $f(-2) = -2$. So the point $(-2, -2)$ is on the graph. By symmetry, the point $(-6, -2)$ (two units to the left of the axis) is on the graph.

$f(x) = x^2 + 8x + 10$

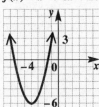

From the graph, we see that the domain is $(-\infty, \infty)$ and the range is $[-6, \infty)$.

25. $y = f(x) = -2x^2 + 4x - 5$

Step 1
Since $a = -2$, the graph opens down and is narrower than the graph of $y = x^2$.

Step 2
From Exercise 7, the vertex is $(1, -3)$. Since the graph opens down, the axis goes through the x-coordinate of the vertex—its equation is $x = 1$.

Step 3
If $x = 0$, $y = -5$, so the y-intercept is $(0, -5)$.
To find the x-intercepts, let $y = 0$.

$$0 = -2x^2 + 4x - 5$$
$$x = \frac{-4 \pm \sqrt{16 - 40}}{2(-2)}$$

The discriminant is negative, so there are no x-intercepts.

Step 4
By symmetry, $(2, -5)$ is also on the graph.

$f(x) = -2x^2 + 4x - 5$

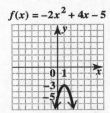

From the graph, we see that the domain is $(-\infty, \infty)$ and the range is $(-\infty, -3]$.

27. $x = (y + 2)^2 + 1 = y^2 + 4y + 5$

The roles of x and y are reversed, so this is a horizontal parabola.

Step 1
The coefficient of y^2 is $1 > 0$, so the graph opens to the right.

Step 2
The vertex can be identified from the given form of the equation. When $y = -2$, $x = 1$, so the vertex is $(1, -2)$. Since the graph opens right, the axis goes through the y-coordinate of the vertex—its equation is $y = -2$.

Step 3

To find the x-intercept, let $y = 0$.
$x = 0^2 + 4(0) + 5 = 5$, so the x-intercept is
$(5, 0)$.

To find the y-intercepts, let $x = 0$.

$$0 = (y + 2)^2 + 1$$
$$-1 = (y + 2)^2$$

Since $(y + 2)^2$ cannot be negative, there are no
y-intercepts.

Step 4

For an additional point on the graph, let $y = -4$
(two units below the axis) to get
$x = (-4 + 2)^2 + 1 = 5$.

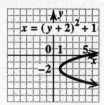

From the graph, we see the domain is $[1, \infty)$ and
the range is $(-\infty, \infty)$.

29. $x = -\frac{1}{5}y^2 + 2y - 4$

The roles of x and y are reversed, so this is a
horizontal parabola.

Step 1

Since $a = -\frac{1}{5} < 0$, the graph opens to the left and
is wider than the graph of $y = x^2$.

Step 2

The y-coordinate of the vertex is

$$\frac{-b}{2a} = \frac{-2}{2\left(-\frac{1}{5}\right)} = \frac{-2}{-\frac{2}{5}} = 5.$$

The x-coordinate of the vertex is

$$-\frac{1}{5}(5)^2 + 2(5) - 4 = -5 + 10 - 4 = 1.$$

Thus, the vertex is $(1, 5)$. Since the graph opens
left, the axis goes through the y-coordinate of the
vertex—its equation is $y = 5$.

Step 3

To find the x-intercept, let $y = 0$.
If $y = 0$, $x = -4$, so the x-intercept is $(-4, 0)$.
To find the y-intercepts, let $x = 0$.

$$0 = -\frac{1}{5}y^2 + 2y - 4$$
$$0 = y^2 - 10y + 20 \quad \textit{Multiply by} -5.$$
$$y = \frac{10 \pm \sqrt{100 - 80}}{2} = \frac{10 \pm \sqrt{20}}{2}$$
$$= \frac{10 \pm 2\sqrt{5}}{2} = 5 \pm \sqrt{5}$$

The y-intercepts are approximately $(0, 7.2)$ and
$(0, 2.8)$.

Step 4

For an additional point on the graph, let $y = 7$
(two units above the axis) to get $x = \frac{1}{5}$. So the
point $\left(\frac{1}{5}, 7\right)$ is on the graph. By symmetry, the
point $\left(\frac{1}{5}, 3\right)$ (two units below the axis) is on the
graph.

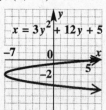

From the graph, we see that the domain is $(-\infty, 1]$
and the range is $(-\infty, \infty)$.

31. $x = 3y^2 + 12y + 5$

The roles of x and y are reversed, so this is a
horizontal parabola.

Step 1

Since $a = 3 > 0$, the graph opens to the right and
is narrower than the graph of $y = x^2$.

Step 2

Use the formula to find the y-value of the vertex.

$$\frac{-b}{2a} = \frac{-12}{2(3)} = -2$$

If $y = -2$, $x = -7$, so the vertex is $(-7, -2)$.
Since the graph opens right, the axis goes through
the y-coordinate of the vertex—its equation is
$y = -2$.

Step 3

If $y = 0$, $x = 5$, so the x-intercept is $(5, 0)$.
To find the y-intercepts, let $x = 0$.

$$0 = 3y^2 + 12y + 5$$
$$y = \frac{-12 \pm \sqrt{144 - 60}}{6} = \frac{-12 \pm \sqrt{84}}{6}$$
$$= \frac{-12 \pm 2\sqrt{21}}{6} = \frac{-6 \pm \sqrt{21}}{3}$$

The y-intercepts are approximately $(0, -0.5)$ and
$(0, -3.5)$.

Step 4

By symmetry, $(5, -4)$ is also on the graph.

From the graph, we see that the domain is $[-7, \infty)$
and the range is $(-\infty, \infty)$.

33. Let $x = $ one number, $40 - x = $ the other number, and $P = $ the product.

$$P = x(40 - x)$$
$$= 40x - x^2 \text{ or } -x^2 + 40x$$

This parabola opens down so the maximum occurs at the vertex.

Here $a = -1$, $b = 40$, and $c = 0$.

$$\frac{-b}{2a} = \frac{-40}{2(-1)} = 20$$

$x = 20$ when the product is a maximum.

Since $x = 20$, $40 - x = 20$, and the two numbers are 20 and 20.

35. Let x represent the length of the two equal sides, and let $280 - 2x$ represent the length of the remaining side (the side parallel to the highway). Then substitute x for L and $280 - 2x$ for W in the formula for the area of a rectangle, $A = LW$.

$$A = x(280 - 2x)$$
$$= 280x - 2x^2 \text{ or } -2x^2 + 280x$$

The maximum area will occur at the vertex.

$$x = \frac{-b}{2a} = \frac{-280}{2(-2)} = 70$$

When $x = 70$, $280 - 2x = 140$, and $A = 9800$. Thus, the dimensions of the lot with maximum area are 140 feet by 70 feet and the maximum area is 9800 square feet.

37. $s(t) = -16t^2 + 32t$

Here, $a = -16 < 0$, so the parabola opens down. The time it takes to reach the maximum height and the maximum height are given by the vertex of the parabola. Use the vertex formula to find that

$$t = \frac{-b}{2a} = \frac{-32}{2(-16)} = \frac{-32}{-32} = 1,$$

and $s(t) = -16(1)^2 + 32(1)$
$$= -16 + 32 = 16.$$

The vertex is $(1, 16)$, so the maximum height is 16 feet which occurs when the time is 1 second. The object hits the ground when $s = 0$.

$$0 = -16t^2 + 32t$$
$$0 = -16t(t - 2)$$

$-16t = 0$ or $t - 2 = 0$
$t = 0$ or $t = 2$

It takes 2 seconds for the object to hit the ground.

39. The graph of the height of the projectile,

$$s(t) = -16t^2 + 64t + 3,$$

is a parabola that opens down since $a = -16 < 0$. The time at which the cork reaches its maximum height and the maximum height are the t and s coordinates of the vertex.

$$t = \frac{-b}{2a} = \frac{-64}{2(-16)} = 2$$
$$s(2) = -16(2)^2 + 64(2) + 3$$
$$= -64 + 128 + 3 = 67$$

The cork reaches a maximum height of 67 feet after 2 seconds.

41. $f(x) = -0.0334x^2 + 0.2351x + 12.79$

(a) Since the graph opens down, the vertex is a maximum.

(b) The x-value of the vertex is given by

$$x = \frac{-b}{2a} = \frac{-0.2351}{2(-0.0334)} \approx 3.519 \approx 3.5$$

The year was $1990 + 3 = 1993$.

$f(3.5) \approx 13.2\%$, which is the maximum percent of births in the U.S. to teenage mothers.

43. $f(x) = -20.57x^2 + 758.9x - 3140$

(a) The coefficient of x^2 is negative because a parabola that models the data must open down.

(b) Use the vertex formula.

$$x = \frac{-b}{2a} = \frac{-758.9}{2(-20.57)} \approx 18.45$$
$$f(18.45) \approx 3860$$

The vertex is approximately $(18.45, 3860)$.

(c) 18 corresponds to 2018, so in 2018 social security assets will reach their maximum value of $3860 billion.

45. The number of people on the plane is $100 - x$ since x is the number of unsold seats. The price per seat is $200 + 4x$.

(a) The total revenue received for the flight is found by multiplying the number of seats by the price per seat. Thus, the revenue is

$$R(x) = (100 - x)(200 + 4x)$$
$$= 20,000 + 200x - 4x^2.$$

(b) Use the formula for the vertex.

$$x = \frac{-b}{2a} = \frac{-200}{2(-4)} = 25$$

$R(25) = 22{,}500$, so the vertex is $(25, 22{,}500)$. $R(0) = 20{,}000$, so the R-intercept is $(0, 20{,}000)$. From the factored form for R, we see that the positive x-intercept is $(100, 0)$. (The factor $200 + 4x$ leads to a negative x-intercept, meaningless in this problem.)

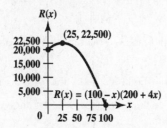

(c) The number of unsold seats x that produce the maximum revenue is 25, the x-value of the vertex.

(d) The maximum revenue is \$22,500, the y-value of the vertex.

47. $f(x) = x^2 - 8x + 18$

$$\frac{-b}{2a} = \frac{-(-8)}{2(1)} = 4$$

$f(4) = 2$, so the vertex is $(4, 2)$, which matches choice **B**.

49. $f(x) = x^2 - 8x + 14$

$$\frac{-b}{2a} = \frac{-(-8)}{2(1)} = 4$$

$f(4) = -2$, so the vertex is $(4, -2)$, which matches choice **A**.

51. (a) $|x - (-p)| = |x + p|$

(b) The focus should have coordinates $(p, 0)$ because the distance from the focus to the origin should equal the distance from the directrix to the origin.

(c) The distance from (x, y) to $(p, 0)$ is

$$\sqrt{(x - p)^2 + (y - 0)^2} = \sqrt{(x - p)^2 + y^2}.$$

(d) Using the results from parts (a) and (c), these distances should be equal.

$$\sqrt{(x - p)^2 + y^2} = |x + p|$$

Square both sides.

$$(x - p)^2 + y^2 = (x + p)^2$$
$$x^2 - 2px + p^2 + y^2 = x^2 + 2px + p^2$$
$$y^2 = 4px$$

53. $[1, 5] \Leftrightarrow 1 \le x \le 5$

55. $(-\infty, 1] \cup [5, \infty) \Leftrightarrow x \le 1$ or $x \ge 5$

57. $-2x + 1 < 4$
$$-2x < 3$$
$$x > -\frac{3}{2}$$

Solution set: $\left(-\frac{3}{2}, \infty\right)$

9.7 Quadratic and Rational Inequalities

1. (a) The x-intercepts determine the solutions of the equation $x^2 - 4x + 3 = 0$. From the graph, the solution set is $\{1, 3\}$.

(b) The x-values of the points on the graph that are *above* the x-axis form the solution set of the inequality $x^2 - 4x + 3 > 0$. From the graph, the solution set is $(-\infty, 1) \cup (3, \infty)$.

(c) The x-values of the points on the graph that are *below* the x-axis form the solution set of the inequality $x^2 - 4x + 3 < 0$. From the graph, the solution set is $(1, 3)$.

3. (a) The x-intercepts determine the solutions of the equation $-2x^2 - x + 15 = 0$. From the graph, the solution set is $\left\{-3, \frac{5}{2}\right\}$.

(b) The x-values of the points on the graph that are *above* the x-axis form the solution set of the inequality $-2x^2 - x + 15 > 0$. From the graph, the solution set for $-2x^2 - x + 15 \ge 0$ is $\left[-3, \frac{5}{2}\right]$.

(c) The x-values of the points on the graph that are *below* the x-axis form the solution set of the inequality $-2x^2 - x + 15 < 0$. From the graph, the solution set for $-2x^2 - x + 15 \le 0$ is $(-\infty, -3] \cup \left[\frac{5}{2}, \infty\right)$.

5. Include the endpoints if the symbol is $\le$ or $\ge$. Exclude the endpoints if the symbol is $<$ or $>$.

7. $(x + 1)(x - 5) > 0$
Solve the equation
$(x + 1)(x - 5) = 0.$

$$x + 1 = 0 \quad \text{or} \quad x - 5 = 0$$
$$x = -1 \quad \text{or} \quad x = 5$$

The numbers -1 and 5 divide a number line into three intervals: A, B, and C.

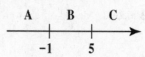

Test a number from each interval in the original inequality.

Interval A: Let $x = -2$.
$$(x + 1)(x - 5) > 0$$
$$(-2 + 1)(-2 - 5) > 0 \qquad ?$$
$$-1(-7) > 0 \qquad ?$$
$$7 > 0 \qquad \textit{True}$$

Interval B: Let $x = 0$.
$$(0 + 1)(0 - 5) > 0 \qquad ?$$
$$-5 > 0 \qquad \textit{False}$$

Interval C: Let $x = 6$.
$$(6 + 1)(6 - 5) > 0 \qquad ?$$
$$7 > 0 \qquad \textit{True}$$

The solution set includes the numbers in Intervals A and C, excluding -1 and 5 because of $>$.
Solution set: $(-\infty, -1) \cup (5, \infty)$

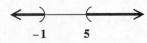

$-1 \qquad 5$

9. $(x + 4)(x - 6) < 0$
Solve the equation
$(x + 4)(x - 6) = 0$.

$$x + 4 = 0 \quad \text{or} \quad x - 6 = 0$$
$$x = -4 \quad \text{or} \qquad x = 6$$

These numbers divide a number line into three intervals: A, B, and C.

$$\begin{array}{c c c}
\textbf{A} & \textbf{B} & \textbf{C} \\
\end{array}$$

$-4 \qquad 6$

Test a number from each interval in the original inequality.

Interval A: Let $x = -5$.
$$(x + 4)(x - 6) < 0$$
$$(-5 + 4)(-5 - 6) < 0 \qquad ?$$
$$-1(-11) < 0 \qquad ?$$
$$11 < 0 \qquad \textit{False}$$

Interval B: Let $x = 0$.
$$4(-6) < 0 \qquad ?$$
$$-24 < 0 \qquad \textit{True}$$

Interval C: Let $x = 7$.
$$(7 + 4)(7 - 6) < 0 \qquad ?$$
$$11(1) < 0 \qquad ?$$
$$11 < 0 \qquad \textit{False}$$

The solution set includes Interval B, where the expression is negative.
Solution set: $(-4, 6)$

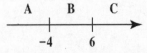

$-4 \qquad\qquad 6$

11. $x^2 - 4x + 3 \geq 0$
Solve the equation
$$x^2 - 4x + 3 = 0.$$
$$(x - 1)(x - 3) = 0$$
$$x - 1 = 0 \quad \text{or} \quad x - 3 = 0$$
$$x = 1 \quad \text{or} \qquad x = 3$$

$$\begin{array}{c c c}
\textbf{A} & \textbf{B} & \textbf{C} \\
\end{array}$$

$1 \qquad 3$

Interval A: Let $x = 0$.
$$3 \geq 0 \qquad \textit{True}$$

Interval B: Let $x = 2$.
$$2^2 - 4(2) + 3 \geq 0 \qquad ?$$
$$-1 \geq 0 \qquad \textit{False}$$

Interval C: Let $x = 4$.
$$4^2 - 4(4) + 3 \geq 0 \qquad ?$$
$$3 \geq 0 \qquad \textit{True}$$

The solution set includes the numbers in Intervals A and C, including 1 and 3 because of $\geq$.
Solution set: $(-\infty, 1] \cup [3, \infty)$

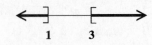

$1 \qquad 3$

13. $10x^2 + 9x \geq 9$
$$10x^2 + 9x - 9 \geq 0$$
Solve the equation
$$10x^2 + 9x - 9 = 0.$$
$$(2x + 3)(5x - 3) = 0$$
$$2x + 3 = 0 \quad \text{or} \quad 5x - 3 = 0$$
$$x = -\tfrac{3}{2} \quad \text{or} \qquad x = \tfrac{3}{5}$$

$$\begin{array}{c c c}
\textbf{A} & \textbf{B} & \textbf{C} \\
\end{array}$$

$-\dfrac{3}{2} \qquad \dfrac{3}{5}$

Test a number from each interval in the original inequality.

Interval A: Let $x = -2$.
$$10(-2)^2 + 9(-2) \geq 9 \qquad ?$$
$$40 - 18 \geq 9 \qquad ?$$
$$22 \geq 9 \qquad \textit{True}$$

Interval B: Let $x = 0$.
$$0 \geq 9 \qquad \textit{False}$$

Interval C: Let $x = 1$.
$$10(1)^2 + 9(1) \geq 9 \qquad ?$$
$$10 + 9 \geq 9 \qquad ?$$
$$19 \geq 9 \qquad \textit{True}$$

continued

The solution set includes the numbers in Intervals A and C, including $-\frac{3}{2}$ and $\frac{3}{5}$ because of $\geq$.
Solution set: $\left(-\infty, -\frac{3}{2}\right] \cup \left[\frac{3}{5}, \infty\right)$

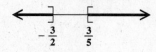

15. $4x^2 - 9 \leq 0$
Solve the equation
        $4x^2 - 9 = 0.$
$(2x + 3)(2x - 3) = 0$

$2x + 3 = 0$ or $2x - 3 = 0$
  $x = -\frac{3}{2}$ or $x = \frac{3}{2}$

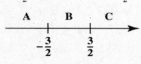

Test a number from each interval in the inequality

        $4x^2 - 9 \leq 0.$

Interval A: Let $x = -2.$
        $4(-2)^2 - 9 \leq 0$?
                $7 \leq 0$ *False*
Interval B: Let $x = 0.$
                $-9 \leq 0$ *True*
Interval C: Let $x = 2.$
        $4(2)^2 - 9 \leq 0$?
                $7 \leq 0$ *False*

The solution set includes Interval B, including the endpoints.
Solution set: $\left[-\frac{3}{2}, \frac{3}{2}\right]$

17. $6x^2 + x \geq 1$
        $6x^2 + x - 1 \geq 0$
Solve the equation
        $6x^2 + x - 1 = 0.$
$(2x + 1)(3x - 1) = 0$

$2x + 1 = 0$ or $3x - 1 = 0$
  $x = -\frac{1}{2}$ or $x = \frac{1}{3}$

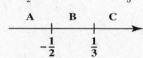

Test a number from each interval in the inequality.

        $6x^2 + x \geq 1.$

Interval A: Let $x = -1.$
        $6(-1)^2 + (-1) \geq 1$?
                $5 \geq 1$ *True*
Interval B: Let $x = 0.$
                $0 \geq 1$ *False*
Interval C: Let $x = 1.$
        $6(1)^2 + 1 \geq 1$?
                $7 \geq 1$ *True*

The solution set includes the numbers in Intervals A and C, including $-\frac{1}{2}$ and $\frac{1}{3}$ because of $\geq$.
Solution set: $\left(-\infty, -\frac{1}{2}\right] \cup \left[\frac{1}{3}, \infty\right)$

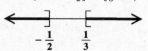

19. $z^2 - 4z \geq 0$
Solve the equation
    $z^2 - 4z = 0.$
$z(z - 4) = 0$

$z = 0$ or $z - 4 = 0$
            $z = 4$

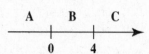

Interval A: Let $z = -1.$
    $(-1)^2 - 4(-1) \geq 0$?
                $5 \geq 0$ *True*
Interval B: Let $z = 2.$
        $2^2 - 4(2) \geq 0$?
                $-4 \geq 0$ *False*
Interval C: Let $z = 5.$
        $5^2 - 4(5) \geq 0$?
                $5 \geq 0$ *True*

The solution set includes the numbers in Intervals A and C, including 0 and 4 because of $\geq$.
Solution set: $(-\infty, 0] \cup [4, \infty)$

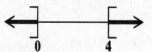

21. $3k^2 - 5k \leq 0$
Solve the equation
    $3k^2 - 5k = 0.$
$k(3k - 5) = 0$

$k = 0$ or $3k - 5 = 0$
            $k = \frac{5}{3}$

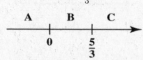

Interval A: Let $k = -1$.
$$3(-1)^2 - 5(-1) \leq 0 \quad ?$$
$$8 \leq 0 \quad \text{False}$$

Interval B: Let $k = 1$.
$$3(1)^2 - 5(1) \leq 0 \quad ?$$
$$-2 \leq 0 \quad \text{True}$$

Interval C: Let $k = 2$.
$$3(2)^2 - 5(2) \leq 0 \quad ?$$
$$2 \leq 0 \quad \text{False}$$

The solution set includes the numbers in Interval B, including the endpoints.

Solution set: $\left[0, \frac{5}{3}\right]$

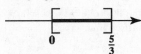

23. $x^2 - 6x + 6 \geq 0$
Solve the equation
$$x^2 - 6x + 6 = 0.$$

Since $x^2 - 6x + 6$ does not factor, let $a = 1$, $b = -6$, and $c = 6$ in the quadratic formula.

$$x = \frac{-(-6) \pm \sqrt{(-6)^2 - 4(1)(6)}}{2(1)}$$
$$= \frac{6 \pm \sqrt{12}}{2} = \frac{6 \pm 2\sqrt{3}}{2}$$
$$= \frac{2(3 \pm \sqrt{3})}{2} = 3 \pm \sqrt{3}$$
$$x = 3 + \sqrt{3} \approx 4.7 \text{ or}$$
$$x = 3 - \sqrt{3} \approx 1.3$$

Test a number from each interval in the inequality
$$x^2 - 6x + 6 \geq 0.$$

Interval A: Let $x = 0$.
$$6 \geq 0 \quad \text{True}$$

Interval B: Let $x = 3$.
$$3^2 - 6(3) + 6 \geq 0 \quad ?$$
$$-3 \geq 0 \quad \text{False}$$

Interval C: Let $x = 5$.
$$5^2 - 6(5) + 6 \geq 0 \quad ?$$
$$1 \geq 0 \quad \text{True}$$

The solution set includes the numbers in Intervals A and C, including $3 - \sqrt{3}$ and $3 + \sqrt{3}$ because of $\geq$.

Solution set: $\left(-\infty, 3 - \sqrt{3}\right] \cup \left[3 + \sqrt{3}, \infty\right)$

25. $(4 - 3x)^2 \geq -2$

Since $(4 - 3x)^2$ is either 0 or positive, $(4 - 3x)^2$ will always be greater than -2. Therefore, the solution set is $(-\infty, \infty)$.

27. $(3x + 5)^2 \leq -4$

Since $(3x + 5)^2$ is never negative, $(3x + 5)^2$ will never be less than or equal to a negative number. Therefore, the solution set is $\emptyset$.

29. $(x - 1)(x - 2)(x - 4) < 0$
The numbers 1, 2, and 4 are solutions of the cubic equation

$$(x - 1)(x - 2)(x - 4) = 0.$$

These numbers divide a number line into four intervals.

Test a number from each interval in the inequality
$$(x - 1)(x - 2)(x - 4) < 0.$$

Interval A: Let $x = 0$.
$$-1(-2)(-4) < 0 \quad ?$$
$$-8 < 0 \quad \text{True}$$

Interval B: Let $x = 1.5$.
$$(1.5 - 1)(1.5 - 2)(1.5 - 4) < 0 \quad ?$$
$$0.5(-0.5)(-2.5) < 0 \quad ?$$
$$0.625 < 0 \quad \text{False}$$

Interval C: Let $x = 3$.
$$(3 - 1)(3 - 2)(3 - 4) < 0 \quad ?$$
$$2(1)(-1) < 0 \quad ?$$
$$-2 < 0 \quad \text{True}$$

Interval D: Let $x = 5$.
$$(5 - 1)(5 - 2)(5 - 4) < 0 \quad ?$$
$$4(3)(1) < 0 \quad ?$$
$$12 < 0 \quad \text{False}$$

The numbers in Intervals A and C, not including 1, 2, or 4, are solutions.
Solution set: $(-\infty, 1) \cup (2, 4)$

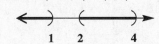

31. $(x - 4)(2x + 3)(3x - 1) \geq 0$

The numbers 4, $-\frac{3}{2}$, and $\frac{1}{3}$ are solutions of the cubic equation

$$(x - 4)(2x + 3)(3x - 1) = 0.$$

These numbers divide a number line into 4 intervals.

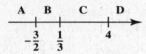

Interval A: Let $x = -2$.

$$-6(-1)(-7) \geq 0 \quad ?$$
$$-42 \geq 0 \qquad \textit{False}$$

Interval B: Let $x = 0$.

$$-4(3)(-1) \geq 0 \quad ?$$
$$12 \geq 0 \qquad \textit{True}$$

Interval C: Let $x = 1$.

$$-3(5)(2) \geq 0 \quad ?$$
$$-30 \geq 0 \qquad \textit{False}$$

Interval D: Let $x = 5$.

$$1(13)(14) \geq 0 \quad ?$$
$$182 \geq 0 \qquad \textit{True}$$

The solution set includes numbers in Intervals B and D, including the endpoints.
Solution set: $\left[-\frac{3}{2}, \frac{1}{3}\right] \cup [4, \infty)$

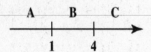

33. $\dfrac{x - 1}{x - 4} > 0$

The number 1 makes the numerator 0, and 4 makes the denominator 0. These two numbers determine three intervals.

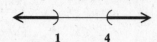

Test a number from each interval in the inequality

$$\frac{x - 1}{x - 4} > 0.$$

Interval A: Let $x = 0$.

$$\frac{0 - 1}{0 - 4} > 0 \quad ?$$
$$\frac{1}{4} > 0 \qquad \textit{True}$$

Interval B: Let $x = 2$.

$$\frac{2 - 1}{2 - 4} > 0 \quad ?$$
$$\frac{1}{-2} > 0 \qquad \textit{False}$$

Interval C: Let $x = 5$.

$$\frac{5 - 1}{5 - 4} > 0 \quad ?$$
$$4 > 0 \qquad \textit{True}$$

The solution set includes numbers in Intervals A and C, excluding endpoints.
Solution set: $(-\infty, 1) \cup (4, \infty)$

35. $\dfrac{2x + 3}{x - 5} \leq 0$

The number $-\frac{3}{2}$ makes the numerator 0, and 5 makes the denominator 0. These two numbers determine three intervals.

Test a number from each interval in the inequality

$$\frac{2x + 3}{x - 5} \leq 0.$$

Interval A: Let $x = -2$.

$$\frac{2(-2) + 3}{(-2) - 5} \leq 0 \quad ?$$
$$\frac{1}{7} \leq 0 \qquad \textit{False}$$

Interval B: Let $x = 0$.

$$\frac{2(0) + 3}{0 - 5} \leq 0 \quad ?$$
$$-\frac{3}{5} \leq 0 \qquad \textit{True}$$

Interval C: Let $x = 6$.

$$\frac{2(6) + 3}{6 - 5} \leq 0 \quad ?$$
$$15 \leq 0 \qquad \textit{False}$$

The solution set includes the points in Interval B. The endpoint 5 is not included since it makes the left side undefined. The endpoint $-\frac{3}{2}$ is included because it makes the left side equal to 0.
Solution set: $\left[-\frac{3}{2}, 5\right)$

37. $\dfrac{8}{x - 2} \geq 2$

Write the inequality so that 0 is on one side.

$$\frac{8}{x - 2} - 2 \geq 0$$
$$\frac{8}{x - 2} - \frac{2(x - 2)}{x - 2} \geq 0$$
$$\frac{8 - 2x + 4}{x - 2} \geq 0$$
$$\frac{-2x + 12}{x - 2} \geq 0$$

The number 6 makes the numerator 0, and 2 makes the denominator 0. These two numbers determine three intervals.

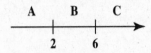

Test a number from each interval in the inequality

$$\frac{8}{x-2} \geq 2.$$

Interval A: Let $x = 0$.

$$\frac{8}{0-2} \geq 2 \quad ?$$

$$-4 \geq 2 \qquad \textit{False}$$

Interval B: Let $x = 3$.

$$\frac{8}{3-2} \geq 2 \quad ?$$

$$8 \geq 2 \qquad \textit{True}$$

Interval C: Let $x = 7$.

$$\frac{8}{7-2} \geq 2 \quad ?$$

$$\frac{8}{5} \geq 2 \qquad \textit{False}$$

The solution set includes numbers in Interval B, including 6 but excluding 2, which makes the fraction undefined.
Solution set: $(2, 6]$

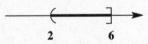

39.
$$\frac{3}{2x-1} < 2$$

Write the inequality so that 0 is on one side.

$$\frac{3}{2x-1} - 2 < 0$$

$$\frac{3}{2x-1} - \frac{2(2x-1)}{2x-1} < 0$$

$$\frac{3-4x+2}{2x-1} < 0$$

$$\frac{-4x+5}{2x-1} < 0$$

The number $\frac{5}{4}$ makes the numerator 0, and $\frac{1}{2}$ makes the denominator 0. These two numbers determine three intervals.

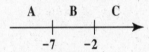

Test a number from each interval in the inequality

$$\frac{3}{2x-1} < 2.$$

Interval A: Let $x = 0$.

$$\frac{3}{2(0)-1} < 2 \quad ?$$

$$-3 < 2 \qquad \textit{True}$$

Interval B: Let $x = 1$.

$$\frac{3}{2(1)-1} < 2 \quad ?$$

$$3 < 2 \qquad \textit{False}$$

Interval C: Let $x = 2$.

$$\frac{3}{2(2)-1} < 2 \quad ?$$

$$1 < 2 \qquad \textit{True}$$

The solution set includes numbers in Intervals A and C, excluding endpoints.
Solution set: $\left(-\infty, \frac{1}{2}\right) \cup \left(\frac{5}{4}, \infty\right)$

41.
$$\frac{x-3}{x+2} \geq 2$$

Write the inequality so that 0 is on one side.

$$\frac{x-3}{x+2} - 2 \geq 0$$

$$\frac{x-3}{x+2} - \frac{2(x+2)}{x+2} \geq 0$$

$$\frac{x-3-2x-4}{x+2} \geq 0$$

$$\frac{-x-7}{x+2} \geq 0$$

The number -7 makes the numerator 0, and -2 makes the denominator 0. These two numbers determine three intervals.

Test a number from each interval in the inequality

$$\frac{x-3}{x+2} \geq 2.$$

Interval A: Let $x = -8$.

$$\frac{-11}{-6} \geq 2 \quad ?$$

$$\frac{11}{6} \geq 2 \qquad \textit{False}$$

Interval B: Let $x = -4$.

$$\frac{-7}{-2} \geq 2 \quad ?$$

$$\frac{7}{2} \geq 2 \qquad \textit{True}$$

Interval C: Let $x = 0$.

$$\frac{-3}{2} \geq 2 \qquad \textit{False}$$

continued

The solution set includes numbers in Interval B, including -7 but excluding -2, which makes the fraction undefined.

Solution set: $[-7, -2)$

43. $\dfrac{x - 8}{x - 4} < 3$

Write the inequality so that 0 is on one side.

$$\dfrac{x - 8}{x - 4} - 3 < 0$$

$$\dfrac{x - 8}{x - 4} - \dfrac{3(x - 4)}{x - 4} < 0$$

$$\dfrac{x - 8 - 3x + 12}{x - 4} < 0$$

$$\dfrac{-2x + 4}{x - 4} < 0$$

The number 2 makes the numerator 0, and 4 makes the denominator 0. These two numbers determine three intervals.

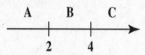

Test a number from each interval in the inequality

$$\dfrac{x - 8}{x - 4} < 3.$$

Interval A: Let $x = 0$.

$$\dfrac{-8}{-4} < 3 \quad ?$$

$$2 < 3 \qquad \textit{True}$$

Interval B: Let $x = 3$.

$$\dfrac{-5}{-1} < 3 \quad ?$$

$$5 < 3 \qquad \textit{False}$$

Interval C: Let $x = 5$.

$$\dfrac{-3}{1} < 3 \qquad \textit{True}$$

The solution set includes numbers in Intervals A and C, excluding endpoints.

Solution set: $(-\infty, 2) \cup (4, \infty)$

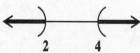

45. $\dfrac{4k}{2k - 1} < k$

Write the inequality so that 0 is on one side.

$$\dfrac{4k}{2k - 1} - k < 0$$

$$\dfrac{4k}{2k - 1} - \dfrac{k(2k - 1)}{2k - 1} < 0$$

$$\dfrac{4k - 2k^2 + k}{2k - 1} < 0$$

$$\dfrac{-2k^2 + 5k}{2k - 1} < 0$$

$$\dfrac{k(-2k + 5)}{2k - 1} < 0$$

The numbers 0 and $\frac{5}{2}$ make the numerator 0, and $\frac{1}{2}$ makes the denominator 0. These three numbers determine four intervals.

Test a number from each interval in the inequality

$$\dfrac{4k}{2k - 1} < k.$$

Interval A: Let $k = -1$.

$$\dfrac{4(-1)}{2(-1) - 1} < -1 \quad ?$$

$$\dfrac{4}{3} < -1 \qquad \textit{False}$$

Interval B: Let $k = \frac{1}{4}$.

$$\dfrac{4\left(\frac{1}{4}\right)}{2\left(\frac{1}{4}\right) - 1} < \dfrac{1}{4} \quad ?$$

$$-2 < \dfrac{1}{4} \qquad \textit{True}$$

Interval C: Let $k = 1$.

$$\dfrac{4(1)}{2(1) - 1} < 1 \quad ?$$

$$4 < 1 \qquad \textit{False}$$

Interval D: Let $k = 3$.

$$\dfrac{4(3)}{2(3) - 1} < 3 \quad ?$$

$$\dfrac{12}{5} < 3 \qquad \textit{True}$$

The solution set includes numbers in Intervals B and D. None of the endpoints are included.

Solution set: $\left(0, \frac{1}{2}\right) \cup \left(\frac{5}{2}, \infty\right)$

47. $\dfrac{2x - 3}{x^2 + 1} \ge 0$

The denominator is positive for all real numbers x, so it has no effect on the solution set for the inequality.

$$2x - 3 \ge 0$$

$$2x \ge 3$$

$$x \ge \dfrac{3}{2}$$

Solution set: $\left[\frac{3}{2}, \infty\right)$

49. $\dfrac{(3x-5)^2}{x+2} > 0$

The numerator is positive for all real numbers x except $x = \frac{5}{3}$, which makes it equal to 0. If we solve the inequality $x + 2 > 0$, then we only have to be sure to exclude $\frac{5}{3}$ from that solution set to determine the solution set of the original inequality.

$$x + 2 > 0$$
$$x > -2$$

Solution set: $\left(-2, \frac{5}{3}\right) \cup \left(\frac{5}{3}, \infty\right)$

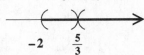

51. $\left\{\left(-3, \frac{1}{8}\right), \left(-2, \frac{1}{4}\right), \left(-1, \frac{1}{2}\right), (0,1), (1,2), (2,4), (3,8)\right\}$

The domain is the set of x-values: $\{-3, -2, -1, 0, 1, 2, 3\}$.
The range is the set of y-values: $\left\{\frac{1}{8}, \frac{1}{4}, \frac{1}{2}, 1, 2, 4, 8\right\}$.

53. Using the vertical line test, we find any vertical line will intersect the graph at most once. This indicates that the graph represents a function.

Chapter 9 Review Exercises

1. $t^2 = 121$
$t = 11 \quad \text{or} \quad t = -11$
Solution set: $\{-11, 11\}$

2. $p^2 = 3$
$p = \sqrt{3} \quad \text{or} \quad p = -\sqrt{3}$
Solution set: $\left\{-\sqrt{3}, \sqrt{3}\right\}$

3. $(2x + 5)^2 = 100$
$2x + 5 = 10 \quad \text{or} \quad 2x + 5 = -10$
$2x = 5 \qquad\qquad 2x = -15$
$x = \frac{5}{2} \quad \text{or} \qquad x = -\frac{15}{2}$
Solution set: $\left\{-\frac{15}{2}, \frac{5}{2}\right\}$

4. $(3k - 2)^2 = -25$
$3k - 2 = \sqrt{-25} \quad \text{or} \quad 3k - 2 = -\sqrt{-25}$
$3k - 2 = 5i \qquad\qquad 3k - 2 = -5i$
$3k = 2 + 5i \qquad\qquad 3k = 2 - 5i$
$k = \dfrac{2 + 5i}{3} \qquad\qquad k = \dfrac{2 - 5i}{3}$
$k = \frac{2}{3} + \frac{5}{3}i \quad \text{or} \qquad k = \frac{2}{3} - \frac{5}{3}i$
Solution set: $\left\{\frac{2}{3} + \frac{5}{3}i, \frac{2}{3} - \frac{5}{3}i\right\}$

5. $x^2 + 4x = 15$
Complete the square.
$\left(\frac{1}{2} \cdot 4\right)^2 = 2^2 = 4$
Add 4 to each side.
$x^2 + 4x + 4 = 15 + 4$
$(x + 2)^2 = 19$
$x + 2 = \sqrt{19} \qquad \text{or} \quad x + 2 = -\sqrt{19}$
$x = -2 + \sqrt{19} \quad \text{or} \qquad x = -2 - \sqrt{19}$
Solution set: $\left\{-2 + \sqrt{19}, -2 - \sqrt{19}\right\}$

6. $2m^2 - 3m = -1$
$m^2 - \frac{3}{2}m = -\frac{1}{2} \qquad\qquad \textit{Divide by 2.}$
Complete the square.
$\left[\frac{1}{2}\left(-\frac{3}{2}\right)\right]^2 = \left(-\frac{3}{4}\right)^2 = \frac{9}{16}$
Add $\frac{9}{16}$ to each side.
$m^2 - \frac{3}{2}m + \frac{9}{16} = -\frac{1}{2} + \frac{9}{16}$
$\left(m - \frac{3}{4}\right)^2 = -\frac{8}{16} + \frac{9}{16}$
$\left(m - \frac{3}{4}\right)^2 = \frac{1}{16}$
$m - \frac{3}{4} = \sqrt{\frac{1}{16}} \quad \text{or} \quad m - \frac{3}{4} = -\sqrt{\frac{1}{16}}$
$m - \frac{3}{4} = \frac{1}{4} \qquad\qquad m - \frac{3}{4} = -\frac{1}{4}$
$m = \frac{3}{4} + \frac{1}{4} \qquad\qquad m = \frac{3}{4} - \frac{1}{4}$
$m = 1 \qquad \text{or} \qquad m = \frac{1}{2}$
Solution set: $\left\{\frac{1}{2}, 1\right\}$

7. By the square root property, the first step should be

$$x = \sqrt{12} \quad \text{or} \quad x = -\sqrt{12}.$$

Solution set: $\left\{-2\sqrt{3}, 2\sqrt{3}\right\}$

8. $16t^2 = d$
$16t^2 = 150 \qquad\qquad \textit{Let d = 150.}$
$t^2 = \frac{150}{16}$
$t = \sqrt{\frac{150}{16}} \qquad\qquad t \geq 0$
$t \approx 3.1$ seconds

It would take about 3.1 seconds for the wallet to fall 150 feet.

9. $x^2 + 5x + 2 = 0$
Here $a = 1$, $b = 5$, and $c = 2$.
$b^2 - 4ac = 5^2 - 4(1)(2)$
$= 25 - 8 = 17$
Since the discriminant is positive, but not a perfect square, there are two distinct irrational number solutions. The answer is **C**.

10.
$$4t^2 = 3 - 4t$$
$$4t^2 + 4t - 3 = 0$$
Here $a = 4$, $b = 4$, and $c = -3$.
$$b^2 - 4ac = 4^2 - 4(4)(-3)$$
$$= 16 + 48$$
$$= 64 \text{ or } 8^2$$
Since the discriminant is positive, and a perfect square, there are two distinct rational number solutions. The answer is **A**.

11.
$$4x^2 = 6x - 8$$
$$4x^2 - 6x + 8 = 0$$
Here $a = 4$, $b = -6$, and $c = 8$.
$$b^2 - 4ac = (-6)^2 - 4(4)(8)$$
$$= 36 - 128 = -92$$
Since the discriminant is negative, there are two distinct nonreal complex number solutions. The answer is **D**.

12. $9z^2 + 30z + 25 = 0$
Here $a = 9$, $b = 30$, and $c = 25$.
$$b^2 - 4ac = 30^2 - 4(9)(25)$$
$$= 900 - 900 = 0$$
Since the discriminant is zero, there is exactly one rational number solution. The answer is **B**.

13. $2x^2 + x - 21 = 0$
Here $a = 2$, $b = 1$, and $c = -21$.
$$x = \frac{-b \pm \sqrt{b^2 - 4ac}}{2a}$$
$$x = \frac{-1 \pm \sqrt{1^2 - 4(2)(-21)}}{2(2)}$$
$$= \frac{-1 \pm \sqrt{1 + 168}}{4}$$
$$= \frac{-1 \pm \sqrt{169}}{4} = \frac{-1 \pm 13}{4}$$
$$x = \frac{-1 + 13}{4} = \frac{12}{4} = 3 \text{ or}$$
$$x = \frac{-1 - 13}{4} = -\frac{14}{4} = -\frac{7}{2}$$
Solution set: $\left\{-\frac{7}{2}, 3\right\}$

14.
$$k^2 + 5k = 7$$
$$k^2 + 5k - 7 = 0$$
Here $a = 1$, $b = 5$, and $c = -7$.
$$k = \frac{-b \pm \sqrt{b^2 - 4ac}}{2a}$$
$$k = \frac{-5 \pm \sqrt{5^2 - 4(1)(-7)}}{2(1)}$$
$$= \frac{-5 \pm \sqrt{25 + 28}}{2}$$
$$= \frac{-5 \pm \sqrt{53}}{2}$$

Solution set: $\left\{\dfrac{-5 + \sqrt{53}}{2}, \dfrac{-5 - \sqrt{53}}{2}\right\}$

15. $(t + 3)(t - 4) = -2$
$$t^2 - t - 12 = -2$$
$$t^2 - t - 10 = 0$$
Here $a = 1$, $b = -1$, and $c = -10$.
$$t = \frac{-b \pm \sqrt{b^2 - 4ac}}{2a}$$
$$t = \frac{-(-1) \pm \sqrt{(-1)^2 - 4(1)(-10)}}{2(1)}$$
$$= \frac{1 \pm \sqrt{1 + 40}}{2} = \frac{1 \pm \sqrt{41}}{2}$$
Solution set: $\left\{\dfrac{1 + \sqrt{41}}{2}, \dfrac{1 - \sqrt{41}}{2}\right\}$

16. $2x^2 + 3x + 4 = 0$
Here $a = 2$, $b = 3$, and $c = 4$.
$$x = \frac{-b \pm \sqrt{b^2 - 4ac}}{2a}$$
$$x = \frac{-3 \pm \sqrt{3^2 - 4(2)(4)}}{2(2)}$$
$$= \frac{-3 \pm \sqrt{9 - 32}}{4}$$
$$= \frac{-3 \pm \sqrt{-23}}{4}$$
$$= \frac{-3 \pm i\sqrt{23}}{4} = -\frac{3}{4} \pm \frac{\sqrt{23}}{4}i$$
Solution set: $\left\{-\frac{3}{4} + \frac{\sqrt{23}}{4}i, -\frac{3}{4} - \frac{\sqrt{23}}{4}i\right\}$

17.
$$3p^2 = 2(2p - 1)$$
$$3p^2 = 4p - 2$$
$$3p^2 - 4p + 2 = 0$$
Here $a = 3$, $b = -4$, and $c = 2$.
$$p = \frac{-b \pm \sqrt{b^2 - 4ac}}{2a}$$
$$p = \frac{-(-4) \pm \sqrt{(-4)^2 - 4(3)(2)}}{2(3)}$$
$$= \frac{4 \pm \sqrt{16 - 24}}{6} = \frac{4 \pm \sqrt{-8}}{6}$$
$$= \frac{4 \pm 2i\sqrt{2}}{6} = \frac{2(2 \pm i\sqrt{2})}{6}$$
$$= \frac{2 \pm i\sqrt{2}}{3} = \frac{2}{3} \pm \frac{\sqrt{2}}{3}i$$
Solution set: $\left\{\frac{2}{3} + \frac{\sqrt{2}}{3}i, \frac{2}{3} - \frac{\sqrt{2}}{3}i\right\}$

18. $m(2m - 7) = 3m^2 + 3$

$2m^2 - 7m = 3m^2 + 3$

$0 = m^2 + 7m + 3$

Here $a = 1$, $b = 7$, and $c = 3$.

$$m = \frac{-b \pm \sqrt{b^2 - 4ac}}{2a}$$

$$m = \frac{-7 \pm \sqrt{7^2 - 4(1)(3)}}{2(1)}$$

$$= \frac{-7 \pm \sqrt{49 - 12}}{2}$$

$$= \frac{-7 \pm \sqrt{37}}{2}$$

Solution set: $\left\{ \frac{-7+\sqrt{37}}{2}, \frac{-7-\sqrt{37}}{2} \right\}$

19. $\dfrac{15}{x} = 2x - 1$

$x \left(\dfrac{15}{x} \right) = x(2x - 1)$ *Multiply by the LCD, x.*

$15 = 2x^2 - x$

$0 = 2x^2 - x - 15$

$0 = (2x + 5)(x - 3)$

$2x + 5 = 0$ or $x - 3 = 0$

$x = -\frac{5}{2}$ or $x = 3$

Check $x = -\frac{5}{2}$: $-6 = -5 - 1$ *True*

Check $x = 3$: $5 = 6 - 1$ *True*

Solution set: $\left\{ -\frac{5}{2}, 3 \right\}$

20. $\dfrac{1}{n} + \dfrac{2}{n + 1} = 2$

$n(n + 1) \left(\dfrac{1}{n} + \dfrac{2}{n + 1} \right)$ *Multiply by the LCD, n(n+1).*

$= n(n + 1) \cdot 2$

$(n + 1) + 2n = 2n^2 + 2n$

$0 = 2n^2 - n - 1$

$0 = (2n + 1)(n - 1)$

$2n + 1 = 0$ or $n - 1 = 0$

$n = -\frac{1}{2}$ or $n = 1$

Check $n = -\frac{1}{2}$: $-2 + 4 = 2$ *True*

Check $n = 1$: $1 + 1 = 2$ *True*

Solution set: $\left\{ -\frac{1}{2}, 1 \right\}$

21. $-2r = \sqrt{\dfrac{48 - 20r}{2}}$

Square both sides.

$$(-2r)^2 = \left(\sqrt{\frac{48 - 20r}{2}} \right)^2$$

$$4r^2 = \frac{48 - 20r}{2}$$

$$4r^2 = 24 - 10r$$

$4r^2 + 10r - 24 = 0$

$2r^2 + 5r - 12 = 0$

$(r + 4)(2r - 3) = 0$

$r + 4 = 0$ or $2r - 3 = 0$

$r = -4$ or $r = \frac{3}{2}$

Check $r = -4$: $8 = \sqrt{64}$ *True*

Check $r = \frac{3}{2}$: $-3 = \sqrt{9}$ *False*

Solution set: $\{-4\}$

22. $8(3x + 5)^2 + 2(3x + 5) - 1 = 0$

Let $u = 3x + 5$. The equation becomes

$$8u^2 + 2u - 1 = 0.$$

$$(2u + 1)(4u - 1) = 0$$

$2u + 1 = 0$ or $4u - 1 = 0$

$u = -\frac{1}{2}$ or $u = \frac{1}{4}$

To find x, substitute $3x + 5$ for u.

$3x + 5 = -\frac{1}{2}$ or $3x + 5 = \frac{1}{4}$

$3x = -\frac{11}{2}$ $3x = -\frac{19}{4}$

$x = -\frac{11}{6}$ or $x = -\frac{19}{12}$

Check $x = -\frac{11}{6}$: $2 - 1 - 1 = 0$ *True*

Check $x = -\frac{19}{12}$: $0.5 + 0.5 - 1 = 0$ *True*

Solution set: $\left\{ -\frac{11}{6}, -\frac{19}{12} \right\}$

23. $2x^{2/3} - x^{1/3} - 28 = 0$

Let $u = x^{1/3}$, so $u^2 = \left(x^{1/3} \right)^2 = x^{2/3}$.

The equation becomes

$$2u^2 - u - 28 = 0.$$

$$(2u + 7)(u - 4) = 0$$

$2u + 7 = 0$ or $u - 4 = 0$

$u = -\frac{7}{2}$ or $u = 4$

To find x, substitute $x^{1/3}$ for u.

$x^{1/3} = -\frac{7}{2}$ or $x^{1/3} = 4$

$\left(x^{1/3} \right)^3 = \left(-\frac{7}{2} \right)^3$ $\left(x^{1/3} \right)^3 = 4^3$

$x = -\frac{343}{8}$ or $x = 64$

Check $x = -\frac{343}{8}$: $24.5 + 3.5 - 28 = 0$ *True*

Check $x = 64$: $32 - 4 - 28 = 0$ *True*

Solution set: $\left\{ -\frac{343}{8}, 64 \right\}$

24. $p^4 - 10p^2 + 9 = 0$

Let $x = p^2$, so $x^2 = p^4$.

$$x^2 - 10x + 9 = 0$$
$$(x - 1)(x - 9) = 0$$

$$x - 1 = 0 \quad \text{or} \quad x - 9 = 0$$
$$x = 1 \quad \text{or} \quad x = 9$$

To find p, substitute p^2 for x.

$$p^2 = 1 \quad \text{or} \quad p^2 = 9$$
$$p = \pm\sqrt{1} \qquad p = \pm\sqrt{9}$$
$$p = \pm 1 \quad \text{or} \quad p = \pm 3$$

Check $p = \pm 1$: $1 - 10 + 9 = 0$ *True*

Check $p = \pm 3$: $81 - 90 + 9 = 0$ *True*

Solution set: $\{-3, -1, 1, 3\}$

25. Let $x =$ Phong's speed.

Make a table. Use $d = rt$, or $t = \dfrac{d}{r}$.

	Distance	Rate	Time
Upstream	20	$x - 3$	$\dfrac{20}{x - 3}$
Downstream	20	$x + 3$	$\dfrac{20}{x + 3}$

Time upstream plus time downstream equals 7 hr.

$$\underset{\downarrow}{\frac{20}{x-3}} \quad \underset{\downarrow}{+} \quad \underset{\downarrow}{\frac{20}{x+3}} \quad \underset{\downarrow}{=} \quad \underset{\downarrow}{7}$$

Multiply each side by the LCD, $(x + 3)(x - 3)$.

$$(x+3)(x-3)\left(\frac{20}{x-3} + \frac{20}{x+3}\right) = (x+3)(x-3)(7)$$
$$20(x+3) + 20(x-3) = 7(x^2 - 9)$$
$$20x + 60 + 20x - 60 = 7x^2 - 63$$
$$0 = 7x^2 - 40x - 63$$
$$0 = (x - 7)(7x + 9)$$

$$x - 7 = 0 \quad \text{or} \quad 7x + 9 = 0$$
$$x = 7 \quad \text{or} \quad x = -\frac{9}{7}$$

Reject $-\frac{9}{7}$ since speed can't be negative. Phong's speed was 7 mph.

26. Let $x =$ Maureen's speed on the trip to pick up Laurie.

Make a chart. Use $d = rt$, or $t = \frac{d}{r}$.

	Distance	Rate	Time
To Laurie	8	x	$\dfrac{8}{x}$
To the Mall	11	$x + 15$	$\dfrac{11}{x + 15}$

Time to pick up Laurie plus time to mall equals 24 min (or 0.4 hr).

$$\frac{8}{x} + \frac{11}{x + 15} = 0.4$$

Multiply each side by the LCD, $x(x + 15)$.

$$x(x+15)\left(\frac{8}{x} + \frac{11}{x+15}\right) = x(x+15)(0.4)$$
$$8(x + 15) + 11x = 0.4x(x + 15)$$
$$8x + 120 + 11x = 0.4x^2 + 6x$$
$$0 = 0.4x^2 - 13x - 120$$

Multiply by 5 to clear the decimal.

$$0 = 2x^2 - 65x - 600$$
$$0 = (x - 40)(2x + 15)$$

$$x - 40 = 0 \quad \text{or} \quad 2x + 15 = 0$$
$$x = 40 \quad \text{or} \quad x = -\frac{15}{2}$$

Speed cannot be negative, so $-\frac{15}{2}$ is not a solution. Maureen's speed on the trip to pick up Laurie was 40 mph.

27. Let $x =$ the amount of time for the old machine alone and

$x - 1 =$ the amount of time for the new machine alone.

Make a chart.

Machine	Rate	Time Together	Fractional Part of the Job Done
Old	$\dfrac{1}{x}$	2	$\dfrac{2}{x}$
New	$\dfrac{1}{x - 1}$	2	$\dfrac{2}{x - 1}$

Part done by old machine plus Part done by new machine = 1 whole job.

$$\frac{2}{x} + \frac{2}{x - 1} = 1$$

Multiply by the LCD, $x(x - 1)$.

$$x(x-1)\left(\frac{2}{x} + \frac{2}{x-1}\right) = x(x-1)\cdot 1$$
$$2(x - 1) + 2x = x^2 - x$$
$$2x - 2 + 2x = x^2 - x$$
$$0 = x^2 - 5x + 2$$

Use the quadratic formula.

$$x = \frac{-b \pm \sqrt{b^2 - 4ac}}{2a}$$
$$x = \frac{-(-5) \pm \sqrt{(-5)^2 - 4(1)(2)}}{2(1)}$$

$$= \frac{5 \pm \sqrt{25-8}}{2} = \frac{5 \pm \sqrt{17}}{2}$$

$$x = \frac{5 + \sqrt{17}}{2} \approx 4.6 \text{ or}$$

$$x = \frac{5 - \sqrt{17}}{2} \approx 0.4$$

Reject 0.4 as the time for the old machine, because that would yield a negative time for the new machine. Thus, the old machine takes about 4.6 hours.

28. Let $x =$ the time for Carter alone and
$x - 1 =$ the time for Greg alone.

Worker	Rate	Time Together	Fractional Part of the Job Done
Carter	$\dfrac{1}{x}$	1.5	$\dfrac{1.5}{x}$
Greg	$\dfrac{1}{x-1}$	1.5	$\dfrac{1.5}{x-1}$

$$\begin{array}{ccccc} \text{Part by} & & \text{part} & & \text{1 whole} \\ \text{Carter} & + & \text{by Greg} & = & \text{job.} \end{array}$$

$$\frac{1.5}{x} + \frac{1.5}{x-1} = 1$$

Multiply by the LCD, $x(x-1)$.

$$x(x-1)\left(\frac{1.5}{x} + \frac{1.5}{x-1}\right) = x(x-1) \cdot 1$$

$$1.5(x-1) + 1.5x = x^2 - x$$

$$1.5x - 1.5 + 1.5x = x^2 - x$$

$$0 = x^2 - 4x + 1.5$$

Use the quadratic formula.

$$x = \frac{-b \pm \sqrt{b^2 - 4ac}}{2a}$$

$$x = \frac{-(-4) \pm \sqrt{(-4)^2 - 4(1)(1.5)}}{2(1)}$$

$$= \frac{4 \pm \sqrt{16-6}}{2} = \frac{4 \pm \sqrt{10}}{2}$$

$$= \frac{4 \pm \sqrt{10}}{2}$$

$$x = \frac{4 + \sqrt{10}}{2} \approx 3.6 \text{ or}$$

$$x = \frac{4 - \sqrt{10}}{2} \approx 0.4$$

Reject 0.4 as Carter's time, because that would make Greg's time negative. Thus, Carter's time alone is about 3.6 hours and Greg's time alone is about $x - 1 = 2.6$ hours.

29. Solve $k = \dfrac{rF}{wv^2}$ for v.

Multiply both sides by v^2, then divide by k.

$$v^2 = \frac{rF}{kw}$$

$$v = \pm \sqrt{\frac{rF}{kw}} = \frac{\pm \sqrt{rF}}{\sqrt{kw}}$$

$$= \frac{\pm \sqrt{rF}}{\sqrt{kw}} \cdot \frac{\sqrt{kw}}{\sqrt{kw}}$$

$$v = \frac{\pm \sqrt{rFkw}}{kw}$$

30. Solve $p = \sqrt{\dfrac{yz}{6}}$ for y.

Square both sides.

$$p^2 = \left(\sqrt{\frac{yz}{6}}\right)^2$$

$$p^2 = \frac{yz}{6}$$

$$\frac{6p^2}{z} = y \quad \text{or} \quad y = \frac{6p^2}{z}$$

31. Solve $mt^2 = 3mt + 6$ for t.

$$mt^2 - 3mt - 6 = 0$$

Use the quadratic formula with $a = m$, $b = -3m$, and $c = -6$.

$$t = \frac{-b \pm \sqrt{b^2 - 4ac}}{2a}$$

$$t = \frac{3m \pm \sqrt{(-3m)^2 - 4(m)(-6)}}{2m}$$

$$= \frac{3m \pm \sqrt{9m^2 + 24m}}{2m}$$

32. Let $x =$ the length of the longer leg;
$\frac{3}{4}x =$ the length of the shorter leg;
$2x - 9 =$ the length of the hypotenuse.

Use the Pythagorean formula.

$$c^2 = a^2 + b^2$$

$$(2x - 9)^2 = x^2 + \left(\tfrac{3}{4}x\right)^2$$

$$4x^2 - 36x + 81 = x^2 + \tfrac{9}{16}x^2$$

$$16\left(4x^2 - 36x + 81\right) = 16\left(x^2 + \tfrac{9}{16}x^2\right)$$

$$64x^2 - 576x + 1296 = 16x^2 + 9x^2$$

$$39x^2 - 576x + 1296 = 0$$

$$13x^2 - 192x + 432 = 0 \qquad \textit{Divide by 3.}$$

$$(13x - 36)(x - 12) = 0$$

$$13x - 36 = 0 \quad \text{or} \quad x - 12 = 0$$

$$x = \tfrac{36}{13} \quad \text{or} \qquad x = 12$$

continued

Reject $x = \frac{36}{13}$ since $2\left(\frac{36}{13}\right) - 9$ is negative.
If $x = 12$, then
$$\tfrac{3}{4}x = \tfrac{3}{4}(12) = 9$$
and
$$2x - 9 = 2(12) - 9 = 15.$$
The lengths of the three sides are 9 feet, 12 feet, and 15 feet.

33. Let $x =$ the amount removed from one dimension.

The area of the square is 256 cm^2, so the length of one side is $\sqrt{256}$ or 16 cm. The dimensions of the new rectangle are $16 + x$ and $16 - x$ cm. The area of the new rectangle is 16 cm^2 less than the area of the square.

$$(16 + x)(16 - x) = 256 - 16$$
$$256 - x^2 = 240$$
$$-x^2 = -16$$
$$x^2 = 16$$
$$x = \pm\sqrt{16} = \pm 4$$

Length cannot be negative, so reject -4. If $x = 4$, then $16 + x = 20$, and $16 - x = 12$. The dimensions are 20 cm by 12 cm.

34. Let $x =$ the width of the border.

Area of mat = length • width
$$352 = (2x + 20)(2x + 14)$$
$$352 = 4x^2 + 68x + 280$$
$$0 = 4x^2 + 68x - 72$$
$$0 = x^2 + 17x - 18$$
$$0 = (x + 18)(x - 1)$$
$$x + 18 = 0 \quad \text{or} \quad x - 1 = 0$$
$$x = -18 \quad \text{or} \quad x = 1$$
Reject the negative answer for length. The mat is 1 inch wide.

35. $f(t) = 100t^2 - 300t$ is the distance of the light from the starting point at t minutes. When the light returns to the starting point, the value of $f(t)$ will be 0.

$$0 = 100t^2 - 300t$$
$$0 = t^2 - 3t \qquad \textit{Divide by 100.}$$
$$0 = t(t - 3)$$
$$t = 0 \quad \text{or} \quad t - 3 = 0$$
$$t = 3$$

Since $t = 0$ represents the starting time, the light will return to the starting point in 3 minutes.

36. $f(t) = -16t^2 + 45t + 400$
$$200 = -16t^2 + 45t + 400 \quad \textit{Let f(t) = 200.}$$
$$0 = -16t^2 + 45t + 200$$
$$0 = 16t^2 - 45t - 200$$

Here $a = 16$, $b = -45$, and $c = -200$.

$$t = \frac{-b \pm \sqrt{b^2 - 4ac}}{2a}$$
$$t = \frac{-(-45) \pm \sqrt{(-45)^2 - 4(16)(-200)}}{2(16)}$$
$$= \frac{45 \pm \sqrt{2025 + 12{,}800}}{32}$$
$$= \frac{45 \pm \sqrt{14{,}825}}{32}$$
$$t = \frac{45 + \sqrt{14{,}825}}{32} \approx 5.2 \quad \text{or}$$
$$t = \frac{45 - \sqrt{14{,}825}}{32} \approx -2.4$$

Reject the negative solution since time cannot be negative. The ball will reach a height of 200 ft above the ground after about 5.2 seconds.

37. $s(t) = -16t^2 + 75t + 407$
$$450 = -16t^2 + 75t + 407 \quad \textit{Let s(t) = 450.}$$
$$0 = -16t^2 + 75t - 43$$

Here $a = -16$, $b = 75$, and $c = -43$.

$$t = \frac{-b \pm \sqrt{b^2 - 4ac}}{2a}$$
$$t = \frac{-75 \pm \sqrt{75^2 - 4(-16)(-43)}}{2(-16)}$$
$$= \frac{-75 \pm \sqrt{5625 - 2752}}{-32}$$
$$= \frac{-75 \pm \sqrt{2873}}{-32}$$
$$t = \frac{-75 + \sqrt{2873}}{-32} \approx 0.7 \quad \text{or}$$
$$t = \frac{-75 - \sqrt{2873}}{-32} \approx 4.0$$

The ball will be 450 ft above the ground after 0.7 second (going up) and again after 4.0 seconds (coming down).

38. Let $p =$ the price.

Demand = Supply
$$\frac{25}{p} = 70p + 15$$
$$p\left(\frac{25}{p}\right) = p(70p + 15) \qquad \textit{Multiply by p.}$$
$$25 = 70p^2 + 15p$$
$$0 = 70p^2 + 15p - 25$$
$$0 = 14p^2 + 3p - 5 \qquad \textit{Divide by 5.}$$
$$0 = (2p - 1)(7p + 5)$$
$$2p - 1 = 0 \quad \text{or} \quad 7p + 5 = 0$$
$$p = \tfrac{1}{2} \quad \text{or} \qquad p = -\tfrac{5}{7}$$

Reject the negative answer for price. Supply and demand are equal when p is $\frac{1}{2}$ dollar or \$0.50.

39. Let $A = 10{,}920.25$, $P = 10{,}000$, and solve for r.

$$A = P(1 + r)^2$$
$$10{,}920.25 = 10{,}000(1 + r)^2$$
$$1.092025 = (1 + r)^2$$

$$1 + r = \sqrt{1.092025} \quad \text{or} \quad 1 + r = -\sqrt{1.092025}$$
$$1 + r = 1.045 \quad \text{or} \quad 1 + r = -1.045$$
$$r = 0.045 \quad \text{or} \quad r = -2.045$$

Reject the negative rate. The required interest rate is $0.045 = 4.5\%$.

40. (a) Use $f(x) = 3.29x^2 - 10.4x + 21.6$ with $x = 11$.

$$f(11) = 3.29(11)^2 - 10.4(11) + 21.6$$
$$f(11) \approx 305$$

The value of 305 is close to the number shown on the graph.

(b) $200 = 3.29x^2 - 10.4x + 21.6$ *Let f(x) = 200.*
$$0 = 3.29x^2 - 10.4x - 178.4$$

Here $a = 3.29$, $b = -10.4$, and $c = -178.4$.

$$x = \frac{-b \pm \sqrt{b^2 - 4ac}}{2a}$$

$$x = \frac{-(-10.4) \pm \sqrt{(-10.4)^2 - 4(3.29)(-178.4)}}{2(3.29)}$$

$$= \frac{10.4 \pm \sqrt{2455.904}}{6.58}$$

$$x = \frac{10.4 + \sqrt{2455.904}}{6.58} \approx 9.1 \quad \text{or}$$

$$x = \frac{10.4 - \sqrt{2455.904}}{6.58} \approx -6.0$$

Since time cannot be negative, the correct answer is $x \approx 9$, which represents 1999. Based on the graph, the number of e-mail boxes reached 200 million in 2000.

41. $y = 6 - 2x^2$
Write in $y = a(x - h)^2 + k$ form as
$y = -2(x - 0)^2 + 6$. The vertex (h, k) is $(0, 6)$.

42. $f(x) = -(x - 1)^2$
Write in $y = a(x - h)^2 + k$ form as
$y = -1(x - 1)^2 + 0$. The vertex (h, k) is $(1, 0)$.

43. $y = (x - 3)^2 + 7$
The equation is in the form $y = a(x - h)^2 + k$, so the vertex (h, k) is $(3, 7)$.

44. $y = f(x) = -3x^2 + 4x - 2$

Use the vertex formula with $a = -3$ and $b = 4$.

The x-coordinate of the vertex is

$$\frac{-b}{2a} = \frac{-4}{2(-3)} = \frac{2}{3}.$$

The y-coordinate of the vertex is

$$f\left(\frac{-b}{2a}\right) = f\left(\frac{2}{3}\right)$$
$$= -3\left(\tfrac{2}{3}\right)^2 + 4\left(\tfrac{2}{3}\right) - 2$$
$$= -\tfrac{4}{3} + \tfrac{8}{3} - 2$$
$$= -\tfrac{2}{3}.$$

The vertex is $\left(\tfrac{2}{3}, -\tfrac{2}{3}\right)$.

45. $x = (y - 3)^2 - 4$

When $y = 3$, $x = -4$, so the vertex is $(-4, 3)$.

46. If the discriminant is negative, there are no x-intercepts.

47. $y = 2(x - 2)^2 - 3$

The graph opens up since $a = 2 > 0$.

The vertex is $(2, -3)$ and the axis is $x = 2$. The domain is $(-\infty, \infty)$. The smallest y-value is -3, so the range is $[-3, \infty)$.

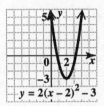

48. $f(x) = -2x^2 + 8x - 5$
Complete the square to find the vertex.

$$f(x) = -2\left(x^2 - 4x\right) - 5$$
$$= -2\left(x^2 - 4x + 4 - 4\right) - 5$$
$$= -2\left(x^2 - 4x + 4\right) - 2(-4) - 5$$
$$= -2(x - 2)^2 + 8 - 5$$
$$= -2(x - 2)^2 + 3$$

The equation is in the form $y = a(x - h)^2 + k$, so the vertex (h, k) is $(2, 3)$ and the axis is $x = 2$. Here, $a = -2 < 0$, so the parabola opens down.

Also, $|a| = |-2| = 2 > 1$, so the graph is narrower than the graph of $y = x^2$. The points $(0, -5)$, $(1, 1)$, and $(3, 1)$ are on the graph.

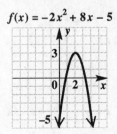

The domain is $(-\infty, \infty)$. The largest y-value is 3, so the range is $(-\infty, 3]$.

49. $x = 2(y+3)^2 - 4$

Since the roles of x and y are reversed, this is a horizontal parabola.

$x = 2[y - (-3)]^2 + (-4)$

The equation is in the form

$$x = a(y - k)^2 + h,$$

so the vertex (h, k) is $(-4, -3)$ and the axis is $y = -3$. Here, $a = 2 > 0$, so the parabola opens to the right and is narrower than the graph of $y = x^2$.

Two other points on the graph are $(4, -1)$ and $(4, -5)$.

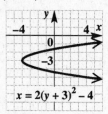

$x = 2(y + 3)^2 - 4$

The smallest x-value is -4, so the domain is $[-4, \infty)$. The range is $(-\infty, \infty)$.

50. $x = -\frac{1}{2}y^2 + 6y - 14$

Since the roles of x and y are reversed, this is a horizontal parabola. Complete the square to find the vertex.

$$\begin{aligned}
x &= -\tfrac{1}{2}y^2 + 6y - 14 \\
&= -\tfrac{1}{2}(y^2 - 12y) - 14 \\
&= -\tfrac{1}{2}(y^2 - 12y + 36 - 36) - 14 \\
&= -\tfrac{1}{2}(y^2 - 12y + 36) - \tfrac{1}{2}(-36) - 14 \\
&= -\tfrac{1}{2}(y - 6)^2 + 18 - 14 \\
x &= -\tfrac{1}{2}(y - 6)^2 + 4
\end{aligned}$$

The equation is in the form $x = a(y - k)^2 + h$, so the vertex (h, k) is $(4, 6)$ and the axis is $y = 6$. Here, $a = -\frac{1}{2} < 0$, so the parabola opens to the left. Also, $|a| = \left|-\frac{1}{2}\right| = \frac{1}{2} < 1$, so the graph is wider than the graph of $y = x^2$. The points $(-14, 0)$, $(2, 4)$, and $(2, 8)$ are on the graph.

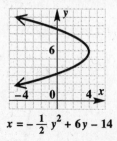

$x = -\frac{1}{2} y^2 + 6y - 14$

The largest x-value is 4, so the domain is $(-\infty, 4]$. The range is $(-\infty, \infty)$.

51. **(a)** Use $ax^2 + bx + c = y$ with $(1, 11.47)$, $(4, 24.45)$, and $(8, 29.78)$.

$$\begin{aligned}
a + b + c &= 11.47 \quad (1) \\
16a + 4b + c &= 24.45 \quad (2) \\
64a + 8b + c &= 29.78 \quad (3)
\end{aligned}$$

(b) Add $-1 \times (1)$ to (2).

$$15a + 3b = 12.98 \quad (4)$$

Add $-1 \times (1)$ to (3).

$$63a + 7b = 18.31 \quad (5)$$

Now eliminate b.

$$\begin{array}{rll}
-105a - 21b &= -90.86 & -7 \times (4) \\
189a + 21b &= 54.93 & 3 \times (5) \\
\hline
84a &= -35.93 & \\
a &= -\frac{35.93}{84} \approx -0.4277 &
\end{array}$$

Use (4) to find b.

$$\begin{aligned}
15a + 3b &= 12.98 \quad (4) \\
15\left(-\tfrac{35.93}{84}\right) + 3b &= 12.98 \\
3b &\approx 19.39607 \\
b &\approx 6.4654
\end{aligned}$$

Use (1) to find c.

$$\begin{aligned}
c &= 11.47 - a - b \\
&\approx 11.47 + 0.4277 - 6.4654 \\
&= 5.4323
\end{aligned}$$

Rounding slightly, we get the quadratic function

$$f(x) = -0.428x^2 + 6.47x + 5.43.$$

(c) For 2000, $x = 2000 - 1995 = 5$.

$$\begin{aligned}
f(5) &= -0.428(5)^2 + 6.47(5) + 5.43 \\
&= 27.08
\end{aligned}$$

From the model, approximate consumer spending for home videos was $27.08, which is slightly higher than the actual value of $25.89.

52. $s(t) = -16t^2 + 160t$

The equation represents a parabola. Since $a = -16 < 0$, the parabola opens down. The time and maximum height occur at the vertex (h, k) of the parabola, given by

$$\left(\frac{-b}{2a}, s\left(\frac{-b}{2a}\right)\right).$$

Using the standard form of the equation, $a = -16$ and $b = 160$, so

$$h = \frac{-b}{2a} = \frac{-160}{2(-16)} = 5,$$

and $k = s(h) = -16(5)^2 + 160(5)$
$$= -400 + 800 = 400.$$

The vertex is $(5, 400)$. The time at which the maximum height is reached is 5 seconds. The maximum height is 400 feet.

53. Let L = the length of the rectangle and W = the width.

The perimeter of the rectangle is 200 m, so

$$2L + 2W = 200$$
$$2W = 200 - 2L$$
$$W = 100 - L.$$

Since the area is length times width, substitute $100 - L$ for W.

$$A = LW$$
$$= L(100 - L)$$
$$= 100L - L^2 \text{ or } -L^2 + 100L$$

Use the vertex formula.

$$L = \frac{-b}{2a} = \frac{-100}{2(-1)} = 50$$

So $L = 50$ meters and
$W = 100 - L = 100 - 50 = 50$ meters.
The maximum area is

$$50 \cdot 50 = 2500 \text{ m}^2.$$

54. $(x - 4)(2x + 3) > 0$
Solve the equation
$(x - 4)(2x + 3) = 0.$
$x - 4 = 0 \quad \text{or} \quad 2x + 3 = 0$
$\quad x = 4 \quad \text{or} \quad x = -\frac{3}{2}$
The numbers $-\frac{3}{2}$ and 4 divide a number line into three intervals.

A	B	C

$\begin{array}{ccc} & & \\ -\frac{3}{2} & 4 & \end{array}$

Test a number from each interval in the inequality

$$(x - 4)(2x + 3) > 0.$$

Interval A: Let $x = -2$.
$\qquad -6(-1) > 0 \qquad ?$
$\qquad\qquad 6 > 0 \qquad\qquad$ *True*

Interval B: Let $x = 0$.
$\qquad -4(3) > 0 \qquad ?$
$\qquad\quad -12 > 0 \qquad\quad$ *False*

Interval C: Let $x = 5$.
$\qquad 1(13) > 0 \qquad ?$
$\qquad\quad 13 > 0 \qquad\qquad$ *True*

The solution set includes numbers in Intervals A and C, excluding endpoints.
Solution set: $\left(-\infty, -\frac{3}{2}\right) \cup (4, \infty)$

$\begin{array}{ccc} & & \\ -\frac{3}{2} & 4 & \end{array}$

55. $\qquad\qquad x^2 + x \le 12$
Solve the equation
$\qquad\qquad x^2 + x = 12.$
$\qquad x^2 + x - 12 = 0$
$(x + 4)(x - 3) = 0$
$x + 4 = 0 \quad \text{or} \quad x - 3 = 0$
$\quad x = -4 \quad \text{or} \qquad x = 3$
The numbers -4 and 3 divide a number line into three intervals.

A	B	C

$\begin{array}{ccc} & & \\ -4 & 3 & \end{array}$

Test a number from each interval in the inequality

$$x^2 + x \le 12.$$

Interval A: Let $x = -5$.
$\qquad 25 - 5 \le 12 \qquad ?$
$\qquad\quad 20 \le 12 \qquad\qquad$ *False*

Interval B: Let $x = 0$.
$\qquad\qquad 0 \le 12 \qquad\qquad$ *True*

Interval C: Let $x = 4$.
$\qquad 16 + 4 \le 12 \qquad ?$
$\qquad\quad 20 \le 12 \qquad\qquad$ *False*

The numbers in Interval B, including -4 and 3, are solutions.
Solution set: $[-4, 3]$

$\begin{array}{ccc} & & \\ -4 & 3 & \end{array}$

56. $(x + 2)(x - 3)(x + 5) \le 0$
The numbers -2, 3, and -5 are solutions of the cubic equation

$$(x + 2)(x - 3)(x + 5) = 0.$$

These numbers divide a number line into four intervals.

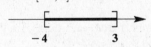

A	B	C	D

$\begin{array}{cccc} & & & \\ -5 & -2 & 3 & \end{array}$

Test a number from each interval in the inequality

$$(x + 2)(x - 3)(x + 5) \le 0.$$

continued

Interval A: Let $x = -6$.
$$(-4)(-9)(-1) \leq 0 \qquad \textit{True}$$

Interval B: Let $x = -3$.
$$(-5)(-6)(2) \leq 0 \qquad \textit{False}$$

Interval C: Let $x = 0$.
$$(2)(-3)(5) \leq 0 \qquad \textit{True}$$

Interval D: Let $x = 4$.
$$(6)(1)(9) \leq 0 \qquad \textit{False}$$

The numbers in Intervals A and C, including -5, -2, and 3, are solutions.
Solution set: $(-\infty, -5] \cup [-2, 3]$

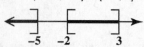

57. $(4m + 3)^2 \leq -4$

The square of a real number is never negative. So, the solution set of this inequality is $\emptyset$.

58. $$\frac{6}{2z - 1} < 2$$

Write the inequality so that 0 is on one side.

$$\frac{6}{2z - 1} - 2 < 0$$

$$\frac{6}{2z - 1} - \frac{2(2z - 1)}{2z - 1} < 0$$

$$\frac{6 - 4z + 2}{2z - 1} < 0$$

$$\frac{-4z + 8}{2z - 1} < 0$$

The number 2 makes the numerator 0, and $\frac{1}{2}$ makes the denominator 0. These two numbers determine three intervals.

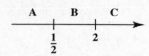

Test a number from each interval in the inequality

$$\frac{6}{2z - 1} < 2.$$

Interval A: Let $z = 0$.
$$-6 < 2 \qquad \textit{True}$$

Interval B: Let $z = 1$.
$$6 < 2 \qquad \textit{False}$$

Interval C: Let $z = 3$.
$$\frac{6}{5} < 2 \qquad \textit{True}$$

The solution set includes numbers in Intervals A and C, excluding endpoints.
Solution set: $\left(-\infty, \frac{1}{2}\right) \cup (2, \infty)$

59. $$\frac{3t + 4}{t - 2} \leq 1$$

Write the inequality so that 0 is on one side.

$$\frac{3t + 4}{t - 2} - 1 \leq 0$$

$$\frac{3t + 4}{t - 2} - \frac{1(t - 2)}{t - 2} \leq 0$$

$$\frac{3t + 4 - t + 2}{t - 2} \leq 0$$

$$\frac{2t + 6}{t - 2} \leq 0$$

The number -3 makes the numerator 0, and 2 makes the denominator 0. These two numbers determine three intervals.

Test a number from each interval in the inequality

$$\frac{3t + 4}{t - 2} \leq 1.$$

Interval A: Let $t = -4$.
$$\frac{-8}{-6} \leq 1 \qquad ?$$
$$\frac{4}{3} \leq 1 \qquad \textit{False}$$

Interval B: Let $t = 0$.
$$\frac{4}{-2} \leq 1 \qquad ?$$
$$-2 \leq 1 \qquad \textit{True}$$

Interval C: Let $t = 3$.
$$\frac{13}{1} \leq 1 \qquad ?$$
$$13 \leq 1 \qquad \textit{False}$$

The numbers in Interval B, including -3 but not 2, are solutions.
Solution set: $[-3, 2)$

60. **[9.4]** Solve $V = r^2 + R^2 h$ for R.
$$V - r^2 = R^2 h$$
$$R^2 h = V - r^2$$
$$R^2 = \frac{V - r^2}{h}$$
$$R = \pm \sqrt{\frac{V - r^2}{h}} = \frac{\pm \sqrt{V - r^2}}{\sqrt{h}}$$
$$= \frac{\pm \sqrt{V - r^2}}{\sqrt{h}} \cdot \frac{\sqrt{h}}{\sqrt{h}}$$
$$= \frac{\pm \sqrt{Vh - r^2 h}}{h}$$

61. **[9.2]** $\quad 3t^2 - 6t = -4$
$$3t^2 - 6t + 4 = 0$$
Use the quadratic formula.

$$t = \frac{-b \pm \sqrt{b^2 - 4ac}}{2a}$$

$$t = \frac{-(-6) \pm \sqrt{(-6)^2 - 4(3)(4)}}{2(3)}$$

$$= \frac{6 \pm \sqrt{-12}}{6} = \frac{6 \pm 2i\sqrt{3}}{6}$$

$$= \frac{2(3 \pm i\sqrt{3})}{6} = \frac{3 \pm i\sqrt{3}}{3} = 1 \pm \frac{\sqrt{3}}{3}i$$

Solution set: $\left\{ 1 + \frac{\sqrt{3}}{3}i, 1 - \frac{\sqrt{3}}{3}i \right\}$

62. **[9.3]**

$$x^4 - 1 = 0$$
$$(x^2 + 1)(x^2 - 1) = 0$$

$x^2 + 1 = 0$	or $x^2 - 1 = 0$
$x^2 = -1$	$x^2 = 1$
$x = \pm\sqrt{-1}$	$x = \pm\sqrt{1}$
$x = \pm i$ or	$x = \pm 1$

Solution set: $\{-i, i, -1, 1\}$

63. **[9.3]** $(x^2 - 2x)^2 = 11(x^2 - 2x) - 24$

Let $u = x^2 - 2x$. The equation becomes

$$u^2 = 11u - 24.$$
$$u^2 - 11u + 24 = 0$$
$$(u - 8)(u - 3) = 0$$

$u - 8 = 0$	or $u - 3 = 0$
$u = 8$ or	$u = 3$

To find x, substitute $x^2 - 2x$ for u.

$x^2 - 2x = 8$	or	$x^2 - 2x = 3$
$x^2 - 2x - 8 = 0$		$x^2 - 2x - 3 = 0$
$(x - 4)(x + 2) = 0$		$(x - 3)(x + 1) = 0$
$x - 4 = 0$ or $x + 2 = 0$		$x - 3 = 0$ or $x + 1 = 0$
$x = 4$ or $x = -2$	or	$x = 3$ or $x = -1$

The proposed solutions all check.
Solution set: $\{-2, -1, 3, 4\}$

64. **[9.7]** $(r - 1)(2r + 3)(r + 6) < 0$

Solve the equation
$$(r - 1)(2r + 3)(r + 6) = 0.$$

$r - 1 = 0$	or	$2r + 3 = 0$	or	$r + 6 = 0$
$r = 1$	or	$r = -\frac{3}{2}$	or	$r = -6$

The numbers -6, $-\frac{3}{2}$, and 1 divide a number line into four intervals.

Test a number from each interval in the inequality

$$(r - 1)(2r + 3)(r + 6) < 0.$$

Interval A:	Let $r = -7$.	
$-8(-11)(-1) < 0$	?	
$-88 < 0$		*True*

Interval B:	Let $r = -2$.	
$-3(-1)(4) < 0$	?	
$12 < 0$		*False*

Interval C:	Let $r = 0$.	
$-1(3)(6) < 0$	?	
$-18 < 0$		*True*

Interval D:	Let $r = 2$.	
$1(7)(8) < 0$	?	
$56 < 0$		*False*

The numbers in Intervals A and C, not including -6, $-\frac{3}{2}$, or 1, are solutions.
Solution set: $(-\infty, -6) \cup \left(-\frac{3}{2}, 1\right)$

65. **[9.3]**

$$2x - \sqrt{x} = 6$$
$$2x - \sqrt{x} - 6 = 0$$

Let $u = \sqrt{x}$, so $u^2 = x$.

$$2u^2 - u - 6 = 0$$
$$(2u + 3)(u - 2) = 0$$

$2u + 3 = 0$	or	$u - 2 = 0$
$u = -\frac{3}{2}$	or	$u = 2$
$\sqrt{x} = -\frac{3}{2}$	or	$\sqrt{x} = 2$

Since $\sqrt{x} \geq 0$, we must have $\sqrt{x} = 2$, or $x = 4$.

Check $x = 4$: $8 - 2 = 6$ *True*
Solution set: $\{4\}$

66. **[9.1]** $(3k + 11)^2 = 7$

$3k + 11 = \sqrt{7}$	or $3k + 11 = -\sqrt{7}$
$3k = -11 + \sqrt{7}$	$3k = -11 - \sqrt{7}$
$k = \dfrac{-11 + \sqrt{7}}{3}$ or	$k = \dfrac{-11 - \sqrt{7}}{3}$

Solution set: $\left\{ \dfrac{-11 + \sqrt{7}}{3}, \dfrac{-11 - \sqrt{7}}{3} \right\}$

67. **[9.4]** Solve $S = \dfrac{Id^2}{k}$ for d.

Multiply both sides by k, then divide by I.

$$\frac{Sk}{I} = d^2$$

$$d = \pm\sqrt{\frac{Sk}{I}} = \frac{\pm\sqrt{Sk}}{\sqrt{I}}$$

$$= \frac{\pm\sqrt{Sk}}{\sqrt{I}} \cdot \frac{\sqrt{I}}{\sqrt{I}}$$

$$d = \frac{\pm\sqrt{SkI}}{I}$$

68. **[9.7]** $(8k - 7)^2 \geq -1$

The square of any real number is always greater than or equal to 0, so any real number satisfies this inequality. Solution set: $(-\infty, \infty)$

69. **[9.3]** $\qquad 6 + \dfrac{15}{s^2} = -\dfrac{19}{s}$

Multiply by the LCD, s^2.

$$s^2\left(6 + \frac{15}{s^2}\right) = s^2\left(-\frac{19}{s}\right)$$

$$6s^2 + 15 = -19s$$

$$6s^2 + 19s + 15 = 0$$

$$(3s + 5)(2s + 3) = 0$$

$$3s + 5 = 0 \quad \text{or} \quad 2s + 3 = 0$$

$$s = -\tfrac{5}{3} \quad \text{or} \qquad s = -\tfrac{3}{2}$$

Check $s = -\tfrac{5}{3}$: $\quad 6 + \tfrac{27}{5} = \tfrac{57}{5} \quad$ *True*

Check $s = -\tfrac{3}{2}$: $\quad 6 + \tfrac{20}{3} = \tfrac{38}{3} \quad$ *True*

Solution set: $\left\{-\tfrac{5}{3}, -\tfrac{3}{2}\right\}$

70. **[9.3]** $\qquad x^4 - 8x^2 = -1$

$$x^4 - 8x^2 + 1 = 0$$

Use the quadratic formula with $a = 1$, $b = -8$, and $c = 1$ to solve for x^2, not x.

$$x^2 = \frac{-b \pm \sqrt{b^2 - 4ac}}{2a}$$

$$x^2 = \frac{-(-8) \pm \sqrt{(-8)^2 - 4(1)(1)}}{2(1)}$$

$$= \frac{8 \pm \sqrt{64 - 4}}{2} = \frac{8 \pm \sqrt{60}}{2}$$

$$= \frac{8 \pm 2\sqrt{15}}{2} = 4 \pm \sqrt{15}$$

Both $4 + \sqrt{15}$ and $4 - \sqrt{15}$ are positive and can be set equal to x^2.

$$x^2 = 4 + \sqrt{15} \qquad \text{or} \quad x^2 = 4 - \sqrt{15}$$

$$x = \pm\sqrt{4 + \sqrt{15}} \qquad x = \pm\sqrt{4 - \sqrt{15}}$$

The proposed solutions all check.

Solution set: $\left\{\sqrt{4 + \sqrt{15}}, -\sqrt{4 + \sqrt{15}},\right.$

$$\left.\sqrt{4 - \sqrt{15}}, -\sqrt{4 - \sqrt{15}}\right\}$$

71. **[9.7]** $\qquad \dfrac{-2}{x + 5} \leq -5$

Write the inequality so that 0 is on one side.

$$\frac{-2}{x + 5} + 5 \leq 0$$

$$\frac{-2}{x + 5} + \frac{5(x + 5)}{x + 5} \leq 0$$

$$\frac{-2 + 5x + 25}{x + 5} \leq 0$$

$$\frac{5x + 23}{x + 5} \leq 0$$

The number $-\tfrac{23}{5}$ makes the numerator 0, and -5 makes the denominator 0. These two numbers determine three intervals.

$$\begin{array}{c c c}
\text{A} & \text{B} & \text{C} \\
\xleftarrow{\hspace{1cm}} \!\!\!\!\!\! + \!\!\!\!\!\!\! \underset{-5}{} \!\!\!\!\!\!\! + \!\!\!\!\!\!\! \underset{-\frac{23}{5}}{} \!\!\!\!\!\!\! \xrightarrow{\hspace{1cm}}
\end{array}$$

Test a number from each interval in the inequality

$$\frac{-2}{x + 5} \leq -5.$$

Interval A: Let $x = -6$.

$$\frac{-2}{-1} \leq -5 \quad ?$$

$$2 \leq -5 \qquad \textit{False}$$

Interval B: Let $x = -\tfrac{24}{5}$.

$$\frac{-2}{\frac{1}{5}} \leq -5 \quad ?$$

$$-10 \leq -5 \qquad \textit{True}$$

Interval C: Let $x = 0$.

$$\frac{-2}{5} \leq -5 \qquad \textit{False}$$

The numbers in Interval B, including $-\tfrac{23}{5}$ but not -5, are solutions.

Solution set: $\left(-5, -\tfrac{23}{5}\right]$

72. **[9.5]** $g(x) = x^2 - 5$ written in the form

$$g(x) = a(x - h)^2 + k \text{ is}$$

$$g(x) = 1(x - 0)^2 + (-5).$$

Here, $h = 0$ and $k = -5$, so the vertex (h, k) is $(0, -5)$. The graph is shifted 5 units down from the graph of $f(x) = x^2$. Since $a = 1 > 0$, the graph opens up. The correct figure is **F**.

73. **[9.5]** $h(x) = -x^2 + 4$

Because the coefficient of x^2 is negative, the parabola opens down. Because $k = 4$, the shift is upward 4 units. The figure that most closely resembles the graph of $h(x)$ is Choice **B**.

74. **[9.5]** $F(x) = (x - 1)^2$ written in the form

$$F(x) = a(x - h)^2 + k \text{ is}$$

$$F(x) = 1(x - 1)^2 + 0.$$

Here, $h = 1$ and $k = 0$, so the vertex (h, k) is $(1, 0)$. The graph is shifted 1 unit to the right of the graph of $f(x) = x^2$. Since $a = 1 > 0$, the graph opens up. The correct figure is **C**.

75. **[9.5]** $G(x) = (x+1)^2$

Because the coefficient of the x^2-term is positive, the graph opens up. Because $h = -1$, the shift is to the left 1 unit. The figure that most closely resembles the graph of $G(x)$ is Choice **A**.

76. **[9.5]** $H(x) = (x-1)^2 + 1$ is written in the form

$$H(x) = a(x-h)^2 + k.$$

Here, $h = 1$ and $k = 1$, so the vertex (h, k) is $(1, 1)$. The graph is shifted 1 unit to the right and 1 unit up from the graph of $f(x) = x^2$. Since $a = 1 > 0$, the graph opens up. The correct figure is **E**.

77. **[9.5]** $K(x) = (x+1)^2 + 1$

Because $h = -1$ and $k = 1$, the shift is to the left 1 unit and up 1 unit. The figure that most closely resembles the graph of $K(x)$ is choice **D**.

78. **[9.6]** $y = f(x) = 4x^2 + 4x - 2$

Complete the square to find the vertex.

$$y = 4\left(x^2 + x\right) - 2$$
$$= 4\left(x^2 + x + \tfrac{1}{4} - \tfrac{1}{4}\right) - 2$$
$$= 4\left(x^2 + x + \tfrac{1}{4}\right) + 4\left(-\tfrac{1}{4}\right) - 2$$
$$= 4\left(x + \tfrac{1}{2}\right)^2 - 1 - 2$$
$$= 4\left(x + \tfrac{1}{2}\right)^2 - 3$$
$$y = 4\left(x + \tfrac{1}{2}\right)^2 - 3$$

The equation is now in the form $y = a(x-h)^2 + k$, so the vertex (h, k) is $\left(-\tfrac{1}{2}, -3\right)$ and the axis is $x = -\tfrac{1}{2}$. Since $a = 4 > 0$, the parabola opens up. Also, $|a| = |4| = 4 > 1$, so the graph is narrower than the graph of $y = x^2$. The points $(-2, 6)$, $(0, -2)$, and $(1, 6)$ are on the graph.

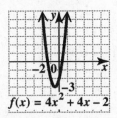

$f(x) = 4x^2 + 4x - 2$

The domain is $(-\infty, \infty)$.
The smallest y-value is -3, so the range is $[-3, \infty)$.

79. **[9.6]** $-0.001x^2 + 0.295x + 0.227$

(a) For 2002, $x = 2002 - 1970 = 32$.
$$f(32) = -0.001(32)^2 + 0.295(32) + 0.227$$
$$= 8.643$$

The nuclear power consumption in the U.S. in 2002 was approximately 8.64 quadrillion Btu.

(b) The result using the model is high.

(c)
$$f(x) = 10$$
$$-0.001x^2 + 0.295x + 0.227 = 10$$
$$-0.001x^2 + 0.295x - 9.773 = 0$$
$$x = \frac{-0.295 \pm \sqrt{(0.295)^2 - 4(-0.001)(-9.773)}}{2(-0.001)}$$
$$= \frac{-0.295 \pm \sqrt{0.047933}}{-0.002}$$
$$\approx 38.03 \quad \text{or} \quad 256.97$$

Rounding down, the year is $1970 + 38 = 2008$. (The other value is outside the practical domain of this model.)

80. **[9.4]** Let $x =$ the speed of the boat in still water. Use $t = d/r$ to make a table.

	Distance	Rate	Time
Upstream	15	$x - 5$	$\dfrac{15}{x-5}$
Downstream	15	$x + 5$	$\dfrac{15}{x+5}$

The total time is 4 hours.

$$\frac{15}{x-5} + \frac{15}{x+5} = 4$$

Multiply by the LCD, $(x+5)(x-5)$.
$$15(x+5) + 15(x-5) = 4(x+5)(x-5)$$
$$15x + 75 + 15x - 75 = 4x^2 - 100$$
$$0 = 4x^2 - 30x - 100$$
$$0 = 2x^2 - 15x - 50$$
$$0 = (2x+5)(x-10)$$

$$2x + 5 = 0 \quad \text{or} \quad x - 10 = 0$$
$$x = -\tfrac{5}{2} \quad \text{or} \quad x = 10$$

The speed must be positive, so check $x = 10$. The 15-mile trip upstream would take 3 hours at 5 mph. The trip downstream would take 1 hour at 15 mph. Thus, the total time would be 4 hours and we have the speed of the boat in still water equal to 10 mph.

81. **[9.4]** Let $x =$ the length of the wire. Use the Pythagorean formula.

$$x^2 = 100^2 + 400^2 \qquad \textit{Height=400.}$$
$$= 10{,}000 + 160{,}000$$
$$= 170{,}000$$
$$x = \pm\sqrt{170{,}000} \approx \pm 412.3$$

Reject the negative solution. The length of the wire is about 412.3 feet.

82. **[9.4]** Let x = the width of the rectangle.

Then $2x - 1$ = the length of the rectangle.

Use the Pythagorean formula.

$$x^2 + (2x - 1)^2 = (2.5)^2$$
$$x^2 + 4x^2 - 4x + 1 = 6.25$$
$$5x^2 - 4x - 5.25 = 0$$

Multiply by 4.

$$20x^2 - 16x - 21 = 0$$
$$(2x - 3)(10x + 7) = 0$$

$$2x - 3 = 0 \quad \text{or} \quad 10x + 7 = 0$$
$$x = \tfrac{3}{2} \quad \text{or} \quad x = -\tfrac{7}{10}$$

Discard the negative solution.
If $x = 1.5$, then

$$2x - 1 = 2(1.5) - 1 = 2.$$

The width of the rectangle is 1.5 cm, and the length is 2 cm.

Chapter 9 Test

1. $t^2 = 54$

$$t = \sqrt{54} \quad \text{or} \quad t = -\sqrt{54}$$
$$t = 3\sqrt{6} \quad \text{or} \quad t = -3\sqrt{6}$$

Solution set: $\left\{ 3\sqrt{6}, -3\sqrt{6} \right\}$

2. $(7x + 3)^2 = 25$

$$7x + 3 = 5 \quad \text{or} \quad 7x + 3 = -5$$
$$7x = 2 \qquad\qquad 7x = -8$$
$$x = \tfrac{2}{7} \quad \text{or} \qquad x = -\tfrac{8}{7}$$

Solution set: $\left\{ -\tfrac{8}{7}, \tfrac{2}{7} \right\}$

3. $2x^2 + 4x = 8$

$$x^2 + 2x = 4 \qquad \textit{Divide by 2.}$$
$$x^2 + 2x + 1 = 4 + 1$$
$$(x + 1)^2 = 5$$

$$x + 1 = \sqrt{5} \qquad \text{or} \quad x + 1 = -\sqrt{5}$$
$$x = -1 + \sqrt{5} \quad \text{or} \qquad x = -1 - \sqrt{5}$$

Solution set: $\left\{ -1 + \sqrt{5}, -1 - \sqrt{5} \right\}$

4. $2x^2 - 3x - 1 = 0$

Here $a = 2$, $b = -3$, and $c = -1$.

$$x = \frac{-b \pm \sqrt{b^2 - 4ac}}{2a}$$
$$x = \frac{-(-3) \pm \sqrt{(-3)^2 - 4(2)(-1)}}{2(2)}$$
$$= \frac{3 \pm \sqrt{17}}{4}$$

Solution set: $\left\{ \dfrac{3 + \sqrt{17}}{4}, \dfrac{3 - \sqrt{17}}{4} \right\}$

5. $3t^2 - 4t = -5$

$$3t^2 - 4t + 5 = 0$$

Here $a = 3$, $b = -4$, and $c = 5$.

$$t = \frac{-b \pm \sqrt{b^2 - 4ac}}{2a}$$
$$t = \frac{-(-4) \pm \sqrt{(-4)^2 - 4(3)(5)}}{2(3)}$$
$$= \frac{4 \pm \sqrt{-44}}{6} = \frac{4 \pm 2i\sqrt{11}}{6}$$
$$= \frac{2\left(2 \pm i\sqrt{11}\right)}{6} = \frac{2 \pm i\sqrt{11}}{3}$$
$$= \frac{2}{3} \pm \frac{\sqrt{11}}{3}i$$

Solution set: $\left\{ \dfrac{2}{3} + \dfrac{\sqrt{11}}{3}i, \dfrac{2}{3} - \dfrac{\sqrt{11}}{3}i \right\}$

6. $$3x = \sqrt{\frac{9x + 2}{2}}$$

Square both sides.

$$9x^2 = \frac{9x + 2}{2}$$
$$18x^2 = 9x + 2$$
$$18x^2 - 9x - 2 = 0$$

As directed, use the quadratic formula with $a = 18$, $b = -9$, and $c = -2$.

$$x = \frac{-b \pm \sqrt{b^2 - 4ac}}{2a}$$
$$x = \frac{-(-9) \pm \sqrt{(-9)^2 - 4(18)(-2)}}{2(18)}$$
$$= \frac{9 \pm \sqrt{225}}{36} = \frac{9 \pm 15}{36}$$
$$x = \frac{9 + 15}{36} = \frac{24}{36} = \frac{2}{3} \text{ or}$$
$$x = \frac{9 - 15}{36} = \frac{-6}{36} = -\frac{1}{6}$$

Check $x = \tfrac{2}{3}$: $\quad 2 = \sqrt{4} \quad$ *True*

Check $x = -\tfrac{1}{6}$: $\quad -\tfrac{1}{2} = \sqrt{\tfrac{1}{4}} \quad$ *False*

Solution set: $\left\{ \tfrac{2}{3} \right\}$

7. If k is a negative number, then $4k$ is also negative, so the equation $x^2 = 4k$ will have two nonreal complex solutions. The answer is **A**.

8. $2x^2 - 8x - 3 = 0$

$$b^2 - 4ac = (-8)^2 - 4(2)(-3)$$
$$= 64 + 24 = 88$$

The discriminant, 88, is positive but not a perfect square, so there will be two distinct irrational number solutions.

9. $3 - \dfrac{16}{x} - \dfrac{12}{x^2} = 0$

Multiply by the LCD, x^2.

$$x^2\left(3 - \frac{16}{x} - \frac{12}{x^2}\right) = x^2 \cdot 0$$
$$3x^2 - 16x - 12 = 0$$
$$(3x + 2)(x - 6) = 0$$

$3x + 2 = 0 \quad$ or $\quad x - 6 = 0$
$\quad\quad x = -\frac{2}{3} \quad$ or $\quad\quad x = 6$

Check $x = -\frac{2}{3}$: $\quad 3 + 24 - 27 = 0 \quad$ *True*

Check $x = 6$: $\quad\quad 3 - \frac{8}{3} - \frac{1}{3} = 0 \quad$ *True*

Solution set: $\left\{-\frac{2}{3}, 6\right\}$

10. $4x^2 + 7x - 3 = 0$

Use the quadratic formula with $a = 4$, $b = 7$, and $c = -3$.

$$x = \frac{-b \pm \sqrt{b^2 - 4ac}}{2a}$$
$$x = \frac{-7 \pm \sqrt{7^2 - 4(4)(-3)}}{2(4)}$$
$$= \frac{-7 \pm \sqrt{97}}{8}$$

Solution set: $\left\{\dfrac{-7 + \sqrt{97}}{8}, \dfrac{-7 - \sqrt{97}}{8}\right\}$

11. $\quad\quad 9x^4 + 4 = 37x^2$

$9x^4 - 37x^2 + 4 = 0$

Let $u = x^2$, so $u^2 = \left(x^2\right)^2 = x^4$.
The equation becomes

$$9u^2 - 37u + 4 = 0.$$
$$(9u - 1)(u - 4) = 0$$

$9u - 1 = 0 \quad$ or $\quad u - 4 = 0$
$\quad u = \frac{1}{9} \quad$ or $\quad\quad u = 4$

To find x, substitute x^2 for u.

$x^2 = \frac{1}{9} \quad\quad$ or $\quad\quad x^2 = 4$

$x = \pm\sqrt{\frac{1}{9}} \quad\quad\quad x = \pm\sqrt{4}$

$x = \pm\frac{1}{3} \quad$ or $\quad x = \pm 2$

Check $x = \pm\frac{1}{3}$: $\quad \frac{1}{9} + 4 = \frac{37}{9} \quad$ *True*

Check $x = \pm 2$: $144 + 4 = 37(4) \quad$ *True*

Solution set: $\left\{-2, -\frac{1}{3}, \frac{1}{3}, 2\right\}$

12. $12 = (2n + 1)^2 + (2n + 1)$

Let $u = 2n + 1$. The equation becomes

$$12 = u^2 + u.$$
$$0 = u^2 + u - 12$$
$$0 = (u + 4)(u - 3)$$

$u + 4 = 0 \quad$ or $\quad u - 3 = 0$
$\quad u = -4 \quad$ or $\quad\quad u = 3$

To find n, substitute $2n + 1$ for u.

$2n + 1 = -4 \quad$ or $\quad 2n + 1 = 3$
$\quad 2n = -5 \quad\quad\quad\quad 2n = 2$
$\quad\quad n = -\frac{5}{2} \quad$ or $\quad\quad n = 1$

Check $n = -\frac{5}{2}$: $\quad 12 = 16 - 4 \quad$ *True*

Check $n = 1$: $\quad\quad 12 = 9 + 3 \quad$ *True*

Solution set: $\left\{-\frac{5}{2}, 1\right\}$

13. Solve $S = 4\pi r^2$ for r.

$$\frac{S}{4\pi} = r^2$$
$$r = \pm\sqrt{\frac{S}{4\pi}} = \frac{\pm\sqrt{S}}{2\sqrt{\pi}}$$
$$= \frac{\pm\sqrt{S}}{2\sqrt{\pi}} \cdot \frac{\sqrt{\pi}}{\sqrt{\pi}}$$
$$r = \frac{\pm\sqrt{\pi S}}{2\pi}$$

14. Let $x =$ Andrew's time alone.
Then $x - 2 =$ Kent's time alone.

Make a table.

	Rate	Time Together	Fractional Part of the Job Done
Andrew	$\dfrac{1}{x}$	5	$\dfrac{5}{x}$
Kent	$\dfrac{1}{x - 2}$	5	$\dfrac{5}{x - 2}$

Part done $\quad$ part done $\quad\quad$ 1 whole
by Andrew plus by Kent equals $\quad$ job.
$\quad\downarrow\quad\quad\quad\quad\downarrow\quad\quad\quad\downarrow\quad\quad\downarrow\quad\quad\downarrow$
$\quad\dfrac{5}{x}\quad + \quad \dfrac{5}{x - 2}\quad = \quad\quad 1$

Multiply both sides by the LCD, $x(x - 2)$.

$$x(x - 2)\left(\frac{5}{x} + \frac{5}{x - 2}\right) = x(x - 2) \cdot 1$$
$$5x - 10 + 5x = x^2 - 2x$$
$$0 = x^2 - 12x + 10$$

Use the quadratic formula with $a = 1$, $b = -12$, and $c = 10$.

continued

$$x = \frac{-b \pm \sqrt{b^2 - 4ac}}{2a}$$

$$x = \frac{-(-12) \pm \sqrt{(-12)^2 - 4(1)(10)}}{2(1)}$$

$$= \frac{12 \pm \sqrt{104}}{2} = \frac{12 \pm 2\sqrt{26}}{2}$$

$$= \frac{2\left(6 \pm \sqrt{26}\right)}{2} = 6 \pm \sqrt{26}$$

$$x = 6 + \sqrt{26} \approx 11.1 \ \text{ or}$$

$$x = 6 - \sqrt{26} \approx 0.9$$

Reject 0.9 for Andrew's time, because that would yield a negative time for Kent. Thus, Andrew's time is about 11.1 hours and Kent's time is $x - 2 \approx 9.1$ hours.

15. Let $x =$ Abby's rate.

Make a table. Use $d = rt$, or $t = \dfrac{d}{r}$.

	Distance	Rate	Time
Upstream	10	$x - 3$	$\dfrac{10}{x-3}$
Downstream	10	$x + 3$	$\dfrac{10}{x+3}$

Time   Time
upstream plus downstream equals 3.5 hr.

$$\underset{\downarrow}{\frac{10}{x-3}} \ + \ \underset{\downarrow}{\frac{10}{x+3}} \ = \ \underset{\downarrow}{\frac{7}{2}}$$

Multiply both sides by the LCD, $2(x+3)(x-3)$.

$$2(x+3)(x-3)\left(\frac{10}{x-3} + \frac{10}{x+3}\right) = 2(x+3)(x-3)\left(\frac{7}{2}\right)$$

$$20(x+3) + 20(x-3) = 7\left(x^2 - 9\right)$$

$$20x + 60 + 20x - 60 = 7x^2 - 63$$

$$0 = 7x^2 - 40x - 63$$

$$0 = (x-7)(7x+9)$$

$$x - 7 = 0 \quad \text{or} \quad 7x + 9 = 0$$

$$x = 7 \quad \text{or} \qquad x = -\tfrac{9}{7}$$

Reject $-\tfrac{9}{7}$ since the rate can't be negative. Abby's rate was 7 mph.

16. Let $x =$ the width of the walk.
The area of the walk is equal to the area of the outer figure minus the area of the pool.

$$152 = (10 + 2x)(24 + 2x) - (24)(10)$$

$$152 = 240 + 68x + 4x^2 - 240$$

$$0 = 4x^2 + 68x - 152$$

$$0 = x^2 + 17x - 38$$

$$0 = (x + 19)(x - 2)$$

$$x + 19 = 0 \qquad \text{or} \quad x - 2 = 0$$

$$x = -19 \quad \text{or} \qquad x = 2$$

Reject -19 since width can't be negative. The walk is 2 feet wide.

17. Let $x =$ the height of the tower. Then
$2x + 2 =$ the distance from the point to the top.

The distance from the base to the point is 30 m. These three segments form a right triangle, so the Pythagorean formula applies.

$$a^2 + b^2 = c^2$$

$$x^2 + 30^2 = (2x + 2)^2$$

$$x^2 + 900 = 4x^2 + 8x + 4$$

$$0 = 3x^2 + 8x - 896$$

$$0 = (x - 16)(3x + 56)$$

$$x - 16 = 0 \qquad \text{or} \quad 3x + 56 = 0$$

$$x = 16 \quad \text{or} \qquad x = -\tfrac{56}{3}$$

Reject $-\tfrac{56}{3}$ since height can't be negative. The tower is 16 m high.

18. $f(x) = a(x - h)^2 + k$
Since $a < 0$, the parabola opens down. Since $h > 0$ and $k < 0$, the x-coordinate is positive and the y-coordinate is negative. Therefore, the vertex is in quadrant IV. The correct graph is **A**.

19. $f(x) = \frac{1}{2}x^2 - 2$
$f(x) = \frac{1}{2}(x - 0)^2 - 2$
The graph is a parabola in $f(x) = a(x - h)^2 + k$ form with vertex (h, k) at $(0, -2)$. The axis is $x = 0$. Since $a = \frac{1}{2} > 0$, the parabola opens up. Also, $|a| = \left|\frac{1}{2}\right| = \frac{1}{2} < 1$, so the graph of the parabola is wider than the graph of $f(x) = x^2$. The points $(2, 0)$ and $(-2, 0)$ are on the graph.

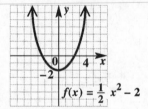

From the graph, we see that the x-values can be any real number, so the domain is $(-\infty, \infty)$. The y-values are greater than or equal to -2, so the range is $[-2, \infty)$.

20. $f(x) = -x^2 + 4x - 1$

The x-coordinate of the vertex is

$$x = \frac{-b}{2a} = \frac{-4}{2(-1)} = 2.$$

The y-coordinate of the vertex is

$$f(2) = -4 + 8 - 1 = 3.$$

The graph is a parabola with vertex (h, k) at $(2, 3)$ and axis $x = 2$. Since $a = -1 < 0$, the parabola opens down.

Also, $|a| = |-1| = 1$, so the graph has the same shape as the graph of $f(x) = x^2$. The points $(0, -1)$ and $(4, -1)$ are on the graph.

$$f(x) = -x^2 + 4x - 1$$

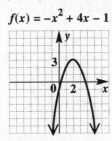

From the graph, we see that the x-values can be any real number, so the domain is $(-\infty, \infty)$.

The y-values are less than or equal to 3, so the range is $(-\infty, 3]$.

21. $x = -(y - 2)^2 + 2$

The equation is in $x = a(y - k)^2 + h$ form. The graph is a horizontal parabola with vertex (h, k) at $(2, 2)$ and axis $y = 2$. Since $a = -1 < 0$, the graph opens to the left. Also, $|a| = |-1| = 1$, so the graph has the same shape as the graph of $y = x^2$. The points $(-2, 0)$ and $(-2, 4)$ are on the same graph.

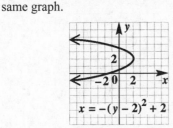

The largest value of x is 2, so the domain is $(-\infty, 2]$.

The y-values can be any real number, so the range is $(-\infty, \infty)$.

22. (a) Use $f(x) = 0.156x^2 - 2.05x + 10.2$ with $x = 11$.

$$f(11) = 0.156(11)^2 - 2.05(11) + 10.2$$
$$\approx 6.5 \text{ (to the nearest tenth)}$$

In 2001, there was a 6.5% increase in tuition.

(b) Find the vertex.

$$\frac{-b}{2a} = \frac{-(-2.05)}{2(0.156)} \approx 6 \text{ (rounding down)}$$

$$f(6) = 0.156(6)^2 - 2.05(6) + 10.2$$
$$\approx 3.5 \text{ (to the nearest tenth)}$$

The minimum tuition increase was 3.5%, which occurred in $1990 + 6 = 1996$.

23. Let $x =$ the width of the lot.
Then $640 - 2x =$ the length of the lot.

Area A is length times width.

$$A = x(640 - 2x)$$
$$A(x) = 640x - 2x^2 = -2x^2 + 640x$$

Use the vertex formula.

$$x = \frac{-b}{2a} = \frac{-640}{2(-2)} = 160$$

$$A(160) = -2(160)^2 + 640(160)$$
$$= -51{,}200 + 102{,}400$$
$$= 51{,}200$$

The graph is a parabola that opens down, so the maximum occurs at the vertex $(160, 51{,}200)$. The maximum area is 51,200 ft^2 if the width x is 160 feet and the length is $640 - 2x = 640 - 2(160) = 320$ feet.

24.
$$2x^2 + 7x > 15$$
$$2x^2 + 7x - 15 > 0$$

Solve the equation
$$2x^2 + 7x - 15 = 0.$$
$$(2x - 3)(x + 5) = 0$$

$$2x - 3 = 0 \quad \text{or} \quad x + 5 = 0$$
$$x = \tfrac{3}{2} \quad \text{or} \quad x = -5$$

The numbers -5 and $\tfrac{3}{2}$ divide a number line into three intervals.

```
        A       B       C
   ─────┼───────┼──────────►
       -5       3
               ─
               2
```

Test a number from each interval in the inequality

$$2x^2 + 7x > 15.$$

Interval A: Let $x = -6$.
$$72 - 42 > 15 \quad ?$$
$$30 > 15 \qquad \text{True}$$

Interval B: Let $x = 0$.
$$0 > 15 \qquad \text{False}$$

Interval C: Let $x = 2$.
$$8 + 14 > 15 \quad ?$$
$$22 > 15 \qquad \text{True}$$

The numbers in Intervals A and C, not including -5 and $\tfrac{3}{2}$, are solutions.

Solution set: $(-\infty, -5) \cup \left(\tfrac{3}{2}, \infty\right)$

```
   ◄──────)      (──────►
         -5       3
                  ─
                  2
```

25. $\dfrac{5}{t-4} \le 1$

Write the inequality so that 0 is on one side.

$$\frac{5}{t-4} - 1 \le 0$$

$$\frac{5}{t-4} - \frac{1(t-4)}{t-4} \le 0$$

$$\frac{5-t+4}{t-4} \le 0$$

$$\frac{-t+9}{t-4} \le 0$$

The number 9 makes the numerator 0, and 4 makes the denominator 0. These two numbers determine three intervals.

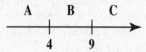

Test a number from each interval in the inequality

$$\frac{5}{t-4} \le 1.$$

Interval A: Let $t = 0$.

$$\frac{5}{-4} \le 1 \qquad True$$

Interval B: Let $t = 7$.

$$\frac{5}{3} \le 1 \qquad False$$

Interval C: Let $t = 10$.

$$\frac{5}{6} \le 1 \qquad True$$

The numbers in Intervals A and C, including 9 but not 4, are solutions.

Solution set: $(-\infty, 4) \cup [9, \infty)$

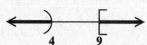

Cumulative Review Exercises
(Chapters 1–9)

1. $S = \left\{ -\frac{7}{3}, -2, -\sqrt{3}, 0, 0.7, \sqrt{12}, \sqrt{-8}, 7, \frac{32}{3} \right\}$

(a) The elements of S that are integers are $-2, 0,$ and 7.

(b) The elements of S that are rational numbers are $-\frac{7}{3}, -2, 0, 0.7, 7,$ and $\frac{32}{3}$.

(c) All the elements of S except $\sqrt{-8}$ are real numbers.

(d) All the elements of S are complex numbers.

2. $|-3| + 8 - |-9| - (-7 + 3) = 3 + 8 - 9 - (-4)$
$$= 3 + 8 - 9 + 4$$
$$= 6$$

3. $2(-3)^2 + (-8)(-5) + (-17)$
$$= 2(9) + 40 - 17$$
$$= 18 + 40 - 17 = 41$$

4. $7 - (4 + 3t) + 2t = -6(t-2) - 5$
$$7 - 4 - 3t + 2t = -6t + 12 - 5$$
$$3 - t = -6t + 7$$
$$5t = 4$$
$$t = \frac{4}{5}$$

Check $t = \frac{4}{5}$: $\frac{11}{5} = \frac{11}{5}$ *True*

Solution set: $\left\{ \frac{4}{5} \right\}$

5. $|6x - 9| = |-4x + 2|$

$6x - 9 = -4x + 2$ or $6x - 9 = -(-4x + 2)$
$10x = 11$ $6x - 9 = 4x - 2$
                           $2x = 7$
$x = \frac{11}{10}$ or $x = \frac{7}{2}$

Check $x = \frac{11}{10}$: $\left|-\frac{24}{10}\right| = \left|-\frac{24}{10}\right|$ *True*

Check $x = \frac{7}{2}$: $|12| = |-12|$ *True*

Solution set: $\left\{ \frac{11}{10}, \frac{7}{2} \right\}$

6. $2x = \sqrt{\dfrac{5x+2}{3}}$

Square both sides.

$$(2x)^2 = \left(\sqrt{\frac{5x+2}{3}} \right)^2$$

$$4x^2 = \frac{5x+2}{3}$$

$$12x^2 = 5x + 2$$

$$12x^2 - 5x - 2 = 0$$

$$(3x - 2)(4x + 1) = 0$$

$3x - 2 = 0$ or $4x + 1 = 0$

$x = \frac{2}{3}$ or $x = -\frac{1}{4}$

Check $x = \frac{2}{3}$: $\frac{4}{3} = \sqrt{\frac{16}{9}}$ *True*

Check $x = -\frac{1}{4}$: $-\frac{1}{2} = \sqrt{\frac{1}{4}}$ *False*

Solution set: $\left\{ \frac{2}{3} \right\}$

7.
$$\frac{3}{x-3} - \frac{2}{x-2} = \frac{3}{x^2 - 5x + 6}$$
$$\frac{3}{x-3} - \frac{2}{x-2} = \frac{3}{(x-3)(x-2)}$$

Multiply by the LCD, $(x-3)(x-2)$.

$$(x-3)(x-2)\left(\frac{3}{x-3} - \frac{2}{x-2}\right)$$
$$= (x-3)(x-2)\left[\frac{3}{(x-3)(x-2)}\right]$$
$$3(x-2) - 2(x-3) = 3$$
$$3x - 6 - 2x + 6 = 3$$
$$x = 3$$

The number 3 is not allowed as a solution since it makes the denominator 0. Solution set: $\emptyset$

8.
$$(r-5)(2r+3) = 1$$
$$2r^2 - 7r - 15 = 1$$
$$2r^2 - 7r - 16 = 0$$

Use the quadratic formula.

$$r = \frac{-b \pm \sqrt{b^2 - 4ac}}{2a}$$
$$r = \frac{-(-7) \pm \sqrt{(-7)^2 - 4(2)(-16)}}{2(2)}$$
$$= \frac{7 \pm \sqrt{49 + 128}}{4} = \frac{7 \pm \sqrt{177}}{4}$$

Solution set: $\left\{\dfrac{7+\sqrt{177}}{4}, \dfrac{7-\sqrt{177}}{4}\right\}$

9.
$$x^4 - 5x^2 + 4 = 0$$

Let $u = x^2$, so $u^2 = \left(x^2\right)^2 = x^4$.
The equation becomes
$$u^2 - 5u + 4 = 0.$$
$$(u-4)(u-1) = 0$$

$$u - 4 = 0 \quad \text{or} \quad u - 1 = 0$$
$$u = 4 \quad \text{or} \quad u = 1$$

To find x, substitute x^2 for u.

$$x^2 = 4 \qquad \text{or} \qquad x^2 = 1$$
$$x = 2 \text{ or } x = -2 \quad \text{or} \quad x = 1 \text{ or } x = -1$$

The proposed solutions check.
Solution set: $\{-2, -1, 1, 2\}$

10.
$$-2x + 4 \le -x + 3$$
$$-x \le -1$$

Multiply by -1, and reverse the direction of the inequality.

$$x \ge 1$$
Solution set: $[1, \infty)$

11.
$$|3x - 7| \le 1$$
$$-1 \le 3x - 7 \le 1$$
$$6 \le 3x \le 8$$
$$2 \le x \le \tfrac{8}{3}$$
Solution set: $\left[2, \tfrac{8}{3}\right]$

12.
$$x^2 - 4x + 3 < 0$$
Solve the equation
$$x^2 - 4x + 3 = 0.$$
$$(x-3)(x-1) = 0$$

$$x - 3 = 0 \quad \text{or} \quad x - 1 = 0$$
$$x = 3 \quad \text{or} \qquad x = 1$$

The numbers 1 and 3 divide a number line into three intervals.

Test a number from each interval in the inequality
$$x^2 - 4x + 3 < 0.$$

Interval A: Let $x = 0$.
$$3 < 0 \qquad \textit{False}$$

Interval B: Let $x = 2$.
$$4 - 8 + 3 < 0 \quad ?$$
$$-1 < 0 \qquad \textit{True}$$

Interval C: Let $x = 4$.
$$16 - 16 + 3 < 0 \quad ?$$
$$3 < 0 \qquad \textit{False}$$

The numbers from Interval B, not including 1 or 3, are solutions.
Solution set: $(1, 3)$

13.
$$\frac{3}{p+2} > 1$$

Write the inequality so that 0 is on one side.

$$\frac{3}{p+2} - 1 > 0$$
$$\frac{3}{p+2} - \frac{1(p+2)}{p+2} > 0$$
$$\frac{3 - p - 2}{p+2} > 0$$
$$\frac{-p+1}{p+2} > 0$$

The number 1 makes the numerator 0, and -2 makes the denominator 0. These two numbers determine three intervals.

continued

Test a number from each interval in the inequality

$$\frac{3}{p+2} > 1.$$

Interval A: Let $p = -4$.

$$\frac{3}{-2} > 1 \qquad \textit{False}$$

Interval B: Let $p = 0$.

$$\frac{3}{2} > 1 \qquad \textit{True}$$

Interval C: Let $p = 2$.

$$\frac{3}{4} > 1 \qquad \textit{False}$$

The numbers from Interval B, not including -2 or 1, are solutions.

Solution set: $(-2, 1)$

14. $4x - 5y = 15$

Draw the line through its intercepts, $\left(\frac{15}{4}, 0\right)$ and $(0, -3)$. The graph passes the vertical line test, so the relation is a function. As with any line that is not horizontal or vertical, the domain and range are both $(-\infty, \infty)$.

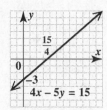

Solve the equation for y.

$$4x - 5y = 15$$
$$-5y = 15 - 4x$$
$$y = \frac{15 - 4x}{-5}, \quad \text{or} \quad \frac{4x - 15}{5}$$

Thus, $f(x) = \frac{4}{5}x - 3$.

15. $4x - 5y < 15$

Draw a dashed line through the points $\left(\frac{15}{4}, 0\right)$ and $(0, -3)$. Check the origin:

$$4(0) - 5(0) < 15 \quad ?$$
$$0 < 15 \qquad \textit{True}$$

Shade the region that contains the origin.

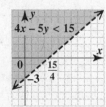

The relation is not a function since for any value of x, there is more than one value of y.

16. $y = -2(x - 1)^2 + 3$

The equation is in $f(x) = a(x - h)^2 + k$ form, so the graph is a parabola with vertex (h, k) at $(1, 3)$. Since $a = -2 < 0$, the parabola opens down. Also $|a| = |-2| = 2 > 1$, so the graph is narrower than the graph of $f(x) = x^2$. The points $(0, 1)$ and $(2, 1)$ are on the graph.

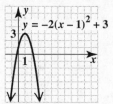

The relation is a function since it passes the vertical line test. The domain is $(-\infty, \infty)$. The largest value of y is 3, so the range is $(-\infty, 3]$. The equation is already solved for y, so using function notation we have

$$f(x) = -2(x - 1)^2 + 3.$$

17. $-2x + 7y = 16$

Solve the equation for y.

$$7y = 2x + 16$$
$$y = \frac{2}{7}x + \frac{16}{7}$$

So the slope is $\frac{2}{7}$ and the y-intercept is $\left(0, \frac{16}{7}\right)$. Let $y = 0$ in $-2x + 7y = 16$ to find the x-intercept.

$$-2x + 7(0) = 16$$
$$-2x = 16$$
$$x = -8$$

The x-intercept is $(-8, 0)$.

18.   **(a)** Solve $5x + 2y = 6$ for y.

$$2y = -5x + 6$$
$$y = -\frac{5}{2}x + 3$$

So the slope of the given line and the desired line is $-\frac{5}{2}$. The required form is

$$y = -\frac{5}{2}x + b.$$
$$-3 = -\frac{5}{2}(2) + b \quad \textit{Let } x=2, \, y=-3.$$
$$-3 = -5 + b$$
$$2 = b$$

The equation is $y = -\frac{5}{2}x + 2$.

(b) The negative reciprocal of the slope in part (a) is

$$-\frac{1}{-\frac{5}{2}} = \frac{2}{5},$$

which is the slope of the line perpendicular to the given line.

So $y = \frac{2}{5}x + b$.

$1 = \frac{2}{5}(-4) + b$ *Let $x = -4$, $y = 1$.*

$1 = -\frac{8}{5} + b$

$\frac{13}{5} = b$

The equation is $y = \frac{2}{5}x + \frac{13}{5}$.

19. **(a)** We are given the y-intercept, $(0, 600)$, so the slope-intercept form of the line is

$$y = mx + 600.$$

Substitute 1340 for y and 6 for x to solve for the slope m.

$$1340 = m(6) + 600$$
$$740 = 6m$$
$$m = \frac{740}{6} = 123.\overline{3}$$

Rounding to the nearest whole number gives us the linear model

$$y = 123x + 600.$$

(b) For 2002, use $x = 7$.

$$y = 123(7) + 600$$
$$= 861 + 600$$
$$= 1461$$

From the model, the approximate sales were $1461 million, which is a little high compared to the table value of $1400 million.

20. No. The graph is a vertical line, which is not the graph of a function by the vertical line test. Also, the only domain value, 5, can have infinitely many range values paired with it.

21. $f(x) = 2(x - 1)^2 - 5$

(a) $f(-2) = 2(-2 - 1)^2 - 5$
$\qquad = 2(-3)^2 - 5$
$\qquad = 2(9) - 5 = 13$

(b) Any value can be substituted for x, so the domain is $(-\infty, \infty)$. The graph of f is a parabola that opens up with vertex $(1, -5)$. The vertex is a minimum point so the y-values are all greater than or equal to -5. Thus, the range is $[-5, \infty)$.

22. $\begin{aligned} 2x - 4y &= 10 \quad (1) \\ 9x + 3y &= 3 \quad (2) \end{aligned}$

Simplify the equations.

$$\begin{aligned} x - 2y &= 5 \quad (3) \; \tfrac{1}{2} \times (1) \\ 3x + y &= 1 \quad (4) \; \tfrac{1}{3} \times (2) \end{aligned}$$

To eliminate y, multiply (4) by 2 and add the result to (3).

$$\begin{array}{rl} x - 2y = 5 & (3) \\ 6x + 2y = 2 & 2 \times (4) \\ \hline 7x = 7 & \\ x = 1 & \end{array}$$

Substitute $x = 1$ into (4).

$$\begin{aligned} 3x + y &= 1 \quad (4) \\ 3(1) + y &= 1 \\ y &= -2 \end{aligned}$$

Solution set: $\{(1, -2)\}$

23. $\begin{aligned} x + y + 2z &= 3 \quad (1) \\ -x + y + z &= -5 \quad (2) \\ 2x + 3y - z &= -8 \quad (3) \end{aligned}$

Eliminate z by adding (2) and (3).

$$\begin{array}{rl} -x + y + z = -5 & (2) \\ 2x + 3y - z = -8 & (3) \\ \hline x + 4y = -13 & (4) \end{array}$$

To get another equation without z, multiply equation (3) by 2 and add the result to equation (1).

$$\begin{array}{rl} x + y + 2z = 3 & (1) \\ 4x + 6y - 2z = -16 & 2 \times (3) \\ \hline 5x + 7y = -13 & (5) \end{array}$$

To eliminate x, multiply (4) by -5 and add the result to (5).

$$\begin{array}{rl} -5x - 20y = 65 & -5 \times (4) \\ 5x + 7y = -13 & (5) \\ \hline -13y = 52 & \\ y = -4 & \end{array}$$

Use (4) to find x.

$$\begin{aligned} x + 4y &= -13 \quad (4) \\ x + 4(-4) &= -13 \\ x - 16 &= -13 \\ x &= 3 \end{aligned}$$

Use (2) to find z.

$$\begin{aligned} -x + y + z &= -5 \quad (2) \\ -3 - 4 + z &= -5 \\ -7 + z &= -5 \\ z &= 2 \end{aligned}$$

Solution set: $\{(3, -4, 2)\}$

24. **(a)** Let x = the amount of sales (in billions) for AOL and y = the amount of sales for Time Warner.

The combined sales were \$34.2 billion, so

$$x + y = 34.2. \quad (1)$$

Sales for AOL were \$.3 billion less than 4 times the sales of Time Warner, so

$$x = 4y - 0.3. \quad (2)$$

(b) Substitute $4y - 0.3$ for x in (1), then solve for y.

$$(4y - 0.3) + y = 34.2$$
$$5y = 34.5$$
$$y = 6.9$$

From (2), $x = 4(6.9) - 0.3 = 27.3$. Thus, AOL had \$27.3 billion in sales and Time Warner had \$6.9 billion in sales.

25. $\left(\dfrac{x^{-3}y^2}{x^5 y^{-2}}\right)^{-1} = \left(x^{-3-5}y^{2-(-2)}\right)^{-1}$

$$= \left(x^{-8}y^4\right)^{-1}$$
$$= x^8 y^{-4}$$
$$= \dfrac{x^8}{y^4}$$

26. $\dfrac{(4x^{-2})^2 (2y^3)}{8x^{-3}y^5} = \dfrac{16x^{-4}(2y^3)}{8x^{-3}y^5}$

$$= \dfrac{4x^{-4}y^3}{x^{-3}y^5}$$
$$= 4x^{-4-(-3)}y^{3-5}$$
$$= 4x^{-1}y^{-2}$$
$$= \dfrac{4}{xy^2}$$

27. $(7x + 4)(2x - 3)$
$$\qquad \textbf{F} \quad \textbf{O} \quad \textbf{I} \quad \textbf{L}$$
$$= 14x^2 - 21x + 8x - 12$$
$$= 14x^2 - 13x - 12$$

28. $\left(\frac{2}{3}t + 9\right)^2 = \left(\frac{2}{3}t\right)^2 + 2\left(\frac{2}{3}t\right)(9) + 9^2$
$$= \frac{4}{9}t^2 + 12t + 81$$

29. $\left(3t^3 + 5t^2 - 8t + 7\right) - \left(6t^3 + 4t - 8\right)$
$$= 3t^3 + 5t^2 - 8t + 7 - 6t^3 - 4t + 8$$
$$= -3t^3 + 5t^2 - 12t + 15$$

30. Divide $4x^3 + 2x^2 - x + 26$ by $x + 2$.

$$
\begin{array}{r}
4x^2 \quad - \quad 6x \quad + \quad 11 \\
x + 2 \overline{\smash{)}\,4x^3 + \quad 2x^2 - \quad x + 26} \\
\underline{4x^3 + \quad 8x^2} \\
-6x^2 - \quad x \\
\underline{-6x^2 - 12x} \\
11x + 26 \\
\underline{11x + 22} \\
4
\end{array}
$$

The answer is

$$4x^2 - 6x + 11 + \dfrac{4}{x + 2}.$$

31. $16x - x^3 = x\left(16 - x^2\right)$
$$= x(4 + x)(4 - x)$$

32. $24m^2 + 2m - 15$

The two integers whose product is $24(-15) = -360$ and whose sum is 2 are 20 and -18.

$$24m^2 + 2m - 15$$
$$= 24m^2 + 20m - 18m - 15$$
$$= 4m(6m + 5) - 3(6m + 5)$$
$$= (6m + 5)(4m - 3)$$

33. $8x^3 + 27y^3$
Use the sum of cubes formula,

$$a^3 + b^3 = (a + b)\left(a^2 - ab + b^2\right),$$

with $a = 2x$ and $b = 3y$.

$$8x^3 + 27y^3$$
$$= (2x + 3y)\left[(2x)^2 - (2x)(3y) + (3y)^2\right]$$
$$= (2x + 3y)\left(4x^2 - 6xy + 9y^2\right)$$

34. $9x^2 - 30xy + 25y^2$
Use the perfect square formula,

$$a^2 - 2ab + b^2 = (a - b)^2,$$

with $a = 3x$ and $b = 5y$.

$$9x^2 - 30xy + 25y^2$$
$$= \left[(3x)^2 - 2(3x)(5y) + (5y)^2\right]$$
$$= (3x - 5y)^2$$

35. $\dfrac{x^2 - 3x - 10}{x^2 + 3x + 2} \cdot \dfrac{x^2 - 2x - 3}{x^2 + 2x - 15}$

$$= \dfrac{(x - 5)(x + 2)}{(x + 2)(x + 1)} \cdot \dfrac{(x - 3)(x + 1)}{(x + 5)(x - 3)}$$

$$= \dfrac{x - 5}{x + 5}$$

36. $\dfrac{3}{2-k} - \dfrac{5}{k} + \dfrac{6}{k^2 - 2k}$

$= \dfrac{3}{2-k} - \dfrac{5}{k} + \dfrac{6}{k(k-2)}$

$= \dfrac{-3}{k-2} - \dfrac{5}{k} + \dfrac{6}{k(k-2)}$

The LCD is $k(k-2)$.

$= \dfrac{-3k}{(k-2)k} - \dfrac{5(k-2)}{k(k-2)} + \dfrac{6}{k(k-2)}$

$= \dfrac{-3k - 5(k-2) + 6}{k(k-2)}$

$= \dfrac{-3k - 5k + 10 + 6}{k(k-2)}$

$= \dfrac{-8k + 16}{k(k-2)}$

$= \dfrac{-8(k-2)}{k(k-2)} = -\dfrac{8}{k}$

37. $\dfrac{\dfrac{r}{s} - \dfrac{s}{r}}{\dfrac{r}{s} + 1}$

Multiply the numerator and denominator by the LCD of all the fractions, rs.

$= \dfrac{\left(\dfrac{r}{s} - \dfrac{s}{r}\right)rs}{\left(\dfrac{r}{s} + 1\right)rs} = \dfrac{r^2 - s^2}{r^2 + rs}$

$= \dfrac{(r-s)(r+s)}{r(r+s)} = \dfrac{r-s}{r}$

38. $\dfrac{1 - x^{-2}y^2}{x^{-1} - x^{-2}y} = \dfrac{1 - \dfrac{y^2}{x^2}}{\dfrac{1}{x} - \dfrac{y}{x^2}}$

Multiply the numerator and denominator by the LCD of all the fractions, x^2.

$= \dfrac{\left(1 - \dfrac{y^2}{x^2}\right)x^2}{\left(\dfrac{1}{x} - \dfrac{y}{x^2}\right)x^2} = \dfrac{x^2 - y^2}{x - y}$

$= \dfrac{(x+y)(x-y)}{x-y} = x + y$

39. $\sqrt[3]{\dfrac{27}{16}} = \dfrac{\sqrt[3]{27}}{\sqrt[3]{16}} = \dfrac{\sqrt[3]{3^3}}{\sqrt[3]{8 \cdot 2}} = \dfrac{3}{2\sqrt[3]{2}}$

$= \dfrac{3 \cdot \sqrt[3]{4}}{2\sqrt[3]{2} \cdot \sqrt[3]{4}} = \dfrac{3\sqrt[3]{4}}{2\sqrt[3]{8}}$

$= \dfrac{3\sqrt[3]{4}}{2 \cdot 2} = \dfrac{3\sqrt[3]{4}}{4}$

40. $\dfrac{2}{\sqrt{7} - \sqrt{5}} = \dfrac{2(\sqrt{7} + \sqrt{5})}{(\sqrt{7} - \sqrt{5})(\sqrt{7} + \sqrt{5})}$

$= \dfrac{2(\sqrt{7} + \sqrt{5})}{7 - 5}$

$= \dfrac{2(\sqrt{7} + \sqrt{5})}{2} = \sqrt{7} + \sqrt{5}$

41. Let $x = $ Tri's rate on the bicycle.
Then $x - 10 = $ Tri's rate while walking.

Make a chart. Use $d = rt$, or $t = \dfrac{d}{r}$.

	Distance	Rate	Time
Bicycle	12	x	$\dfrac{12}{x}$
Walking	8	$x - 10$	$\dfrac{8}{x-10}$

$$\underset{\text{the bicycle}}{\text{Tri's time on}} + \underset{\text{walking}}{\text{Tri's time}} = \underset{\text{hours.}}{5}$$

$$\dfrac{12}{x} + \dfrac{8}{x-10} = 5$$

Multiply by the LCD, $x(x - 10)$.

$x(x-10)\left(\dfrac{12}{x} + \dfrac{8}{x-10}\right) = x(x-10) \cdot 5$

$12(x-10) + 8x = 5x(x-10)$

$12x - 120 + 8x = 5x^2 - 50x$

$0 = 5x^2 - 70x + 120$

$0 = x^2 - 14x + 24$

$0 = (x-12)(x-2)$

$x - 12 = 0 \quad$ or $\quad x - 2 = 0$

$x = 12 \quad$ or $\qquad x = 2$

Reject 2 for Tri's bicycle speed, since it would yield a negative walking speed. Thus, his bicycle speed was 12 mph, and his walking speed was $x - 10 = 2$ mph.

42. Let $\quad x =$ the distance traveled by
the southbound car and

$\quad 2x - 38 =$ the distance traveled by
the eastbound car.

Since the cars are traveling at right angles with one another, the Pythagorean formula can be applied.

$$a^2 + b^2 = c^2$$
$$x^2 + (2x - 38)^2 = 95^2$$
$$x^2 + 4x^2 - 152x + 1444 = 9025$$
$$5x^2 - 152x - 7581 = 0$$

$$x = \frac{-(-152) \pm \sqrt{(-152)^2 - 4(5)(-7581)}}{2(5)}$$

$$= \frac{152 \pm \sqrt{174{,}724}}{10} = \frac{152 \pm 418}{10}$$

Thus, $x = \dfrac{152 + 418}{10} = 57$ (the other value is negative). The southbound car traveled 57 miles, and the eastbound car traveled
$2x - 38 = 2(57) - 38 = 76$ miles.

43. Since 31% of the people in the United States curse the ATM, the number of people in the United States who curse the ATM in a sample of 4000 is

$$0.31(4000) = 1240.$$

44. $0.24(4000) = 960$

45. $0.33(4000) = 1320$

46. German cursers: $0.53(4000) = 2120$
United States thankers: $0.22(4000) = 880$

The difference is

$$2120 - 880 = 1240.$$

CHAPTER 10 INVERSE, EXPONENTIAL, AND LOGARITHMIC FUNCTIONS

10.1 Inverse Functions

1. This function is not one-to-one because both France and the United States are paired with the same trans fat percentage, 11. Also both Hungary and Poland are paired with the same trans fat percentage, 8.

3. The function in the table that pairs a city with a distance is a one-to-one function because for each city there is one distance and each distance has only one city to which it is paired.

   If the distance from Indianapolis to Denver had 1 mile added to it, it would be $1058 + 1 = 1059$ mi, the same as the distance from Los Angeles to Denver. In this case, one distance would have two cities to which it is paired, and the function would not be one-to-one.

5. If a function is made up of ordered pairs in such a way that the same y-value appears in a correspondence with two different x-values, then the function is not one-to-one. Choice **B**

7. All of the graphs pass the vertical line test, so they all represent functions. The graph in choice **A** is the only one that passes the horizontal line test, so it is the one-to-one function.

9. $\{(3, 6), (2, 10), (5, 12)\}$ is a one-to-one function, since each x-value corresponds to only one y-value and each y-value corresponds to only one x-value. To find the inverse, interchange x and y in each ordered pair. The inverse is

   $$\{(6, 3), (10, 2), (12, 5)\}.$$

11. $\{(-1, 3), (2, 7), (4, 3), (5, 8)\}$ is not a one-to-one function. The ordered pairs $(-1, 3)$ and $(4, 3)$ have the same y-value for two different x-values.

13. The graph of $f(x) = 2x + 4$ is a nonvertical, nonhorizontal line. By the horizontal line test, $f(x)$ is a one-to-one function. To find the inverse, replace $f(x)$ with y.

    $$y = 2x + 4$$

    Interchange x and y.

    $$x = 2y + 4$$

    Solve for y.

    $$2y = x - 4$$
    $$y = \frac{x - 4}{2}$$

Replace y with $f^{-1}(x)$.

$$f^{-1}(x) = \frac{x - 4}{2}, \text{ or } f^{-1}(x) = \frac{1}{2}x - 2$$

15. Write $g(x) = \sqrt{x - 3}$ as $y = \sqrt{x - 3}$.

    Since $x \geq 3$, $y \geq 0$. The graph of g is half of a horizontal parabola that opens to the right. The graph passes the horizontal line test, so g is one-to-one. To find the inverse, interchange x and y to get

    $$x = \sqrt{y - 3}.$$

    Note that now $y \geq 3$, so $x \geq 0$.
    Solve for y by squaring both sides.

    $$x^2 = y - 3$$
    $$x^2 + 3 = y$$

    Replace y with $g^{-1}(x)$.
    $$g^{-1}(x) = x^2 + 3, x \geq 0$$

17. $f(x) = 3x^2 + 2$ is not a one-to-one function because two x-values, such as 1 and -1, both have the same y-value, in this case 5. The graph of this function is a vertical parabola which does not pass the horizontal line test.

19. The graph of $f(x) = x^3 - 4$ is the graph of $g(x) = x^3$ shifted down 4 units. (Recall that $g(x) = x^3$ is the elongated S-shaped curve.) The graph of f passes the horizontal line test, so f is one-to-one.

    Replace $f(x)$ with y.
    $$y = x^3 - 4$$
    Interchange x and y.
    $$x = y^3 - 4$$
    Solve for y.
    $$x + 4 = y^3$$
    Take the cube root of each side.
    $$\sqrt[3]{x + 4} = y$$
    Replace y with $f^{-1}(x)$.
    $$f^{-1}(x) = \sqrt[3]{x + 4}$$

In Exercises 21–24, $f(x) = 2^x$ is a one-to-one function.

21. **(a)** To find $f(3)$, substitute 3 for x.

    $f(x) = 2^x$, so $f(3) = 2^3 = 8$.

    (b) Since f is one-to-one and $f(3) = 8$, it follows that $f^{-1}(8) = 3$.

23. **(a)** To find $f(0)$, substitute 0 for x.

    $f(x) = 2^x$, so $f(0) = 2^0 = 1$.

    (b) Since f is one-to-one and $f(0) = 1$, it follows that $f^{-1}(1) = 0$.

25. **(a)** The function is one-to-one since any horizontal line intersects the graph at most once.

(b) In the graph, the two points marked on the line are $(-1, 5)$ and $(2, -1)$. Interchange x and y in each ordered pair to get $(5, -1)$ and $(-1, 2)$. Plot these points, then draw a dashed line through them to obtain the graph of the inverse function.

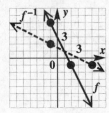

27. **(a)** The function is not one-to-one since there are horizontal lines that intersect the graph more than once. For example, the line $y = 1$ intersects the graph twice.

29. **(a)** The function is one-to-one since any horizontal line intersects the graph at most once.

(b) In the graph, the four points marked on the curve are $(-4, 2)$, $(-1, 1)$, $(1, -1)$, and $(4, -2)$. Interchange x and y in each ordered pair to get $(2, -4)$, $(1, -1)$, $(-1, 1)$, and $(-2, 4)$. Plot these points, then draw a dashed curve (symmetric to the original graph about the line $y = x$) through them to obtain the graph of the inverse.

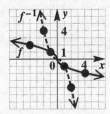

31. $f(x) = 2x - 1$ or $y = 2x - 1$
The graph is a line through $(-2, -5)$, $(0, -1)$, and $(3, 5)$. Plot these points and draw the solid line through them. Then the inverse will be a line through $(-5, -2)$, $(-1, 0)$, and $(5, 3)$. Plot these points and draw the dashed line through them.

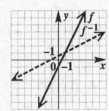

33. $g(x) = -4x$ or $y = -4x$
The graph is a line through $(0, 0)$ and $(1, -4)$. For the inverse, interchange x and y in each ordered pair to get the points $(0, 0)$ and $(-4, 1)$. Draw a dashed line through these points to obtain the graph of the inverse function.

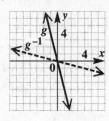

35. $f(x) = \sqrt{x}$, $x \geq 0$
Complete the table of values.

x	$f(x)$
0	0
1	1
4	2

Plot these points and connect them with a solid smooth curve.
Since $f(x)$ is one-to-one, make a table of values for $f^{-1}(x)$ by interchanging x and y.

x	$f^{-1}(x)$
0	0
1	1
2	4

Plot these points and connect them with a dashed smooth curve.

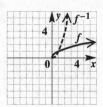

37. $f(x) = x^3 - 2$
Complete the table of values.

x	$f(x)$
-1	-3
0	-2
1	-1
2	6

Plot these points and connect them with a solid smooth curve.
Make a table of values for f^{-1}.

x	$f^{-1}(x)$
-3	-1
-2	0
-1	1
6	2

Plot these points and connect them with a dashed smooth curve.

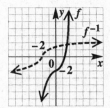

39.
$$f(x) = 4x - 5$$
Replace $f(x)$ with y.
$$y = 4x - 5$$
Interchange x and y.
$$x = 4y - 5$$
Solve for y.
$$x + 5 = 4y$$
$$\frac{x + 5}{4} = y$$
Replace y with $f^{-1}(x)$.
$$\frac{x + 5}{4} = f^{-1}(x),$$
$$\text{or} \quad f^{-1}(x) = \frac{1}{4}x + \frac{5}{4}$$

40. Insert each number in the inverse function found in Exercise 39,
$$f^{-1}(x) = \frac{x + 5}{4}.$$
$$f^{-1}(47) = \frac{47 + 5}{4} = \frac{52}{4} = 13 = \text{M},$$
$$f^{-1}(95) = \frac{95 + 5}{4} = \frac{100}{4} = 25 = \text{Y},$$
and so on.

The decoded message is as follows:
My graphing calculator is the greatest thing since sliced bread.

41. A one-to-one code is essential to this process because if the code is not one-to-one, an encoded number would refer to two different letters.

42. Answers will vary according to the student's name. For example, Jane Doe is encoded as follows:

1004 5 2748 129 68 3379 129.

43.
$$Y_1 = f(x) = 2x - 7$$
Replace $f(x)$ with y.
$$y = 2x - 7$$
Interchange x and y.
$$x = 2y - 7$$
Solve for y.
$$x + 7 = 2y$$
$$\frac{x + 7}{2} = y$$
Replace y with $f^{-1}(x)$.
$$\frac{x + 7}{2} = f^{-1}(x) = Y_2$$

Now graph Y_1 and Y_2.

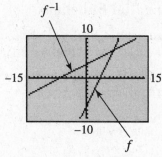

45.
$$Y_1 = f(x) = x^3 + 5$$
Replace $f(x)$ with y.
$$y = x^3 + 5$$
Interchange x and y.
$$x = y^3 + 5$$
Solve for y.
$$x - 5 = y^3$$
Take the cube root of each side.
$$\sqrt[3]{x - 5} = y$$
Replace y with $f^{-1}(x)$.
$$\sqrt[3]{x - 5} = f^{-1}(x) = Y_2$$

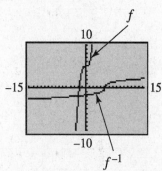

47.
$$Y_1 = X^2 + 3X + 4$$
Graph Y_1 and its inverse in the same square window on a graphing calculator. On a TI-83, graph Y_1 and then enter

$$\text{DrawInv } Y_1$$

on the home screen. DrawInv is choice 8 under the DRAW menu. Y_1 is choice 1 under VARS, Y-VARS, Function.

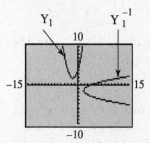

49. $f(x) = 4^x$, so $f(3) = 4^3 = 64$.

51. $f(x) = 4^x$, so $f\left(\frac{1}{2}\right) = 4^{1/2} = \sqrt{4} = 2$.

53. $f(x) = 4^x$, so $f(2.73) = 4^{2.73} \approx 44.02$.

10.2 Exponential Functions

1. Since the graph of $F(x) = a^x$ always contains the point $(0, 1)$, the correct response is **C**.

3. Since the graph of $F(x) = a^x$ always approaches the x-axis, the correct response is **A**.

5. $f(x) = 3^x$
Make a table of values.
$$f(-2) = 3^{-2} = \frac{1}{3^2} = \frac{1}{9},$$
$$f(-1) = 3^{-1} = \frac{1}{3^1} = \frac{1}{3}, \text{ and so on.}$$

x	-2	-1	0	1	2
$f(x)$	$\frac{1}{9}$	$\frac{1}{3}$	1	3	9

Plot the points from the table and draw a smooth curve through them.

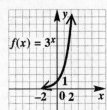

7. $g(x) = \left(\frac{1}{3}\right)^x$
Make a table of values.
$$g(-2) = \left(\frac{1}{3}\right)^{-2} = \left(\frac{3}{1}\right)^2 = 9,$$
$$g(-1) = \left(\frac{1}{3}\right)^{-1} = \left(\frac{3}{1}\right)^1 = 3, \text{ and so on.}$$

x	-2	-1	0	1	2
$g(x)$	9	3	1	$\frac{1}{3}$	$\frac{1}{9}$

Plot the points from the table and draw a smooth curve through them.

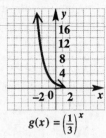
$$g(x) = \left(\frac{1}{3}\right)^x$$

9. $y = 4^{-x}$
This equation can be rewritten as
$$y = (4^{-1})^x = \left(\frac{1}{4}\right)^x,$$
which shows that it is *falling* from left to right.
Make a table of values.

x	-2	-1	0	1	2
y	16	4	1	$\frac{1}{4}$	$\frac{1}{16}$

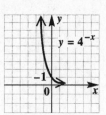

11. $y = 2^{2x-2}$
Make a table of values. It will help to find values for $2x - 2$ before you find y.

x	-2	-1	0	1	2	3
$2x - 2$	-6	-4	-2	0	2	4
y	$\frac{1}{64}$	$\frac{1}{16}$	$\frac{1}{4}$	1	4	16

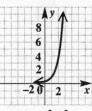
$$y = 2^{2x-2}$$

13. **(a)** For an exponential function defined by $f(x) = a^x$, if $a > 1$, the graph <u>rises</u> from left to right. (See Example 1, $f(x) = 2^x$, in your text.) If $0 < a < 1$, the graph <u>falls</u> from left to right. (See Example 2, $g(x) = \left(\frac{1}{2}\right)^x = 2^{-x}$, in your text.)

(b) An exponential function defined by $f(x) = a^x$ is one-to-one and has an inverse, since each value of $f(x)$ corresponds to one and only one value of x.

15.
$$6^x = 36$$
Write each side as a power of 6.
$$6^x = 6^2$$
For $a > 0$ and $a \neq 1$, if $a^x = a^y$, then $x = y$.
Set the exponents equal to each other.
$$x = 2$$
Check $x = 2$: $6^2 = 36$ *True*
The solution set is $\{2\}$.

17.
$$100^x = 1000$$
Write each side as a power of 10.
$$\left(10^2\right)^x = 10^3$$
$$10^{2x} = 10^3$$
For $a > 0$ and $a \neq 1$, if $a^x = a^y$, then $x = y$.
Set the exponents equal to each other.
$$2x = 3$$
$$x = \frac{3}{2}$$
Check $x = \frac{3}{2}$: $100^{3/2} = 1000$ *True*
The solution set is $\left\{\frac{3}{2}\right\}$.

19. $16^{2x+1} = 64^{x+3}$

Write each side as a power of 4.

$(4^2)^{2x+1} = (4^3)^{x+3}$

$4^{4x+2} = 4^{3x+9}$

Set the exponents equal.

$4x + 2 = 3x + 9$

$x = 7$

Check $x = 7$: $16^{15} = 64^{10}$ *True*

The solution set is $\{7\}$.

21. $5^x = \frac{1}{125}$

$5^x = \left(\frac{1}{5}\right)^3$

Write each side as a power of 5.

$5^x = 5^{-3}$

Set the exponents equal.

$x = -3$

Check $x = -3$: $5^{-3} = \frac{1}{125}$ *True*

The solution set is $\{-3\}$.

23. $5^x = 0.2$

$5^x = \frac{2}{10} = \frac{1}{5}$

Write each side as a power of 5.

$5^x = 5^{-1}$

Set the exponents equal.

$x = -1$

Check $x = -1$: $5^{-1} = 0.2$ *True*

The solution set is $\{-1\}$.

25. $\left(\frac{3}{2}\right)^x = \frac{8}{27}$

$\left(\frac{3}{2}\right)^x = \left(\frac{2}{3}\right)^3$

Write each side as a power of $\frac{3}{2}$.

$\left(\frac{3}{2}\right)^x = \left(\frac{3}{2}\right)^{-3}$

Set the exponents equal.

$x = -3$

Check $x = -3$: $\left(\frac{3}{2}\right)^{-3} = \frac{8}{27}$ *True*

The solution set is $\{-3\}$.

27. $12^{2.6} \approx 639.545$

29. $0.5^{3.921} \approx 0.066$

31. $2.718^{2.5} \approx 12.179$

33. **(a)** The increase for the exponential-type curve in the year 2000 is about 0.5°C.

(b) The increase for the linear graph in the year 2000 is about 0.35°C.

35. **(a)** The increase for the exponential-type curve in the year 2020 is about 1.6°C.

(b) The increase for the linear graph in the year 2020 is about 0.5°C.

37. $f(x) = 220{,}717(1.0217)^{-x}$

(a) 1970 corresponds to $x = 0$.

$f(0) = 220{,}717(1.0217)^0$

$= 220{,}717(1) = 220{,}717$

The answer has units in thousands of tons.

(b) 1995 corresponds to $x = 25$.

$f(25) = 220{,}717(1.0217)^{-25} \approx 129{,}048$

(c) 2002 corresponds to $x = 32$.

$f(32) = 220{,}717(1.0217)^{-32} \approx 111{,}042$

The actual amount, 112,049 thousand tons, is greater than the 111,042 thousand tons that the model provides.

39. $V(t) = 5000(2)^{-0.15t}$

(a) The original value is found when $t = 0$.

$V(0) = 5000(2)^{-0.15(0)}$

$= 5000(2)^0$

$= 5000(1) = 5000$

The original value is $5000.

(b) The value after 5 years is found when $t = 5$.

$V(5) = 5000(2)^{-0.15(5)}$

$= 5000(2)^{-0.75} \approx 2973.02$

The value after 5 years is about $2973.

(c) The value after 10 years is found when $t = 10$.

$V(10) = 5000(2)^{-0.15(10)}$

$= 5000(2)^{-1.5} \approx 1767.77$

The value after 10 years is about $1768.

(d) Use the results of parts (a) – (c) to make a table of values.

t	0	5	10
$V(t)$	5000	2973	1768

Plot the points from the table and draw a smooth curve through them.

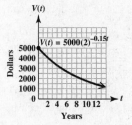

41. $V(t) = 5000(2)^{-0.15t}$

$2500 = 5000(2)^{-0.15t}$ *Let V(t) = 2500.*

$\frac{1}{2} = (2)^{-0.15t}$ *Divide by 5000.*

$2^{-1} = 2^{-0.15t}$

$-1 = -0.15t$ *Equate exponents.*

$t = \dfrac{-1}{-0.15} \approx 6.67$

The value of the machine will be $2500 in approximately 6.67 years after it was purchased.

43. $16 = 2 \cdot 2 \cdot 2 \cdot 2 = 2^4$, so $\square = 4$.

45. $2^0 = 1$, so $\square = 0$.

10.3 Logarithmic Functions

1. (a) $\log_{1/3} 3 = -1$ is equivalent to $\left(\frac{1}{3}\right)^{-1} = 3$. **(B)**

(b) $\log_5 1 = 0$ is equivalent to $5^0 = 1$. **(E)**

(c) $\log_2 \sqrt{2} = \frac{1}{2}$ is equivalent to $2^{1/2} = \sqrt{2}$. **(D)**

(d) $\log_{10} 1000 = 3$ is equivalent to $10^3 = 1000$. **(F)**

(e) $\log_8 \sqrt[3]{8} = \frac{1}{3}$ is equivalent to $8^{1/3} = \sqrt[3]{8}$ **(A)**

(f) $\log_4 4 = 1$ is equivalent to $4^1 = 4$. **(C)**

3. The base is 4, the exponent (logarithm) is 5, and the number is 1024, so $4^5 = 1024$ becomes $\log_4 1024 = 5$ in logarithmic form.

5. $\frac{1}{2}$ is the base and -3 is the exponent, so $\left(\frac{1}{2}\right)^{-3} = 8$ becomes $\log_{1/2} 8 = -3$ in logarithmic form.

7. The base is 10, the exponent (logarithm) is -3, and the number is 0.001, so $10^{-3} = 0.001$ becomes $\log_{10} 0.001 = -3$ in logarithmic form.

9. $\sqrt[4]{625} = 625^{1/4} = 5$

The base is 625, the exponent (logarithm) is $\frac{1}{4}$, and the number is 5, so $\sqrt[4]{625} = 5$ becomes $\log_{625} 5 = \frac{1}{4}$ in logarithmic form.

11. In $\log_4 64 = 3$, 4 is the base and 3 is the logarithm (exponent), so $\log_4 64 = 3$ becomes $4^3 = 64$ in exponential form.

13. In $\log_{10} \frac{1}{10,000} = -4$, the base is 10, the logarithm (exponent) is -4, and the number is $\frac{1}{10,000}$, so $\log_{10} \frac{1}{10,000} = -4$ becomes $10^{-4} = \frac{1}{10,000}$ in exponential form.

15. In $\log_6 1 = 0$, 6 is the base and 0 is the logarithm (exponent), so $\log_6 1 = 0$ becomes $6^0 = 1$ in exponential form.

17. In $\log_9 3 = \frac{1}{2}$, the base is 9, the logarithm (exponent) is $\frac{1}{2}$, and the number is 3, so $\log_9 3 = \frac{1}{2}$ becomes $9^{1/2} = 3$ in exponential form.

19. Use the properties of logarithms,

$$\log_b b = 1 \quad \text{and} \quad \log_b 1 = 0,$$

for $b > 0$, $b \neq 1$.

(a) $\log_8 8 = 1$ **(C)**

(b) $\log_{16} 1 = 0$ **(B)**

(c) $\log_{0.3} 1 = 0$ **(B)**

(d) $\log_{\sqrt{7}} \sqrt{7} = 1$ **(C)**

21.
$$x = \log_{27} 3$$

Write in exponential form.

$$27^x = 3$$

Write each side as a power of 3.

$$(3^3)^x = 3$$
$$3^{3x} = 3^1$$

Set the exponents equal.

$$3x = 1$$
$$x = \frac{1}{3}$$

Check $x = \frac{1}{3}$: $\frac{1}{3} = \log_{27} 3$ since $27^{1/3} = 3$.

The solution set is $\left\{\frac{1}{3}\right\}$.

23. $\log_x 9 = \frac{1}{2}$

Change to exponential form.

$$x^{1/2} = 9$$
$$(x^{1/2})^2 = 9^2 \quad \text{Square.}$$
$$x^1 = 81$$
$$x = 81$$

$x = 81$ is an acceptable base since it is a positive number (not equal to 1).

Check $x = 81$: $81^{1/2} = 9$ *True*

The solution set is $\{81\}$.

25.
$$\log_x 125 = -3$$

Write in exponential form.

$$x^{-3} = 125$$
$$\frac{1}{x^3} = 125$$
$$1 = 125(x^3)$$
$$\frac{1}{125} = x^3$$

Take the cube root of each side.

$$\sqrt[3]{\frac{1}{125}} = \sqrt[3]{x^3}$$
$$x = \sqrt[3]{\frac{1}{5^3}} = \frac{1}{5}$$

$x = \frac{1}{5}$ is an acceptable base since it is a positive number (not equal to 1).

The solution set is $\left\{\frac{1}{5}\right\}$.

27. $\log_{12} x = 0$

Write in exponential form.

$$12^0 = x$$
$$1 = x$$

The argument (the input of the logarithm) must be a positive number, so $x = 1$ is acceptable.

The solution set is $\{1\}$.

29. $\log_x x = 1$

Write in exponential form.

$$x^1 = x$$

This equation is true for all the numbers x that are allowed as the base of a logarithm; that is, all positive numbers x, $x \neq 1$.

The solution set is $\{x \mid x > 0,\ x \neq 1\}$.

31. $\log_x \dfrac{1}{25} = -2$

Write in exponential form.

$$x^{-2} = \frac{1}{25}$$
$$\frac{1}{x^2} = \frac{1}{25}$$
$$x^2 = 25 \qquad \textit{Denominators must be equal.}$$
$$x = \pm 5$$

Reject $x = -5$ since the base of a logarithm must be positive and not equal to 1.

The solution set is $\{5\}$.

33. $\log_8 32 = x$

$$8^x = 32 \qquad \textit{Exponential form}$$

Write each side as a power of 2.

$$\left(2^3\right)^x = 2^5$$
$$2^{3x} = 2^5$$
$$3x = 5 \qquad \textit{Equate exponents.}$$
$$x = \frac{5}{3}$$

Check $x = \frac{5}{3}$: $\log_8 32 = \frac{5}{3}$ since $8^{5/3} = 2^5 = 32$.

The solution set is $\left\{\frac{5}{3}\right\}$.

35. $\log_\pi \pi^4 = x$

$$\pi^x = \pi^4 \qquad \textit{Exponential form}$$
$$x = 4 \qquad \textit{Equate exponents.}$$

Check $x = 4$: $\log_\pi \pi^4 = 4$ since $\pi^4 = \pi^4$.

The solution set is $\{4\}$.

37. $\log_6 \sqrt{216} = x$

$$\log_6 216^{1/2} = x \qquad \textit{Equivalent form}$$
$$6^x = 216^{1/2} \qquad \textit{Exponential form}$$
$$6^x = \left(6^3\right)^{1/2} \qquad \textit{Same base}$$
$$6^x = 6^{3/2}$$
$$x = \frac{3}{2} \qquad \textit{Equate exponents.}$$

Check $x = \frac{3}{2}$:

$$\log_6 \sqrt{216} = \frac{3}{2} \text{ since } 6^{3/2} = \sqrt{6^3} = \sqrt{216}.$$

The solution set is $\left\{\frac{3}{2}\right\}$.

39. $\log_4 (2x + 4) = 3$

$$2x + 4 = 4^3 \qquad \textit{Exponential form}$$
$$2x = 64 - 4$$
$$2x = 60$$
$$x = 30$$

Check $x = 30$:

$$\log_4 (2 \cdot 30 + 4) = \log_4 64 = 3.$$

The solution set is $\{30\}$.

41. $$y = \log_3 x$$

Change to exponential form.

$$3^y = x$$

Refer to Section 10.2, Exercise 5, for the graph of $f(x) = 3^x$. Since $y = \log_3 x$ (or $3^y = x$) is the inverse of $f(x) = y = 3^x$, its graph is symmetric about the line $y = x$ to the graph of $f(x) = 3^x$. The graph can be plotted by reversing the ordered pairs in the table of values belonging to $f(x) = 3^x$.

x	$\frac{1}{9}$	$\frac{1}{3}$	1	3	9
y	-2	-1	0	1	2

Plot the points, and draw a smooth curve through them.

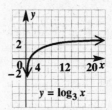

43. $y = \log_{1/3} x$

Change to exponential form.

$$\left(\tfrac{1}{3}\right)^y = x$$

Refer to Section 10.2, Exercise 7, for the graph of $g(x) = \left(\frac{1}{3}\right)^x$. Since $y = \log_{1/3} x$ (or $\left(\frac{1}{3}\right)^y = x$) is the inverse of $y = \left(\frac{1}{3}\right)^x$, its graph is symmetric about the line $y = x$ to the graph of $y = \left(\frac{1}{3}\right)^x$. The graph can be plotted by reversing the ordered pairs in the table of values belonging to $g(x) = \left(\frac{1}{3}\right)^x$.

x	9	3	1	$\frac{1}{3}$	$\frac{1}{9}$
y	-2	-1	0	1	2

Plot the points, and draw a smooth curve through them.

continued

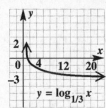

$y = \log_{1/3} x$

45. The number 1 is not used as a base for a logarithmic function since the function would look like $x = 1^y$ in exponential form. Then, for any real value of y, the statement $1 = 1$ would always be the result since every power of 1 is equal to 1.

47. The range of $F(x) = a^x$ is the domain of $G(x) = \log_a x$, that is, $\underline{(0, \infty)}$.
The domain of $F(x) = a^x$ is the range of $G(x) = \log_a x$, that is, $\underline{(-\infty, \infty)}$.

49. The values of t are on the horizontal axis, and the values of $f(t)$ are on the vertical axis. Read the value of $f(t)$ from the graph for the given value of t. At $t = 0$, $f(0) = 8$.

51. To find $f(60)$, find 60 on the t-axis, then go up to the graph and across to the $f(t)$ axis to read the value of $f(60)$. At $t = 60$, $f(60) = 24$.

53. $f(x) = 3800 + 585 \log_2 x$

(a) $x = 1982 - 1980 = 2$

$$f(2) = 3800 + 585 \log_2 2$$
$$= 3800 + 585(1)$$
$$= 4385$$

The model gives an approximate withdrawal of 4385 billion ft^3 of natural gas from crude oil wells in the United States for 1982.

(b) $x = 1988 - 1980 = 8$

$$f(8) = 3800 + 585 \log_2 8$$
$$= 3800 + 585(3)$$
$$= 5555$$

The model gives an approximate withdrawal of 5555 billion ft^3 of natural gas from crude oil wells in the United States for 1988.

(c) $x = 1996 - 1980 = 16$

$$f(16) = 3800 + 585 \log_2 16$$
$$= 3800 + 585(4)$$
$$= 6140$$

The model gives an approximate withdrawal of 6140 billion ft^3 of natural gas from crude oil wells in the United States for 1996.

55. $S(t) = 100 + 30 \log_3 (2t + 1)$

(a) $S(1) = 100 + 30 \log_3 (2 \cdot 1 + 1)$
$$= 100 + 30 \log_3 (3)$$
$$= 100 + 30(1) = 130$$

After 1 year, the sales were 130 thousand units.

(b) $S(13) = 100 + 30 \log_3 (2 \cdot 13 + 1)$
$$= 100 + 30 \log_3 (27)$$
$$= 100 + 30(3) = 190$$

After 13 years, the sales were 190 thousand units.

(c) Make a table of values, plot the points they represent, and draw a smooth curve through them.

To make the table, find values of t such that $2t + 1 = 3^k$, where $k = 0, 1, 2, 3, 4$.

k	0	1	2	3	4
3^k	1	3	9	27	81
$2t + 1$	1	3	9	27	81
$2t$	0	2	8	26	80
t	0	1	4	13	40
$S(t)$	100	130	160	190	220

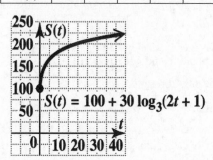

$S(t) = 100 + 30 \log_3 (2t + 1)$

57. $R = \log_{10} \dfrac{x}{x_0}$

Change to exponential form.

$$10^R = \frac{x}{x_0}, \text{ so } x = x_0 \, 10^R.$$

Let $R = 6.7$ for the Northridge earthquake, with intensity x_1.

$$x_1 = x_0 10^{6.7}$$

Let $R = 7.3$ for the Landers earthquake, with intensity x_2.

$$x_2 = x_0 10^{7.3}$$

The ratio of x_2 to x_1 is

$$\frac{x_2}{x_1} = \frac{x_0 10^{7.3}}{x_0 10^{6.7}} = 10^{0.6} \approx 3.98.$$

The Landers earthquake was about 4 times more powerful than the Northridge earthquake.

59. $g(x) = \log_3 x$

On a TI-83, assign 3^x to Y_1. Then enter

$$\text{DrawInv } Y_1$$

on the home screen to obtain the figure that follows. (See Exercise 47 in Section 10.1 for TI-83 specifics.)

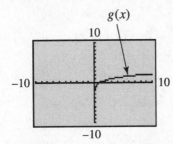

61. $g(x) = \log_{1/3} x$

Assign (1/3)^x to Y_1 and enter $\text{DrawInv } Y_1$.

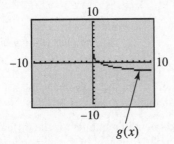

63. $4^7 \cdot 4^2 = 4^{7+2} = 4^9$

65. $\dfrac{5^{-3}}{5^8} = 5^{-3-8} = 5^{-11} = \dfrac{1}{5^{11}}$

67. $\left(9^3\right)^{-2} = 9^{3(-2)} = 9^{-6} = \dfrac{1}{9^6}$

10.4 Properties of Logarithms

1. By the product rule,

$$\log_{10}(3 \cdot 4) = \log_{10} 3 + \log_{10} 4.$$

3. By a special property (see page 656 in the text),

$$3^{\log_3 4} = 4.$$

5. By a special property (see page 656 in the text),

$$\log_3 3^4 = 4.$$

7. Use the product rule for logarithms.

$$\log_7(4 \cdot 5) = \log_7 4 + \log_7 5$$

9. Use the quotient rule for logarithms.

$$\log_5 \tfrac{8}{3} = \log_5 8 - \log_5 3$$

11. Use the power rule for logarithms.

$$\log_4 6^2 = 2 \log_4 6$$

13. $\log_3 \dfrac{\sqrt[3]{4}}{x^2 y} = \log_3 \dfrac{4^{1/3}}{x^2 y}$

Use the quotient rule for logarithms.

$$= \log_3 4^{1/3} - \log_3 \left(x^2 y\right)$$

Use the product rule for logarithms.

$$= \log_3 4^{1/3} - \left(\log_3 x^2 + \log_3 y\right)$$

$$= \log_3 4^{1/3} - \log_3 x^2 - \log_3 y$$

Use the power rule for logarithms.

$$= \tfrac{1}{3} \log_3 4 - 2 \log_3 x - \log_3 y$$

15. $\log_3 \sqrt{\dfrac{xy}{5}}$

$$= \log_3 \left(\dfrac{xy}{5}\right)^{1/2}$$

$$= \tfrac{1}{2} \log_3 \left(\dfrac{xy}{5}\right) \qquad \textit{Power rule}$$

$$= \tfrac{1}{2} \left[\log_3 (xy) - \log_3 5\right] \qquad \textit{Quotient rule}$$

$$= \tfrac{1}{2} \left(\log_3 x + \log_3 y - \log_3 5\right) \qquad \textit{Product rule}$$

$$= \tfrac{1}{2} \log_3 x + \tfrac{1}{2} \log_3 y - \tfrac{1}{2} \log_3 5$$

17. $\log_2 \dfrac{\sqrt[3]{x} \cdot \sqrt[5]{y}}{r^2}$

$$= \log_2 \dfrac{x^{1/3} y^{1/5}}{r^2}$$

$$= \log_2 \left(x^{1/3} y^{1/5}\right) - \log_2 r^2 \qquad \textit{Quotient rule}$$

$$= \log_2 x^{1/3} + \log_2 y^{1/5} - \log_2 r^2 \quad \textit{Product rule}$$

$$= \tfrac{1}{3} \log_2 x + \tfrac{1}{5} \log_2 y - 2 \log_2 r \quad \textit{Power rule}$$

19. The distributive property tells us that the *product* $a(x + y)$ equals the sum $ax + ay$. In the notation $\log_a (x + y)$, the parentheses do not indicate multiplication. They indicate that $x + y$ is the result of raising a to some power.

21. By the product rule for logarithms,

$$\log_b x + \log_b y = \log_b xy.$$

23. By the quotient rule for logarithms,

$$\log_a m - \log_a n = \log_a \dfrac{m}{n}.$$

25. $\left(\log_a r - \log_a s\right) + 3 \log_a t$

Use the quotient and power rules for logarithms.

$$= \log_a \dfrac{r}{s} + \log_a t^3$$

$$= \log_a \dfrac{rt^3}{s} \qquad\qquad \textit{Product rule}$$

27. $3 \log_a 5 - 4 \log_a 3$

$= \log_a 5^3 - \log_a 3^4$ *Power rule*

$= \log_a \dfrac{5^3}{3^4}$ *Quotient rule*

$= \log_a \dfrac{125}{81}$

29. $\log_{10} (x + 3) + \log_{10} (x - 3)$

$= \log_{10} (x + 3)(x - 3)$ *Product rule*

$= \log_{10} \left(x^2 - 9\right)$

31. By the power rule for logarithms,

$3 \log_p x + \frac{1}{2} \log_p y - \frac{3}{2} \log_p z - 3 \log_p a$

$= \log_p x^3 + \log_p y^{1/2} - \log_p z^{3/2} - \log_p a^3$

Group the terms into sums.

$= (\log_p x^3 + \log_p y^{1/2}) - (\log_p z^{3/2} + \log_p a^3)$

$= \log_p x^3 y^{1/2} - \log_p z^{3/2} a^3$ *Product rule*

$= \log_p \dfrac{x^3 y^{1/2}}{z^{3/2} a^3}$ *Quotient rule*

In Exercises 33–44, $\log_{10} 2 \approx 0.3010$ and $\log_{10} 9 \approx 0.9542$.

33. By the product rule for logarithms,

$\log_{10} 18 = \log_{10} (2 \cdot 9)$

$= \log_{10} 2 + \log_{10} 9$

$\approx 0.3010 + 0.9542$

$= 1.2552.$

35. By the quotient rule for logarithms,

$\log_{10} \frac{2}{9} = \log_{10} 2 - \log_{10} 9$

$\approx 0.3010 - 0.9542$

$= -0.6532.$

37. By the product and power rules for logarithms,

$\log_{10} 36 = \log_{10} 2^2 \cdot 9$

$= 2 \log_{10} 2 + \log_{10} 9$

$\approx 2(0.3010) + 0.9542$

$= 1.5562.$

39. $\log_{10} 3 = \log_{10} 9^{1/2}$ *Rename 3*

$= \frac{1}{2} \log_{10} 9$ *Power rule*

$\approx \frac{1}{2}(0.9542)$

$= 0.4771$

41. $\log_{10} \sqrt[4]{9} = \log_{10} 9^{1/4}$

$= \frac{1}{4} \log_{10} 9$ *Power rule*

$\approx \frac{1}{4}(0.9542)$

$= 0.23855 \approx 0.2386$

43. $\log_{10} 9^5 = 5 \log_{10} 9$ *Power rule*

$\approx 5(0.9542)$

$= 4.7710$

45. LS $= \log_2 (8 + 32) = \log_2 40$

RS $= \log_2 8 + \log_2 32 = \log_2 (8 \cdot 32)$

$= \log_2 256$

LS $\neq$ RS, so the statement is *false*.

47. $\log_3 7 + \log_3 7^{-1} = \log_3 7 + (-1) \log_3 7$

$= 0$

The statement is *true*.

49. $\log_6 60 - \log_6 10 = \log_6 \frac{60}{10}$

$= \log_6 6 = 1$

The statement is *true*.

51. $\dfrac{\log_{10} 7}{\log_{10} 14} \stackrel{?}{=} \dfrac{1}{2}$

$2 \log_{10} 7 \stackrel{?}{=} 1 \log_{10} (7 \cdot 2)$

Cross products are equal

$2 \log_{10} 7 \stackrel{?}{=} \log_{10} 7 + \log_{10} 2$

$\log_{10} 7 \stackrel{?}{=} \log_{10} 2$ *Subtract $\log_{10} 7$.*

The statement is *false*.

53. The exponent of a quotient is the difference between the exponent of the numerator and the exponent of the denominator.

55. $\log_2 8 - \log_2 4 = \log_2 \frac{8}{4}$

$= \log_2 2 = 1$

57. $10^4 = 10{,}000$ becomes $\log_{10} 10{,}000 = 4$.

59. $10^{-2} = 0.01$ becomes $\log_{10} 0.01 = -2$.

61. $\log_{10} 1 = 0$ becomes $10^0 = 1$.

10.5 Common and Natural Logarithms

1. Since $\log x = \log_{10} x$, the base is 10. The correct response is **C**.

3. $10^0 = 1$ and $10^1 = 10$, so $\log 1 = 0$ and $\log 10 = 1$. Thus, the value of $\log 5.6$ must lie between 0 and 1. The correct response is **C**.

5. $\log 10^{19.2} = \log_{10} 10^{19.2} = 19.2$ by the special property, $\log_b b^x = x$.

7. To four decimal places,

$\log 43 \approx 1.6335.$

9. $\log 328.4 \approx 2.5164$

11. $\log 0.0326 \approx -1.4868$

13. $\log\left(4.76 \times 10^9\right) \approx 9.6776$
On a TI-83, enter

LOG 4.76 2nd EE 9).

15. $\ln 7.84 \approx 2.0592$

17. $\ln 0.0556 \approx -2.8896$

19. $\ln 388.1 \approx 5.9613$

21. $\ln\left(8.59 \times e^2\right) \approx 4.1506$
On a TI-83, enter

LN 8.59 X 2nd e^x 2)).

23. $\ln 10 \approx 2.3026$

25. **(a)** $\log 356.8 \approx 2.552\,424\,846$

(b) $\log 35.68 \approx 1.552\,424\,846$

(c) $\log 3.568 \approx 0.552\,424\,846$

(d) The whole number part of the answers (2, 1, or 0) varies, whereas the decimal part (0.552 424 846) remains the same, indicating that the whole number part corresponds to the placement of the decimal point and the decimal part corresponds to the digits 3, 5, 6, and 8.

27. When you try to find $\log\left(-1\right)$ on a calculator, an error message is displayed. This is because the domain of the logarithmic function is $(0, \infty)$; -1 is not in the domain.

29. $\text{pH} = -\log\left[H_3O^+\right]$
$= -\log\left(2.5 \times 10^{-2}\right) \approx 1.6$
Since the pH is less than 3.0, the wetland is classified as a *bog*.

31. Ammonia has a hydronium ion concentration of 2.5×10^{-12}.
$\text{pH} = -\log\left[H_3O^+\right]$
$\text{pH} = -\log\left(2.5 \times 10^{-12}\right) \approx 11.6$

33. Grapes have a hydronium ion concentration of 5.0×10^{-5}.
$\text{pH} = -\log\left[H_3O^+\right]$
$\text{pH} = -\log\left(5.0 \times 10^{-5}\right) \approx 4.3$

35. Human blood plasma has a pH of 7.4.
$\text{pH} = -\log\left[H_3O^+\right]$
$7.4 = -\log\left[H_3O^+\right]$
$\log_{10}\left[H_3O^+\right] = -7.4$
$\left[H_3O^+\right] = 10^{-7.4} \approx 4.0 \times 10^{-8}$

37. Spinach has a pH value of 5.4.
$\text{pH} = -\log\left[H_3O^+\right]$
$5.4 = -\log\left[H_3O^+\right]$
$\log_{10}\left[H_3O^+\right] = -5.4$
$\left[H_3O^+\right] = 10^{-5.4} \approx 4.0 \times 10^{-6}$

39. $D = 10\log\left(\dfrac{I}{I_0}\right)$

(a) $D = 10\log\left(\dfrac{5.012 \times 10^{10} I_0}{I_0}\right)$
$= 10\log\left(5.012 \times 10^{10}\right) \approx 107$

The average decibel level for *Spider-Man 2* is about 107 dB.

(b) $D = 10\log\left(\dfrac{10^{10} I_0}{I_0}\right)$
$= 10\log 10^{10} = 10 \cdot 10 = 100$

The average decibel level for *Finding Nemo* is 100 dB.

(c) $D = 10\log\left(\dfrac{6,310,000,000\,I_0}{I_0}\right)$
$= 10\log\left(6,310,000,000\right) \approx 98$

The average decibel level for *Saving Private Ryan* is about 98 dB.

41. $N(r) = -5000\ln r$

(a) 85% (or 0.85)

$N(0.85) = -5000\ln 0.85 \approx 813 \approx 800$

The number of years since the split for 85% is about 800 years.

(b) 35% (or 0.35)

$N(0.35) = -5000\ln 0.35 \approx 5249 \approx 5200$

The number of years since the split for 35% is about 5200 years.

(c) 10% (or 0.10)

$N(0.10) = -5000\ln 0.10 \approx 11,513 \approx 11,500$

The number of years since the split for 10% is about 11,500 years.

43. $f(x) = -1317 + 304\ln x$

(a) $x = 1998 - 1900 = 98$

$f(98) = -1317 + 304\ln(98) \approx 77$

The prediction for 1998 is 77%.

(b) $f(x) = 50$
$$50 = -1317 + 304 \ln x$$
$$1367 = 304 \ln x$$
$$\ln x = \tfrac{1367}{304}$$
$$e^{\ln x} = e^{1367/304}$$
$$x \approx 89.72$$

The outpatient surgeries reached 50% in 1989.

45. $T = -0.642 - 189 \ln (1 - p)$

(a) $T = -0.642 - 189 \ln (1 - 0.25)$
$$= -0.642 - 189 \ln (0.75) \approx 53.73$$

About $54 per ton will reduce emissions by 25%.

(b) If $p = 0$, then $\ln (1 - p) = \ln 1 = 0$, so T would be negative. If $p = 1$, then $\ln (1 - p) = \ln 0$, but the domain of $\ln x$ is $(0, \infty)$.

47.   The change-of-base rule is
$$\log_a x = \frac{\log_b x}{\log_b a}.$$

Use common logarithms ($b = 10$).
$$\log_3 12 = \frac{\log_{10} 12}{\log_{10} 3} = \frac{\log 12}{\log 3} \approx 2.2619$$

49.   Use natural logarithms ($b = e$).
$$\log_5 3 = \frac{\log_e 3}{\log_e 5} = \frac{\ln 3}{\ln 5} \approx 0.6826$$

51. $\log_3 \sqrt{2} = \dfrac{\ln \sqrt{2}}{\ln 3} \approx 0.3155$

53. $\log_\pi e = \dfrac{\ln e}{\ln \pi} = \dfrac{1}{\ln \pi} \approx 0.8736$

55. $\log_e 12 = \dfrac{\ln 12}{\ln e} = \dfrac{\ln 12}{1} \approx 2.4849$

57.   Let $m =$ the number of letters in your first name and $n =$ the number of letters in your last name.

Answers will vary, but suppose the name is Paul Bunyan, with $m = 4$ and $n = 6$.

(a) $\log_m n = \log_4 6$ is the exponent to which 4 must be raised in order to obtain 6.

(b) Use the change-of-base rule.
$$\log_4 6 = \frac{\log 6}{\log 4}$$
$$\approx 1.292\,481\,25$$

(c) Here, $m = 4$. Use the power key ($y^x, x^y, \wedge$) on your calculator.
$$4^{1.292\,481\,25} \approx 6$$

The result is 6, the value of n.

59. $f(x) = 3800 + 585 \log_2 x$

$x = 2003 - 1980 = 23$
$$f(23) = 3800 + 585 \log_2 23$$
$$= 3800 + 585 \left(\frac{\log 23}{\log 2} \right)$$
$$\approx 6446$$

The model gives an approximate withdrawal of 6446 billion ft^3 of natural gas from crude oil wells in the United States for 2003.

61.   To graph $g(x) = \log_3 x$, assign either $\dfrac{\log X}{\log 3}$ or $\dfrac{\ln X}{\ln 3}$ to Y_1.

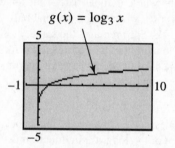

$$g(x) = \log_3 x$$

63.   To graph $g(x) = \log_{1/3} x$, assign either $\dfrac{\log X}{\log \frac{1}{3}}$ or $\dfrac{\ln X}{\ln \frac{1}{3}}$ to Y_1.

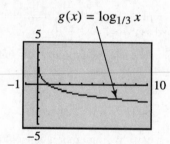

$$g(x) = \log_{1/3} x$$

65. $4^{2x} = 8^{3x+1}$
$$\left(2^2 \right)^{2x} = \left(2^3 \right)^{3x+1}$$
$$2^{4x} = 2^{9x+3}$$
$$4x = 9x + 3$$
$$-3 = 5x$$
$$x = -\tfrac{3}{5}$$

The solution set is $\left\{ -\tfrac{3}{5} \right\}$.

67. $\log_3 (x + 4) = 2$
$$x + 4 = 3^2$$
$$x + 4 = 9$$
$$x = 5$$

The solution set is $\{5\}$.

69. $\log_{1/2} 8 = x$

$\left(\frac{1}{2}\right)^x = 8$

$\left(2^{-1}\right)^x = 2^3$

$2^{-x} = 2^3$

$-x = 3$

$x = -3$

The solution set is $\{-3\}$.

71. $\log (x + 2) + \log (x - 3)$

$= \log (x + 2)(x - 3), \text{ or } \log \left(x^2 - x - 6\right)$

10.6 Exponential and Logarithmic Equations; Further Applications

1. $5^x = 125$

$\log 5^x = \log 125$

2. $x \log 5 = \log 125$

3. $\dfrac{x \log 5}{\log 5} = \dfrac{\log 125}{\log 5}$

$x = \dfrac{\log 125}{\log 5}$

4. $\dfrac{\log 125}{\log 5} = 3 \text{ (from calculator)}$

The solution set is $\{3\}$.

5. $7^x = 5$

Take the logarithm of each side.

$\log 7^x = \log 5$

Use the power rule for logarithms.

$x \log 7 = \log 5$

$x = \dfrac{\log 5}{\log 7} \approx 0.827$

The solution set is $\{0.827\}$.

7. $9^{-x+2} = 13$

$\log 9^{-x+2} = \log 13$

$(-x + 2) \log 9 = \log 13 \ (*)$

$-x \log 9 + 2 \log 9 = \log 13$

$-x \log 9 = \log 13 - 2 \log 9$

$x \log 9 = 2 \log 9 - \log 13$

$x = \dfrac{2 \log 9 - \log 13}{\log 9}$

≈ 0.833

$(*)$ Alternative solution steps:

$(-x + 2) \log 9 = \log 13$

$-x + 2 = \dfrac{\log 13}{\log 9}$

$2 - \dfrac{\log 13}{\log 9} = x$

The solution set is $\{0.833\}$.

9. $3^{2x} = 14$

$\log 3^{2x} = \log 14$

$2x \log 3 = \log 14$

$x = \dfrac{\log 14}{2 \log 3} \approx 1.201$

The solution set is $\{1.201\}$.

11. $2^{x+3} = 5^x$

$\log 2^{x+3} = \log 5^x$

$(x + 3) \log 2 = x \log 5$

$x \log 2 + 3 \log 2 = x \log 5$

Get x-terms on one side.

$x \log 2 - x \log 5 = -3 \log 2$

$x (\log 2 - \log 5) = -3 \log 2 \quad \text{Factor out } x.$

$x = \dfrac{-3 \log 2}{\log 2 - \log 5}$

≈ 2.269

The solution set is $\{2.269\}$.

13. $2^{x+3} = 3^{x-4}$

$\log 2^{x+3} = \log 3^{x-4}$

$(x + 3) \log 2 = (x - 4) \log 3$

$x \log 2 + 3 \log 2 = x \log 3 - 4 \log 3$

$\qquad\qquad\qquad\qquad \textit{Distributive property}$

Get x-terms on one side.

$x \log 2 - x \log 3 = -3 \log 2 - 4 \log 3$

Factor out x.

$x(\log 2 - \log 3) = -3 \log 2 - 4 \log 3$

$x = \dfrac{-3 \log 2 - 4 \log 3}{\log 2 - \log 3}$

≈ 15.967

The solution set is $\{15.967\}$.

15. $e^{0.012x} = 23$

$\ln e^{0.012x} = \ln 23$

$0.012x \, (\ln e) = \ln 23$

$0.012x = \ln 23 \quad \textit{ln e = 1}$

$x = \dfrac{\ln 23}{0.012} \approx 261.291$

The solution set is $\{261.291\}$.

17.
$$e^{-0.205x} = 9$$
$$\ln e^{-0.205x} = \ln 9$$
$$-0.205x \, (\ln e) = \ln 9$$
$$-0.205x = \ln 9 \qquad \textit{ln e = 1}$$
$$x = \frac{\ln 9}{-0.205} \approx -10.718$$

The solution set is $\{-10.718\}$.

19.
$$\ln e^{3x} = 9$$
$$3x = 9$$
$$x = 3$$

The solution set is $\{3\}$.

21.
$$\ln e^{0.45x} = \sqrt{7}$$
$$0.45x = \sqrt{7}$$
$$x = \frac{\sqrt{7}}{0.45} \approx 5.879$$

The solution set is $\{5.879\}$.

23. Let's try Exercise 14.
$$e^{0.006x} = 30$$
$$\log e^{0.006x} = \log 30$$
$$0.006x \, (\log e) = \log 30$$
$$x = \frac{\log 30}{0.006 \log e} \approx 566.866$$

The natural logarithm is a better choice because $\ln e = 1$, whereas $\log e$ needs to be calculated.

25.
$$\log_3 (6x + 5) = 2$$
$$6x + 5 = 3^2 \qquad \textit{Exponential form}$$
$$6x + 5 = 9$$
$$6x = 4$$
$$x = \frac{4}{6} = \frac{2}{3}$$

Check $x = \frac{2}{3}$: $\log_3 9 = \log_3 3^2 = 2$

The solution set is $\left\{\frac{2}{3}\right\}$.

27.
$$\log_2 (2x - 1) = 5$$
$$2x - 1 = 2^5 \quad \textit{Exponential form}$$
$$2x - 1 = 32$$
$$2x = 33$$
$$x = \frac{33}{2}$$

Check $x = \frac{33}{2}$: $\log_2 32 = \log_2 2^5 = 5$

The solution set is $\left\{\frac{33}{2}\right\}$.

29.
$$\log_7 (x + 1)^3 = 2$$
$$(x + 1)^3 = 7^2 \qquad \textit{Exponential form}$$
$$x + 1 = \sqrt[3]{49} \qquad \textit{Cube root}$$
$$x = -1 + \sqrt[3]{49}$$

Check $x = -1 + \sqrt[3]{49}$: $\log_7 49 = \log_7 7^2 = 2$

The solution set is $\left\{-1 + \sqrt[3]{49}\right\}$.

31. 2 cannot be a solution because $\log (2 - 3) = \log (-1)$, and -1 is not in the domain of $\log x$.

33.
$$\log (6x + 1) = \log 3$$
$$6x + 1 = 3 \qquad \textit{Property 4}$$
$$6x = 2$$
$$x = \frac{2}{6} = \frac{1}{3}$$

Check $x = \frac{1}{3}$: $\log (2 + 1) = \log 3 \quad \textit{True}$

The solution set is $\left\{\frac{1}{3}\right\}$.

35.
$$\log_5 (3t + 2) - \log_5 t = \log_5 4$$
$$\log_5 \frac{3t + 2}{t} = \log_5 4$$
$$\frac{3t + 2}{t} = 4$$
$$3t + 2 = 4t$$
$$2 = t$$

Check $t = 2$: $\log_5 8 - \log_5 2 = \log_5 \frac{8}{2} = \log_5 4$

The solution set is $\{2\}$.

37.
$$\log 4x - \log (x - 3) = \log 2$$
$$\log \frac{4x}{x - 3} = \log 2$$
$$\frac{4x}{x - 3} = 2$$
$$4x = 2(x - 3)$$
$$4x = 2x - 6$$
$$2x = -6$$
$$x = -3$$

Reject $x = -3$, because $4x = -12$, which yields an equation in which the logarithm of a negative number must be found.

The solution set is $\emptyset$.

39.
$$\log_2 x + \log_2 (x - 7) = 3$$
$$\log_2 [x(x - 7)] = 3$$
$$x(x - 7) = 2^3 \qquad \textit{Exponential form}$$
$$x^2 - 7x = 8$$
$$x^2 - 7x - 8 = 0$$
$$(x - 8)(x + 1) = 0$$
$$x - 8 = 0 \quad \text{or} \quad x + 1 = 0$$
$$x = 8 \quad \text{or} \qquad x = -1$$

Reject $x = -1$, because it yields an equation in which the logarithm of a negative number must be found.

Check $x = 8$: $\log_2 8 + \log_2 1 = \log_2 2^3 + 0 = 3$

The solution set is $\{8\}$.

41. $\log 5x - \log (2x - 1) = \log 4$

$$\log \frac{5x}{2x - 1} = \log 4$$

$$\frac{5x}{2x - 1} = 4$$

$$5x = 8x - 4$$

$$4 = 3x$$

$$\tfrac{4}{3} = x$$

Check $x = \tfrac{4}{3}$: $\log \frac{20}{3} - \log \frac{5}{3} = \log \frac{20/3}{5/3} = \log 4$

The solution set is $\left\{ \tfrac{4}{3} \right\}$.

43. $\log_2 x + \log_2 (x - 6) = 4$

$$\log_2 [x(x - 6)] = 4$$

$$x(x - 6) = 2^4 \quad \textit{Exponential form}$$

$$x^2 - 6x = 16$$

$$x^2 - 6x - 16 = 0$$

$$(x - 8)(x + 2) = 0$$

$$x - 8 = 0 \quad \text{or} \quad x + 2 = 0$$

$$x = 8 \quad \text{or} \quad x = -2$$

Reject $x = -2$, because it yields an equation in which the logarithm of a negative number must be found.

Check $x = 8$:

$$\log_2 8 + \log_2 2 = \log_2 16 = \log_2 2^4 = 4$$

The solution set is $\{8\}$.

45. (a) Use the formula $A = P\left(1 + \dfrac{r}{n}\right)^{nt}$ with $P = 2000$, $r = 0.04$, $n = 4$, and $t = 6$.

$$A = 2000\left(1 + \tfrac{0.04}{4}\right)^{4 \cdot 6}$$

$$= 2000(1.01)^{24} \approx 2539.47$$

The account will contain $2539.47.

(b) $\quad 3000 = 2000\left(1 + \tfrac{0.04}{4}\right)^{4t}$

$$\frac{3000}{2000} = (1.01)^{4t}$$

$$\log \left(\tfrac{3}{2}\right) = \log (1.01)^{4t}$$

$$\log \left(\tfrac{3}{2}\right) = 4t \log (1.01)$$

$$t = \frac{\log \left(\tfrac{3}{2}\right)}{4 \log (1.01)} \approx 10.2$$

It will take about 10.2 years for the account to grow to $3000.

47. (a) Use the formula $A = Pe^{rt}$ with $P = 4000$, $r = 0.035$, and $t = 6$.

$$A = 4000e^{(0.035)(6)}$$

$$= 4000e^{0.21} \approx 4934.71$$

There will be $4934.71 in the account.

(b) If the initial amount doubles, then $A = 2P$, or $8000.

$$8000 = 4000e^{0.035t}$$

$$2 = e^{0.035t} \qquad \textit{Divide by 4000.}$$

$$\ln 2 = \ln e^{0.035t}$$

$$\ln 2 = 0.035t$$

$$t = \frac{\ln 2}{0.035} \approx 19.8$$

The initial amount will double in about 19.8 years.

49. Use $A = P\left(1 + \dfrac{r}{n}\right)^{nt}$ with $P = 5000$, $r = 0.07$, and $t = 12$.

(a) If the interest is compounded annually, $n = 1$.

$$A = 5000\left(1 + \tfrac{0.07}{1}\right)^{1 \cdot 12}$$

$$= 5000(1.07)^{12} \approx 11{,}260.96$$

There will be $11,260.96 in the account.

(b) If the interest is compounded semiannually, $n = 2$.

$$A = 5000\left(1 + \tfrac{0.07}{2}\right)^{2 \cdot 12}$$

$$= 5000(1.035)^{24} \approx 11{,}416.64$$

There will be $11,416.64 in the account.

(c) If the interest is compounded quarterly, $n = 4$.

$$A = 5000\left(1 + \tfrac{0.07}{4}\right)^{4 \cdot 12}$$

$$= 5000(1.0175)^{48} \approx 11{,}497.99$$

There will be $11,497.99 in the account.

(d) If the interest is compounded daily, $n = 365$.

$$A = 5000\left(1 + \tfrac{0.07}{365}\right)^{365 \cdot 12}$$

$$\approx 11{,}580.90$$

There will be $11,580.90 in the account.

(e) Use the continuous compound interest formula.

$$A = Pe^{rt}$$

$$A = 5000e^{0.07(12)}$$

$$= 5000e^{0.84} \approx 11{,}581.83$$

There will be $11,581.83 in the account.

51. In the continuous compound interest formula, let $A = 1850$, $r = 0.065$, and $t = 40$.

$$A = Pe^{rt}$$

$$1850 = Pe^{0.065(40)}$$

$$1850 = Pe^{2.6}$$

$$P = \frac{1850}{e^{2.6}} \approx 137.41$$

Deposit $137.41 today.

53. $f(x) = 15.80e^{0.0708x}$

(a) 1980 corresponds to $x = 0$.

$$f(0) = 15.80e^{0.0708(0)}$$
$$= 15.80(1) = 15.8$$

The approximate volume of materials recovered from municipal solid waste collections in the United States in 1980 was 15.8 million tons.

(b) 1985 corresponds to $x = 5$.

$$f(5) \approx 22.5 \text{ million tons}$$

(c) 1995 corresponds to $x = 15$.

$$f(15) \approx 45.7 \text{ million tons}$$

(d) 2003 corresponds to $x = 23$.

$$f(23) \approx 80.5 \text{ million tons}$$

55. $B(x) = 27{,}190e^{0.0448x}$
2004 corresponds to $x = 9$.
$B(9) = 27{,}190e^{0.0448(9)}$
$\approx 40{,}693$
The approximate value of consumer expenditures for 2004 was 40,693 million dollars.

57. $A(t) = 2.00e^{-0.053t}$

(a) Let $t = 4$.

$$A(4) = 2.00e^{-0.053(4)}$$
$$= 2.00e^{-0.212}$$
$$\approx 1.62$$

About 1.62 grams would be present.

(b) $A(10) = 2.00e^{-0.053(10)}$
$$\approx 1.18$$

About 1.18 grams would be present.

(c) $A(20) = 2.00e^{-0.053(20)}$
$$\approx 0.69$$

About 0.69 grams would be present.

(d) The initial amount is the amount $A(t)$ present at time $t = 0$.

$$A(0) = 2.00e^{-0.053(0)}$$
$$= 2.00e^0 = 2.00(1) = 2.00$$

Initially, 2.00 grams were present.

59. (a) Find $A(t) = 400e^{-0.032t}$ when $t = 25$.

$$A(25) = 400e^{-0.032(25)}$$
$$= 400e^{-0.8} \approx 179.73$$

About 179.73 grams of lead will be left.

(b) Use $A(t) = 400e^{-0.032t}$, with
$A(t) = \frac{1}{2}(400) = 200$.

$$200 = 400e^{-0.032t}$$
$$\frac{200}{400} = e^{-0.032t}$$
$$0.5 = e^{-0.032t}$$
$$\ln 0.5 = \ln e^{-0.032t}$$
$$\ln 0.5 = -0.032t(\ln e)$$
$$t = \frac{\ln 0.5}{-0.032} \approx 21.66$$

It would take about 21.66 years for the sample to decay to half its original amount.

61.
$$f(x) = 15.80e^{0.0708x}$$
$$100 = 15.8e^{0.0708x}$$
$$e^{0.0708x} = \frac{100}{15.8}$$
$$\ln e^{0.0708x} = \ln \frac{100}{15.8}$$
$$0.0708x (\ln e) = \ln \frac{100}{15.8}$$
$$0.0708x = \ln \frac{100}{15.8}$$
$$x = \frac{\ln \frac{100}{15.8}}{0.0708} \approx 26$$

Since $x = 0$ corresponds to 1980, $x = 26$ corresponds to 2006.

63. $f(t) = 300e^{0.4t}$
The original number of ants, at $t = 0$, is 300.
Replace $f(t)$ with 600, (double the number of ants), and solve for t.

$$600 = 300e^{0.4t}$$
$$2 = e^{0.4t}$$
$$\ln 2 = \ln e^{0.4t}$$
$$\ln 2 = 0.4t (\ln e)$$
$$\ln 2 = 0.4t$$
$$t = \frac{\ln 2}{0.4} \approx 1.733$$

It will take about 1.733 days for the number of ants to double.

65. (a) The expression $\frac{1}{x}$ in the base cannot be evaluated since division by 0 is not defined.

(b) When X = 100,000, $Y_1 \approx 2.7183$. The decimal approximation appears to be close to the decimal approximation for e, 2.71828....

(c) $\left(1 + \dfrac{1}{1{,}000{,}000}\right)^{1{,}000{,}000} \approx 2.718280469$

$$e = e^1 \approx 2.718281828$$

The two values differ in the sixth decimal place.

(d) As the values of x approach infinity, the value of $\left(1 + \dfrac{1}{x}\right)^x$ approaches $\underline{e}$.

67. $f(x) = 2x^2$

The graph of f is narrower than the graph of $y = x^2$.

x	0	± 1	± 2
y	0	2	8

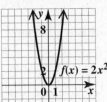

69. $f(x) = (x + 1)^2$

The graph of f is the graph of $y = x^2$ shifted *left* 1 unit.

x	-3	-2	-1	0	1
y	4	1	0	1	4

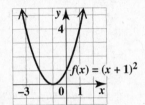

Chapter 10 Review Exercises

1. Since a horizontal line intersects the graph in two points, the function is not one-to-one.

2. Since every horizontal line intersects the graph in no more than one point, the function is one-to-one.

3. This function is not one-to-one because two sodas in the list have 41 mg of caffeine.

4. The function $f(x) = -3x + 7$ is a linear function. By the horizontal line test, it is a one-to-one function. To find the inverse, replace $f(x)$ with y.

$$y = -3x + 7$$

Interchange x and y.

$$x = -3y + 7$$

Solve for y.

$$x - 7 = -3y$$

$$\frac{x - 7}{-3} = y \quad \text{or} \quad \frac{7 - x}{3} = y$$

Replace y with $f^{-1}(x)$.

$$f^{-1}(x) = \frac{x - 7}{-3}, \quad \text{or} \quad f^{-1}(x) = -\frac{1}{3}x + \frac{7}{3}$$

5. $f(x) = \sqrt[3]{6x - 4}$

The function is one-to-one since each $f(x)$-value corresponds to exactly one x-value.

To find the inverse, replace $f(x)$ with y.

$$y = \sqrt[3]{6x - 4}$$

Cube both sides.

$$y^3 = 6x - 4$$

Interchange x and y.

$$x^3 = 6y - 4$$

Solve for y.

$$x^3 + 4 = 6y$$

$$\frac{x^3 + 4}{6} = y$$

Replace y with $f^{-1}(x)$.

$$\frac{x^3 + 4}{6} = f^{-1}(x)$$

6. $f(x) = -x^2 + 3$

This is an equation of a vertical parabola which opens down.

Since a horizontal line will intersect the graph in two points, the function is not one-to-one.

7. The graph is a linear function through $(0, 1)$ and $(3, 0)$. The graph of $f^{-1}(x)$ will include the points $(1, 0)$ and $(0, 3)$, found by interchanging x and y. Plot these points, and draw a straight line through them.

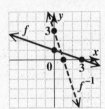

8. The graph is a curve through $(1, 2)$, $(0, 1)$, and $\left(-1, \frac{1}{2}\right)$. Interchange x and y to get $(2, 1)$, $(1, 0)$, and $\left(\frac{1}{2}, -1\right)$, which are on the graph of $f^{-1}(x)$. Plot these points, and draw a smooth curve through them.

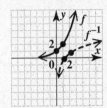

9. $f(x) = 3^x$

Make a table of values.

x	-2	-1	0	1	2
$f(x)$	$\frac{1}{9}$	$\frac{1}{3}$	1	3	9

Plot the points from the table and draw a smooth curve through them.

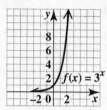

10. $f(x) = \left(\frac{1}{3}\right)^x$

Make a table of values.

x	-2	-1	0	1	2
$f(x)$	9	3	1	$\frac{1}{3}$	$\frac{1}{9}$

Plot the points from the table and draw a smooth curve through them.

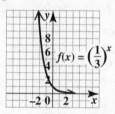

11. $y = 3^{x+1}$

Make a table of values.

x	-3	-2	-1	0	1
y	$\frac{1}{9}$	$\frac{1}{3}$	1	3	9

Plot the points from the table and draw a smooth curve through them.

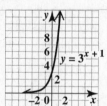

12. $y = 2^{2x+3}$

Make a table of values.

x	-2	$-\frac{3}{2}$	-1	0	$\frac{1}{2}$
y	$\frac{1}{2}$	1	2	8	16

Plot the points from the table and draw a smooth curve through them.

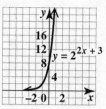

13. $4^{3x} = 8^{x+4}$

Write each side as a power of 2.

$$\left(2^2\right)^{3x} = \left(2^3\right)^{x+4}$$
$$2^{6x} = 2^{3x+12}$$
$$6x = 3x + 12 \quad \textit{Equate exponents.}$$
$$3x = 12$$
$$x = 4$$

Check $x = 4$: $4^{12} = 8^8 \quad$ *True*

The solution set is $\{4\}$.

14. $\left(\frac{1}{27}\right)^{x-1} = 9^{2x}$

$$\left[\left(\frac{1}{3}\right)^3\right]^{x-1} = \left(3^2\right)^{2x}$$

Write each side as a power of 3.

$$\left(3^{-3}\right)^{x-1} = \left(3^2\right)^{2x}$$
$$3^{-3x+3} = 3^{4x}$$
$$-3x + 3 = 4x \quad \textit{Equate exponents.}$$
$$3 = 7x$$
$$\frac{3}{7} = x$$

Check $x = \frac{3}{7}$: $\left(\frac{1}{27}\right)^{-4/7} = 9^{6/7} \quad$ *True*

The solution set is $\left\{\frac{3}{7}\right\}$.

15. $W(x) = 7.77(1.059)^x$

(a) $x = 1985 - 1980 = 5$

$$W(5) = 7.77(1.059)^5$$
$$\approx 10.3$$

The approximate amount of plastic waste in 1985 was 10.3 million tons.

(b) $x = 1995 - 1980 = 15$

$W(15) \approx 18.4$ million tons

(c) $x = 2000 - 1980 = 20$

$W(20) \approx 24.5$ million tons

16. $g(x) = \log_3 x$

Replace $g(x)$ with y, and write in exponential form.

$$y = \log_3 x$$
$$3^y = x$$

Make a table of values. Since $x = 3^y$ is the inverse of $f(x) = y = 3^x$ in Exercise 9, simply reverse the ordered pairs in the table of values belonging to $f(x) = 3^x$.

x	$\frac{1}{9}$	$\frac{1}{3}$	1	3	9
y	-2	-1	0	1	2

Plot the points from the table and draw a smooth curve through them.

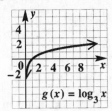

$g(x) = \log_3 x$

17. $g(x) = \log_{1/3} x$

Replace $g(x)$ with y, and write in exponential form.

$$y = \log_{1/3} x$$

$$\left(\tfrac{1}{3}\right)^y = x$$

Make a table of values. Since $x = \left(\tfrac{1}{3}\right)^y$ is the inverse of $f(x) = y = \left(\tfrac{1}{3}\right)^x$ in Exercise 10, simply reverse the ordered pairs in the table of values belonging to $f(x) = \left(\tfrac{1}{3}\right)^x$.

x	9	3	1	$\frac{1}{3}$	$\frac{1}{9}$
y	-2	-1	0	1	2

Plot the points from the table and draw a smooth curve through them.

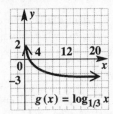

$g(x) = \log_{1/3} x$

18. $\log_8 64 = x$

$\quad 8^x = 64$ *Exponential form*

Write each side as a power of 8.

$\quad 8^x = 8^2$

$\quad\quad x = 2$ *Equate exponents.*

The solution set is $\{2\}$.

19. $\log_2 \sqrt{8} = x$

$\quad 2^x = \sqrt{8}$ *Exponential form*

$\quad 2^x = 8^{1/2}$

Write each side as a power of 2.

$\quad 2^x = \left(2^3\right)^{1/2}$

$\quad 2^x = 2^{3/2}$

$\quad\quad x = \tfrac{3}{2}$ *Equate exponents.*

The solution set is $\left\{\tfrac{3}{2}\right\}$.

20. $\log_x \left(\dfrac{1}{49}\right) = -2$

$\quad x^{-2} = \dfrac{1}{49}$ *Exponential form*

$\quad \dfrac{1}{x^2} = \dfrac{1}{49}$

$\quad\quad x^2 = 49$

$\quad\quad x = \pm 7$

Since x is the base, we cannot have a negative number.

The solution set is $\{7\}$.

21. $\log_4 x = \tfrac{3}{2}$

$\quad x = 4^{3/2}$ *Exponential form*

$\quad x = \left(\sqrt{4}\right)^3 = 2^3 = 8$

The solution set is $\{8\}$.

22. $\log_k 4 = 1$

$\quad k^1 = 4$ *Exponential form*

$\quad k = 4$

The solution set is $\{4\}$.

23. $\log_b b^2 = 2$

$\quad b^2 = b^2$ *Exponential form*

This is an identity. Thus, b can be any real number, $b > 0$ and $b \neq 1$.

The solution set is $\{b \mid b > 0, \ b \neq 1\}$.

24. $\log_b a$ is the exponent to which b must be raised to obtain a.

25. From Exercise 24,

$$b^{\log_b a} = a.$$

26. $S(x) = 100 \log_2 (x + 2)$

(a) When $x = 6$,

$$S(6) = 100 \log_2 (6 + 2)$$

$$= 100(3) = 300.$$

After 6 weeks the sales were 300 thousand dollars or \$300,000.

(b) To graph the function, make a table of values that includes the ordered pair from above.

x	0	2	6
$S(x)$	100	200	300

Plot the ordered pairs and draw the graph through them.

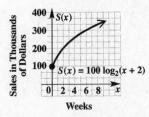

$S(x) = 100 \log_2(x + 2)$

Weeks

27. $\log_2 3xy^2$

$\quad = \log_2 3 + \log_2 x + \log_2 y^2 \quad$ *Product rule*

$\quad = \log_2 3 + \log_2 x + 2\log_2 y \quad$ *Power rule*

28. $\log_4 \dfrac{\sqrt{x} \cdot w^2}{z}$

$\quad = \log_4 \left(\sqrt{x} \cdot w^2\right) - \log_4 z \quad$ *Quotient rule*

$\quad = \log_4 x^{1/2} + \log_4 w^2 - \log_4 z \quad$ *Product rule*

$\quad = \frac{1}{2}\log_4 x + 2\log_4 w - \log_4 z \quad$ *Power rule*

29. $\log_b 3 + \log_b x - 2\log_b y$

Use the product and power rules for logarithms.

$\quad = \log_b (3 \cdot x) - \log_b y^2$

$\quad = \log_b \dfrac{3x}{y^2} \qquad\qquad$ *Quotient rule*

30. $\log_3 (x + 7) - \log_3 (4x + 6)$

$\quad = \log_3 \left(\dfrac{x + 7}{4x + 6}\right) \qquad$ *Quotient rule*

31. $\log 28.9 \approx 1.4609$

32. $\log 0.257 \approx -0.5901$

33. $\ln 28.9 \approx 3.3638$

34. $\ln 0.257 \approx -1.3587$

35. $\log_{16} 13 = \dfrac{\log 13}{\log 16} \approx 0.9251$

36. $\log_4 12 = \dfrac{\log 12}{\log 4} \approx 1.7925$

37. Milk has a hydronium ion concentration of 4.0×10^{-7}.

$\text{pH} = -\log\left[\text{H}_3\text{O}^+\right]$

$\text{pH} = -\log\left(4.0 \times 10^{-7}\right) \approx 6.4$

38. Crackers have a hydronium ion concentration of 3.8×10^{-9}.

$\text{pH} = -\log\left[\text{H}_3\text{O}^+\right]$

$\text{pH} = -\log\left(3.8 \times 10^{-9}\right) \approx 8.4$

39. Orange juice has a pH of 4.6.

$\qquad \text{pH} = -\log\left[\text{H}_3\text{O}^+\right]$

$\qquad 4.6 = -\log\left[\text{H}_3\text{O}^+\right]$

$\log_{10}\left[\text{H}_3\text{O}^+\right] = -4.6$

$\qquad \left[\text{H}_3\text{O}^+\right] = 10^{-4.6} \approx 2.5 \times 10^{-5}$

40. $Q(t) = 500e^{-0.05t}$

(a) Let $t = 0$.

$Q(0) = 500e^{-0.05(0)}$

$\qquad = 500e^0 = 500(1) = 500$

There are 500 grams.

(b) Let $t = 4$.

$Q(4) = 500e^{-0.05(4)}$

$\qquad = 500e^{-0.2} \approx 409.4$

There will be about 409 grams in 4 days.

41. $t(r) = \dfrac{\ln 2}{\ln (1 + r)}$

(a) $4\% = 0.04$; $t(0.04) = \dfrac{\ln 2}{\ln (1 + 0.04)} \approx 18$

At 4%, it would take about 18 years.

(b) $6\% = 0.06$; $t(0.06) = \dfrac{\ln 2}{\ln (1 + 0.06)} \approx 12$

At 6%, it would take about 12 years.

(c) $10\% = 0.10$; $t(0.10) = \dfrac{\ln 2}{\ln (1 + 0.10)} \approx 7$

At 10%, it would take about 7 years.

(d) $12\% = 0.12$; $t(0.12) = \dfrac{\ln 2}{\ln (1 + 0.12)} \approx 6$

At 12%, it would take about 6 years.

(e) Each comparison shows approximately the same number. For example, in part (a) the doubling time is 18 yr (rounded) and $\frac{72}{4} = 18$.

Thus, the formula $t = \dfrac{72}{100r}$ (called the *rule of 72*) is an excellent approximation of the doubling time formula. (It is used by bankers for that purpose.)

42. $\qquad 3^x = 9.42$

$\qquad \log 3^x = \log 9.42$

$\qquad x\log 3 = \log 9.42$

$\qquad\qquad x = \dfrac{\log 9.42}{\log 3} \approx 2.042$

Check $x = 2.042$: $3^{2.042} \approx 9.425$

The solution set is $\{2.042\}$.

43. $\qquad\qquad 2^{x-1} = 15$

$\qquad\quad \log 2^{x-1} = \log 15$

$\qquad (x - 1)\log 2 = \log 15$

$\qquad\qquad x - 1 = \dfrac{\log 15}{\log 2}$

$\qquad\qquad\qquad x = \dfrac{\log 15}{\log 2} + 1 \approx 4.907$

Check $x = 4.907$: $2^{4.907} \approx 15.0$

The solution set is $\{4.907\}$.

44. $e^{0.06x} = 3$

Take base e logarithms on both sides.

$\ln e^{0.06x} = \ln 3$

$0.06x \ln e = \ln 3$

$0.06x = \ln 3 \qquad\qquad ln\ e = 1$

$x = \dfrac{\ln 3}{0.06} \approx 18.310$

Check $x = 18.310$: $e^{1.0986} \approx 3.0$

The solution set is $\{18.310\}$.

45. $\log_3 (9x + 8) = 2$

$9x + 8 = 3^2 \quad$ *Exponential form*

$9x + 8 = 9$

$9x = 1$

$x = \tfrac{1}{9}$

Check $x = \tfrac{1}{9}$: $\log_3 9 = \log_3 3^2 = 2$

The solution set is $\left\{\tfrac{1}{9}\right\}$.

46. $\log_5 (y + 6)^3 = 2$

Change to exponential form.

$(y + 6)^3 = 5^2$

$(y + 6)^3 = 25$

Take the cube root of each side.

$y + 6 = \sqrt[3]{25}$

$y = \sqrt[3]{25} - 6$

Check $y = -6 + \sqrt[3]{25}$: $\log_5 25 = \log_5 5^2 = 2$

The solution set is $\left\{-6 + \sqrt[3]{25}\right\}$.

47. $\log_3 (p + 2) - \log_3 p = \log_3 2$

$\log_3 \dfrac{p + 2}{p} = \log_3 2 \quad$ *Quotient rule*

$\dfrac{p + 2}{p} = 2 \qquad\qquad$ *Property 4*

$p + 2 = 2p$

$2 = p$

Check $p = 2$: $\log_3 4 - \log_3 2 = \log_3 \tfrac{4}{2} = \log_3 2$

The solution set is $\{2\}$.

48. $\log (2x + 3) = 1 + \log x$

$\log (2x + 3) - \log x = 1$

$\log_{10} \dfrac{2x + 3}{x} = 1 \quad$ *Quotient rule*

$10^1 = \dfrac{2x + 3}{x} \quad$ *Exponential form*

$10x = 2x + 3$

$8x = 3$

$x = \tfrac{3}{8}$

Check $x = \tfrac{3}{8}$:

LS $= \log \left(\tfrac{3}{4} + 3\right) = \log \tfrac{15}{4}$

RS $= \log 10 + \log \tfrac{3}{8} = \log \tfrac{30}{8} = \log \tfrac{15}{4}$

The solution set is $\left\{\tfrac{3}{8}\right\}$.

49. $\log_4 x + \log_4 (8 - x) = 2$

$\log_4 [x(8 - x)] = 2 \quad$ *Product rule*

$x(8 - x) = 4^2 \quad$ *Exponential form*

$8x - x^2 = 16$

$x^2 - 8x + 16 = 0$

$(x - 4)(x - 4) = 0$

$x - 4 = 0$

$x = 4$

Check $x = 4$: $\log_4 4 + \log_4 4 = 1 + 1 = 2$

The solution set is $\{4\}$.

50. $\log_2 x + \log_2 (x + 15) = \log_2 16$

$\log_2 [x(x + 15)] = \log_2 16 \quad$ *Product rule*

$x^2 + 15x = 16 \qquad$ *Property 4*

$x^2 + 15x - 16 = 0$

$(x + 16)(x - 1) = 0$

$x + 16 = 0 \qquad$ or $\quad x - 1 = 0$

$x = -16 \quad$ or $\qquad x = 1$

Reject $x = -16$, because it yields an equation in which the logarithm of a negative number must be found.

Check $x = 1$:

$\log_2 1 + \log_2 16 = 0 + \log_2 16 = \log_2 16$

The solution set is $\{1\}$.

51. When the power rule was applied in the second step, the domain was changed from $\{x \mid x \neq 0\}$ to $\{x \mid x > 0\}$. Instead of using the power rule for logarithms, we can change the original equation to the exponential form $x^2 = 10^2$ and get $x = \pm 10$. As you can see in the erroneous solution, the valid solution -10 was "lost." The solution set is $\{\pm 10\}$.

52. $A = P\left(1 + \dfrac{r}{n}\right)^{nt}$

Let $P = 20{,}000$, $r = 0.07$, and $t = 5$. For $n = 4$ (quarterly compounding),

$$A = 20{,}000\left(1 + \dfrac{0.07}{4}\right)^{4 \cdot 5} \approx 28{,}295.56.$$

There will be $28,295.56 in the account after 5 years.

53. In the continuous compounding formula, let $P = 10{,}000$, $r = 0.06$, and $t = 3$.

$$A = Pe^{rt}$$
$$A = 10{,}000e^{0.06(3)} \approx 11{,}972.17$$

There will be $11,972.17 in the account after 3 years.

54. Use $A = P\left(1 + \dfrac{r}{n}\right)^{nt}$.

Plan A:
Let $P = 1000$, $r = 0.04$, $n = 4$, and $t = 3$.

$$A = 1000\left(1 + \frac{0.04}{4}\right)^{4 \cdot 3} \approx 1126.83$$

Plan B:
Let $P = 1000$, $r = 0.039$, $n = 12$, and $t = 3$.

$$A = 1000\left(1 + \frac{0.039}{12}\right)^{12 \cdot 3} \approx 1123.91$$

Plan A is the better plan by $2.92.

55. Let $Q(t) = \frac{1}{2}(500)$ to find the half-life of the radioactive substance.

$$Q(t) = 500e^{-0.05t}$$
$$\tfrac{1}{2}(500) = 500e^{-0.05t}$$
$$0.5 = e^{-0.05t}$$
$$\ln 0.5 = \ln e^{-0.05t}$$
$$\ln 0.5 = -0.05t\,(\ln e)$$
$$-0.05t = \ln 0.5$$
$$t = \frac{\ln 0.5}{-0.05} \approx 13.9$$

The half-life is about 13.9 days.

56. $S = C(1 - r)^n$

(a) Let $C = 30{,}000$, $r = 0.15$, and $n = 12$.

$$S = 30{,}000(1 - 0.15)^{12}$$
$$= 30{,}000(0.85)^{12} \approx 4267$$

The scrap value is about $4267.

(b) Let $S = \frac{1}{2}C$ and $n = 6$.

$$S = C(1 - r)^n$$
$$\tfrac{1}{2}C = C(1 - r)^6$$
$$0.5 = (1 - r)^6 \;(*)$$
$$\ln 0.5 = \ln (1 - r)^6$$
$$\ln 0.5 = 6\ln (1 - r)$$
$$\ln (1 - r) = \frac{\ln 0.5}{6}$$
$$\ln (1 - r) \approx -0.1155$$
$$1 - r = e^{-0.1155}$$
$$1 - r \approx 0.89$$
$$r = 0.11$$

The rate is approximately 11%.

$(*)$ Alternative solution steps without logarithms:

$$0.5 = (1 - r)^6$$
$$\sqrt[6]{0.5} = 1 - r$$
$$r = 1 - \sqrt[6]{0.5} \approx 0.11$$

Note that $1 - r$ must be positive, so $\pm \sqrt[6]{0.5}$ is not needed.

57. $N(r) = -5000 \ln r$
Replace $N(r)$ with 2000, and solve for r.

$$2000 = -5000 \ln r$$
$$\ln r = -\tfrac{2000}{5000}$$
$$\log_e r = -0.4$$

Change to exponential form and approximate.

$$r = e^{-0.4} \approx 0.67$$

About 67% of the words are common to both of the evolving languages.

58. A. Solve $7^x = 23$ by using the power rule with common logarithms.

$$7^x = 23$$
$$\log 7^x = \log 23$$
$$x \log 7 = \log 23$$
$$x = \frac{\log 23}{\log 7}$$

B. Solve $7^x = 23$ by using the power rule with natural logarithms.

$$7^x = 23$$
$$\ln 7^x = \ln 23$$
$$x \ln 7 = \ln 23$$
$$x = \frac{\ln 23}{\ln 7}$$

C. Use the change-of-base rule with the solution from **A.**

$$x = \frac{\log 23}{\log 7} = \log_7 23$$

D. $x = \dfrac{\log 23}{\log 7} \neq \log_{23} 7$

The answer is **D**.

59. [10.4] $\log_2 128 = \log_2 2^7 = 7$, by a special property of logarithms.

60. [10.4] By a special property of logarithms,
$$5^{\log_5 36} = 36.$$

61. [10.5] $e^{\ln 4} = 4$ since $e^{\ln x} = x$.

62. [10.5] $10^{\log e} = e$ since $10^{\log x} = x$.

63. [10.4] $\log_3 3^{-5} = -5$ since $\log_a a^x = x$.

64. [10.5] $\ln e^{5.4} = 5.4$ since $\ln e^x = x$.

65. **[10.6]** $\log_3 (x + 9) = 4$

$\qquad x + 9 = 3^4$ *Exponential form*

$\qquad x + 9 = 81$

$\qquad x = 72$

Check $x = 72$: $\log_3 81 = \log_3 3^4 = 4$

The solution set is $\{72\}$.

66. **[10.6]** $\ln e^x = 3$

$\qquad x = 3$

The solution set is $\{3\}$.

67. **[10.3]** $\log_x \frac{1}{81} = 2$

$\qquad x^2 = \frac{1}{81}$ *Exponential form*

$\qquad x^2 = \left(\frac{1}{9}\right)^2$

$\qquad x = \pm \frac{1}{9}$ *Square root property*

The base x cannot be negative.

Check $x = \frac{1}{9}$: $\log_{1/9} \frac{1}{81} = 2$ since $\left(\frac{1}{9}\right)^2 = \frac{1}{81}$

The solution set is $\left\{\frac{1}{9}\right\}$.

68. **[10.2]** $27^x = 81$

Write each side as a power of 3.

$\qquad \left(3^3\right)^x = 3^4$

$\qquad 3^{3x} = 3^4$

$\qquad 3x = 4$ *Equate exponents.*

$\qquad x = \frac{4}{3}$

Check $x = \frac{4}{3}$: $27^{4/3} = 3^4 = 81$

The solution set is $\left\{\frac{4}{3}\right\}$.

69. **[10.2]** $2^{2x-3} = 8$

Write each side as a power of 2.

$\qquad 2^{2x-3} = 2^3$

$\qquad 2x - 3 = 3$ *Equate exponents.*

$\qquad 2x = 6$

$\qquad x = 3$

Check $x = 3$: $2^3 = 8$ *True*

The solution set is $\{3\}$.

70. **[10.2]** $5^{x+2} = 25^{2x+1}$

$\qquad 5^{x+2} = \left(5^2\right)^{2x+1}$ *Equal bases*

$\qquad 5^{x+2} = 5^{4x+2}$

$\qquad x + 2 = 4x + 2$ *Equate exponents.*

$\qquad 0 = 3x$

$\qquad 0 = x$

Check $x = 0$: $5^2 = 25^1$ *True*

The solution set is $\{0\}$.

71. **[10.6]**

$\log_3 (x + 1) - \log_3 x = 2$

$\qquad \log_3 \frac{x+1}{x} = 2$ *Quotient rule*

$\qquad \frac{x+1}{x} = 3^2$ *Exponential form*

$\qquad 9x = x + 1$

$\qquad 8x = 1$

$\qquad x = \frac{1}{8}$

Check $x = \frac{1}{8}$:

$\log_3 \frac{9}{8} - \log_3 \frac{1}{8} = \log_3 9 = \log_3 3^2 = 2$

The solution set is $\left\{\frac{1}{8}\right\}$.

72. **[10.6]** $\log (3x - 1) = \log 10$

$\qquad 3x - 1 = 10$

$\qquad 3x = 11$

$\qquad x = \frac{11}{3}$

Check $x = \frac{11}{3}$: $\log (11 - 1) = \log 10$ *True*

The solution set is $\left\{\frac{11}{3}\right\}$.

73. **[10.6]** $\ln \left(x^2 + 3x + 4\right) = \ln 2$

$\qquad x^2 + 3x + 4 = 2$

$\qquad x^2 + 3x + 2 = 0$

$\qquad (x + 2)(x + 1) = 0$

$x + 2 = 0 \quad$ or $\quad x + 1 = 0$

$x = -2 \quad$ or $\qquad x = -1$

Check $x = -2$: $\ln (4 - 6 + 4) = \ln 2$ *True*
Check $x = -1$: $\ln (1 - 3 + 4) = \ln 2$ *True*

The solution set is $\{-2, -1\}$.

74. **[10.6] (a)** $\log (2x + 3) = \log x + 1$

$\qquad \log (2x + 3) - \log x = 1$

$\qquad \log_{10} \frac{2x+3}{x} = 1$ *Quotient rule*

$\qquad \frac{2x+3}{x} = 10^1$ *Exponential form*

$\qquad 2x + 3 = 10x$

$\qquad 3 = 8x$

$\qquad \frac{3}{8} = x$

Check $x = \frac{3}{8}$:

LS $= \log \left(\frac{3}{4} + 3\right) = \log \frac{15}{4}$

RS $= \log \frac{3}{8} + \log 10 = \log \frac{30}{8} = \log \frac{15}{4}$

The solution set is $\left\{\frac{3}{8}\right\}$.

(b) From the graph, the x-value of the x-intercept is 0.375, the decimal equivalent of $\frac{3}{8}$. Note that the solutions of $Y_1 - Y_2 = 0$ are the same as the solutions of $Y_1 = Y_2$.

75. **[10.6]** $x = 2000 - 1980 = 20$

$$R(x) = 10.001e^{0.0521x}$$
$$R(20) = 10.001e^{0.0521(20)}$$
$$\approx 28.35$$

In 2000, about 28.35% of municipal solid waste was recovered.

76. **[10.5]** **(a)** There are 90 of one species and 10 of another, so

$$p_1 = \frac{90}{100} = 0.9 \quad \text{and} \quad p_2 = \frac{10}{100} = 0.1.$$

Thus, the index of diversity is

$$-(p_1 \ln p_1 + p_2 \ln p_2)$$
$$= -(0.9 \ln 0.9 + 0.1 \ln 0.1)$$
$$\approx 0.325.$$

(b) There are 60 of one species and 40 of another, so

$$p_1 = \frac{60}{100} = 0.6 \quad \text{and} \quad p_2 = \frac{40}{100} = 0.4.$$

Thus, the index of diversity is

$$-(p_1 \ln p_1 + p_2 \ln p_2)$$
$$= -(0.6 \ln 0.6 + 0.4 \ln 0.4)$$
$$\approx 0.673.$$

Chapter 10 Test

1. **(a)** $f(x) = x^2 + 9$

This function is not one-to-one. The graph of $f(x)$ is a vertical parabola. A horizontal line will intersect the graph more than once.

(b) This function is one-to-one. A horizontal line will not intersect the graph in more than one point.

2. $$f(x) = \sqrt[3]{x + 7}$$
Replace $f(x)$ with y.
$$y = \sqrt[3]{x + 7}$$
Interchange x and y.
$$x = \sqrt[3]{y + 7}$$
Solve for y.
$$x^3 = y + 7$$
$$x^3 - 7 = y$$
Replace y with $f^{-1}(x)$.
$$f^{-1}(x) = x^3 - 7$$

3. By the horizontal line test, $f(x)$ is a one-to-one function and has an inverse. Choose some points on the graph of $f(x)$, such as $(4, 0)$, $(3, -1)$, and $(0, -2)$. To graph the inverse, interchange the x- and y-values to get $(0, 4)$, $(-1, 3)$, and $(-2, 0)$. Plot these points and draw a smooth curve through them.

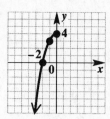

4. $f(x) = 6^x$
Make a table of values.

x	-2	-1	0	1
$f(x)$	$\frac{1}{36}$	$\frac{1}{6}$	1	6

Plot these points and draw a smooth exponential curve through them.

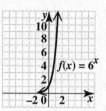

5. $g(x) = \log_6 x$
Make a table of values.

Powers of 6	6^{-2}	6^{-1}	6^0	6^1
x	$\frac{1}{36}$	$\frac{1}{6}$	1	6
$g(x)$	-2	-1	0	1

Plot these points and draw a smooth logarithmic curve through them.

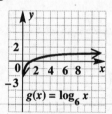

6. $y = 6^x$ and $y = \log_6 x$ are inverse functions. To use the graph from Exercise 4 to obtain the graph of the function in Exercise 5, interchange the x- and y-coordinates of the ordered pairs $\left(-2, \frac{1}{36}\right)$, $\left(-1, \frac{1}{6}\right)$, $(0, 1)$, and $(1, 6)$ to get $\left(\frac{1}{36}, -2\right)$, $\left(\frac{1}{6}, -1\right)$, $(1, 0)$, and $(6, 1)$. Plot these points and draw a smooth logarithmic curve through them.

7. $$5^x = \frac{1}{625}$$
$$5^x = \left(\frac{1}{5}\right)^4$$
Write each side as a power of 5.
$$5^x = 5^{-4}$$
$$x = -4 \qquad \textit{Equate exponents.}$$
The solution set is $\{-4\}$.

8.
$$2^{3x-7} = 8^{2x+2}$$

Write each side as a power of 2.
$$2^{3x-7} = \left(2^3\right)^{2x+2}$$
$$3x - 7 = 3(2x + 2) \quad \textit{Equate exponents.}$$
$$3x - 7 = 6x + 6$$
$$-13 = 3x$$
$$-\frac{13}{3} = x$$

Check $x = -\frac{13}{3}$:
$$\text{LS} = 2^{-13-7} = 2^{-20}$$
$$\text{RS} = 8^{-20/3} = \left(2^3\right)^{-20/3} = 2^{-20}$$

The solution set is $\left\{-\frac{13}{3}\right\}$.

9. $f(x) = 26.7e^{0.023x}$

(a) 2010: $x = 2010 - 1995 = 15$
$$f(15) = 26.7e^{0.023(15)} \approx 37.7$$

The U.S. Hispanic population estimate for 2010 is 37.7 million.

(b) 2015: $x = 2015 - 1995 = 20$
$$f(20) = 26.7e^{0.023(20)} \approx 42.3$$

The U.S. Hispanic population estimate for 2015 is 42.3 million.

10. The base is 4, the exponent (logarithm) is -2, and the number is 0.0625, so $4^{-2} = 0.0625$ becomes $\log_4 0.0625 = -2$ in logarithmic form.

11. The base is 7, the logarithm (exponent) is 2, and the number is 49, so $\log_7 49 = 2$ becomes $7^2 = 49$ in exponential form.

12. $\log_{1/2} x = -5$
$$x = \left(\tfrac{1}{2}\right)^{-5} \quad \textit{Exponential form}$$
$$x = \left(\tfrac{2}{1}\right)^5 = 32$$

The argument (the input of the logarithm) must be a positive number, so $x = 32$ is acceptable.

Check $x = 32$:
$$\log_{1/2} 32 = -5 \text{ since } \left(\tfrac{1}{2}\right)^{-5} = 2^5 = 32$$

The solution set is $\{32\}$.

13.
$$x = \log_9 3$$
$$9^x = 3 \quad \textit{Exponential form}$$
Write each side as a power of 3.
$$\left(3^2\right)^x = 3$$
$$3^{2x} = 3^1$$
$$2x = 1 \quad \textit{Equate exponents.}$$
$$x = \tfrac{1}{2}$$

Check $x = \tfrac{1}{2}$:
$$\tfrac{1}{2} = \log_9 3 \text{ since } 9^{1/2} = \sqrt{9} = 3$$

The solution set is $\left\{\tfrac{1}{2}\right\}$.

14. $\log_x 16 = 4$
$$x^4 = 16 \quad \textit{Exponential form}$$
$$x^2 = \pm 4 \quad \textit{Square root property}$$
Reject -4 since $x^2 \geq 0$.
$$x^2 = 4$$
$$x = \pm 2 \quad \textit{Square root property}$$
Reject -2 since the base cannot be negative.

Check $x = 2$: $\log_2 16 = 4$ since $2^4 = 16$

The solution set is $\{2\}$.

15. The value of $\log_2 32$ is $\underline{5}$. This means that if we raise $\underline{2}$ to the $\underline{5\text{th}}$ power, the result is $\underline{32}$.

16. $\log_3 x^2 y$
$$= \log_3 x^2 + \log_3 y \quad \textit{Product rule}$$
$$= 2\log_3 x + \log_3 y \quad \textit{Power rule}$$

17. $\log_5 \left(\dfrac{\sqrt{x}}{yz}\right)$
$$= \log_5 \sqrt{x} - \log_5 yz \qquad \textit{Quotient rule}$$
$$= \log_5 x^{1/2} - \left(\log_5 y + \log_5 z\right) \textit{Product rule}$$
$$= \tfrac{1}{2}\log_5 x - \log_5 y - \log_5 z \quad \textit{Power rule}$$

18. $3\log_b s - \log_b t$
$$= \log_b s^3 - \log_b t \quad \textit{Power rule}$$
$$= \log_b \frac{s^3}{t} \qquad \textit{Quotient rule}$$

19. $\tfrac{1}{4}\log_b r + 2\log_b s - \tfrac{2}{3}\log_b t$

Use the power rule for logarithms.
$$= \log_b r^{1/4} + \log_b s^2 - \log_b t^{2/3}$$

Use the product and quotient rules for logarithms.
$$= \log_b \frac{r^{1/4}s^2}{t^{2/3}}$$

20. **(a)** $\log 23.1 \approx 1.3636$

(b) $\ln 0.82 \approx -0.1985$

21. **(a)** $\log_3 19 = \dfrac{\log_{10} 19}{\log_{10} 3} = \dfrac{\log 19}{\log 3}$

(b) $\log_3 19 = \dfrac{\log_e 19}{\log_e 3} = \dfrac{\ln 19}{\ln 3}$

(c) The four-decimal-place approximation of either fraction is 2.6801.

22.
$$3^x = 78$$
$$\ln 3^x = \ln 78$$
$$x \ln 3 = \ln 78 \qquad \textit{Power rule}$$
$$x = \frac{\ln 78}{\ln 3} \approx 3.9656$$
Check $x = 3.9656$: $3^{3.9656} \approx 78.0$
The solution set is $\{3.9656\}$.

23. $\log_8 (x + 5) + \log_8 (x - 2) = 1$
Use the product rule for logarithms.
$$\log_8 [(x + 5)(x - 2)] = 1$$
$$(x + 5)(x - 2) = 8^1 \qquad \textit{Exp. form}$$
$$x^2 + 3x - 10 = 8$$
$$x^2 + 3x - 18 = 0$$
$$(x + 6)(x - 3) = 0$$
$$x + 6 = 0 \quad \text{or} \quad x - 3 = 0$$
$$x = -6 \quad \text{or} \quad x = 3$$

Reject $x = -6$, because $x + 5 = -1$, which yields an equation in which the logarithm of a negative number must be found.

Check $x = 3$: $\log_8 8 + \log_8 1 = 1 + 0 = 1$
The solution set is $\{3\}$.

24. $A = P\left(1 + \dfrac{r}{n}\right)^{nt}$

$$A = 10{,}000\left(1 + \frac{0.045}{4}\right)^{4 \cdot 5} \approx 12{,}507.51$$

$10,000 invested at 4.5% annual interest, compounded quarterly, will increase to $12,507.51 in 5 years.

25. $A = Pe^{rt}$

(a) $A = 15{,}000e^{0.05(5)} \approx 19{,}260.38$
There will be $19,260.38 in the account.

(b) Let $A = 2(15{,}000)$ and solve for t.
$$2(15{,}000) = 15{,}000e^{0.05t}$$
$$2 = e^{0.05t}$$
$$\ln 2 = \ln e^{0.05t}$$
$$\ln 2 = 0.05t \,(\ln e)$$
$$0.05t = \ln 2$$
$$t = \frac{\ln 2}{0.05} \approx 13.9$$

The principal will double in about 13.9 years.

Cumulative Review Exercises (Chapters 1–10)

For Exercises 1–4,
$$S = \left\{ -\tfrac{9}{4}, -2, -\sqrt{2}, 0, 0.6, \sqrt{11}, \sqrt{-8}, 6, \tfrac{30}{3} \right\}.$$

1. The integers are -2, 0, 6, and $\frac{30}{3}$ (or 10).

2. The rational numbers are $-\frac{9}{4}$, -2, 0, 0.6, 6, and $\frac{30}{3}$ (or 10). Each can be expressed as a quotient of two integers.

3. The irrational numbers are $-\sqrt{2}$ and $\sqrt{11}$.

4. All are real numbers except $\sqrt{-8}$.

5. $|-8| + 6 - |-2| - (-6 + 2)$
$$= 8 + 6 - 2 - (-4)$$
$$= 14 - 2 + 4 = 16$$

6. $-12 - |-3| - 7 - |-5|$
$$= -12 - 3 - 7 - 5 = -27$$

7. $2(-5) + (-8)(4) - (-3)$
$$= -10 - 32 + 3 = -39$$

8. $7 - (3 + 4a) + 2a = -5(a - 1) - 3$
$$7 - 3 - 4a + 2a = -5a + 5 - 3$$
$$4 - 2a = -5a + 2$$
$$3a = -2$$
$$a = -\tfrac{2}{3}$$

The solution set is $\left\{ -\frac{2}{3} \right\}$.

9. $2m + 2 \leq 5m - 1$
$$-3m \leq -3$$
Divide by -3; reverse the inequality.
$$m \geq 1$$

The solution set is $[1, \infty)$.

10. $|2x - 5| = 9$

$2x - 5 = 9 \quad$ or $\quad 2x - 5 = -9$
$$2x = 14 \qquad\qquad 2x = -4$$
$$x = 7 \quad \text{or} \qquad x = -2$$

The solution set is $\{-2, 7\}$.

11. $|3p| - 4 = 12$
$$|3p| = 16$$

$3p = 16 \quad$ or $\quad 3p = -16$
$$p = \tfrac{16}{3} \quad \text{or} \quad p = -\tfrac{16}{3}$$

The solution set is $\left\{ \pm \frac{16}{3} \right\}$.

12. $|3k - 8| \leq 1$
$$-1 \leq 3k - 8 \leq 1$$
$$7 \leq 3k \leq 9$$
$$\tfrac{7}{3} \leq k \leq 3$$

The solution set is $\left[\frac{7}{3}, 3 \right]$.

13. $|4m + 2| > 10$

$$4m + 2 > 10 \quad \text{or} \quad 4m + 2 < -10$$
$$4m > 8 \qquad\qquad 4m < -12$$
$$m > 2 \quad \text{or} \qquad m < -3$$

The solution set is $(-\infty, -3) \cup (2, \infty)$.

14. $5x + 2y = 10$

Find the x- and y-intercepts. To find the x-intercept, let $y = 0$.

$$5x + 2(0) = 10$$
$$5x = 10$$
$$x = 2$$

The x-intercept is $(2, 0)$.
To find the y-intercept, let $x = 0$.

$$5(0) + 2y = 10$$
$$2y = 10$$
$$y = 5$$

The y-intercept is $(0, 5)$.
Plot the intercepts and draw the line through them.

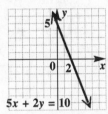

15. $-4x + y \leq 5$

Graph the line $-4x + y = 5$, which has intercepts $(0, 5)$ and $\left(-\frac{5}{4}, 0\right)$, as a solid line because the inequality involves $\leq$. Test $(0, 0)$, which yields $0 \leq 5$, a true statement. Shade the region that includes $(0, 0)$.

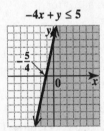

16. **(a)** Yes, this graph is the graph of a function because it passes the vertical line test.

(b) $(x_1, y_1) = (2000, 61{,}327)$ and $(x_2, y_2) = (2003, 56{,}250)$.

$$m = \frac{y_2 - y_1}{x_2 - x_1} = \frac{56{,}250 - 61{,}327}{2003 - 2000}$$
$$= \frac{-5077}{3} = -1692.\overline{3}$$

The slope of the line in the graph is about -1692 and can be interpreted as follows: The number of

U.S. travelers to international countries decreased by approximately 1692 thousand per year during 2000-2003.

17. Through $(5, -1)$; parallel to $3x - 4y = 12$
Find the slope of

$$3x - 4y = 12$$
$$-4y = -3x + 12$$
$$y = \tfrac{3}{4}x - 3.$$

The slope is $\frac{3}{4}$, so a line parallel to it also has slope $\frac{3}{4}$. Let $m = \frac{3}{4}$ and $(x_1, y_1) = (5, -1)$ in the point-slope form.

$$y - y_1 = m(x - x_1)$$
$$y - (-1) = \tfrac{3}{4}(x - 5)$$
$$y + 1 = \tfrac{3}{4}x - \tfrac{15}{4}$$
$$y = \tfrac{3}{4}x - \tfrac{19}{4}$$

18. $5x - 3y = 14 \quad (1)$
$2x + 5y = 18 \quad (2)$

Multiply equation (1) by 5 and equation (2) by 3. Then add the results.

$$\begin{array}{rcll} 25x - 15y &=& 70 & 5 \times (1) \\ 6x + 15y &=& 54 & 3 \times (2) \\ \hline 31x &=& 124 & \textit{Add} \\ x &=& 4 & \end{array}$$

Substitute 4 for x in equation (1) to find y.

$$5x - 3y = 14 \quad (1)$$
$$5(4) - 3y = 14$$
$$20 - 3y = 14$$
$$-3y = -6$$
$$y = 2$$

The solution set is $\{(4, 2)\}$.

19. $2x - 7y = 8 \quad (1)$
$4x - 14y = 3 \quad (2)$

Multiply equation (1) by -2 and add the result to equation (2).

$$\begin{array}{rcll} -4x + 14y &=& -16 & -2 \times (1) \\ 4x - 14y &=& 3 & (2) \\ \hline 0 &=& -13 & \textit{Add} \end{array}$$

The statement $0 = -13$ is false, so the solution set is $\emptyset$.

20.
$$\begin{aligned} x + 2y + 3z &= 11 \quad (1) \\ 3x - y + z &= 8 \quad (2) \\ 2x + 2y - 3z &= -12 \quad (3) \end{aligned}$$

To eliminate z, add equations (1) and (3).

$$\begin{array}{rrrrl} x &+ 2y &+ 3z &= &11 \quad (1) \\ 2x &+ 2y &- 3z &= &-12 \quad (3) \\ \hline 3x &+ 4y & &= &-1 \quad (4) \end{array}$$

To eliminate z again, multiply equation (2) by 3 and add the result to equation (3).

$$\begin{array}{rrrrll} 9x &- 3y &+ 3z &= &24 & 3 \times (2) \\ 2x &+ 2y &- 3z &= &-12 & (3) \\ \hline 11x &- y & &= &12 & (5) \end{array}$$

Multiply equation (5) by 4 and add the result to equation (4).

$$\begin{array}{rrrll} 44x &- 4y &= &48 & 4 \times (5) \\ 3x &+ 4y &= &-1 & (4) \\ \hline 47x & &= &47 & \\ & x &= &1 & \end{array}$$

Substitute 1 for x in equation (5) to find y.

$$\begin{aligned} 11x - y &= 12 \quad (5) \\ 11(1) - y &= 12 \\ 11 - y &= 12 \\ -y &= 1 \\ y &= -1 \end{aligned}$$

Substitute 1 for x and -1 for y in equation (2) to find z.

$$\begin{aligned} 3x - y + z &= 8 \quad (2) \\ 3(1) - (-1) + z &= 8 \\ 3 + 1 + z &= 8 \\ 4 + z &= 8 \\ z &= 4 \end{aligned}$$

The solution set is $\{(1, -1, 4)\}$.

21. Let $x =$ the amount of candy at \$1.00 per pound.

	Number of Pounds	Price per Pound	Value
First Candy	x	\$1.00	$1.00x$
Second Candy	10	\$1.96	$1.96(10)$
Mixture	$x + 10$	\$1.60	$1.60(x + 10)$

The sum of the values of each candy must equal the value of the mixture.

$$1.00x + 1.96(10) = 1.60(x + 10)$$

Multiply by 10 to clear the decimals.

$$\begin{aligned} 10x + 196 &= 16(x + 10) \\ 10x + 196 &= 16x + 160 \\ 36 &= 6x \\ 6 &= x \end{aligned}$$

Use 6 pounds of the \$1.00 candy.

22.
$$\begin{aligned} (2p + 3)(3p - 1) &= 6p^2 - 2p + 9p - 3 \\ &= 6p^2 + 7p - 3 \end{aligned}$$

23.
$$\begin{aligned} (4k - 3)^2 &= (4k)^2 - 2(4k)(3) + 3^2 \\ &= 16k^2 - 24k + 9 \end{aligned}$$

24.
$$\begin{aligned} \left(3m^3 + 2m^2 - 5m\right) - \left(8m^3 + 2m - 4\right) \\ = 3m^3 + 2m^2 - 5m - 8m^3 - 2m + 4 \\ = 3m^3 - 8m^3 + 2m^2 - 5m - 2m + 4 \\ = -5m^3 + 2m^2 - 7m + 4 \end{aligned}$$

25.

$$\begin{array}{r} 2t^3 + 5t^2 - 3t + 4 \\ 3t + 1 \overline{\smash{)}\, 6t^4 + 17t^3 - 4t^2 + 9t + 4} \\ \underline{6t^4 + 2t^3} \\ 15t^3 - 4t^2 \\ \underline{15t^3 + 5t^2} \\ -9t^2 + 9t \\ \underline{-9t^2 - 3t} \\ 12t + 4 \\ \underline{12t + 4} \\ 0 \end{array}$$

The quotient is $2t^3 + 5t^2 - 3t + 4$.

26. $8x + x^3$
Factor out the GCF, x.

$$8x + x^3 = x(8 + x^2)$$

27. $24y^2 - 7y - 6$
Factor by trial and error.

$$24y^2 - 7y - 6 = (8y + 3)(3y - 2)$$

28. $5z^3 - 19z^2 - 4z$
Factor out the GCF, z, and then factor by trial and error.

$$\begin{aligned} 5z^3 - 19z^2 - 4z &= z\left(5z^2 - 19z - 4\right) \\ &= z(5z + 1)(z - 4) \end{aligned}$$

29. $16a^2 - 25b^4$

Use the difference of squares formula,

$$x^2 - y^2 = (x + y)(x - y),$$

where $x = 4a$ and $y = 5b^2$.

$$16a^2 - 25b^4 = (4a + 5b^2)(4a - 5b^2)$$

30. $8c^3 + d^3$

Use the sum of cubes formula,

$$x^3 + y^3 = (x + y)(x^2 - xy + y^2),$$

where $x = 2c$ and $y = d$.

$$8c^3 + d^3 = (2c + d)(4c^2 - 2cd + d^2)$$

31. $16r^2 + 56rq + 49q^2$
$$= (4r)^2 + 2(4r)(7q) + (7q)^2$$

Use the perfect square formula,

$$x^2 + 2xy + y^2 = (x + y)^2,$$

where $x = 4r$ and $y = 7q$.

$$16r^2 + 56rq + 49q^2 = (4r + 7q)^2$$

32. $\dfrac{(5p^3)^4 (-3p^7)}{2p^2(4p^4)} = \dfrac{(5^4p^{12})(-3p^7)}{8p^6}$

$$= \dfrac{(625)(-3)p^{19}}{8p^6}$$

$$= -\dfrac{1875p^{13}}{8}$$

33. $\dfrac{x^2 - 9}{x^2 + 7x + 12} \div \dfrac{x - 3}{x + 5}$

Multiply by the reciprocal.

$$= \dfrac{x^2 - 9}{x^2 + 7x + 12} \cdot \dfrac{x + 5}{x - 3}$$

$$= \dfrac{(x + 3)(x - 3)}{(x + 3)(x + 4)} \cdot \dfrac{(x + 5)}{(x - 3)} \quad \text{\textit{Factor}}$$

$$= \dfrac{x + 5}{x + 4}$$

34. $\dfrac{2}{k + 3} - \dfrac{5}{k - 2}$

The LCD is $(k + 3)(k - 2)$.

$$= \dfrac{2(k - 2)}{(k + 3)(k - 2)} - \dfrac{5(k + 3)}{(k - 2)(k + 3)}$$

$$= \dfrac{2k - 4 - 5k - 15}{(k + 3)(k - 2)}$$

$$= \dfrac{-3k - 19}{(k + 3)(k - 2)}$$

35. $\dfrac{3}{p^2 - 4p} - \dfrac{4}{p^2 + 2p}$

$$= \dfrac{3}{p(p - 4)} - \dfrac{4}{p(p + 2)}$$

The LCD is $p(p - 4)(p + 2)$.

$$= \dfrac{3(p + 2)}{p(p - 4)(p + 2)} - \dfrac{4(p - 4)}{p(p + 2)(p - 4)}$$

$$= \dfrac{3p + 6 - 4p + 16}{p(p - 4)(p + 2)}$$

$$= \dfrac{22 - p}{p(p - 4)(p + 2)}$$

36. $\sqrt{288} = \sqrt{144 \cdot 2} = \sqrt{144}\sqrt{2} = 12\sqrt{2}$

37. $2\sqrt{32} - 5\sqrt{98} = 2\sqrt{16 \cdot 2} - 5\sqrt{49 \cdot 2}$

$$= 2 \cdot 4\sqrt{2} - 5 \cdot 7\sqrt{2}$$

$$= 8\sqrt{2} - 35\sqrt{2}$$

$$= -27\sqrt{2}$$

38. $\sqrt{2x + 1} - \sqrt{x} = 1$

$$\sqrt{2x + 1} = 1 + \sqrt{x}$$

$$\left(\sqrt{2x + 1}\right)^2 = \left(1 + \sqrt{x}\right)^2$$

$$2x + 1 = 1 + 2\sqrt{x} + x$$

$$x = 2\sqrt{x}$$

$$(x)^2 = \left(2\sqrt{x}\right)^2$$

$$x^2 = 4x$$

$$x^2 - 4x = 0$$

$$x(x - 4) = 0$$

$$x = 0 \quad \text{or} \quad x = 4$$

Check $x = 0$: $\quad \sqrt{1} - \sqrt{0} = 1 \quad$ *True*

Check $x = 4$: $\quad \sqrt{9} - \sqrt{4} = 1 \quad$ *True*

The solution set is $\{0, 4\}$.

39. $(5 + 4i)(5 - 4i) = 5^2 - (4i)^2$

$$= 25 - 16i^2$$

$$= 25 - 16(-1)$$

$$= 25 + 16 = 41$$

40. $3x^2 - x - 1 = 0$

Here $a = 3$, $b = -1$, and $c = -1$.

Use the quadratic formula.

$$x = \dfrac{-b \pm \sqrt{b^2 - 4ac}}{2a}$$

$$x = \dfrac{-(-1) \pm \sqrt{(-1)^2 - 4(3)(-1)}}{2(3)}$$

$$= \dfrac{1 \pm \sqrt{1 + 12}}{6} = \dfrac{1 \pm \sqrt{13}}{6}$$

The solution set is $\left\{ \dfrac{1 + \sqrt{13}}{6}, \dfrac{1 - \sqrt{13}}{6} \right\}$.

41. $k^2 + 2k - 8 > 0$

Solve the equation

$$k^2 + 2k - 8 = 0.$$

$$(k + 4)(k - 2) = 0$$

$$k + 4 = 0 \quad \text{or} \quad k - 2 = 0$$

$$k = -4 \quad \text{or} \quad k = 2$$

The numbers -4 and 2 divide a number line into three intervals.

continued

A   B   C

$$-4 \qquad 2$$

Test a number from each interval in the inequality

$$k^2 + 2k - 8 > 0.$$

Interval A: Let $k = -5$.
$$25 - 10 - 8 > 0 \qquad ?$$
$$7 > 0 \qquad \textit{True}$$

Interval B: Let $k = 0$.
$$-8 > 0 \qquad \textit{False}$$

Interval C: Let $k = 3$.
$$9 + 6 - 8 > 0 \qquad ?$$
$$7 > 0 \qquad \textit{True}$$

The numbers in Intervals A and C, not including -4 or 2 because of $>$, are solutions. The solution set is $(-\infty, -4) \cup (2, \infty)$.

42. $x^4 - 5x^2 + 4 = 0$

Let $u = x^2$, so $u^2 = \left(x^2\right)^2 = x^4$.

$$u^2 - 5u + 4 = 0$$
$$(u - 1)(u - 4) = 0$$
$$u - 1 = 0 \quad \text{or} \quad u - 4 = 0$$
$$u = 1 \quad \text{or} \qquad u = 4$$

To find x, substitute x^2 for u.

$$x^2 = 1 \qquad \text{or} \quad x^2 = 4$$
$$x = \pm 1 \quad \text{or} \qquad x = \pm 2$$

The solution set is $\{\pm 1, \pm 2\}$.

43.   Let $x =$ one of the numbers.
Then $300 - x =$ the other number.

The product of the two numbers is given by

$$P = x(300 - x).$$

Writing this equation in standard form gives us

$$P = -x^2 + 300x.$$

Finding the maximum of the product is the same as finding the vertex of the graph of P. The x-value of the vertex is

$$x = -\frac{b}{2a} = -\frac{300}{2(-1)} = 150.$$

If x is 150, then $300 - x$ must also be 150. The two numbers are 150 and 150 and the product is $150 \cdot 150 = 22{,}500$.

44. $f(x) = \frac{1}{3}(x - 1)^2 + 2$ is in $f(x) = a(x - h)^2 + k$ form. The graph is a vertical parabola with vertex (h, k) at $(1, 2)$. Since $a = \frac{1}{3} > 0$, the graph opens up. Also, $|a| = \left|\frac{1}{3}\right| = \frac{1}{3} < 1$, so the graph is wider than the graph of $f(x) = x^2$. The points $\left(0, 2\frac{1}{3}\right)$, $(-2, 5)$, and $(4, 5)$ are also on the graph.

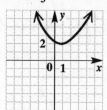

$$f(x) = \tfrac{1}{3}(x - 1)^2 + 2$$

45. $f(x) = 2^x$
Make a table of values.

x	-2	-1	0	1	2
$f(x)$	$\frac{1}{4}$	$\frac{1}{2}$	1	2	4

Plot the ordered pairs from the table, and draw a smooth exponential curve through the points.

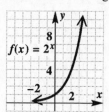

$$f(x) = 2^x$$

46.
$$5^{x+3} = \left(\tfrac{1}{25}\right)^{3x+2}$$
$$5^{x+3} = \left[\left(\tfrac{1}{5}\right)^2\right]^{3x+2}$$

Write each side to the power of 5.

$$5^{x+3} = \left(5^{-2}\right)^{(3x+2)}$$
$$5^{x+3} = 5^{-2(3x+2)}$$
$$x + 3 = -2(3x + 2)$$
$$x + 3 = -6x - 4$$
$$7x = -7$$
$$x = -1$$

Check $x = -1$: $5^2 = \left(\tfrac{1}{25}\right)^{-1}$ *True*
The solution set is $\{-1\}$.

47. $f(x) = \log_3 x$

Make a table of values.

Powers of 3	3^{-2}	3^{-1}	3^0	3^1	3^2
x	$\frac{1}{9}$	$\frac{1}{3}$	1	3	9
y	-2	-1	0	1	2

Plot the ordered pairs and draw a smooth logarithmic curve through the points.

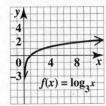

48. $\log_2 81 = \log_2 9^2 = 2\log_2 9$

$\approx 2(3.1699) = 6.3398$

49. $\log \dfrac{x^3 \sqrt{y}}{z}$

$= \log \dfrac{x^3 y^{1/2}}{z}$

$= \log\left(x^3 y^{1/2}\right) - \log z \qquad$ *Quotient rule*

$= \log x^3 + \log y^{1/2} - \log z \quad$ *Product rule*

$= 3\log x + \frac{1}{2}\log y - \log z \quad$ *Power rule*

50. $B(t) = 25{,}000 e^{0.2t}$

(a) At noon, $t = 0$.

$B(0) = 25{,}000 e^{0.2(0)}$

$= 25{,}000 e^0 = 25{,}000(1) = 25{,}000$

25,000 bacteria are present at noon.

(b) At 1 P.M., $t = 1$.

$B(1) = 25{,}000 e^{0.2(1)} \approx 30{,}535$

About 30,500 bacteria are present at 1 P.M.

(c) At 2 P.M., $t = 2$.

$B(2) = 25{,}000 e^{0.2(2)} \approx 37{,}296$

About 37,300 bacteria are present at 2 P.M.

(d) The population doubles when $B(t) = 50{,}000$.

$50{,}000 = 25{,}000 e^{0.2t}$

$2 = e^{0.2t}$

$\ln 2 = \ln e^{0.2t}$

$\ln 2 = 0.2t \ln e$

$t = \dfrac{\ln 2}{0.2} \approx 3.5$

The population will double in about 3.5 hours, or at about 3:30 P.M.

CHAPTER 11 NONLINEAR FUNCTIONS, CONIC SECTIONS, AND NONLINEAR SYSTEMS

11.1 Additional Graphs of Functions

1. For the reciprocal function defined by $f(x) = 1/x$, <u>0</u> is the only real number not in the domain since division by 0 is undefined.

3. The lowest point on the graph of $f(x) = |x|$ has coordinates (<u>0</u> , <u>0</u>).

5. $f(x) = |x - 2| + 2$

The graph of this function has its "vertex" at $(2, 2)$, so the correct graph is **B**.

7. $f(x) = |x - 2| - 2$

The graph of this function has its "vertex" at $(2, -2)$, so the correct graph is **A**.

9. $f(x) = |x + 1|$

Since x can be any real number, the domain is $(-\infty, \infty)$.

The value of y is always greater than or equal to 0, so the range is $[0, \infty)$.

The graph of $y = |x + 1|$ looks like the graph of the absolute value function $y = |x|$, but the graph is translated 1 unit to the left. The x-value of its "vertex" is obtained by setting $x + 1 = 0$ and solving for x:

$$x + 1 = 0$$
$$x = -1.$$

Since the corresponding y-value is 0, the "vertex" is $(-1, 0)$. Some additional points are $(-3, 2)$, $(-2, 1)$, $(0, 1)$, and $(1, 2)$.

11. $f(x) = \dfrac{1}{x} + 1$

The graph of this function is similar to the graph of $g(x) = \dfrac{1}{x}$, except that each point is translated 1 unit upward. Just as with $g(x) = \dfrac{1}{x}$, $x = 0$ is the vertical asymptote, but this graph has $y = 1$ as its horizontal asymptote.

The domain is all real numbers except 0, that is, $(-\infty, 0) \cup (0, \infty)$.

The range is all real numbers except 1, that is, $(-\infty, 1) \cup (1, \infty)$.

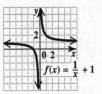

13. $f(x) = \sqrt{x - 2}$

The graph is found by shifting the graph of $y = \sqrt{x}$ two units to the right. The following table of ordered pairs gives some specific points the graph passes through.

x	2	3	6
y	0	1	2

The domain of the function is $[2, \infty)$ and its range is $[0, \infty)$.

15. $f(x) = \dfrac{1}{x - 2}$

This is the graph of the reciprocal function, $g(x) = \dfrac{1}{x}$, shifted 2 units to the right. Since $x \neq 2$ (or a denominator of 0 results), the line $x = 2$ is a vertical asymptote. Since $\dfrac{1}{x - 2} \neq 0$, the line $y = 0$ is a horizontal asymptote.

x	-1	0	1	$\frac{3}{2}$
y	$-\frac{1}{3}$	$-\frac{1}{2}$	-1	-2

x	$\frac{5}{2}$	3	4	5
y	2	1	$\frac{1}{2}$	$\frac{1}{3}$

The domain of the function is $(-\infty, 2) \cup (2, \infty)$ and its range is $(-\infty, 0) \cup (0, \infty)$.

17. $f(x) = \sqrt{x + 3} - 3$

The graph is found by shifting the graph of $y = \sqrt{x}$ three units to the left and three units down. The following table of ordered pairs gives some specific points the graph passes through.

x	-3	-2	1
y	-3	-2	-1

$f(x) = \sqrt{x + 3} - 3$

The domain of the function is $[-3, \infty)$ and its range is $[-3, \infty)$.

In Exercises 18–25, the answer is the greatest integer that is less than or equal to the number in the greatest integer symbol.

19. $[\![3]\!] = 3 \quad (3 \le 3)$

21. $[\![\frac{1}{2}]\!] = 0 \quad (0 \le \frac{1}{2})$

23. $[\![-14]\!] = -14 \quad (-14 \le -14)$

25. $[\![-10.1]\!] = -11 \quad (-11 \le -10.1)$

27. $f(x) = [\![x - 3]\!]$

This is the graph of the greatest integer function, $g(x) = [\![x]\!]$, shifted 3 units to the right.

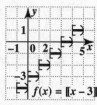

$f(x) = [\![x - 3]\!]$

29. For any portion of the first ounce, the cost will be one 39¢ stamp. If the weight exceeds one ounce (up to two ounces), an additional 24¢ stamp is required. The following table summarizes the weight of a letter, x, and the number of stamps required, $p(x)$, on the interval $(0, 5]$.

x	$(0, 1]$	$(1, 2]$	$(2, 3]$	$(3, 4]$	$(4, 5]$
$p(x)$	1	2	3	4	5

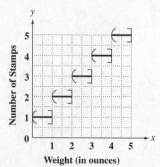

In Exercises 31–34, use the distance formula

$$d = \sqrt{(x_2 - x_1)^2 + (y_2 - y_1)^2}.$$

31. $(2, -1)$ and $(4, 3)$

$$\begin{aligned} d &= \sqrt{(4 - 2)^2 + [3 - (-1)]^2} \\ &= \sqrt{2^2 + 4^2} = \sqrt{4 + 16} \\ &= \sqrt{20} = \sqrt{4} \cdot \sqrt{5} = 2\sqrt{5} \end{aligned}$$

33. (x, y) and $(-2, 5)$

$$\begin{aligned} d &= \sqrt{[x - (-2)]^2 + (y - 5)^2} \\ &= \sqrt{(x + 2)^2 + (y - 5)^2} \end{aligned}$$

11.2 The Circle and the Ellipse

1. **(a)** $x^2 + y^2 = 25$ can be written in the center-radius form as

$$(x - 0)^2 + (y - 0)^2 = 5^2.$$

The center is the point $(0, 0)$.

(b) The radius is 5.

(c) The x-intercepts are $(5, 0)$ and $(-5, 0)$. The y-intercepts are $(0, 5)$ and $(0, -5)$.

$x^2 + y^2 = 25$

3. $(x - 3)^2 + (y - 2)^2 = 25$ is an equation of a circle with center $(3, 2)$ and radius 5, choice **B**.

5. $(x + 3)^2 + (y - 2)^2 = 25$ is an equation of a circle with center $(-3, 2)$ and radius 5, choice **D**.

7. Center: $(-4, 3)$; radius: 2

Substitute $h = -4$, $k = 3$, and $r = 2$ in the center-radius form of the equation of a circle.

$$\begin{aligned} (x - h)^2 + (y - k)^2 &= r^2 \\ [x - (-4)]^2 + (y - 3)^2 &= 2^2 \\ (x + 4)^2 + (y - 3)^2 &= 4 \end{aligned}$$

9. Center: $(-8, -5)$; radius: $\sqrt{5}$

Substitute $h = -8$, $k = -5$, and $r = \sqrt{5}$ in the center-radius form of the equation of a circle.

$$\begin{aligned} (x - h)^2 + (y - k)^2 &= r^2 \\ [x - (-8)]^2 + [y - (-5)]^2 &= \left(\sqrt{5}\right)^2 \\ (x + 8)^2 + (y + 5)^2 &= 5 \end{aligned}$$

11. $\qquad x^2 + y^2 + 4x + 6y + 9 = 0$

Rewrite the equation keeping only the variable terms on the left and grouping the x-terms and y-terms.

$$x^2 + 4x + y^2 + 6y = -9$$

Complete both squares on the left, and add the same constants to the right.

$$\left(x^2 + 4x + \underline{4}\right) + \left(y^2 + 6y + \underline{9}\right) = -9 + \underline{4} + \underline{9}$$
$$(x + 2)^2 + (y + 3)^2 = 4$$

From the form $(x - h)^2 + (y - k)^2 = r^2$, we have $h = -2$, $k = -3$, and $r = 2$. The center is $(-2, -3)$, and the radius r is 2.

13. $\quad x^2 + y^2 + 10x - 14y - 7 = 0$

$$\left(x^2 + 10x \quad\right) + \left(y^2 - 14y \quad\right) = 7$$
$$\left(x^2 + 10x + \underline{25}\right) + \left(y^2 - 14y + \underline{49}\right)$$
$$= 7 + \underline{25} + \underline{49}$$
$$(x + 5)^2 + (y - 7)^2 = 81$$

The center is $(-5, 7)$, and the radius is $\sqrt{81} = 9$.

15. $\qquad 3x^2 + 3y^2 - 12x - 24y + 12 = 0$

$$3\left(x^2 - 4x \quad\right) + 3\left(y^2 - 8y \quad\right) = -12$$

Divide by 3.

$$\left(x^2 - 4x \quad\right) + \left(y^2 - 8y \quad\right) = -4$$
$$\left(x^2 - 4x + \underline{4}\right) + \left(y^2 - 8y + \underline{16}\right)$$
$$= -4 + \underline{4} + \underline{16}$$
$$(x - 2)^2 + (y - 4)^2 = 16$$

The center is $(2, 4)$, and the radius is $\sqrt{16} = 4$.

17. $\qquad\qquad x^2 + y^2 = 9$

$$(x - 0)^2 + (y - 0)^2 = 3^2$$

Here, $h = 0$, $k = 0$, and $r = 3$, so the graph is a circle with center $(0, 0)$ and radius 3.

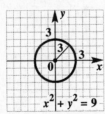

$$x^2 + y^2 = 9$$

19. $\qquad\qquad 2y^2 = 10 - 2x^2$

$$2x^2 + 2y^2 = 10$$
$$x^2 + y^2 = 5 \qquad \textit{Divide by 2.}$$
$$(x - 0)^2 + (y - 0)^2 = \left(\sqrt{5}\right)^2$$

Here, $h = 0$, $k = 0$, and $r = \sqrt{5} \approx 2.2$, so the graph is a circle with center $(0, 0)$ and radius $\sqrt{5}$.

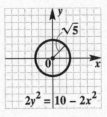

$$2y^2 = 10 - 2x^2$$

21. $(x + 3)^2 + (y - 2)^2 = 9$

Here, $h = -3$, $k = 2$, and $r = \sqrt{9} = 3$. The graph is a circle with center $(-3, 2)$ and radius 3.

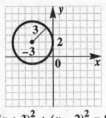

$$(x + 3)^2 + (y - 2)^2 = 9$$

23. $\qquad\quad x^2 + y^2 - 4x - 6y + 9 = 0$

$$\left(x^2 - 4x \quad\right) + \left(y^2 - 6y \quad\right) = -9$$

Complete the square for each variable.

$$\left(x^2 - 4x + \underline{4}\right) + \left(y^2 - 6y + \underline{9}\right)$$
$$= -9 + \underline{4} + \underline{9}$$
$$(x - 2)^2 + (y - 3)^2 = 4$$

Here, $h = 2$, $k = 3$, and $r = \sqrt{4} = 2$. The graph is a circle with center $(2, 3)$ and radius 2.

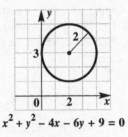

$$x^2 + y^2 - 4x - 6y + 9 = 0$$

25. $\quad x^2 + y^2 + 6x - 6y + 9 = 0$

$$\left(x^2 + 6x \quad\right) + \left(y^2 - 6y \quad\right) = -9$$
$$\left(x^2 + 6x + \underline{9}\right) + \left(y^2 - 6y + \underline{9}\right) = -9 + \underline{9} + \underline{9}$$
$$(x + 3)^2 + (y - 3)^2 = 9$$

The center is $(-3, 3)$, and the radius is $\sqrt{9} = 3$.

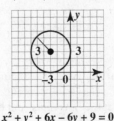

$$x^2 + y^2 + 6x - 6y + 9 = 0$$

27. This method works because the pencil is always the same distance from the fastened end. The fastened end works as the center, and the length of the string from the fastened end to the pencil is the radius.

29. The equation $\dfrac{x^2}{9} + \dfrac{y^2}{25} = 1$ is of the form

$\dfrac{x^2}{a^2} + \dfrac{y^2}{b^2} = 1$. The graph is an ellipse with

$a^2 = 9$ and $b^2 = 25$, so $a = 3$ and $b = 5$. The x-intercepts are $(3, 0)$ and $(-3, 0)$. The y-intercepts are $(0, 5)$ and $(0, -5)$. Plot the intercepts, and draw the ellipse through them.

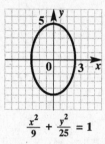

$$\dfrac{x^2}{9} + \dfrac{y^2}{25} = 1$$

31. $\dfrac{x^2}{36} = 1 - \dfrac{y^2}{16}$

$\dfrac{x^2}{36} + \dfrac{y^2}{16} = 1$ is in the form $\dfrac{x^2}{a^2} + \dfrac{y^2}{b^2} = 1$.

The graph is an ellipse with $a^2 = 36$ and $b^2 = 16$, so $a = 6$ and $b = 4$. The x-intercepts are $(6, 0)$ and $(-6, 0)$. The y-intercepts are $(0, 4)$ and $(0, -4)$. Plot the intercepts, and draw the ellipse through them.

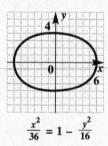

$$\dfrac{x^2}{36} = 1 - \dfrac{y^2}{16}$$

33. $\dfrac{y^2}{25} = 1 - \dfrac{x^2}{49}$

$\dfrac{x^2}{49} + \dfrac{y^2}{25} = 1$ is in the form $\dfrac{x^2}{a^2} + \dfrac{y^2}{b^2} = 1$.

The graph is an ellipse with $a^2 = 49$ and $b^2 = 25$, so $a = 7$ and $b = 5$. The x-intercepts are $(7, 0)$ and $(-7, 0)$. The y-intercepts are $(0, 5)$ and $(0, -5)$. Plot the intercepts, and draw the ellipse through them.

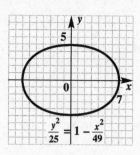

$$\dfrac{y^2}{25} = 1 - \dfrac{x^2}{49}$$

35. $\dfrac{x^2}{16} + \dfrac{y^2}{4} = 1$ is in the form

$$\dfrac{x^2}{a^2} + \dfrac{y^2}{b^2} = 1,$$

so $a = 4$ and $b = 2$. Its x-intercepts are $(4, 0)$ and $(-4, 0)$, and its y-intercepts are $(0, 2)$ and $(0, -2)$. Plot the intercepts, and draw the ellipse through them.

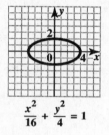

$$\dfrac{x^2}{16} + \dfrac{y^2}{4} = 1$$

37. $\dfrac{(x+1)^2}{64} + \dfrac{(y-2)^2}{49} = 1$

This equation is of the form

$$\dfrac{(x-h)^2}{a^2} + \dfrac{(y-k)^2}{b^2} = 1,$$

so the center of the ellipse is at $(-1, 2)$. Since $a^2 = 64$, $a = 8$. Since $b^2 = 49$, $b = 7$. Add ± 8 to -1, and add ± 7 to 2 to find the points $(7, 2)$, $(-9, 2)$, $(-1, 9)$, and $(-1, -5)$. Plot the points, and draw the ellipse through them.

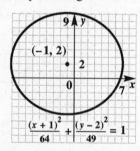

$$\dfrac{(x+1)^2}{64} + \dfrac{(y-2)^2}{49} = 1$$

39. $\dfrac{(x-2)^2}{16} + \dfrac{(y-1)^2}{9} = 1$

The center of the ellipse is at $(2, 1)$. Since $a^2 = 16$, $a = 4$. Since $b^2 = 9$, $b = 3$. Add ± 4 to 2, and add ± 3 to 1 to find the points $(6, 1)$, $(-2, 1)$, $(2, 4)$, and $(2, -2)$. Plot the points, and draw the ellipse through them.

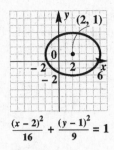

$$\frac{(x-2)^2}{16} + \frac{(y-1)^2}{9} = 1$$

41. By the vertical line test the set is not a function, because a vertical line may intersect the graph of an ellipse in two points.

43.
$$(x+2)^2 + (y-4)^2 = 16$$
$$(y-4)^2 = 16 - (x+2)^2$$

Take the square root of each side.

$$y - 4 = \pm\sqrt{16 - (x+2)^2}$$
$$y = 4 \pm \sqrt{16 - (x+2)^2}$$

Therefore, the two functions used to obtain the graph were

$$y_1 = 4 + \sqrt{16 - (x+2)^2} \quad \text{and}$$
$$y_2 = 4 - \sqrt{16 - (x+2)^2}.$$

45.
$$x^2 + y^2 = 36$$
$$y^2 = 36 - x^2$$

Take the square root of both sides.

$$y = \pm\sqrt{36 - x^2}$$

Therefore,

$$y_1 = \sqrt{36 - x^2} \quad \text{and} \quad y_2 = -\sqrt{36 - x^2}.$$

Use these two functions to obtain the graph.

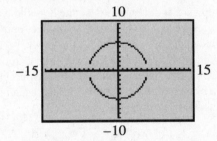

47.
$$\frac{x^2}{16} + \frac{y^2}{4} = 1$$
$$\frac{y^2}{4} = 1 - \frac{x^2}{16}$$
$$y^2 = 4\left(1 - \frac{x^2}{16}\right)$$

Take the square root of both sides.

$$y = \pm 2\sqrt{1 - \frac{x^2}{16}}$$

Therefore,

$$y_1 = 2\sqrt{1 - \frac{x^2}{16}} \quad \text{and} \quad y_2 = -2\sqrt{1 - \frac{x^2}{16}}.$$

Use these two functions to obtain the graph.

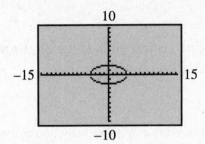

49.
$$\frac{x^2}{36} + \frac{y^2}{9} = 1$$
$$c^2 = a^2 - b^2 = 36 - 9 = 27, \quad \text{so}$$
$$c = \sqrt{27} = 3\sqrt{3}.$$

The kidney stone and the source of the beam must be placed $3\sqrt{3}$ units from the center of the ellipse.

51. **(a)** $100x^2 + 324y^2 = 32{,}400$

$$\frac{x^2}{324} + \frac{y^2}{100} = 1 \qquad \textit{Divide by 32,400.}$$
$$\frac{x^2}{18^2} + \frac{y^2}{10^2} = 1$$

The height in the center is the y-coordinate of the positive y-intercept. The height is 10 meters.

(b) The width of the ellipse is the distance between the x-intercepts, $(-18, 0)$ and $(18, 0)$. The width across the bottom of the arch is $18 + 18 = 36$ meters.

53.
$$\frac{x^2}{141.7^2} + \frac{y^2}{141.1^2} = 1$$

(a) $c^2 = a^2 - b^2$, so

$$c = \sqrt{a^2 - b^2} = \sqrt{141.7^2 - 141.1^2}$$
$$= \sqrt{169.68} \approx 13.0$$

From the figure, the *apogee* is
$a + c = 141.7 + 13.0 = 154.7$ million miles.

(b) The *perigee* is $a - c = 141.7 - 13.0 = 128.7$ million miles.

55. Plot the points $(3, 4)$, $(-3, 4)$, $(3, -4)$, and $(-3, -4)$.

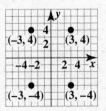

57. $4x + 3y = 12$

If $y = 0$, then $4x = 12$, so $x = 3$ and $(3, 0)$ is the
x-intercept.
If $x = 0$, then $3y = 12$, so $y = 4$ and $(0, 4)$ is the
y-intercept.

11.3 The Hyperbola and Functions Defined by Radicals

1. $\dfrac{x^2}{25} + \dfrac{y^2}{9} = 1$

This is the standard form for the equation of an
ellipse with x-intercepts $(5, 0)$ and $(-5, 0)$ and
y-intercepts $(0, 3)$ and $(0, -3)$. This is graph **C**.

3. $\dfrac{x^2}{9} - \dfrac{y^2}{25} = 1$

This is the standard form for the equation of a
hyperbola that opens left and right. Its
x-intercepts are $(3, 0)$ and $(-3, 0)$. This is graph
D.

5. If the equation of a hyperbola is in standard form
(that is, equal to one), the hyperbola would open
to the left and right if the x^2-term was positive. It
would open up and down if the y^2-term was
positive.

7. The equation $\dfrac{x^2}{16} - \dfrac{y^2}{9} = 1$ is in the form

$\dfrac{x^2}{a^2} - \dfrac{y^2}{b^2} = 1$. The graph is a hyperbola with
$a = 4$ and $b = 3$. The x-intercepts are $(4, 0)$ and
$(-4, 0)$. There are no y-intercepts. The vertices
of the fundamental rectangle are $(4, 3)$, $(4, -3)$,
$(-4, -3)$, and $(-4, 3)$. Extend the diagonals of
the rectangle through these points to get the
asymptotes. Graph a branch of the hyperbola
through each intercept and approaching the
asymptotes.

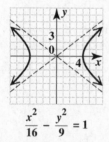

$$\frac{x^2}{16} - \frac{y^2}{9} = 1$$

9. $\dfrac{y^2}{4} - \dfrac{x^2}{25} = 1$ is in the form $\dfrac{y^2}{b^2} - \dfrac{x^2}{a^2} = 1$, where
$b = 2$ and $a = 5$. The branches open upward and
downward, so the hyperbola has y-intercepts $(0, 2)$
and $(0, -2)$. The fundamental rectangle has
vertices $(5, 2)$, $(5, -2)$, $(-5, -2)$, and $(-5, 2)$.

Sketch the extended diagonals, which are the
asymptotes of this hyperbola. Graph a branch of
the hyperbola through each intercept and
approaching the asymptotes.

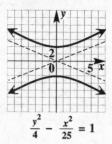

$$\frac{y^2}{4} - \frac{x^2}{25} = 1$$

11. $\dfrac{x^2}{25} - \dfrac{y^2}{36} = 1$ is a hyperbola with $a = 5$ and
$b = 6$. The x-intercepts are $(5, 0)$ and $(-5, 0)$.
There are no y-intercepts. To sketch the graph,
draw the extended diagonals of the fundamental
rectangle with vertices $(5, 6)$, $(5, -6)$, $(-5, -6)$
and $(-5, 6)$. Graph a branch of the hyperbola
through each intercept and approaching the
asymptotes.

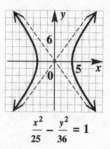

$$\frac{x^2}{25} - \frac{y^2}{36} = 1$$

13. $\dfrac{y^2}{16} - \dfrac{x^2}{16} = 1$ is in the form $\dfrac{y^2}{b^2} - \dfrac{x^2}{a^2} = 1$,
where $b = 4$ and $a = 4$. The branches open
upward and downward, so the hyperbola has
y-intercepts $(0, 4)$ and $(0, -4)$. The fundamental
rectangle has vertices $(4, 4)$, $(4, -4)$, $(-4, -4)$,
and $(-4, 4)$. Sketch the extended diagonals,
which are the asymptotes of this hyperbola. Graph
a branch of the hyperbola through each intercept
and approaching the asymptotes.

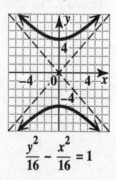

$$\frac{y^2}{16} - \frac{x^2}{16} = 1$$

15. $x^2 - y^2 = 16$

$\dfrac{x^2}{16} - \dfrac{y^2}{16} = 1$ *Divide by 16.*

This equation is in the form $\dfrac{x^2}{a^2} - \dfrac{y^2}{b^2} = 1$ with $a = 4$ and $b = 4$. The graph is a hyperbola with x-intercepts $(4, 0)$ and $(-4, 0)$ and no y-intercepts. One asymptote passes through $(4, 4)$ and $(-4, -4)$. The other asymptote passes through $(-4, 4)$ and $(4, -4)$. Sketch the graph through the intercepts and approaching the asymptotes.

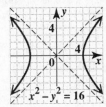

17. $4x^2 + y^2 = 16$

$\dfrac{x^2}{4} + \dfrac{y^2}{16} = 1$ *Divide by 16.*

This equation is in the form $\dfrac{x^2}{a^2} + \dfrac{y^2}{b^2} = 1$ with $a = 2$ and $b = 4$. The graph is an ellipse. The x-intercepts $(2, 0)$ and $(-2, 0)$. The y-intercepts are $(0, 4)$ and $(0, -4)$. Plot the intercepts and draw the ellipse through them.

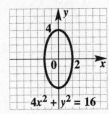

19. $x^2 - 2y = 0$

$x^2 = 2y$

$\frac{1}{2}x^2 = y$

The equation is in the form

$$f(x) = a(x - h)^2 + k,$$

with $a = \frac{1}{2}$, $h = 0$, and $k = 0$. The graph is a parabola that opens up and is wider than the graph of $f(x) = x^2$ because $|a| = \frac{1}{2} < 1$. The vertex is at $(0, 0)$.

x	1	2	3
y	$\frac{1}{2}$	2	4.5

Use symmetry about $x = 0$ to draw the graph.

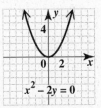

21. $9x^2 = 144 + 16y^2$

$9x^2 - 16y^2 = 144$

$\dfrac{x^2}{16} - \dfrac{y^2}{9} = 1$ *Divide by 144.*

The equation is a hyperbola in the form $\dfrac{x^2}{a^2} - \dfrac{y^2}{b^2} = 1$ with $a = 4$ and $b = 3$. The x-intercepts are $(4, 0)$ and $(-4, 0)$. There are no y-intercepts. To sketch the graph, draw the extended diagonals of the fundamental rectangle with vertices $(4, 3)$, $(4, -3)$, $(-4, -3)$ and $(-4, 3)$. These are the asymptotes. Graph a branch of the hyperbola through each intercept approaching the asymptotes.

23. $y^2 = 4 + x^2$

$y^2 - x^2 = 4$

$\dfrac{y^2}{4} - \dfrac{x^2}{4} = 1$ *Divide by 4.*

The graph is a hyperbola with y-intercepts $(0, 2)$ and $(0, -2)$. One asymptote passes through $(2, 2)$ and $(-2, -2)$. The other asymptote passes through $(-2, 2)$ and $(2, -2)$.

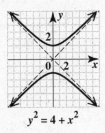

25. $f(x) = \sqrt{16 - x^2}$

Replace $f(x)$ with y and square both sides to get the equation

$$y^2 = 16 - x^2 \quad \text{or} \quad x^2 + y^2 = 16.$$

This is the graph of a circle with center $(0, 0)$ and radius 4. Since $f(x)$, or y, represents a principal square root in the original equation, $f(x)$ must be nonnegative. This restricts the graph to the upper half of the circle.

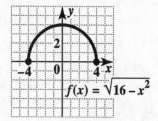

The domain is $[-4, 4]$, and the range is $[0, 4]$.

27. $\qquad\qquad f(x) = -\sqrt{36 - x^2}$

Replace $f(x)$ with y, and square both sides of the equation.

$$y = -\sqrt{36 - x^2}$$
$$y^2 = 36 - x^2$$
$$x^2 + y^2 = 36$$

This is a circle centered at the origin with radius $\sqrt{36} = 6$. Since $f(x)$, or y, represents a nonpositive square root in the original equation, $f(x)$ must be nonpositive. This restricts the graph to the bottom half of the circle.

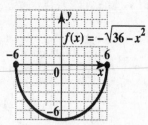

The domain is $[-6, 6]$, and the range is $[-6, 0]$.

29. $\qquad\qquad \dfrac{y}{3} = \sqrt{1 + \dfrac{x^2}{9}}$

Square both sides.

$$\frac{y^2}{9} = 1 + \frac{x^2}{9}$$
$$\frac{y^2}{9} - \frac{x^2}{9} = 1$$

This is a hyperbola opening up and down with y-intercepts $(0, 3)$ and $(0, -3)$. The four points $(3, 3)$, $(3, -3)$, $(-3, 3)$, and $(-3, -3)$ are the vertices of the rectangle that determine the asymptotes. Since $f(x)$, or y, represents a square root in the original equation, $f(x)$ must be

nonnegative. This restricts the graph to the upper half of the hyperbola.

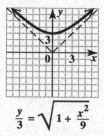

$$\frac{y}{3} = \sqrt{1 + \frac{x^2}{9}}$$

The domain is $(-\infty, \infty)$, and the range is $[3, \infty)$.

31. $\dfrac{(x - 2)^2}{4} - \dfrac{(y + 1)^2}{9} = 1$

is a hyperbola centered at $(2, -1)$, with $a = 2$ and $b = 3$. The x-intercepts are $(2 \pm 2, -1)$ or $(4, -1)$ and $(0, -1)$. The asymptotes are the extended diagonals of the rectangle with vertices $(2, 3)$, $(2, -3)$, $(-2, -3)$ and $(-2, 3)$ shifted 2 units right and 1 unit down, or $(4, 2)$, $(4, -4)$, $(0, -4)$ and $(0, 2)$. Draw the hyperbola.

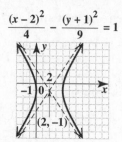

33. $\dfrac{y^2}{36} - \dfrac{(x - 2)^2}{49} = 1$

is a hyperbola centered at $(2, 0)$ with $a = 7$, and $b = 6$. The asymptotes are the extended diagonals of the rectangle with vertices $(7, 6)$, $(7, -6)$, $(-7, -6)$, and $(-7, 6)$ shifted right 2 units, or $(9, 6)$, $(9, -6)$, $(-5, -6)$, and $(-5, 6)$. Draw the hyperbola.

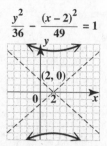

35. (a) $400x^2 - 625y^2 = 250{,}000$

$$\frac{x^2}{625} - \frac{y^2}{400} = 1 \qquad \textit{Divide by 250,000.}$$
$$\frac{x^2}{25^2} - \frac{y^2}{20^2} = 1$$

The x-intercepts are $(25, 0)$ and $(-25, 0)$. The distance between the buildings is the distance between the x-intercepts. The buildings are $25 + 25 = 50$ meters apart at their closest point.

(b) At $x = 50$, $y = \dfrac{d}{2}$, so $d = 2y$.

$$400(50)^2 - 625y^2 = 250{,}000$$
$$1{,}000{,}000 - 625y^2 = 250{,}000$$
$$-625y^2 = -750{,}000$$
$$y^2 = 1200$$
$$y = \sqrt{1200}$$

The distance d is $2\sqrt{1200} \approx 69.3$ meters.

37. $\dfrac{x^2}{9} - y^2 = 1$

$$-y^2 = 1 - \dfrac{x^2}{9}$$

$$y^2 = \dfrac{x^2}{9} - 1 \qquad \textit{Multiply by } -1.$$

Take the square root of both sides.

$$y = \pm\sqrt{\dfrac{x^2}{9} - 1}$$

The two functions used to obtain the graph were

$$y_1 = \sqrt{\dfrac{x^2}{9} - 1} \quad \text{and} \quad y_2 = -\sqrt{\dfrac{x^2}{9} - 1}.$$

39. $\dfrac{x^2}{25} - \dfrac{y^2}{49} = 1$

$$-\dfrac{y^2}{49} = 1 - \dfrac{x^2}{25}$$

$$\dfrac{y^2}{49} = \dfrac{x^2}{25} - 1 \qquad \textit{Multiply by } -1.$$

$$y^2 = 49\left(\dfrac{x^2}{25} - 1\right)$$

Take the square root of both sides.

$$y = \pm 7\sqrt{\dfrac{x^2}{25} - 1}$$

To obtain the graph, use the two functions

$$y_1 = 7\sqrt{\dfrac{x^2}{25} - 1} \quad \text{and} \quad y_2 = -7\sqrt{\dfrac{x^2}{25} - 1}.$$

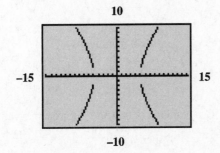

41. $y^2 - 9x^2 = 9$

$$y^2 = 9 + 9x^2$$
$$y^2 = 9\left(1 + x^2\right)$$

Take the square root of both sides.

$$y = \pm 3\sqrt{1 + x^2}$$

To obtain the graphs, use the two functions

$$y_1 = 3\sqrt{1 + x^2} \quad \text{and} \quad y_2 = -3\sqrt{1 + x^2}.$$

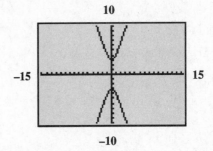

43. $2x + y = 13$ (1)
$y = 3x + 3$ (2)

Substitute $3x + 3$ for y in (1).

$$2x + (3x + 3) = 13$$
$$5x + 3 = 13$$
$$5x = 10$$
$$x = 2$$

Substitute 2 for x in (2).

$$y = 3(2) + 3 = 9$$

The solution set is $\{(2, 9)\}$.

45. $9x + 2y = 10$ (1)
$x - y = -5$ (2)

Solve (2) for x.

$$x = y - 5 \quad (3)$$

Substitute $y - 5$ for x in (1).

$$9(y - 5) + 2y = 10$$
$$9y - 45 + 2y = 10$$
$$11y = 55$$
$$y = 5$$

Substitute 5 for y in (3).

$$x = 5 - 5 = 0$$

The solution set is $\{(0, 5)\}$.

47. $2x^4 - 5x^2 - 3 = 0$

Let $u = x^2$, so that $u^2 = x^4$.

$2u^2 - 5u - 3 = 0$

$(2u + 1)(u - 3) = 0$

$2u + 1 = 0$ or $u - 3 = 0$

$u = -\frac{1}{2}$ or $u = 3$

$x^2 = -\frac{1}{2}$ or $x^2 = 3$

$x = \pm\sqrt{-\frac{2}{4}}$ or $x = \pm\sqrt{3}$

$x = \pm\frac{\sqrt{2}}{2}i$ or $x = \pm\sqrt{3}$

The solution set is $\left\{-\sqrt{3}, \sqrt{3}, -\frac{\sqrt{2}}{2}i, \frac{\sqrt{2}}{2}i\right\}$.

49. $x^4 - 7x^2 + 12 = 0$

Let $u = x^2$, so that $u^2 = x^4$.

$u^2 - 7u + 12 = 0$

$(u - 4)(u - 3) = 0$

$u - 4 = 0$ or $u - 3 = 0$

$u = 4$ or $u = 3$

$x^2 = 4$ or $x^2 = 3$

$x = \pm\sqrt{4}$ or $x = \pm\sqrt{3}$

$x = \pm 2$ or $x = \pm\sqrt{3}$

The solution set is $\left\{-2, -\sqrt{3}, \sqrt{3}, 2\right\}$.

11.4 Nonlinear Systems of Equations

1. Substitute $x - 1$ for y in the first equation. Then solve for x. Find the corresponding y-values by substituting back into $y = x - 1$. In the first equation, both variables are squared and in the second, both variables are to the first power, so the elimination method is not appropriate.

3. The line intersects the ellipse in exactly one point, so there is *one* point in the solution set of the system.

5. The line does not intersect the hyperbola, so there are no points in the solution set of the system.

7. A line and a circle; no points

Draw any circle, and then draw a line that does not cross the circle.

9. A line and a hyperbola; one point

The line is tangent to the hyperbola.

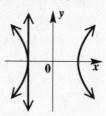

11. A circle and an ellipse; four points

Draw any ellipse, and then draw a circle with the same center whose radius is just large enough so that there are four points of intersection. (If the radius of the circle is too large or too small, there may be fewer points of intersection.)

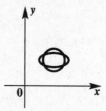

13. A parabola and an ellipse; four points

Draw any parabola, and then draw an ellipse large enough so that there are four points of intersection. (If the ellipse is too large or too small, there may be fewer points of intersection.)

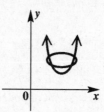

15. $y = 4x^2 - x$ (1)

$y = x$ (2)

Substitute x for y in equation (1).

$$y = 4x^2 - x \quad (1)$$
$$x = 4x^2 - x$$
$$0 = 4x^2 - 2x$$
$$0 = 2x(2x - 1)$$

$2x = 0$ or $2x - 1 = 0$

$x = 0$ or $x = \frac{1}{2}$

Use equation (2) to find y for each x-value.

If $x = 0$, then $y = 0$.

If $x = \frac{1}{2}$, then $y = \frac{1}{2}$.

The solution set is $\left\{(0, 0), \left(\frac{1}{2}, \frac{1}{2}\right)\right\}$.

17.
$$y = x^2 + 6x + 9 \quad (1)$$
$$x + y = 3 \quad\quad\quad (2)$$

Substitute $x^2 + 6x + 9$ for y in equation (2).

$$x + y = 3 \quad (2)$$
$$x + (x^2 + 6x + 9) = 3$$
$$x^2 + 7x + 9 = 3$$
$$x^2 + 7x + 6 = 0$$
$$(x + 6)(x + 1) = 0$$

$$x + 6 = 0 \quad \text{or} \quad x + 1 = 0$$
$$x = -6 \quad \text{or} \quad\quad x = -1$$

Substitute these values for x in equation (2) and solve for y.

If $x = -6$, then
$$x + y = 3 \quad (2)$$
$$-6 + y = 3$$
$$y = 9.$$

If $x = -1$, then
$$x + y = 3 \quad (2)$$
$$-1 + y = 3$$
$$y = 4.$$

The solution set is $\{(-6, 9), (-1, 4)\}$.

19.
$$x^2 + y^2 = 2 \quad (1)$$
$$2x + y = 1 \quad (2)$$

Solve equation (2) for y.

$$y = 1 - 2x \quad (3)$$

Substitute $1 - 2x$ for y in equation (1).

$$x^2 + y^2 = 2 \quad (1)$$
$$x^2 + (1 - 2x)^2 = 2$$
$$x^2 + 1 - 4x + 4x^2 = 2$$
$$5x^2 - 4x - 1 = 0$$
$$(5x + 1)(x - 1) = 0$$

$$5x + 1 = 0 \quad \text{or} \quad x - 1 = 0$$
$$x = -\tfrac{1}{5} \quad \text{or} \quad\quad x = 1$$

Use equation (3) to find y for each x-value.

If $x = -\frac{1}{5}$, then
$$y = 1 - 2\left(-\tfrac{1}{5}\right) = 1 + \tfrac{2}{5} = \tfrac{7}{5}.$$

If $x = 1$, then
$$y = 1 - 2(1) = -1.$$

The solution set is $\left\{\left(-\tfrac{1}{5}, \tfrac{7}{5}\right), (1, -1)\right\}$.

21.
$$xy = 4 \quad\quad\quad (1)$$
$$3x + 2y = -10 \quad (2)$$

Solve equation (1) for y to get $y = \dfrac{4}{x}$.

Substitute $\dfrac{4}{x}$ for y in equation (2) to find x.

$$3x + 2y = -10 \quad (2)$$
$$3x + 2\left(\frac{4}{x}\right) = -10$$

Multiply by the LCD, x.

$$3x^2 + 8 = -10x$$
$$3x^2 + 10x + 8 = 0$$
$$(3x + 4)(x + 2) = 0$$

$$3x + 4 = 0 \quad \text{or} \quad x + 2 = 0$$
$$x = -\tfrac{4}{3} \quad \text{or} \quad\quad x = -2$$

Since $y = \dfrac{4}{x}$, if $x = -\dfrac{4}{3}$, then $y = \dfrac{4}{-\frac{4}{3}} = -3$.

If $x = -2$, then $y = \frac{4}{-2} = -2$.

The solution set is $\left\{(-2, -2), \left(-\tfrac{4}{3}, -3\right)\right\}$.

23.
$$xy = -3 \quad (1)$$
$$x + y = -2 \quad (2)$$

Solve equation (2) for y.

$$y = -x - 2 \quad (3)$$

Substitute $-x - 2$ for y in equation (1).

$$xy = -3 \quad (1)$$
$$x(-x - 2) = -3$$
$$-x^2 - 2x = -3$$
$$-x^2 - 2x + 3 = 0$$
$$x^2 + 2x - 3 = 0 \quad\quad \textit{Multiply by -1.}$$
$$(x + 3)(x - 1) = 0$$

$$x + 3 = 0 \quad \text{or} \quad x - 1 = 0$$
$$x = -3 \quad \text{or} \quad\quad x = 1$$

Use equation (3) to find y for each x-value.

If $x = -3$, then
$$y = -(-3) - 2 = 1.$$

If $x = 1$, then
$$y = -(1) - 2 = -3.$$

The solution set is $\{(-3, 1), (1, -3)\}$.

25. $y = 3x^2 + 6x \qquad (1)$

$y = x^2 - x - 6 \qquad (2)$

Substitute $x^2 - x - 6$ for y in equation (1) to find x.

$$y = 3x^2 + 6x \qquad (1)$$
$$x^2 - x - 6 = 3x^2 + 6x$$
$$0 = 2x^2 + 7x + 6$$
$$0 = (2x + 3)(x + 2)$$

$$2x + 3 = 0 \quad \text{or} \quad x + 2 = 0$$
$$x = -\tfrac{3}{2} \quad \text{or} \qquad x = -2$$

Substitute $-\tfrac{3}{2}$ for x in equation (1) to find y.

$$y = 3x^2 + 6x \qquad (1)$$
$$y = 3\left(-\tfrac{3}{2}\right)^2 + 6\left(-\tfrac{3}{2}\right)$$
$$= 3\left(\tfrac{9}{4}\right) + 6\left(-\tfrac{6}{4}\right)$$
$$= \tfrac{27}{4} - \tfrac{36}{4} = -\tfrac{9}{4}$$

Substitute -2 for x in equation (1) to find y.

$$y = 3x^2 + 6x \qquad (1)$$
$$y = 3(-2)^2 + 6(-2)$$
$$= 12 - 12 = 0$$

The solution set is $\left\{\left(-\tfrac{3}{2}, -\tfrac{9}{4}\right), (-2, 0)\right\}$.

27. $2x^2 - y^2 = 6 \qquad (1)$

$\qquad y = x^2 - 3 \quad (2)$

Substitute $x^2 - 3$ for y in equation (1).

$$2x^2 - y^2 = 6 \quad (1)$$
$$2x^2 - \left(x^2 - 3\right)^2 = 6$$
$$2x^2 - \left(x^4 - 6x^2 + 9\right) = 6$$
$$-x^4 + 8x^2 - 9 = 6$$
$$-x^4 + 8x^2 - 15 = 0$$
$$x^4 - 8x^2 + 15 = 0 \qquad \textit{Multiply by } -1.$$

Let $z = x^2$, so $z^2 = x^4$.

$$z^2 - 8z + 15 = 0$$
$$(z - 3)(z - 5) = 0$$

$$z - 3 = 0 \quad \text{or} \quad z - 5 = 0$$
$$z = 3 \quad \text{or} \qquad z = 5$$

Since $z = x^2$,

$$x^2 = 3 \quad \text{or} \quad x^2 = 5.$$

Use the square root property.

$$x = \pm\sqrt{3} \quad \text{or} \quad x = \pm\sqrt{5}$$

Use equation (2) to find y for each x-value.

If $x = \sqrt{3}$ or $-\sqrt{3}$, then

$$y = \left(\pm\sqrt{3}\right)^2 - 3 = 0.$$

(Note: We could substitute 3 for x^2 in (2) to obtain the same result.)

If $x = \sqrt{5}$ or $-\sqrt{5}$, then

$$y = \left(\pm\sqrt{5}\right)^2 - 3 = 2.$$

The solution set is

$$\left\{\left(-\sqrt{3}, 0\right), \left(\sqrt{3}, 0\right), \left(-\sqrt{5}, 2\right), \left(\sqrt{5}, 2\right)\right\}.$$

29. $x^2 - xy + y^2 = 0 \qquad (1)$

$\quad x - 2y = 1 \qquad\qquad (2)$

Solve equation (2) for x.

$$x = 2y + 1 \qquad (3)$$

Substitute $2y + 1$ for x in equation (1).

$$(2y + 1)^2 - (2y + 1)y + y^2 = 0$$
$$4y^2 + 4y + 1 - 2y^2 - y + y^2 = 0$$
$$3y^2 + 3y + 1 = 0$$

Use the quadratic formula with $a = 3$, $b = 3$, and $c = 1$.

$$y = \frac{-b \pm \sqrt{b^2 - 4ac}}{2a}$$
$$y = \frac{-3 \pm \sqrt{9 - 12}}{6}$$
$$= \frac{-3 \pm \sqrt{-3}}{6} = \frac{-3 \pm i\sqrt{3}}{6}$$

Use equation (3) to solve for x.

$$x = 2y + 1 \qquad\qquad\qquad (3)$$
$$= 2\left(\frac{-3 \pm i\sqrt{3}}{6}\right) + 1$$
$$= \frac{-3 \pm i\sqrt{3}}{3} + \frac{3}{3}$$
$$= \pm\frac{i\sqrt{3}}{3}$$

The solution set is

$$\left\{\left(\tfrac{\sqrt{3}}{3}i, -\tfrac{1}{2} + \tfrac{\sqrt{3}}{6}i\right), \left(-\tfrac{\sqrt{3}}{3}i, -\tfrac{1}{2} - \tfrac{\sqrt{3}}{6}i\right)\right\}.$$

31. $3x^2 + 2y^2 = 12 \quad (1)$

$\quad x^2 + 2y^2 = 4 \quad (2)$

Multiply equation (2) by -1 and add the result to equation (1).

$$
\begin{array}{rcll}
3x^2 + 2y^2 &=& 12 & (1) \\
-x^2 - 2y^2 &=& -4 & -1 \times (2) \\
\hline
2x^2 &=& 8 & \\
x^2 &=& 4 & \\
x &=& \pm 2 &
\end{array}
$$

Substitute ± 2 for x in equation (2) to find y.

$$x^2 + 2y^2 = 4 \quad (2)$$
$$(\pm 2)^2 + 2y^2 = 4$$
$$4 + 2y^2 = 4$$
$$2y^2 = 0$$
$$y^2 = 0$$
$$y = 0$$

The solution set is $\{(-2, 0), (2, 0)\}$.

33. $2x^2 + 3y^2 = 6 \quad (1)$
$\quad\;\; x^2 + 3y^2 = 3 \quad (2)$

Multiply equation (2) by -1 and add the result to equation (1).

$$2x^2 + 3y^2 = \quad\;\; 6 \quad (1)$$
$$\underline{-x^2 - 3y^2 = \quad\; -3} \qquad -1 \times (2)$$
$$x^2 \qquad\quad = \qquad 3$$
$$x = \pm\sqrt{3}$$

Substitute $\pm\sqrt{3}$ for x in equation (2).

$$x^2 + 3y^2 = 3 \quad (2)$$
$$\left(\pm\sqrt{3}\right)^2 + 3y^2 = 3$$
$$3 + 3y^2 = 3$$
$$y^2 = 0$$
$$y = 0$$

The solution set is $\left\{\left(\sqrt{3}, 0\right), \left(-\sqrt{3}, 0\right)\right\}$.

35. $2x^2 + \quad y^2 = 28 \quad (1)$
$\quad\;\; 4x^2 - 5y^2 = 28 \quad (2)$

Multiply (1) by 5 and add the result to equation (2).

$$10x^2 + 5y^2 = 140 \qquad 5 \times (1)$$
$$\underline{\;\; 4x^2 - 5y^2 = \quad 28} \qquad (2)$$
$$14x^2 \qquad\quad = 168$$
$$x^2 = \quad 12$$
$$x = \pm\sqrt{12} = \pm 2\sqrt{3}$$

Let $x = \pm 2\sqrt{3}$ in (1).

$$2\left(\pm 2\sqrt{3}\right)^2 + y^2 = 28$$
$$2(12) + y^2 = 28$$
$$y^2 = 4$$
$$y = \pm 2$$

The solution set is $\left\{\left(2\sqrt{3}, 2\right), \left(2\sqrt{3}, -2\right),\right.$
$\qquad\qquad\qquad \left.\left(-2\sqrt{3}, 2\right), \left(-2\sqrt{3}, -2\right)\right\}$.

37. $2x^2 = \;\; 8 - 2y^2 \quad (1)$
$\quad\;\; 3x^2 = 24 - 4y^2 \quad (2)$

Multiply equation (1) by -2 and add the result to equation (2).

$$-4x^2 = -16 + 4y^2 \qquad -2 \times (1)$$
$$\underline{\;\; 3x^2 = \quad 24 - 4y^2} \qquad (2)$$
$$-x^2 = \qquad 8$$
$$x^2 = \quad -8$$
$$x = \pm\sqrt{-8} = \pm 2i\sqrt{2}$$

Substitute $\pm 2i\sqrt{2}$ for x in equation (1).

$$2x^2 = 8 - 2y^2 \qquad\qquad (1)$$
$$2\left(\pm 2i\sqrt{2}\right)^2 = 8 - 2y^2$$
$$2(-8) = 8 - 2y^2$$
$$-16 = 8 - 2y^2$$
$$2y^2 = 24$$
$$y^2 = 12$$
$$y = \pm\sqrt{12} = \pm 2\sqrt{3}$$

Since $2i\sqrt{2}$ can be paired with either $2\sqrt{3}$ or $-2\sqrt{3}$ and $-2i\sqrt{2}$ can be paired with either $2\sqrt{3}$ or $-2\sqrt{3}$, there are four possible solutions.

The solution set is

$$\left\{\left(-2i\sqrt{2}, -2\sqrt{3}\right), \left(-2i\sqrt{2}, 2\sqrt{3}\right),\right.$$
$$\left.\left(2i\sqrt{2}, -2\sqrt{3}\right), \left(2i\sqrt{2}, 2\sqrt{3}\right)\right\}.$$

39. $x^2 + xy + y^2 = 15 \quad (1)$
$\qquad\;\; x^2 + y^2 = 10 \quad (2)$

Multiply equation (2) by -1 and add the result to equation (1).

$$x^2 + xy + y^2 = \quad 15 \quad (1)$$
$$\underline{-x^2 \qquad - y^2 = -10} \qquad -1 \times (2)$$
$$xy \qquad\quad = \quad 5$$
$$y = \frac{5}{x}$$

Substitute $\dfrac{5}{x}$ for y in equation (2).

$$x^2 + y^2 = 10 \qquad (2)$$
$$x^2 + \left(\frac{5}{x}\right)^2 = 10$$
$$x^2 + \frac{25}{x^2} = 10$$
$$x^4 + 25 = 10x^2 \quad \textit{Multiply by } x^2.$$
$$x^4 - 10x^2 + 25 = 0$$

continued

Let $z = x^2$, so $z^2 = x^4$.

$$z^2 - 10z + 25 = 0$$
$$(z - 5)^2 = 0$$
$$z - 5 = 0$$
$$z = 5$$

Since $z = x^2$,

$$x^2 = 5, \text{ and } x = \pm\sqrt{5}.$$

Using the equation $y = \dfrac{5}{x}$, we get the following.

If $x = -\sqrt{5}$, then

$$y = \frac{5}{-\sqrt{5}} = \frac{5 \cdot \sqrt{5}}{-\sqrt{5} \cdot \sqrt{5}} = \frac{5\sqrt{5}}{-5} = -\sqrt{5}.$$

Similarly, if $x = \sqrt{5}$, then $y = \sqrt{5}$.

The solution set is $\left\{\left(-\sqrt{5}, -\sqrt{5}\right), \left(\sqrt{5}, \sqrt{5}\right)\right\}$.

41.
$$3x^2 + 2xy - 3y^2 = 5 \quad (1)$$
$$-x^2 - 3xy + y^2 = 3 \quad (2)$$

Multiply equation (2) by 3 and add the result to equation (1).

$$
\begin{array}{rcll}
3x^2 + 2xy - 3y^2 &=& 5 & (1) \\
-3x^2 - 9xy + 3y^2 &=& 9 & 3 \times (2) \\
\hline
-7xy &=& 14 & \\
x &=& \dfrac{14}{-7y} = -\dfrac{2}{y} &
\end{array}
$$

Substitute $-\dfrac{2}{y}$ for x in equation (2).

$$-x^2 - 3xy + y^2 = 3 \qquad (2)$$
$$-\left(-\frac{2}{y}\right)^2 - 3\left(-\frac{2}{y}\right)y + y^2 = 3$$
$$-\left(\frac{4}{y^2}\right) + 6 + y^2 = 3$$
$$y^2 + 3 - \frac{4}{y^2} = 0$$
$$y^4 + 3y^2 - 4 = 0 \quad \textit{Multiply by } y^2.$$
$$(y^2 + 4)(y^2 - 1) = 0$$

$$
\begin{array}{lll}
y^2 + 4 = 0 & \text{or} & y^2 - 1 = 0 \\
y^2 = -4 & & y^2 = 1 \\
y = \pm 2i & \text{or} & y = \pm 1
\end{array}
$$

Since $x = -\dfrac{2}{y}$, substitute these values for y to find the values of x.

If $y = 2i$, then

$$x = -\frac{2}{2i} = -\frac{1}{i} = -\frac{1}{i} \cdot \frac{i}{i} = \frac{-i}{-1} = i.$$

If $y = -2i$, then

$$x = -\frac{2}{-2i} = \frac{1}{i} = \frac{1}{i} \cdot \frac{i}{i} = \frac{i}{-1} = -i.$$

If $y = 1$, then

$$x = -\frac{2}{1} = -2.$$

If $y = -1$, then

$$x = -\frac{2}{-1} = 2.$$

The solution set is
$$\{(i, 2i), (-i, -2i), (2, -1), (-2, 1)\}.$$

43.
$$xy = -6 \quad (1)$$
$$x + y = -1 \quad (2)$$

Solve both equations for y.

$$y = -\frac{6}{x} \quad \text{and} \quad y = -x - 1$$

Graph

$$Y_1 = -\frac{6}{X} \quad \text{and} \quad Y_2 = -X - 1$$

on a graphing calculator to obtain the solution set $\{(2, -3), (-3, 2)\}$.

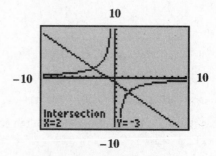

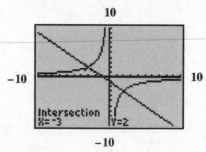

Now solve the system algebraically using substitution.

$$xy = -6 \quad (1)$$
$$x + y = -1 \quad (2)$$

Solve equation (2) for y.

$$y = -1 - x \qquad (3)$$

Substitute $-1 - x$ for y in equation (1).

$$x(-1 - x) = -6$$
$$-x - x^2 = -6$$
$$0 = x^2 + x - 6$$
$$0 = (x + 3)(x - 2)$$

$$x + 3 = 0 \quad \text{or} \quad x - 2 = 0$$
$$x = -3 \quad \text{or} \quad x = 2$$

Substitute these values for x in (3) to find y.

If $x = -3$, then $y = -1 - (-3) = 2$.

If $x = 2$, then $y = -1 - 2 = -3$.

The solution set, $\{(2, -3), (-3, 2)\}$, is the same as that obtained using a graphing calculator.

45. Let W = the width, and
L = the length.

The formula for the area of a rectangle is $LW = A$, so

$$LW = 84. \quad (1)$$

The perimeter of a rectangle is given by $2L + 2W = P$, so

$$2L + 2W = 38. \quad (2)$$

Solve equation (2) for L to get

$$L = 19 - W. \quad (3)$$

Substitute $19 - W$ for L in equation (1).

$$LW = 84 \quad (1)$$
$$(19 - W)W = 84$$
$$19W - W^2 = 84$$
$$-W^2 + 19W - 84 = 0$$
$$W^2 - 19W + 84 = 0 \qquad \textit{Multiply by } -1.$$
$$(W - 7)(W - 12) = 0$$

$$W - 7 = 0 \quad \text{or} \quad W - 12 = 0$$
$$W = 7 \quad \text{or} \quad W = 12$$

Using equation (3), with $W = 7$,

$$L = 19 - 7 = 12.$$

If $W = 12$, then $L = 7$, which are the same two numbers. Length must be greater than width, so the length is 12 feet and the width is 7 feet.

47.
$$px = 16 \qquad (1)$$
$$p = 10x + 12 \quad (2)$$

Substitute $10x + 12$ for p in equation (1).

$$px = 16 \quad (1)$$
$$(10x + 12)x = 16$$
$$10x^2 + 12x = 16$$
$$10x^2 + 12x - 16 = 0$$
$$5x^2 + 6x - 8 = 0$$
$$(5x - 4)(x + 2) = 0$$

$$5x - 4 = 0 \quad \text{or} \quad x + 2 = 0$$
$$x = \tfrac{4}{5} \quad \text{or} \quad x = -2$$

Since x cannot be negative, eliminate -2 as a value of x. Substitute $\frac{4}{5}$ for x in equation (2) to find p.

$$p = 10x + 12 \qquad (2)$$
$$p = 10\left(\tfrac{4}{5}\right) + 12$$
$$= 8 + 12 = 20$$

Since the demand x is in thousands,

$$\text{demand} = \tfrac{4}{5}(1000) = 800.$$

The equilibrium price is \$20. The supply/demand at that price is 800 calculators.

49. $2x - y \leq 4$

Draw a solid line through the intercepts $(2, 0)$ and $(0, -4)$. Checking $(0, 0)$ gives us $0 \leq 4$, a true statement, so shade the side containing $(0, 0)$.

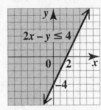

51. $-5x + 3y \leq 15$

Draw a solid line through the intercepts $(-3, 0)$ and $(0, 5)$. Checking $(0, 0)$ gives us $0 \leq 15$, a true statement, so shade the side containing $(0, 0)$.

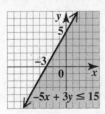

11.5 Second-Degree Inequalities and Systems of Inequalities

1. To graph the solution set of a nonlinear inequality, first graph the corresponding equality. This graph should be a dashed curve for $<$ or $>$ inequalities or a solid curve for $\leq$ or $\geq$ inequalities. Next, decide which region to shade by substituting any point not on the boundary (usually $(0,0)$ is the easiest) into the inequality. If the statement is true, then shade that area. If the statement is false, then shade the other area.

3. $x^2 + y^2 < 25$
$y > -2$

The boundary, $x^2 + y^2 = 25$, is a circle (dashed because of the $<$ sign) with center $(0,0)$ and radius 5. When $(0,0)$ is tested, a true statement, $0 < 25$, results, so the inside of the circle is shaded. The graph of $y = -2$ is a horizontal line through $(0,-2)$ with shading above the dashed line, since $y > -2$. The correct answer is **C**.

5. $y \geq x^2 + 4$

This is an inequality whose boundary is a solid parabola, opening up, with vertex $(0,4)$. The inside of the parabola is shaded since $(0,0)$ makes the inequality false. This is graph **B**.

7. $y < x^2 + 4$

This is an inequality whose boundary is a dashed parabola, opening up, with vertex $(0,4)$. The outside of the parabola is shaded since $(0,0)$ makes the inequality true. This is graph **A**.

9. $y^2 > 4 + x^2$
$y^2 - x^2 > 4$
$\dfrac{y^2}{4} - \dfrac{x^2}{4} > 1$

The boundary, $\dfrac{y^2}{4} - \dfrac{x^2}{4} = 1$, is a hyperbola with y-intercepts $(0,2)$ and $(0,-2)$ and asymptotes formed by the extended diagonals of the fundamental rectangle with vertices at $(2,2)$, $(2,-2)$, $(-2,-2)$, and $(-2,2)$. The hyperbola has dashed branches because of $>$. Test $(0,0)$.

$$0^2 > 4 + 0^2 \quad ?$$
$$0 > 4 \qquad \textit{False}$$

Shade the sides of the hyperbola that do not contain $(0,0)$. These are the regions inside the branches of the hyperbola.

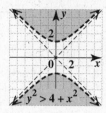

11. $y \geq x^2 - 2$
$y \geq (x - 0)^2 - 2$

Graph the solid vertical parabola $y = x^2 - 2$ with vertex $(0,-2)$. Two other points on the parabola are $(2,2)$ and $(-2,2)$. Test a point not on the parabola, say $(0,0)$, in $y \geq x^2 - 2$ to get $0 \geq -2$, a true statement. Shade that portion of the graph that contains the point $(0,0)$. This is the region inside the parabola.

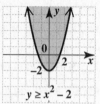

13. $2y^2 \geq 8 - x^2$
$x^2 + 2y^2 \geq 8$
$\dfrac{x^2}{8} + \dfrac{y^2}{4} \geq 1$

The boundary, $\dfrac{x^2}{8} + \dfrac{y^2}{4} = 1$, is the ellipse with intercepts $(2\sqrt{2}, 0)$, $(-2\sqrt{2}, 0)$, $(0, 2)$, and $(0, -2)$, drawn as a solid curve because of $\geq$. Test $(0,0)$.

$$2(0)^2 \geq 8 - 0^2 \quad ?$$
$$0 \geq 8 \qquad \textit{False}$$

Shade the region of the ellipse that does not contain $(0,0)$. This is the region outside the ellipse.

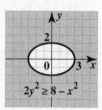

15. $y \leq x^2 + 4x + 2$

Graph the solid vertical parabola $y = x^2 + 4x + 2$. Use the vertex formula $x = \dfrac{-b}{2a}$ to obtain the vertex $(-2, -2)$. Two other points on the parabola are $(0, 2)$ and $(1, 7)$. Test a point not on the parabola, say $(0,0)$, in

$y \leq x^2 + 4x + 2$ to get $0 \leq 2$, a true statement. Shade outside the parabola, since this region contains $(0, 0)$.

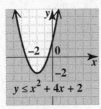

17. $9x^2 > 16y^2 + 144$

$9x^2 - 16y^2 > 144$

$\dfrac{x^2}{16} - \dfrac{y^2}{9} > 1$

The boundary, $\dfrac{x^2}{16} - \dfrac{y^2}{9} = 1$, is a hyperbola with x-intercepts $(4, 0)$ and $(-4, 0)$ and asymptotes formed by the extended diagonals of the fundamental rectangle with vertices at $(4, 3)$, $(4, -3)$, $(-4, -3)$, and $(-4, 3)$. The hyperbola has dashed branches because of $>$. Test $(0, 0)$.

$9(0)^2 > 16(0)^2 + 144$?

$0 > 144$ *False*

Shade the sides of the hyperbola that do not contain $(0, 0)$. These are the regions inside the branches of the hyperbola.

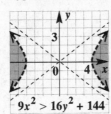

19. $x^2 - 4 \geq -4y^2$

$x^2 + 4y^2 \geq 4$

$\dfrac{x^2}{4} + \dfrac{y^2}{1} \geq 1$

Graph the solid ellipse $\dfrac{x^2}{4} + \dfrac{y^2}{1} = 1$ through the x-intercepts $(2, 0)$ and $(-2, 0)$ and y-intercepts $(0, 1)$ and $(0, -1)$. Test a point not on the ellipse, say $(0, 0)$, in $x^2 - 4 \geq -4y^2$ to get $-4 \geq 0$, a false statement. Shade outside the ellipse, since this region does *not* include $(0, 0)$.

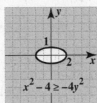

21. $x \leq -y^2 + 6y - 7$

Complete the square to find the vertex.

$x = -y^2 + 6y - 7$

$\quad = -(y^2 - 6y) - 7$

$\quad = -(y^2 - 6y + 9) + 9 - 7$

$x = -(y - 3)^2 + 2$

The boundary, $x = -(y - 3)^2 + 2$, is a solid horizontal parabola with vertex at $(2, 3)$ that opens to the left. Test $(0, 0)$.

$0 \leq -0^2 + 6(0) - 7$?

$0 \leq -7$ *False*

Shade the region of the parabola that does not contain $(0, 0)$. This is the region inside the parabola.

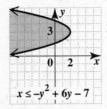

23. $2x + 5y < 10$

$\quad x - 2y < 4$

Graph $2x + 5y = 10$ as a dashed line through $(5, 0)$ and $(0, 2)$. Test $(0, 0)$.

$2x + 5y < 10$

$2(0) + 5(0) < 10$?

$0 < 10$ *True*

Shade the region containing $(0, 0)$. Graph $x - 2y = 4$ as a dashed line through $(4, 0)$ and $(0, -2)$. Test $(0, 0)$.

$x - 2y < 4$

$0 - 2(0) < 4$?

$0 < 4$ *True*

Shade the region containing $(0, 0)$. The graph of the system is the intersection of the two shaded regions.

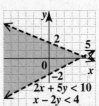

25. $5x - 3y \leq 15$
  $4x + y \geq 4$

The boundary, $5x - 3y = 15$, is a solid line with intercepts $(3, 0)$ and $(0, -5)$. Test $(0, 0)$.

$$5(0) - 3(0) \leq 15 \ ?$$
$$0 \leq 15 \ \textit{True}$$

Shade the side of the line that contains $(0, 0)$.
The boundary, $4x + y = 4$, is a solid line with intercepts $(1, 0)$ and $(0, 4)$. Test $(0, 0)$.

$$4(0) + 0 \geq 4 \ ?$$
$$0 \geq 4 \ \textit{False}$$

Shade the side of the line that does not contain $(0, 0)$.
The graph of the system is the intersection of the two shaded regions.

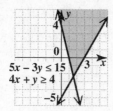

27. $x \leq 5$
  $y \leq 4$

Graph $x = 5$ as a solid vertical line through $(5, 0)$.
Since $x \leq 5$, shade the left side of the line.
Graph $y = 4$ as a solid horizontal line through $(0, 4)$. Since $y \leq 4$, shade below the line.
The graph of the system is the intersection of the two shaded regions.

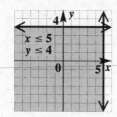

29. $y > \ x^2 - 4$
  $y < -x^2 + 3$

The boundary, $y = x^2 - 4$, is a dashed parabola with vertex $(0, -4)$ that opens up. Test $(0, 0)$.

$$0 > 0^2 - 4 \ ?$$
$$0 > -4 \qquad \textit{True}$$

Shade the side of the parabola that contains $(0, 0)$.
This is the region inside the parabola.
The boundary, $y = -x^2 + 3$, is a dashed parabola with vertex $(0, 3)$ that opens down. Test $(0, 0)$.

$$0 < -0^2 + 3 \ ?$$
$$0 < 3 \qquad \textit{True}$$

Shade the side of the parabola that contains $(0, 0)$.
This is the region inside the parabola.
The graph of the system is the intersection of the two shaded regions.

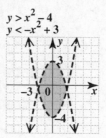

31. $x^2 + y^2 \geq 4 \ \ (1)$
  $x + y \leq 5 \ \ (2)$
  $x \geq 0 \ \ (3)$
  $y \geq 0 \ \ (4)$

The graph of (1) is the region outside the solid circle boundary $x^2 + y^2 = 4$ with center $(0, 0)$ and radius 2.
The graph of (2) is the region below the solid line $x + y = 5$ with intercepts $(5, 0)$ and $(0, 5)$.
The intersection of the graphs of (3) and (4) is quadrant I where both x and y are positive. The positive x- and y-axes are included.
The graph of the system is the intersection of the graphs of (1) and (2) in quadrant I.

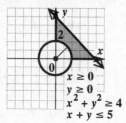

33. $y \leq -x^2$
  $y \geq x - 3$
  $y \leq -1$
  $x < 1$

The boundary, $y = -x^2$, is a solid parabola with vertex at $(0, 0)$ that opens down. Test $(0, -1)$.

$$-1 \leq -(0)^2 \ ?$$
$$-1 \leq 0 \qquad \textit{True}$$

Shade the side of the parabola that contains $(0, -1)$. This is the region inside the parabola.
The boundary, $y = x - 3$, is a solid line with intercepts $(3, 0)$ and $(0, -3)$. Test $(0, 0)$.

$$0 \geq 0 - 3 \ ?$$
$$0 \geq -3 \qquad \textit{True}$$

Shade the side of the line that contains $(0, 0)$.
For $y \leq -1$, shade below the solid horizontal line $y = -1$.

For $x < 1$, shade to the left of the dashed vertical line $x = 1$.

The intersection of the four shaded regions is the graph of the system.

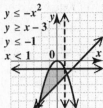

35. $x^2 + y^2 > 36, x \geq 0$

This is a circle of radius 6 centered at the origin. The graph is a dashed curve. Since $(0, 0)$ does not satisfy the inequality, the shading will be outside the circle. By including the restriction $x \geq 0$, we consider only the shading to the right of, and including, the y-axis.

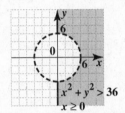

37. $x < y^2 - 3, x < 0$

Consider the equation.

$$x = y^2 - 3, \text{ or}$$
$$x = (y - 0)^2 - 3.$$

This is a parabola with vertex $(-3, 0)$, opening to the right having the same shape as $x = y^2$. The graph is a dashed curve. The shading will be outside the parabola, since $(0, 0)$ does not satisfy the inequality. By including the restriction $x < 0$, we consider only the shading to the left of, but not including, the y-axis. The y-axis is a dashed line.

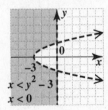

39. $4x^2 - y^2 > 16, x < 0$

Consider the equation

$$4x^2 - y^2 = 16, \text{ or}$$
$$\frac{x^2}{4} - \frac{y^2}{16} = 1.$$

This is a hyperbola with x-intercepts $(2, 0)$ and $(-2, 0)$. The graph is a dashed curve. The shading will be to the left of the left branch and to the right of the right branch of the hyperbola, since

$(0, 0)$ does not satisfy the inequality. By including the restriction $x < 0$, we consider only the shading to the left of the y-axis.

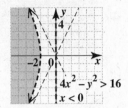

41. $x^2 + 4y^2 \geq 1, x \geq 0, y \geq 0$

Consider the equation

$$x^2 + 4y^2 = 1, \text{ or}$$
$$\frac{x^2}{1} + \frac{y^2}{\frac{1}{4}} = 1.$$

The graph is a solid ellipse with x-intercepts $(1, 0)$ and $(-1, 0)$ and y-intercepts $\left(0, \frac{1}{2}\right)$ and $\left(0, -\frac{1}{2}\right)$. The shading will be outside the ellipse, since $(0, 0)$ does not satisfy the inequality. By including the restrictions $x \geq 0$ and $y \geq 0$, we consider only the shading in quadrant I and portions of the x- and y-axis.

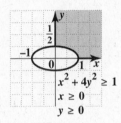

43. $y \geq x - 3$
$y \leq -x + 4$

The graphs of both inequalities include the points on the lines as part of the solution because of $\geq$ and $\leq$.

To produce the graph of the system on the TI-83, make the following Y-assignments:

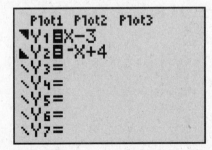

To get the upper and lower darkened triangles to the left of Y_1 and Y_2, simply place the cursor in that spot and press $\boxed{\text{ENTER}}$ to cycle through the graphing choices.

continued

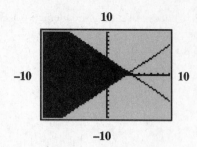

45. $y < x^2 + 4x + 4$

$y > -3$

The graphs do *not* include the points on the parabola or the line as part of the solution because of $<$ and $>$.

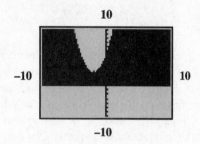

47. $\dfrac{n+5}{n}$

(a) $[n = 1]$ $\dfrac{1+5}{1} = \dfrac{6}{1} = 6$

(b) $[n = 2]$ $\dfrac{2+5}{2} = \dfrac{7}{2}$

(c) $[n = 3]$ $\dfrac{3+5}{3} = \dfrac{8}{3}$

(d) $[n = 4]$ $\dfrac{4+5}{4} = \dfrac{9}{4}$

49. $n^2 - n$

(a) $[n = 1]$ $1^2 - 1 = 1 - 1 = 0$

(b) $[n = 2]$ $2^2 - 2 = 4 - 2 = 2$

(c) $[n = 3]$ $3^2 - 3 = 9 - 3 = 6$

(d) $[n = 4]$ $4^2 - 4 = 16 - 4 = 12$

Chapter 11 Review Exercises

1. $f(x) = |x + 4|$

This is the graph of the absolute value function $g(x) = |x|$ shifted 4 units to the left (since $x + 4 = 0$ if $x = -4$).

x	y
-6	2
-5	1
-4	0
-3	1
-2	2

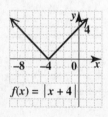

2. $f(x) = \dfrac{1}{x - 4}$

This is the graph of the reciprocal function, $g(x) = \dfrac{1}{x}$, shifted 4 units to the right. Since $x \neq 4$ (or a denominator of 0 results), the line $x = 4$ is a vertical asymptote. Since $\dfrac{1}{x - 4} \neq 0$, the line $y = 0$ is a horizontal asymptote.

x	y
2	$-\frac{1}{2}$
3	-1
$3\frac{1}{2}$	-2
$4\frac{1}{2}$	2
5	1
6	$\frac{1}{2}$

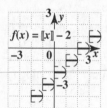

3. $f(x) = \sqrt{x} + 3$

This is the graph of the square root function, $g(x) = \sqrt{x}$, shifted 3 units upward.

x	y
0	3
1	4
4	5

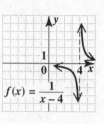

4. $f(x) = [\![x]\!] - 2$

This is the graph of the greatest integer function, $g(x) = [\![x]\!]$, shifted down 2 units.

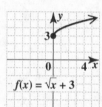

5. Center $(-2, 4)$, $r = 3$
Here $h = -2$, $k = 4$, and $r = 3$, so an equation of the circle is

$$(x - h)^2 + (y - k)^2 = r^2$$
$$[x - (-2)]^2 + (y - 4)^2 = 3^2$$
$$(x + 2)^2 + (y - 4)^2 = 9.$$

6. Center $(-1, -3)$, $r = 5$
Here $h = -1$, $k = -3$, and $r = 5$, so an equation of the circle is

$$(x - h)^2 + (y - k)^2 = r^2$$
$$[x - (-1)]^2 + [y - (-3)]^2 = 5^2$$
$$(x + 1)^2 + (y + 3)^2 = 25.$$

7. Center $(4, 2)$, $r = 6$
Here $h = 4$, $k = 2$, and $r = 6$, so an equation of the circle is
$$(x - h)^2 + (y - k)^2 = r^2$$
$$(x - 4)^2 + (y - 2)^2 = 6^2$$
$$(x - 4)^2 + (y - 2)^2 = 36.$$

8. $x^2 + y^2 + 6x - 4y - 3 = 0$
Write the equation in center-radius form,
$$(x - h)^2 + (y - k)^2 = r^2,$$
by completing the squares on x and y.
$$\left(x^2 + 6x \quad\right) + \left(y^2 - 4y \quad\right) = 3$$
$$\left(x^2 + 6x + \underline{9}\right) + \left(y^2 - 4y + \underline{4}\right)$$
$$= 3 + \underline{9} + \underline{4}$$
$$(x + 3)^2 + (y - 2)^2 = 16$$
$$[x - (-3)]^2 + (y - 2)^2 = 16$$
The circle has center (h, k) at $(-3, 2)$ and radius $\sqrt{16} = 4$.

9. $x^2 + y^2 - 8x - 2y + 13 = 0$
Write the equation in center-radius form,
$$(x - h)^2 + (y - k)^2 = r^2,$$
by completing the squares on x and y.
$$\left(x^2 - 8x \quad\right) + \left(y^2 - 2y \quad\right) = -13$$
$$\left(x^2 - 8x + \underline{16}\right) + \left(y^2 - 2y + \underline{1}\right)$$
$$= -13 + \underline{16} + \underline{1}$$
$$(x - 4)^2 + (y - 1)^2 = 4$$
The circle has center (h, k) at $(4, 1)$ and radius $\sqrt{4} = 2$.

10. $2x^2 + 2y^2 + 4x + 20y = -34$
$$x^2 + y^2 + 2x + 10y = -17$$
Write the equation in center-radius form,
$$(x - h)^2 + (y - k)^2 = r^2,$$
by completing the squares on x and y.
$$\left(x^2 + 2x \quad\right) + \left(y^2 + 10y \quad\right) = -17$$
$$\left(x^2 + 2x + \underline{1}\right) + \left(y^2 + 10y + \underline{25}\right)$$
$$= -17 + \underline{1} + \underline{25}$$
$$(x + 1)^2 + (y + 5)^2 = 9$$
$$[x - (-1)]^2 + [y - (-5)]^2 = 9$$
The circle has center (h, k) at $(-1, -5)$ and radius $\sqrt{9} = 3$.

11. $4x^2 + 4y^2 - 24x + 16y = 48$
$$x^2 + y^2 - 6x + 4y = 12$$
Write the equation in center-radius form,
$$(x - h)^2 + (y - k)^2 = r^2,$$
by completing the squares on x and y.
$$\left(x^2 - 6x \quad\right) + \left(y^2 + 4y \quad\right) = 12$$
$$\left(x^2 - 6x + \underline{9}\right) + \left(y^2 + 4y + \underline{4}\right)$$
$$= 12 + \underline{9} + \underline{4}$$
$$(x - 3)^2 + (y + 2)^2 = 25$$
$$(x - 3)^2 + [y - (-2)]^2 = 25$$
The circle has center (h, k) at $(3, -2)$ and radius $\sqrt{25} = 5$.

12. $$x^2 + y^2 = 16$$
$$(x - 0)^2 + (y - 0)^2 = 4^2$$
This is a circle with center $(0, 0)$ and radius 4.

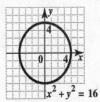

13. $\dfrac{x^2}{16} + \dfrac{y^2}{9} = 1$ is in $\dfrac{x^2}{a^2} + \dfrac{y^2}{b^2} = 1$ form with $a = 4$ and $b = 3$. The graph is an ellipse with x-intercepts $(4, 0)$ and $(-4, 0)$ and y-intercepts $(0, 3)$ and $(0, -3)$. Plot the intercepts, and draw the ellipse through them.

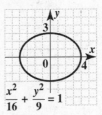

14. $\dfrac{x^2}{49} + \dfrac{y^2}{25} = 1$ is in $\dfrac{x^2}{a^2} + \dfrac{y^2}{b^2} = 1$ form with $a = 7$ and $b = 5$. The graph is an ellipse with x-intercepts $(7, 0)$ and $(-7, 0)$ and y-intercepts $(0, 5)$ and $(0, -5)$. Plot the intercepts, and draw the ellipse through them.

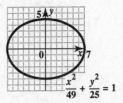

15. The total distance on the horizontal axis is $160 + 16,000 = 16,160$ km. This represents $2a$, so $a = \frac{1}{2}(16,160) = 8080$. The distance from Earth to the center of the ellipse is

$$8080 - 160 = 7920,$$

which is the value of c. From Exercise 50 in Section 11.2, we know that $c^2 = a^2 - b^2$, so

$$b^2 = a^2 - c^2.$$
$$b^2 = 8080^2 - 7920^2$$
$$= 2,560,000$$

Thus, $b = \sqrt{2,560,000} = 1600$ and the equation is

$$\frac{x^2}{8080^2} + \frac{y^2}{1600^2} = 1$$

or $\dfrac{x^2}{65,286,400} + \dfrac{y^2}{2,560,000} = 1.$

16. **(a)** The distance between the foci is $2c$, where c can be found using the relationship

$$c = \sqrt{a^2 - b^2}.$$
$$= \sqrt{310^2 - (513/2)^2}$$
$$\approx 174.1 \text{ feet}$$

So the distance is about 348.2 feet.

(b) The approximate circumference of the Roman Colosseum is

$$C \approx 2\pi \sqrt{\frac{a^2 + b^2}{2}}.$$
$$= 2\pi \sqrt{\frac{310^2 + (513/2)^2}{2}}$$
$$\approx 1787.6 \text{ feet}$$

17. $\dfrac{x^2}{16} - \dfrac{y^2}{25} = 1$ is in $\dfrac{x^2}{a^2} - \dfrac{y^2}{b^2} = 1$ form with $a = 4$ and $b = 5$. The graph is a hyperbola with x-intercepts $(4, 0)$ and $(-4, 0)$ and asymptotes that are the extended diagonals of the rectangle with vertices $(4, 5)$, $(4, -5)$, $(-4, -5)$, and $(-4, 5)$. Graph a branch of the hyperbola through each intercept approaching the asymptotes.

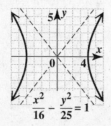

18. $\dfrac{y^2}{25} - \dfrac{x^2}{4} = 1$ is in $\dfrac{y^2}{b^2} - \dfrac{x^2}{a^2} = 1$ form with $a = 2$ and $b = 5$. The graph is a hyperbola with y-intercepts $(0, 5)$ and $(0, -5)$ and asymptotes that are the extended diagonals of the rectangle with vertices $(2, 5)$, $(2, -5)$, $(-2, -5)$, and $(-2, 5)$. Graph a branch of the hyperbola through each intercept approaching the asymptotes.

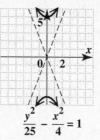

19. $$f(x) = -\sqrt{16 - x^2}$$
Replace $f(x)$ with y.
$$y = -\sqrt{16 - x^2}$$
Square both sides.
$$y^2 = 16 - x^2$$
$$x^2 + y^2 = 16$$

This equation is the graph of a circle with center $(0, 0)$ and radius 4. Since $f(x)$ represents a nonpositive square root, $f(x)$ is nonpositive and its graph is the lower half of the circle.

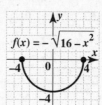

20. $$x^2 + y^2 = 64$$
$$(x - 0)^2 + (y - 0)^2 = 8^2$$

The last equation is in $(x - h)^2 + (y - k)^2 = r^2$ form. The graph is a *circle*.

21. $$y = 2x^2 - 3$$
$$y = 2(x - 0)^2 - 3$$

The last equation is in $y = a(x - h)^2 + k$ form. The graph is a *parabola*.

22. $$y^2 = 2x^2 - 8$$
$$2x^2 - y^2 = 8$$
$$\frac{x^2}{4} - \frac{y^2}{8} = 1$$

The last equation is in $\dfrac{x^2}{a^2} - \dfrac{y^2}{b^2} = 1$ form, so the graph is a *hyperbola*.

23.
$$y^2 = 8 - 2x^2$$
$$2x^2 + y^2 = 8$$
$$\frac{x^2}{4} + \frac{y^2}{8} = 1$$

The last equation is in $\frac{x^2}{a^2} + \frac{y^2}{b^2} = 1$ form, so the graph is an *ellipse*.

24.
$$x = y^2 + 4$$
$$x = (y - 0)^2 + 4$$

The last equation is in $x = a(y - k)^2 + h$ form, so the graph is a *parabola*.

25.
$$x^2 - y^2 = 64$$
$$\frac{x^2}{64} - \frac{y^2}{64} = 1$$

The last equation is in $\frac{x^2}{a^2} - \frac{y^2}{b^2} = 1$ form, so the graph is a *hyperbola*.

26. A hyperbola is defined as the set of all points in a plane such that the absolute value of the *difference* of the distances from two fixed points (called *foci*) is constant. The hyperbola shown in the text will have an equation of the form

$$\frac{x^2}{a^2} - \frac{y^2}{b^2} = 1. \quad (1)$$

The constant difference for this hyperbola is

$$|d_1 - d_2| = |80 - 30| = 50.$$

Let Q be the point on the right branch of the hyperbola that is on $\overline{MS}$. Let $x = \overline{MQ}$ and $y = \overline{QS}$. Since $\overline{MS} = 100$ and $\overline{MQ} - \overline{QS}$ must be 50, we have the following system.

$$
\begin{array}{rcll}
x + y &=& 100 & (2) \\
x - y &=& 50 & \\
\hline
2x &=& 150 & \text{Add.} \\
x &=& 75 &
\end{array}
$$

From (2), we see that $y = 25$. The distance from the center of the hyperbola, C, to S, is 50, so $a = \overline{CQ} = 50 - 25 = 25$.
To find b^2, we'll find a point on the hyperbola, substitute for a, x, and y, and then solve for b^2. Let P have coordinates (d, e). If we draw a perpendicular line from P to $\overline{MS}$, we see two right triangles, PQM and PQS. Using the Pythagorean formula, we get the following system of equations.

$$
\begin{array}{rcll}
(50 + d)^2 + e^2 &=& 80^2 & \\
(50 - d)^2 + e^2 &=& 30^2 & (3) \\
\hline
(2500 + 100d + d^2) - (2500 - 100d + d^2) & & & \\
= 6400 - 900 & & & \text{Subtract.} \\
200d = 5500 & & & \\
d = 27.5 & & &
\end{array}
$$

From (3) with $d = 27.5$, we get $e^2 = 30^2 - 22.5^2$, so $e = \sqrt{393.75}$. Now substitute 25 for a, 27.5 for x, and $\sqrt{393.75}$ for y in (1) to solve for b^2.

$$\frac{27.5^2}{25^2} - \frac{393.75}{b^2} = 1$$
$$\frac{756.25}{625} - \frac{625}{625} = \frac{393.75}{b^2}$$
$$\frac{131.25}{625} = \frac{393.75}{b^2}$$
$$\frac{625}{131.25} = \frac{b^2}{393.75}$$
$$b^2 = \frac{625}{131.25}(393.75)$$
$$b^2 = 1875$$

Thus, equation (1) becomes

$$\frac{x^2}{625} - \frac{y^2}{1875} = 1.$$

27.
$$
\begin{array}{ll}
2y = 3x - x^2 & (1) \\
x + 2y = -12 & (2)
\end{array}
$$

Substitute $3x - x^2$ for $2y$ in equation (2).

$$
\begin{array}{rl}
x + 2y = -12 & (2) \\
x + (3x - x^2) = -12 & \\
-x^2 + 4x + 12 = 0 & \\
x^2 - 4x - 12 = 0 & \\
(x - 6)(x + 2) = 0 &
\end{array}
$$

$$x - 6 = 0 \quad \text{or} \quad x + 2 = 0$$
$$x = 6 \quad \text{or} \quad x = -2$$

Substitute these values for x in equation (2) to find y.

If $x = 6$, then
$$
\begin{array}{rl}
x + 2y = -12 & (2) \\
6 + 2y = -12 & \\
2y = -18 & \\
y = -9. &
\end{array}
$$

If $x = -2$, then
$$
\begin{array}{rl}
x + 2y = -12 & (2) \\
-2 + 2y = -12 & \\
2y = -10 & \\
y = -5. &
\end{array}
$$

The solution set is $\{(6, -9), (-2, -5)\}$.

28.
$$y + 1 = x^2 + 2x \quad (1)$$
$$y + 2x = 4 \quad (2)$$

Solve equation (2) for y.

$$y = 4 - 2x \quad (3)$$

Substitute $4 - 2x$ for y in (1).

$$(4 - 2x) + 1 = x^2 + 2x$$
$$0 = x^2 + 4x - 5$$
$$0 = (x + 5)(x - 1)$$

$$x + 5 = 0 \quad \text{or} \quad x - 1 = 0$$
$$x = -5 \quad \text{or} \quad x = 1$$

Substitute these values for x in equation (3) to find y.

If $x = -5$, then $y = 4 - 2(-5) = 14$.
If $x = 1$, then $y = 4 - 2(1) = 2$.

The solution set is $\{(1, 2), (-5, 14)\}$.

29.
$$x^2 + 3y^2 = 28 \quad (1)$$
$$y - x = -2 \quad (2)$$

Solve equation (2) for y.

$$y = x - 2$$

Substitute $x - 2$ for y in equation (1).

$$x^2 + 3y^2 = 28 \quad (1)$$
$$x^2 + 3(x - 2)^2 = 28$$
$$x^2 + 3(x^2 - 4x + 4) - 28 = 0$$
$$x^2 + 3x^2 - 12x + 12 - 28 = 0$$
$$4x^2 - 12x - 16 = 0$$
$$4(x^2 - 3x - 4) = 0$$
$$4(x - 4)(x + 1) = 0$$

$$x - 4 = 0 \quad \text{or} \quad x + 1 = 0$$
$$x = 4 \quad \text{or} \quad x = -1$$

Since $y = x - 2$, if $x = 4$, then $y = 4 - 2 = 2$.

If $x = -1$, then $y = -1 - 2 = -3$.

The solution set is $\{(4, 2), (-1, -3)\}$.

30.
$$xy = 8 \quad (1)$$
$$x - 2y = 6 \quad (2)$$

Solve equation (2) for x.

$$x = 2y + 6 \quad (3)$$

Substitute $2y + 6$ for x in equation (1) to find y.

$$xy = 8 \quad (1)$$
$$(2y + 6)y = 8$$
$$2y^2 + 6y - 8 = 0$$
$$2(y^2 + 3y - 4) = 0$$
$$2(y + 4)(y - 1) = 0$$

$$y + 4 = 0 \quad \text{or} \quad y - 1 = 0$$
$$y = -4 \quad \text{or} \quad y = 1$$

Substitute these values for y in equation (3) to find x.
If $y = -4$, then $x = 2(-4) + 6 = -2$.
If $y = 1$, then $x = 2(1) + 6 = 8$.

The solution set is $\{(-2, -4), (8, 1)\}$.

31.
$$x^2 + y^2 = 6 \quad (1)$$
$$x^2 - 2y^2 = -6 \quad (2)$$

Multiply equation (2) by -1 and add the result to equation (1).

$$
\begin{array}{rl}
x^2 + y^2 = 6 & (1) \\
-x^2 + 2y^2 = 6 & -1 \times (2) \\
\hline
3y^2 = 12 & \\
y^2 = 4 &
\end{array}
$$

$$y = 2 \quad \text{or} \quad y = -2$$

Substitute these values for y in equation (1) to find x.
If $y = \pm 2$, then

$$x^2 + y^2 = 6 \quad (1)$$
$$x^2 + (\pm 2)^2 = 6$$
$$x^2 + 4 = 6$$
$$x^2 = 2.$$

$$x = \sqrt{2} \quad \text{or} \quad x = -\sqrt{2}$$

Since each value of x can be paired with each value of y, there are four points and the solution set is
$$\left\{ \left(\sqrt{2}, 2\right), \left(-\sqrt{2}, 2\right), \left(\sqrt{2}, -2\right), \left(-\sqrt{2}, -2\right) \right\}.$$

32.
$$3x^2 - 2y^2 = 12 \quad (1)$$
$$x^2 + 4y^2 = 18 \quad (2)$$

Multiply equation (1) by 2 and add the result to equation (2).

$$
\begin{array}{rl}
6x^2 - 4y^2 = 24 & 2 \times (1) \\
x^2 + 4y^2 = 18 & (2) \\
\hline
7x^2 = 42 & \\
x^2 = 6 &
\end{array}
$$

$$x = \sqrt{6} \quad \text{or} \quad x = -\sqrt{6}$$

Substitute these values for x in equation (2) to find y.
If $x = \pm\sqrt{6}$, then

$$x^2 + 4y^2 = 18. \quad (2)$$
$$\left(\pm\sqrt{6}\right)^2 + 4y^2 = 18$$
$$6 + 4y^2 = 18$$
$$4y^2 = 12$$
$$y^2 = 3$$

$y = \sqrt{3}$ or $y = -\sqrt{3}$

The solution set is

$$\left\{ (\sqrt{6}, \sqrt{3}), (\sqrt{6}, -\sqrt{3}), \right.$$
$$\left. (-\sqrt{6}, \sqrt{3}), (-\sqrt{6}, -\sqrt{3}) \right\}.$$

33. A circle and a line can intersect in zero, one, or two points, so zero, one, or two solutions are possible.

34. A parabola and a hyperbola can intersect in zero, one, two, three, or four points, so zero, one, two, three, or four solutions are possible.

35. $9x^2 \geq 16y^2 + 144$

$9x^2 - 16y^2 \geq 144$

$\dfrac{x^2}{16} - \dfrac{y^2}{9} \geq 1$

The boundary, $\dfrac{x^2}{16} - \dfrac{y^2}{9} = 1$, is a solid hyperbola with x-intercepts $(4, 0)$ and $(-4, 0)$. The asymptotes are the extended diagonals of the rectangle with vertices $(4, 3)$, $(4, -3)$, $(-4, -3)$, and $(-4, 3)$. Test $(0, 0)$.

$9(0)^2 \geq 16(0)^2 + 144$?

$0 \geq 144$ *False*

Shade the sides of the hyperbola that do not contain $(0, 0)$. These are the regions inside the branches of the hyperbola.

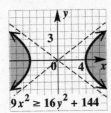

36. $4x^2 + y^2 \geq 16$

$\dfrac{x^2}{4} + \dfrac{y^2}{16} \geq 1$

The boundary, $\dfrac{x^2}{4} + \dfrac{y^2}{16} = 1$, is a solid ellipse with intercepts $(2, 0)$, $(-2, 0)$, $(0, 4)$, and $(0, -4)$. Test $(0, 0)$.

$4(0)^2 + 0^2 \geq 16$?

$0 \geq 16$ *False*

Shade the side of ellipse that does not contain $(0, 0)$. This is the region outside the ellipse.

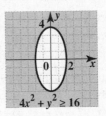

37. $y < -(x + 2)^2 + 1$

The boundary, $y = -(x + 2)^2 + 1$, is a dashed vertical parabola with vertex $(-2, 1)$. Since $a = -1 < 0$, the parabola opens down. Also, $|a| = |-1| = 1$, so the graph has the same shape as the graph of $y = x^2$. Test $(0, 0)$.

$0 < -(0 + 2)^2 + 1$?

$0 < -(4) + 1$?

$0 < -3$ *False*

Shade the side of the parabola that does not contain $(0, 0)$. This is the region inside the parabola.

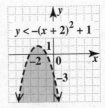

38. $2x + 5y \leq 10$
         $3x - y \leq 6$

The boundary, $2x + 5y = 10$ is a solid line with intercepts $(5, 0)$ and $(0, 2)$. Test $(0, 0)$.

$2(0) + 5(0) \leq 10$?

$0 \leq 10$ *True*

Shade the side of the line that contains $(0, 0)$. The boundary, $3x - y = 6$, is a solid line with intercepts $(2, 0)$ and $(0, -6)$. Test $(0, 0)$.

$3(0) - 0 \leq 6$?

$0 \leq 6$ *True*

Shade the side of the line that contains $(0, 0)$. The graph of the system is the intersection of the two shaded regions.

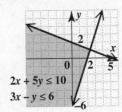

39.
$$|x| \leq 2$$
$$|y| > 1$$
$$4x^2 + 9y^2 \leq 36$$

The equation of the boundary, $|x| = 2$, can be written as

$$x = -2 \quad \text{or} \quad x = 2.$$

The graph is these two solid vertical lines. Since $|0| \leq 2$ is true, the region between the lines, containing $(0, 0)$, is shaded.

The boundary, $|y| = 1$, consists of the two dashed horizontal lines $y = 1$ and $y = -1$. Since $|0| > 1$ is false, the regions above $y = 1$ and below $y = -1$, not containing $(0, 0)$, are shaded.

The boundary given by

$$4x^2 + 9y^2 = 36$$
$$\text{or} \quad \frac{x^2}{9} + \frac{y^2}{4} = 1$$

is graphed as a solid ellipse with intercepts $(3, 0)$, $(-3, 0)$, $(0, 2)$, and $(0, -2)$. Test $(0, 0)$.

$$4(0)^2 + 9(0)^2 \leq 36 \quad ?$$
$$0 \leq 36 \quad \textit{True}$$

The region inside the ellipse, containing $(0, 0)$, is shaded.

The graph of the system consists of the regions that include the common points of the three shaded regions.

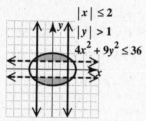

40.
$$9x^2 \leq 4y^2 + 36$$
$$x^2 + y^2 \leq 16$$

The equation of the first boundary is

$$9x^2 = 4y^2 + 36$$
$$9x^2 - 4y^2 = 36$$
$$\frac{x^2}{4} - \frac{y^2}{9} = 1.$$

The graph is a solid hyperbola with x-intercepts $(2, 0)$ and $(-2, 0)$. The asymptotes are the extended diagonals of the rectangle with vertices $(2, 3)$, $(2, -3)$, $(-2, -3)$, and $(-2, 3)$. Test $(0, 0)$.

$$9(0)^2 \leq 4(0)^2 + 36 \quad ?$$
$$0 \leq 36 \quad \textit{True}$$

Shade the region between the branches of the hyperbola that contains $(0, 0)$.

The equation of the second boundary is $x^2 + y^2 = 16$. This is a solid circle with center $(0, 0)$ and radius 4. Test $(0, 0)$.

$$0^2 + 0^2 \leq 16 \quad ?$$
$$0 \leq 16 \quad \textit{True}$$

Shade the region inside the circle.

The graph of the system is the intersection of the shaded regions which is between the two branches of the hyperbola and inside the circle.

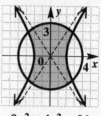

41. **[11.2]** $\frac{x^2}{64} + \frac{y^2}{25} = 1$ is in $\frac{x^2}{a^2} + \frac{y^2}{b^2} = 1$ form with $a = 8$ and $b = 5$. The graph is an ellipse with intercepts $(8, 0)$, $(-8, 0)$, $(0, 5)$, and $(0, -5)$. Plot the intercepts, and draw the ellipse through them.

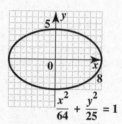

42. **[11.3]**
$$\frac{y^2}{4} - 1 = \frac{x^2}{9}$$
$$\frac{y^2}{4} - \frac{x^2}{9} = 1$$

The equation is in $\frac{y^2}{b^2} - \frac{x^2}{a^2} = 1$ form with $a = 3$ and $b = 2$. The graph is a hyperbola with y-intercepts $(0, 2)$ and $(0, -2)$ and asymptotes that are the extended diagonals of the rectangle with vertices $(3, 2)$, $(3, -2)$, $(-3, -2)$, and $(-3, 2)$. Draw a branch of the hyperbola through each intercept and approaching the asymptotes.

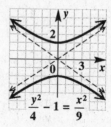

43. **[11.2]** $x^2 + y^2 = 25$ is in $(x-h)^2 + (y-k)^2 = r^2$ form. The graph is a circle with center at $(0,0)$ and radius 5.

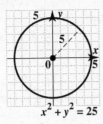

44. **[11.2]** $x^2 + 9y^2 = 9$ in $\dfrac{x^2}{a^2} + \dfrac{y^2}{b^2} = 1$ form is

$\dfrac{x^2}{9} + \dfrac{y^2}{1} = 1$ with $a = 3$ and $b = 1$. The graph is an ellipse with x-intercepts $(3,0)$ and $(-3,0)$ and y-intercepts $(0,1)$ and $(0,-1)$. Plot the intercepts, and draw the ellipse through them.

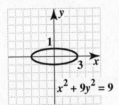

45. **[11.3]** $x^2 - 9y^2 = 9$

$$\dfrac{x^2}{9} - \dfrac{y^2}{1} = 1$$

The equation is in the $\dfrac{x^2}{a^2} - \dfrac{y^2}{b^2} = 1$ form with $a = 3$ and $b = 1$. The graph is a hyperbola with x-intercepts $(3,0)$ and $(-3,0)$ and asymptotes that are the extended diagonals of the rectangle with vertices $(3,1)$, $(3,-1)$, and $(-3,-1)$, and $(-3,1)$. Graph a branch of the hyperbola through each intercept and approaching the asymptotes.

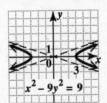

46. **[11.3]** $f(x) = \sqrt{4-x}$

Replace $f(x)$ with y.

$$y = \sqrt{4-x}$$

Square both sides.

$$y^2 = 4 - x$$
$$x = -y^2 + 4$$
$$x = -1(y-0)^2 + 4$$

This equation is the graph of a horizontal parabola with vertex $(4,0)$. Since $a = -1 < 0$, the graph opens to the left. Also, $|a| = |-1| = 1$, so the graph has the same shape as the graph of $y = x^2$.

The points $(0,2)$ and $(3,1)$ are on the graph. Since $f(x)$ represents a square root, $f(x)$ is nonnegative and its graph is the upper half of the parabola.

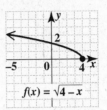

47. **[11.5]** $3x + 2y \geq 0$

$y \leq 4$

$x \leq 4$

The boundary $3x + 2y = 0$ is a solid line through $(0,0)$ and $(2,-3)$. Test $(0,1)$.

$$3(0) + 2(1) \geq 0 \quad ?$$
$$2 \geq 0 \quad \textit{True}$$

Shade the side of the line that contains $(0,1)$. The boundary $y = 4$ is a solid horizontal line through $(0,4)$. Since $y \leq 4$, shade below the line.

The boundary $x = 4$ is a solid vertical line through $(4,0)$. Since $x \leq 4$, shade the region to the left of the line.

The graph of the system is the intersection of the three shaded regions.

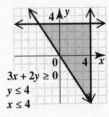

48. **[11.5]** $4y > 3x - 12$

$x^2 < 16 - y^2$

The boundary $4y = 3x - 12$ is a dashed line with intercepts $(0,-3)$ and $(4,0)$. Test $(0,0)$.

$$4(0) > 3(0) - 12 \quad ?$$
$$0 > -12 \quad \textit{True}$$

Shade the side of the line that contains $(0,0)$. The boundary $x^2 = 16 - y^2$, or $x^2 + y^2 = 16$, is a dashed circle with center at $(0,0)$ and radius 4. Test $(0,0)$.

$$0^2 < 16 - 0^2 \quad ?$$
$$0 < 16 \quad \textit{True}$$

Shade the region inside the circle. The graph of the system is the intersection of the two shaded regions.

continued

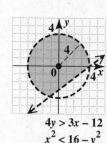

4y > 3x − 12
$x^2 < 16 - y^2$

Chapter 11 Test

1. **(a)** $f(x) = \sqrt{x - 2}$ is the graph of $g(x) = \sqrt{x}$ shifted 2 units right. The graph is **C**.

    (b) $f(x) = \sqrt{x + 2}$ is the graph of $g(x) = \sqrt{x}$ shifted 2 units left. The graph is **A**.

    (c) $f(x) = \sqrt{x} + 2$ is the graph of $g(x) = \sqrt{x}$ shifted 2 units up. The graph is **D**.

    (d) $f(x) = \sqrt{x} - 2$ is the graph of $g(x) = \sqrt{x}$ shifted 2 units down. The graph is **B**.

2. $f(x) = |x - 3| + 4$

    This is the graph of the absolute value function, $g(x) = |x|$, shifted 3 units to the right (since $x - 3 = 0$ if $x = 3$) and 4 units upward (because of the $+4$). Its lowest point is at $(3, 4)$.

x	0	1	2	3	4	5	6
y	7	6	5	4	5	6	7

    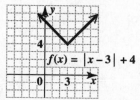

    $f(x) = |x - 3| + 4$

3. $(x - 2)^2 + (y + 3)^2 = 16$
    $(x - 2)^2 + [y - (-3)]^2 = 4^2$

    The graph is a circle with center $(2, -3)$ and radius 4.

    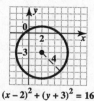

    $(x - 2)^2 + (y + 3)^2 = 16$

4. $x^2 + y^2 + 8x - 2y = 8$

    To find the center and radius, complete the squares on x and y.

    $\left(x^2 + 8x \quad\right) + \left(y^2 - 2y \quad\right) = 8$
    $\left(x^2 + 8x + \underline{16}\right) + \left(y^2 - 2y + \underline{1}\right) = 8 + \underline{16} + \underline{1}$
    $(x + 4)^2 + (y - 1)^2 = 25$

    The graph is a circle with center $(-4, 1)$ and radius $\sqrt{25} = 5$.

5. $$f(x) = \sqrt{9 - x^2}$$
    Replace $f(x)$ with y.
    $$y = \sqrt{9 - x^2}$$
    Square both sides.
    $$y^2 = 9 - x^2$$
    $$x^2 + y^2 = 9$$

    The graph of $x^2 + y^2 = 9$ is a circle of radius $\sqrt{9} = 3$ centered at the origin. Since $f(x)$ is nonnegative, only the top half of the circle is graphed.

    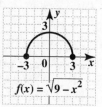

    $f(x) = \sqrt{9 - x^2}$

6. $4x^2 + 9y^2 = 36$
    $$\frac{x^2}{9} + \frac{y^2}{4} = 1$$

    The equation is in $\dfrac{x^2}{a^2} + \dfrac{y^2}{b^2} = 1$ form with $a = 3$ and $b = 2$. The graph is an ellipse with intercepts $(3, 0)$, $(-3, 0)$, $(0, 2)$, and $(0, -2)$. Plot these intercepts, and draw the ellipse through them.

    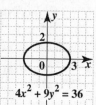

    $4x^2 + 9y^2 = 36$

7. $16y^2 - 4x^2 = 64$
    $$\frac{y^2}{4} - \frac{x^2}{16} = 1$$

    The equation is in $\dfrac{y^2}{b^2} - \dfrac{x^2}{a^2} = 1$ form with $a = 4$ and $b = 2$. The graph is a hyperbola with y-intercepts $(0, 2)$ and $(0, -2)$ and asymptotes that are the extended diagonals of the rectangle with vertices $(4, 2)$, $(4, -2)$, $(-4, -2)$, and $(-4, 2)$.

    Draw a branch of the hyperbola through each intercept and approaching the asymptotes.

    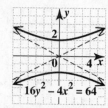

    $16y^2 - 4x^2 = 64$

8.
$$\frac{y}{2} = -\sqrt{1 - \frac{x^2}{9}}$$

Square both sides.

$$\frac{y^2}{4} = 1 - \frac{x^2}{9}$$

$$\frac{x^2}{9} + \frac{y^2}{4} = 1$$

This is an ellipse with x-intercepts $(3, 0)$ and $(-3, 0)$ and y-intercepts $(0, 2)$ and $(0, -2)$. Since y represents a negative square root in the original equation, y must be nonpositive. This restricts the graph to the lower half of the ellipse.

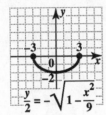

9. $6x^2 + 4y^2 = 12$

We have the *sum* of squares with different coefficients equal to a positive number, so this is an equation of an *ellipse*.

10.
$$16x^2 = 144 + 9y^2$$
$$16x^2 - 9y^2 = 144$$

We have the *difference* of squares equal to a positive number, so this is an equation of a *hyperbola*.

11. $4y^2 + 4x = 9$
$$4x = -4y^2 + 9$$

We have an x-term and a y^2-term, so this is an equation of a horizontal *parabola*.

12. $2x - y = 9$ (1)
     $xy = 5$ (2)

Solve equation (1) for y.

$$y = 2x - 9 \quad (3)$$

Substitute $2x - 9$ for y in equation (2).

$$xy = 5 \quad (2)$$
$$x(2x - 9) = 5$$
$$2x^2 - 9x = 5$$
$$2x^2 - 9x - 5 = 0$$
$$(2x + 1)(x - 5) = 0$$

$$2x + 1 = 0 \quad \text{or} \quad x - 5 = 0$$
$$x = -\tfrac{1}{2} \quad \text{or} \quad x = 5$$

Substitute these values for x in equation (3) to find y.

If $x = -\frac{1}{2}$, then $y = 2\left(-\frac{1}{2}\right) - 9 = -10$.

If $x = 5$, then $y = 2(5) - 9 = 1$.

The solution set is $\left\{ \left(-\frac{1}{2}, -10\right), (5, 1) \right\}$.

13. $x - 4 = 3y$ (1)
    $x^2 + y^2 = 8$ (2)

Solve equation (1) for x.

$$x = 3y + 4$$

Substitute $3y + 4$ for x in equation (2).

$$x^2 + y^2 = 8 \quad (2)$$
$$(3y + 4)^2 + y^2 = 8$$
$$9y^2 + 24y + 16 + y^2 = 8$$
$$10y^2 + 24y + 8 = 0$$
$$2(5y^2 + 12y + 4) = 0$$
$$2(5y + 2)(y + 2) = 0$$

$$5y + 2 = 0 \quad \text{or} \quad y + 2 = 0$$
$$y = -\tfrac{2}{5} \quad \text{or} \quad y = -2$$

Since $x = 3y + 4$, substitute these values for y to find x.

If $y = -\frac{2}{5}$, then
$$x = 3\left(-\tfrac{2}{5}\right) + 4 = -\tfrac{6}{5} + 4 = \tfrac{14}{5}.$$

If $y = -2$, then $x = 3(-2) + 4 = -2$.

The solution set is $\left\{ (-2, -2), \left(\frac{14}{5}, -\frac{2}{5}\right) \right\}$.

14. $x^2 + y^2 = 25$ (1)
    $x^2 - 2y^2 = 16$ (2)

Multiply equation (1) by 2 and add the result to equation (2).

$$
\begin{array}{rll}
2x^2 + 2y^2 &= 50 & 2 \times (1) \\
x^2 - 2y^2 &= 16 & (2) \\
\hline
3x^2 &= 66 & \\
x^2 &= 22 &
\end{array}
$$

$$x = \sqrt{22} \quad \text{or} \quad x = -\sqrt{22}$$

Substitute 22 for x^2 in equation (1).

$$x^2 + y^2 = 25 \quad (1)$$
$$22 + y^2 = 25$$
$$y^2 = 3$$

$$y = \sqrt{3} \quad \text{or} \quad y = -\sqrt{3}$$

The solution set is
$$\left\{ \left(\sqrt{22}, \sqrt{3}\right), \left(\sqrt{22}, -\sqrt{3}\right), \right.$$
$$\left. \left(-\sqrt{22}, \sqrt{3}\right), \left(-\sqrt{22}, -\sqrt{3}\right) \right\}.$$

15. $y < x^2 - 2$

The boundary, $y = x^2 - 2$, is a dashed parabola in $y = a(x - h)^2 + k$ form with vertex (h, k) at $(0, -2)$. Since $a = 1 > 0$, the parabola opens up. It also has the same shape as $y = x^2$. The points $(2, 2)$ and $(-2, 2)$ are on the graph. Test $(0, 0)$.

$$0 \le (0)^2 - 2 \quad ?$$
$$0 \le -2 \qquad \textit{False}$$

Shade the side of the parabola that does not contain $(0, 0)$. This is the region outside the parabola.

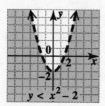

16. $x^2 + 25y^2 \le 25$
$x^2 + y^2 \le 9$

The first boundary, $\dfrac{x^2}{25} + \dfrac{y^2}{1} = 1$, is a solid ellipse with intercepts $(5, 0)$, $(-5, 0)$, $(0, 1)$, and $(0, -1)$. Test $(0, 0)$.

$$0^2 + 25 \cdot 0^2 \le 25 \quad ?$$
$$0 \le 25 \qquad \textit{True}$$

Shade the region inside the ellipse.
The second boundary, $x^2 + y^2 = 9$, is a solid circle with center $(0, 0)$ and radius 3. Test $(0, 0)$.

$$0^2 + 0^2 \le 9 \quad ?$$
$$0 \le 9 \qquad \textit{True}$$

Shade the region inside the circle. The solution of the system is the intersection of the two shaded regions.

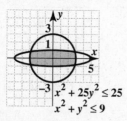

Cumulative Review Exercises (Chapters 1–11)

1.
$$-10 + |-5| - |3| + 4$$
$$= -10 + 5 - 3 + 4$$
$$= -5 - 3 + 4$$
$$= -8 + 4 = -4$$

2.
$$4 - (2x + 3) + x = 5x - 3$$
$$4 - 2x - 3 + x = 5x - 3$$
$$-x + 1 = 5x - 3$$
$$-6x = -4$$
$$x = \tfrac{2}{3}$$

The solution set is $\left\{ \tfrac{2}{3} \right\}$.

3.
$$-4k + 7 \ge 6k + 1$$
$$-10k \ge -6$$

Divide by -10; reverse the direction of the inequality.

$$k \le \tfrac{-6}{-10}$$
$$k \le \tfrac{3}{5}$$

The solution set is $\left(-\infty, \tfrac{3}{5}\right]$.

4.
$$|5m| - 6 = 14$$
$$|5m| = 20$$

$$5m = 20 \quad \text{or} \quad 5m = -20$$
$$m = 4 \quad \text{or} \quad m = -4$$

The solution set is $\{-4, 4\}$.

5. $|2p - 5| > 15$

$$2p - 5 > 15 \quad \text{or} \quad 2p - 5 < -15$$
$$2p > 20 \qquad\qquad 2p < -10$$
$$p > 10 \quad \text{or} \qquad p < -5$$

The solution set is $(-\infty, -5) \cup (10, \infty)$.

6. Let $(x_1, y_1) = (2, 5)$ and $(x_2, y_2) = (-4, 1)$.

$$m = \frac{y_2 - y_1}{x_2 - x_1} = \frac{1 - 5}{-4 - 2} = \frac{-4}{-6} = \frac{2}{3}$$

7. Through $(-3, -2)$; perpendicular to $2x - 3y = 7$

Write $2x - 3y = 7$ in slope-intercept form.

$$-3y = -2x + 7$$
$$y = \tfrac{2}{3}x - \tfrac{7}{3}$$

The slope is $\tfrac{2}{3}$. Perpendicular lines have slopes that are negative reciprocals of each other, so a line perpendicular to the given line will have slope $-\tfrac{3}{2}$. Let $m = -\tfrac{3}{2}$ and $(x_1, y_1) = (-3, -2)$ in the point-slope form.

$$y - y_1 = m(x - x_1)$$
$$y - (-2) = -\tfrac{3}{2}[x - (-3)]$$
$$y + 2 = -\tfrac{3}{2}(x + 3)$$

Multiply by 2 to clear the fraction.

$$2y + 4 = -3(x + 3)$$
$$2y + 4 = -3x - 9$$
$$3x + 2y = -13$$

8.
$$3x - y = 12 \quad (1)$$
$$2x + 3y = -3 \quad (2)$$

Multiply equation (1) by 3 and add the result to equation (2).

$$
\begin{array}{r}
9x - 3y = 36 \qquad 3 \times (1) \\
2x + 3y = -3 \quad (2) \\
\hline
11x \qquad = 33 \\
x = 3
\end{array}
$$

Substitute 3 for x in equation (1) to find y.

$$3x - y = 12 \quad (1)$$
$$3(3) - y = 12$$
$$9 - y = 12$$
$$-y = 3$$
$$y = -3$$

The solution set is $\{(3, -3)\}$.

9.
$$x + y - 2z = 9 \quad (1)$$
$$2x + y + z = 7 \quad (2)$$
$$3x - y - z = 13 \quad (3)$$

Add equation (2) and equation (3).

$$
\begin{array}{r}
2x + y + z = 7 \quad (2) \\
3x - y - z = 13 \quad (3) \\
\hline
5x \qquad = 20 \\
x = 4
\end{array}
$$

Multiply equation (1) by -1 and add the result to equation (2).

$$
\begin{array}{r}
-x - y + 2z = -9 \\
2x + y + z = 7 \quad (2) \\
\hline
x \qquad + 3z = -2 \quad (4)
\end{array}
$$

Substitute 4 for x in equation (4) to find z.

$$x + 3z = -2 \quad (4)$$
$$4 + 3z = -2$$
$$3z = -6$$
$$z = -2$$

Substitute 4 for x and -2 for z in equation (2) to find y.

$$2x + y + z = 7 \quad (2)$$
$$2(4) + y - 2 = 7$$
$$y + 6 = 7$$
$$y = 1$$

The solution set is $\{(4, 1, -2)\}$.

10.
$$xy = -5 \quad (1)$$
$$2x + y = 3 \quad (2)$$

Solve equation (2) for y.

$$y = -2x + 3 \quad (3)$$

Substitute $-2x + 3$ for y in equation (1).

$$xy = -5 \quad (1)$$
$$x(-2x + 3) = -5$$
$$-2x^2 + 3x = -5$$
$$-2x^2 + 3x + 5 = 0$$
$$2x^2 - 3x - 5 = 0$$
$$(2x - 5)(x + 1) = 0$$

$$2x - 5 = 0 \quad \text{or} \quad x + 1 = 0$$
$$x = \tfrac{5}{2} \quad \text{or} \qquad x = -1$$

Substitute these values for x in equation (3) to find y.

If $x = \tfrac{5}{2}$, then $y = -2\left(\tfrac{5}{2}\right) + 3 = -2$.

If $x = -1$, then $y = -2(-1) + 3 = 5$.

The solution set is $\left\{(-1, 5), \left(\tfrac{5}{2}, -2\right)\right\}$.

11. Let $s = $ Al's speed, $2s = $ Bev's speed, $t = $ Bev's time, and $t + \tfrac{1}{2} = $ Al's time.

	Distance	Rate	Time
Al	20	s	$t + \tfrac{1}{2}$
Bev	20	$2s$	t

Since $d = rt$, the system of equations is

$$s\left(t + \tfrac{1}{2}\right) = 20 \quad (1)$$
$$2st = 20. \quad (2)$$

Solve equation (2) for s.

$$2st = 20 \qquad\qquad (2)$$
$$s = \frac{20}{2t} = \frac{10}{t}$$

Substitute $\dfrac{10}{t}$ for s in equation (1) to find t.

$$s\left(t + \tfrac{1}{2}\right) = 20 \qquad\qquad (1)$$
$$\frac{10}{t}\left(t + \frac{1}{2}\right) = 20$$
$$10 + \frac{5}{t} = 20$$

Multiply each term by the LCD, t.

$$10t + 5 = 20t$$
$$5 = 10t$$
$$t = \tfrac{5}{10} = \tfrac{1}{2}$$

continued

Since $s = \dfrac{10}{t}$ and $t = \dfrac{1}{2}$,

$$s = \frac{10}{\frac{1}{2}} = 20.$$

So Al's speed was 20 mph and Bev's speed was $2 \cdot 20 = 40$ mph.

Another solution:

$$st + \tfrac{1}{2}s = 20 \quad (1)$$
$$2st = 20 \quad (2)$$

Multiply equation (1) by -2 and add the result to equation (2).

$$
\begin{array}{rl}
-2st - \quad s = -40 & -2 \times (1) \\
2st \qquad\;\; = \;\;20 & (2) \\
\hline
-s = -20 & \\
s = \;\;\;20 &
\end{array}
$$

Substitute 20 for s in equation (2) to find $t = \tfrac{1}{2}$.

12. $f(x) = 0.07x + 135$
$f(2000) = 0.07(2000) + 135 = 275$

The weekly fee is $275 if $2000 is taken in for the week.

13. $(5y - 3)^2 = (5y)^2 - 2(5y)3 + 3^2$
$= 25y^2 - 30y + 9$

14. $(2r + 7)(6r - 1)$
$= 12r^2 - 2r + 42r - 7$
$= 12r^2 + 40r - 7$

15. $\dfrac{8x^4 - 4x^3 + 2x^2 + 13x + 8}{2x + 1}$

$$
\begin{array}{r}
4x^3 - 4x^2 + 3x + 5 \\
2x + 1 \,\overline{\smash{\big)}\, 8x^4 - 4x^3 + 2x^2 + 13x + 8} \\
\underline{8x^4 + 4x^3} \qquad\qquad\qquad\quad \\
-8x^3 + 2x^2 \qquad\qquad \\
\underline{-8x^3 - 4x^2} \qquad\qquad \\
6x^2 + 13x \qquad \\
\underline{6x^2 + 3x} \qquad \\
10x + 8 \\
\underline{10x + 5} \\
3
\end{array}
$$

The answer is

$$4x^3 - 4x^2 + 3x + 5 + \frac{3}{2x + 1}.$$

16. $12x^2 - 7x - 10 = (4x - 5)(3x + 2)$

17. $2y^4 + 5y^2 - 3$

Let $p = y^2$, so $p^2 = y^4$.

$$
\begin{aligned}
2y^4 + 5y^2 - 3 &= 2p^2 + 5p - 3 \\
&= (2p - 1)(p + 3)
\end{aligned}
$$

Now substitute y^2 for p.

$$= (2y^2 - 1)(y^2 + 3)$$

18. $z^4 - 1 = (z^2 + 1)(z^2 - 1)$
$= (z^2 + 1)(z + 1)(z - 1)$

19. $a^3 - 27b^3 = a^3 - (3b)^3$
$= (a - 3b)(a^2 + 3ab + 9b^2)$

20. $\dfrac{5x - 15}{24} \cdot \dfrac{64}{3x - 9} = \dfrac{5(x - 3)}{3 \cdot 8} \cdot \dfrac{8 \cdot 8}{3(x - 3)}$

$$= \frac{5 \cdot 8}{3 \cdot 3} = \frac{40}{9}$$

21. $\dfrac{y^2 - 4}{y^2 - y - 6} \div \dfrac{y^2 - 2y}{y - 1}$

Multiply by the reciprocal.

$$= \frac{y^2 - 4}{y^2 - y - 6} \cdot \frac{y - 1}{y^2 - 2y}$$

Factor and simplify.

$$= \frac{(y + 2)(y - 2)}{(y - 3)(y + 2)} \cdot \frac{(y - 1)}{y(y - 2)}$$

$$= \frac{y - 1}{y(y - 3)}$$

22. $\dfrac{5}{c + 5} - \dfrac{2}{c + 3}$

The LCD is $(c + 5)(c + 3)$.

$$= \frac{5(c + 3)}{(c + 5)(c + 3)} - \frac{2(c + 5)}{(c + 3)(c + 5)}$$

$$= \frac{5c + 15 - 2c - 10}{(c + 5)(c + 3)}$$

$$= \frac{3c + 5}{(c + 5)(c + 3)}$$

23. $\dfrac{p}{p^2 + p} + \dfrac{1}{p^2 + p} = \dfrac{p + 1}{p^2 + p}$

$$= \frac{p + 1}{p(p + 1)} = \frac{1}{p}$$

24. Let x = the time to do the job working together.

Make a chart.

Worker	Rate	Time Together	Fractional Part of the Job Done
Kareem	$\dfrac{1}{3}$	x	$\dfrac{x}{3}$
Jamal	$\dfrac{1}{2}$	x	$\dfrac{x}{2}$

Part done by Kareem plus part done by Jamal equals 1 whole job.

$$\frac{x}{3} + \frac{x}{2} = 1$$

Multiply by the LCD, 6.

$$6\left(\frac{x}{3} + \frac{x}{2}\right) = 6 \cdot 1$$
$$2x + 3x = 6$$
$$5x = 6$$
$$x = \frac{6}{5} \text{ or } 1\tfrac{1}{5}$$

It takes $\frac{6}{5}$ or $1\tfrac{1}{5}$ hours to do the job together.

25. $\left(\frac{4}{3}\right)^{-1} = \left(\frac{3}{4}\right)^{1} = \frac{3}{4}$

26.
$$\frac{(2a)^{-2}a^4}{a^{-3}} = \frac{2^{-2}a^{-2}a^4}{a^{-3}} = \frac{2^{-2}a^2}{a^{-3}}$$
$$= \frac{a^2 a^3}{2^2} = \frac{a^5}{4}$$

27.
$$4\sqrt[3]{16} - 2\sqrt[3]{54} = 4\sqrt[3]{8 \cdot 2} - 2\sqrt[3]{27 \cdot 2}$$
$$= 4 \cdot 2\sqrt[3]{2} - 2 \cdot 3\sqrt[3]{2}$$
$$= 8\sqrt[3]{2} - 6\sqrt[3]{2} = 2\sqrt[3]{2}$$

28.
$$\frac{3\sqrt{5x}}{\sqrt{2x}} = \frac{3\sqrt{5x} \cdot \sqrt{2x}}{\sqrt{2x} \cdot \sqrt{2x}} = \frac{3\sqrt{10x^2}}{2x}$$
$$= \frac{3x\sqrt{10}}{2x} = \frac{3\sqrt{10}}{2}$$

29. $\dfrac{5 + 3i}{2 - i}$

Multiply the numerator and denominator by the conjugate of the denominator.

$$= \frac{(5 + 3i)(2 + i)}{(2 - i)(2 + i)}$$
$$= \frac{10 + 5i + 6i + 3i^2}{4 - i^2}$$
$$= \frac{10 + 11i + 3(-1)}{4 - (-1)}$$
$$= \frac{7 + 11i}{5} = \frac{7}{5} + \frac{11}{5}i$$

30. $2\sqrt{k} = \sqrt{5k + 3}$

$$4k = 5k + 3 \qquad \textit{Square both sides.}$$
$$-k = 3$$
$$k = -3$$

Since k must be nonnegative so that $\sqrt{k}$ is a real number, -3 cannot be a solution. The solution set is $\emptyset$.

31.
$$10q^2 + 13q = 3$$
$$10q^2 + 13q - 3 = 0$$
$$(5q - 1)(2q + 3) = 0$$
$$5q - 1 = 0 \quad \text{or} \quad 2q + 3 = 0$$
$$q = \tfrac{1}{5} \quad \text{or} \qquad q = -\tfrac{3}{2}$$

The solution set is $\left\{\frac{1}{5}, -\frac{3}{2}\right\}$.

32. $(4x - 1)^2 = 8$
$$4x - 1 = \pm\sqrt{8}$$
$$4x = 1 \pm 2\sqrt{2}$$
$$x = \frac{1 \pm 2\sqrt{2}}{4}$$

The solution set is $\left\{\frac{1 + 2\sqrt{2}}{4}, \frac{1 - 2\sqrt{2}}{4}\right\}$.

33. $3k^2 - 3k - 2 = 0$

Use the quadratic formula with $a = 3$, $b = -3$, and $c = -2$.

$$k = \frac{-b \pm \sqrt{b^2 - 4ac}}{2a}$$
$$k = \frac{-(-3) \pm \sqrt{(-3)^2 - 4(3)(-2)}}{2(3)}$$
$$= \frac{3 \pm \sqrt{9 + 24}}{6} = \frac{3 \pm \sqrt{33}}{6}$$

The solution set is $\left\{\frac{3 + \sqrt{33}}{6}, \frac{3 - \sqrt{33}}{6}\right\}$.

34. $2(x^2 - 3)^2 - 5(x^2 - 3) = 12$

Let $u = (x^2 - 3)$.

$$2u^2 - 5u = 12$$
$$2u^2 - 5u - 12 = 0$$
$$(2u + 3)(u - 4) = 0$$
$$2u + 3 = 0 \quad \text{or} \quad u - 4 = 0$$
$$u = -\tfrac{3}{2} \quad \text{or} \qquad u = 4$$

Substitute $x^2 - 3$ for u to find x.

If $u = -\frac{3}{2}$, then

$$x^2 - 3 = -\tfrac{3}{2}$$
$$x^2 = \tfrac{3}{2}$$
$$x = \pm\sqrt{\frac{3}{2}} = \pm\frac{\sqrt{3}}{\sqrt{2}} \cdot \frac{\sqrt{2}}{\sqrt{2}} = \pm\frac{\sqrt{6}}{2}.$$

If $u = 4$, then

$$x^2 - 3 = 4$$
$$x^2 = 7$$
$$x = \pm\sqrt{7}.$$

The solution set is $\left\{-\frac{\sqrt{6}}{2}, \frac{\sqrt{6}}{2}, -\sqrt{7}, \sqrt{7}\right\}$.

35. Solve $F = \dfrac{kwv^2}{r}$ for v.

$$Fr = kwv^2$$
$$v^2 = \frac{Fr}{kw}$$

Take the square root of each side.

$$v = \pm\sqrt{\frac{Fr}{kw}} = \frac{\pm\sqrt{Fr}}{\sqrt{kw}} \cdot \frac{\sqrt{kw}}{\sqrt{kw}} = \frac{\pm\sqrt{Frkw}}{kw}$$

36. $f(x) = y = x^3 + 4$

Interchange x and y and solve for y.

$$x = y^3 + 4$$
$$y^3 = x - 4$$
$$y = \sqrt[3]{x - 4}$$
$$f^{-1}(x) = \sqrt[3]{x - 4}$$

37. $a^{\log_a x} = x$, so $3^{\log_3 4} = 4$.

38. $e^{\ln x} = x$, so $e^{\ln 7} = 7$.

39. $2\log(3x + 7) - \log 4 = \log(3x + 7)^2 - \log 4$

$$= \log\frac{(3x + 7)^2}{4}$$

40. $\log(x + 2) + \log(x - 1) = 1$

$$\log_{10}[(x + 2)(x - 1)] = 1$$
$$(x + 2)(x - 1) = 10^1$$
$$x^2 + x - 2 = 10$$
$$x^2 + x - 12 = 0$$
$$(x + 4)(x - 3) = 0$$
$$x + 4 = 0 \quad \text{or} \quad x - 3 = 0$$
$$x = -4 \quad \text{or} \quad x = 3$$

The original equation is undefined when $x = -4$.

The solution set is $\{3\}$.

41. $P = 10,000, r = 0.05, t = 4$

(a) $A = P\left(1 + \dfrac{r}{n}\right)^{nt}$

$$= 10,000\left(1 + \frac{0.05}{4}\right)^{4 \cdot 4} \quad \textit{Let } n = 4.$$
$$= 10,000(1.0125)^{16}$$
$$\approx 12,198.90$$

There will be \$12,198.90 in the account.

(b) $A = Pe^{rt}$

$$= 10,000e^{(0.05)(4)}$$
$$\approx 12,214.03$$

There will be \$12,214.03 in the account.

42. $y = 1.38(1.65)^x$

$$y = 1.38(1.65)^5 \qquad \textit{Let x = 5.}$$
$$\approx 16.9 \text{ billion dollars}$$

The sales are estimated to be \$16.9 billion in 2000.

43. $2003 - 1995 = 8$

$$y = 1.38(1.65)^8 \qquad \textit{Let x = 8.}$$
$$\approx 75.8 \text{ billion dollars}$$

The sales are estimated to be \$75.8 billion in 2003.

44. $f(x) = |x - 3|$

The function is defined for any value of x, so the domain is $(-\infty, \infty)$. The result of the absolute value is nonnegative, so the range is $[0, \infty)$.

45. $f(x) = -3x + 5$

The equation is in slope-intercept form, so the y-intercept is $(0, 5)$ and $m = -3$ or $\frac{-3}{1}$.

Plot $(0, 5)$. From $(0, 5)$, move down 3 units and right 1 unit. Draw the line through these two points.

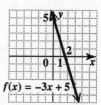

$f(x) = -3x + 5$

46. $f(x) = -2(x - 1)^2 + 3$

The graph is a parabola that has been shifted 1 unit to the right and 3 units upward from $(0, 0)$, so its vertex is at $(1, 3)$. Since $a = -2 < 0$, the parabola opens down. Also $|a| = |-2| = 2 > 1$, so the graph is narrower than the graph of $f(x) = x^2$. The points $(0, 1)$ and $(2, 1)$ are on the graph.

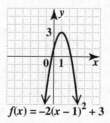

$f(x) = -2(x - 1)^2 + 3$

47. $\dfrac{x^2}{25} + \dfrac{y^2}{16} \le 1$

The boundary, $\dfrac{x^2}{25} + \dfrac{y^2}{16} = 1$, is a solid ellipse in

$\dfrac{x^2}{a^2} + \dfrac{y^2}{b^2} = 1$ form with intercepts $(5, 0)$, $(-5, 0)$,

$(0, 4)$, and $(0, -4)$. Test $(0, 0)$.

$$\dfrac{0^2}{25} + \dfrac{0^2}{16} \le 1 \quad ?$$
$$0 \le 1 \qquad True$$

Shade the region inside the ellipse.

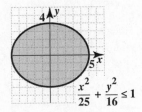

48. $f(x) = \sqrt{x - 2}$

This is the graph of $g(x) = \sqrt{x}$ shifted 2 units right.

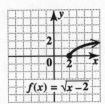

49. $\dfrac{x^2}{4} - \dfrac{y^2}{16} = 1$ is in $\dfrac{x^2}{a^2} - \dfrac{y^2}{b^2} = 1$ form.

The graph is a hyperbola with x-intercepts $(2, 0)$ and $(-2, 0)$ and asymptotes that are the extended diagonals of the rectangle with vertices $(2, 4)$, $(2, -4)$, $(-2, -4)$, and $(-2, 4)$. Draw a branch of the hyperbola through each intercept approaching the asymptotes.

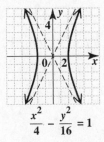

50. $f(x) = 3^x$

The graph of f is an increasing exponential.

x	-1	0	1	2
$f(x)$	$\frac{1}{3}$	1	3	9

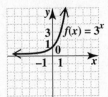

CHAPTER 12 SEQUENCES AND SERIES

12.1 Sequences and Series

1. $a_n = n + 1$
To get a_1, the first term, replace n with 1.

$$a_1 = 1 + 1 = 2$$

To get a_2, the second term, replace n with 2.

$$a_2 = 2 + 1 = 3$$

To get a_3, the third term, replace n with 3.

$$a_3 = 3 + 1 = 4$$

To get a_4, the fourth term, replace n with 4.

$$a_4 = 4 + 1 = 5$$

To get a_5, the fifth term, replace n with 5.

$$a_5 = 5 + 1 = 6$$

Answer: $2, 3, 4, 5, 6$

3. $a_n = \dfrac{n+3}{n}$
To get a_1, the first term, replace n with 1.

$$a_1 = \frac{1+3}{1} = \frac{4}{1} = 4$$

To get a_2, the second term, replace n with 2.

$$a_2 = \frac{2+3}{2} = \frac{5}{2}$$

To get a_3, the third term, replace n with 3.

$$a_3 = \frac{3+3}{3} = \frac{6}{3} = 2$$

To get a_4, the fourth term, replace n with 4.

$$a_4 = \frac{4+3}{4} = \frac{7}{4}$$

To get a_5, the fifth term, replace n with 5.

$$a_5 = \frac{5+3}{5} = \frac{8}{5}$$

Answer: $4, \frac{5}{2}, 2, \frac{7}{4}, \frac{8}{5}$

5. $a_n = 3^n$

$$a_1 = 3^1 = 3$$
$$a_2 = 3^2 = 9$$
$$a_3 = 3^3 = 27$$
$$a_4 = 3^4 = 81$$
$$a_5 = 3^5 = 243$$

Answer: $3, 9, 27, 81, 243$

7. $a_n = \dfrac{1}{n^2}$

$$a_1 = \frac{1}{1^2} = 1$$
$$a_2 = \frac{1}{2^2} = \frac{1}{4}$$
$$a_3 = \frac{1}{3^2} = \frac{1}{9}$$
$$a_4 = \frac{1}{4^2} = \frac{1}{16}$$
$$a_5 = \frac{1}{5^2} = \frac{1}{25}$$

Answer: $1, \frac{1}{4}, \frac{1}{9}, \frac{1}{16}, \frac{1}{25}$

9. $a_n = (-1)^n$

$$a_1 = (-1)^1 = -1$$
$$a_2 = (-1)^2 = 1$$
$$a_3 = (-1)^3 = -1$$
$$a_4 = (-1)^4 = 1$$
$$a_5 = (-1)^5 = -1$$

Answer: $-1, 1, -1, 1, -1$

11. $a_n = -9n + 2$
To find a_8, replace n with 8.

$$a_8 = -9(8) + 2 = -72 + 2 = -70$$

13. $a_n = \dfrac{3n+7}{2n-5}$

$$a_{14} = \frac{3(14)+7}{2(14)-5} = \frac{42+7}{28-5} = \frac{49}{23}$$

15. $a_n = (n+1)(2n+3)$

$$a_8 = (8+1)[2(8)+3]$$
$$= 9(19)$$
$$= 171$$

17. $4, 8, 12, 16, \ldots$ can be written as
$4 \cdot 1, 4 \cdot 2, 4 \cdot 3, 4 \cdot 4, \ldots$, so $a_n = 4n$.

19. $\frac{1}{3}, \frac{1}{9}, \frac{1}{27}, \frac{1}{81}, \ldots$ can be written as
$\dfrac{1}{3^1}, \dfrac{1}{3^2}, \dfrac{1}{3^3}, \dfrac{1}{3^4}, \ldots$, so $a_n = \dfrac{1}{3^n}$.

21. Make a table as follows:

Month	Interest	Payment	Unpaid balance
0			1000
1	$1000(0.01) = 10$	$100 + 10 = 110$	$1000 - 100 = 900$
2	$900(0.01) = 9$	$100 + 9 = 109$	$900 - 100 = 800$
3	$800(0.01) = 8$	$100 + 8 = 108$	$800 - 100 = 700$
4	$700(0.01) = 7$	$100 + 7 = 107$	$700 - 100 = 600$
5	$600(0.01) = 6$	$100 + 6 = 106$	$600 - 100 = 500$
6	$500(0.01) = 5$	$100 + 5 = 105$	$500 - 100 = 400$

The payments are \$110, \$109, \$108, \$107, \$106, and \$105; the unpaid balance is \$400.

23. When new, the car is worth $20,000.
Let $a_n =$ the value of the car after the nth year.
The car retains $\frac{4}{5}$ of its value each year.

$$a_1 = \tfrac{4}{5}(20{,}000) = \$16{,}000$$
(value after the first year)
$$a_2 = \tfrac{4}{5}(16{,}000) = \$12{,}800$$
$$a_3 = \tfrac{4}{5}(12{,}800) = \$10{,}240$$
$$a_4 = \tfrac{4}{5}(10{,}240) = \$8192$$
$$a_5 = \tfrac{4}{5}(8192) = \$6553.60$$

The value of the car after 5 years is about $6554.

25. $\displaystyle\sum_{i=1}^{5}(i+3)$

$= (1+3) + (2+3) + (3+3) + (4+3) + (5+3)$
Let i = 1 to 5 and add terms.
$= 4 + 5 + 6 + 7 + 8$
$= 30$

27. $\displaystyle\sum_{i=1}^{3}(i^2 + 2)$

$= (1^2 + 2) + (2^2 + 2) + (3^2 + 2)$
Let i = 1 to 3 and add terms.
$= 3 + 6 + 11$
$= 20$

29. $\displaystyle\sum_{i=1}^{6}(-1)^i$

$= (-1)^1 + (-1)^2 + (-1)^3 + (-1)^4$
$\quad + (-1)^5 + (-1)^6$
$= -1 + 1 - 1 + 1 - 1 + 1$
$= 0$

31. $\displaystyle\sum_{i=3}^{7}(i-3)(i+2)$

$= (3-3)(3+2) + (4-3)(4+2)$
$\quad + (5-3)(5+2) + (6-3)(6+2)$
$\quad + (7-3)(7+2)$
$= 0(5) + 1(6) + 2(7) + 3(8) + 4(9)$
$= 0 + 6 + 14 + 24 + 36$
$= 80$

33. $\displaystyle\sum_{i=1}^{5} 2x \cdot i$

$= 2x \cdot 1 + 2x \cdot 2 + 2x \cdot 3 + 2x \cdot 4$
$\quad + 2x \cdot 5$
$= 2x + 4x + 6x + 8x + 10x$

35. $\displaystyle\sum_{i=1}^{5} i \cdot x^i$

$= 1 \cdot x^1 + 2 \cdot x^2 + 3 \cdot x^3 + 4 \cdot x^4 + 5 \cdot x^5$
$= x + 2x^2 + 3x^3 + 4x^4 + 5x^5$

37. $3 + 4 + 5 + 6 + 7$
$= (1+2) + (2+2) + (3+2)$
$\quad + (4+2) + (5+2)$
$= \displaystyle\sum_{i=1}^{5}(i+2)$

39. $\frac{1}{2} + \frac{1}{3} + \frac{1}{4} + \frac{1}{5} + \frac{1}{6}$

$= \dfrac{1}{1+1} + \dfrac{1}{2+1} + \dfrac{1}{3+1} + \dfrac{1}{4+1}$
$\quad + \dfrac{1}{5+1}$
$= \displaystyle\sum_{i=1}^{5}\dfrac{1}{i+1}$

41. $1 + 4 + 9 + 16 + 25$
$= 1^2 + 2^2 + 3^2 + 4^2 + 5^2$
$= \displaystyle\sum_{i=1}^{5} i^2$

43. The similarities are that both are defined by the same linear expression and that points satisfying both lie in a straight line. The difference is that the domain of f consists of all real numbers, but the domain of the sequence is $\{1, 2, 3, \ldots\}$. An example of a similarity is that $f(1) = 6$ and $a_1 = 6$. An example of a difference is that $f\left(\frac{3}{2}\right) = 7$, but $a_{3/2}$ is not defined.

45. A sequence is a list of terms in a specific order, while a series is the indicated sum of the terms of a sequence.

47. $\overline{x} = \dfrac{\displaystyle\sum_{i=1}^{n} x_i}{n} = \dfrac{\displaystyle\sum_{i=1}^{7} x_i}{7}$

$= \dfrac{8 + 11 + 14 + 9 + 3 + 6 + 8}{7} = \dfrac{59}{7}$

49. $\overline{x} = \dfrac{5 + 9 + 8 + 2 + 4 + 7 + 3 + 2}{8} = \dfrac{40}{8} = 5$

51. $\overline{x} = \dfrac{8155 + 8305 + 8244 + 8126 + 8044}{5}$

$= \dfrac{40{,}874}{5} = 8174.8$

The average number of funds available for this five-year period was about 8175.

53. $a + 3d = 12$ (1)
$a + 8d = 22$ (2)

Multiply (1) by -1 and add the result to (2).

$-a - 3d = -12 \quad -1 \times (1)$
$\underline{a + 8d = \quad 22}$
$\qquad 5d = \quad 10 \quad$ *Add.*
$\qquad d = \quad 2$

Substitute 2 for d in (1).

$$a + 3d = 12 \quad (1)$$
$$a + 3(2) = 12$$
$$a + 6 = 12$$
$$a = 6$$

Thus, $a = 6$ and $d = 2$.

55. Given $a = -2$, $n = 5$, and $d = 3$,

$$a + (n-1)d = -2 + (5-1)3$$
$$= -2 + (4)3$$
$$= -2 + 12$$
$$= 10.$$

12.2 Arithmetic Sequences

1. An arithmetic sequence is a sequence (list) of numbers in a specific order such that there is a common difference between any two successive terms. For example, the sequence $1, 5, 9, 13, \ldots$ is arithmetic with difference $d = 5 - 1 = 9 - 5 = 13 - 9 = 4$. As another example, $2, -1, -4, -7, \ldots$ is an arithmetic sequence with $d = -3$.

3. $1, 2, 3, 4, 5, \ldots$

d is the difference between any two adjacent terms. Choose the terms 3 and 2.

$$d = 3 - 2 = 1$$

The terms 2 and 1 would give

$$d = 2 - 1 = 1,$$

the same result. Therefore, the common difference is $d = 1$.

Note: You should find the difference for all pairs of adjacent terms to determine if the sequence is arithmetic.

5. $2, -4, 6, -8, 10, -12, \ldots$

The difference between the first two terms is $-4 - 2 = -6$, but the difference between the second and third terms is $6 - (-4) = 10$. The differences are not the same so the sequence is *not arithmetic*.

7. $-10, -5, 0, 5, 10, \ldots$

Choose the terms 10 and 5, and find the difference.

$$d = 10 - 5 = 5$$

The terms -5 and -10 would give

$$d = -5 - (-10) = -5 + 10 = 5,$$

the same result. Therefore, the common difference is $d = 5$.

9. $a_1 = 5$
$a_2 = a_1 + d = 5 + 4 = 9$
$a_3 = a_2 + d = 9 + 4 = 13$
$a_4 = a_3 + d = 13 + 4 = 17$
$a_5 = a_4 + d = 17 + 4 = 21$

11. $a_1 = -2$
$a_2 = a_1 + d = -2 + (-4) = -6$
$a_3 = a_2 + d = -6 + (-4) = -10$
$a_4 = a_3 + d = -10 + (-4) = -14$
$a_5 = a_4 + d = -14 + (-4) = -18$

13. $a_1 = 2$, $d = 5$
$a_n = a_1 + (n-1)d$
$\quad = 2 + (n-1)5$
$\quad = 2 + 5n - 5$
$\quad = 5n - 3$

15. $3, \frac{15}{4}, \frac{9}{2}, \frac{21}{4}, \ldots$

To find d, subtract any two adjacent terms.

$$d = \frac{15}{4} - 3 = \frac{15}{4} - \frac{12}{4} = \frac{3}{4}$$

The first term is $a_1 = 3$. Now find a_n.

$$a_n = a_1 + (n-1)d$$
$$= 3 + (n-1)\left(\frac{3}{4}\right)$$
$$= 3 + \frac{3}{4}n - \frac{3}{4}$$
$$= \frac{3}{4}n + \frac{9}{4}$$

17. $-3, 0, 3, \ldots$

To find d, subtract any two adjacent terms.

$$d = 0 - (-3) = 3$$

The first term is $a_1 = -3$. Now find a_n.

$$a_n = a_1 + (n-1)d$$
$$= -3 + (n-1)3$$
$$= -3 + 3n - 3$$
$$= 3n - 6$$

19. Given $a_1 = 4$ and $d = 3$, find a_{25}.

$$a_n = a_1 + (n-1)d$$
$$a_{25} = 4 + (25-1)3$$
$$= 4 + 72$$
$$= 76$$

21. Given $2, 4, 6, \ldots$, find a_{24}.

Here, $a_1 = 2$ and $d = 4 - 2 = 2$.

$$a_n = a_1 + (n-1)d$$
$$a_{24} = 2 + (24-1)2$$
$$= 2 + 46$$
$$= 48$$

23. Given $a_{12} = -45$ and $a_{10} = -37$, find a_1.

Use $a_n = a_1 + (n-1)d$ to write a system of equations.

$$a_{12} = a_1 + (12-1)d$$
$$-45 = a_1 + 11d \qquad (1)$$
$$a_{10} = a_1 + (10-1)d$$
$$-37 = a_1 + 9d \qquad (2)$$

To eliminate d, multiply equation (1) by -9 and equation (2) by 11. Then add the results.

$$\begin{array}{rll} 405 = & -9a_1 - 99d & -9 \times (1) \\ \underline{-407 = } & \underline{11a_1 + 99d} & 11 \times (2) \\ -2 = & 2a_1 & \\ -1 = & a_1 & \end{array}$$

25. $3, 5, 7, \ldots, 33$

Let n represent the number of terms in the sequence. So, $a_n = 33$, $a_1 = 3$, and $d = 5 - 3 = 2$.

$$a_n = a_1 + (n-1)d$$
$$33 = 3 + (n-1)2$$
$$33 = 3 + 2n - 2$$
$$33 = 2n + 1$$
$$32 = 2n$$
$$n = 16$$

The sequence has 16 terms.

27. $\frac{3}{4}, 3, \frac{21}{4}, \ldots, 12$

Let n represent the number of terms in the sequence. So, $a_n = 12$, $a_1 = \frac{3}{4}$, and $d = 3 - \frac{3}{4} = \frac{9}{4}$.

$$a_n = a_1 + (n-1)d$$
$$12 = \frac{3}{4} + (n-1)\left(\frac{9}{4}\right)$$
$$\frac{45}{4} = (n-1)\left(\frac{9}{4}\right)$$
$$5 = n - 1 \qquad \textit{Multiply by } \frac{4}{9}.$$
$$6 = n$$

The sequence has 6 terms.

29. n represents the number of terms.

31. Find S_6 given $a_1 = 6$, $d = 3$, and $n = 6$.

$$S_n = \frac{n}{2}\big[2a_1 + (n-1)d\big]$$
$$S_6 = \frac{6}{2}\big[2 \cdot 6 + (6-1)3\big]$$
$$= 3(12 + 5 \cdot 3)$$
$$= 3(27)$$
$$= 81$$

33. Find S_6 given $a_1 = 7$, $d = -3$, and $n = 6$.

$$S_n = \frac{n}{2}\big[2a_1 + (n-1)d\big]$$
$$S_6 = \frac{6}{2}\big[2 \cdot 7 + (6-1)(-3)\big]$$
$$= 3\big[14 + 5(-3)\big]$$

$$= 3(14 - 15)$$
$$= 3(-1)$$
$$= -3$$

35. Find S_6 given $a_n = 4 + 3n$.

Find the first and last terms.

$$a_1 = 4 + 3(1) = 7$$
$$a_6 = 4 + 3(6) = 22$$

Now find the sum.

$$S_n = \frac{n}{2}(a_1 + a_n)$$
$$S_6 = \frac{6}{2}(7 + 22)$$
$$= 3(29)$$
$$= 87$$

37. $\displaystyle\sum_{i=1}^{10} (8i - 5)$

$$a_n = 8n - 5$$
$$a_1 = 8(1) - 5 = 3$$
$$a_{10} = 8(10) - 5 = 75$$

Use $S_n = \dfrac{n}{2}(a_1 + a_n)$ with $n = 10$, $a_1 = 3$, and $a_{10} = 75$.

$$S_{10} = \frac{10}{2}(3 + 75)$$
$$= 5(78)$$
$$= 390$$

39. $\displaystyle\sum_{i=1}^{20} (2i - 5)$

$$a_n = 2n - 5$$
$$a_1 = 2(1) - 5 = -3$$
$$a_{20} = 2(20) - 5 = 35$$

Use $S_n = \dfrac{n}{2}(a_1 + a_n)$ with $n = 20$, $a_1 = -3$, and $a_{20} = 35$.

$$S_{20} = \frac{20}{2}(-3 + 35)$$
$$= 10(32)$$
$$= 320$$

41. $\displaystyle\sum_{i=1}^{250} i$

Here, $a_n = n$, $a_1 = 1$, and $a_{250} = 250$. Use $S_n = \dfrac{n}{2}(a_1 + a_n)$ with $n = 250$, $a_1 = 1$, and $a_{250} = 250$.

$$S_{250} = \frac{250}{2}(1 + 250)$$
$$= 125(251)$$
$$= 31,375$$

43. The sequence is 1, 2, 3, ..., 30.

$$S_n = \frac{n}{2}(a_1 + a_n)$$
$$S_{30} = \frac{30}{2}(1 + 30)$$
$$= 15(31)$$
$$= 465$$

The account will have $465 deposited in it over the entire month.

45. Your salaries at six-month intervals form an arithmetic sequence with $a_1 = 1600$ and $d = 50$. Since your salary is increased every 6 months, or $2(5) = 10$ times, after 5 years your salary will equal the term a_{11}.

$$a_n = a_1 + (n-1)d$$
$$a_{11} = 1600 + (11-1)50$$
$$= 1600 + 500$$
$$= 2100$$

Your salary will be $2100/month.

47. Given the sequence 20, 22, 24, ..., for 25 terms, find a_{25}. Here, $a_1 = 20$, $d = 22 - 20 = 2$, and $n = 25$.

$$a_n = a_1 + (n-1)d$$
$$a_{25} = 20 + (25-1)2$$
$$= 20 + 48$$
$$= 68$$

There are 68 seats in the last row. Now find S_{25}.

$$S_n = \frac{n}{2}(a_1 + a_n)$$
$$S_{25} = \frac{25}{2}(20 + 68)$$
$$= \frac{25}{2}(88)$$
$$= 25(44)$$
$$= 1100$$

There are 1100 seats in the section.

49. Given the sequence 35, 31, 27, ..., can the sequence end in 1? If not, find the last positive value. If the sequence ends in 1, we can find n, a whole number.

$$d = 31 - 35 = -4$$
$$a_n = a_1 + (n-1)d$$
$$1 = 35 + (n-1)(-4)$$
$$1 = 35 - 4n + 4$$
$$-38 = -4n$$
$$9.5 = n$$

Since n is not a whole number, the sequence cannot end in 1. The largest n possible is $n = 9$.

$$a_n = a_1 + (n-1)d$$
$$a_9 = 35 + (9-1)(-4)$$
$$= 35 - 32 = 3$$

She can build 9 rows. There are 3 blocks in the last row.

51. $a = 2, r = 3, n = 2$

$$ar^n = 2(3)^2 = 2(9) = 18$$

53. $a = 4, r = \frac{1}{2}, n = 3$

$$ar^n = 4\left(\frac{1}{2}\right)^3 = 4\left(\frac{1}{8}\right) = \frac{1}{2}$$

12.3 Geometric Sequences

1. A geometric sequence is an ordered list of numbers such that each term after the first is obtained by multiplying the previous term by a constant, r, called the common ratio. For example, if the first term is 3 and $r = 4$, then the sequence is 3, 12, 48, 192, ... If the first term is 2 and $r = -1$, then the sequence is 2, -2, 2, -2, ...

3. 4, 8, 16, 32, ...

To find r, choose any two adjacent terms and divide the second one by the first one.

$$r = \frac{8}{4} = 2$$

Notice that any two other adjacent terms could have been used with the same result. The common ratio is $r = 2$.

Note: You should find the ratio for all pairs of adjacent terms to determine if the sequence is geometric.

5. $\frac{1}{3}, \frac{2}{3}, \frac{3}{3}, \frac{4}{3}, \frac{5}{3}, \cdots$

Choose any two adjacent terms and divide the second by the first.

$$r = \frac{\frac{2}{3}}{\frac{1}{3}} = \frac{2}{3} \cdot 3 = 2$$

Confirm this result with any two other adjacent terms.

$$r = \frac{\frac{3}{3}}{\frac{2}{3}} = 1 \cdot \frac{3}{2} = \frac{3}{2}$$

Since $2 \neq \frac{3}{2}$, the ratios are not the same. The sequence is *not geometric*.

7. 1, -3, 9, -27, 81, ...

[1st and 2nd] $r = \frac{-3}{1} = -3$
[2nd and 3rd] $r = \frac{9}{-3} = -3$

The common ratio is $r = -3$.

9. $1, -\frac{1}{2}, \frac{1}{4}, -\frac{1}{8}, \frac{1}{16}, \cdots$

[1st and 2nd] $r = \frac{-\frac{1}{2}}{1} = -\frac{1}{2}$

[2nd and 3rd] $r = \frac{\frac{1}{4}}{-\frac{1}{2}} = \frac{1}{4} \cdot (-2) = -\frac{1}{2}$

The common ratio is $r = -\frac{1}{2}$.

11. Find a general term for 5, 10, ...

First, find r.

$$r = \tfrac{10}{5} = 2$$

Use $a_1 = 5$ and $r = 2$ to find a_n.

$$a_n = a_1 r^{n-1}$$
$$a_n = 5(2)^{n-1}$$

13. Find a general term for $\frac{1}{9}, \frac{1}{3}, \ldots$.

Here, $a_1 = \frac{1}{9}$. Find r.

$$r = \frac{\frac{1}{3}}{\frac{1}{9}} = \frac{1}{3} \cdot \frac{9}{1} = 3$$

Now find a_n.

$$a_n = a_1 r^{n-1}$$
$$a_n = \frac{1}{9}(3)^{n-1}, \quad \text{or} \quad a_n = \frac{3^{n-1}}{9}$$

15. Find a general term for $10, -2, \ldots$.

Here, $a_1 = 10$. Find r.

$$r = \frac{-2}{10} = -\frac{1}{5}$$

Now find a_n.

$$a_n = a_1 r^{n-1}$$
$$a_n = 10\left(-\frac{1}{5}\right)^{n-1}$$

17. Substitute $a_1 = 2$, $r = 5$, and $n = 10$ in the nth-term formula to find a_{10}.

$$a_n = a_1 r^{n-1}$$
$$a_{10} = a_1(r)^{10-1}$$
$$= 2(5)^9 = 3,906,250$$

19. Given $\frac{1}{2}, \frac{1}{6}, \frac{1}{18}, \ldots$, find a_{12}.

First find the common ratio.

$$r = \frac{\frac{1}{6}}{\frac{1}{2}} = \frac{1}{6} \cdot 2 = \frac{1}{3}$$

Substitute $a_1 = \frac{1}{2}$, $r = \frac{1}{3}$, and $n = 12$ in the nth-term formula.

$$a_n = a_1 r^{n-1}$$
$$a_{12} = a_1(r)^{12-1}$$
$$= \frac{1}{2}\left(\frac{1}{3}\right)^{11}$$

21. Given $a_3 = \frac{1}{2}$ and $a_7 = \frac{1}{32}$, find a_{25}.

Find a_1 and r using the general term $a_n = a_1 r^{n-1}$.

$$a_3 = a_1 r^{3-1}$$
$$\frac{1}{2} = a_1 r^2 \quad (1)$$
$$a_7 = a_1 r^{7-1}$$
$$\frac{1}{32} = a_1 r^6 \quad (2)$$

Solve (1) for a_1.

$$a_1 = \frac{1}{2r^2}$$

Substitute $\frac{1}{2r^2}$ for a_1 in (2).

$$\frac{1}{32} = \frac{1}{2r^2} r^6$$
$$\frac{1}{16} = r^4$$
$$r^2 = \pm \frac{1}{4}$$

Since r^2 is positive,

$$r^2 = \tfrac{1}{4} \quad \text{and} \quad r = \pm \tfrac{1}{2}.$$

Substitute $\frac{1}{4}$ for r^2 in (1).

$$\tfrac{1}{2} = a_1\left(\tfrac{1}{4}\right)$$
$$2 = a_1$$

Use $a_1 = 2$ and $r = \frac{1}{2}$ (or $-\frac{1}{2}$) to find a_{25}.

$$a_{25} = a_1(r)^{25-1}$$
$$= 2\left(\tfrac{1}{2}\right)^{24}$$
$$= \frac{1}{2^{23}}$$

23. $a_1 = 2$, $r = 3$; use $a_n = a_1 r^{n-1}$.

$$a_2 = a_1 r^1 = 2(3) = 6$$
$$a_3 = a_1 r^2 = 2(3)^2 = 18$$
$$a_4 = a_1 r^3 = 2(3)^3 = 54$$
$$a_5 = a_1 r^4 = 2(3)^4 = 162$$

Answer: 2, 6, 18, 54, 162

25. $a_1 = 5$, $r = -\frac{1}{5}$; use $a_n = a_1 r^{n-1}$.

$$a_2 = a_1 r^1 = 5\left(-\tfrac{1}{5}\right) = -1$$
$$a_3 = a_1 r^2 = 5\left(-\tfrac{1}{5}\right)^2 = \tfrac{1}{5}$$
$$a_4 = a_1 r^3 = 5\left(-\tfrac{1}{5}\right)^3 = -\tfrac{1}{25}$$
$$a_5 = a_1 r^4 = 5\left(-\tfrac{1}{5}\right)^4 = \tfrac{1}{125}$$

Answer: $5, -1, \frac{1}{5}, -\frac{1}{25}, \frac{1}{125}$

27. $\frac{1}{3}, \frac{1}{9}, \frac{1}{27}, \frac{1}{81}, \frac{1}{243}$

Here, $a_1 = \frac{1}{3}$, $n = 5$, and

$$r = \frac{\frac{1}{9}}{\frac{1}{3}} = \frac{1}{9} \cdot 3 = \frac{1}{3}.$$

$$S_n = \frac{a_1(r^n - 1)}{r - 1}$$
$$S_5 = \frac{\frac{1}{3}\left[\left(\frac{1}{3}\right)^5 - 1\right]}{\frac{1}{3} - 1}$$

$$= \frac{\frac{1}{3}\left(\frac{1}{243} - 1\right)}{-\frac{2}{3}}$$

$$= \frac{\frac{1}{3}\left(-\frac{242}{243}\right)}{-\frac{2}{3}} = \frac{121}{243}$$

29. $-\frac{4}{3}, -\frac{4}{9}, -\frac{4}{27}, -\frac{4}{81}, -\frac{4}{243}, -\frac{4}{729}$

Here, $a_1 = -\frac{4}{3}$, $n = 6$, and

$$r = \frac{-\frac{4}{9}}{-\frac{4}{3}} = -\frac{4}{9} \cdot \left(-\frac{3}{4}\right) = \frac{1}{3}.$$

$$S_n = \frac{a_1(r^n - 1)}{r - 1}$$

$$S_6 = \frac{-\frac{4}{3}\left[\left(\frac{1}{3}\right)^6 - 1\right]}{\frac{1}{3} - 1}$$

$$= \frac{-\frac{4}{3}\left(\frac{1}{729} - 1\right)}{-\frac{2}{3}}$$

$$= \frac{-\frac{4}{3}\left(-\frac{728}{729}\right)}{-\frac{2}{3}} = -\frac{1456}{729} \approx -1.997$$

31. $\sum_{i=1}^{7} 4\left(\frac{2}{5}\right)^i$

Use $a_1 = 4\left(\frac{2}{5}\right) = \frac{8}{5}$, $n = 7$, and $r = \frac{2}{5}$.

$$S_n = \frac{a_1(r^n - 1)}{r - 1}$$

$$S_7 = \frac{\frac{8}{5}\left[\left(\frac{2}{5}\right)^7 - 1\right]}{\frac{2}{5} - 1}$$

$$= \frac{\frac{8}{5}\left[\left(\frac{2}{5}\right)^7 - 1\right]}{-\frac{3}{5}}$$

$$= -\frac{8}{3}\left[\left(\frac{2}{5}\right)^7 - 1\right] \approx 2.662$$

33. $\sum_{i=1}^{10} (-2)\left(\frac{3}{5}\right)^i$

Use $a_1 = (-2)\left(\frac{3}{5}\right) = -\frac{6}{5}$, $n = 10$, and $r = \frac{3}{5}$.

$$S_n = \frac{a_1(r^n - 1)}{r - 1}$$

$$S_{10} = \frac{-\frac{6}{5}\left[\left(\frac{3}{5}\right)^{10} - 1\right]}{\frac{3}{5} - 1}$$

$$= \frac{-\frac{6}{5}\left[\left(\frac{3}{5}\right)^{10} - 1\right]}{-\frac{2}{5}}$$

$$= 3\left[\left(\frac{3}{5}\right)^{10} - 1\right] \approx -2.982$$

35. There are 22 deposits, so $n = 22$.

$$S = R\left[\frac{(1+i)^n - 1}{i}\right]$$

$$= 1000\left[\frac{(1 + 0.065)^{22} - 1}{0.065}\right]$$

$$= 46,101.64$$

There will be $46,101.64 in the account.

37. Quarterly deposits for 10 years give us $n = 4 \cdot 10 = 40$. The interest rate per period is $i = \frac{0.07}{4} = 0.0175$.

$$S = R\left[\frac{(1+i)^n - 1}{i}\right]$$

$$= 1200\left[\frac{(1 + 0.0175)^{40} - 1}{0.0175}\right]$$

$$= 68,680.96$$

We now use the compound interest formula to determine the value of this money after 5 more years.

$$A = P\left(1 + \frac{r}{n}\right)^{nt}$$

$$= 68,680.96\left(1 + \frac{0.09}{12}\right)^{12(5)}$$

$$= 107,532.48$$

The woman is also saving $300 per month, so we use the annuity formula to determine that value.

$$S = R\left[\frac{(1+i)^n - 1}{i}\right]$$

$$= 300\left[\frac{\left(1 + \frac{0.09}{12}\right)^{12(5)} - 1}{\frac{0.09}{12}}\right]$$

$$= 22,627.24$$

Adding $22,627.24 to $107,532.48 gives a total of $130,159.72 in the account.

39. Find the sum if $a_1 = 6$ and $r = \frac{1}{3}$.
Since $|r| < 1$, the sum exists.

$$S = \frac{a_1}{1 - r} = \frac{6}{1 - \frac{1}{3}} = \frac{6}{\frac{2}{3}} = 6 \cdot \frac{3}{2} = 9$$

41. Find the sum if $a_1 = 1000$ and $r = -\frac{1}{10}$.
Since $|r| < 1$, the sum exists.

$$S = \frac{a_1}{1 - r} = \frac{1000}{1 - \left(-\frac{1}{10}\right)} = \frac{1000}{\frac{11}{10}}$$

$$= 1000 \cdot \frac{10}{11} = \frac{10,000}{11}$$

43. $\sum_{i=1}^{\infty} \frac{9}{8}\left(-\frac{2}{3}\right)^i$

$a_1 = \frac{9}{8}\left(-\frac{2}{3}\right)^1 = -\frac{3}{4}$ and $r = -\frac{2}{3}$.
Since $|r| < 1$, the sum exists.

$$S = \frac{a_1}{1 - r} = \frac{-\frac{3}{4}}{1 - \left(-\frac{2}{3}\right)} = \frac{-\frac{3}{4}}{\frac{5}{3}}$$

$$= -\frac{3}{4} \cdot \frac{3}{5} = -\frac{9}{20}$$

45. $\sum_{i=1}^{\infty} \frac{12}{5}\left(\frac{5}{4}\right)^i$

Since $|r| = \frac{5}{4} > 1$, the sum *does not exist*.

47. The ball is dropped from a height of 10 feet and will rebound $\frac{3}{5}$ of its original height.

Let a_n = the ball's height on the nth rebound.

$$a_1 = 10 \quad \text{and} \quad r = \frac{3}{5}.$$

Since we must find the height after the fourth bounce, $n = 5$ (since a_1 is the starting point). Use $a_n = a_1 r^{n-1}$.

$$a_5 = 10\left(\frac{3}{5}\right)^{5-1} = 10\left(\frac{3}{5}\right)^4 \approx 1.3$$

The ball will rebound approximately 1.3 feet after the fourth bounce.

49. This exercise can be modeled by a geometric sequence with $a_1 = 256$ and $r = \frac{1}{2}$. First we need to find n so that $a_n = 32$.

$$32 = a_1 r^{n-1}$$
$$32 = 256\left(\frac{1}{2}\right)^{n-1}$$
$$\frac{1}{8} = \left(\frac{1}{2}\right)^{n-1}$$
$$\left(\frac{1}{2}\right)^3 = \left(\frac{1}{2}\right)^{n-1}$$
$$3 = n - 1$$
$$4 = n$$

Since n is 4, this means that 32 grams will be present on the day which corresponds to the 4th term of the sequence. That would be on day 3. To find what is left after the tenth day, we need to find a_{11} since we started with a_1.

$$a_{11} = a_1 r^{11-1} = 256\left(\frac{1}{2}\right)^{10} = \frac{256}{1024} = \frac{1}{4}$$

There will be $\frac{1}{4}$ gram of the substance after 10 days.

51. **(a)** Here, $a_1 = 1.1$ billion and $r = 106\% = 1.06$. Since we must find the consumption after 5 years, $n = 6$ (since a_1 is the starting point).

$$a_6 = a_1 r^{n-1}$$
$$a_6 = 1.1(1.06)^{6-1}$$
$$= 1.1(1.06)^5 \approx 1.5$$

The community will use about 1.5 billion units 5 years from now.

(b) If consumption doubles, then the consumption would be $2a_1$.

$$2a_1 = a_1(1.06)^{n-1}$$
$$2 = (1.06)^{n-1}$$
$$\ln 2 = \ln(1.06)^{n-1}$$
$$\ln 2 = (n-1)\ln(1.06)$$
$$n - 1 = \frac{\ln 2}{\ln 1.06}$$
$$n - 1 \approx 12$$
$$n \approx 13$$

Since n is about 13, that would represent the 13th term of the sequence, which represents about 12 years after the start.

53. Since the machine depreciates by $\frac{1}{4}$ of its value, it retains $1 - \frac{1}{4} = \frac{3}{4}$ of its value. Since the cost of the machine new is \$50,000, $a_1 = 50,000$. We want the value after 8 years so since the original cost is a_1, we need to find a_9.

$$a_9 = a_1 r^{9-1} = 50,000\left(\frac{3}{4}\right)^8 \approx 5006$$

The machine's value after 8 years is about \$5000.

55. $\frac{1}{3} = 0.33333\ldots$

56. $\frac{2}{3} = 0.66666\ldots$

57.
$$\begin{array}{r} 0.33333\ldots \\ + 0.66666\ldots \\ \hline 0.99999\ldots \end{array}$$

58. $S = \dfrac{a_1}{1-r} = \dfrac{0.9}{1-0.1} = \dfrac{0.9}{0.9} = 1$

Therefore, $0.99999\ldots = 1$.

59. From Exercise 58, $0.99999\ldots = 1$, so choice **B** is correct.

60. $0.49999\ldots = 0.4 + 0.09999\ldots$
$$= \frac{4}{10} + \frac{1}{10}(0.9999\ldots)$$
$$= \frac{4}{10} + \frac{1}{10}(1)$$
$$= \frac{5}{10} = \frac{1}{2}$$

61. $(3x + 2y)^2 = (3x)^2 + 2(3x)(2y) + (2y)^2$
$$= 9x^2 + 12xy + 4y^2$$

63. $(a - b)^2 = a^2 - 2ab + b^2$
$(a - b)^3$
$$= (a - b)^2(a - b)^1$$
$$= \left(a^2 - 2ab + b^2\right)(a - b)$$
$$= a^3 - 2a^2b + ab^2 - a^2b + 2ab^2 - b^3$$
$$= a^3 - 3a^2b + 3ab^2 - b^3$$

12.4   The Binomial Theorem

1. $6! = 6 \cdot 5 \cdot 4 \cdot 3 \cdot 2 \cdot 1 = 720$

3. $\dfrac{6!}{4!\,2!} = \dfrac{6 \cdot 5 \cdot 4 \cdot 3 \cdot 2 \cdot 1}{(4 \cdot 3 \cdot 2 \cdot 1)(2 \cdot 1)} = \dfrac{6 \cdot 5}{2 \cdot 1} = 15$

5. ${}_6C_2 = \dfrac{6!}{2!\,(6-2)!} = \dfrac{6!}{2!\,4!} = 15,$

by Exercise 3.

7. $\dfrac{4!}{0!\,4!} = \dfrac{4!}{(1)(4!)} = \dfrac{1}{1} = 1$

9. $4! \cdot 5 = (4 \cdot 3 \cdot 2 \cdot 1) \cdot 5$
$$= 24 \cdot 5 = 120$$

11. $_{13}C_{11} = \dfrac{13!}{11!\,(13-11)!} = \dfrac{13!}{11!\,2!}$

$\qquad = \dfrac{13 \cdot 12}{2 \cdot 1} \qquad$ *11! cancels*

$\qquad = 13 \cdot 6 = 78$

13. $(m+n)^4$

$\qquad = m^4 + \dfrac{4!}{3!\,1!}m^3n^1 + \dfrac{4!}{2!\,2!}m^2n^2$

$\qquad\quad + \dfrac{4!}{1!\,3!}m^1n^3 + n^4$

$\qquad = m^4 + 4m^3n + 6m^2n^2 + 4mn^3 + n^4$

15. $(a-b)^5$

$\qquad = \left[a + (-b)\right]^5$

$\qquad = a^5 + \dfrac{5!}{4!\,1!}a^4(-b)^1 + \dfrac{5!}{3!\,2!}a^3(-b)^2$

$\qquad\quad + \dfrac{5!}{2!\,3!}a^2(-b)^3 + \dfrac{5!}{1!\,4!}a^1(-b)^4 + (-b)^5$

$\qquad = a^5 - 5a^4b + 10a^3b^2 - 10a^2b^3 + 5ab^4 - b^5$

17. $(2x+3)^3$

$\qquad = (2x)^3 + \dfrac{3!}{2!\,1!}(2x)^2(3)^1 + \dfrac{3!}{1!\,2!}(2x)(3)^2$

$\qquad\quad + (3)^3$

$\qquad = 8x^3 + 36x^2 + 54x + 27$

19. $\left(\dfrac{x}{2} - y\right)^4$

$\qquad = \left[\left(\dfrac{x}{2}\right) + (-y)\right]^4$

$\qquad = \left(\dfrac{x}{2}\right)^4 + \dfrac{4!}{3!\,1!}\left(\dfrac{x}{2}\right)^3(-y)^1 + \dfrac{4!}{2!\,2!}\left(\dfrac{x}{2}\right)^2(-y)^2$

$\qquad\quad + \dfrac{4!}{1!\,3!}\left(\dfrac{x}{2}\right)^1(-y)^3 + (-y)^4$

$\qquad = \dfrac{x^4}{16} - \dfrac{x^3y}{2} + \dfrac{3x^2y^2}{2} - 2xy^3 + y^4$

21. $(mx - n^2)^3$

$\qquad = \left[mx + (-n^2)\right]^3$

$\qquad = (mx)^3 + \dfrac{3!}{2!\,1!}(mx)^2(-n^2)^1$

$\qquad\quad + \dfrac{3!}{1!\,2!}(mx)^1(-n^2)^2 + (-n^2)^3$

$\qquad = m^3x^3 - 3m^2n^2x^2 + 3mn^4x - n^6$

23. $(r+2s)^{12}$

$\qquad = r^{12} + \dfrac{12!}{11!\,1!}r^{11}(2s)^1 + \dfrac{12!}{10!\,2!}r^{10}(2s)^2$

$\qquad\quad + \dfrac{12!}{9!\,3!}r^9(2s)^3 + \cdots$

The first four terms are

$\qquad r^{12} + 24r^{11}s + 264r^{10}s^2 + 1760r^9s^3.$

25. $(3x - y)^{14}$

$\qquad = \left[3x + (-y)\right]^{14}$

$\qquad = (3x)^{14} + \dfrac{14!}{13!\,1!}(3x)^{13}(-y)^1$

$\qquad\quad + \dfrac{14!}{12!\,2!}(3x)^{12}(-y)^2$

$\qquad\quad + \dfrac{14!}{11!\,3!}(3x)^{11}(-y)^3 + \cdots$

The first four terms are

$\qquad 3^{14}x^{14} - 14\left(3^{13}\right)x^{13}y + 91\left(3^{12}\right)x^{12}y^2$

$\qquad - 364\left(3^{11}\right)x^{11}y^3.$

27. $(t^2 + u^2)^{10}$

$\qquad = \left(t^2\right)^{10} + \dfrac{10!}{9!\,1!}\left(t^2\right)^9\left(u^2\right)^1$

$\qquad\quad + \dfrac{10!}{8!\,2!}\left(t^2\right)^8\left(u^2\right)^2$

$\qquad\quad + \dfrac{10!}{7!\,3!}\left(t^2\right)^7\left(u^2\right)^3 + \cdots$

The first four terms are

$\qquad t^{20} + 10t^{18}u^2 + 45t^{16}u^4 + 120t^{14}u^6.$

29. The rth term of the expansion of $(x+y)^n$ is

$$\dfrac{n!}{\left[n - (r-1)\right]!\,(r-1)!}(x)^{n-(r-1)}(y)^{r-1}.$$

Start with the exponent on y, which is 1 less than the term number r. In this case, we are looking for the fourth term, so $r = 4$ and $r - 1 = 3$. Thus, the fourth term of $(2m + n)^{10}$ is

$$\dfrac{10!}{(10-3)!\,3!}(2m)^{10-3}(n)^3$$

$$= \dfrac{10!}{7!\,3!}2^7m^7n^3$$

$$= 120\left(2^7\right)m^7n^3.$$

31. The seventh term of $\left(x + \dfrac{y}{2}\right)^8$ is

$$\dfrac{8!}{(8-6)!\,6!}(x)^{8-6}\left(\dfrac{y}{2}\right)^6$$

$$= \dfrac{8!}{6!\,2!}x^2\,\dfrac{y^6}{2^6}$$

$$= \dfrac{7x^2y^6}{16}.$$

33. The third term of $(k-1)^9$ is

$$\dfrac{9!}{(9-2)!\,2!}k^{9-2}(-1)^2 = \dfrac{9!}{7!\,2!}k^7 = 36k^7.$$

35. The expansion of $(x^2 - 2y)^6$ has seven terms, so the middle term is the fourth. The fourth term of $(x^2 - 2y)^6$ is

$$\frac{6!}{(6-3)!\,3!}(x^2)^{6-3}(-2y)^3$$
$$= \frac{6!}{3!\,3!}(x^2)^3(-8y^3)$$
$$= 20x^6(-8y^3) = -160x^6y^3.$$

37. The term of the expansion of $(3x^3 - 4y^2)^5$ with x^9y^4 in it is the term with $(3x^3)^3(-4y^2)^2$, since $(x^3)^3(y^2)^2 = x^9y^4$. The term is

$$\frac{5!}{3!\,2!}(3x^3)^3(-4y^2)^2$$
$$= 10(27x^9)(16y^4) = 4320x^9y^4.$$

Chapter 12 Review Exercises

1. $a_n = 2n - 3$
$a_1 = 2(1) - 3 = -1$
$a_2 = 2(2) - 3 = 1$
$a_3 = 2(3) - 3 = 3$
$a_4 = 2(4) - 3 = 5$

Answer: $-1, 1, 3, 5$

2. $a_n = \dfrac{n-1}{n}$
$a_1 = \dfrac{1-1}{1} = 0$
$a_2 = \dfrac{2-1}{2} = \dfrac{1}{2}$
$a_3 = \dfrac{3-1}{3} = \dfrac{2}{3}$
$a_4 = \dfrac{4-1}{4} = \dfrac{3}{4}$

Answer: $0, \frac{1}{2}, \frac{2}{3}, \frac{3}{4}$

3. $a_n = n^2$
$a_1 = (1)^2 = 1$
$a_2 = (2)^2 = 4$
$a_3 = (3)^2 = 9$
$a_4 = (4)^2 = 16$

Answer: $1, 4, 9, 16$

4. $a_n = \left(\frac{1}{2}\right)^n$
$a_1 = \left(\frac{1}{2}\right)^1 = \frac{1}{2}$
$a_2 = \left(\frac{1}{2}\right)^2 = \frac{1}{4}$
$a_3 = \left(\frac{1}{2}\right)^3 = \frac{1}{8}$
$a_4 = \left(\frac{1}{2}\right)^4 = \frac{1}{16}$

Answer: $\frac{1}{2}, \frac{1}{4}, \frac{1}{8}, \frac{1}{16}$

5. $a_n = (n+1)(n-1)$
$a_1 = (1+1)(1-1) = 2(0) = 0$
$a_2 = (2+1)(2-1) = 3(1) = 3$
$a_3 = (3+1)(3-1) = 4(2) = 8$
$a_4 = (4+1)(4-1) = 5(3) = 15$

Answer: $0, 3, 8, 15$

6. $\displaystyle\sum_{i=1}^{5} i^2 x$
$= 1^2x + 2^2x + 3^2x + 4^2x + 5^2x$
$= x + 4x + 9x + 16x + 25x$

7. $\displaystyle\sum_{i=1}^{6} (i+1)x^i$
$= (1+1)x^1 + (2+1)x^2 + (3+1)x^3$
$\quad + (4+1)x^4 + (5+1)x^5 + (6+1)x^6$
$= 2x + 3x^2 + 4x^3 + 5x^4 + 6x^5 + 7x^6$

8. $\displaystyle\sum_{i=1}^{4} (i+2)$
$= (1+2) + (2+2) + (3+2) + (4+2)$
$= 3 + 4 + 5 + 6$
$= 18$

9. $\displaystyle\sum_{i=1}^{6} 2^i$
$= 2^1 + 2^2 + 2^3 + 2^4 + 2^5 + 2^6$
$= 2 + 4 + 8 + 16 + 32 + 64$
$= 126$

10. $\displaystyle\sum_{i=4}^{7} \frac{i}{i+1}$
$= \frac{4}{4+1} + \frac{5}{5+1} + \frac{6}{6+1} + \frac{7}{7+1}$
$= \frac{4}{5} + \frac{5}{6} + \frac{6}{7} + \frac{7}{8} \qquad LCD = 2^3 \cdot 3 \cdot 5 \cdot 7 = 840$
$= \frac{672}{840} + \frac{700}{840} + \frac{720}{840} + \frac{735}{840} = \frac{2827}{840}$

11. $\bar{x} = \dfrac{2535 + 2478 + 2342 + 2078 + 2662}{5}$

$\bar{x} = \dfrac{12{,}095}{5} = 2419$

The average mutual fund retirement assets were 2419 billion dollars for the five given years.

12. $2, 5, 8, 11, \ldots$ is an *arithmetic* sequence with

$$d = 5 - 2 = 3.$$

Note: You should find the difference for all pairs of adjacent terms to determine if the sequence is arithmetic.

13. $-6, -2, 2, 6, 10, \ldots$ is an *arithmetic* sequence with

$$d = -2 - (-6) = 4.$$

14. $\frac{2}{3}, -\frac{1}{3}, \frac{1}{6}, -\frac{1}{12}, \ldots$ is a *geometric* sequence with

$$r = \frac{-\frac{1}{3}}{\frac{2}{3}} = -\frac{1}{3} \cdot \frac{3}{2} = -\frac{1}{2}.$$

Note: You should find the ratio for all pairs of adjacent terms to determine if the sequence is geometric.

15. $-1, 1, -1, 1, -1, \ldots$ is a *geometric* sequence with

$$r = \frac{1}{-1} = -1.$$

16. $64, 32, 8, \frac{1}{2}, \ldots$

Find two differences:
$32 - 64 = -32$ and $8 - 32 = -24$, so the sequence is not arithmetic.

Find two ratios:
$\frac{32}{64} \neq \frac{8}{32}$, so the sequence is not geometric.

Therefore, the sequence is *neither*.

17. $64, 32, 16, 8, \ldots$ is a *geometric* sequence with

$$r = \frac{32}{64} = \frac{1}{2}.$$

18. $10, 8, 6, 4, \ldots$ is an *arithmetic* sequence with

$$d = 8 - 10 = -2.$$

19. Given $a_1 = -2$ and $d = 5$, find a_{16}.

$$\begin{aligned} a_n &= a_1 + (n-1)d \\ a_{16} &= -2 + (16-1)5 \\ &= -2 + 15(5) \\ &= -2 + 75 = 73 \end{aligned}$$

20. Given $a_6 = 12$ and $a_8 = 18$, find a_{25}.

$$\begin{aligned} a_n &= a_1 + (n-1)d \\ a_6 &= a_1 + 5d, \quad \text{so} \quad 12 = a_1 + 5d \quad (1) \\ a_8 &= a_1 + 7d, \quad \text{so} \quad 18 = a_1 + 7d \quad (2) \end{aligned}$$

Multiply equation (1) by -1 and add the result to equation (2).

$$\begin{array}{ll} -12 = -a_1 - 5d & -1 \times (1) \\ \underline{18 = a_1 + 7d} & (2) \\ 6 = 2d \\ 3 = d \end{array}$$

To find a_1, substitute $d = 3$ in equation (1).

$$\begin{aligned} 12 &= a_1 + 5d \quad (1) \\ 12 &= a_1 + 5(3) \\ 12 &= a_1 + 15 \\ -3 &= a_1 \end{aligned}$$

Use $a_1 = -3$ and $d = 3$ to find a_{25}.

$$\begin{aligned} a_n &= a_1 + (n-1)d \\ a_{25} &= -3 + (25-1)3 \\ &= -3 + 24(3) \\ &= -3 + 72 = 69 \end{aligned}$$

21. $a_1 = -4$, $d = -5$

$$\begin{aligned} a_n &= a_1 + (n-1)d \\ a_n &= -4 + (n-1)(-5) \\ &= -4 - 5n + 5 \\ &= -5n + 1 \end{aligned}$$

22. $6, 3, 0, -3, \ldots$

To get the general term, a_n, first find d.
$d = 3 - 6 = -3$

$$\begin{aligned} a_n &= a_1 + (n-1)d \\ a_n &= 6 + (n-1)(-3) \\ &= 6 - 3n + 3 \\ &= -3n + 9 \end{aligned}$$

23. $7, 10, 13, \ldots, 49$

Here, $a_1 = 7$ and $d = 10 - 7 = 3$.
Now find n, the number of terms.

$$\begin{aligned} a_n &= a_1 + (n-1)d \\ 49 &= 7 + (n-1)(3) \\ 42 &= 3(n-1) \\ 14 &= n - 1 \qquad \textit{Divide by 3.} \\ 15 &= n \end{aligned}$$

There are 15 terms in this sequence.

24. $5, 1, -3, \ldots, -79$

Here, $a_1 = 5$ and $d = 1 - 5 = -4$.
Now find n, the number of terms.

$$\begin{aligned} a_n &= a_1 + (n-1)d \\ -79 &= 5 + (n-1)(-4) \\ -79 &= 9 - 4n \\ -88 &= -4n \\ n &= 22 \end{aligned}$$

There are 22 terms in this sequence.

25. Find S_8 if $a_1 = -2$ and $d = 6$. Find a_8 first.

$$\begin{aligned} a_8 &= a_1 + (8-1)d \\ &= -2 + 7(6) \\ &= -2 + 42 = 40 \end{aligned}$$

Now find the sum.

$$\begin{aligned} S_n &= \frac{n}{2}(a_1 + a_n) \\ S_8 &= \frac{8}{2}(-2 + 40) \\ &= 4(38) = 152 \end{aligned}$$

26. Find S_8 if $a_n = -2 + 5n$.
Find the first and last terms.

$$a_1 = -2 + 5(1) = 3$$
$$a_8 = -2 + 5(8) = 38$$

Now find the sum.

$$S_n = \frac{n}{2}(a_1 + a_n)$$
$$S_8 = \frac{8}{2}(3 + 38)$$
$$= 4(41)$$
$$= 164$$

27. Find the general term for the geometric sequence $-1, -4, \ldots$.

$$a_1 = -1 \quad \text{and} \quad r = \frac{-4}{-1} = 4.$$
$$a_n = a_1 r^{n-1}$$
$$a_n = -1(4)^{n-1}$$

28. $\frac{2}{3}, \frac{2}{15}, \ldots$

$$a_1 = \frac{2}{3} \quad \text{and} \quad r = \frac{\frac{2}{15}}{\frac{2}{3}} = \frac{2}{15} \cdot \frac{3}{2} = \frac{1}{5}.$$
$$a_n = a_1 r^{n-1}$$
$$a_n = \frac{2}{3}\left(\frac{1}{5}\right)^{n-1}$$

29. Find a_{11} for $2, -6, 18, \ldots$.

$$a_1 = 2 \quad \text{and} \quad r = \frac{-6}{2} = -3.$$
$$a_n = a_1 r^{n-1}$$
$$a_{11} = 2(-3)^{11-1}$$
$$= 2(-3)^{10} = 118{,}098$$

30. Given $a_3 = 20$ and $a_5 = 80$, find a_{10}.
$$a_n = a_1 r^{n-1}$$
For a_3, $\qquad a_3 = a_1 r^{3-1}$
$$20 = a_1 r^2.$$
For a_5, $\qquad a_5 = a_1 r^{5-1}$
$$80 = a_1 r^4.$$
The ratio of a_5 to a_3 is

$$\frac{80}{20} = \frac{a_1 r^4}{a_1 r^2}$$
$$4 = r^2$$
$$r = \pm 2.$$

Since $20 = a_1 r^2$, and $r^2 = 4$,
$$20 = a_1(4)$$
$$5 = a_1.$$
Now find a_{10}.
$$a_n = a_1 r^{n-1}$$
$$a_{10} = 5(\pm 2)^{10-1}$$
$$= 5(\pm 2)^9$$

Two answers are possible for a_{10}:

$$5(2)^9 = 2560 \quad \text{or} \quad 5(-2)^9 = -2560.$$

31. $\displaystyle\sum_{i=1}^{5} \left(\frac{1}{4}\right)^i$
$a_1 = \frac{1}{4}$, $r = \frac{1}{4}$, and $n = 5$.

$$S_n = \frac{a_1(1 - r^n)}{1 - r}$$
$$S_5 = \frac{\frac{1}{4}\left[1 - \left(\frac{1}{4}\right)^5\right]}{1 - \frac{1}{4}}$$
$$= \frac{\frac{1}{4}\left(1 - \frac{1}{1024}\right)}{\frac{3}{4}}$$
$$= \frac{1}{3}\left(\frac{1023}{1024}\right) = \frac{341}{1024}$$

32. $\displaystyle\sum_{i=1}^{8} \frac{3}{4}(-1)^i$
$a_1 = -\frac{3}{4}$, $r = -1$, and $n = 8$.

$$S_n = \frac{a_1(1 - r^n)}{1 - r}$$
$$S_8 = \frac{-\frac{3}{4}\left[1 - (-1)^8\right]}{1 - (-1)}$$
$$= \frac{-\frac{3}{4}(1 - 1)}{2}$$
$$= \frac{-\frac{3}{4}(0)}{2} = 0$$

33. $\displaystyle\sum_{i=1}^{\infty} 4\left(\frac{1}{5}\right)^i$
The terms are the terms of an infinite geometric sequence with $a_1 = \frac{4}{5}$ and $r = \frac{1}{5}$.

$$S = \frac{a_1}{1 - r} = \frac{\frac{4}{5}}{1 - \frac{1}{5}} = \frac{\frac{4}{5}}{\frac{4}{5}} = 1$$

34. $\displaystyle\sum_{i=1}^{\infty} 2(3)^i$
The terms are the terms of an infinite geometric sequence with $a_1 = 6$ and $r = 3$.

$$S = \frac{a_1}{1 - r} \quad \text{if} \quad |r| < 1,$$

but $r = 3$ so S does not exist, and, thus, the sum does not exist.

35. $(2p - q)^5$
$$= [2p + (-q)]^5$$
$$= (2p)^5 + \frac{5!}{4!\,1!}(2p)^4(-q)^1$$
$$+ \frac{5!}{3!\,2!}(2p)^3(-q)^2 + \frac{5!}{2!\,3!}(2p)^2(-q)^3$$
$$+ \frac{5!}{1!\,4!}(2p)^1(-q)^4 + (-q)^5$$
$$= 32p^5 + 5(16p^4)(-q) + 10(8p^3)q^2$$
$$+ 10(4p^2)(-q^3) + 5(2p)q^4 - q^5$$
$$= 32p^5 - 80p^4q + 80p^3q^2 - 40p^2q^3$$
$$+ 10pq^4 - q^5$$

36. $(x^2 + 3y)^4$

$$= (x^2)^4 + \frac{4!}{3!\,1!}(x^2)^3(3y)^1$$
$$+ \frac{4!}{2!\,2!}(x^2)^2(3y)^2$$
$$+ \frac{4!}{1!\,3!}(x^2)^1(3y)^3 + (3y)^4$$
$$= x^8 + 4(x^6)(3y) + 6(x^4)(9y^2)$$
$$+ 4(x^2)(27y^3) + 81y^4$$
$$= x^8 + 12x^6y + 54x^4y^2 + 108x^2y^3$$
$$+ 81y^4$$

37. $\left(\sqrt{m} + \sqrt{n}\right)^4$

$$= \left(\sqrt{m}\right)^4 + \frac{4!}{3!\,1!}\left(\sqrt{m}\right)^3\left(\sqrt{n}\right)^1$$
$$+ \frac{4!}{2!\,2!}\left(\sqrt{m}\right)^2\left(\sqrt{n}\right)^2$$
$$+ \frac{4!}{1!\,3!}\left(\sqrt{m}\right)^1\left(\sqrt{n}\right)^3 + \left(\sqrt{n}\right)^4$$
$$= m^2 + 4(m\sqrt{m})(\sqrt{n}) + 6(m)(n)$$
$$+ 4(\sqrt{m})(n\sqrt{n}) + n^2$$
$$= m^2 + 4m\sqrt{mn} + 6mn + 4n\sqrt{mn} + n^2$$

38. The fourth term ($r = 4$, so $r - 1 = 3$) of $(3a + 2b)^{19}$ is

$$\frac{19!}{16!\,3!}(3a)^{16}(2b)^3$$
$$= 969(3)^{16}(a^{16})(8)b^3$$
$$= 7752(3)^{16}a^{16}b^3.$$

39. The twenty-third term ($r = 23$, so $r - 1 = 22$) of $(-2k + 3)^{25}$ is

$$= \frac{25!}{3!\,22!}(-2k)^3(3)^{22}$$
$$= -18{,}400(3)^{22}k^3.$$

40. **[12.2]** The arithmetic sequence $1, 7, 13, \ldots$ has $a_1 = 1$ and $d = 7 - 1 = 6$. First find a_{40}.

$$a_n = a_1 + (n-1)d$$
$$a_{40} = 1 + (40 - 1)6$$
$$= 1 + (39)6$$
$$= 1 + 234 = 235$$

Now find the tenth term so that we can find the sum of the first 10 terms.

$$a_{10} = 1 + (10 - 1)6$$
$$= 1 + (9)6 = 55$$

Now find the sum.

$$S_n = \frac{n}{2}(a_1 + a_n)$$
$$S_{10} = \frac{10}{2}(1 + 55)$$
$$= 5(56) = 280$$

41. **[12.3]** The geometric sequence $-3, 6, -12, \ldots$ has $a_1 = -3$ and $r = \frac{6}{-3} = -2$. First find a_{10}.

$$a_n = a_1 r^{n-1}$$
$$a_{10} = -3(-2)^9$$
$$= -3(-512) = 1536$$

Now find the sum of the first ten terms.

$$S_n = \frac{a_1(1 - r^n)}{1 - r}$$
$$S_{10} = \frac{-3[1 - (-2)^{10}]}{1 - (-2)}$$
$$= \frac{-3(1 - 1024)}{3}$$
$$= -(-1023) = 1023$$

42. **[12.3]** The geometric sequence has $a_1 = 1$ and $r = -3$. First find a_9.

$$a_n = a_1 r^{n-1}$$
$$a_9 = 1(-3)^{9-1}$$
$$= (-3)^8 = 6561$$

Now find the sum of the first ten terms.

$$S_n = \frac{a_1(r^n - 1)}{r - 1}$$
$$S_{10} = \frac{1[(-3)^{10} - 1]}{-3 - 1}$$
$$= -\tfrac{1}{4}(3^{10} - 1)$$
$$= -\tfrac{1}{4}(59{,}049 - 1)$$
$$= -14{,}762$$

43. **[12.2]** The arithmetic sequence with $a_1 = -4$ and $d = 3$ is $-4, -1, 2, 5, \ldots$. First find a_{15}.

$$a_n = a_1 + (n-1)d$$
$$a_{15} = -4 + (15 - 1)3$$
$$= -4 + 42 = 38$$

Now find the sum of the first ten terms.

$$S_n = \frac{n}{2}[2a_1 + (n-1)d]$$
$$S_{10} = \frac{10}{2}[2(-4) + (10 - 1)3]$$
$$= 5(-8 + 27)$$
$$= 5(19) = 95$$

44. **[12.2]** $2, 7, 12, \ldots$
This is an arithmetic sequence with $a_1 = 2$ and $d = 7 - 2 = 5$.

$$a_n = a_1 + (n-1)d$$
$$a_n = 2 + (n-1)5$$
$$= 2 + 5n - 5$$
$$= 5n - 3$$

45. **[12.3]** $2, 8, 32, \ldots$
This is a geometric sequence with $a_1 = 2$ and $r = \frac{8}{2} = 4$.

$$a_n = a_1 r^{n-1}$$
$$a_n = 2(4)^{n-1}$$

46. **[12.3]** $27, 9, 3, \ldots$
This is a geometric sequence with $a_1 = 27$ and $r = \frac{9}{27} = \frac{1}{3}$.

$$a_n = a_1 r^{n-1}$$
$$a_n = 27\left(\frac{1}{3}\right)^{n-1}$$

47. **[12.2]** $12, 9, 6, \ldots$
This is an arithmetic sequence with $a_1 = 12$ and $d = -3$.

$$a_n = a_1 + (n-1)d$$
$$a_n = 12 + (n-1)(-3)$$
$$= 12 - 3n + 3$$
$$= -3n + 15$$

48. **[12.2]** The distances traveled in successive seconds are

$$3, 7, 11, 15, 19, \ldots.$$

This is an arithmetic sequence with $a_1 = 3$ and $d = 4$. Since we know a_1 and d, we'll use the second formula for the sum of an arithmetic sequence with $S_n = 210$.

$$S_n = \frac{n}{2}\left[2a_1 + (n-1)d\right]$$
$$210 = \frac{n}{2}\left[2(3) + (n-1)4\right]$$
$$420 = n(6 + 4n - 4)$$
$$420 = 6n + 4n^2 - 4n$$
$$0 = 4n^2 + 2n - 420$$
$$0 = 2n^2 + n - 210$$
$$0 = (2n + 21)(n - 10)$$
$$2n + 21 = 0 \quad \text{or} \quad n - 10 = 0$$
$$n = -\frac{21}{2} \quad \text{or} \quad n = 10$$

Discard $-\frac{21}{2}$ since time cannot be negative. It takes her 10 seconds.

49. **[12.3]** Use the formula for the future value of an ordinary annuity with $R = 672$, $i = \frac{0.06}{4} = 0.015$, and $n = 7(4) = 28$.

$$S = R\left[\frac{(1+i)^n - 1}{i}\right]$$
$$S = 672\left[\frac{(1+0.015)^{28} - 1}{0.015}\right]$$
$$= 23,171.55$$

The future value of the annuity is $23,171.55.

50. **[12.1]** Since $100\% - 3\% = 97\% = 0.97$, the population after 1 year is $0.97(50,000)$, after 2 years is $0.97\left[0.97(50,000)\right]$ or $(0.97)^2(50,000)$, and after n years is $(0.97)^n(50,000)$. After 6 years, the population is

$$(0.97)^6(50,000) \approx 41,649 \approx 42,000.$$

51. **[12.1]** $\left(\frac{1}{2}\right)^n$ is left after n strokes. So $\left(\frac{1}{2}\right)^7 = \frac{1}{128} = 0.0078125$ is left after 7 strokes.

52. **[12.3]** **(a)** We can write the repeating decimal number $0.55555\ldots$ as an infinite geometric sequence as follows:

$$\frac{5}{10} + \frac{5}{10}\left(\frac{1}{10}\right) + \frac{5}{10}\left(\frac{1}{10}\right)^2 + \frac{5}{10}\left(\frac{1}{10}\right)^3 + \cdots$$

(b) The common ratio r is

$$\frac{\frac{5}{10}\left(\frac{1}{10}\right)}{\frac{5}{10}} = \frac{1}{10}.$$

(c) Since $|r| < 1$, the sum exists.

$$S = \frac{a_1}{1-r} = \frac{\frac{5}{10}}{1 - \frac{1}{10}} = \frac{\frac{5}{10}}{\frac{9}{10}} = \frac{5}{9}$$

53. **[12.3]** No, the sum cannot be found, because $r = 2$ and this value of r does not satisfy $|r| < 1$.

54. **[12.3]** No, the terms must be successive, such as the first and second or the second and third.

Chapter 12 Test

1. $a_n = (-1)^n + 1$
$a_1 = (-1)^1 + 1 = 0$
$a_2 = (-1)^2 + 1 = 1 + 1 = 2$
$a_3 = (-1)^3 + 1 = -1 + 1 = 0$
$a_4 = (-1)^4 + 1 = 1 + 1 = 2$
$a_5 = (-1)^5 + 1 = -1 + 1 = 0$

Answer: $0, 2, 0, 2, 0$

2. $a_1 = 4$, $d = 2$
$a_2 = a_1 + d = 4 + 2 = 6$
$a_3 = a_2 + d = 6 + 2 = 8$
$a_4 = a_3 + d = 8 + 2 = 10$
$a_5 = a_4 + d = 10 + 2 = 12$

Answer: $4, 6, 8, 10, 12$

3. $a_4 = 6$, $r = \frac{1}{2}$
First find a_1.

$$a_n = a_1 r^{n-1}$$
$$a_4 = a_1\left(\frac{1}{2}\right)^{4-1}$$
$$6 = a_1\left(\frac{1}{8}\right)$$
$$a_1 = 48$$

Now find the remaining terms.

$$a_2 = \tfrac{1}{2}a_1 = \tfrac{1}{2}(48) = 24$$
$$a_3 = \tfrac{1}{2}a_2 = \tfrac{1}{2}(24) = 12$$
$$a_4 = \tfrac{1}{2}a_3 = \tfrac{1}{2}(12) = 6$$
$$a_5 = \tfrac{1}{2}a_4 = \tfrac{1}{2}(6) = 3$$

Answer: 48, 24, 12, 6, 3

4. Given $a_1 = 6$ and $d = -2$, find a_4.

$$a_n = a_1 + (n-1)d$$
$$a_4 = a_1 + (4-1)d$$
$$= 6 + (3)(-2)$$
$$= 6 - 6 = 0$$

5. Given $a_5 = 16$ and $a_7 = 9$, find a_4.
This is a geometric sequence, so

$$a_6 = a_5 r \quad \text{and} \quad a_7 = a_6 r.$$

From the last equation, $a_6 = \dfrac{a_7}{r}$, so by
substitution,

$$a_5 r = \frac{a_7}{r}.$$

Multiply by r and divide by a_5 to get
$$r^2 = \frac{a_7}{a_5}$$
$$r^2 = \frac{9}{16}$$
$$r = \pm\sqrt{\frac{9}{16}} = \pm\frac{3}{4}$$

Since $a_5 = a_4 r$, we know that $a_4 = \dfrac{16}{r}$.
Substituting each value of r in the last equation
gives us two values of a_4.

$$a_4 = \frac{16}{\frac{3}{4}} = \frac{64}{3} \quad \text{or} \quad a_4 = \frac{16}{-\frac{3}{4}} = -\frac{64}{3}$$

6. Given the arithmetic sequence with $a_2 = 12$ and
$a_3 = 15$, find S_5. First find d.

$$d = a_3 - a_2 = 15 - 12 = 3$$

Now find the first and fifth terms.

$$a_1 = a_2 - 3 = 12 - 3 = 9$$
$$a_5 = a_4 + 3 = a_3 + 6 = 15 + 6 = 21$$

Now find the sum of the first five terms.

$$S_n = \frac{n}{2}(a_1 + a_n)$$
$$S_5 = \frac{5}{2}(9 + 21)$$
$$= \frac{5}{2}(30) = 75$$

7. Given the geometric sequence with $a_5 = 4$ and
$a_7 = 1$, find S_5.

$$r^2 = \frac{a_7}{a_5} = \frac{1}{4}, \text{ so } r = \frac{1}{2} \text{ or } r = -\frac{1}{2}.$$

Use $a_7 = a_1 r^6$ to get $1 = a_1\left(\pm\frac{1}{2}\right)^6$, and so
$a_1 = 64$.

Use $S_n = \dfrac{a_1(r^n - 1)}{r - 1}$ or $S_n = \dfrac{a_1(1 - r^n)}{1 - r}$.

$$S_5 = \frac{64\left[1 - \left(\frac{1}{2}\right)^5\right]}{1 - \frac{1}{2}} \quad \text{or} \quad S_5 = \frac{64\left[1 - \left(-\frac{1}{2}\right)^5\right]}{1 - \left(-\frac{1}{2}\right)}$$

$$= \frac{64}{\frac{1}{2}}\left(1 - \frac{1}{32}\right) \qquad = \frac{64}{\frac{3}{2}}\left(1 + \frac{1}{32}\right)$$

$$= 128\left(\tfrac{31}{32}\right) \qquad\qquad = \frac{128}{3}\left(\tfrac{33}{32}\right)$$

$$= 124 \qquad\qquad\qquad = 44$$

8. $\overline{x} = \dfrac{\text{total}}{5}$, where total $=$

$$71{,}911 + 72{,}458 + 73{,}527 + 74{,}638 + 76{,}579.$$

Thus, $\overline{x} = \dfrac{369{,}113}{5} = 73{,}822.6 \approx 73{,}823.$

The average number of banks for this five-year
period was 73,823.

9. $S = R\left[\dfrac{(1+i)^n - 1}{i}\right]$

$$= 4000\left[\frac{\left(1 + \frac{0.06}{4}\right)^{4(7)} - 1}{\frac{0.06}{4}}\right]$$

$$= 4000\left[\frac{(1.015)^{28} - 1}{0.015}\right]$$

$$\approx 137{,}925.91$$

The account will have \$137,925.91 at the end of
this term.

10. An infinite geometric series has a sum if $|r| < 1$,
where r is the common ratio.

11. $\displaystyle\sum_{i=1}^{5}(2i + 8)$

$$= [2(1) + 8] + [2(2) + 8] + [2(3) + 8]$$
$$+ [2(4) + 8] + [2(5) + 8]$$
$$= 10 + 12 + 14 + 16 + 18$$
$$= 70$$

12. $\displaystyle\sum_{i=1}^{6}(3i - 5)$

Find the first and sixth terms.

$$a_1 = 3(1) - 5 = -2$$
$$a_6 = 3(6) - 5 = 13$$

Now find the sum of the first six terms.

$$S_6 = \frac{n}{2}(a_1 + a_6)$$
$$S_6 = \frac{6}{2}(-2 + 13)$$
$$= 3(11) = 33$$

13. $\displaystyle\sum_{i=1}^{500} i$

Use the formula $S_{500} = \dfrac{n}{2}(a_1 + a_{500})$ with $a_1 = 1$
and $a_{500} = 500$.

$$S_{500} = \frac{500}{2}(1 + 500)$$
$$= 250(501) = 125{,}250$$

14. $\displaystyle\sum_{i=1}^{3} \frac{1}{2}(4^i) = \frac{1}{2}\sum_{i=1}^{3}(4^i)$
$$= \frac{1}{2}(4^1 + 4^2 + 4^3)$$
$$= \frac{1}{2}(4 + 16 + 64)$$
$$= \frac{1}{2}(84) = 42$$

15. $\displaystyle\sum_{i=1}^{\infty} \left(\frac{1}{4}\right)^i$

This is an infinite geometric series with $a_1 = \frac{1}{4}$
and $r = \frac{1}{4}$.
Since $|r| = \frac{1}{4} < 1$, the sum exists.

$$S = \frac{a_1}{1 - r} = \frac{\frac{1}{4}}{1 - \frac{1}{4}} = \frac{\frac{1}{4}}{\frac{3}{4}} = \frac{1}{4} \cdot \frac{4}{3} = \frac{1}{3}$$

16. $\displaystyle\sum_{i=1}^{\infty} 6\left(\frac{3}{2}\right)^i$

This is an infinite geometric series with
$a_1 = 6\left(\frac{3}{2}\right)^1 = 9$ and $r = \frac{3}{2}$.
Since $|r| = \frac{3}{2} > 1$, the sum does not exist.

17. $8! = 8 \cdot 7 \cdot 6 \cdot 5 \cdot 4 \cdot 3 \cdot 2 \cdot 1$
$$= 40{,}320$$

18. By definition, $0! = 1$.

19. $\dfrac{6!}{4!\, 2!} = \dfrac{6 \cdot 5 \cdot 4 \cdot 3 \cdot 2 \cdot 1}{(4 \cdot 3 \cdot 2 \cdot 1)(2 \cdot 1)} = \dfrac{6 \cdot 5}{2 \cdot 1} = 15$

20. $_{12}C_{10} = \dfrac{12!}{10!\,(12 - 10)!} = \dfrac{12 \cdot 11 \cdot 10!}{10!\,(2!)}$
$$= \frac{12 \cdot 11}{2} = 66$$

21. $(3k - 5)^4$
$$= (3k)^4 + \frac{4!}{3!\, 1!}(3k)^3(-5)^1$$
$$+ \frac{4!}{2!\, 2!}(3k)^2(-5)^2 + \frac{4!}{1!\, 3!}(3k)^1(-5)^3$$
$$+ (-5)^4$$
$$= 81k^4 - 4\left(27k^3\right)(5) + 6\left(9k^2\right)(25)$$
$$- 4(3k)(125) + 625$$
$$= 81k^4 - 540k^3 + 1350k^2 - 1500k + 625$$

22. The fifth term ($r = 5$, so $r - 1 = 4$) of
$\left(2x - \dfrac{y}{3}\right)^{12}$ is

$$\frac{12!}{(12 - 4)!\, 4!}(2x)^{12-4}\left(-\frac{y}{3}\right)^4$$
$$= \frac{12!}{8!\, 4!}(2x)^8\left(-\frac{y}{3}\right)^4$$
$$= \frac{14{,}080 x^8 y^4}{9}.$$

23. The amounts of unpaid balance during 15 months
form an arithmetic sequence

$$300, 280, 260, \ldots, 40, 20,$$

which is the sequence with $n = 15$, $a_1 = 300$, and
$a_{15} = 20$. Find the sum of these balances.

$$S_n = \frac{n}{2}(a_1 + a_n)$$
$$S_{15} = \frac{15}{2}(300 + 20)$$
$$= \frac{15}{2}(320) = 2400$$

Since 1% interest is paid on this total, the interest
paid is 1% of $2400 or $24. The sewing machine
cost $300 (paid monthly at $20), so the total cost
is $300 + $24 = $324.

24. The weekly populations form a geometric
sequence with $a_1 = 20$ and $r = 3$ since the colony
begins with 20 insects and triples each week. Find
the general term of this geometric sequence.

$$a_n = a_1 r^{n-1}$$
$$a_n = 20(3)^{n-1}$$

We're assuming that from the beginning of July to
the end of September is 12 weeks, so find a_{12}.

$$a_n = 20(3)^{n-1}$$
$$a_{12} = 20(3)^{11}$$

At the end of September, $20(3)^{11} = 3{,}542{,}940$
insects will be present in the colony.

Cumulative Review Exercises (Chapters 1–12)

1. $|-7| + 6 - |-10| - (-8 + 3)$
$$= 7 + 6 - 10 - (-5)$$
$$= 13 - 10 + 5 = 8$$

2. $-15 - |-4| - 10 - |-6|$
$$= -15 - 4 - 10 - 6 = -35$$

3. $4(-6) + (-8)(5) - (-9)$
$$= -24 - 40 + 9 = -55$$

In Exercises 4–7, let

$$P = \left\{-\frac{8}{3}, 10, 0, \sqrt{13}, -\sqrt{3}, \frac{45}{15}, \sqrt{-7}, 0.82, -3\right\}.$$

4. The integers are 10, 0, $\frac{45}{15}$ (or 3), and -3.

5. The rational numbers are

$$-\frac{8}{3}, 10, 0, \frac{45}{15} \text{ (or 3)}, 0.82, \text{ and } -3.$$

6. The irrational numbers are $\sqrt{13}$ and $-\sqrt{3}$.

7. All are real numbers except $\sqrt{-7}$.

8.
$$9 - (5 + 3a) + 5a = -4(a - 3) - 7$$
$$9 - 5 - 3a + 5a = -4a + 12 - 7$$
$$4 + 2a = -4a + 5$$
$$6a = 1$$
$$a = \frac{1}{6}$$

The solution set is $\left\{\frac{1}{6}\right\}$.

9.
$$7m + 18 \le 9m - 2$$
$$-2m \le -20$$
Divide by -2; reverse the direction of the inequality symbol.
$$m \ge 10$$
The solution set is $[10, \infty)$.

10. $|4x - 3| = 21$
$$4x - 3 = 21 \quad \text{or} \quad 4x - 3 = -21$$
$$4x = 24 \qquad\qquad 4x = -18$$
$$x = 6 \quad \text{or} \quad x = -\frac{18}{4} = -\frac{9}{2}$$
The solution set is $\left\{-\frac{9}{2}, 6\right\}$.

11.
$$\frac{x + 3}{12} - \frac{x - 3}{6} = 0$$
Multiply by the LCD, 12.
$$12\left(\frac{x + 3}{12} - \frac{x - 3}{6}\right) = 12(0)$$
$$x + 3 - 2(x - 3) = 0$$
$$x + 3 - 2x + 6 = 0$$
$$9 - x = 0$$
$$9 = x$$

Check $x = 9$: $1 - 1 = 0$ *True*
The solution set is $\{9\}$.

12. $2x > 8 \quad \text{or} \quad -3x > 9$
$$x > 4 \quad \text{or} \quad x < -3$$

The solution set is $(-\infty, -3) \cup (4, \infty)$.

13. $|2m - 5| \ge 11$
$$2m - 5 \ge 11 \quad \text{or} \quad 2m - 5 \le -11$$
$$2m \ge 16 \qquad\qquad 2m \le -6$$
$$m \ge 8 \quad \text{or} \qquad m \le -3$$
The solution set is $(-\infty, -3] \cup [8, \infty)$.

14. Let $(x_1, y_1) = (4, -5)$
and $(x_2, y_2) = (-12, -17)$. Then

$$m = \frac{y_2 - y_1}{x_2 - x_1} = \frac{-17 - (-5)}{-12 - 4} = \frac{-12}{-16} = \frac{3}{4}.$$
The slope is $\frac{3}{4}$.

15. To find the equation of the line through $(-2, 10)$ and parallel to $3x + y = 7$, find the slope of
$$3x + y = 7$$
$$y = -3x + 7.$$
The slope is -3, so a line parallel to it also has slope -3. Use $m = -3$ and $(x_1, y_1) = (-2, 10)$ in the point-slope form.

$$y - y_1 = m(x - x_1)$$
$$y - 10 = -3[x - (-2)]$$
$$y - 10 = -3(x + 2)$$

Write in standard form.

$$y - 10 = -3x - 6$$
$$3x + y = 4$$

Alternative solution: The line must be of the form $3x + y = k$ since it is parallel to $3x + y = 7$. Substitute -2 for x and 10 for y to find k.
$$3(-2) + 10 = k$$
$$4 = k$$
The equation is $3x + y = 4$.

16. $x - 3y = 6$

Find the x- and y-intercepts. To find the x-intercept, let $y = 0$.
$$x - 3(0) = 6$$
$$x = 6$$
The x-intercept is $(6, 0)$.

To find the y-intercept, let $x = 0$.
$$0 - 3y = 6$$
$$y = -2$$
The y-intercept is $(0, -2)$.
Plot the intercepts and draw the line through them.

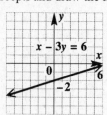

17. $4x - y < 4$

Graph the line $4x - y = 4$, which has intercepts $(0, -4)$ and $(1, 0)$, as a dashed line because the inequality involves $<$. Test $(0, 0)$, which yields $0 < 4$, a true statement. Shade the region on the side of the line that includes $(0, 0)$.

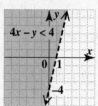

18. $\{(-3, 2), (-2, 6), (0, 4), (1, 2), (2, 6)\}$

(a) The set of ordered pairs is a function since every first coordinate is paired with a unique second coordinate.

(b) The domain is the set of first coordinates, that is,

$$\{-3, -2, 0, 1, 2\}.$$

(c) The range is the set of second coordinates, that is,

$$\{2, 6, 4\}.$$

19. $2x + 5y = -19$ (1)
$-3x + 2y = -19$ (2)

To eliminate x, multiply equation (1) by 3 and equation (2) by 2. Then add the results.

$$
\begin{array}{rcl}
6x + 15y & = & -57 \quad\quad 3 \times (1) \\
-6x + 4y & = & -38 \quad\quad 2 \times (2) \\
\hline
19y & = & -95 \\
y & = & -5
\end{array}
$$

Substitute -5 for y in equation (1) to find x.

$$
\begin{array}{rcl}
2x + 5y & = & -19 \quad\quad (1) \\
2x + 5(-5) & = & -19 \\
2x - 25 & = & -19 \\
2x & = & 6 \\
x & = & 3
\end{array}
$$

The solution set is $\{(3, -5)\}$.

20. $y = 5x + 3$ (1)
$2x + 3y = -8$ (2)

From equation (1), substitute $5x + 3$ for y in equation (2). Then solve for x.

$$
\begin{array}{rcl}
2x + 3(5x + 3) & = & -8 \\
2x + 15x + 9 & = & -8 \\
17x & = & -17 \\
x & = & -1
\end{array}
$$

From (1), $y = 5(-1) + 3 = -2$.

The solution $(-1, -2)$ checks.

The solution set is $\{(-1, -2)\}$.

21. $\begin{array}{rrrcr} x & + 2y & + z & = & 8 \\ 2x & - y & + 3z & = & 15 \\ -x & + 3y & - 3z & = & -11 \end{array}$

Write the augmented matrix.

$$
\left[\begin{array}{rrr|r}
1 & 2 & 1 & 8 \\
2 & -1 & 3 & 15 \\
-1 & 3 & -3 & -11
\end{array}\right]
$$

$$
\left[\begin{array}{rrr|r}
1 & 2 & 1 & 8 \\
0 & -5 & 1 & -1 \\
0 & 5 & -2 & -3
\end{array}\right]
\quad
\begin{array}{l}
-2R_1 + R_2 \\
R_1 + R_3
\end{array}
$$

$$
\left[\begin{array}{rrr|r}
1 & 2 & 1 & 8 \\
0 & 1 & -\frac{1}{5} & \frac{1}{5} \\
0 & 5 & -2 & -3
\end{array}\right]
\quad
-\frac{1}{5}R_2
$$

$$
\left[\begin{array}{rrr|r}
1 & 2 & 1 & 8 \\
0 & 1 & -\frac{1}{5} & \frac{1}{5} \\
0 & 0 & -1 & -4
\end{array}\right]
\quad
-5R_2 + R_3
$$

$$
\left[\begin{array}{rrr|r}
1 & 2 & 1 & 8 \\
0 & 1 & -\frac{1}{5} & \frac{1}{5} \\
0 & 0 & 1 & 4
\end{array}\right]
\quad
-R_3
$$

This matrix gives the system

$$
\begin{array}{rcl}
x + 2y + z & = & 8 \\
y - \frac{1}{5}z & = & \frac{1}{5} \\
z & = & 4.
\end{array}
$$

Substitute $z = 4$ in the second equation.

$$
\begin{array}{rcl}
y - \frac{1}{5}(4) & = & \frac{1}{5} \\
y & = & \frac{1}{5} + \frac{4}{5} = 1
\end{array}
$$

Substitute $y = 1$ and $z = 4$ in the first equation.

$$
\begin{array}{rcl}
x + 2(1) + 4 & = & 8 \\
x + 6 & = & 8 \\
x & = & 2
\end{array}
$$

The solution set is $\{(2, 1, 4)\}$.

22. Let $x =$ the number of pounds of \$3 per pound nuts.

	Number of Pounds	Price per Pound	Value
\$3/lb nuts	x	3	$3x$
\$4.25/lb nuts	8	4.25	4.25(8)
Mixture	$x + 8$	4	$4(x + 8)$

The last column gives the equation.

$$
\begin{array}{rcl}
3x + 4.25(8) & = & 4(x + 8) \\
3x + 34 & = & 4x + 32 \\
2 & = & x
\end{array}
$$

Use 2 pounds of the \$3 per pound nuts.

23. $(4p + 2)(5p - 3)$
　　　F　　O　　I　　L
$= 20p^2 - 12p + 10p - 6$
$= 20p^2 - 2p - 6$

24. $(3k - 7)^2 = (3k)^2 - 2(3k)(7) + 7^2$
　　　　$= 9k^2 - 42k + 49$

25. $(2m^3 - 3m^2 + 8m) - (7m^3 + 5m - 8)$
$= 2m^3 - 7m^3 - 3m^2 + 8m - 5m + 8$
$= -5m^3 - 3m^2 + 3m + 8$

26.

$$
\begin{array}{r}
2t^3 + 3t^2 - 4t + 2 \\
3t - 2\overline{)6t^4 + 5t^3 - 18t^2 + 14t - 1} \\
\underline{6t^4 - 4t^3} \\
9t^3 - 18t^2 \\
\underline{9t^3 - 6t^2} \\
-12t^2 + 14t \\
\underline{-12t^2 + 8t} \\
6t - 1 \\
\underline{6t - 4} \\
3 \quad \text{Remainder}
\end{array}
$$

Answer: $2t^3 + 3t^2 - 4t + 2 + \dfrac{3}{3t - 2}$

27. $7x + x^3 = x(7 + x^2)$

28. $14y^2 + 13y - 12$
Look for two integers whose product is
$(14)(-12) = -168$ and whose sum is 13. The
required numbers are 21 and -8.

$$14y^2 + 13y - 12$$
$$= 14y^2 + 21y - 8y - 12$$
$$= 7y(2y + 3) - 4(2y + 3)$$
$$= (2y + 3)(7y - 4)$$

29. $6z^3 + 5z^2 - 4z = z(6z^2 + 5z - 4)$
$\qquad\qquad\qquad\quad = z(3z + 4)(2z - 1)$

30. $49a^4 - 9b^2 = (7a^2)^2 - (3b)^2$
$\qquad\qquad\quad = (7a^2 + 3b)(7a^2 - 3b)$

31. $c^3 + 27d^3 = c^3 + (3d)^3$
$\qquad\qquad\quad = (c + 3d)(c^2 - 3cd + 9d^2)$

32. $64r^2 + 48rq + 9q^2$
$= (8r)^2 + 2(8r)(3q) + (3q)^2$
$= (8r + 3q)^2$

33. $\qquad 2x^2 + x = 10$
$\qquad 2x^2 + x - 10 = 0$
$\qquad (2x + 5)(x - 2) = 0$

$\qquad 2x + 5 = 0 \quad$ or $\quad x - 2 = 0$
$\qquad\quad x = -\frac{5}{2} \quad$ or $\qquad x = 2$

Check $x = -\frac{5}{2}$: $\frac{25}{2} - \frac{5}{2} = 10$ *True*
Check $x = 2$: $8 + 2 = 10$ *True*

The solution set is $\left\{-\frac{5}{2}, 2\right\}$.

34. $\qquad k^2 - k - 6 \le 0$

Solve the equation

$\qquad\qquad k^2 - k - 6 = 0.$
$\qquad\qquad (k - 3)(k + 2) = 0$

$\qquad k - 3 = 0 \quad$ or $\quad k + 2 = 0$
$\qquad\quad k = 3 \quad$ or $\qquad k = -2$

The numbers -2 and 3 divide a number line into
three intervals.

$$
\begin{array}{ccc}
\text{A} & \text{B} & \text{C} \\
\end{array}
$$

Test a number from each interval in the original
inequality.

$\qquad\qquad k^2 - k - 6 \le 0$
Interval A: Let $k = -3$.
$(-3)^2 - (-3) - 6 \le 0$?
$\qquad\qquad\qquad 6 \le 0$ *False*

Interval B: Let $k = 0$.
$\qquad 0^2 - 0 - 6 \le 0$?
$\qquad\qquad\quad -6 \le 0$ *True*

Interval C: Let $k = 4$.
$\qquad 4^2 - 4 - 6 \le 0$?
$\qquad\qquad\quad 6 \le 0$ *False*

The numbers in Interval B, including the
endpoints -2 and 3 because of $\le$, are solutions.

The solution set is $[-2, 3]$.

35. $\left(\frac{2}{3}\right)^{-2} = \left(\frac{3}{2}\right)^2 = \frac{3}{2} \cdot \frac{3}{2} = \frac{9}{4}$

36. $\dfrac{(3p^2)^3(-2p^6)}{4p^3(5p^7)} = \dfrac{3^3 p^6(-2)p^6}{20p^{10}}$

$\qquad\qquad\qquad = \dfrac{-54p^{12}}{20p^{10}}$

$\qquad\qquad\qquad = -\dfrac{27}{10}p^{12-10}$

$\qquad\qquad\qquad = -\dfrac{27p^2}{10}$

37. $f(x) = \dfrac{2}{x^2 - 81} = \dfrac{2}{(x + 9)(x - 9)}$

The domain of f is the set of all real numbers
excluding ± 9 since division by 0 is not defined.
In interval notation, the domain is

$$(-\infty, -9) \cup (-9, 9) \cup (9, \infty).$$

We can also write this as

$$\{x \mid x \neq -9, 9\}.$$

38. $\dfrac{x^2 - 16}{x^2 + 2x - 8} \div \dfrac{x - 4}{x + 7}$

$\qquad = \dfrac{x^2 - 16}{x^2 + 2x - 8} \cdot \dfrac{x + 7}{x - 4}$

$\qquad = \dfrac{(x + 4)(x - 4)(x + 7)}{(x + 4)(x - 2)(x - 4)}$

$\qquad = \dfrac{x + 7}{x - 2}$

39. $\dfrac{5}{p^2 + 3p} - \dfrac{2}{p^2 - 4p}$

$= \dfrac{5}{p(p+3)} - \dfrac{2}{p(p-4)}$

The LCD is $p(p+3)(p-4)$.

$= \dfrac{5(p-4)}{p(p+3)(p-4)} - \dfrac{2(p+3)}{p(p-4)(p+3)}$

$= \dfrac{5p - 20 - 2p - 6}{p(p+3)(p-4)}$

$= \dfrac{3p - 26}{p(p+3)(p-4)}$

40. $\dfrac{4}{x-3} - \dfrac{6}{x+3} = \dfrac{24}{x^2 - 9}$

$\dfrac{4}{x-3} - \dfrac{6}{x+3} = \dfrac{24}{(x+3)(x-3)}$

Multiply by the LCD, $(x+3)(x-3)$. $(x \neq \pm 3)$

$4(x+3) - 6(x-3) = 24$
$4x + 12 - 6x + 18 = 24$
$-2x + 30 = 24$
$-2x = -6$
$x = 3$

But $x \neq 3$.
The solution set is $\emptyset$.

41. $6x^2 + 5x = 8$

$6x^2 + 5x - 8 = 0$

Use the quadratic formula with $a = 6$, $b = 5$, and $c = -8$.

$x = \dfrac{-b \pm \sqrt{b^2 - 4ac}}{2a}$

$x = \dfrac{-5 \pm \sqrt{5^2 - 4(6)(-8)}}{2(6)}$

$= \dfrac{-5 \pm \sqrt{25 + 192}}{12}$

$= \dfrac{-5 \pm \sqrt{217}}{12}$

Solution set: $\left\{ \dfrac{-5 + \sqrt{217}}{12}, \dfrac{-5 - \sqrt{217}}{12} \right\}$

42. $\sqrt{3x - 2} = x$

$3x - 2 = x^2$ *Square.*

$0 = x^2 - 3x + 2$

$0 = (x-1)(x-2)$

$x - 1 = 0$ or $x - 2 = 0$
$x = 1$ or $x = 2$

Check $x = 1$: $\sqrt{1} = 1$ *True*
Check $x = 2$: $\sqrt{4} = 2$ *True*

The solution set is $\{1, 2\}$.

43. $5\sqrt{72} - 4\sqrt{50} = 5\sqrt{36 \cdot 2} - 4\sqrt{25 \cdot 2}$

$= 5 \cdot 6\sqrt{2} - 4 \cdot 5\sqrt{2}$

$= 30\sqrt{2} - 20\sqrt{2}$

$= 10\sqrt{2}$

44. $(8 + 3i)(8 - 3i) = 8^2 - (3i)^2$

$= 64 - 9i^2$

$= 64 - 9(-1)$

$= 64 + 9 = 73$

45. The graph of $f(x) = 9x + 5$ is a line. To find the inverse, replace $f(x)$ with y.

$y = 9x + 5$

Interchange x and y.

$x = 9y + 5$

Solve for y.

$x - 5 = 9y$

$\dfrac{x - 5}{9} = y$

Replace y with $f^{-1}(x)$.

$f^{-1}(x) = \dfrac{x - 5}{9}$,   or $f^{-1}(x) = \dfrac{1}{9}x - \dfrac{5}{9}$

46. Graph $g(x) = \left(\frac{1}{3}\right)^x$.

Make a table of values.

x	-2	-1	0	1	2
$g(x)$	9	3	1	$\frac{1}{3}$	$\frac{1}{9}$

Plot these points, and draw a smooth decreasing exponential curve through them.

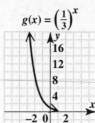

$$g(x) = \left(\frac{1}{3}\right)^x$$

47. $3^{2x-1} = 81$

$3^{2x-1} = 3^4$

$2x - 1 = 4$ *Equate exponents.*

$2x = 5$

$x = \frac{5}{2}$

Check $x = \frac{5}{2}$: $3^{5-1} = 3^4 = 81$

The solution set is $\left\{ \frac{5}{2} \right\}$.

48. Graph $y = \log_{1/3} x$.

Change to exponential form.

$$\left(\tfrac{1}{3}\right)^y = x$$

This is the inverse of the graph of $g(x) = y = \left(\tfrac{1}{3}\right)^x$ in Exercise 46. To find points on the graph, interchange the x- and y-values in the table.

x	9	3	1	$\frac{1}{3}$	$\frac{1}{9}$
y	-2	-1	0	1	2

Plot these points, and draw a smooth decreasing logarithmic curve through them.

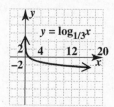

49. $\log_8 x + \log_8 (x + 2) = 1$

Use the product rule for logarithms.

$$\log_8 x(x + 2) = 1$$

Change to exponential form.

$$x(x + 2) = 8^1$$
$$x^2 + 2x - 8 = 0$$
$$(x + 4)(x - 2) = 0$$
$$x + 4 = 0 \quad \text{or} \quad x - 2 = 0$$
$$x = -4 \quad \text{or} \quad x = 2$$

$x \neq -4$ because $\log_8 (-4)$ does not exist.

Check $x = 2$: $\log_8 2 + \log_8 4 = \log_8 (2 \cdot 4) = 1$

The solution set is $\{2\}$.

50. $f(x) = 2(x - 2)^2 - 3$ is in

$f(x) = a(x - h)^2 + k$ form.

The graph is a vertical parabola with vertex (h, k) at $(2, -3)$. Since $a = 2 > 0$, the graph opens up. Also, $|a| = |2| = 2 > 1$, so the graph is narrower than the graph of $f(x) = x^2$. The points $(0, 5)$ and $(4, 5)$ are on the graph.

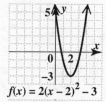

51. $\dfrac{x^2}{9} + \dfrac{y^2}{25} = 1$ is in $\dfrac{x^2}{a^2} + \dfrac{y^2}{b^2} = 1$ form with $a = 3$ and $b = 5$. The graph is an ellipse centered at $(0, 0)$ with x-intercepts $(3, 0)$ and $(-3, 0)$ and y-intercepts $(0, 5)$ and $(0, -5)$. Plot the intercepts and draw the ellipse through them.

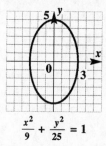

52. $x^2 - y^2 = 9$

$$\dfrac{x^2}{9} - \dfrac{y^2}{9} = 1$$

The graph is a hyperbola centered at $(0, 0)$ with x-intercepts $(3, 0)$ and $(-3, 0)$. The asymptotes are $y = \pm x$. Draw the right and left branches through the intercepts and approaching the asymptotes.

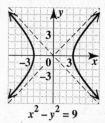

53. $\quad xy = -5 \quad (1)$
$2x + y = 3 \quad (2)$

Solve equation (2) for y.

$$y = -2x + 3 \quad (3)$$

Substitute $-2x + 3$ for y in equation (1).

$$xy = -5 \quad (1)$$
$$x(-2x + 3) = -5$$
$$-2x^2 + 3x = -5$$
$$-2x^2 + 3x + 5 = 0$$
$$2x^2 - 3x - 5 = 0$$
$$(2x - 5)(x + 1) = 0$$
$$2x - 5 = 0 \quad \text{or} \quad x + 1 = 0$$
$$x = \tfrac{5}{2} \quad \text{or} \quad x = -1$$

Substitute these values for x in equation (3) to find y.

If $x = \tfrac{5}{2}$, then $y = -2\left(\tfrac{5}{2}\right) + 3 = -2$.

If $x = -1$, then $y = -2(-1) + 3 = 5$.

The solution set is $\left\{ (-1, 5), \left(\tfrac{5}{2}, -2\right) \right\}$.

54. Center at $(-5, 12)$; radius 9

Use the equation of a circle with $h = -5$, $k = 12$, and $r = 9$.

$$(x - h)^2 + (y - k)^2 = r^2$$
$$[x - (-5)]^2 + (y - 12)^2 = 9^2$$
$$(x + 5)^2 + (y - 12)^2 = 81$$

55. $a_n = 5n - 12$

$a_1 = 5(1) - 12 = 5 - 12 = -7$
$a_2 = 5(2) - 12 = 10 - 12 = -2$
$a_3 = 5(3) - 12 = 15 - 12 = 3$
$a_4 = 5(4) - 12 = 20 - 12 = 8$
$a_5 = 5(5) - 12 = 25 - 12 = 13$

Answer: $-7, -2, 3, 8, 13$

56. **(a)** $a_1 = 8$, $d = 2$

$$S_n = \frac{n}{2}\left[2a_1 + (n-1)d\right]$$
$$S_6 = \frac{6}{2}\left[2(8) + (6-1)2\right]$$
$$= 3(16 + 10) = 3(26) = 78$$

(b) $15 - 6 + \frac{12}{5} - \frac{24}{25} + \cdots$

This is an infinite geometric series with $a_1 = 15$ and $r = \frac{-6}{15} = -\frac{2}{5}$. The sum is

$$S = \frac{a_1}{1 - r} = \frac{15}{1 - \left(-\frac{2}{5}\right)}$$
$$= \frac{15}{\frac{7}{5}} = 15 \cdot \frac{5}{7} = \frac{75}{7}.$$

57. $\sum\limits_{i=1}^{4} 3i = 3\sum\limits_{i=1}^{4} i = 3(1 + 2 + 3 + 4)$

$$= 3(10) = 30$$

58. $9! = 9 \cdot 8 \cdot 7 \cdot 6 \cdot 5 \cdot 4 \cdot 3 \cdot 2 \cdot 1$

$$= 362,880$$

59. $(2a - 1)^5$

$$= (2a)^5 + \frac{5!}{4!\,1!}(2a)^4(-1)^1$$
$$+ \frac{5!}{3!\,2!}(2a)^3(-1)^2 + \frac{5!}{2!\,3!}(2a)^2(-1)^3$$
$$+ \frac{5!}{1!\,4!}(2a)^1(-1)^4 + (-1)^5$$
$$= 32a^5 + 5(16a^4)(-1) + 10(8a^3)(1)$$
$$+ 10(4a^2)(-1) + 5(2a)(1) + (-1)$$
$$= 32a^5 - 80a^4 + 80a^3 - 40a^2 + 10a - 1$$

60. The fourth term $(r = 4$, so $r - 1 = 3)$ of $\left(3x^4 - \frac{1}{2}y^2\right)^5$ is

$$\frac{5!}{(5-3)!\,3!}(3x^4)^{5-3}\left(-\frac{1}{2}y^2\right)^3$$
$$= \frac{5!}{2!\,3!}(3x^4)^2\left(-\frac{1}{2}\right)^3(y^2)^3$$
$$= 10(9x^8)\left(-\frac{1}{8}\right)y^6$$
$$= -\frac{45x^8y^6}{4}.$$

APPENDIX B DETERMINANTS AND CRAMER'S RULE

1. **(a)** *True*, a matrix is an array of numbers, while a determinant is a single number.

(b) *True*, a square matrix has the same number of rows as columns.

(c) *False*, the determinant $\begin{vmatrix} a & b \\ c & d \end{vmatrix}$ is equal to $ad - bc$.

(d) The value $\begin{vmatrix} 0 & 0 \\ x & y \end{vmatrix}$ is zero for any replacements of x and y.

$$\begin{vmatrix} 0 & 0 \\ x & y \end{vmatrix} = 0(y) - 0(x) = 0$$

No matter what replacements are used for x and y, the value of the determinant is zero since both x and y are being multiplied by zero. The statement is *true*.

3. $\begin{vmatrix} -2 & 5 \\ -1 & 4 \end{vmatrix} = -2(4) - 5(-1)$

$$= -8 + 5 = -3$$

5. $\begin{vmatrix} 1 & -2 \\ 7 & 0 \end{vmatrix} = 1(0) - (-2)7$

$$= 0 + 14 = 14$$

7. $\begin{vmatrix} 0 & 4 \\ 0 & 4 \end{vmatrix} = 0(4) - 4(0)$

$$= 0 - 0 = 0$$

9. $\begin{vmatrix} -1 & 2 & 4 \\ -3 & -2 & -3 \\ 2 & -1 & 5 \end{vmatrix}$ Expand by minors about the first column.

$$= -1\begin{vmatrix} -2 & -3 \\ -1 & 5 \end{vmatrix} - (-3)\begin{vmatrix} 2 & 4 \\ -1 & 5 \end{vmatrix}$$
$$+ 2\begin{vmatrix} 2 & 4 \\ -2 & -3 \end{vmatrix}$$
$$= -1[-2(5) - (-3)(-1)]$$
$$+ 3[2(5) - 4(-1)] + 2[2(-3) - 4(-2)]$$
$$= -1(-13) + 3(14) + 2(2)$$
$$= 13 + 42 + 4 = 59$$

11. $\begin{vmatrix} 1 & 0 & -2 \\ 0 & 2 & 3 \\ 1 & 0 & 5 \end{vmatrix}$ There are two 0s in column 2. We'll expand about that column since there is only 1 minor to evaluate.

$$= -0\begin{vmatrix} 0 & 3 \\ 1 & 5 \end{vmatrix} + 2\begin{vmatrix} 1 & -2 \\ 1 & 5 \end{vmatrix} - 0\begin{vmatrix} 1 & -2 \\ 0 & 3 \end{vmatrix}$$
$$= 0 + 2[1(5) - (-2)(1)] - 0$$
$$= 2[5 + 2] = 2(7) = 14$$

13. Multiply the upper left and lower right entries. Then multiply the upper right and lower left entries. Subtract the second product from the first to obtain the determinant. For example,

$$\begin{vmatrix} 4 & 2 \\ 7 & 1 \end{vmatrix} = 4 \cdot 1 - 2 \cdot 7$$
$$= 4 - 14 = -10.$$

15. $\begin{vmatrix} 3 & -1 & 2 \\ 1 & 5 & -2 \\ 0 & 2 & 0 \end{vmatrix}$ Expand about row 3.

$$= 0 - 2\begin{vmatrix} 3 & 2 \\ 1 & -2 \end{vmatrix} + 0$$
$$= -2[3(-2) - 2(1)]$$
$$= -2[-6 - 2] = -2[-8] = 16$$

17. $\begin{vmatrix} 0 & 0 & 3 \\ 4 & 0 & -2 \\ 2 & -1 & 3 \end{vmatrix}$ Expand about row 1.

$$= 0 - 0 + 3\begin{vmatrix} 4 & 0 \\ 2 & -1 \end{vmatrix}$$
$$= 3[4(-1) - 0(2)]$$
$$= 3(-4) = -12$$

19. $\begin{vmatrix} 1 & 1 & 2 \\ 5 & 5 & 7 \\ 3 & 3 & 1 \end{vmatrix}$ Expand about row 1.

$$= 1\begin{vmatrix} 5 & 7 \\ 3 & 1 \end{vmatrix} - 1\begin{vmatrix} 5 & 7 \\ 3 & 1 \end{vmatrix} + 2\begin{vmatrix} 5 & 5 \\ 3 & 3 \end{vmatrix}$$
$$= 1[5(1) - 7(3)] - 1[5(1) - 7(3)]$$
$$+ 2[5(3) - 5(3)]$$
$$= 1(-16) - 1(-16) + 2(0)$$
$$= -16 + 16 + 0 = 0$$

21. $x = \dfrac{D_x}{D} = \dfrac{-43}{-43} = 1$

$y = \dfrac{D_y}{D} = \dfrac{0}{-43} = 0$

$z = \dfrac{D_z}{D} = \dfrac{43}{-43} = -1$

The solution set is $\{(1, 0, -1)\}$.

23. $5x + 2y = -3$
$4x - 3y = -30$

$$D = \begin{vmatrix} 5 & 2 \\ 4 & -3 \end{vmatrix} = -15 - 8 = -23$$

$$D_x = \begin{vmatrix} -3 & 2 \\ -30 & -3 \end{vmatrix} = 9 + 60 = 69$$

$$D_y = \begin{vmatrix} 5 & -3 \\ 4 & -30 \end{vmatrix} = -150 + 12 = -138$$

$$x = \frac{D_x}{D} = \frac{69}{-23} = -3; \; y = \frac{D_y}{D} = \frac{-138}{-23} = 6$$

The solution set is $\{(-3, 6)\}$.

25. $3x - y = 9$
$2x + 5y = 8$

$$D = \begin{vmatrix} 3 & -1 \\ 2 & 5 \end{vmatrix} = 15 + 2 = 17$$

$$D_x = \begin{vmatrix} 9 & -1 \\ 8 & 5 \end{vmatrix} = 45 + 8 = 53$$

$$D_y = \begin{vmatrix} 3 & 9 \\ 2 & 8 \end{vmatrix} = 24 - 18 = 6$$

$$x = \frac{D_x}{D} = \frac{53}{17}; y = \frac{D_y}{D} = \frac{6}{17}$$

The solution set is $\left\{ \left(\frac{53}{17}, \frac{6}{17} \right) \right\}$.

27. $4x + 5y = 6$
$7x + 8y = 9$

$$D = \begin{vmatrix} 4 & 5 \\ 7 & 8 \end{vmatrix} = 32 - 35 = -3$$

$$D_x = \begin{vmatrix} 6 & 5 \\ 9 & 8 \end{vmatrix} = 48 - 45 = 3$$

$$D_y = \begin{vmatrix} 4 & 6 \\ 7 & 9 \end{vmatrix} = 36 - 42 = -6$$

$$x = \frac{D_x}{D} = \frac{3}{-3} = -1; y = \frac{D_y}{D} = \frac{-6}{-3} = 2$$

The solution set is $\{(-1, 2)\}$.

29. $x - y + 6z = 19$
$3x + 3y - z = 1$
$x + 9y + 2z = -19$

$$D = \begin{vmatrix} 1 & -1 & 6 \\ 3 & 3 & -1 \\ 1 & 9 & 2 \end{vmatrix} \quad \text{Expand about column 1.}$$

$$= 1 \begin{vmatrix} 3 & -1 \\ 9 & 2 \end{vmatrix} - 3 \begin{vmatrix} -1 & 6 \\ 9 & 2 \end{vmatrix} + 1 \begin{vmatrix} -1 & 6 \\ 3 & -1 \end{vmatrix}$$

$$= 1(6 + 9) - 3(-2 - 54) + 1(1 - 18)$$

$$= 15 + 168 - 17 = 166$$

$$D_x = \begin{vmatrix} 19 & -1 & 6 \\ 1 & 3 & -1 \\ -19 & 9 & 2 \end{vmatrix} \quad \text{Expand about column 3.}$$

$$= 6 \begin{vmatrix} 1 & 3 \\ -19 & 9 \end{vmatrix} - (-1) \begin{vmatrix} 19 & -1 \\ -19 & 9 \end{vmatrix}$$

$$+ 2 \begin{vmatrix} 19 & -1 \\ 1 & 3 \end{vmatrix}$$

$$= 6(9 + 57) + 1(171 - 19) + 2(57 + 1)$$

$$= 396 + 152 + 116 = 664$$

$$D_y = \begin{vmatrix} 1 & 19 & 6 \\ 3 & 1 & -1 \\ 1 & -19 & 2 \end{vmatrix} \quad \text{Expand about column 1.}$$

$$= 1 \begin{vmatrix} 1 & -1 \\ -19 & 2 \end{vmatrix} - 3 \begin{vmatrix} 19 & 6 \\ -19 & 2 \end{vmatrix} + 1 \begin{vmatrix} 19 & 6 \\ 1 & -1 \end{vmatrix}$$

$$= 1(2 - 19) - 3(38 + 114) + 1(-19 - 6)$$

$$= -17 - 456 - 25 = -498$$

$$D_z = \begin{vmatrix} 1 & -1 & 19 \\ 3 & 3 & 1 \\ 1 & 9 & -19 \end{vmatrix} \quad \text{Expand about column 1.}$$

$$= 1 \begin{vmatrix} 3 & 1 \\ 9 & -19 \end{vmatrix} - 3 \begin{vmatrix} -1 & 19 \\ 9 & -19 \end{vmatrix} + 1 \begin{vmatrix} -1 & 19 \\ 3 & 1 \end{vmatrix}$$

$$= 1(-57 - 9) - 3(19 - 171) + 1(-1 - 57)$$

$$= -66 + 456 - 58 = 332$$

$$x = \frac{D_x}{D} = \frac{664}{166} = 4; y = \frac{D_y}{D} = \frac{-498}{166} = -3$$

$$z = \frac{D_z}{D} = \frac{332}{166} = 2$$

The solution set is $\{(4, -3, 2)\}$.

31. $7x + y - z = 4$
$2x - 3y + z = 2$
$-6x + 9y - 3z = -6$

$$D = \begin{vmatrix} 7 & 1 & -1 \\ 2 & -3 & 1 \\ -6 & 9 & -3 \end{vmatrix} \quad \text{Expand about column 3.}$$

$$= -1 \begin{vmatrix} 2 & -3 \\ -6 & 9 \end{vmatrix} - 1 \begin{vmatrix} 7 & 1 \\ -6 & 9 \end{vmatrix} - 3 \begin{vmatrix} 7 & 1 \\ 2 & -3 \end{vmatrix}$$

$$= -1(18 - 18) - 1(63 + 6) - 3(-21 - 2)$$

$$= 0 - 69 + 69 = 0$$

Because $D = 0$, Cramer's rule does not apply.

33. $-x + 2y = 4$
$3x + y = -5$
$2x + z = -1$

$$D = \begin{vmatrix} -1 & 2 & 0 \\ 3 & 1 & 0 \\ 2 & 0 & 1 \end{vmatrix} \quad \text{Expand about column 3.}$$

$$= 0 - 0 + 1 \begin{vmatrix} -1 & 2 \\ 3 & 1 \end{vmatrix}$$

$$= 1(-1 - 6) = -7$$

$$D_x = \begin{vmatrix} 4 & 2 & 0 \\ -5 & 1 & 0 \\ -1 & 0 & 1 \end{vmatrix} \quad \text{Expand about column 3.}$$

$$= 0 - 0 + 1 \begin{vmatrix} 4 & 2 \\ -5 & 1 \end{vmatrix}$$

$$= 1(4 + 10) = 14$$

$$D_y = \begin{vmatrix} -1 & 4 & 0 \\ 3 & -5 & 0 \\ 2 & -1 & 1 \end{vmatrix} \quad \text{Expand about column 3.}$$

$$= 0 - 0 + 1 \begin{vmatrix} -1 & 4 \\ 3 & -5 \end{vmatrix}$$

$$= 1(5 - 12) = -7$$

$$D_z = \begin{vmatrix} -1 & 2 & 4 \\ 3 & 1 & -5 \\ 2 & 0 & -1 \end{vmatrix}$$ Expand about row 3.

$$= 2 \begin{vmatrix} 2 & 4 \\ 1 & -5 \end{vmatrix} - 0 - 1 \begin{vmatrix} -1 & 2 \\ 3 & 1 \end{vmatrix}$$

$$= 2(-10 - 4) - 1(-1 - 6)$$

$$= -28 + 7 = -21$$

$$x = \frac{D_x}{D} = \frac{14}{-7} = -2; y = \frac{D_y}{D} = \frac{-7}{-7} = 1$$

$$z = \frac{D_z}{D} = \frac{-21}{-7} = 3$$

The solution set is $\{(-2, 1, 3)\}$.

35.
$$\begin{aligned} -5x - y &= -10 \\ 3x + 2y + z &= -3 \\ -y - 2z &= -13 \end{aligned}$$

$$D = \begin{vmatrix} -5 & -1 & 0 \\ 3 & 2 & 1 \\ 0 & -1 & -2 \end{vmatrix}$$ Expand about column 3.

$$= 0 - 1 \begin{vmatrix} -5 & -1 \\ 0 & -1 \end{vmatrix} - 2 \begin{vmatrix} -5 & -1 \\ 3 & 2 \end{vmatrix}$$

$$= -1(5 - 0) - 2(-10 + 3)$$

$$= -5 + 14 = 9$$

$$D_x = \begin{vmatrix} -10 & -1 & 0 \\ -3 & 2 & 1 \\ -13 & -1 & -2 \end{vmatrix}$$ Expand about column 3.

$$= 0 - 1 \begin{vmatrix} -10 & -1 \\ -13 & -1 \end{vmatrix} - 2 \begin{vmatrix} -10 & -1 \\ -3 & 2 \end{vmatrix}$$

$$= -1(10 - 13) - 2(-20 - 3)$$

$$= 3 + 46 = 49$$

$$D_y = \begin{vmatrix} -5 & -10 & 0 \\ 3 & -3 & 1 \\ 0 & -13 & -2 \end{vmatrix}$$ Expand about column 3.

$$= 0 - 1 \begin{vmatrix} -5 & -10 \\ 0 & -13 \end{vmatrix} - 2 \begin{vmatrix} -5 & -10 \\ 3 & -3 \end{vmatrix}$$

$$= -1(65) - 2(15 + 30)$$

$$= -65 - 90 = -155$$

$$D_z = \begin{vmatrix} -5 & -1 & -10 \\ 3 & 2 & -3 \\ 0 & -1 & -13 \end{vmatrix}$$ Expand about row 3.

$$= 0 - (-1) \begin{vmatrix} -5 & -10 \\ 3 & -3 \end{vmatrix} - 13 \begin{vmatrix} -5 & -1 \\ 3 & 2 \end{vmatrix}$$

$$= 1(15 + 30) - 13(-10 + 3)$$

$$= 45 + 91 = 136$$

$$x = \frac{D_x}{D} = \frac{49}{9}; y = \frac{D_y}{D} = \frac{-155}{9} = -\frac{155}{9}$$

$$z = \frac{D_z}{D} = \frac{136}{9}$$

The solution set is $\left\{ \left(\frac{49}{9}, -\frac{155}{9}, \frac{136}{9} \right) \right\}$.

37. $\begin{vmatrix} 4 & x \\ 2 & 3 \end{vmatrix} = 8$

Evaluate the determinant.

$$\begin{vmatrix} 4 & x \\ 2 & 3 \end{vmatrix} = 4(3) - x(2)$$

$$= 12 - 2x$$

Solve the equation.

$$\begin{aligned} 12 - 2x &= 8 \\ -2x &= -4 \\ x &= 2 \end{aligned}$$

The solution set is $\{2\}$.

39. $\begin{vmatrix} x & 4 \\ x & -3 \end{vmatrix} = 0$

Evaluate the determinant.

$$\begin{vmatrix} x & 4 \\ x & -3 \end{vmatrix} = x(-3) - 4x$$

$$= -3x - 4x = -7x$$

Solve the equation.

$$\begin{aligned} -7x &= 0 \\ x &= 0 \end{aligned}$$

The solution set is $\{0\}$.

APPENDIX C SYNTHETIC DIVISION

1. Synthetic division provides a quick, easy way to divide a polynomial by a binomial of the form $x - k$.

3. $\dfrac{x^2 - 6x + 5}{x - 1}$

$$
\begin{array}{r|rrr}
1 & 1 & -6 & 5 \\
  &   & 1 & -5 \\
\hline
  & 1 & -5 & 0
\end{array}
$$
$\leftarrow$ *Coefficients of numerator*

Write the answer from the bottom row.

$$\underset{x}{\downarrow} \quad \underset{-5}{\downarrow}$$

Answer: $x - 5$

5. $\dfrac{4m^2 + 19m - 5}{m + 5}$

$m + 5 = m - (-5)$, so use -5.

$$
\begin{array}{r|rrr}
-5 & 4 & 19 & -5 \\
   &   & -20 & 5 \\
\hline
   & 4 & -1 & 0
\end{array}
$$

Answer: $4m - 1$

7. $\dfrac{2a^2 + 8a + 13}{a + 2}$

$a + 2 = a - (-2)$, so use -2.

$$
\begin{array}{r|rrr}
-2 & 2 & 8 & 13 \\
   &   & -4 & -8 \\
\hline
   & 2 & 4 & 5
\end{array}
$$
$\leftarrow$ *Remainder*

The quotient polynomial is $2a + 4$ and the remainder is 5.

Answer: $2a + 4 + \dfrac{5}{a + 2}$

9. $(p^2 - 3p + 5) \div (p + 1)$

$$
\begin{array}{r|rrr}
-1 & 1 & -3 & 5 \\
   &   & -1 & 4 \\
\hline
   & 1 & -4 & 9
\end{array}
$$

Answer: $p - 4 + \dfrac{9}{p + 1}$

11. $\dfrac{4a^3 - 3a^2 + 2a - 3}{a - 1}$

$$
\begin{array}{r|rrrr}
1 & 4 & -3 & 2 & -3 \\
  &   & 4 & 1 & 3 \\
\hline
  & 4 & 1 & 3 & 0
\end{array}
$$

Answer: $4a^2 + a + 3$

13. $(x^5 - 2x^3 + 3x^2 - 4x - 2) \div (x - 2)$

Insert 0 for the missing x^4-term.

$$
\begin{array}{r|rrrrrr}
2 & 1 & 0 & -2 & 3 & -4 & -2 \\
  &   & 2 & 4 & 4 & 14 & 20 \\
\hline
  & 1 & 2 & 2 & 7 & 10 & 18
\end{array}
$$
$\leftarrow$ *Remainder*

Answer: $x^4 + 2x^3 + 2x^2 + 7x + 10 + \dfrac{18}{x - 2}$

15. $(-4r^6 - 3r^5 - 3r^4 + 5r^3 - 6r^2 + 3r + 3) \div (r - 1)$

$$
\begin{array}{r|rrrrrrr}
1 & -4 & -3 & -3 & 5 & -6 & 3 & 3 \\
  &   & -4 & -7 & -10 & -5 & -11 & -8 \\
\hline
  & -4 & -7 & -10 & -5 & -11 & -8 & -5
\end{array}
$$
$\leftarrow$ *Remainder*

Answer:

$-4r^5 - 7r^4 - 10r^3 - 5r^2 - 11r - 8 + \dfrac{-5}{r - 1}$

17. $(-3y^5 + 2y^4 - 5y^3 - 6y^2 - 1) \div (y + 2)$

Insert 0 for the missing y-term.

$$
\begin{array}{r|rrrrrr}
-2 & -3 & 2 & -5 & -6 & 0 & -1 \\
   &   & 6 & -16 & 42 & -72 & 144 \\
\hline
   & -3 & 8 & -21 & 36 & -72 & 143
\end{array}
$$
$\leftarrow$ *Remainder*

Answer:

$-3y^4 + 8y^3 - 21y^2 + 36y - 72 + \dfrac{143}{y + 2}$

19. $\dfrac{y^3 + 1}{y - 1} = \dfrac{y^3 + 0y^2 + 0y + 1}{y - 1}$

$$
\begin{array}{r|rrrr}
1 & 1 & 0 & 0 & 1 \\
  &   & 1 & 1 & 1 \\
\hline
  & 1 & 1 & 1 & 2
\end{array}
$$
$\leftarrow$ *Remainder*

Answer: $y^2 + y + 1 + \dfrac{2}{y - 1}$

21. $P(x) = 2x^3 - 4x^2 + 5x - 3; k = 2$

To find $P(2)$, divide the polynomial by $x - 2$. $P(2)$ will be the remainder.

$$
\begin{array}{r|rrrr}
2 & 2 & -4 & 5 & -3 \\
  &   & 4 & 0 & 10 \\
\hline
  & 2 & 0 & 5 & 7
\end{array}
$$
$\leftarrow$ *Remainder*

By the remainder theorem, $P(2) = 7$.

23. $P(r) = -r^3 - 5r^2 - 4r - 2; k = -4$

Divide by $r + 4$. The remainder is equal to $P(-4)$.

$$
\begin{array}{r|rrrr}
-4 & -1 & -5 & -4 & -2 \\
   &   & 4 & 4 & 0 \\
\hline
   & -1 & -1 & 0 & -2
\end{array}
$$
$\leftarrow$ *Remainder*

By the remainder theorem, $P(-4) = -2$.

25. $P(y) = 2y^3 - 4y^2 + 5y - 33; \ k = 3$

Divide by $y - 3$. The remainder is equal to $P(3)$.

$$3 \ \overline{\left)\ \begin{array}{rrrr} 2 & -4 & 5 & -33 \\ & 6 & 6 & 33 \end{array}\right.}$$
$$\begin{array}{rrrr} 2 & 2 & 11 & 0 \end{array} \leftarrow \textit{Remainder}$$

By the remainder theorem, $P(3) = 0$.

27. By the remainder theorem, a zero remainder means that $P(k) = 0$; that is, k is a number that makes $P(x) = 0$.

29. Is $x = -2$ a solution of

$$x^3 - 2x^2 - 3x + 10 = 0?$$

To decide whether -2 is a solution to the given equation, divide the polynomial by $x + 2$.

$$-2 \ \overline{\left)\ \begin{array}{rrrr} 1 & -2 & -3 & 10 \\ & -2 & 8 & -10 \end{array}\right.}$$
$$\begin{array}{rrrr} 1 & -4 & 5 & 0 \end{array} \leftarrow \textit{Remainder}$$

Since the remainder is 0, -2 is a solution of the equation.

31. Is $m = -2$ a solution of

$$m^4 + 2m^3 - 3m^2 + 8m - 8 = 0?$$

To decide whether -2 is a solution to the given equation, divide the polynomial by $m + 2$.

$$-2 \ \overline{\left)\ \begin{array}{rrrrr} 1 & 2 & -3 & 8 & -8 \\ & -2 & 0 & 6 & -28 \end{array}\right.}$$
$$\begin{array}{rrrrr} 1 & 0 & -3 & 14 & -36 \end{array} \leftarrow \textit{Remainder}$$

Since the remainder is not 0, -2 is not a solution of the equation.

33. Is $a = -2$ a solution of

$$3a^3 + 2a^2 - 2a + 11 = 0?$$

$$-2 \ \overline{\left)\ \begin{array}{rrrr} 3 & 2 & -2 & 11 \\ & -6 & 8 & -12 \end{array}\right.}$$
$$\begin{array}{rrrr} 3 & -4 & 6 & -1 \end{array} \leftarrow \textit{Remainder}$$

Since the remainder is not 0, -2 is not a solution of the equation.

35. Is $x = -3$ a solution of

$$2x^3 - x^2 - 13x + 24 = 0?$$

$$-3 \ \overline{\left)\ \begin{array}{rrrr} 2 & -1 & -13 & 24 \\ & -6 & 21 & -24 \end{array}\right.}$$
$$\begin{array}{rrrr} 2 & -7 & 8 & 0 \end{array} \leftarrow \textit{Remainder}$$

Since the remainder is 0, -3 is a solution of the equation.

In Exercises 37–41,

$$P(x) = 2x^2 + 5x - 12.$$

37. Factor $P(x)$.

$$2x^2 + 5x - 12 = (2x - 3)(x + 4)$$

38. Solve $P(x) = 0$.

$$2x^2 + 5x - 12 = 0$$
$$(2x - 3)(x + 4) = 0$$

$$2x - 3 = 0 \quad \text{or} \quad x + 4 = 0$$
$$2x = 3 \qquad\qquad x = -4$$
$$x = \tfrac{3}{2}$$

The solution set is $\left\{ -4, \tfrac{3}{2} \right\}$.

39.
$$\begin{aligned} P(-4) &= 2(-4)^2 + 5(-4) - 12 \\ &= 2(16) - 20 - 12 \\ &= 32 - 20 - 12 = 0 \end{aligned}$$

$$\begin{aligned} P\left(\tfrac{3}{2}\right) &= 2\left(\tfrac{3}{2}\right)^2 + 5\left(\tfrac{3}{2}\right) - 12 \\ &= 2\left(\tfrac{9}{4}\right) + \tfrac{15}{2} - 12 \\ &= \tfrac{9}{2} + \tfrac{15}{2} - \tfrac{24}{2} = 0 \end{aligned}$$

40. If $P(a) = 0$, then $x - \underline{\ a\ }$ is a factor of $P(x)$.

41. $Q(x) = 3x^3 - 4x^2 - 17x + 6$
$$\begin{aligned} Q(3) &= 3(3)^3 - 4(3)^2 - 17(3) + 6 \\ &= 81 - 36 - 51 + 6 = 0 \end{aligned}$$

Since $Q(3) = 0$, $x - 3$ is a factor of $Q(x)$. To check, use synthetic division to see if 3 is a solution of the equation.

$$3 \ \overline{\left)\ \begin{array}{rrrr} 3 & -4 & -17 & 6 \\ & 9 & 15 & -6 \end{array}\right.}$$
$$\begin{array}{rrrr} 3 & 5 & -2 & 0 \end{array}$$

Therefore, $x - 3$ is a factor of the polynomial and

$$3x^3 - 4x^2 - 17x + 6 = (x - 3)(3x^2 + 5x - 2)$$
$$Q(x) = (x - 3)(3x - 1)(x + 2).$$

43. From the graph, it appears that $x = 3$ is a solution of the equation

$$x^3 - x^2 - 21x + 45 = 0.$$

Check this with synthetic division.

$$3 \ \overline{\left)\ \begin{array}{rrrr} 1 & -1 & -21 & 45 \\ & 3 & 6 & -45 \end{array}\right.}$$
$$\begin{array}{rrrr} 1 & 2 & -15 & 0 \end{array} \leftarrow \textit{Remainder}$$

Since the remainder is 0, 3 is a solution of the equation.

45. From the graph, it appears that $x = -1$ is a solution of the equation

$$x^3 + 3x^2 - 13x - 15 = 0.$$

Check this with synthetic division.

$$-1 \ \overline{\left)\ \begin{array}{rrrr} 1 & 3 & -13 & -15 \\ & -1 & -2 & 15 \end{array}\right.}$$
$$\begin{array}{rrrr} 1 & 2 & -15 & 0 \end{array} \leftarrow \textit{Remainder}$$

Since the remainder is 0, -1 is a solution of the equation.